THE BASAL
GANGLIA VI

ADVANCES IN BEHAVIORAL BIOLOGY

Editorial Board

Jan Bures	Institute of Physiology, Prague, Czech Republic
Irwin Kopin	National Institute of Mental Health, Bethesda, Maryland
Bruce McEwen	Rockefeller University, New York, New York
Karl Pribram	Radford University, Radford, Virginia
Jay Rosenblatt	Rutgers University, Newark, New Jersey
Lawrence Weiskranz	University of Oxford, Oxford, England

Recent Volumes in This Series

Volume 39   THE BASAL GANGLIA III
Edited by Giorgio Bernardi, Malcolm B. Carpenter, Gaetano Di Chiara, Micaela Morelli, and Paolo Stanzione

Volume 40   TREATMENT OF DEMENTIAS: A New Generation of Progress
Edited by Edwin M. Meyer, James W. Simpkins, Jyunji Yamamoto, and Fulton T. Crews

Volume 41   THE BASAL GANGLIA IV: New Ideas and Data on Structure and Function
Edited by Gérard Percheron, John S. McKenzie, and Jean Féger

Volume 42   CALLOSAL AGENESIS: A Natural Split Brain?
Edited by Maryse Lassonde and Malcolm A. Jeeves

Volume 43   NEUROTRANSMITTERS IN THE HUMAN BRAIN
Edited by David J. Tracey, George Paxinos, and Jonathan Stone

Volume 44   ALZHEIMER'S AND PARKINSON'S DISEASES: Recent Developments
Edited by Israel Hanin, Mitsuo Yoshia, and Abraham Fisher

Volume 45   EPILEPSY AND THE CORPUS CALLOSUM 2
Edited by Alexander G. Reeves and David W. Roberts

Volume 46   BIOLOGY AND PHYSIOLOGY OF THE BLOOD–BRAIN BARRIER: Transport, Cellular Interactions, and Brain Pathologies
Edited by Pierre-Olivier Couraud and Daniel Scherman

Volume 47   THE BASAL GANGLIA V
Edited by Chihoto Ohye, Minoru Kimura, and John S. McKenzie

Volume 48   KINDLING 5
Edited by Michael E. Corcoran and Solomon Moshé

Volume 49   PROGRESS IN ALZHEIMER'S AND PARKINSON'S DISEASES
Edited by Abraham Fisher, Israel Hanin, and Mitsuo Yoshida

Volume 50   NEUROPSYCHOLOGY OF CHILDHOOD EPILEPSY
Edited by Isabelle Jambaqué, Maryse Lassonde, and Olivier Dulac

Volume 51   MAPPING THE PROGRESS OF ALZHEIMER'S AND PARKINSON'S DISEASE
Edited by Yoshikuni Mizuno, Abraham Fisher, and Israel Hanin

Volume 52   THE BASAL GANGLIA VII
Edited by Louise F. B. Nicholson and Richard L. M. Faull

Volume 53   CATECHOLAMINE RESEARCH: From Molecular Insights to Clinical Medicine
Edited by Toshiharu Nagatsu, Toshitaka Nabeshima, Richard McCarty, and David S. Goldstein

Volume 54   THE BASAL GANGLIA VI
Edited by Ann M. Graybiel, Mahlon R. Delong, and Stephen T. Kitai

---

A Continuation Order Plan is available for this series. A continuation order will bring delivery of each new volume immediately upon publication. Volumes are billed only upon actual shipment. For further information please contact the publisher.

THE BASAL GANGLIA VI

Edited by

Ann M. Graybiel
Massachusetts Institute of Technology
Cambridge, Massachusetts, USA

Mahlon R. Delong
Emory University
Atlanta, Georgia, USA

and

Stephen T. Kitai
University of Tennessee Memphis
Memphis, Tennessee, USA

Kluwer Academic / Plenum Publishers
New York, Boston, Dordrecht, London, Moscow

Library of Congress Cataloging-in-Publication Data

International Basal Ganglia Society. Symposium (6th: 1998: Brewster, Mass.)
 The basal ganglia VI/edited by Ann M. Graybiel, Mahlon R. Delong and Stephen T. Kitai.
   p. ; cm.
 Includes bibliographical references and index.
 ISBN 0-306-47499-9
   1. Basal ganglia—Congresses.   I. Title: Basal ganglia 6.   II. Title: Basal ganglia six.   III. Graybiel, A. M. (Ann M.), 1942–   IV. DeLong, Mahlon R.   V. Kitai, Stephen T.   VI. Title.
   [DNLM: 1. Basal Ganglia Diseases—therapy—Congresses. 2. Basal Ganglia—pathology—Congresses.   3. Basal Ganglia—physiology—Congresses.   WL 307 I605b 2003]
 QP383.3.I58 1998
 616.8′33—dc21

2002043275

Proceedings of the 6th Triennial Meeting of the International Basal Ganglia Society, Brewster, Massachusetts, October 15–18, 1998

ISBN 0-306-47499-9

©2003 Kluwer Academic/Plenum Publishers, New York
233 Spring Street, New York, New York 10013

http://www.wkap.com

10 9 8 7 6 5 4 3 2 1

A C.I.P. record for this book is available from the Library of Congress

All rights reserved

No part of this book may be reproduced, stored in a retrieval system, or transmitted in any form or by any means, electronic, mechanical, photocopying, microfilming, recording, or otherwise, without written permission from the Publisher, with the exception of any material supplied specifically for the purpose of being entered and executed on a computer system, for exclusive use by the purchaser of the work

Printed in the United States of America

INTERNATIONAL BASAL GANGLIA SOCIETY

OFFICERS (1995-1998)

PRESIDENT
Ann M. Graybiel, USA

SECRETARY
Mahlon R. DeLong, USA

TREASURER
Stephen T. Kitai, USA

PRESIDENT ELECT
Richard L. Faull, New Zealand

SECRETARY ELECT
Louise F. Nicholson, New Zealand

PAST PRESIDENT
Chihiro Ohye, Japan

COUNCIL
Paul J. Bedard, Canada
Hagai Bergman, Israel
Paolo Calabresi, Italy
Alexander Cools, Netherlands
Ivan Divac, Denmark
Piers Emson, UK
Minoru Kimura, Japan

HONORARY MEMBERS
Denise Albe-Fessard
Malcolm B. Carpenter
John S. McKenzie
Hirotaro Narabayashi

IBAGS VI

The 7th triennial meeting of the International Basal Ganglia Society was held at
Brewster, Massachusetts,
USA
October 15th to 18th, 1998

Special Lecturer   Steven Kitai

Program Committee

Elizabeth Abercrombie
Marie-Francoise Chesselet
Mike Levine
Jack Penney
Jim Tepper
Judy Walters
Charlie Wilson
Mike Zigmond

The following organizations and institutions are gratefully acknowledged:

Allergan
Medtronics
Pharmacia and UpJohn
SmithKline Beecham

PREFACE

This volume, the sixth in the IBAGS series, summarizes major contributions in clinical and basic research on the basal ganglia. The sixth meeting of the Society was held on Cape Cod, in the state of Massachusetts, USA, in October, 1998. Altogether 16 countries were represented by 227 participants. This volume contains papers contributed by participants.

The focus of the sixth triennial IBAGS meeting, and of this volume, was to bring together leaders in basic and clinical science to address two sets of still-persisting questions in the field. The first set focuses on the functions of the basal ganglia in health and disease: What are the core functions of the basal ganglia and cortico-basal ganglia loops? How are these core functions disrupted in disorders affecting the basal ganglia? How do we account for the broad range of behaviors affected by basal ganglia disorders and for the increasing evidence that the basal ganglia influence cognitive as well as motor functions? These issues are addressed in the first five sections of the current volume, which summarize advances in the study of basal ganglia disorders based on studies in humans (Section 1), new results obtained with experimental animal models of basal ganglia disorders (Section 2), results of experiments on information coding in the basal ganglia (Section 3) and new information about functions of the basal ganglia related to learning and adaptive motor control (Section 4). The second group of questions concerns the neural mechanisms that underpin these functions. Section 5 reviews new insights into the anatomy of the basal ganglia based on new methodologies. Section 6 focuses on experiments aimed at identifying the actions of dopamine in the basal ganglia. Section 7 summarizes new results on neurotransmitter functions in the basal ganglia, emphasizing glutamate and GABA, but also including non-classical neuromodulators such as adenosine and cannabinoids.

The volume opens with a special tribute to Professor Hiro Narabayashi, a pioneer in clinical treatment of basal ganglia disorders and a true hero in the field of Parkinson's disease. His renowned talent in bringing care to patients was combined with an insatiable curiosity about neural mechanisms and a delight in new discoveries. He was the mentor of many of us in the field, either directly or indirectly, and a giant of clinical medicine. He was, above all, a luminous, caring individual who faced life with the highest principles and yet with gentle humility and wonderful humor. Hiro Narabayashi died 18 March 2001. We will miss him deeply.

Ann M. Graybiel
Mahlon R. Delong
Stephen T. Kitai

SPECIAL ACKNOWLEDGEMENTS

The editors would like to express special appreciation to John S. Kinnebrew, Albert S. Wang, Brandy Baker, and Clark Brayton Jr. for their exceptional efforts in the editorial and production process for this volume.

A MEMORIAL LETTER FOR PROFESSOR HIROTARO NARABAYASHI WHO DIED IN THIS SPRING

It is greatly to be regretted that we mourn over the death of Prof. Hirotaro Narabayashi, a member of the International Basal Ganglia Society, and an honorary member of the Juntendo Medical Society at the time of his death. I, as a neurobiologist, was myself fascinated with Dr. Narabayashi's work in the Department of Neurology (Juntendo University School of Medicine). I hope that with this introduction to his academic achievement, we may celebrate him instead of mourning his loss.

Prof. Narabayashi entered the department of Psychiatry at the University of Tokyo, headed by Prof. Yushi Uchimura, after graduating from the University of Tokyo in 1947. Soon after (in 1949), when he was only 27 years old, he started independently, for the first time in the world, a stereotaxic surgery for neuro-psychiatric diseases. He made an apparatus for the surgery by himself. He performed the pallidotomy by a procaine-oil blocking method in patients with Parkinson's disease, with an athetotic type of cerebral palsy characterized by hypertonic muscular activity, and with diseases characterized by involuntary movements including tremor and athethosis. The pallidotomy procedure produced a clear positive response for both the rigidity and the tremor. Up to the present time, there has been a dramatic and long-lasting benefit for the patients who came to Prof. Narabayashi with symptoms including involuntary movements induced by dysfunction of the basal ganglia. Prof. Narabayashi was among the several creators of stereotaxic brain surgery. Soon after World War II, most passed away or retired. Prof. Narabayashi was an exception. He continued to stay in active service for surgery, finally finishing his work at age 77, in April, 2000, only because of the onset of age-related macular degeneration. His patients and colleagues alike were saddened to hear of his retirement, as was Prof. Narabayashi.

Stereotaxic surgery is a paradoxical treatment, because the surgeon makes a lesion to deep-lying intact regions of the brain in order to make behavior improve. Prof. Narabayashi understood that this method can contribute to major developments in anatomy, neurophysiology and neuropharmacology - especially for motor functions. In fact, many doctors gathered in his clinic, called Shin-kuri (which comes from 'neurological clinic' in Japanese). They were young psychiatrists, neurologists, famous neurophysiologists - including Prof. Hiroshi

Shimazu (Univ. of Tokyo) — and also anatomists. Further, many researchers from abroad came as observers and students; the clinic received such visitors almost continuously. It was as if the clinic was an ideal center for clinical research. Later, Prof. Narabayashi became interested in, and changed, the target of the stereotaxic surgeries that he performed from pallidotomy to thalamotomy, and changed from the procaine-oil blocking method to high-frequency electrocoagulation. He also introduced unit recording in the patient before the thalamotomy or pallidotomy. By the use of these recordings, the precision and safety of the surgery dramatically improved. Furthermore, Prof. Narabayashi found that a key site for relieving rigidity exists in the VL nucleus of the thalamus and that a key point for relieving tremor was in the Vim nucleus; and he found that these two sites were located adjacent to each other. This was a great discovery for sensorimotor neurophysiology as well as for neurology.

The main object of Narabayashi's stereotaxic surgery was to treat Parkinson's disease. In 1960, more than ten years after the development of stereotaxy, L-Dopa therapy to counter dopamine deficiency in Parkinson's disease was found to be dramatically effective. This discovery was recognized by the award of the Nobel Prize. After the introduction of L-Dopa therapy, stereotaxic surgery became a less popular method for treating Parkinson's disease in the Western world. Dr Narabayashi led the popularization of L-Dopa therapy in Japan, and at the same time continued to improve stereotaxic surgical methods for treating Parkinson's disease. Finally, he gained a near-perfect method for the surgery, which he continued to perform himself. In his clinic, large numbers of patients came from all over the world to receive treatment. Prof. Narabayashi also traveled all over the world to teach his method of stereotaxic surgery to neurosurgeons and neurologists.

Recently, it became clear that long-term L-Dopa therapy has serious problems. In some countries, a new therapy was developed. This treatment was to have a neonatal midbrain transplanted into the striatum of the Parkinson's patient and, more recently, other dopamine-releasing cell types. Finally, in late 1980's, because of problems with the implantation therapy, pallidotomy was reconsidered in Sweden as a treatment of choice, and this surgical method spread again all over the world. This was a revival of Narabayashi's original idea of pallidotomy. It is easy for doctors to change the therapies they use. But Prof. Narabayashi continued to use and to perfect his method for over 50 years, because he had a sense of duty toward his patients. Prof. Narabayashi had an extremely positive attitude, free and flexible ideas, and was interdisciplinary and internationally-minded. His character teaches us a lesson. Prof. Narabayashi was a truly dynamic person. Now he has finished his life and lives calmly in peace with his darling wife in the other world. May his soul rest in peace!

Hisamasa Imai, M.D.
Professor of Neurology
Editor of Juntendo Medical Journal

CONTENTS

SECTION I. STUDIES IN HUMANS

1. What has Stereotactic Surgery Taught Us About Basal Ganglia Function? 3
David J. Brooks

2. Pathophysiology of the Internal Segment of the Globus Pallidus in Parkinson's Disease: What Have We Learned from Surgery? 15
Andres M. Lozano, William D. Hutchison, Ron Levy, Anthony E. Lang, and Jonathan O. Dostrovsky

3. Apomorphine Responses in Globus Pallidus and Subthalamus of Parkinsonian Patients Undergoing Stereotaxic Neurosurgery: Lack of Evidence for the So-Called Indirect Pathway 23
A. Stefani, A. Bassi, A. Peppe, P. Mazzone, M. G. Altibrandi, G. Gattoni, G. Bernardi, and P. Stanzione

4. Neuronal Activity of GPe in Parkinson's Disease: Is It Really Hypoactive? 33
Chihiro Ohye, Masafumi Hirato, Akio Takahashi, and Katsushige Watanabe

5. Acute Effects of Levodopa and Pallidotomy on Bimanual Repetitive Arm Movements in Patients With Parkinson's Disease 43
R. Levy, W.D. Hutchison, A.E. Lang, A.M. Lozano, and J.O. Dostrovsky

6. Tremor in Parkinson's Disease: A Simplification in Network Dynamics? 51
Anne Beuter and Roderick Edwards

7. Visuo-spatial Attention in Non-demented Patients with Parkinson's Disease: An Event-related Potential Study 61
Constantin Potagas, Nguyen Bathien, and André Hugelin

8. Neuronal Intranuclear Inclusions in Neostriatal Striosomes and Matrix in Huntington's Disease 83
John C. Hedreen

9. Probing Striatal Function in Obsessive Compulsive Disorder: Neuroimaging Studies of Implicit Sequence Learning 87
Scott L. Rauch

SECTION II. ANIMAL MODELS

10. Evolution of the Multiunit Activity of the Basal Ganglia in the Course of Dynamic Experimental Parkinsonism 97
Erwan Bezard, Thoams Boraud, Bernard Bioulac, and Christian E. Gross

11. Evidence for Neuronal Dysfunction in a Mouse Model of Early Stages of Huntington's Disease 107
M.-F. Chesselet, M.S. Levine, C. Cepeda, L. Menalled, and H. Zanjani

12. Effects of GDNF on Nigrostriatal Dopamine Neurons: Experience from Rodent Models of Parkinson's Disease 117
Carl Rosenblad, Deniz Kirik and Anders Björklund

13. A Role for the Vesicular Monoamine Transporter (VMAT2) in Parkinson's Disease 131
Dwight C. German and Patricia K. Sonsalla

14. Quantitative Analysis and Behavioural Correlates of Lesioning of the Subthalamic Nucleus in the Hemiparkinsonian Marmoset 139
J.M. Henderson, L.E. Annet, E.M. Torres, and S.B. Dunnett

15. Effects of Reversible Blockade of Pedunculopontine Tegmental Nucleus on Voluntary Arm Movement in Monkey 151
Masura Matsumura and Katsuhige Watanabe

16. Differential Reductions in Dopaminergic Innervation of the Motor-related Areas of the Frontal Lobe in MPTP-treated Monkeys 159
Yoshio Yamaji, Masaru Matsumura, Jun Kojima, Hironobu Tokuno, Atsushi Nambu, Masahiko Inase, Hisamasa Imai, and Masahiko Takada

17. Control of Epileptic Seizures: Another Function for the Basal Ganglia? 169
Colin Deransart, Véronique Riban, Laurent Vercueil, Karine Nail-Boucherie, Christian Marescaux, and Antoine Depaulis

SECTION III. INFORMATION CODING IN THE BASAL GANGLIA

18. Surround Inhibition in the Basal Ganglia: A Brief Review 181
Michel Filion

19. Surround Inhibition in the Basal Ganglia 187
Jeffery R. Wickens

20. Information Processing in the Cortico-Striato-Nigral Circuits of the Rat Basal
    Ganglia: Anatomical and Neurophysiological Aspects 199
S. Charpier, P. Mailly, S. Mahon, A. Menetrey, M.J. Besson and J.M. Deniau

21. The Control of Spiking by Synaptic Input in Striatal and Pallidal Neurons 209
Dieter Jaeger

22. Excitatory Cortical Inputs to Pallidal Neurons Through the Cortico-Subthalamo-
    Pallidal Hyperdirect Pathway in the Monkey 217
Atsushi Nambu, Hironobu Tokuno, Ikuma Hamada, Hitoshi Kita, Michiko Imanishi,
    Toshikazu Akazawa, Yoko Ikeuchi, and Naomi Hasegawa

23. TANs, PANs, and STANs 225
Ben D. Bennett and Charles J. Wilson

24. Dopamine and Ensemble Coding in the Striatum and Nucleus Accumbens: A
    Coincidence Detection Mechanism 237
Patricio O'Donnell

25. Neuronal Firing Patterns in the Subthalamic Nucleus: Effects of Dopamine
    Receptor Stimulation on Multisecond Oscillations 245
Kelly A. Allers, Deborah S. Kreiss, and Judith R. Walters

SECTION IV. LEARNING FUNCTIONS OF THE BASAL GANGLIA AND ADAPTIVE MOTOR CONTROL

26. Selection and the Basal Ganglia: A Role for Dopamine 257
P. Redgrave, T. Prescott, and K. Gurney

27. Movement Inhibition and Next Sensory State Prediction in the Basal Ganglia 267
Amanda Bischoff-Grethe, Michael G. Crowley, and Michael A. Arbib

28. Basal Ganglia Neural Coding of Natural Action Sequences 279
J. Wayne Aldridge and Kent C. Berridge

29. Activity of the Putamenal Neurons in the Monkey During the Sequential Stages of the Behavioural Task 289
B.F. Tolkunov, A.A. Orlov, S.V. Afanas'ev, and E.V. Selezneva

30. Learning-Selective Changes in Activity of the Basal Ganglia 297
Masahiko Inase, Bao-Ming Li, Ichiro Takashima, and Toshio Iijima

31. Cognitive Decision Processes and Functional Characteristics of the Basal Ganglia Reward System 303
Moti Shatner, Gali Havazelet-Heimer, Aeyal Raz and Hagai Bergman

32. A Partial Dopamine Lesion Impairs Performance on a Procedural Learning Task: Implications for Parkinson's Disease 311
Julie L. Fudge, David D. Song, and Suzanne N. Haber

33. Neurochemical Evidence that Mesolimbic Noradrenaline Directs Mesolimbic Dopamine, Implying That Noradrenaline, Like Dopamine, Plays a Key Role in Goal-directed and Motivational Behavior 323
A.R. Cools and T. Tuinstra

34. Tonically Active Neurons in the Striatum of the Monkey Rapidly Signal a Switch in Behavioral Set 335
Traci M. Thomas and Michael D. Crutcher

35. Tonically Active Neurons in the Monkey Striatum Are Sensitive to Sensory Events in a Manner That Reflects Their Predictability in Time 347
Paul Apicella, Sabrina Ravel, Pierangelo Sardo, and Eric Legallet

SECTION V. NEW INSIGHTS INTO THE ANATOMY OF THE BASAL GANGLIA

36. Distribution of Substantia Nigra Pars Compacta Neurons With Respect to Pars Reticulata Striato-Nigral Afferences: Computer-assisted Three-dimensional Reconstructions 359
B. Banrezes, P. Andrey, A. Menetrey, P. Mailly, J.-M. Deniau, and Y. Maurin

37. The Immunocytochemical Localization of Tyrosine Hydroxylase in the Human Striatum and Substantia Nigra: A Postmortem Ultrastructural Study 369
Rosalinda C. Roberts

38. Dendritic Changes in Medium Spiny Neurons of the Weaver Striatum: A Golgi Study 379
Diane E. Smith, Beverly Glover, and C'Lita Henry

39. Early Corticostriatal Projections and Development of Striatal Patch/Matrix Organization 385
Abigail Snyder-Keller, Yili Lin, and David J. Graber

40. Re-evaluation of Markers for the Patch-Matrix Organization of the Rat Striatum: Core and Shell in the Striatum 393
Richard E. Harlan, Monique Guillot, and Meredith M. Garcia

41. Physiological and Morphological Classification of Single Neurons in Barrel Cortex With Axons in Neostriatum 399
E.A.M. Hutton, A.K. Wright, and G.W. Arbuthnott

42. The Enkephalin-expressing Cells of the Rodent Globus Pallidus 411
J.F. Marshall, B.R. Hoover, and J.J. Schuller

43. Corticostriatal Projections From the Cingulate Motor Areas in the Macaque Monkey 419
Masahiko Takada, Ikuma Hamada, Hiornobu Tokuno, Masahiko Inase, yumi Ito, Naomi Hasegawa, Yoko Ikeuchi, Michiko Imanishi, Toshikazu Akazawa, Nobuhiko Hatanaka, and Atsuhi Nambu

44. Superficial and Deep Thalamo-Cortical Projections From the Oral Part of the Ventral Lateral Thalamic Nucleus (VLo) Receiving Inputs to the Internal Pallidal Segment (GPi) and Cerebellar Dentate Nucleus in the Macaque Monkey 429
Katsuma Nakano, Tetsuro Kayahara, and Iiji Nagaoka

45. Morphological and Electrophysiological Studies of Substantia Nigra, Tegmental Pedunculopontine Nucleus and Subthalamus in Organotypic Co-Culture 437
S.T. Kitai, N, Ichinohe, J. Rohrbacher, and B. Teng

46. Distribution of the Basal Ganglia and Cerebellar Projections to the Rodent Motor Thalamus 455
S. T. Sakai and I. Grofovà

SECTION VI. THE ACTIONS OF DOPAMINE IN THE BASAL GANGLIA

47. Effects of Dopamine Receptor Stimulation on Single Unit Activity in the Basal Ganglia 465
Judith R. Walters, David N. Ruskin, Kelly A. Allers, and Debra A. Bergstron

48. Immunocytochemical Characterization of Catecholaminergic Neurons in the Rat Striatum Following Dopamine Depleting Lesions 477
S. Totterdell and G.E. Meredith

49. Actions of Dopamine on the Rat Striatal Cholinergic Interneurons 489
Toshihiko Aosaki

50. Effect of Different Dopaminergic Agonists on the Activity of Pallidal Neurons in the Normal Monkey 499
Thomas Boraud, Erwan Bezard, Christelle Imbert, Bernard Bioulac, and Christian E. Gross

51. D_2 Dopamine Receptor-deficient Mutant Mice: New Tools to Tease Apart Receptor Subtype Electrophysiology 507
M. S. Levine, C. Cepeda, R. S. Hurst, M. A. Ariano, M. J. Low, and D. K. Grandy

52. Antipsychotic Drug-induced Muscle Rigidity and D_2 Receptor Occupancy in the Basal Ganglia of the Rat 519
Kim M. Hemsley and Ann D. Crocker

53. Differential Modulation of Single-Unit Activity in the Striatum of Freely Behaving Rats by D1 and D2 Dopamine Receptors 527
George V. Rebec and Eugene A. Kiyatkin

54. The Relationship of Dopamine Agonist-induced Rotation to Firing Rate Changes in the Basal Ganglia Output Nuclei 537
David N. Ruskin, Debra A. Bergstrom, and Judith R. Walters

55. Firing Patterns of Single Nucleus Accumbens Neurons During Intravenous Cocaine Self-Administration Sessions 547
Laura L. Peoples

SECTION VII. NEUROTRANSMITTER FUNCTIONS IN THE BASAL GANGLIA

56. Immediate Early Gene (IEG) Induction in the Basal Ganglia upon Electrical Stimulation of the Cerebral Cortex: Involvement of the MAPkinase Pathway IEG Induction upon Cortico-Striatal Stimulation 559
M.J. Besson, V. Sgambato, P. Vanhoutte, M. Rogard, C. Pages, A.M. Thierry, N. Maurice, J.M. Deniau, and J. Caboche

57. Subsynaptic Localization of Group I Metabotropic Glutamate Receptors in the Basal Ganglia 567
Yoland Smith, Maryse Paquet, Jesse E. Hanson, and George W. Hubert

58. Localization and Physiological Roles of Metabotropic Glutamate Receptors in the Indirect Pathway 581
Michael J. Marino, Stefania R. Bradley, Hazar Awad, Marion Wittmann, and P. Jeffrey Conn

59. Nicotine Affects Striatal Glutamatergic Function in 6-OHDA Lesioned Rats 589
Charles K. Meshul, Cynthia Allen, and Tom S. Kay

60. Subcellular and Subsynaptic Localization of Glutamate Transporters in the
    Monkey Basal Ganglia 599
Ali Charara, Maryse Paquet, Jeffrey D. Rothstein, and Yoland Smith

61. Glutamate-Dopamine Interactions in Striatum and Nucleus Accumbens of
    the Conscious Rat during Aging 615
F. Mora, A. Del Arco, and G. Segovia

62. The Immunohistochemical Localisation of $GABA_A$ Receptor Subunits in the
    Human Striatum 623
H.J. Waldvogel, W.M.C. van Roon-Mom, and R.L.M. Faull

63. Synaptic Localization of $GABA_A$ Receptor Subunits in the Basal Ganglia of the
    Rat 631
F. Fujiyama, M.-M. Fritschy, F.A. Stephenson, and J.P. Bolam

64. Afferent Control of Nigral Dopaminergic Neurons: The Role of GABAergic
    Inputs 641
James M. Tepper, Pau Celada, Yuji Iribe, and Carlos A. Paladini

65. The Role of Striatal Adenosine A_{2A} Receptors in Motor Control of Rats 653
Wolfgang Hauber, Jens Nagel, Partic Neuscheler, and Michael Koch

66. A Novel Neuromodulator in Basal Ganglia 661
M. Clara Sañudo-Peña

Section I

STUDIES IN HUMANS

WHAT HAS STEREOTACTIC SURGERY TAUGHT US ABOUT BASAL GANGLIA FUNCTION ?

David J. Brooks*

1. INTRODUCTION

The basal ganglia and cortex are thought to communicate via a series of anatomically and functionally segregated looped circuits which link the striatum with the frontal association areas via pallidal and thalamic nuclei (Alexander et al., 1986). A 'motor' loop consists of motor cortex projecting to dorsal putamen, and then to posteroventral globus pallidus (GP), ventrolateral thalamus, and finally supplementary motor area (SMA) which, in turn, projects back to motor cortex. A 'dorsal prefrontal' (DLPFC) loop connects this area to dorsal head of caudate, anterodorsal pallidum, and ventroanterior thalamus which then projects back to DLPFC. A 'limbic' circuit connects ventral striatum to the anterior cingulate cortex while an 'orbitofrontal' loop links this area to the ventromedial caudate, dorsomedial pallidum and anterior thalamic nuclei.

Current models for basal ganglia connectivity postulate that the striatum influences the internal portion of the globus pallidus (GPi) via distinct "direct" and "indirect" pathways (Penney, Jr. and Young, 1986; Alexander et al., 1990; Crossman, 1990). The direct striatal-GPi pathway is monosynaptic and acts to inhibit the neurons of the GPi. Loss of striatal dopamine in PD is believed to result in underactivity of this direct pathway and hence overactivity of inhibitory GPi output to ventral thalamus. In contrast, the striatal projections to lateral globus pallidus (GPe) constituting the first arm of the indirect pathway become overactive in PD. These act to inhibit GPe which in turn leads to disinhibition of the subthalamic nucleus (STN). Since this nucleus normally excites the neurones of the GPM via glutamatergic projections, disinhibition of STN in PD also leads to overactivity of GPi output projections. Thus, the overall effect of loss of striatal dopamine in Parkinson's disease is to disinhibit GPi inhibitory output to ventral thalamus via both the monosynapic "direct" and the polysynaptic "indirect" striatal-pallidal pathways.

In a review of the functional effects of thalamotomy and pallidotomy performed in patients with Parkinson's disease (PD) prior to the advent of computer-guided stereotactic approaches, Marsden and Obeso (Marsden and Obeso, 1994) recently highlighted the

---

* David J. Brooks, MRC Cyclotron Unit, Hammersmith Hospital, London, UK

apparently paradoxical functional outcomes: First, ventral thalamotomy is associated with relief of tremor and rigidity with little change in the level of akinesia of PD patients despite destruction of facilitatory projections to motor cortical areas. Second, pallidotomy dramatically improves levodopa-induced dyskinesias in PD (whereas the above models of basal ganglia connectivity would predict that dyskinesias should be worsened) but is associated with only mild improvements in bradykinesia and rigidity. These authors also noted that otherwise intact subjects who sustained bilateral pallidal lesions due to anoxic or toxic damage were generally able to function, at least superficially, at an acceptable level. The same was true for ventral thalamic stroke patients. Mink and Thatch (Mink and Thach, 1991) have reported that monkeys given injections of kainic acid or the GABA agonist muscimol into the internal pallidum show a combination of both dystonic and slowed limb movements though no apparent impairment of reaction time. Marsden and Obeso (Marsden and Obeso, 1994) felt that some of these observations could be rationalised if the basal ganglia were regarded as part of an extensive parallel processing system involved in focussing and filtering programmes that originate from cortical areas. Basal ganglia or ventral thalamic damage would then fail to interfere with automatic behaviour as adaptive processes should allow non-basal ganglia dependent cirtcuitary, such as cerebellar – lateral parietal - lateral premotor cortical loops, to be abnormally recruited. H ^{15}O PET activation data exists in support of this viewpoint, Parkinson's disease patients showing abnormally raised activation of lateral premotor and parietal areas during automatic performance of sequential finger movements (Samuel et al., 1997a). Marsden and Obeso (Marsden and Obeso, 1994) also hypothesised that the basal ganglia play a primary role in alerting the cortex to novel or rewarding circumstances allowing it to switch from one motor programme to another as appropriate. They predicted that Parkinson's disease patients would be expected to show impairment in learning new skills and be less able to react to rewarding situations and indeed this is our experience (I. Goerendt, A.D. Lawrence, and D.J. Brooks–unpublished observations). Marsden and Obeso (Marsden and Obeso, 1994), however, were unable to produce a satisfactory explanation as to why pallidotomy was so effective for relieving dyskinesias.

Since the time of Marsden and Obeso's review there have been a number of clinical trials evaluating the functional effects of thermocoagulation lesions and high-frequency electrical stimulators in Parkinson's disease placed stereotactically under computerised tomographic guidance. In this chapter the functional effects of these procedures is summarised and insight that they may provide into basal ganglia function is reviewed. Before proceeding to do this, however, four caveats need to be made:

1. Both lesion and stimulator therapeutic approaches in PD have targeted the "motor" rather than "association" areas of the basal ganglia and so conclusions concerning the higher cognitive role of the basal ganglia cannot necessarily be drawn from the clinical outcome of these patients.
2. High frequency electrical stimulators may have excitatory as well as inhibitory effects on target projections.
3. Lesion models *per se* have inherent limitations – one cannot necessarily ascribe a primary function to a structure from the deficit caused when it is ablated if it is part of a distributed parallel processing system.

4. No randomised control trials comparing the efficacy and safety of stereotactic surgery in PD with best medical practice or bilateral pallidal and STN lesioning / stimulation have, to date, been reported.

2. CLINICAL OBSERVATIONS

2.1 Unilateral Posteroventral Pallidotomy

This procedure has generally been limited to advanced PD patients with severe fluctuating responses to levodopa and associated dyskinesias who had previously enjoyed a significant and sustained response to this agent. The rationale underlying pallidotomy is that the loss of striatal dopamine in PD is believed to be associated with reduced inhibition of the medial globus pallidus (GPM) by both the direct and indirect striatal pathways. This results in an excessive inhibitory output from GPM to the ventral thalamus and in turn its excitatory projections to SMA and prefrontal cortex (Penney, Jr. and Young,1986; Alexander et al.,1990; Crossman, 1990). Intra-operative single-cell recordings in PD patients have confirmed the presence of abnormal burst firing of the GPM at rest at an increased level (Hutchison et al., 1994; Sterio et al., 1994). It is, therefore, argued that by lesioning the motor GPM this excessive inhibition is removed so facilitating volitional and sequential movements in PD.

Following unilateral posteroventral pallidotomy the positive clinical outcomes include (Samuel et al., 1998; Scott et al., 1998; Lang et al., 1997; Baron et al., 1996; Dogali et al., 1995; Ondo et al., 1998):

1. A marked (70-90%) improvement in severity of contralateral dyskinesias.
2. A more moderate (30%) improvement in contralateral rigidity and bradykinesia in the "off" state associated with a 50% decrease, often after a delay of some days, of contralateral tremor score on the Unified Parkinson's Disease Rating Scale (UPDRS).
3. A moderate (30%) improvement in the Activities of Daily Living (ADL) score on the UPDRS allowing some patients with advanced and disabling disease to regain functional independence.
4. An increase in "on" time after levodopa challenges.

These improvements in parkinsonism and drug induced dyskinesias have been maintained for at least two years but little longer term follow-up data are available.

Functions that are not improved in PD after unilateral posteroventral pallidotomy include:

1. "Off" ipsilateral bradykinesia and tremor - though rigidity improves marginally.
2. "On" contralateral and ipsilateral UPDRS limb scores.
3. Gait and freezing – indeed it is the impression of some centres that freezing continues to deteriorate despite the improvement in dyskinesias.
4. Bulbar dysfunction - such as pallalalia and dysphagia.
5. It is usually not possible to reduce the levodopa requirement – indeed abolition of dyskinesias allows some patients to succesfully increase their daily intake.

It has also been noted that those parkinsonian patients thought to have atypical disease with a poor initial levodopa response do not respond well to pallidotomy. This may reflect the presence of superadded striatal or pallidal degeneration.

2.2 Bilateral Posteroventral Pallidal Stimulation

A major difficulty with pallidotomy is that the operative risk of bulbar dysfunction (worsening dysarthria and dysphagia) increases significantly to around 10% when bilateral procedures are attempted (Hariz and DeSalles, 1997). Some PD patients have also noted impairment of short term recall. For this reason bilateral posteroventral pallidal high frequency electrical stimulation was developed in order to induce a more selective pallidal conduction block. Again, patients selected for surgery have invariably had advanced disease with associated motor fluctuations and dopa dyskinesias.

The benefits from bilateral posteroventral pallidal stimulation or unilateral pallidotomy with contralateral pallidal stimulation include (Ghika et al., 1998; Galvez-Jiminez et al., 1998; Krack et al., 1998; Bejjani et al., 1997):

1. Bilateral near abolition of levodopa associated dyskinesias.
2. Bilateral 30% improvement in "off" bradykinesia, rigidity, and tremor UPDRS scores
3. Improved gait and freezing in "off"
4. Improved activities of daily living
5. Increased "on" time after levodopa

Comparisons concerning the relative effects of unilateral and bilateral pallidal stimulation have not been reported although unilateral stimulation appears to produce similar effects to unilateral pallidotomy. As with pallidotomy, the follow up is short and limited to 1-2 years.

Functions that have been reported not to respond to bilateral pallidal stimulation include:

1. "On" bradykinesia and tremor (rigidity improves marginally)
2. "On" UPDRS limb, gait, and ADL scores
3. "On" freezing
4. Speech and swallowing
5. Levodopa requirement – this may increase over time.

As pallidal stimulation affects a far smaller volume of tissue than pallidotomy it has been possible to investigate the relative functional effects of stimulating ventrally and dorsally (Krack et al., 1998; Bejjani et al., 1997). Ventral stimulation appears to be most effective for abolishing levodopa-induced dyskinesias and improving rigidity in the "off" state but has only a marginal effect on bradykinesia and impairs relief of bradykinesia by levodopa. In contrast, more dorsal stimulation is effective for improving bradykinesia but can cause levodopa-induced dyskinesias to worsen. As posteroventral pallidotomy lesions both ventral and more dorsal territories it improves both bradykinesia and dyskinesia though, being ventrally placed, is most effective for the latter. As with pallidotomy, bilateral pallidal stimulation does not appear to help parkinsonian patients with atypical syndromes.

2.3 Bilateral Subthalamic Nucleus (STN) Stimulation

While the excessive inhibitory output of the GPi in PD can be directly reduced by medial pallidotomy, an alternative approach is to reduce the excitatory input to GPi from the subthalamic nucleus via high-frequency electrical stimulation (Limousin et al., 1995; Limousin et al., 1998; Kumar et al., 1998). As with pallidotomy, to date this procedure has been restricted to advanced PD patients with severe motor fluctuations with or without dyskinesias who had previously enjoyed a significant and sustained response to levodopa.

Functions that have been reported to respond to STN stimulation include:

1. "Off" bradykinesia, rigidity, and tremor by 60-80% on the UPDRS
2. "Off" dystonia, gait, and freezing
3. Activities of daily living
4. Levodopa reqirement halved
5. Dyskinesias moderately improve probably secondary to levodopa reduction. If levodopa levels are left unchanged they may increase in severity.
6. "On" time increases by 70%

Patients have been followed for up to 3 years and these improvements appear to be stable though some centers have noted worsening of freezing.

Functions that do not improve after bilateral STN stimulation include:

1. "On" bradykinesia and activities of daily living
2. "On" freezing

Interestingly, some patients complain that if they stop taking levodopa when their stimulators are working they lose the euphoria associated with being "on" despite their improved mobility when "off". Recently a short report suggesting that bilateral subthalamotomy can be performed safely in PD and has similar efficacy to STN stimulation has appeared (Gill and Heywood, 1997).

3. FUNCTIONAL IMAGING AND STEREOTACTIC SURGERY

3.1 Pallidotomy

There have been three PET reports on the effects of medial pallidotomy in PD on regional brain function. The first studied changes in resting glucose metabolism with [18]FDG PET before and after pallidotomy in eight PD patients (Eidelberg et al., 1996). Post-operatively, pallidotomy was found to have increased the resting metabolism of primary motor cortex, lateral premotor cortex, and dorsolateral prefrontal cortex, but surprisingly not that of SMA. Preoperative levels of lentiform nucleus metabolism correlated with post-operative clinical outcome. Principal component analysis showed that pre-operatively the PD patient group had an abnormal pattern of relatively increased lentiform nucleus and reduced premotor resting metabolism. Medial pallidotomy acted to normalise this aberrant pattern. The degree of change in expression of this pattern of covariance post-operatively correlated with the improvement in clinical disability. Surprisingly, post-operative clinical outcome and resting metabolic changes did not correlate with the size of the pallidotomy lesion as measured volumetrically with MRI.

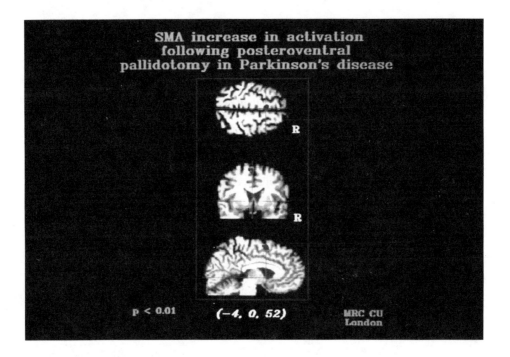

Figure 1a. MRI surface renderings of the cerebral hemispheres in standard stereotactic space for a group of six PD patients before and after right medial pallidotomy with the PET activation rCBF increases superimposed. The darker gray areas show voxels in the rostral SMA and dorsal prefrontal cortex that are significantly more activated (p<0.001) after pallidotomy while patients perform paced joystick movements in freely chosen directions. (Picture courtesy of M. Samuel)

$H_2^{15}O$ PET activation studies with PD patients scanned after cessation of dopaminergic medication for twelve hours have shown that the supplementary motor area (SMA) and dorsolateral prefrontal cortex (DLPFC) blood flow increases are both attenuated during performance of paced joystick movements in freely chosen directions (Playford et al., 1992). This impairment of SMA and DLPFC function in PD is postulated to be a consequence of excessive inhibitory output from GPi to thalamus and frontal cortex secondary to the loss of striatal dopamine (Penney, Jr. and Young, 1986; Alexander et al., 1990; Crossman, 1990). An

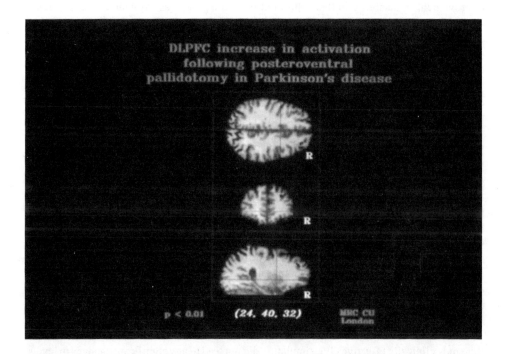

Figure 1b. MRI surface renderings of the cerebral hemispheres in standard stereotactic space for a group of six PD patients before and after right medial pallidotomy with the PET activation rCBF increases superimposed. The darker gray areas show voxels in the rostral SMA and dorsal prefrontal cortex that are significantly more activated (p<0.001) after pallidotomy while patients perform paced joystick movements in freely chosen directions. (Picture courtesy of M. Samuel)

inability to normally activate SMA and DLPFC in PD may well explain the difficulties that these patients experience when preparing and planning volitional movements or performing sequences of actions (Benecke et al., 1987; Dick et al., 1989). A PET activation study from our unit reported significantly increased activation of SMA, lateral premotor cortex, and dorsal prefrontal cortex in a single PD case while off medication after undergoing unilateral pallidotomy (Ceballos-Baumann et al., 1994). This patient was studied while performing paced joystick movements in freely selected directions and post-operatively showed an improvement in both response time and the number of completed joystick movements. This study was subsequently extended to involve a group of six PD patients (Samuel et al., 1997b). As a group, these patients showed significant improvements in dyskinesia score (75%), "off" contralateral wrist rigidity (83%) and contralateral bradykinesia (56%) when assessed with the UPDRS after pallidotomy. There were associated significant increases in levels of SMA and right DLPFC activation on joystick movement (Fig. 1a and 1b).

A second PET activation pallidotomy study involved six PD patients who were scanned off medication at rest and then while reaching at three second intervals out to grasp different lighted targets arranged in a row (Grafton et al., 1995). Pallidotomy resulted in no overall improvement in disability in this cohort but the motor task was associated with increased levels of caudal SMA and ventral lateral premotor activation after surgery. Despite the lack of clinical improvement, levels of SMA and lateral premotor activation before and after pallidotomy correlated with levels of performance in this patient group. As this was an externally cued reaching task it was not designed to activate prefrontal areas.

3.2 Pallidal Stimulation

There have been two PET reports on the effects of high freqency electrical pallidal stimulation on regional brain function. In the first study (Davis et al., 1997) levels of resting rCBF were measured with the stimulator switched off, switched on at a sub-effective intensity, and switched on at a clinically effective intensity. Effective posteroventral GPi stimulation improved contralateral bradykinesia and rigidity in eight of the nine PD patients when they were assessed off medication and this was associated with increased resting SMA and putamen / external pallidal rCBF. Stimulation of the GPi at a lower intensity did not lead to clinical improvement or increase SMA rCBF. The authors concluded that decreased akinesia in PD following pallidal stimulation resulted from increased SMA activation though their study design was strictly resting rather than activation. They felt that the increased levels of putamen / external pallidal rCBF during stimulation reflected antidromic conduction. The second PET study measured changes in activation associated with moving a joystick in freely chosen directions in six PD patients while off medication with the GPi stimulator switched off and then again when the stimulator was switched on (Limousin et al., 1997). In this study clinically effective GPi stimulation did not lead to any significant changes in levels of SMA or DLPFC activation.

3.3 Subthalamic Stimulation

To date, there have been two reports of H ^{15}O PET activation fingings in PD patients before and after STN stimulation. In both studies levels of rCBF were measured in patients with PD when off medication during performance of paced joystick movements in freely selected directions. The first study (Ceballos-Baumann et al., 1997) reported relative increases in activation of rostral SMA lateral premotor, and dorsolateral prefrontal cortex during movement when the STN stimulator was switched on in a group of seven PD patients, findings similar to that reported following pallidotomy. In contrast to pallidotomy, however, STN stimulation was also associated with reduced motor cortex activation and led to increases in resting thalamic and decreases in resting motor cortex and caudal SMA rCBF. The second study (Limousin et al., 1997) also found that STN stimulation in their six PD cases led to increased SMA and DLPFC and decreased motor cortex activation. The mechanism causing the decreased motor cortex activation remains unclear.

4. SYNTHESIS

Using an elegant approach with attenuated neurotropic viruses to track anterograde and retrograde basal ganglia – cortical connectivity, Middleton and Strick (Middleton,1997) have been able to demonstrate that the most posteroventral Gpi projects mainly to ventral premotor (PMC) and sensorimotor cortex (SMC) via the ventrolateral thalmic nuclei whereas those GPi neurons situated more anterodorsally project mainly to SMA. Posteroventral pallidotomy would be expected to lesion all these projections given the 1 cm size of the lesion whereas posteroventral pallidal stimulation could be targeted to one set of projections or the other. Given this, and the observation that posteroventral GPi stimulation predominantly relieves levodopa-induced dyskinesias while more anterodorsal stimulation relieves akinesia and can worsen dyskinesia (Krack et al., 1998), it has been hypothesised that dyskinesias may be mediated via overactivity of posteroventral GPi – SMC projections while akinesia results from underactivity of more anterodorsal GPi – SMA projections.

There is some functional imaging evidence in support of this hypothesis: Samuel and co-workers found that during performance of paced joystick movements in freely chosen directions with eyes closed posteroventral pallidotomy improved SMA activation in PD patients when "off" along with akinesia (Samuel et al., 1997b). DLPFC activation also unexpectedly increased suggesting that association pallidal projections may have been affected along with motor pallidum by the surgery. The task in Samuel's study was not reliant on visual cueing. When such a reaching task was employed by Grafton et al. pallidotomy also improved levels of lateral as well as mesial premotor function during performance (Grafton et al., 1995).

More recently Piccini and co-workers have examined the functional correlates of dyskinesias (Piccini et al., 1997). They compared patterns of resting regional cerebral blood flow (rCBF) in PD patients after receiving levodopa before and after focal limb dyskinesias occurred. Dyskinesias were indeed associated with relatively increased lentiform nucleus and lateral premotor cortex rCBF, however, mesial premotor rCBF was also increased.

It would seem, therefore, that the attractive model proposed above to explain the selective relief of dyskinesias and akinesia by posteroventral and more anterodorsal pallidal stimulation, respectively, may be over simplistic. Functional imaging findings suggest that both lateral and mesial premotor area underfunction is relevant to akinesia and overactivity to dyskinesia generation.

Compared with pallidotomy / pallidal stimulation, bilateral STN stimulation appears to be more effective against parkinsonism but primarily relieves dyskinesias via a levodopa sparing effect. It also appears to be more effective for axial symptoms such as festination of gait. Neither bilateral pallidal or STN stimulation helps freezing while in the "on" condition.

One explanation for the dramatic effects of STN stimulation on akinesia may be provided by Parent who has reported that the STN sends facilitatory projections predominantly to dorsal rather than ventral GPi. Stimulating motor STN would, therefore, preferentially inhibit dorsal GPi – SMA projections relieving akinesia but potentially worsen levodopa-induced dyskinesias. In contrast, posteroventral pallidotomy or pallidal stimulation primarily targets the dyskinesia generating posteroventral GPi – SMC projections.

The STN also sends direct projections to the pedunculopontine nuclei (PPN) and the substantia nigra reticulata (SNr). The PPN is thought to be responsible for mediating postural stability and lesions of this nucleus result in impairment of gait in primates. The SNr mediates coordination of eye and head position. Given this, it is reasonable to postulate that the superior effect of STN relative to pallidal stimulation on axial symptoms such as gait festination relates to direct facilitation of PPN and SNr output which has become impaired in the presence of dopamine depletion.

5. CONCLUSIONS

1. The basal ganglia receive widespread cortical input and probably act to filter and focus programmes arising from these areas. As they are part of a distibuted parallel processing system the brain can adapt to the presence of pallidal lesions in otherwise intact subjects so leaving automatic performance unimpaired. Subjects with pallidal lesions, however, would be predicted to show impaired skill acquisition and responses to rewarding situations.
2. The effects of stereotactic pallidal stimulation in PD suggest that levodopa-induced dyskinesias may be mediated via overactivity of posteroventral GPi – SMC projections while akinesia results from underactivity of more anterodorsal GPi – SMA projections. Functional imaging studies, however, suggest that this division may be over simplistic.
3. Bilateral STN stimulation is more effective than posteroventral pallidal stimulation for relieving akinesia as STN neurons project mainly to the dorsal GPi. Conversely, bilateral posteroventral pallidal stimulation is most effective for amelioration of levodopa-induced dyskinesias.
4. The STN sends direct projections to GPi, the pedunculopontine nuclei, and the SNr whereas GPi projects to ventral thalamus. As the PPN and SNr play a major role in controlling posture and gait bilateral STN stimulation has a more dramatic effect on axial symptoms in PD than posteroventral pallidotomy or pallidal stimulation.

6. REFERENCES

Alexander, G.E., Crutcher, M.D., and Delong, M.R., 1990, Basal ganglia thalamo-cortical circuits: parallel substrates for motor, oculomotor, "prefrontal" and "limbic" functions. *Progress in Brain Research* **85**:119-146.

Alexander, G.E., Delong, M.R., and Strick, P.L. 1986, Parallel organization of functionally segregated circuits linking basal ganglia and cortex. *Ann. Rev. Neurosci.* **9**:357-381.

Baron, M.S., Vitek, J.L., Bakay, R.A.E. *et al.*, 1996, Treatment of advanced Parkinson's disease by posterior GPi pallidotomy: 1-year results of a pilot study. *Ann.Neurol.* **40**:355-366.

Bejjani, B., Damier, P., Arnulf, I. *et al.*, 1987, Pallidal stimulation for Parkinson's disease. Two targets? *Neurology* **49**:1564-1569.

Benecke, R., Rothwell, J.C., Dick, J.P., Day, B.L., and Marsden, C.D. 1987, Disturbance of sequential movements in patients with Parkinson's disease. *Brain* **110**:361-379.

Ceballos-Baumann, A.O., Bartenstein, P., Von Falkenhayn, I. *et al.*, 1997, Parkinson's disease ON and OFF subthalamic nucleus stimulation: A PET activation study. *Neurology* **48** Supp 2:A250.

Ceballos-Baumann, A.O., Obeso, J.A., Delong, M.R. et al., 1994, Functional reafferentation of striatal-frontal connections after posteroventral pallidotomy in Parkinson's disease. *Lancet* **344**:814.

Crossman, A.R., 1990, A hypothesis on the pathophysiological mechanisms that underlie levodopa- or dopamine agonist-induced dyskinesia in Parkinson's disease: implications for future strategies in treatment. *Movement Disorders* **5**:100-108.

Davis, K.D., Taub, E., Houle, S. et al., 1997, Globus pallidus stimulation activates the cortical motor system during alleviation of parkinsonian symptoms. *Nature Medicine* **3**:671-674.

Dick, J.P.R., Rothwell, J.C., Day, B.L. et al., 1989, The Bereitschaftspotential is abnormal in Parkinson's Disease. *Brain* **112**:233-244.

Dogali, M., Fazzini, E., Kolodny, E. et al., 1995, Stereotaxic ventral pallidotomy for Parkinson's disease. *Neurology* **45**:753-761.

Eidelberg, D., Moeller, J.R., Ishikawa, T. et al., 1996, Regional metabolic correlates of surgical outcome following unilateral pallidotomy for Parkinson's disease. *Ann.Neurol.* **39**:450-459.

Galvez-Jiminez, N., Lozano, A., Tasker, R., Duff, J., Hutchison, W., and Lang, A.E., 1998, Pallidal stimulation in Parkinson's disease patients with a prior unilateral pallidotomy. *Can.J.Neurol.Sci.* **25**:300-305.

Ghika, J., Villemure, J.G., Fankhauser, H., Favre, J., Assal, G., and Ghika-Schmid, F., 1998, Efficiency and safety of bilateral contemporaneous pallidal stimulation (deep brain stimulation) in levodopa-responsive patients with Parkinson's disease with severe motor fluctuations: a 2-year follow-up review. *J.Neurosurg.* **89**:713-7.18.

Gill, S.S., and Heywood, P., 1997, Bilateral dorsolateral subthalamotomy for advanced Parkinson's disease. *Lancet* **350**:1224.

Grafton, S.T., Waters, C., Sutton, J., Lew, M.F., and Couldwell, W., 1995, Pallidotomy increases activity of motor association cortex in Parkinson's disease - a positron emission tomographic study. *Ann.Neurol.* **37**:776-783.

Hariz, M.I., and DeSalles, A.A., 1997, The side effects and complications of posteroventral pallidotomy. *Acta Neurochir. Suppl. (Wien)* **68**:42-48.

Hutchison, W.D., Lozano, A.M., Davis, K.D., Saint Cyr, J.A., Lang, A.E., and Dostrovsky, J.O., 1994, Differential neuronal activity in segments of globus pallidus in Parkinson's disease patients. *NeuroReport* **5**:1533-1537.

Krack, P., Pollak, P., Limousin, P. et al., 1998, Opposite motor effects of pallidal stimulation in Parkinson's disease. *Ann.Neurol.* **43**: 180-192.

Kumar, R., Lozano, A.M., Kim, Y.J. et al., 1998, Double-blind evaluation of subthalamic nucleus deep brain stimulation in advanced Parkinson's disease. *Neurology* **51**: 850-855.

Lang, A.E., Lozano, A.M., Montgomery, E., Duff, J., Tasker, R., and Hutchinson, W., 1997, Posteroventral medial pallidotomy in advanced Parkinson's disease. *New Eng.J.Med.* **337**:1036-1042.

Limousin, P., Greene, J., Polak, P., Rothwell, J.C., Benabid, A.L., and Frackowiak, R.S.J., 1997, Positron emission tomography (PET) study of modulation of cerebral activity by subthalamic nucleus (STN) and internal globus pallidus (GPi) stimulation in Parkinson's disease. *Neurology* **48** Supp 2:A249.

Limousin, P., Krack, P., Pollak, P. et al., 1998, Electrical stimulation of the subthalamic nucleus in advanced Parkinson's disease. *N.Eng.J.Med.* **339**:1105-1111.

Limousin, P., Pollak, P., Benazzouz, A., et al., 1995, Effect on parkinsonian signs and symptoms of bilateral subthalamic nucleus stimulation. *Lancet* **345**:91-95.

Marsden, C.D., and Obeso, J.A., 1994, The functions of the basal ganglia and the paradox of stereotactic surgery in Parkinson's disease. *Brain* **117**:877-897.

Middleton, F.A.S.P.L., 1997, New concepts about the organization of basal ganglia output. *Adv.Neurol.* **74**:57-68.

Mink, J.W., and Thach, W.T., 1991, Basal ganglia motor control. III. Pallidal ablation: Normal reaction time, muscle cocontraction, and slow movement. *J.Neurophysiol.* **65**:330-351.

Ondo, W.G., Jankovic, J., Lai, E.C. *et al.*, 1998, Assessment of motor function after stereotactic pallidotomy. *Neurology* **50**:266-270.

Penney, Jr., J.B. and Young, A.B., 1986, Striatal inhomogeneities and basal ganglia function. *Movement Disorders* **1**:3-15.

Piccini, P., Boecker, H., Weeks, R.A., and Brooks, D.J., 1997, Dyskinesia correlated regional blood flow changes in Parkinson's disease. *Neurology* **48** Supp 2:A327.

Playford, E.D., Jenkins, I.H., Passingham, R.E., Nutt, J., Frackowiak, R.S.J., and Brooks, D.J., 1992, Impaired mesial frontal and putamen activation in Parkinson's disease: a PET study. *Ann.Neurol.* **32**:151-161.

Samuel, M., Caputo, E., Brooks, D.J. *et al.*, 1998, A study of medial pallidotomy for Parkinson's disease: Clinical outcome, MRI location and complications. *Brain* **121**:59-75.

Samuel, M., Ceballos-Baumann, A.O., Blin, J., Uema, T., Boecker, H., and Brooks, D.J., 1997a, Evidence for lateral premotor and parietal overactivity in Parkinson's disease during sequential and bimanual movements: A PET study. *Brain* **120**: 963-976.

Samuel, M., Ceballos-Baumann, A.O., Turjanski, N. *et al.*, 1997b, Pallidotomy in Parkinson's disease increases SMA and prefrontal activation during performance of volitional movements: An $H_2^{15}O$ PET study. *Brain* **120**:1301-1313.

Scott, R., Gregory, R., Hines, N. *et al.*, 1998, Neuropsychological, neurological, and functional outcome following pallidotomy for Parkinson's disease: A consecutive series of eight simultaneous and twelve unilateral procedures. *Brain*, **121**:659-671.

Sterio, D., Beric, A., Dogali, M., Fazzini, E., Alfaro, G., and Devinsky, O., 1994, Neurophysiological properties of pallidal neurons in Parkinson's disease. *Ann.Neurol.* **35**:586-591.

PATHOPHYSIOLOGY OF THE INTERNAL SEGMENT OF THE GLOBUS PALLIDUS IN PARKINSON'S DISEASE

What have we learned from surgery?

Andres M. Lozano, William D. Hutchison, Ron Levy, Anthony E. Lang, and Jonathan O. Dostrovsky*

Key words: Parkinson's disease, pallidotomy, deep brain stimulation, apomorphine

Abstract: Although the internal segment of the globus pallidus (GPi) has been a neurosurgical target for some 50 years, only recently has a detailed analysis of the clinical effects of interventions in this region been made. This is largely the result of new surgical options for the large number of patients that remain poorly managed after many years of anti-parkinson medication and advances in stereotactic functional neurosurgery including imaging. A short review is given of microelectrode techniques used for targeting the GPi for pallidotomy and deep brain stimulation (DBS) procedures. Analysis of neuronal firing rates suggest that GPi outflow is hyperactive in unmedicated PD patients. This is further supported by observations that the administration of apomorphine reduces the firing rate of single neurons recorded in the GPi of PD patients. Finally, an overview is provided of the clinical outcome of these procedures and what they suggest about the pathophysiology of GPi and its role in PD symptomatology.

1. INTRODUCTION

The globus pallidus has been used as a target for surgical intervention for Parkinson's disease (PD) for approximately half a century.[1,2] Its selection as a target was at first empirical,

---

* A.M. Lozano and W. D. Hutchison, Division of Neurosurgery, The Toronto Western Hospital and Department of Surgery, University of Toronto; A.E. Lang, Division of Neurology, The Toronto Western Hospital;. A.E. Lang, Departmen of, Medicine; R. Levy and J.O. Dostrovsky, Department of Physiology, University of Toronto, Ontario, Canada

based on surgical trials that observed the clinical effects of lesions in a variety of cortical and subcortical targets in patients with movement disorders. There is now, however, a strong rationale for surgical intervention in the globus pallidus in patients with PD who continue to be disabled despite available medications. This has come about largely through observations in experimental animal models of parkinsonism. Microelectrode recordings in these animals have shown that the parkinsonian state is associated with a large increase in the spontaneous rate of discharge in the internal segment of the globus pallidus (GPi) and the subthalamic nucleus (STN).[3,4] Current models of the pathophysiology of disease suggest that in the dopamine deficiency state, the enhanced driving from the overactive STN neurons coupled with a diminution in direct striatal inhibitory influences contribute to the heightened activity in GPi.[5] The increased GABAergic output from GPi is believed to interfere with the function of brainstem and thalamocortical motor pathways leading to the disruption and poverty of movement, which characterizes parkinsonism. The surgical strategy is to eliminate the abnormal activity in GPi, either through lesioning (pallidotomy) or the application of chronic electrical deep brain stimulation (GPi-DBS). Here we review some of the physiological attributes of pallidal neurons recorded in PD patients undergoing stereotactic neurosurgery including their response to the dopamine agonist apomorphine and briefly discuss the effects of surgery.

2. RECORDINGS FROM PALLIDUM IN PATIENTS WITH PD

2.1 GPe

The patterns of neuronal discharge in the pallidal complex are shown in Figure 1. In our stereotactic procedures recordings usually start within the external nucleus of the globus pallidus (GPe).[6] In GPe there are cells with large amplitude spikes and tonic ongoing activity. Two 'signature' types have been identified in GPe based on both monkey and human neuronal recordings.[7] The first type is commonly encountered and these neurons have a slow frequency of discharge (about 30-50 Hz) interspersed with pauses termed SFD-P. The second type comprises a smaller group of neurons and they have lower discharge frequencies (10-20Hz) and bursts, or LFD-B neurons. There is clearly a range of firing rates in GPe and not all neurons can be classified neatly into these two groups, but the identification of these types is characteristic of the GPe. Neurons in GPe may also respond to limb or orofacial movements, and these responses can be excitatory or inhibitory.

2.2 White Matter Laminae and Border Cells

Between GPe and GPi is a white matter lamina carrying fibres projecting to and from pallidum. The recordings in this region reveal few if any action potentials and can thus be distinguished from the cellular regions above and below. At the margins of this region with the internal and external segments are cells with regular discharges at 20-40 Hz with longer spike durations (compared to pallidal cells) referred to as border cells. Border cells (Bor in Fig. l) are thought to be cholinergic and to have similar features to those of the cortically-projecting neurons of the nucleus basalis of Meynert.[8] Another lamina exists within GPi and

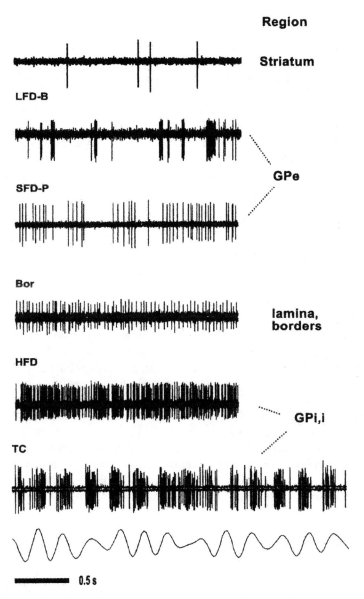

Figure 1. Recordings of identified cell types in electrode trajectories targeting the GPi in surgery for PD. LFD-B, low frequency discharge with bursts; SFD-P, slow frequency discharge with pauses; Bor, border cell; HFD, high frequency discharge. Below the tremor cell (TC) record is an accelerometer trace from the back of the contralateral hand. Reproduced with permission from Ref. 11.

is referred to as the lamina incompleta, which divides this nucleus into GPi,e and GPi,i and can also contain border cells.

2.3 GPi

Neuronal action potentials in GPi have larger amplitudes and higher spontaneous firing rates than those in GPe. They discharge in an irregular fashion in the range of 60-100 Hz, and are referred to as high frequency discharge neurons, or HFD.[7] In patients with tremor, neurons may be found with periodic oscillations in firing rate at the same frequency as the tremor. These are colloquially referred to as "tremor cells" (TC in Fig. 1). About 20% of neurons examined alter their firing rate in response to passive or active movements of limbs or orofacial structures.[6] Responses to voluntary movements are frequently a brief pause in activity at the start of a movement sequence or inhibition during the whole duration of a movement sequence.[9] Below the GPi the neural recordings become quiet again as the electrode tip passes out of the cellular region.

2.4 Optic Tract and Internal Capsule

Injury to the optic tract and internal capsule, which lie adjacent to GPi, must be avoided. In most cases, an increase in noise in the high frequency band can be discerned signalling the entry into optic tract.[6] Passing electrical current through the electrode at this point normally evokes visual sensations known as phosphenes. These are commonly referred to as stars, or bright lights and occur in a band or wedge shaped portion of the contralateral visual field. More rarely scotomata are reported. It is also good routine practice to record the background axonal responses to a Strobe light. The latency of these flash-evoked potentials is in the range of 35-45 ms. This technique can be very useful in rare instances where phosphenes are not reported by the patient with electrical stimulation at locations that are nonetheless suspected to be within the optic tract. This may be a technical problem if there is a "shorting-out" of the high voltage required for stimulation and insufficient current density to stimulate the optic tract.

Posterior and medial to the pallidum is the internal capsule, which is always identified using microstimulation in our intra-operative mapping sessions. Currents from ≤ 100 µA delivered through the electrode tip produce tetanic contraction of a body part. Microstimulation of capsule may also produce arrest or reduction of tremor.

3. APOMORPHINE STUDIES

The basal firing rate for GPi neurons in non-human primates is approximately 60-70Hz.[4] The rate of firing of GPi neurons in MPTP treated parkinsonian animals was found to be elevated by approximately 22%.[4] The average firing rate of neurons in ventral GPi of patients with PD is about 83 Hz, which is very similar to that found in MPTP monkeys.[7] Because of the lack of a control population, we do not know if they are increased relative to normal humans or whether decreasing the GPi discharge rate can improve parkinsonian symptoms. In experimental animals, the administration of the non-selective dopamine agonist apomorphine has been shown to reduce the firing rate of GPi neurons.[10] We injected apomorphine (2-4 mg)

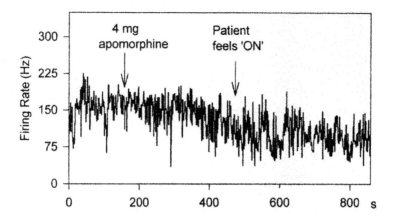

Figure 2. Apomorphine decreases the firing rate of a GPi neuron in a PD patient

subcutaneously into patients off their usual anti-parkinsonian medications while they were undergoing pallidal procedures to test whether this influenced GPi neuronal activity. Apomorphine produced improvement in the parkinsonism and was associated with a striking reduction in the spontaneous discharge rate of GPi neurons in the PD patients. Firing rates diminished approximately 50% with a latency of approximately 5 minutes and a duration of 30 minutes (see Fig. 2). These findings suggest that GPi neuronal firing rates are indeed elevated in unmedicated PD patients and that a reduction in GPi firing with dopaminergic drugs is a likely mechanism of the anti-parkinsonian action.

4. EFFECTS OF SURGERY

Pallidotomy and pallidal DBS both improve the major clinical manifestations of PD. However the location of the lesion or chronic DBS within GPi has profound differential effect on symptoms. In an analysis of 33 patients having high resolution 3 dimensional magnetic resonance imaging reconstructions of their pallidotomy lesions, Gross et al.[11] found that the elimination of L-dopa-induced dyskinesia and the decrease in rigidity was seen after anteromedial GPi lesions. Bradykinesia, rigidity, akinesia, gait and posture and drug-induced dyskinesias improved with central lesions and tremor was most effectively treated at the posterolateral region of GPi. Further, differential effects of pallidal stimulation are obtained depending on the position of the stimulating electrode in the pallidal complex. Two groups have reported that electrical stimulation through the most ventral contacts of quadripolar DBS electrodes eliminates dyskinesias and produces an anti-L-dopa effect with respect to bradykinesia but not rigidity. Electrical stimulation through the dorsal contacts (in possibly dorsal GPi or ventral GPe) had a pro-dyskinetic effect with improvements in parkinsonism (i.e., effects similar to L-dopa).[12,13] The results were explained by different output projections through the ansa lenticularis and thalamic fasciculus. A pro-dyskinetic effect of electrical stimulation in GPe is consistent with observations of dyskinesias in normal monkeys following focal inhibition with micro-injections of muscimol.[14]

These observations are consistent with the hypothesis that different regions in GPi subserve specific parkinsonian motor symptoms. This is not surprising given the segregated nature of basal ganglia circuits and the neuroanatomical evidence that sub-territories within GPi project through a thalamic relay to influence separate and distinct cortical fields. Anatomical studies in primates[15] have revealed that the dorsal motor GPi relates to the supplementary motor cortex while the central portion of the motor GPi projects to primary motor cortex and the ventral motor GPi area to premotor cortex.

5. MECHANISM OF IMPROVEMENT

Although the exact mechanism through which destruction or electrical stimulation of GPi leads to improvement in the major manifestations of PD is not clear, reduction of the hyperactive basal ganglia output of GPi appears to be involved. A perplexing observation is that lesioning or chronic electrical stimulation of GPi (and indeed the subthalamic nucleus and the motor thalamus) is devoid in the majority of cases, of any overt disturbances in motor function. Recent positron emission tomography (PET) studies suggest that the depressed activity in the supplementary motor area of the cortex in PD patients is reversed when the akinesia is treated with pallidotomy or pallidal DBS.[16,17] These functional imaging data together with the observation of the similar nature and magnitude of improvement with lesioning or stimulating GPi [18,19] suggest that the two procedures may have a similar mechanism. A number of pathological patterns of activity of pallidal neurons are now being described in patients with chorea, ballism, and dystonia[2,6,20] and in patients experiencing dyskinesias after the administration of dopaminergic agents. These conditions respond to pallidotomy or GPi DBS. This suggests that the suppression of an abnormal basal ganglia output in the pathological GPi resulting from pallidotomy or pallidal DBS restores motor function that is preferable to that which occurs under the influence of the disrupted signals generated in a diseased GPi.

6. REFERENCES

1. Meyers, R., 1942, The modification of alternating tremors, rigidity and festination by surgery of the basal ganglia. *A.Research Nerv.& Ment.Dis.* **20**:602-665.
2. Meyers, R., 1942, Surgical interruption of the pallidofugal fibres. Its effect on the syndrome of paralysis agitans and technical considerations in its application. *N.Y.State J.Med.* **42**:317-325.
3. Bergman, H., Wichmann, T., Karmon, B., et al., 1994, The primate subthalamic nucleus. II. Neuronal activity in the MPTP model of parkinsonism. *J.Neurophysiol.* **72**:507-520.
4. Filion, M., and Tremblay, L., 1991, Abnormal spontaneous activity of globus pallidus neurons in monkeys with MPTP-induced parkinsonism. *Brain Res.* **547**:142-151.
5. DeLong, M.R., 1990, Primate models of movement disorders of basal ganglia origin. *Trends Neurosci.* **13**:281-285.
6. Lozano, A.M., Hutchison, W.D., Kiss, Z.H.T., et al., 1996, Methods for microelectrode-guided posteroventral pallidotomy. *J.Neurosurg.* **84**:194-202
7. Hutchison, W.D., Lozano, C.A., Davis, K.D., et al., 1994, Differential neuronal activity in segments of globus pallidus in Parkinson's disease patients. *Neuroreport* **5**:1533-1537.
8. DeLong, M.R.. 1971, Activity of pallidal neurons during movement. *J.Neurophysiol.* **34**:414-427.

9. Hutchison, W.D., Lozano, A.M., Lang, A.E., et al., 1996, Cognitive modulation of globus pallidus (GP) neurons in patients with Parkinson's disease (PD). *Soc.Neurosci.Abstr.* **22**: 414.
10. Filion, M., Tremblay, L., and Bédard, P.J., 1991, Effects of dopamine agonists on the spontaneous activity of globus pallidus neurons in monkeys with MPTP-induced parkinsonism. *Brain Res.* **547**:152-161.
11. Gross, R.E., Lombardi, W.J., Lang, A.E., et al., 1999, Relationship of lesion location to clinical outcome following microelectrode-guided pallidotomy for Parkinson's disease. *Brain* **122**:405-416.
12. Bejjani, B.P., Damier, P., Arnulf, I., et al., 1998, Deep brain stimulation in Parkinson's disease: opposite effects of stimulation in the pallidum. *Mov Disord.* **13**:969-970.
13. Hutchison, W.D., 1998, Microelectrode techniques and findings of globus pallidus, in Krauss, J.K., Grossman, R.G., Jankovic,, J. (eds): Pallidal surgery for the treatment of Parkinson's disease and movement disorders. Philadelphia: Lippincott-Raven, pp. 135-152.
14. Lozano, A.M., Lang,A.E., Galvez-Jimenez, N., et al., 1995, Effect of GPi pallidotomy on motor function in Parkinson's disease. *Lancet* **346**:1383-1387.
15. Hoover, J.E., and Strick, P.L., 1993, Multiple output channels in the basal ganglia. *Science* **259**:819-821.
16. Ceballos-Baumann, A.O., Obeso, J.A., Vitek, J.L., et al., 1994, Restoration of thalamocortical activity after posteroventral pallidotomy in Parkinson's disease [letter]. *Lancet* **344** :814.
17. Davis, K.D., Taub, E., Houle, S., et al., 1997, Globus pallidus stimulation activates the cortical motor system during allieviation of parkinsonian symptoms. *Nature Medicine* **3**:671-674.
18. Krack, P., Pollak, P., Limousin, P., et al., 1998, Opposite motor effects of pallidal stimulation in Parkinson's disease. *Ann.Neurol.* **43**:180-192.
19. Lozano, A.M., Kumar, R., Gross, R.E., et al., 1997, Globus pallidus internus pallidotomy for generalized dystonia. *Mov.Dis.* **12**:865-870.
20. Vitek, J.L., Chockkan, V., Zhang, J.Y., et al., 1999, Neuronal activity in the basal ganglia in patients with generalized dystonia and hemiballismus. *Ann.Neurol.* **46**:22-35.

APOMORPHINE RESPONSES IN GLOBUS PALLIDUS AND SUBTHALAMUS OF PARKINSONIAN PATIENTS UNDERGOING STEREOTAXIC NEUROSURGERY: LACK OF EVIDENCE FOR THE SO-CALLED INDIRECT PATHWAY

A.Stefani, A. Bassi, A. Peppe, M. Pierantozzi, S. Frasca, L. Brusa, G. Gattoni, G. Bernardi, and P. Stanzione*

Abstract: Changes in the electrophysiological firing properties of the external pallidus (GPe) and the subthalamic nucleus (STN) have been highlighted as crucial hallmarks of Parkinson's disease. Increase in the firing discharge of STN glutamatergic neurons, putatively driven by pallidal hypo-activity, was suggested to underlie the progressive development of the disease. We have challenged this typical view, based on the so-called "indirect pathway", by recording from both structures of parkinsonian patients undergoing deep brain stimulation. In GPe, the firing frequency (from 17 to 43 Hz) and the firing pattern (random or train/burstlike) were rather variable. Surprisingly, the administration of the dopaminergic agonist apomorphine, at doses capable to ameliorate hypokinetic symptoms without producing involuntary movements, was followed by negligible electrophysiological changes. In particular, no evidence of a pronounced, long-lasting and clinically related increase of excitability emerged. A modest decrease in the GPe "burst index" was, however, detected in 3 out of 6 GPe. On the other hand, the firing discharge of the STN of parkinsonian patients in *off*-state was significantly augmented if compared to the subthalamic units recorded in the subthalamus of a patient affected by essential tremor. In addition, apomorphine consistently reduced the firing discharge only of parkinsonian subthalamus. Peculiar features of the apomorphine responses in the parkinsonian subthalamus were the relative fastness of the observed inhibition (significant at 6-10 minutes from sc. administration) and the slowness and incompleteness of electrophysiological recovery, despite the vanishing of the clinical benefit. Following apomorphine concentrations that induced dyskinesias, a movement-related increase of peak firing discharge occurred, obscuring the inhibitory response. Our results demonstrate

---

*A. Stefani, G. Gattoni, G. Bernardi, M. Pierantozzi, S. Frasca, L. Brusa and P. Stanzione, Clinica Neurologica, Univ. di Tor Vergata, Rome, Italy, A. Bassi, A. Peppe, G. Bernardi, and P. Stanzione, IRCCS Ospedale S. Lucia, Rome,Italy. P.

that some part of the acclaimed models of basal ganglia circuitry is obsolete. The re-balancing of subthalamic excitability, when patients benefit from dopaminergic therapy, may not be primarily attributed to an increased GABAergic input from the external pallidus, but, instead, to changes in STN intrinsic firing properties and/or modulation of glutamatergic inputs.

1. INTRODUCTION

In the early nineties, it was proposed a model of basal ganglia (BG) circuitry (Albin et al., 1989; Bergmann et al., 1990; DeLong, 1990) mostly based upon the well known finding that lesions of subthalamus (STN) may produce ballistic movements, the latter interpreted as a clinical condition opposite to Parkinson's disease (PD) akinesia. In order to support the physio-pathological role of STN in altering the muscular tone, a regulatory influence of external pallidus (GPe) on STN was supposed (Bergman et al., 1990). In turn, STN firing activity should regulate, in balance with concomitant putaminal inputs, the GPi inhibitory control of thalamic nuclei. Although more recent findings in PD patients confirmed the strict relation between GPi electrophysiological hyperactivity and clinical status (Stefani et al., 1997; Stanzione et al., 1998), the whole model of the so-called "indirect pathway" (Putamen-GPe-STN-GPi) has been extensively questioned (Chesselet and Delfs, 1997; Parent and Cicchetti, 1998). As a matter of fact, lesioning of GPe produces rather modest changes in the firing activity of the STN-GPi pathway (Hassani *et al* 1996). More importantly, neither in chronic parkinsonism nor in MPTP monkeys were found reduced, in GPe, the biochemical markers (as GAD or cytochrome oxydase) of the excitability of GABAergic neurons (Chesselet & Delfs 1997; Levy *et al* 1997; Smith *et al* 1998). Therefore, the putative "hypoactivity" of GPe as a contributing factor to PD hypokinesia is doubtful. What tells us the extracellular recordings from those nuclei in mammalian rendered "parkinsonian»? So far, inconclusive results, given that the "hypoactivity" of GPe (pioneering reports by Filion *et al* 1991 and by Miller & DeLong) tends to be a) transient and reversible and b) associated with a bursting discharge (implying the co-occurrence of more excitatory inputs, Filion *et al* 1991). Interestingly, the resurgent stereotactic approach to PD is providing an unique opportunity to verify the presence and the prominence of any clear-cut modifications of GPe and STN firing activity and, in addition, to match those with precisely assessed motor behaviors and clinical signs (Benabid *et al* 1994; Gross *et al* 1997; Limousin *et al* 1995, 1998; Siegfried & Lippitz 1994).

Our recent papers showed the importance to record from human BG following prolonged wash-out from therapy (up to 2 weeks). Otherwise, when a simple CAPIT protocol is applied, most extracellular findings could be attributed to long-lasting effect of levodopa-centered therapy and not to the pathological hallmarks of PD (Lang & Lozano, 1998; Stefani et al., 1997). Thus, we have detailed the apomorphine-induced reduction of excitability of human GPi and its time-locked correlation to the drug-mediated improvement of PD signs (Stanzione et al 1998; Stefani et al 1997). Further, we have provided evidence on the desensitization of both clinical and electrophysiological responses inside GP, since following the administration of two closely spaced apomorphine doses both clinical and electrophysiological response declined (Stefani et al 1999). Moreover, by collecting microdialysis samples from GPe and STN before and following drug concentrations capable to ensure brief but unequivocal benefits, it was possible to establish that the extracellular concentrations of GABA and glutamate were substantially unaltered. Again, this finding weakened the presumptive major alterations in fast endogenous transmission as suggested by some early oversimplification of the "parallel model theory". Here, we maintain our procedural approach in extending the analysis in the GPe and STN of human PD patients. Furthermore, we have recorded from the STN of a patient affected by essential tremor (and selected for DBS in thalamic nuclei). As a consequence, we may compare the STN firing properties between two clinical conditions, characterized by a negligible or a crucial impairment of the dopaminergic transmission.

2. METHODS

The patients of this report are 10 PD patients, with mean disease duration of 11-years. They were selected for stereotactic implantation of permanent stimulating electrodes in the GPi or STN in consideration of a very poor control of clinical symptoms, with on-off fluctuations and marked levodopa-induced involuntary movements in on-state. Before stereotactic neurosurgery, they underwent a prolonged washout period of 12-18 days in order to eliminate long-lasting levodopa effect. In order to minimize discomfort, patients were treated with daily repeated physiotherapy sessions and low doses of benzodiazepine were administered. The apomorphine dose not-producing dyskinetic movements was carefully established in each patients by repeated pre-operative challenge test doses. Surgical procedures and electrode positioning in the GPi of each hemisphere were done in two operative sessions by modifications of the techniques originally described (Benabid et al 1994), as detailed elsewhere (Stefani et al 1997). Five or two electrode trajectories were directed, respectively, to the center of GPi or STN. Along each trajectory, a tungsten electrode (.8-1.4 MOhms, FHC Inc, Bowdoinham, ME) was positioned allowing single unit recordings and electrical stimulation. The latter was performed by an external stimulator (Grass 8800). The multiple traces were simultaneously amplified by a DAM-8 (WPI), displayed on multi-channel oscilloscopes, digitized at 20 kHz, and, if relevant, stored on PC for on- and off-line analysis (LabView). GPe single units were identified by the electrode deepness (+6/+1 with respect to the intercommissural line) and by their electrical properties (train-like mode or regular discharge, however interrupted by sequences of electrical silence, see Fig. 1B). Frequency histograms were done by positioning level discriminators to count single unit activity. Spike frequency was expressed as spike/second. Comparison of the firing frequency in different conditions (pre- vs. post-apomorphine) was always performed by calculating mean and standard deviation (SD) of the values measured during 30 consecutive seconds in each of the experimental conditions. Comparisons were performed only among data obtained from the same single unit. Significance was assessed by Student t-test corrected by Bonferroni method for multiple comparison. Pearson's regression line analysis was used to calculate the slope of apomorphine-induced firing decrease (see Stefani et al 1999 for details). Rigidity was assessed before and during surgery according to the U.P.D.R.S (items N° 5 of Section III) and akynesia was expressed as the number of finger tapping in 30 s. This study was approved by the University of Tor Vergata ethic committee and the patients gave their written consent.

3. RESULTS

Figure 1 illustrates an exemplary response to apomorphine of GPe multiunit recordings. Noticeably, the GPe firing discharge is characterized by the occurrence of a "train/silence" pattern (1B). Following the sc. administration of 2-mg apomorphine (capable to provide a transitory amelioration of rigidity and akinesia, but no involuntary movements), it was possible to detect a transient but very modest increase in excitability, followed by a slight but prolonged reduction in firing frequency (Fig. 1A). Neither the decrease nor the increases of excitability, however, were statistically significant when compared to physiological alterations under placebo (data not shown). Analogous responses (substantially, the lack of a consistent apomorphine-mediated effect on GPe firing discharge) were confirmed in all our observations (6 out of 6 recordings > 1 hr), as detailed in table I. More intriguingly, in the GPe of 3 patients, we have observed a significant change in patterning, in that the clinical improvement was

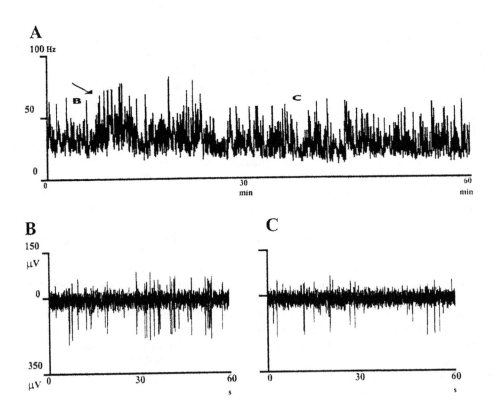

Figure 1: Apomorphine fails to produce large modifications of GPe excitability. In A: the firing frequency vs. time histogram follows the administration of the dopaminergic agonist. A brief, but transient and not significant excitation early takes place (arrow). A slight inhibition then occurs. The only substantial change is the modulation of the bursting firing mode, as revealed by selected traces (compare C, steady state effect, to B, control).

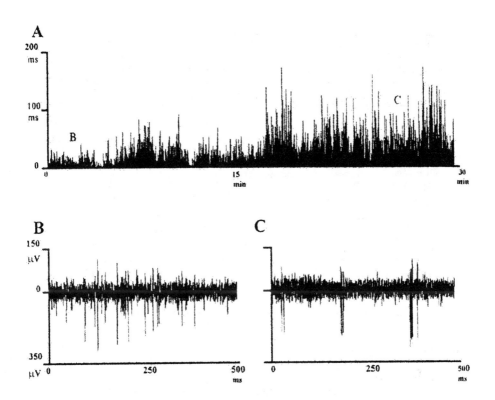

Figure 2: Apomorphine strongly reduces STN excitability. In A, the histogram shows the changes in the inter-spike interval of STN multiunit following the administration of the dopaminergic agonist. B, C: selected traces before (B) and during apomorphine (C) demonstrating how the clinical amelioration correlates with the recovery of prolonged electrophysiological pauses.

paralleled by the decrease of the bursting mode. The latter was manifested in the decrement of the number of "trains" as well as in the reduction of the number of spikes/each train (Fig. 1B, C).

Consistent responses to apomorphine were observed in STN. As emphasized by the exemplary histogram of Fig. 2A, we may detect an early increase of inter-spike interval, confirmed, at about 15 min, by a steady-state "plateau" increment, the latter suggestive of a pronounced drug-mediated reduction of excitability. Similar results were observed in all our patients, despite a surprising variability of the response (table I). Selected spike traces are shown below (1B, C). Noticeably, as apomorphine produces a clear amelioration of hypokinetic symptoms (table I), the firing discharge becomes less randomly hyperactive and the "physiological silent periods" emerge. These findings are reinforced by the observations that negligible modifications of the firing frequency/patterning were provided by either the administration of placebo in PD patients or by the administration of apomorphine in a patient affected by ET (data not shown).

Table I: Summary of electrophysiological and clinical findings (selected items) in 10 PD patients.

		Electro-physiological change under apomorphine (at 20 min)	Electro-physiological recovery after apomorphine (at 1 hour)	Clinical Improvement (upper limb rigidity before and after apomorphine)	Firing Pattern (before and after apomorphine)
GPe	1	- 5.6 %	98 %	4-->2	Burst --> random
	2	-3.1 %	91 %	3-->1	Burst --> random
	3	-8.7 %	78 %	3-->1	Burst --> random
	4	- 11.5 %	87 %	3-->2	Negligible
	5	+ 1.5 %	82 %	3-->2	Negligible
	6	- 2.0 %	97 %	4-->1	Negligible
STN	1	- 38.9 %	75 %	4-->½	Negligible
	2	- 21.7 %	84 %	3-->1	Increased pauses
	3	-15.5 %	92 %	3-->2	Negligible
	4	- 63.2 %	52 %	4-->2	Increased pauses

4. DISCUSSION

This study was undertaken to investigate the electrophysiological changes and the responses to apomorphine in the GPe and STN of PD patients in well-assessed clinical conditions. In particular, we aimed at evaluating the basic pattern in *off* and *on state*, by recording, respectively, after prolonged washout from dopamine-centered therapy or after systemic administration of apomorphine. Our findings clearly show that GPe is relatively insensitive to the dopaminergic agonist, at least when administered at subdyskinetic doses. In fact, negligible are the changes in the nucleus excitability when PD patients experience a good relief from rigidity and akinesia. The only, though modest, modification observed was a slight apomorphine-induced modulation of the bursting discharge mode. The lack of substantial changes in firing activity of GPe is surprisingly at add with the findings presented by Hutchison and co-authors (1997), as they claimed a consistent dopamine-favored increased of the GPe firing activity during amelioration of hypokinetic signs. Such discrepancy, however, may be explained by the differential approaches utilized. In particular, the brief suspension from levodopa (12 hours, CAPIT), as performed by Toronto's group (Hutchinson *et al* 1997) is probably insufficient to provide an unequivocal condition of dopamine deprivation, given the persistence of plastic and/or compensatory mechanisms in the first days off-therapy. Although not always tolerated, a prolonged washout is crucial, in our opinion, to better guarantee a reliable identification of the stimulation targets and a correct interpretation of electrophysiological results. Therefore, we propose that one of the most traditional assumptions of the so-called indirect pathway –*the reduced GABA release from GPe to STN as a major factor in inducing STN hyperexcitability-* should be largely revised.

On the other hand, the data collected in STN confirm that the STN firing discharge, in human PD, is increased and modulated by the administration of dopaminergic agonist (Limousin et al 1998). In line with the results described in PD experimental models, apomorphine had the ability to decrease STN excitability. Interestingly, we had the opportunity to validate the specificity of the apomorphine response, since it was negligible in a non-PD patient affected by ET. As predictable, the apomorphine-induced changes of STN discharge were consistent with the clinical improvement. Yet, i) the reduction in STN firing discharge was rather variable and, in 2 out of 4 patients, even quite modest (see table I); ii) the reduction of STN excitability might precede (or even not strictly correlate with) the clinical amelioration and, finally, iii) the changes in STN firing properties may persist well after the end of the clinical action of apomorphine, and complete recovery of the firing discharge was not obtained. These findings are in striking contrast with the responses of PD GPi, where a) the apomorphine-mediated reduction of excitability is time-locked with the drug-induced amelioration of hypokinetic symptoms and b) reliable recovery are observed (Fedele et al., 2001; Stefani et al 1997, 1999). The described difference might take place for several reasons. At first, we may consider the possibility that apomorphine activates different subgroups of receptors (with distinct distribution) in STN and GPi. For instance, dopamine D1 receptors on glutamatergic axon terminals impinging STN were recently described, thus supporting the possibility of an "early" presynaptic-like effect of apomorphine. The latter might turn out to influence the incoming glutamatergic inputs. An alternative possibility might regard concomitant, although transient changes in GPe firing properties. We have documented the existence of early and brief "excitatory" modifications in GPe discharge following apomorphine administrations. Yet, this effect was very infrequent (2 out of 6 GPe) and brief, thus supporting only transitory augmentations of GABA release towards STN. Finally, the observed changes in human PD STN might depend upon long-lasting modifications of the intrinsic, intranuclear excitability asides from any modulation of pallidal excitability.

A major issue, which this report has marginally approached, concerns the pattern of discharge. In particular, different groups have addressed the relative prevalence of bursting versus tonic firing discharge in the normal or PD STN. Conclusive evidence has not emerged. In the MPTP-treated green monkeys (Bergman et al 1994), it was described a slight increase of the bursting firing mode; similar findings were reported by Hassani in 6-OHDA-lesioned rats. On the contrary, Kris and others recently showed that the "burstiness" of STN single units in a rodent model of PD is reduced (Kreiss et al 1997). Technical (diverse methods to evaluate the firing patterning) or methodological aspects may explain this discrepancy. Because patterning is one of the main neuronal feature in determining efficacy of transmission, thus, a correct knowledge of firing properties in disease vs. normal conditions is critical in validating putative new therapies or further understanding the psychopathology of PD. Our preliminary data support the proposition that, if bursting mode is a pathological condition in GPe, on the contrary the recovery of a bursting discharge in STN matches with drug effectiveness. Detailed analysis of these aspects, by utilizing a refined version of the Kanaoke-Vitek method (1996) is actually under scrutiny on a larger sample of GPe and STN neurons.

5. ACKNOWLEDGEMENTS

This work was supported by Telethon Onlus Grant n° EC0998 to PS and Min.Sanità Grant tp PS and AS

6. REFERENCES

Albin, R.L., Young, A.B., and Penny, J.B., 1989, The functional anatomy of the basal ganglia. *Trends Neurosci* **12**: 366-375.

Benabid, A.L., Pollak, L., Gross, C., Hoffman, D., Benazzouz, A., Gao, D.M., Laurent, G., Gentil, M., and Perret, J., 1994, Acute and long-term effects of subthalamic nucleus stimulation in Parkinson's disease. *Stereotaxic Funct Neurosurg.* **62**(1-4): 76-84.

Bergmann, H., Wichmann, T., and DeLong, M.R., 1990, Reversal of experimental parkinsonism by lesions of the subthalamic nucleus. *Science* **249**(4975): 1436-1438.

Chesselet, M.-F., and Delfs, J.M., 1996, Basal ganglia and movement disorders: an update. *TINS* **19**: 417-422.

DeLong, M.R., 1990, Primate models of movement disorders of basal ganglia origin. *TINS* **13**: 281-285.

Filion, M., Tremblay, L., and Bedard, P.J., 1991, Effects of dopamine agonists on the spontaneous activity of globus pallidus neurons in monkeys with MPTP-induced parkinsonism. *Brain Res* **547**: 152-161.

Fedele E., Mazzone P., Stefani A., Bassi A., Ansaldo MA, Raiteri M., Altibrandi MH, Pierantozzi M., Giacomini P., Bernardi G., and P. Stanzione, 2001, Microdialysis in Parkinonian patient basal ganglia: acute apomorphine-induced clinical and electrophysiological effects not paralleled by changes in the release of neuroactive amino acids. *Exp. Neurol.* **167**: 356-365.

Gross, C., Rougier, A., Guehl, D., Boraud, T., Julien, J., and Bioulac, B., 1997, High-frequency stimulation of the globus pallidus internalis in Parkinson's disease: a study of seven cases. *J. Neurosurg.* **87**: 491-498.

Hassani, O.K., Mouroux, M., and Feger, J, 1996, Increased subthalamic neuronal activity after nigral dopaminergic lesion independent of disinhibition via the globus pallidus. *Neuroscience* **72**: 105-115.

Hutchison, W.D., Levy, R., and Dostrovsky, J.O, 1997, Effects of apomorphine on globus pallidus neurons in parkinsonian patients. *Ann. Neurol.* **42**: 767-775.

Kaneoke, Y. and Vitek, J.L., 1996, Burst and oscillation as disparate neuronal properties. *J. Neurosci. Meth* **68**: 211-223.

Kreiss, D.S., Mastropietro, C.W., Rawji, S.S., and Walters, J.R., 1997, The response of subthalamic nucleus neurons to dopamine receptor stimulation in a rodent model of Parkinson's disease. *J Neurosci* **17**: 6807-6819.

Lang, A., and Lozano, A.M., 1998, Parkinson's disease. First of two parts. *N. Engl. J. Med.* **339**(15):1044-1053.

Levy, R., Hazrat, L.N., Vila, M., Hassani, O.K., Mouroux, M., Ruberg, M., Asensi, H., Agid, Y., Feger, J.,Obeso, J.A., Parent, A., and Hyrsch, E.C., 1997, Re-evaluation of the functional anatomy of the basal ganglia in normal and parkinsonian states. *Neuroscience* **76**: 335-343.

Limousin, P., Pollak, P., Benazzouz, A., Hoffman, D., Le Bas, J.F., Broussolle, E., Perret, J.E., and Benabid, A.L., 1995, Effect on parkinsonian signs and symptoms of bilateral subthalamic nucleus stimulation. *Lancet* **345**: 91-95.

Limousin, P., Krack, P., Pollak, P., Benazzouz, A., Arduin, C., Hoffmann, D., and Benabid, A.L., 1998, Electrical stimulation of the subthalamic nucleus in advanced Parkinson's disease. *N. Engl. J. Med.* **339**(16)1105-1111.

Miller, W.C., and DeLong, M.R., 1987, Altered tonic activity of neurons in the globus pallidus and subthalamic nucleus in the primate MPTP model of parkinsonism. In *The Basal Ganglia II Structure and Function.* (M.B. Carpenter and A. Jayaraman, eds.), Advances in Behavioral Biology, Vol 32. Plenum Press, New York, pp 415-427.

Parent, A., and Cicchetti, F., 1998, The current model of basal banglia organization under scrutiny. *Mov. Dis.* **13(2)**: 199-202.

Siegfried, J., and Lippitz, B., 1994, Bilateral chronic electrostimulation of ventroposterolateral pallidum a new therapeutic approach for alleviating all parkinsonian symptoms. *Neurosurgery* **35**: 1126-1130.

Smith, Y., Bevan, D., Shink, E., and Bolam J.P., 1998, Microcircuitry of the direct and indirect pathways of the basal ganglia. *Neuroscience.* **86(2)**:353-387.

Stanzione, P., Mazzone, P., Peppe, A., Pierantozzi, M., Stefani, A., Bassi, A., and Bernardi G., 1998, Antiparkinsonian and anti-LID effects can be obtained by stimulating the same site within the GPi in Parkinson's disease. *Neurology* **51**: 1776-1777.

Stefani, A., Stanzione, P, Bassi, A, Mazzone, P., Vangelista, T., and G. Bernardi., 1997, Effects of increasing doses of apomorphine during stereotaxic neurosurgery in Parkinson's disease: clinical score and globus pallidus activity. *J. Neural Transm.* **104**: 895-904.

Stefani, A., Mazzone, P., Bassi, A., Bernardi, G., Altibrandi, M.G., Peppe, A., Pierantozzi, M. and Stanzione, P, 1999, Electrophysiological and clinical desensitization to apomorphine administration in parkinsonian patients undergoing stereotaxic neurosurgery. *Exp. Neurol.* **156**: 209-213.

NEURONAL ACTIVITY OF GPe IN PARKINSON'S DISEASE

Is it really hypoactive?

Chihiro Ohye, Masafumi Hirato, Akio Takahashi and Katsushige Watanabe*

1. INTRODUCTION

Revived stereotactic pallidotomy (Laitinen et al., 1992) is now probably one of the most prevailing stereotactic methods for the treatment of Parkinsonian motor symptoms such as rigidity, akinesia and tremor. According to recent advancements in neuroscience (Crossman, 1987), the subthalamic nucleus (STN), one of the members of the so-called indirect pathway of the basal ganglia circuit, exerts excitatory input over the internal segment of the globus pallidus (GPi). In cases with Parkinson's disease (PD), because of the lack of inhibitory influence from the external segment of the globus pallidus (GPe) to STN, there would be an excess excitation from STN to GPi, the latter sending a strong inhibitory message to the thalamic ventrolateral nucleus which causes several motor disturbances characteristic of PD (DeLong, 1990). If that is the case, the neuronal activity of GPe in PD should be hypoactive.

To test this hypothesis, the neuronal activity of GPe and GPi were directly recorded during the course of stereotactic pallidotomy for PD, and a quantitative analysis was carried out.

2. SUBJECTS AND METHODS

From our series of pallidotomy, early 20 cases with rigid or akinetic type of PD with minimal tremor if any, were subjected to this study (Table 1). Tremor types of PD were excluded because much stable effect is accomplished by ventrointermediate (Vim) thalamotomy (Ohye, 1998a).

---

* C. Ohye, Functional and Gamma Knife Surgery Center, Hidaka Hospital, Takasaki, Gunma, Japan. M. Hirato, A. Takahashi, and K. Watanabe, Department of Neurosurgery, Gunma Univeristy School of Medicine, Gunma, Japan.

Table 1. Patients list of pallidotomy, with the maximum spontaneous activity of GPe and GPi respectively, and the results of operation.

NO	NAME	AGE	SEX	SYMPTOMS	Duration(y)	OPE DATE	SIDE	GPe		GPi	RESULT
1	S.Y.	66	M	akinesia, rigid, tremor	4	940808	R	130(+4)	<	260(-2)	○
2	I.I.	57	F	akinesia, rigid, DID	32	941018	L	20(+4)	<	150(-4)	○
3	Y.T.	46	F	akinesia, rigid, tremor	4	941020	L	150(+3)	<	180(-2)	○
4	T.T.	66	F	akinesia, rigid, tremor	5	941124	R	70(+3)	=	80(-3)*	×
5	M.Y.	44	M	akinesia, tremor, DID	30	941201	L	150(+3)	<	300(-2)*	○
6	Y.I.	48	F	akinesia, rigid, DID	10	950202	L	200(+3)	<	230(-4)	△
7	Y.O.	58	F	akinesia, tremor	10	950216	L	140(+3)	<	170(-3)	△
8	Y.M.	76	F	akinesia, tremor	5	950302	L	120(+3)	=	120(-2)	○
9	M.T.	61	M	akinesia, rigid, tremor	20	950406	R	220(+6)	>	130(-4)	○
10	N.K.	64	F	akinesia, rigid	15	950420	L	110(+4)	>	70(-2)	○
11	M.Y.	54	F	akinesia, tremor	1.5	950803	R	100(+3)	=	120(-3)	△
12	N.K.	72	F	akinesia, tremor	10	951005	L	110(+3)	=	90(-4)*	△
13	S.F.	59	M	rigid	8	951207	L	180(+4)	>	150(-3)	○
14	H.I.	61	M	akinesia, rigid	6	960125	L	100(+4)	<	150(-2)	×
15	Y.E.	53	F	akinesia, DID	15	960328	L	50(+4)	<	80(-3)	△
16	Y.H.	37	M	akinesial rigid	12	960530	L	150(+7)	=	160(-2)*	○
17	C.I.	62	F	akinesia, rigid, tremor	12	970821	L	110(+3)	=	100(-3)	△
18	K.S.	68	F	akinesia, DID	32	971127	L	70(+4)	<	100(-2)	△
19	E.I.	58	M	akinesia, rigid, tremor	6	980209	R	150(+4)	=	130(-3)	○
20	A.T.	47	M	akinesia, DID	12	980216	R	80(+6)	<	130(0)	○

Principal methods of stereotactic pallidotomy are essentially the same as thalamotomy with microrecording (Ohye, 1998a). Briefly, the patient is placed under local anesthesia and Leksell's stereotactic apparatus is fixed on the skull. After making a short straight incision in the scalp over the frontal area, a burr hole is opened and ventriculography is performed with a small amount of room air and a radioopaque substance. Then, the anterior and posterior commissures are read on the X-ray film, a tentative target point for introduction of the microelectrodes is chosen at 2 mm anterior to the midpoint of the intercommissural (IC) line, 20~23 mm lateral to the midsagittal line on the level of the IC line. To determine and verify the safety and proper tentative target point, preoperative MRI and CT images give us very useful information because the GPi in basal ganglia is visible. Therefore, especially for the lateral coordinate, the border between the internal capsule and the posteroventral GPi is precisely measured and 2~3 mm lateral from the border is chosen.

The recording electrode is a bipolar concentric needle type with an outer diameter of 0.4 mm, a tip of approximately 10 µm, an interpolar distance of 0.1~0.2 mm, and an electrical resistance about 100 Kohm. Two electrodes are used as a pair, set parallel to the midsagittal plane. Since Leksell's stereotactic apparatus uses two coagulation needles, the recording electrodes take the place of the coagulation needles. Conveniently, the recording electrode just enters the position of needle stylet. Usually, the center position of the guide hole is used for the reference electrode (anterior electrode, oriented to the tentative target of zero point) and the posterior hole, (3 mm apart) is used for the posterior electrode. After setting the

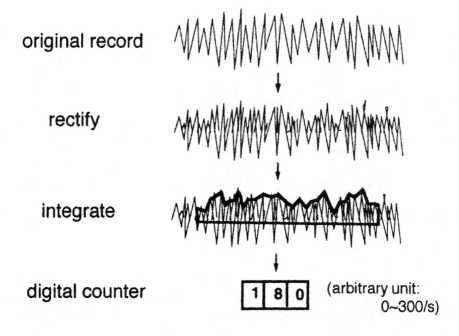

Fig. 1. Method of quantitative assay (arbitrary unit) of the depth electrical activity. Original record (spike discharge) is rectified, electrically integrated and shown on the digital meter by an arbitrary unit (180 in this example).

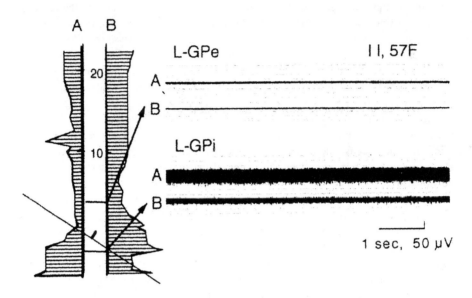

Fig. 2. An example of quantitative assay of the GP activity. Left: Line graph showing the profile of spontaneous activity (GPe<GPi) along the trajectory toward GPi. A pair of recording electrodes was introduced in parallel to the midsagittal plane, 3 mm apart (two vertical lines, A and B). An oblique line depicts level of the intercommissural line. Right: Examples of the spontaneous activity at two different points along the trajectory shown on the left line graph.

electrodes, they are introduced slowly into the brain using a micromanipulator which is driven by a step motor controlled by a remote switch. Speed and step are chosen as required, and the distance from zero (tentative target) to the tip of electrode is always automatically shown on the digital counter in micron steps. Spontaneous electrical activity is recorded continuously from the cortical surface to the tentative target point or to a deeper point, if necessary. Depth electrical activity recorded simultaneously by a pair of electrodes is led to an oscilloscope, then through an electrically integrating device, the amount of spontaneous background basic electrical activity is shown as seconds on an electric digital meter. For this, an arbitrary unit is used which offers the basis for a quantitative assay to compare general activity at different deep structures (Ohye, 1998b). The principle of the integrating system, and its application in making a line histogram for pallidotomy are shown in Fig. 1 and Fig. 2, respectively.

In addition to the spontaneous background activity mentioned above, the unitary or distinguishable multi-unitary spike discharge(s) is selected with the aid of the micromanipulator. Whenever an isolated stable spike discharge is found, passive and/or active limb movement, several natural stimuli are given to test if the pattern of spike discharge is modified. Using a light touch, tap, compression or muscle stretch, voluntary contraction on the contra- and ipsi-lateral extremities or face area are tested.

Depth activity together with related EMG are recorded on a thermorecorder and magnetic tape for later analysis.

One day before the operation, all operative procedures, especially the process and importance of microrecording, are fully explained to the patient and his (or her) family to obtain informed consent.

3. RESULTS

Gradually introducing a pair of electrodes, the background activity changes according to the different depth structures. Usually, from the cortical surface, the electrodes pass through cortical gray matter, white matter, putamen, GPe, GPi, in this order. As each depth structure exhibits characteristic electrical activity, it is not difficult to identify the position of the electrode tip, referring to the standard atlas (Hassler, 1977). Thus, the white matter is almost electrically silent, the putamen shows random spike train on fast oscillation of around 20 Hz just like caudate neurons, GPe exhibits somewhat irregular long bursts interrupted by a short pause, and GPi is characterized by a high frequency (sometimes over 200 Hz) continuous train of spikes.

Comparing the activities recorded by a pair of electrodes, it is usually the case that, in our approach, the activity in GP area from the posterior electrode is less than that from the

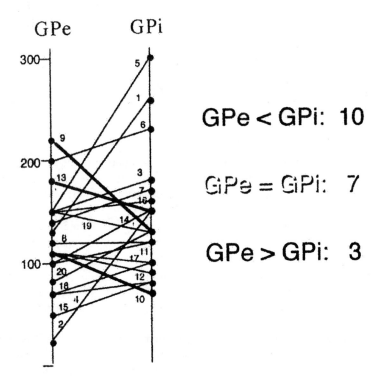

Fig. 3. In each individual case (20 cases), the maximum value of GPe and GPi, respectively, are plotted, and the points from the same patient are connected.

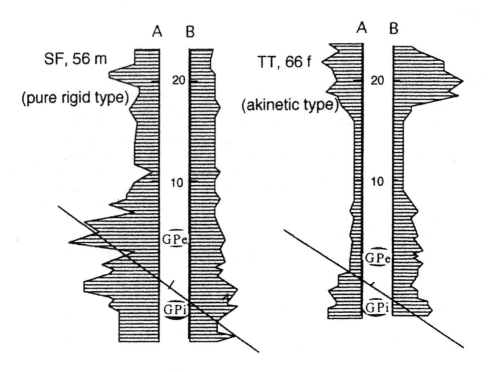

Fig. 4. Two particular profiles of Parkinson's disease. Left: A case with pure rigidity (Case 13). In this case, GPe nd GPi activities are almost equally, very high. Right: A case with severe akinesia and rigidity (Case 4), showing reduced spontaneous activity in both segments of GP all through the trajectory.

anterior electrode. This may be due to the geographical relation of the posterior tracking passing the area close to the border of internal capsule. In a few cases, the relation is reversed and the activity in the anterior electrode is less than that in the posterior electrode. Nevertheless, one of the electrodes penetrates GPi after GPe, engendering comparison of both activities. When changes of the spontaneous activity are plotted along the track to make a line histogram, it is easy to understand the distribution of spontaneous background activity. As expected, activity in GPi is usually more than that in Gpe (10 cases in 20), but in 7 cases it was almost equal and in 3 cases it was less.

So we take the maximum value of the spontaneous activity from supposed GPe (about 4 mm above zero) and GPi (about 3 mm below zero), respectively, and make a diagram as shown in Fig. 3.

Among our series of 20 cases, two particularly impressive cases will be illustrated (Fig. 4). In Case 4, a 66-year-old female with marked akinesia and rigidity, the spontaneous activity of both GPe and GPi, are equally very low, less than 100 in our measurement. Although spontaneous activity was low, in supposed GPi, four motor responses were found (two related to upper limb movement, two related to lower limb). A radiofrequency lesion was made including this GP point which resulted in, unfortunately, no remarkable improvement of symptoms. In this case, the PET scan before surgery revealed a decreased uptake of glucose (FDG) over the entire frontal cortical area, demonstrating a typical aged PD profile as described elsewhere (Ohye, 1991; Hirato et al., 1997). In Case 13, a 59-year-old male with pure rigidity, marked increase of both GPe and GPi activities (150 and 180, respectively) were seen. In this case, six motor related responses were recorded (five related to upper limb, one related to lower limb). Coagulation including this point resulted in marked reduction of his rigidity and concomitant improvement of the movement.

Responses to passive and active movement of the limb or face, either contra or ipsilateral side, manifested by the increase of the GPi spike discharge were found in every case. Distribution of such movement related cells were essentially the same as the previous findings (Ohye et al., 1994; Ohye, 1995; Ohye et al., 1996). Bilateral responses were found in about 40 % of the responsible cells. It is our strategy to make a therapeutic lesion including such motor related zone in the posteroventral part of GPi.

The overall clinical course after surgery is satisfactory. Reduced rigidity and improved akinesia to some extent, are observed. However, we are not yet sure of the long term effect. We have an impression that the immediate favorable effect may decrease within a year or so.

4. DISCUSSION

Despite the fact that microrecording has been used widely for stereotactic pallidotomy (Dogali et al., 1994; Sterio et al., 1994; Tsao et al., 1998; Vitek et al., 1998; Gross et al., 1999) their interest was mainly focused on the neuronal response of GP, and practically no study was conducted to compare global activity of both GPe and GPi in PD. Under such conditions, in this study, we measured directly and quantitatively the spontaneous background activity of both segments of the globus pallidus in PD during the course of stereotactic pallidotomy. It was revealed that in half of the cases (10/20), GPi is more active than GPe, as expected from the current concept of pathophysiology of PD (DeLong, 1990). However, in the rest of the cases, the spontaneous activity of both segments of globus pallidus is almost equal in 7 cases and moreover, GPi is less active than GPe in 3 cases. The results suggest that the widely accepted hypothesis of the basal ganglia circuit and its disease state in PD, for example, is not always true as criticized by Parent and Cicchetti (1998).

We tried to correlate these three different profiles of the spotaneous background activity with patients' core symptoms, duration of disease and drug used before surgery, but there was no positive relation.

In order to compare the spontaneous background activity as accurately as possible, the problem of data sampling (in this case, the amount of spontaneous activity in GP) should be considered. In this study, a pair of recording electrodes were used. The electrodes were introduced toward a certain definite point in every case, but certainly, there was always an individual difference as seen by the relative difference of activity recorded by anterior and posterior electrodes. This difference is partly due to the geographical difference where the electrode passes through the part of GP, if not reflecting the disease state. Usually the recording was conducted only once, and we took a maximum value (see Table 1 and Fig.3) in supposed GPe and GPi, respectively, regardless of anterior or posterior tracking. Thus, in the limited condition during operation, the comparison was made in reasonable condition.

5. CONCLUSION

1. In Parkinson's disease of akinetic and/or rigid type, spontaneous background activity in both segments of GP is measured quantitatively during the course of stereotactic pallidotomy
2. Generally, the activity of GPi is higher than that of Gpe (10 cases).
3. However, there are cases with different profiles: GPe=GPi (7 cases), GPe>GPi (3 cases)
4. No particular clinical symptom can be correlated to the different GPe-GPi profile.
5. Results of pallidotomy were better in cases with hyperactive GPi.
6. In contrast, results of pallidotomy were rather poor in cases with hypoactive GPi. Moreover, in these cases, metabolic rate of glucose was much reduced in the prefrontal area.

6. REFERENCES

Crossman, A.R., 1987, Primate models of dyskinesia: the experimental approach to the study of basal ganglia related involuntary movement disorders. *Neuroscience* 21:1.

Dogali, M., Beric,A., Sterio, D., Eidelberg, D., Fazzini, E., Takikawa, S., Samelson, D.R., Devinsky, O., and Kolodny, E.H., 1994, Anatomic and physiological considerations in pallidotomy for Parkinson's disease. *Stereotac Func Neurosurg,* **62**:53.

DeLong, M.R., 1990, Primate model of movement disorders of basal ganglia origin, *TINS* 13:281.

Gross, R. E., Lombardi, W.J., Lang, A. E, Duff, J., Hutchison, W. D., SaintCyr, J.A., Tasker, R.R., and Lozano, A.M., 1999, Relationship of lesion location to clinical outcome following microelectrode-guided pallidotomy for Parkinson's disease. *Brain* **122**:405.

Hassler, R., 1977, Architectonic organization of the thalamic nucleus. in: *Atlas for Stereotaxy of The Human Brain.* Ed 2, G. Schaltenbrandt. and W. Wahren, eds., G. Thicme, Stuttgart.

Hirato, M., Ohye, C., Takahashi, A., Negishi, M., and Shibasaki, T., 1997, Study on the function of the basal ganglia and frontal cortex using depth microrecording and PET scan in relation to the outcome of pallidotomy for the treatment of rigid akinesia type Parkinson's disease. *Stereotac Func Neurosurgery,* **60**:86.

Leitinen, L.V., Bergenheim, A.T., and Mariz, M.I., 1992, Leksell's posteroventral pallidotomy in the treatment of parkinson's disease. *J. Neurosurg.* **76**: 53.

Ohye, C., 1991, Positron emission tomographic study in Parkinson's disease—Rigid vs tremor type. in: *Parkinson's Disease. From Clinical Aspects to Molecular Basis.* T. Nagatsu, H. Narabayashi, M. Yoshida, eds., Springer, Wien- New York.

Ohye, C., 1995, Activity of the pallidal neurons related to voluntary and involuntary movements in humans. in: *Functions of Cortico-Basal Ganglia Loop.* M. Kimura. and A. Graybiel, eds., Springer Verlag, Tokyo.

Ohye, C: 1998a, Thalamotomy for Parkinson's disease and other types of tremor. Part 1, Historical background and technique. in *Textbook of Stereotactic and Functional Neurosurgery.* PL Gildenberg, RR Tasker, eds., MacGraw-Hills, New York.

Ohye, C, 1998b, Neural Noise Recording in Functional Neurosurgery. in *Textbook of Stereotactic andFunctional Neurosurgery.* PL Gildenberg, RR Tasker, eds., MacGraw-Hill, New York.

Ohye, C., Hirato, M., Takahashi, A., Watanabe K., and Murata, H., 1996, Physiological study of the pallidal neurons in Parkinson's disease. in *The Basal Ganglia V,* C. Ohye, M. Kimura, J.S. McKenzie, eds., Plenum, New York-London.

Ohye, C., Hirato, M., Kawashima, Y., Hayase, N., and Takahashi, A., 1994, Neuronal activity of the human basal ganglia in parkinsonism compared to other motor disorders. in : The *Basal Ganglia* IV. G. Percheron, J.S. McKenzie, J. Feger, eds., Plenum, New York-London.

Parent, A., and Cicchetti, F., 1998, The current model of basal ganglia organization under scrutiny. *Mov Disord.* **13**:199.

Tsao, K., Wilkinson, S., Overman, J., Koller, W.C., Batnitzzky, S., Gordon, M.A., 1998, Pallidotomy lesion locations: significance of microelectrode refinement. *Neurosurgery* **43**:506.

Sterio, D., Beric, A., Dogali, M., Fazzini, E., Alfaro, G., and Devinsky, O., 1994,. Neurophysiological properties of pallidal neurons in Parkinson's disease. *Ann Neurol.* **35**:586.

Vitek, J.L., Bakay, R.A., Hashimoto, T., Kaneko, Y., Mewes, K., Zhang, J.Y., Rye, D., Stair, P., Baron, M., Turner, R., and DeLong, M.R., 1998, Microelectrode-guided pallidotomy: technical approach and its application in medically intractable Parkinson's disease. *J.Neurosurg.* **88**: 1027.

ACUTE EFFECTS OF LEVODOPA AND PALLIDOTOMY ON BIMANUAL REPETITIVE ARM MOVEMENTS IN PATIENTS WITH PARKINSON'S DISEASE

R. Levy, W.D. Hutchison, A.E. Lang, A.M. Lozano, and J.O. Dostrovsky*

Key words: Parkinson's disease, simultaneous movements, pallidotomy, levodopa

Abstract: Patients with Parkinson's disease (PD) exhibit difficulty in performing simultaneous or sequential movements. This inability is thought to be due in part to a loss of dopaminergic innervation of the striatum and subsequent overactivity in the internal globus pallidus (GPi) resulting in inhibition of the motor thalamus and associated cortical motor areas. The performance of two groups of PD patients on a wrist pronation-supination task (WPS) was examined before and after levodopa (n=10) administration or acutely before and after unilateral pallidotomy (n=9). WPS was carried out both alone and with simultaneous button pressing (bimanual task). The amplitude and frequency of WPS was averaged over 3 X 10 s trials in each case. The administration of levodopa caused significant improvement in WPS amplitude when performed in isolation (mean=35%) and during concomitant button tapping (mean=93%) (2 way ANOVA, $p<0.05$). No significant improvement due to levodopa was observed in the frequency of WPS movements. Pallidotomy increased the amplitude of contralateral WPS alone by 28% (2 way ANOVA, $p<0.05$), and by 61% in the bimanual task (2 way ANOVA, $p<0.01$). Pallidotomy did not change WPS frequency either alone or in the bimanual task. In both groups there was a greater improvement in the amplitude of WPS in the bimanual task than in WPS alone (paired-T, $p<0.05$). These findings suggest that pallidotomy improves complex motor tasks to a greater extent than more simple motor tasks, similar to the effects produced by levodopa therapy.

---

* R. Levy, W.D. Hutchison, and J.O. Dostrovsky, Dept. of Physiology, Faculty of Medicine, University of Toronto, Toronto, Ontario, Canada M5S 1 A8. W.D. Hutchison and A.M. Lozano, M5S 1 A8, Dept. of Surgery and A.E. Lang, Dept. of Medicine, Faculty of Medicine, University of Toronto; W.D. Hutchison, A.M. Lozano, A.E. Lang, and J.O. Dostrovsky, Toronto Western Research Institute, The Toronto Western Hospital, 399 Bathurst St., Toronto, Ontario, Canada M5T 2S8

1. INTRODUCTION

Patients with Parkinson's disease (PD) exhibit difficulty in performing concurrent movements (1,2,3,4,5) especially when these simultaneous movements are composed of different temporal or spatial components (6,7). The performance of a bimanual simultaneous motor task in PD patients is thought to be degraded due to the impaired ability to shift attention between two motor tasks (5,8). This deficiency in shifting attention has been associated with the depressed functioning of the dopaminergic system (9). Furthermore, it has been shown that the administration of levodopa causes a greater improvement in the performance of complex simultaneous or sequential movements versus simple single movements (10).

Ablation of the posteroventral GPi (pallidotomy) relieves bradykinesia, tremor, and especially drug induced dyskinesias in PD patients (11,12,13). The motor deficits of PD are associated with a hyperactivity of the inhibitory GABAergic pallidofugal outflow neurons of the GPi in the 1-methyl-4-phenyl-2,3,6-tetra-hydro-pyridine (MPTP) non-human primate model (14,15). Hyperactivity in the GPi has also been demonstrated in humans with PD (16). Pallidotomy is believed to cause a direct *reduction* of GPi hyperactivity thereby releasing the inhibition of brainstem, thalamic, and cortical motor areas (17). The release of the motor thalamus and cortical motor areas such as the supplementary motor area is thought to underlie the improvement in akinesia and bradykinesia and the execution of complex motor tasks (18). Consistent with this hypothesis is the finding that dopamine agonists decrease the overall tonic discharge in the internal segment of the globus pallidus (19).

However, the effectiveness of pallidotomy remains a paradox since GPi lesions that eliminate most of the pallidal motor output would be predicted to worsen voluntary movements (20). It is therefore conceivable that complex movements could be worsened initially to a greater extent than simple movements and that improvement occurs only after the motor system "learns" to adapt to the new network conditions. The goal of this study was to compare the acute effects of levodopa and pallidotomy on the ability of PD patients to perform single and simultaneous tasks.

2. METHODS

2.1 Subjects

The patient group in which the effects of levodopa was studied consisted of 10 PD patients. Eight of these patients were enrolled in a clinical trial at the movement disorders clinic in the Toronto Western Hospital and had an indwelling cerebral catheter implanted into the third ventricle for subsequent infusion of glial-derived neurotrophic factor (GDNF). The other 2 patients were involved in pre-operative assessments before procedures to insert deep brain stimulating electrodes within the subthalamic nuclei. The group consisted of 3 females and 7 males and at the time of the test had a mean age of 58.7 years (±7.4 SD). The average duration of the disease was 11.6 years (±4.7 SD) and all had PD for at least 6 years.

The pallidotomy subjects consisted of 9 patients with idiopathic PD undergoing unilateral pallidotomy for the treatment of their motor symptoms. Pallidotomy involved lesioning the posteroventral GPi on the side opposite the worst motor symptoms. When symptoms were symmetrical, the GPi opposite the dominant hand was lesioned. There were 7 males and 2 females, with a mean age of 59.9 years (±8.7 SD). The average duration of the disease was 12 years (±3.9 SD) and all had PD for at least 7 years. Five patients had the operation contralateral to their dominant hand and seven were right hand dominant. Patients were studied in the practically defined OFF state 12-14 hours after the last oral dose of levodopa. All tests were conducted on the surgical bed with the patient's head constrained in the stereotactic frame that was attached to the bed. The patients were tested after the functional mapping was complete about 3 minutes before the lesion (pre-lesion), and within 3 minutes immediately following the lesion (post-lesion). Patients gave informed consent.

2.2 Apparatus and Tasks

The ability of the patients to perform an isolated task versus a bilateral task was compared by quantifying motor performance using a repetitive self-paced wrist pronation-supination (WPS) movement as the primary task. The secondary task (or "interfering" task) was button tapping by the opposite hand. The performance of button tapping and WPS was assessed using the apparatus shown in the Figure (part A, left side). Patients held a WPS handle (90° to forearm) fixed to a shaft coaxial with wrist rotation. To prevent upper arm movement, the patient's elbow was maintained in a fixed position against the side of the body. The rotational shaft was connected to a potentiometer producing an analogue voltage signal proportional to the angle of rotation. In the simultaneous task, patients tapped a 10 mm high momentary button (25 mm x 25 mm surface area) placed 30 cm lateral (left/right) to the WPS shaft. All data were collected at a 500 Hz sampling rate (CED-1401 Spike2, Cambridge Electronic Design). The apparatus was positioned approximately 20 cm above the patient's chest during the intra-operative testing and was positioned in front of the seated patients during the L-dopa testing.

Unilateral or isolated task: Subjects were instructed to move as fast and through as wide an angle as possible for 3 trials of 10 seconds, all separated by 10 second resting periods. The start and stop times were cued by an audio beep. The time and magnitude of the peak to peak WPS excursions were obtained from the WPS voltage traces. Bilateral task: Subjects were instructed to tap the button as fast as possible with the opposite hand starting 5 seconds before WPS began and continue over the duration of WPS (i.e. 15 seconds in total). Typical traces of both the WPS and button tapping are shown in the Figure (part A, right side).

2.3 Data Analysis

The data were grouped with respect to being contralateral or ipsilateral to the lesion side of the brain for the case of the pallidotomy patients and grouped into right and left sides in the levodopa patients. WPS amplitude was calculated as the peak to peak WPS amplitude (degrees) and WPS frequency was calculated as the reciprocal of the period for one complete pronation-supination cycle. The mean WPS amplitude and frequency per cycle were computed and subsequently averaged over all 3 trials. Patient group means were calculated by averaging the raw performance scores. The improvement of any single measure due to the effect of pallidotomy was calculated by dividing each patient's post-lesion score by their pre-lesion score. This normalised value was expressed as a percent improvement.

Statistical analysis was carried out using paired student t-tests or 2 factor repeated measures ANOVA (body side x time) followed by a Student Newman Keuls all pair-wise comparison test. Statistical significance was assigned at $p < 0.05$ (i.e. $a = 0.05$). In cases of non-normality, the Wilcoxon Signed Rank Test was employed.

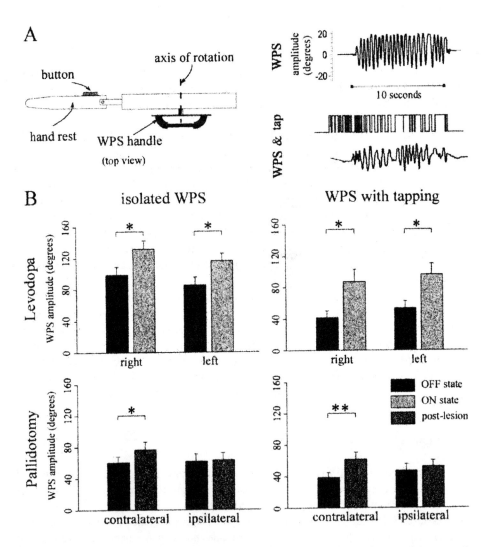

Figure. 1: A: Apparatus (left) and an example of WPS and WPS plus tapping task traces from one patient during the non-medicated state (right). B: Levodopa and pallidotomy effects on WPS amplitude during the isolated and simultaneous tasks (*$p<0.05$, **$p<0.01$).

3. RESULTS

3.1 Effects of Levodopa

There were no significant differences in the amplitude or frequencies of right and left WPS movements during the OFF state (Figure part B, upper left panel). The administration of levodopa increased the amplitude of WPS movements of both the right and left wrists by 34% and 36% respectively (2-factor repeated measures ANOVA, $p<0.05$). There were no changes in WPS frequency.

When WPS was performed concurrently with button tapping of the opposite hand, there was a significant reduction of the WPS movement amplitudes when performed in isolation. During the OFF period testing, there was an average decrease of 50% in WPS amplitude when tapping with the opposite hand was performed (paired t-test, $p<0.01$). WPS amplitude during concomitant button tapping was significantly increased by the administration of levodopa. Right WPS amplitude increased by 107% and left WPS amplitude increased by 80% (2-factor repeated measures ANOVA, $p<0.05$) (Figure part B, upper right panel). There were no changes in WPS frequency due to concomitant button tapping or the effect of levodopa. When compared to the isolated WPS movements performed during the ON period, the reduction of WPS amplitude due to tapping was 26% (paired t-test, $p<0.01$). This was significantly less than the degradation of WPS amplitude by tapping during the OFF period (paired t-test, $p<0.01$).

3.2 Effects of Pallidotomy

For all patients, pre-lesion WPS amplitudes and frequencies of the ipsilateral versus the contralateral side were not statistically different. Pallidotomy resulted in an immediate 28% increase in the group mean contralateral WPS amplitude (2 factor ANOVA, $p<0.05$) (Figure part B, lower left panel). There was no significant improvement in contralateral WPS frequency (Paired t-test, $p=0.14$). There was no detectable effect of pallidotomy on either amplitude or frequency on the body side ipsilateral to the lesion.

Before the lesion, concurrent button tapping had the effect of reducing WPS amplitude by 37% on the side contralateral to the lesion (Paired t-test, $p<0.01$) and 26% on the side ipsilateral to the lesion when compared to the respective isolated WPS amplitudes. This effect was greatest on the contralateral side reflecting that the pallidotomy was performed contralateral to the worst side. Immediately following the lesion, the amplitude of WPS movements (when performed simultaneously with button tapping) improved by 61% (2 factor ANOVA, $p<0.01$) (Figure part B, lower right panel). There was no significant improvement in contralateral WPS frequency. There was no significant effect of pallidotomy on ipsilateral WPS amplitude or ipsilateral WPS frequency. Following pallidotomy the decrease in contralateral WPS amplitude caused by concomitant button tapping was only 20% (Paired t-test, $p<0.01$). For the body side ipsilateral to the lesion, the decrease in WPS amplitude due to tapping improved to 17% (Paired t-test, $p<0.05$) following pallidotomy. The improvement in the ability to maintain WPS amplitude in spite of button tapping was significantly greater for the contralateral side versus the ipsilateral side (Paired t-test, $p<0.01$).

4. DISCUSSION

This study demonstrates that pallidotomy causes an immediate improvement in the performance of a repetitive motor task and a greater improvement in the performance of a simultaneous task. These results are consistent with the improvements seen in the performance of PD patients due to levodopa using the same task. This study supports the notion that a pallidal lesion releases cortical areas in a similar fashion to levodopa. For both patient groups, the greatest effect observed was an increase in the amplitude of WPS movements during the simultaneous tasks. In pallidotomy patients, this occurred contralateral to the lesion. WPS frequency was not affected. These findings are consistent with the hypothesis that many of the deficits observed in PD can be modeled in terms of scaling the amplitude of intended movements (21,22) and the suppression of antagonistic automatic movements which would tend to interfere with the intended movements. One caveat is that the two patient groups are not directly comparable since the severity of illness of patients enrolled in the GDNF infusion study was less than that of the patients receiving the pallidal lesion.

The degradation of WPS amplitude caused by concomitant button tapping was also significantly reduced in both groups, indicating that movements became more autonomous following either intervention. These results support similar findings of Oliveira *et al* (23) in which patients performed opposition of the thumb and forefinger with a reduced amplitude when combined with a secondary lexical decision task. Morris *et al* (24) demonstrated that following levodopa, PD patients walked with a larger stride length over all cadence rates tested. Both the effects of pallidotomy and the administration of levodopa suggest that these changes are likely the result of release of cortical structures such as the supplementary motor area (SMA). Eidelberg *et al* (25) showed that post pallidotomy bilateral increases in SMA metabolism correlated significantly with improvement in contralateral and ipsilateral motor performance. They also demonstrated that following pallidotomy the improvement in contralateral limb motor performance correlated with declines in thalamic metabolism and increases in lateral frontal metabolism. Furthermore, deep brain stimulation of the globus pallidus in parkinsonian patients has been shown to activate the SMA (26). It has also been shown using positron emission tomography that the impaired activation of the SMA in PD patients can be reversed by administration of apomorphine, a non selective D1/D2 receptor agonist (27,28). A release of other motor systems is also probable because the observed changes due to pallidotomy occurred within several minutes following the lesion; any synaptic reorganization would be less possible during this time span.

Since both groups of PD patients improved to a greater extent on the simultaneous task versus the isolated task, this suggests that the simultaneous task, or any sufficiently complex task, is a more sensitive measure of PD dysfunction than a simple task. The use of such tasks could improve the localisation of lesions or placement and programming of deep brain stimulating electrodes. A minimal improvement in WPS and tapping performance immediately following a lesion might suggest that the size of the lesion should be increased or it's location modified, especially in patients whose performance on the tasks is poor.

5. ACKNOWLEDGEMENTS

We would like to thank the patients for their participation. This study was supported by the Parkinson Foundation of Canada.

6. REFERENCES

1. Schwab R.S., Chafetz M.E., and Wang, B., 1954, Control of two simultaneous voluntary motor acts in normals and in parkinsonism. *Arch of Neurol and Psych* **72**: 591-598.
2. Benecke, R., Rothwell, J.C., Dick, J.P., Day, B.L., and Marsden, C.D., 1986, Performance of simultaneous movements in patients with Parkinson's disease. *Brain* **109**: 739-757.
3. Benecke, R., Rothwell, J.C., Dick, J.P., Day, B.L., and Marsden, C.D., 1987a, Disturbance of sequential movements in patients with Parkinson's disease. *Brain* **110**: 361-379.
4. Shimizu, N., Yoshida, M., and Nagatsuka, Y., 1987, Disturbance of two simultaneous motor acts in patients with parkinsonism and cerebellar ataxia. *Adv.Neurol.* **45**:367-70: 367-370.
5. Horstink, M.W., Berger, H.J., van Spaendonck, K.P., van den Bercken, J.H., and Cools, A.R., 1990, Bimanual simultaneous motor performance and impaired ability to shift attention in Parkinson's disease. *J.Neurol.Neurosurg.Psychiatry* **53**: 685-690.
6. Stelmach, G.E. and Worringham, C.J., 1988, The control of bimanual aiming movements in Parkinson's disease. *J.Neurol.Neurosurg.Psychiatry* **51**: 223-231.
7. Johnson, K.A., Cunnington, R., Bradshaw, J.L., Phillips, J.G., Iansek, R., and Rogers, M.A., 1998, Bimanual co-ordination in Parkinson's disease. *Brain* **121**: 743-753.
8. Flowers, K.A. and Robertson, C., 1995, The effect of Parkinson's disease on the ability to maintain a mental set. *J.Neurol.Neurosurg.Psychiatry* **48**: 517-529.
9. Cools, A.R., van den Bercken, J.H., Horstink, M.W., van Spaendonck, K.P., and Berger, H.J., 1984, Cognitive and motor shifting aptitude disorder in Parkinson's disease. *J.Neurol.Neurosurg.Psychiatry* **47**: 443-453.
10. Benecke, R., Rothwell, J.C., Dick, J.P., Day, B.L., and Marsden, C.D., 1987b, Simple and complex movements off and on treatment in patients with Parkinson's disease. *J.Neurol.Neurosurg.Psychiatry* **50**: 296-303.
11. Lang, A.E., Lozano, A.M., Montgomery, E., Duff, J., Tasker, R., and Hutchinson, W., 1997, Posteroventral medial pallidotomy in advanced Parkinson's disease. *N.Engl.J.Med.* **337**: 1036-1042.
12. Dogali, M., Fazzini, E., Kolodny, E., Eidelberg, D., Sterio, D., Devinsky, O., and Beric, A., 1995, Stereotactic ventral pallidotomy for Parkinson's disease. *Neurology* **45**: 753-761.
13. Baron, M.S., Vitek, J.L., Bakay, R.A., Green, J., Kaneoke, Y., Hashimoto, T., Turner, R.S., Woodard, J.L., Cole, S.A., McDonald, W.M., and DeLong, M.R., 1996, Treatment of advanced Parkinson's disease by posterior GPi pallidotomy: 1-year results of a pilot study. *Ann.Neurol.* **40**: 355-366.
14. Alexander, G.E. and Crutcher, M.D., 1990, Functional architecture of basal ganglia circuits: neural substrates of parallel processing. *Trends Neurosci.* **13**: 266-271.
15. Filion, M. and Tremblay, L., 1991, Abnormal spontaneous activity of globus pallidus neurons in monkeys with MPTP-induced parkinsonism. *Brain Res.* **547**: 142-151.
16. Hutchison, W.D., Lozano, A.M., Davis, K.D., Saint-Cyr, J.A., Lang, A.E., and Dostrovsky, J.O., 1994, Differential neuronal activity in segments of globus pallidus in Parkinson's disease patients. *Neuroreport* **5**: 1533-1537.
17. Wichmann, T. and DeLong, M.R., 1996, Functional and pathophysiological models of the basal ganglia. *Curr.Opin.Neurobiol.* **6**: 751-758.
18. Ceballos-Baumann, A.O., Obeso, J.A., Vitek, J.L., DeLong, M.R., Bakay, R., Linazasoro, G., and Brooks, D.J., 1994, Restoration of thalamocortical activity after posteroventral pallidotomy in Parkinson's disease. *Lancet* **344**: 814.
19. Hutchison, W.D., Levy, R, Dostrovsky, J.O., Lozano, A.M., and Lang, A.E., 1997, Effects of apomorphine on globus pallidus neurons in parkinsonian patients. *Ann.Neurol.* **42**: 767-775.
20. Marsden, C.D. and Obeso, J.A., 1994, The functions of the basal ganglia and the paradox of stereotaxic surgery in Parkinson's disease. *Brain* **117**: 877-897.

21. Georgopoulos, A.P., DeLong, M.R., and Crutcher, M.D., 1983, Relations between parameters of step-tracking movements and single cell discharge in the globus pallidus and subthalamic nucleus of the behaving monkey. *J.Neurosci.* **3**: 1586-1598.
22. Iansek, R., Bradshaw, J., Phillips, J., Cunnington, R., and Morris, M.E., 1995, Interaction of the basal ganglia and supplementary motor area in the elaboration of a movement. In Glencross D. and Pillon, B. eds. Motor control and sensory motor integration. Amsterdam, Elsevier, 49-60.
23. Oliveira, R.M., Gurd, J.M., Nixon, P., Marshall, J.C., and Passingham, R.E., 1998, Hypometria in Parkinson's disease: automatic versus controlled processing. *Mov Disord.* **13**: 422-427.
24. Morris, M., Iansek, R., Matyas, T., and Summers, J., 1998, Abnormalities in the stride length-cadence relation in parkinsonian gait. *Mov Disord.* **13**: 61-69.
25. Eidelberg, D., Moeller, J.R., Ishikawa, T., Dhawan, V., Spetsieris, P., Silbersweig, D., Stern, E., Woods, R.P., Fazzini, E., Dogali, M., and Beric, A., 1996, Regional metabolic correlates of surgical outcome following unilateral pallidotomy for Parkinson's disease. *Ann.Neurol.* **39**: 450-459.
26. Davis, K.D., Taub E., Houle S., Lang A.E., Dostrovsky J.O., Tasker R.R., Lozano A.M., 1997, Globus pallidus stimulation activates the cortical motor system during alleviation of parkinsonian symptoms. *Nat. Medicine.* **3**(6): 671-674.
27. Rascol, O., Sabatini, U., Chollet, F., Celsis, P., Montastruc, J.L., Marc-Vergnes, J.P., and Rascol, A., 1992, Supplementary and primary sensory motor area activity in Parkinson's disease. Regional cerebral blood flow changes during finger movements and effects of apomorphine. *Arch.Neurol.* **49**: 144-148.
28. Jenkins, I.H., Fernandez, W., Playford, E.D., Lees, A.J., Frackowiak, R.S., Passingham, R.E., and Brooks, D.J., 1992, Impaired activation of the supplementary motor area in Parkinson's disease is reversed when akinesia is treated with apomorphine. *Ann.Neurol.* **32**: 749-757.

TREMOR IN PARKINSON'S DISEASE
A simplification in network dynamics?

Anne Beuter and Roderick Edwards*

Key words: Tremor, Parkinson's disease, neural networks

Abstract: Normal physiological tremor tends to be irregular while tremor in Parkinson's Disease (PD) shows more regular oscillations. We explore this observation using richly connected inhibitory neural networks under a change of parameter that corresponds to the synaptic efficacy of dopaminergic neurons in the nigro-striatal pathway, and show that transitions from irregular to periodic dynamics are common in such systems. The weakening of these connections leads to a reduction in the number of units that effectively drive the dynamics and thus to simpler behavior. The multiple interconnecting loops of the brain's motor circuitry which involve many inhibitory connections, and the multiplicity of structures involved in the production of symptoms in PD suggest that such a neural network model is appropriate. Furthermore, fixed points that can occur in such networks are suggestive of akinesia in PD. This model is consistent with the view that normal physiological systems can be regulated by robust and richly connected feedback networks with complex dynamics, and that loss of complexity in the feedback structure due to disease leads to more orderly behavior.

1. INTRODUCTION AND HYPOTHESIS

It has been proposed that parkinsonian tremor is produced either by the activity of an intrinsic thalamic pacemaker (central hypothesis) or by the oscillation of an unstable long loop reflex arc (peripheral hypothesis). Recently Zirh et al (1) studied the pattern of interspike intervals occurring within bursts recorded in patients with PD and found that the spike trains of the majority of the recorded cells was not consistent with either hypothesis. These results suggest that another oscillatory process is involved. We propose that the mechanism by

---

* A. Beuter, Institut de Biologie, Laboratoire de Physiologie, Université de Montpellier 1, 4 boulevard Henri IV, 34060, Montpellier Cedex 1, France. R. Edwards, Department of Mathematics and Statistics, University of Victoria, PO Box 3045, Victoria, BC Canada V8W 3P4

The Basal Ganglia VI
Edited by Graybiel *et al.*, Kluwer Academic/Plenum Publishers, 2002

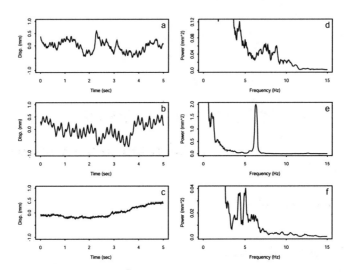

Figure 1:. Displacement recordings of postural tremor sampled at 200 Hz. On the left are 5 s segments of the time series; on the right are corresponding power spectra for the full 29.5 s recordings (estimated by smoothing with an 11-point Daniell filter). The large amount of power at low frequencies arising from the drift in finger position is not shown. Note that the scales on the power spectra are different. (a,d) Normal physiological tremor in a control subject. It is irregular and has a broad spectrum. The frequency of normal physiological tremor is usually cited as being between 8 and 12 Hz, but in displacement (rather than acceleration) lower frequencies are relatively enhanced and the power between about 6 and 11 Hz seen here is typical. (b,e) Parkinsonian tremor, with a frequency of 6.3 Hz, though the amplitude is relatively small compared to many patients with PD. (c,f) Unusually low amplitude tremor (compared to control subjects) of another patient with PD known to have rigidity, as shown by a neurological exam, and difficulty with initiating and sustaining rapid alternating movements, as shown by other motor tests. Although the amplitude is very low, there does seem to be some oscillation at about 5 Hz.

which tremor and other parkinsonian signs arise is one of a simplification in a dynamical process that is always active, but changes when parameters such as the amount of dopamine available in the nigro-striatal pathway decreases. We hypothesise that the onset of a regular oscillation in PD is a change in dynamical regime of a network involving several brain structures from a normally aperiodic one to a more regular one as the parameter corresponding to dopamine efficacy decreases. This implies that tremor in PD and normal physiological tremor are produced by the same motor circuitry, but operating in different parameter ranges. Figure 1 shows the displacement recordings and power spectra of postural tremor in a control subject and in two patients with PD who have low amplitude tremor.

2. MODELING APPROACH

We ask whether simple models that share some basic properties with the network of brain structures involved in the motor circuits show the type of dynamical changes in behavior observed in PD when a similar parameter is changed. We do not argue that our particular model is exactly right, but that the types of behavior observed are **generic** in systems of this type. It is clear that the relevant motor circuitry involves a network of multiple interacting loops in which many of the connections are inhibitory. Though in principle this could apply on a fine scale, where the units of the network are individual neurons, we have in mind a coarse description, in which anatomical structures (or functional substructures of them) are considered as individual units (e.g. Vim in thalamus) and a network that includes all the relevant structures for (in)voluntary motor control, including those in the basal ganglia thalamocortical loops, the cerebellar loops and possibly descending pathways and feedback from the periphery.

We expect the input-output (response) function for a unit to be one that sums inputs and that saturates: beyond a certain level of stimulation the unit does not become more active, and similarly, below a certain level it does not become less active. To the extent that cells within a structure simply act in parallel, they may be lumped into a single unit having properties like those of a single neuron. Though this is a simplification, it is a reasonable first approximation. Therefore, standard Hopfield-type neural network models with sigmoidal response functions and inhibitory connections are an appropriate framework for this investigation (2). We have chosen to make all our connections inhibitory, but essentially the same results are found if excitatory connections are included, as long as inhibition is abundant. For simulations we make the further simplification of replacing the steep sigmoid response functions by binary step functions. The resulting equations are piecewise linear and easily integrated. These piecewise-linear neural networks form a subset of a larger class of Boolean networks not restricted by the Hopfield model requirement that the combined effect of the inputs to a unit be a linear combination. For details, see (3).

3. PERIODIC AND APERIODIC BEHAVIOR IN NEURAL NETWORKS

Hopfield-type neural networks (coupled systems of ordinary differential equations) can be expressed as

$$(1) \; dy_i/dt = -y_i + \Sigma_j w_{ij} g(y_j) - \tau_i, \; i = 1,\ldots,N,$$

where w_{ij} represents the 'synaptic efficacy' or connection weight of neuron j acting on neuron i, τ_i is the threshold level of neuron i, y_i represents the neuron membrane potential (above threshold), and g is the response function of a neuron to its input. In order to look at the typical effect of reducing the strengths of outputs of a group of units in an inhibitory network, we generated random networks in the form of Equation 1 with the binary response function ($g(y_i) = 0$ if $y_i<0$; $g(y_i) = 1$ if $y_i>0$), having N units, each unit receiving inputs from K others. That is, for each unit, we randomly selected K of the remaining N-1 units to provide inputs to it, and set the corresponding connection weight to -1, all other connections being 0. The choice of input units was further restricted to preclude 2-loops, that is, if unit i has an input to unit j, we do not allow an input from unit j to unit i. All thresholds were set close to 1.5-K. See Edwards et al. (3) for details. Trajectories can be computed in terms of points at which units switch state (change sign) (4). We integrated networks in this way from random initial conditions until the trajectory converged to a fixed point or periodic cycle or for 304000 steps (switchings) whichever came first. In the process we kept track of the sequence of units that switched as well as the times of these switchings. Periodicity was determined by

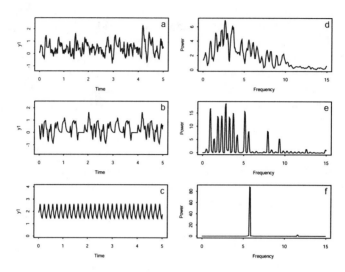

Figure 2: Interpolated time series from one unit of each of 3 random networks (N=50, K=10, d=8) with decreasing values of α after transients have died away. On the left are segments of the time series; on the right are the corresponding power spectra, based on 6000 data points (estimated by smoothing with an 11-point Daniell filter). The time scale is arbitrary and has been divided by 30 for the plots, so that the frequencies in the spectra are similar to those in the tremor examples. (a,d) α = 1, aperiodic, all units are switching; (b,e) α = 0.5, period = 64.30 time units. (c,f) α = 0.2, period 5.18 time units.

looking for at least five consecutive repetitions of a sequence of switching units (up to 2000 steps long) and checking for convergence of the period of this cycle (i.e. the last two circuits taking the same amount of time within 10^{-12} time units).

The matrix of connection weights, $W=(w_{ij})$, was adjusted by reducing the entries in the first few (d) columns by a factor α <= 1. This corresponds to weakening the output of these d network units. We tried 100 random networks with each of three different choices of N, K and d, and in each network 21 values of the parameter. Figure 2 shows typical behaviors for decreasing parameter values. In general, as α decreases, the behavior becomes more regular. Some units become 'fixed' in the 'on' or 'off' state and the number of fixed units tends to increase as the parameter decreases. The fraction of fixed units is usually much larger than the fraction of units weakened — the dynamical effect of the damage is widespread. The fraction of aperiodic networks generally decreases with the parameter, with a dip at 0.5. The proportion of fixed points increases slightly as α decreases with a big jump at 0.25. The average period of the periodic solutions does not seem to vary in any simple way with the parameter, except that at around α = 0.5 we tend to have very long periods. At this parameter value, there seems to be a dearth of aperiodic networks but these are replaced by an abundance of very long-period ones. There may be a smaller but similar effect at α = 0.75. In general, in these examples, when α > 0.5 we tend to find aperiodic behavior, but if it is periodic, the period tends to be short or intermediate in length, and fixed points are rare. For 0.25< α <0.5, there are few aperiodic networks, still not too many fixed points, and we usually get periodic

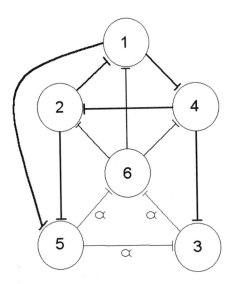

Figure 3: The connection structure of the example 6-unit network. All connections are inhibitory with weight 1 except the three labeled α, which are weakened as α decreases.

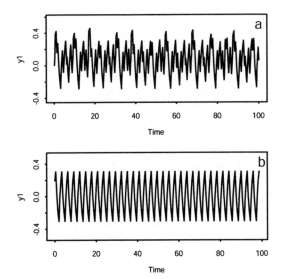

Figure 4: Behavior of one unit of the 6-unit network when (a) $\alpha = 1.0$, aperiodic, and (b) $\alpha = 0.7$, period = 2.88727 time units.

behavior of intermediate length. When $\alpha < 0.25$ we find more fixed points, short period oscillations and very few aperiodic networks.

4. A SIMPLE EXAMPLE

We wish to address the question of how networks of complex inhibitory loops can show the type of transition from irregular to regular behavior that we see in PD by means of weakening of synaptic connections. An analysis of this process is presented in (3) but it can be illustrated by the following example. Consider the 6-unit inhibitory network (Equation 1 and Figure 3) with connection matrix:

$$W = \begin{bmatrix} 0 & -1 & 0 & 0 & 0 & -1 \\ 0 & 0 & 0 & -1 & 0 & -1 \\ 0 & 0 & 0 & -1 & -\alpha & 0 \\ -1 & 0 & 0 & 0 & 0 & -1 \\ -1 & -1 & 0 & 0 & 0 & 0 \\ 0 & 0 & -\alpha & 0 & -\alpha & 0 \end{bmatrix}$$

and

$$\tau_i = -3/2, i=1,...,6.$$

The parameter α was varied. This network, though simple, still shares some features with the one thought to be involved in the production of tremor in PD. In particular, there are several units that participate in more than one inhibitory loop each. The parameter α represents the modulation of outputs from two of the units (the third and fifth) which might be considered as synaptic efficacies (in a global sense) of signals emanating from neurons in these units.

This network also changes from irregular to regular behavior as α decreases. When $\alpha = 1$, its behavior seems irregular as shown in Figure 4a. However, if $\alpha < 3/4$ it can be shown that the behavior must be periodic (Figure 4b). In this case, the sixth unit, which receives inputs only from the third and fifth, receives a total (inhibitory) input of at most $2\alpha < 1.5$ so that

$$\Lambda_6 \equiv w_{6,3} g(y_3) + w_{6,5} g(y_5) - \tau_6 \geq 1.5 - 2\alpha > 0$$

The equation driving the sixth unit at any time is

$$dy_6/dt = -y_6 + \Lambda_6$$

where $\Lambda_6 = 1.5$, $1.5 - \alpha$, or $1.5 - 2\alpha$, depending on the signs of y_3 and y_5 (i.e., on which orthant [the n-dimensional analog of a quadrant in 2 dimensions] of phase space the trajectory is in). Since Λ_6 is always positive, in a finite time y_6 will become positive, and from then on, $g(y_6)$ will be fixed at 1. In other words, the system can be reduced to 5 dimensions with input from unit 6 incorporated into τ_1, τ_2, τ_4. The equation for unit 1, for example, becomes

$$dy_1/dt = -y_1 - g(y_2) - g(y_6) + 3/2 = -y_1 - g(y_2) + 1/2$$

The network has reached a 'boundary' in parameter space, where unit 6 ceases to contribute to the dynamics of the system. Furthermore, under these circumstances, units 1, 2 and 4 receive a constant input from unit 6 and behave as an isolated network, i.e., the resulting 5 x 5 W matrix is reducible. Amongst themselves, these three units now form the simple inhibitory cycle of Glass and Pasternack (4), which is known to oscillate periodically.

Units 3 and 5 are then driven by the periodic oscillation of units 1, 2 and 4 and follow the same rhythm. Even for α somewhat larger than 3/4 the network has this same behavior.

In the random networks, the number of fixed units increased as the parameter decreased suggesting that the same type of dynamical simplification is occurring as in the 6-unit example. The phenomenon can be understood in terms of the approach to boundaries of the central region in parameter space, where even before leaving the central region, the probability of a unit becoming fixed in the 'on' or 'off' state (though not fixed in level of internal activity) increases. The number of units contributing to the dynamics (and therefore the effective dimension of the dynamics) decreases as synaptic efficacies in parts of the network are weakened. This is a probabilistic statement: it clearly depends to some extent on the particular structure and the particular trajectory.

5. PHYSIOLOGICAL PLAUSIBILITY

It is evident that the basal ganglia motor circuitry is a network which we attempt to mimic with the 6-unit example (above). It is composed of two primary input structures (striatum and subthalamic nucleus), two primary output structures (globus pallidus internal segment and substantia nigra pars reticulata) and two intrinsic nuclei (globus pallidus external part, substantia nigra pars compacta) (5). The input structures receive excitatory input from most areas of the cerebral cortex (striatum) and from motor areas of the frontal lobe (subthalamic nucleus). The output structures receive excitatory input from the subthalamic nucleus and inhibitory input from the striatum and send inhibitory input to the motor areas of the thalamus and brain stem. Excitatory input is then sent back from nuclei of the thalamus to the cerebral cortex. The internal nuclei modulate the action of the input and output structures via mostly inhibitory influences. For example, the substantia nigra pars compacta receives inhibitory input from the striatum and other inputs which have been difficult to assess. It sends dopaminergic input to the striatum, the subthalamic nucleus and the globus pallidus (internal segment). Thus it acts both on input and output structures of the basal ganglia but its action has not been well characterized.

We have seen that several interacting loops, many of which are inhibitory are involved in motor function (schematics of these may be found in, e.g., (6) and (7)). Inhibitory feedback is known to produce oscillation. Thus, it seems plausible that PD is a 'dynamical disease' in the sense that it arises from normal tremor via bifurcations in a dynamical process. This was previously suggested by Beuter and Vasilakos (8), where the reversibility of symptoms in PD by pharmacological and electrical interventions were also argued to support the idea, though the model presented there was different.

The normal dynamical regime in the network discussed above should be one of irregular activity (Figure 1). Everyone has a normal physiological tremor that is usually quite small in amplitude and irregular in character, involving what seems to be random firing of motoneurons. Of course, amplitude generally increases when tremor develops in PD, though the regular oscillations can sometimes be seen even when amplitude is not abnormally large (9).

Tremor in subjects with PD can undergo changes over periods of minutes or seconds. The higher amplitude regular oscillations can appear and disappear suddenly or gradually (10,11,12). This suggests that the parameters determining the dynamical regime are also fluctuating, or that additional inputs are being given to the network from other areas (voluntary commands or emotional state), or that other systemic variables such as the ballistocardiogram are modifying the dynamics via peripheral feedback (8).

6. DISCUSSION AND CONCLUSION

Recent reviews have underlined the necessity of network models for understanding basal ganglia functions. The model presented here, though somewhat abstracted from the details of connections of the basal ganglia and related structures, responds to this need. A number of recent network models of the basal ganglia and motor function (13,14,15,16,17) have concentrated on voluntary movement and learning, whereas we tackle an involuntary movement (i.e., tremor) for the first time from a network perspective. In this context, too, we take the approach that basal ganglia functions are understandable as part of a larger motor network.

We provide a plausible mechanism for dynamical changes in the motor circuitry of the brain that lead to regular tremor and possibly even akinesia in PD. The basic observation is that as synaptic efficacies of a group of units in the system are weakened (as for the dopaminergic neurons of the SNc in PD), then dynamical simplification can take place. Irregular activity in the network can become regular periodic oscillation or can stop altogether, going to a fixed state. Aspects of the model that seem necessary for transitions from irregular to regular behavior are: the presence of inhibition, which is needed to prevent approach to a fixed point; several interconnecting loops; saturation of activity in units. These conditions are not very restrictive.

The irregular regime in the model resembles recordings of normal physiological tremor at the periphery. The periodic regime looks increasingly like the regular oscillations of tremor in PD as the period shortens. Long period behavior still looks irregular over short intervals. It is estimated that the symptoms of PD emerge only when 80-90% of the cells in the SNc are damaged. It could be argued that our simulations show a similar robustness in that short period behavior, that really looks like tremor, becomes common only when the α parameter is decreased below about 0.25.

Recordings of finger position in patients with PD who have rigidity or akinesia but not the typical tremor may sometimes resemble fixed point behavior, as much as can be expected in such a complex physiological system. Also, the difficulty in initiating movement in PD is suggestive of the necessity to provide a strong external input to a dynamical system stuck in a stable fixed state. Interestingly, the neural network model of Borrett et al. (15) for parkinsonian bradykinesia also associates akinesia with a fixed point of its dynamics. The presence of transitions in patients with PD between regular and irregular tremor suggests that a fluctuating parameter or an external input is causing a bifurcation in the dynamics of the motor circuit network, and this is understandable in terms of the model.

We know that other structures besides the SNc and other neurotransmitters besides dopamine are affected in PD. The model still ignores many of these complexities. It also does not specify the structures (and their number) which are represented by the units of the model. A weakness of this model is that it does not yet reproduce the increase in amplitude that is usually associated with harmonic tremor in PD. These facts, however, do not undermine the observation that the features of our network produce results resembling the symptoms observed in PD. Further work is under way to improve the model and to determine how oscillations in the network are expressed at the periphery.

The implication of this model that normal physiological tremor is the output of an aperiodic but deterministic dynamical system (and there is evidence for 'chaos', strictly speaking, in models of this type) is provocative. It has been argued that normal hand tremor is a result of uncorrelated firing of motor neurons driving a damped linear oscillator (18). Our model suggests that this uncorrelated firing may be deterministic. This does not rule out the possibility that other influences could be at work simultaneously in normal physiological tremor, including stochastic ones, but it does suggest that there is an aperiodic deterministic component to it. It is different from other potential models of the transition such as a Hopf bifurcation from a fixed point (normal) to a periodic orbit (parkinsonian). Transitions from fixed point to periodic behavior also occur in our network model and it might be suggested that these transitions better reflect the appearance of parkinsonian tremor. However, such transitions correspond to an increase in the synaptic efficacies, rather than a decrease, and though the increase in amplitude in parkinsonian tremor would be accounted for, other factors such as akinesia would not and normal physiological tremor would then have to originate elsewhere.

7. ACKNOWLEDGEMENTS

This work was supported by grants from FCAR (Quebec) and NCERC (Canada).

8. REFERENCES

1. Zirh, T.A., Lenz, F.A., Reich, S.G., and Dougherty, P.M., 1998, Patterns of bursting occurring in thalamic cells during parkinsonian tremor. *Neuroscience* **53**: 1, 107-121.
2. Hopfield, J. J., 1984, Neurons with graded response have collective computational properties like those of two-state neurons. *Proc. Natl. Acad. Sci. USA* **81**: 3088-3092.
3. Edwards, R., Beuter, A. and Glass, L., 1999, Parkinsonian tremor and simplification in network dynamics. *Bull. Math. Biol.* **61**:157-177.
4. Glass, L. and Pasternack, J.S., 1978, Stable oscillations in mathematical models of biological control systems. *J. Math. Biol.* **6**: 207-223.
5. Mink, J.W., 1996, The basal ganglia: focused selection and inhibition of competing motor programs. *Progress in Neurobiology* **50**: 381-425.
6. Graybiel, A.M., 1991, Basal ganglia — input, neural activity, and relation to the cortex. *Curr. Opin. Neurobiol.* **1**: 644-651.
7. Elble, R. J., 1996, Central mechanisms of tremor. *J. Clinical Neurophysiology* **13**: 133-144.
8. Beuter, A. and Vasilakos, K., 1995a, Tremor: Is Parkinson's disease a dynamical disease?. *Chaos*, **5**: 35-42.
9. Beuter, A. and Edwards, R., 1999. Evaluating frequency domain characteristics to discriminate physiologic and parkinsonian tremors. *J. of Clinical Neurophysiology* **16**: 484-494.
10. Gurfinkel, V. S. and Osovets, S.M., 1973, Mechanism of generation of oscillations in the tremor form of parkinsonism. *Biofizika* **18**: 731-738.
11. Beuter, A. and Vasilakos, K., 1995b, Fluctuations in tremor and respiration in patients with Parkinson's disease. *Parkinsonism and Related Disorders*, **1**: 103-111.
12. Edwards, R. and Beuter, A., 1996, How to handle intermittent phenomena in time series analysis: A case study of neuromotor deficit. *Brain and Cognition* **32**: 262-266.
13. Kwan, H. C., Yeap, T.H., Jiang, B.C., and Borrett, D., 1990, Neural network control of simple limb movements *Can. J. Physiol. Pharmacol.* **68**: 126-130.

14. Mitchell, I. J., Brotchie, J.M., Brown, G.D.A., and Crossman, A.R., 1991, Modeling the functional organization of the basal ganglia: A parallel distributed processing approach. *Movement Disorders* **6**, 189-204.
15. Borrett, D. S., Yeap, T.H., and Kwan, H.C., 1993, Neural networks and Parkinson's disease. *Canadian Journal of Neurological Sciences* **20**: 107-113.
16. Contreras-Vidal, J. L. and Stelmach, G.E., 1995, A neural model of basal ganglia-thalamocortical relations in normal and parkinsonian movement. *Biol. Cybern*, **73**: 467-476.
17. Suri, R. E., Albani, C. and Glattfelder, A.H., 1997, A dynamic model of motor basal ganglia functions. *Biol. Cybern* **76**: 451-458.
18. Gantert, C., Honerkamp, J., and Timmer, J., 1992, Analyzing the dynamics of hand tremor time series. *Biol. Cybern.* **66**, 479-484.

VISUO-SPATIAL ATTENTION IN NON-DEMENTED PATIENTS WITH PARKINSON'S DISEASE
An event-related potential study

Constantin Potagas, Nguyen Bathien and André Hugelin*

Key words: Event-related potentials, Parkinson's disease, visual tasks, cognitive demands, P300.

Summary: The P3(00) event-related potentials (ERPs) elicited with visual stimuli were assessed in 53 non-demented patients with Parkinson's disease (PD) and 20 age-matched normal controls. PD patients were classified as having normal cognitive function (NC) (n = 39) or patients with cognitive deficit (CD) (n = 14), defined by basic psychometric data. We used visual tasks with various cognitive difficulties: simple visual tasks with central or displaced targets (SV/ct and SV/dt), visual oddball tasks with central or displaced targets (VO/ct and VO/dt). P3 changes in patient groups were related to the cognitive demands.

The amplitude of P3 was larger for the NC group in the VO/ct task and for the CD group in the SV/dt and VO/ct tasks. It was lower than contols for CD patients for the VO/dt task. Estimates of cognitive load in each task by the difference waves (ERPs to dt minus ERPs to ct) demonstrated that DLP (difference late positivity, component related to P3) was negative for the CD group in the VO task, whereas DLP was positive and greater than the controls in both the NC and CD groups in the SV task. Prolongation of P3 latency for patient groups was also related to task demands. There were significant differences between patients and controls for NC patients in the SV/dt, VO/ct and VO/dt tasks, and for CD patients in all four sets of experimental conditions. The results suggest that the amplitude of P3 reflects the use of the attentional resources available for the task, and P3 latency, the time required for cognitive processing.

---

* C. Potogas, N. Bathien, and A. Hugelin, Centre R. Garcin, Service de Neurologie, Paris, France

1. INTRODUCTION

The feature-integration theory of visual attention (Treisman, 1986) gave spatial processing a central integrative role in visual perception. Feature targets can be identified without correctly localized, while feature-conjunctions are correctly perceived only when attention is focused on their location (Treisman, 1982). The P3 (00) component of the event-related potentials (ERPs) is related to cognitive processing (Donchin et al., 1978 ; Hansch et al., 1982). Recording of ERPs has been widely used to study cognitive function in patients with Parkinson's disease (PD) (Ebmeyer, 1992). The P3 component is commonly elicited in the oddball paradigm, which requires subjects to detect a target stimulus occurring less frequently than the non-target or standard stimulus. P3 is a long-latency positive component with its maximum amplitude at parieto-central scalp locations. The latency of P3 peak is a measure of stimulus duration evaluation processes, involving for example encoding, recognition and classification (Kutas et al., 1977 ; McCarthy and Donchin, 1981). The amplitude of P3 is an index of the perceptual/cognitive resources devoted to processing the stimulus (Israel et al., 1980 ; Wickens et al., 1983). In a previous study (Bange and Bathien, 1998), we showed that the P3 ERP may be an index of the contribution of the central processing to psychomotor retardation in depressed patients.

Changes in P3 amplitude and latency in PD have been reported in numerous ERP studies, and seem to depend upon disease severity. Green et al. (1996) found that P3 amplitude was larger in mildly affected PD patients, with no change in P3 latency. Studies reporting prolonged P3 latency have included patients with more advanced disease (Hansch et al., 1982 ; Stanzione et al., 1991). It has also been suggested that prolongation of P3 latency occur only in PD patients who are demented (Goodin and Aminoff, 1987 ; Tachibana et al., 1992). If P3 abnormalities were to be reliably demonstrated in PD, they could reflect changes in cognitive processing. Therefore it is of interest to record P3 amplitude and latency as a function of task demands. Presumably, P3 data would deteriorate for more challenging tasks exceeding the available resources or in more severely affected patients with cognitive dysfunction.

The relationship between cognitive function or the P3 component and antiparkinsonian medication is still unclear. Some authors have found an effect of such medication on cognitive function assessed by the P3 component (Riklan et al., 1976 ; Prasher and Findley, 1991 ; Stanzione et al., 1991), whereas others found no cognitive changes or alterations in P3 latency as a result of l-Dopa or anticholinergic therapy (Rafal et al.,1984 ; Goodin and Aminoff, 1987 ; Tachibana et al. , 1992). It has been suggested that the effect of antiparkinsonian therapy upon cognitive function is small or non-existent.

The effect of patient age on P3 changes should also be investigated, because an effect of normal ageing has been demonstrated using both oddball and non-oddball paradigms (Snyder and Hillyard, 1979 ; Kutas et al., 1994). P3 latency increases systematically with advancing age and the function is linear (Anderer et al., 1996). The results for P3 amplitude are less consistent in that changes in amplitude are associated with a shift in its scalp distribution (Polich, 1991 ; Friedman , 1995). One study has reported no effect of age on P3 amplitude and latency in PD (Green et al., 1996).

The aim of this study was to investigate P3 related to visual task performance in PD patients and attempt to address the critical issues arising from previous studies. Firstly, to obtain less heterogenous patient samples, non-demented PD patients were classified as having normal cognitive function (NC) or cognitive deficit (CD), defined by basic psychometric examination. Secondly, to examine the effects of task demands on P3 changes, the chosen tasks included a simple visual detection task (SV task) and a standard visual oddball task (VO task). The cognitive difficulty of each task was manipulated using visual targets presented at a predictable location in the fovea (centre targets) or targets at unpredictable location in the eccentric visual field (displaced targets). Thirdly, the relationship between P3 amplitude and latency with age were assessed by regression analysis. To eliminate the age shift effect, P3 amplitude and latency were also assessed using data adjusted to an age of 65 years. Thus this study examines the relationship between changes in P3 and cognitive function in PD patients and determines the contribution of age to the effect. The cognitive load on P3 in each task was investigated using the difference wave technique.

2. SUBJECTS AND METHODS

2.1 Subjects

A group of 53 patients with idiopathic Parkinson's disease (PD) attending the outpatient clinic (Centre R.Garcin - Neurologie) and 20 normal control subjects were studied. All gave informed consent. The patients selected had no history of thalamotomy or any other previous or concurrent neurological disease. Patients who could not understand the behavioural tasks were not included in the study. Dementia was assessed using the neurologist's mental status examination and clinically demented patients were excluded from this study. All patients were evaluated with the Mini-Mental State Examination (Folstein et al., 1975): a result over 26/30 was required. Each patient underwent complete neurological evaluation prior to electrophysiological and neuropsychological testing. Tasks were selected to minimise motor requirements and they tested: word fluency (Binois - Pichot test; Binois and Pichot, 1959); visual perception and organisation (Benton Visual Retention Test -form F) (Benton et al., 1983); abstract reasoning (Raven's Progressive Matrices Test-PM 47)(Raven, 1960); and memory (Logical Memory Subtest of the Wechsler Memory Scale) (Wechsler, 1981). Individual results were rated by a clinical psychologist (E. Debrandt). Performance was classified as pathological when there was a significant difference between the actual performance and normative data, estimated on the basis of the patient's education and A patient was considered to have a cognitive deficit when he had pathological occupation.performances on three or more of the five tests. PD patients were thus classified as having cognitive deficit (CD patients), or normal cognitive function (NC patients).

Table 1 shows the clinical characteristics of the 53 PD patients; of these 14 patients met the criteria for cognitive dysfunction. 2 of the 14 CD patients were receiving l-Dopa plus trihexiphenidyl, and 4 were receiving L-Dopa alone. One NC patient was treated with L-Dopa and trihexiphenidyl and 10 were on L-Dopa alone. Other patients were taking dopaminergic medication plus other medication, including amantadine, bromocriptine, lisuride and piribedil (Table 1).

The controls were 20 subjects, matched for age and socio-economic status, with no clinical evidence of neurological or psychiatric disorders (mean age 65.7 [8.4] years, range 55-82 years; 7 women and 13 men). There was no significant difference in the ages of the patients and controls [$F (2, 69) = 2.75$, NS]. All subjects described themselves as right-handed.

2.2 Stimuli and Procedures

Each subject was seated in a comfortable chair with neck support in a dimly lit room. All stimuli were presented on a 22 inch video monitor 80 cm in front of the subject's eyes. Subjects were asked to focus on a plus symbol continously present on the monitor screen. Eye fixation was checked during a practice session of the experiment, and eye position was continously monitored by electro-oculogram (EOG) recording.

Visual stimuli consisted of vertical bars (0.5° X 1.5°) flashed for 200 msec at random intervals. The central target (ct) was in a fixed position to the near right of the fixation point (plus symbol) in the centre of the screen, in the binocular zone of the visual field. There were 24 possible locations in the eccentric visual field in which the displaced target (dt) could appear. The most peripheral target was separated from the fixation point by 12°. Inter-stimulus-intervals (ISIs) were varied randomly between 800 ms and 3 s (rectangular distribution).

Target presentations were organized according to two paradigms. For the single-stimulus paradigm, the simple visual task (SV task), the subject was asked to respond as fast as he/she could by pressing a button with the preferred hand when he/she detected the visual stimulus. There were two SV tasks: one with central targets and one with displaced targets. For the oddball paradigm, the visual oddball task (VO task), the two stimuli were a target bar and a non-target square. Non-target squares (1.5° X 1.5°) were flashed for 200 ms to the near left of the fixation point, with the plus symbol in the centre of the screen. In the VO task with central targets, the target stimuli were vertical bars (0.5° X 1.5°) to the near right of the fixation point. In the VO task with displaced targets, the target stimuli were randomly distributed among the 24 possible locations in the eccentric visual field. Stimuli were delivered randomly and ISIs varied between 800 ms to 3 s. Eighty percent of the stimuli were non-target and twenty percent were target. The subject was instructed to respond selectively to the stimulus target by pressing a button as in the SV task.

The experiment consisted of 4 experimental blocks of 30 trials which were the SV task with central targets, the SV task with displaced targets, the VO task with central targets and the VO task with displaced targets, performed successively. Each block of trials was initiated by the experimenter. Subjects were given feedback (correct/incorrect) at the end of each trial. The 4 experimental tasks (SV/ct, SV/dt, VO/ct and VO/dt) were given randomly to avoid learning effect.

There was a practice run before the experimental session to ensure that each subject understood and was able to perform the task. Electroencephalogram (EEG) and electro-oculogram (EOG) were monitored on-line, and subjects were informed if eye movements occurred and were encouraged to maintain eye fixation on the centre of the screen.

2.3 EEG Recording and Averaging

The EEG was recorded using Ag/AgCl disk electrodes placed on the midline frontal (Fz), central (Cz) and parietal (Pz) sites, with one electrode below the outer canthus of the left eye for monitoring eye movements. All electrodes were referred to linked ear lobes. Electrode impedance was kept below 5 kilo-ohms. All signals were fed through a Nihon-Kohden Neuropack 8 system. The EEG was amplified (bandpass 0.03 - 30 Hz), digitized on-line at a sampling rate of 200 Hz over 2000 ms, beginning 400 ms before the stimulus. A computer controlled stimulus generation and timing, and recorded subject reponses and reaction time (RT).

ERPs were averaged separately for each task, and for each of the rare and frequent stimuli of the VO tasks. Trials contaminated by eye blinks or movements (excessive muscle activity) were rejected by a computer algorithm during averaging; less than 10 % of the trials were lost due to such artefacts for all age groups. ERP averaging was performed on-line and there were 30 averaged trials for each task. Data were coded and stored for off-line analysis.

2.4 Data Analysis

ERPs were quantified by computer analysis of peak measures (latency and amplitude of maximum negative or positive deflection within a specific time window) and mean amplitude within the time window, all referred to a baseline voltage averaged over a 400 ms interval before the stimulus. Activities at latencies greater than 800 ms were considered to be non-specific post-task activity and were not measured. The measurement window for components was based on inspection of grand-averaged and averaged wave forms for each subject. For the SV tasks (SV with central or displaced targets), the N1 component was identified as the maximum negative peak between 125 and 250 ms, and the peak of the late positive component (defined as P3) as the maximum positive peak between 250 and 700 ms. Measurement of the P2 and N2 components was too unreliable to be of use. For the VO task, the peak amplitude and latency of the N2 component were measured, N2 being defined as a negative peak between 200 and 400 ms of the response to rare stimuli. P3 was identified as the largest positive peak between 280 and 700 ms.

Data from SV and VO tasks were analysed separately. Amplitude and latency values were subjected to a mixed design analysis of variance (ANOVA) using the 3 subject groups (Controls, NC and CD patients) as the between-subject variable, and the target (central or displaced) and electrode site (Fz, Cz, Pz) as the within-subject factors. Repeated measures with greater than 1 degree of freedom were evaluated with P values adjusted using the Greenhouse - Geisser epsilon correction factor. Post-hoc analyses using the Tukey Honest test (Tukey HSD) were performed in some cases, depending on the significance of the main effects and interactions in the initial analysis. Linear regression analysis was performed for the P3 amplitudes and latencies with age as the independent measure. A significance level of $P < 0.05$ was adopted for all statistical comparisons.

To clarify ERP changes associated with displaced target processing, difference waves were also calculated by substracting the averaged ERPs to central targets from ERPs to displaced targets. The analysis was confined to the digitising epoch (i.e 1600 ms after stimulus onset) and performed on 2 components of the difference waves: DN (130 to 270 ms) and DLP (300 to 600 ms) for the SV and VO tasks.

3. RESULTS

3.1 Behaviour Measurements

The mean percentage of task performance errors within subject groups was lesser than 5% for the various tasks. A 2-way ANOVA (group x accuracy) was used to analyze the task accuracy (percentage of targets correctly identified) for both tasks with the two types of target. There was no main effect of " group " (all at p > 0.24). Thus all subjects were able to perform the tasks equally well in all experimental conditions. Table 2 gives the mean reaction times (RTs) for all three groups and the results of the 2-way ANOVA (group x target) for the simple visual (SV) and visual oddball (VO) tasks. The RT of both tasks was higher for displaced targets [$F(1, 69) = 92.28; p < 0.001$]. There was also a significant main effect of " group ". Post hoc testing (Tukey HSD) showed that the RTs of the patient groups were significantly longer than those of the controls for the SV tasks. The RTs of the CD group only were longer in the VO tasks. ANOVA also demonstrated a significant group x target interaction (Table 2). The results showed that the groups did not react similarly to the SV and VO tasks.

3.2 P3 Component

Grand average ERPs from the 73 subjects divided into 2 patient groups (NC and CD patient) and controls are shown superimposed for the four sets of experimental conditions (Fig 1). For all subjects, the averaged ERP curves have similar morphology, including N1 and P3 components in the SV task (N2 was often difficult to identify), and N1, N2 and P3 components in the VO task. The P3 component generally peaked at 300 to 600 ms and had its maximum scalp distribution at the Pz site. 2-way ANOVA (target x site) was used to analyze P3 amplitude and latency in the SV and VO tasks for the three groups, separately. P3 amplitude varied in response to targets as a function of electrode location in the two tasks, being largest at the Pz electrode site (all at $p < 0.001$). There was no significant difference in P3 latency as a function of electrode location.

3.2.1. Effects of Task Demands (Central or Displaced Target)

The P3 component at the Pz site was evaluated with a group X target ANOVA for the SV and VO tasks, separately. Table 3 includes target P3 peak amplitude and latency for each group and task. ANOVA showed that there was no main effect of " group " in the SV and VO tasks for P3 amplitude. However, post hoc testing (Tukey HSD) showed that P3 amplitude in the SV/dt task was larger for CD patients ($p = 0.029$) and in the VO/ct task for both NC ($p < 0.001$) and CD ($p = 0.012$) patients, with a smaller P3 amplitude for CD patients ($p = 0.002$) in the VO/dt task. The main effect of " target " was significant for the SV task [$F(1, 69) = 35.75; p < 0.001$], but not for the VO task [$F(1, 67) = 0.01$; NS]. In the VO task, the "target" effect was significant for controls ($p = 0.006$), but not for NC patients. P3 amplitude of CD patients in the VO/dt task was lower than that for CD patients in the with VO/ct task ($p = 0.005$). This pattern of changes was shown by a significant group X target interaction [$F(2, 67) = 13.617 ; p < 0.001$].

P3/Pz latency varied as a function of task and target, and there was a significant difference between groups [for the SV task: $F(2, 69) = 13.10, p < 0.001$; for the VO task: $F(2, 69) = 28.78, p < 0.001$] (Fig 1). Post hoc tests showed that P3 latency was consistently higher for CD patients in the SV/ct task ($p = 0.002$), and for both NC ($p = 0.023$) and CD ($p < 0.001$) patients in the SV/dt task. In the VO tasks, P3 latency was longer for both types of target for both patient groups (NC patients: $p < 0.001$ for ct and $p = 0.027$ for dt; CD patients: $p < 0.001$ in all

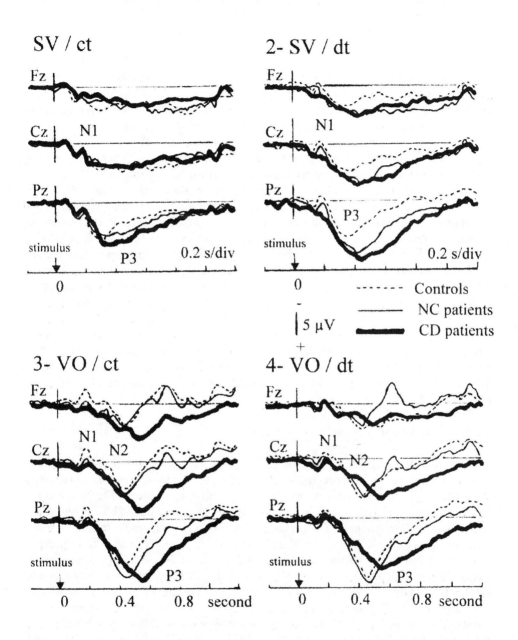

Figure 1. Upper row (1,2): superimposition of group average ERPs in the simple visual task with central/displaced targets (SV/ct and SV/dt) for each subject group (Controls, NC: patients with normal cognitive function, CD: patients with cognitive deficit). Note the change in P3 amplitude and latency in SV/dt for the patient groups. Lower row (3,4): group average ERPs in the visual oddball task (VO/ct, VO/dt) for each subject group. The CD group had a longer P3 latency than did the controls for the VO/ct task, and a reduced P3 amplitude for the VO/dt task.

cases). The main effect of "target" was significant for the SV task [$F(1, 69) = 25.16, p < 0.001$] but not for the VO task [$F(1, 69) = 3.74, p = 0.056$].

3.2.2. Effects of Age

The relationship between and P3 amplitude and latency were assessed by regression analysis. Separate regression analyses of age on P3 amplitude and latency at the Pz site of the three groups were performed for both tasks. In normal controls, there was no significant correlation between age and P3 amplitude in the SV task for both ct and dt conditions. However there were significant and negative correlations in the VO task (ct: $r = -0.60$, $p = 0.004$; dt: $r = -0.50$, $p = 0.024$) and lower P3 amplitude was associated with advancing age. The age/P3 amplitude correlations in both tasks were not significant for either NC or CD patients.

For P3 latency, the age effect also differed between patients and controls. In normal controls, P3 latency was gradually delayed with increasing age during the SV/ct and dt and VO/ct and dt tasks. The relationship between P3 latency and age was linear. The age/latency correlations were between 0.48 and 0.56 (SV/ct task: $r = 0.55$, $p = 0.010$; SV/dt task: $r = 0.48$, $p = 0.030$; VO/ct: $r = 0.56$, $p = 0.009$; VO/dt: $r = 0.55$, $p = 0.011$). The slopes were 2.2 ms/year (ct) and 1.9 ms/year (dt) for the SV task, and 1.7 ms/year (ct) and 3.3 ms/year (dt) for the VO task.

The age/P3 latency correlation in PD patients was only significant for NC patients (SV/ct task: $r = 0.59$, $p < 0.001$; SV/dt task: $r = 0.50$, $p = 0.001$; VO/ct task: $r = 0.63$, $p < 0.001$; VO/dt task: $r = 0.63$, $p < 0.001$). Correlations were not significant for the CD patients ($r < 0.14$ for the four sets of experimental conditions). 7 of the NC patients (19 %) and 11 (28 %) had longer P3 latencies than the controls for the SV/ct and SV/dt tasks. P3 latency was longer for 18 of the NC patients (46 %) in the VO/ct task and 10 (25 %) in the VO/dt task. P3 latencies for more than half the CD patients were above the 95 % confidence limits in the four sets of experimental conditions (Fig 2).

3.2.3. P3 Component Adjusted to Age 65 Years

To eliminate the age shift effect, the P3 amplitude and latency data of the three groups were adjusted for age, using the regression slope of the age/P3 relationship in normal controls. The P3 latency of each subject at the Pz site was adjusted to age 65 years. For P3 amplitude, data from the VO task only were adjusted. Fig 3 shows the mean adjusted P3 amplitudes and latencies of the three groups for the two tasks. ANOVA gave similar results to those obtained from analysis of the raw P3 amplitude and latency data. There was a significant group X target effect for P3 amplitude in the VO task [$F(2, 68) = 14.70, p < 0.001$] mainly due to changes in the P3 amplitude of NC and CD patients (Fig 3). Post hoc testing showed that P3 amplitude was larger for the NC and CD group than for the controls in the VO/ct task (NC patient: $p = 0.021$; CD patient: $p < 0.001$). The difference between patients and controls was not significant in the VO/dt task. However, whereas there was no difference in P3 amplitude for the two target types for NC patients, P3 amplitude was significantly lower for CD patients in the VO/dt task ($p = 0.002$).

For P3 latency, there was a significant main effect of "group" for both SV and VO tasks [SV task: $F(2, 69) = 9.71, p < 0.001$; VO task: $F(2, 70) = 25.14, p < 0.001$]. Post hoc tests indicated that P3 latency of CD patients was longer in both SV and VO tasks (SV/ct: $p = 0.008$; SV/dt: $p < 0.001$; VO/ct and VO/dt: $p < 0.001$). P3 latency of NC patients was longer in the SV/

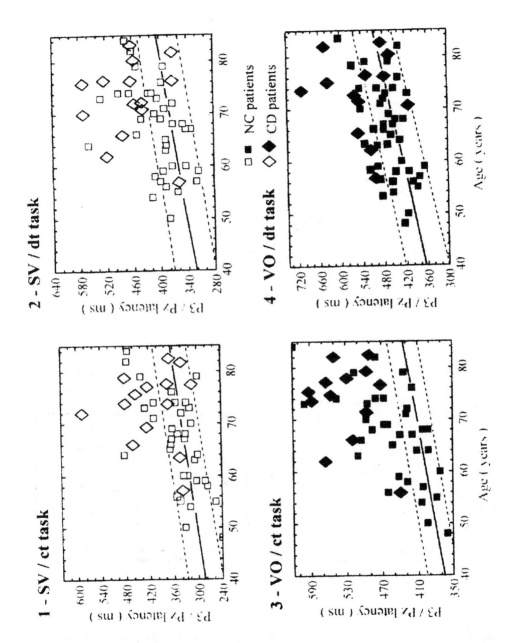

Figure 2. P3 latency data for the NC and CD patients plotted against age in the SV (1,2) and VO (3,4) tasks. The normal regression line is indicated with dotted lines representing 95 % confidence limits based on the normal control subjects. Seven and 11 patients of the NC group (n = 39) had a prolonged P3 latency for the SV/ct and SV/dt tasks respectively, 18 and 10 NC patients for the VO/ct and VO/dt tasks. P3 latencies of more than half CD patients were above the confidence limits in the four tasks.

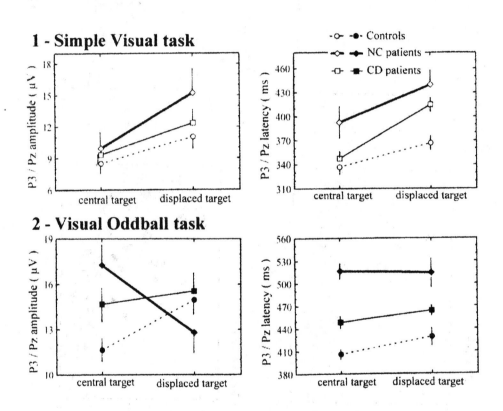

Figure 3. P3 amplitude and latency (mean +/- SE) at Pz electrode site adjusted to an age of 65 years for the three groups in the SV and VO tasks. The amplitude was significantly larger for CD patients in the SV/dt task ($p = 0.029$), and for both NC ($p = 0.021$) and CD ($p < 0.001$) patients in the VO/ct task. It was lower ($p = 0.002$) for CD patients in the VO/dt task than in the VO/ct task. For P3 latency, there was signficant prolongation in the SV/ct task for CD patients ($p < 0.001$), and in the SV/dt, VO/ct and VO/dt tasks for both patient groups.

dt task (p = 0.005). In the VO task, P3 latency was longer in the NC patients than controls for both target types (VO/ct: p = 0.001; VO/dt: p = 0.018).

3.3 Other ERP Measures

The amplitudes and latencies of the ERP components preceding P3 (N1 and N 2) were compared for the groups. As these components differed from individual to individual, their parameters were better determined from individual ERP measurements than from grand mean ERPs. 2-way ANOVA (group x target) was performed to assess the main effect of "group" on the amplitude and latency of the N1 component for the SV task and of N1 and N2 for the VO task at the Cz midline electrode. There were no statistically significant differences between groups for N1 amplitude and latency in the two tasks [N1/Cz amplitude, SV task: $F(2, 69) = 0.08$, NS; VO task: $F(2, 65) = 2.45$, NS; N1/Cz latency, SV task: $F(2, 69) = 0.74$, NS; VO task: $F(2, 65) = 2.36$, NS]. For N1 amplitude, ANOVA showed that there was a significant main effect of "target" in the SV task [$F(1, 69) = 5.85, p = 0.018$] mainly due to a larger N1 amplitude in the SV/dt task. This pattern of change was not observed in the VO task [$F(1, 65) = 0.02, p = 0.868$]. There was no main effect of "group" nor any significant effect of "target" for N2 amplitude and latency in the VO task [N2/Cz amplitude: $F(2, 45) = 0.98$, NS; N2/Cz latency: $F(2, 45) = 2.30$, NS].

3.4 Difference Wave ERPs

For further analysis of P3 modulations by the two tasks with various demand levels, difference waves were calculated by subtracting the averaged ERPs to ct from ERPs to dt for each subject. Fig 4 shows grand mean difference waves obtained from the SV and VO tasks for the three groups. 2-way ANOVAs (group x electrode site and group x task) were used to analyze the amplitudes and latencies of the difference wave components.

3.4.1. DN Component

The first robust component was the DN (difference negativity), a negative deflection. In the measurement window (130 to 270 ms) it was mainly generated by the N1 component in the SV task and by the N1 and N2 components in the VO task: DN latencies were longer in the VO task (Table 4). There was no main effect of "group" nor any significant group on task interaction for DN amplitude and latency. DN amplitude of the three groups was similar in the anterior-posterior distribution across the midline electrode sites.

3.4.2. DLP component

Following the DN component was a large positive deflection, DLP (difference late positivity). The DLP amplitude of controls at the midline electrodes in the SV and VO tasks was at a maximum at the Pz electrode [$F(2, 72) = 11.17, p < 0.001$, epsilon = 0.91]. There was no main effect of "electrode site" for DLP latency.

2-way ANOVAs (group X task) were performed on the amplitude and latency of the DLP component at Pz (Table 4). For DLP latency, there was a significant main effect of "group". Post hoc testing of the group x task interaction showed that the DLP latency of CD patients

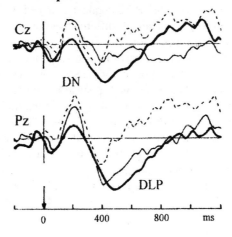
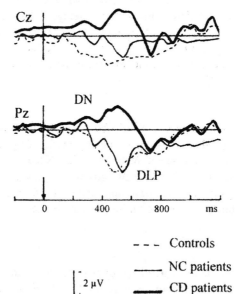

Figure 4. Superimposition of group average ERPs in the SV (1) and VO (2) tasks for each subject group after subtraction. ERPs to center targets were subtracted from ERPs to displaced targets. Note the more positive deflection of DLP (difference late positivity) for both patient groups in the SV task and the negative DLP deflection for CD patients in the VO task.

was longer than for controls in the SV task (p < 0.001). This latency was significantly longer than that of the controls for both NC (p = 0.006) and CD (p < 0.001) patients in the VO task.

For DLP amplitude, ANOVA showed there was a significant group x task interaction, mainly due to CD patient DLP modulations in the VO task. The DLP amplitude of CD patients was positive in the SV task and negative in the VO task (Fig 4). Post hoc tests showed that the DLP amplitude of both NC and CD patients was significantly higher in the SV task. In the VO task, DLP amplitude was negative for CD patients. In contrast, it was positive for NC patients in the VO task, with an amplitude that did not differ significantly from that of the controls (Table 4).

4. DISCUSSION

This investigation has demonstrated differences in P3 amplitude and latency between non-demented PD patients and controls in the performance of simple visual detection and discrimination tasks. In both patient groups, P3 changes occurred as the cognitive load of the task increased, and changes in P3 amplitude seemed to reflect the patient's ability to mobilize sufficient attention resources to perform the task. Before considering the implications of these results for cognitive dysfunction, the factors that could have confounding effects will be discussed. These are the cognitive effects of the SV and VO tasks with central and displaced targets.

4.1. SV and VO Tasks

In the VO task, subjects are presented with two classes of stimuli in a pseudo-random distribution. The target or "oddball" stimuli occur unexpectedly with an *a-priori* probability (20 %) lower than that of the standard or non-target stimulus. The SV task is a variant of the oddball paradigm in which a single stimulus is presented. The task is simpler than the VO task because the subject is required to react to every stimulus. It has been demonstrated that both procedures engage the same neural and cognitive mechanisms to elicit the P3 component (Polich *et al.*, 1994).

Visuospatial attention has been studied extensively in paradigms that use a warning cue to indicate the probable location of a subsequent target stimulus. Accelerated responses (benefits) to accurately cued targets and slower responses (costs) to falsely cued targets can be measured and may reflect the action of the spatial orienting mechanism (Posner, 1980 ; Downing, 1988 ; Berlucchi et al., 1989 ; Müller and Rabbitt, 1989). Egly and Homa (1984) demonstrated that cuing improves the subject's identification and location of a displaced letter-target from the fixation point, a central target. The displaced letter appears at one of eight locations on one of the three rings surrounding the fixation point. Invalid cuing produces costs comparable in magnitude to the benefits. RT studies have reported stronger facilitation for peripheral than for central cues (Müller and Rabbitt, 1989 ; Yamaguchi *et al.*, 1994). In this study, uncued trials were used for the SV/dt and VO/dt tasks and prior information regarding the location of the displaced target was not given. Displaced targets in each task corresponded to conditions with ambiguously located targets. The subject's performance in identifying the displaced target may reflect his ability to assign additional processing capacity to the given item. Our RT data support this hypothesis (Table 2). RTs to displaced targets were significantly longer than those to central targets in the SV and VO tasks. Thus, the task with displaced targets is more demanding than the one with central targets. We presume that in this procedure the four ordered tasks (SV/ct and SV/dt, VO/ct and VO/dt) represent an increasing heriarchy of cognitive difficulty, and include an increasing number of cognitive components, whereas the motor requirements are kept constant.

4.2. P3 Amplitude Modulations

P3 amplitude is an index of the perceptual/cognitive resources devoted to processing the stimulus (Israel et al., 1980 ; Wickens et al., 1983). Using paradigms which changed the cognitive demands of the visual tasks, these results confirm that there is an increase in P3 amplitude for the more demanding VO task in control subjects. They indicate that changes in P3 amplitude differ in both patient groups from those of age-matched controls. P3 amplitude increases were recorded in NC patients performing the VO/ct task, and in CD patients performing the SV/dt and VO/ct tasks. P3 amplitude reduction was observed in CD patients for the VO/dt task.

P3 amplitude modulations in PD are widely reported in the literature. Most studies have either not reported amplitude analysis (Goodin and Aminoff, 1987 ; O'Donnell et al., 1987 ; Stanzione et al., 1991) or did not include a control group for comparison (Pang et al., 1990). Others have reported no change in P3 amplitude in severely affected patients (Hansch et al., 1982 ; Graham et al., 1990). There is one study that supports to some extent these results: Green et al. (1996) recorded an increase in P3 amplitude in auditory modality for patients with mild PD but no neuropsychological deficit. They suggested that these patients were still able to mobilize sufficient attentional resources to perform some cognitive tasks well.

This interpretation is consistent with neuropsychological findings showing changes in the attentional resources of PD patients (Brown and Marsden, 1991 ; Cooper and Sagar, 1993). PD patients perform less well than controls in demanding mental processes. Brown and Marsden (1991) have proposed " a resource depletion hypothesis " suggesting that PD patients have fewer attention resources to allocate to tasks. Our finding of larger P3 amplitude in NC and CD patients performing the VO/ct task does not necessarily conflict with this hypothesis. Our results also demonstrate that P3 amplitude for CD patients was abnormally large in the SV/dt task and small in the VO/dt task (Table 3). Green et al. (1996) described a positive correlation between P3 amplitude and scores on the more attention-demanding tests for PD patients without neuropsychological deficit. They suggested that mildly affected PD patients may still be able to perform well in neuropsychological tests whose demands are within the available attention resources. In such conditions, the four tasks used (SV/ct, SV/dt, VO/ct and VO/dt) did not exceed the total resources available to NC patients and they were devoting more attentional resources to perform the VO/ct task than the controls. The same mechanism may explain the changes in P3 amplitude for CD patients in the SV/dt and VO/dt tasks. The reduction in P3 amplitude in the VO/dt task for CD patients may reflect insufficient attentional resources in this condition for these patients. The NC patients may still have sufficient attentional resources to perform the VO/dt task.

The design of the tasks with two difficulty levels (ct and dt conditions) in each paradigm makes it possible to analyze attention-resource allocation in task performance. The DLP component of the difference waves reflects the difference in P3 potentials between dt and ct conditions in which there are various difficulty levels for target detection and discrimination.

The larger DLPs obtained in the SV task for NC and CD patients (Fig. 4) indicate that these patients mobilize more attentional resources than the controls. The amplitude ratio P3 dt/P3 ct was positive for both patient groups, but the difference in P3 amplitude was not significant for the controls. DLP for CD patients was negative in the VO task, reflecting lower P3 amplitude in the VO/dt task than in the VO/ct task (Fig 3). DLP in the VO task was always positive for NC patients and controls. The results suggest that CD patients experience greater " resource depletion " than NC patients.

An alternative explanation for the abnormal P3 amplitude in PD is that P3 modulations in patients reflect motivation rather than brain dysfunction. Individual motivation has been shown to affect P3 in normal subjects (Begleiter et al., 1983). This possibility can be excluded because of the protocol design. Visual tasks were designed to ensure that the average ERP was not contaminated by responses to targets which had not been identified as such by the subject, and with objective measures of the task performance. Both patients and controls were highly and equally motivated to perform well, and their accuracy did not differ.

4.2. Prolongation of P3 Latency

P3 latency is a measure of the duration of stimulus evaluation processes such as encoding, recognition and classification (Kutas et al., 1977 ; McCarthy and Donchin, 1981). However, individual measurements do not only depend on the cognitive ability of the individual subject, but also on several factors (Polich and Kok, 1995 ; Friedman, 1995). The subject's age is one of the most important factors. P3 latency increases with normal aging, regardless of the stimulus modality. The degree to which P3 latency increases differs from study to study, but is between 1 and 2 ms per year (inferred from the slope of P3 latency on age) in the oddball paradigm (Anderer et al., 1996). The age-related change in P3 latency is also found in the visual non-oddball paradigm (Kutas et al., 1994). Our results for control subjects are consistent with previous reports. In this study, the age-related increase in P3 latency for controls was between 1.9 ms/year in the SV/dt task (non-oddball paradigm) and 3.3 ms/year in the VO/dt task (oddball paradigm).

P3 latency prolongation in PD is commonly reported. Previous studies suggest that changes in P3 latency in PD patients may be related to disease severity: there is no change in mildly affected patients (Goodin and Aminoff, 1987 ; Prasher and Findley, 1991 ; Tachibana et al., 1992 ; Green et al., 1996), and P3 latency is prolonged in patients with more advanced desease (Hansch et al., 1982 ; Pang et al., 1990 ; Stanzione et al., 1991 ; Rumbach et al., 1993).The results of this study show that P3 latency from NC and CD patients was differentially influenced by visuals tasks with a hierarchical cognitive difficulty. Whereas there was prolongation of P3 latency for both NC and CD patients in the VO/ct and VO/dt tasks, P3 latency was only prolonged in the SV/dt task for NC patients, and in both SV/ct and SV/dt tasks for CD patients (Table 3). This phenomenon was reproduced when each subject's data was adjusted to age 65 years using the regression slope of the age/P3 latency relationship in normal controls. Thus, the change in P3 latency in PD patients is not linked to the normal decrease in cognition function with age, but is instead associated with disease severity.

The cognitive load effect on P3 latency, as measured by the dt/ct ratio was different for the SV and VO tasks. P3 latency was more prolonged in the SV/dt task than the SV/ct task for both NC and CD patients, but there was no difference in P3 latency under dt and ct conditions for all three subject groups (Table 3, Fig.3). The results are supported by findings from RT studies using classic simple and choice reaction time paradigms in P3. This was expected because P3 latency can be correlated with RT when tasks that induce P3 are combined with the requirement to make the motor response to the targets for RT measures (Ritter et al., 1972 ; Kutas et al., 1977 ; Roth et al., 1978). Goodrich et al. (1989) found prolonged RTs of simple reaction time paradigms in PD that they interpreted as an impaired-demanding process involved in rapid demanding. Zimmerman et al. (1992) demonstrated slowing of cognitive processes in choice reaction time paradigm in PD by investigating the decision time (overall response time minus movement time in a simple reaction task). If Goodrich's interpretation (1989) is correct, then PD patients (NC and CD) may still be able to generate attention-demanding processes in the SV task. The time demands of various cognitive processes for the VO task in both NC and CD patients exceeds the total resources available.

The results show that P3 amplitude and latency may be sensitive indicators of the change in cognitive function that occurs in PD. P3 changes were related to the cognitive demands of the visual tasks and the degree of cognitive deficit.

5. TABLES

Table 1. Clinical characteristics of patients with Parkinson's disease

	NC patients	CD patients
No of subjects	39	14
Sex F/M	7/32	5/9
Age (years) mean (SD)	66.2 (8.9)	72.0 (7.0)
range	48-82	57-82
Duration of disease		
(years) mean (SD)	7.7 (4.3)	8.1 (5.7)
range	2-15	2-19
Stage (Hoehn - Yahr scale) (n)		
I	5	0
II	16	5
III	16	6
IV	2	3
V	0	0
Antiparkinsonian therapy (n)		
Dopa	10	4
Amantadine	1	0
Dopa + trihexyphenidyl	1	2
Dopa + bromocriptine	7	2
Dopa + lisuride	2	0
Dopa + piribedil	14	3
Dopa + amantadine		
+ trihexyphenidyl	1	1
Dopa + amantadine		
+ bromocriptine	2	0
Dopa + amantadine		
+ lisuride	0	2
Dopa + bromocriptine		
+ piribedil	1	0

NC: normal cognitive function; CD: cognitive deficit
Dopa: with dopa-decarboxylase inhibitors

Table 2. Reaction times (in ms) in the simple visual and visual oddball tasks with the 2 target types recorded for the subject groups

Group	Control	NC patient	CD patient	2-way ANOVA factors	F	df
Simple visual task				group(1)	47.44***	2/69
center target	273 (9.3)	314 (8.5)**	403 (22.7)***	target(2)	92.28***	1/69
displaced target	318 (8.9)	362 (8.5)**	506 (16.9)***	1 X 2	6.35*	2/69
Visual oddball task				group(1)	61.83***	2/69
center target	378 (7.3)	402 (7.0)	543 (24.3)***	target(2)	60.82***	2/69
displaced target	423 (8.1)	441 (7.5)	604 (24.2)***	1 X 2	1.13	2/69

Means and SE are shown. Probabilities relative to controls for post hoc tests are indicated.
* = p<0.05, ** = p<0.01, *** =p<0.001.

Table 3. P3 / Pz amplitude and latency as a function of group and target presentation for the Simple visual task and the Visual oddball task.

Group	Control	NC patient	CD patient	2-way ANOVA factors	F	df
P3 amplitude (μV)						
Simple visual task				group(1)	.86	2/69
center target	8.6 (.6)	9.3 (.8)	9.9 (1.5)	target(2)	35.76***	1/69
displaced target	11.0 (1.0)	12.3 (1.2)	**15.2 (2.2)***	1 X 2	1.25	2/69
Visual oddball task						
center target	11.6 (1.0)	**15.3 (1.0)****	**15.1 (1.6)****	group(1)	1.66	2/67
displaced target	15.0 (1.1)	15.8 (1.1)	**11.0 (1.6)****	target(2)	.01	1/67
				1 X 2	13.61***	2/67
P3 latency (ms)						
Simple visual task				group(1)	13.10***	2/69
center target	336 (8.7)	350 (9.9)	**402 (17.1)***	target(2)	25.16***	1/69
displaced target	356 (9.7)	**401 (9.7)***	**454 (18.1)*****	1 X 2	1.83	2/69
Visual oddball task						
center target	406 (6.2)	**450 (9.7)***	**529 (12.0)*****	group(1)	28.61***	2/70
displaced target	425 (9.2)	**457 (6.7)***	**535 (20.7)*****	target(2)	3.74*	1/70
				1 X 2	.66	2/70

Means and SE are shown. Probabilities for post hoc tests of the interaction between group x target obtained in the ANOVA are indicated as compared to controls.
* = p<0.05, ** = p<0.01, *** = p<0.001.

Table 4. Difference wave ERPs: DN and DLP amplitude
and latency at Pz electrode as a function of group and task

	SV task	VO task	2-way ANOVA		
Group			Factor	F	df
DN amplitude (μV)					
Control	4.3 (0.4)	2.8 (0.5)	Group	0.51	2/60
Parkinson's disease					
NC patient	4.0 (0.6)	3.2 (0.6)	Task	0.01	1/60
CD patient	1.5 (0.7)	3.7 (1.1)	Group x task	2.72	2/60
DN latency (ms)					
Control	202 (8.2)	253 (14.3)	Group	1.41	2/57
Parkinson's disease					
NC patient	204 (8.7)	246 (11.2)	Task	9.43**	1/57
CD patient	237 (14.7)	264 (19.5)	Group x task	0.22	2/57
DLP amplitude (μV)					
Control	3.5 (0.2)	7.1 (1.4)	Group	11.76***	2/67
Parkinson's disease					
NC patient	**7.3 (0.3)***	5.7 (0.7)	Task	12.41***	1/67
CD patient	**7.9 (0.8)***	**-2.4 (1.9)***	Group x task	14.09***	2/67
DLP latency (ms)					
Control	415 (15.1)	462 (16.4)	Group	10.69***	2/68
Parkinson's disease					
NC patient	403 (9.4)	**516 (8.5)**	Task	100.10***	1/68
CD patient	**472 (13.9)***	**569 (17.4)***	Group x task	6.53**	2/68

Means and SE are shown. Probabilities relative to controls for post hoc tests of
the interaction between group x task obtained in the ANOVA are indicated.
* = $p<0.05$, ** = $p<0.01$, *** = $p<0.001$.

6. ACKNOWLEDGEMENTS

This work was supported by INSERM and C.H.Ste Anne. We thank M. Benkherrat for software development, E. Debrandt for neuropsychological assessment and Julie Knight for editorial help. We are also grateful to all the patients and their consultant Dr P. Rondot and Dr M. Ziegler for participating in this study.

7. REFERENCES

Anderer, P., Semlitsch, H.V., and Saletu, B., 1996, Multichannel auditory event-related potentials: effects of normal aging on the scalp distribution of N1, P2, N2 and P300 latencies and amplitudes. *Electroencephalogr. Clin. Neurophysiol.* **99**: 458-472.

Bange F., and Bathien N., 1998, Visual cognitive dysfunction in depression : an event-related potential study. *Electroencephalogr. Clin. Neurophysiol.* **108** : 472-481.

Begleiter, H., Porjesz, B., Chou, C.L., and Aunon, J.I., 1983, P3 and stimulus incentive.*Psychophysiology* **20**: 95-101.

Benton, A., Hamsher, K., Varney, N., and Spreen, O., 1983,Contributions to neuropsychological assessment: a clinical manual. Oxford University Press, New York, 1983.

Berlucchi, G., Tassinari, G., Marzi, C.A., Di Stefano, M., 1989, Spatial distribution of the inhibitory effect of peripheral non-informative cues on simple reaction time to non-fixated visual targets. *Neuropsychologia* **27**: 201-221.

Binois, R., and Pichot, P., 1959, Le test du vocabulaire. Editions du Centre de Psychologie appliquée, Paris, Section 1, 1959.

Brown, R.G., and Marsden, C.D., 1986, Visuospatial function in Parkinson's disease. *Brain* **109**: 987-1002.

Cooper, J.A., and Sagar, H.J., 1993, Incidental and intentional recall in Parkinson's disease: an account based on diminished attentional ressources. *J. Clin. Exp. Neuropsychol.* **15**: 713-731.

Donchin, E., Ritter, W., and McCallum, W.C., 1978, Cognitive psychophysiology: the endogenous components of the ERP. In: Callaway E, Tueting P & Koslow SH, eds, Event-related potentials in man, Academic Press, NY, 1978, pp 349-411.

Downing, C.J., 1988, Expectancy and visual-spatial attention: effects on perceptual quality. *J. Exp. Psychol. Hum. Percept. Perform.* **14**: 188-202.

Ebmeyer, K., 1992, A quantitative method for the assessment of overall effects from a number of similar electrophysiological studies: description and application to event-related potentials in Parkinson's disease. *Electroencephalogr. Clin. Neurophysiol.* **84**: 440-446.

Egly, R., and Homa, D., 1984, Sensitization of the visual field. *J. Exp. Psychol. Hum.Percept.Perform.* **10**: 778-793.

Folstein, M.L., Folstein, C.E., and McHuch, P.R., 1975, Mini-mental state: a pratical method for grading the cognitive state of patients for the clinician. *J. Psychiat.Res.* **12**: 189-198.

Friedman, D., 1995, Cognition in the elderly: an event-related potential perpective. In: Boller F & Grafman J, eds, Handbook of Neuropsychology, Elsevier Science B.V., Amsterdam, vol. **10**, pp 213-240.

Goodin, D., and Aminoff, M.J., 1987, Electrophysiological differences between demented and non-demented patients with Parkinson's disease. *Ann Neurol.* **21**: 90-94.

Goodrich, S., Henderson, L., and Kennard, C. On the existence of an attention-demanding process peculiar to simple reaction time: Converging evidence from Parkinson's disease. *Cognitive Neuropsychology* **6**: 309-331.

Graham, J.S., Yiannikis, C., Gordon, E., Coyle, S., and Morris, J.G.L., 1990, P300 event-related potentials in de-novo Parkinson's disease. *Clin. Exp. Neurol.* 1990. **27**: 89-98.

Green, J., Woodard ,J.L., Sirockman, B.E., Zakers, G.O., Maier, C.L., Green, R., and Watts, R.L., 1996, Event-related potential P3 change in mild Parkinson's disease. *Movement Dis.* **11**: 32-42.

Hansch, E.C., Syndulko, K., Cohen, S.N., Goldberg, Z.I., Potvin, A.R., and Tortelotte, W.W., 1982, Cognition in Parkinson's disease: an event-related potential perspective.*Ann. Neurol.* **11**: 599-607.

Israel, J.B., Wickens, C.D., Chesney, G.L., and Donchin, E., 1980, The event-related potential as an index of display-monitoring workload. *Hum Factors* **22**: 211-224.

Kutas, M., McCarthy, G., and Donchin, E., 1977, Augmenting mental chronometry: the P300 is a measure of stimulus evaluation time. *Science* **197**: 792-795.

Kutas, M., Iragui, V., and Hillyard, S.A., 1994, Effects of aging on event-related brain potentials (ERPs) in a visual detection task. *Electroencephalogr. Clin. Neurophysiol.* **92**: 126-139.

McCarthy, G., and Donchin, E., 1981, A metric for thought: a comparison of P300 latency and reaction time. *Science* **211**: 77-80.

Müller, H.J., and Rabbitt, P.M., 1989, Reflexive and voluntary orienting of visual attention: time course of activation and resistance to interruption. *J. Exp. Psychol. Hum. Percept. Perform.* **15**: 315-330.

O'Donnell, B.F., Squires, N.K., Martz, M.J., Chen, J.R., and Phay, A.J., 1987, Evoked potential changes and neuropsychological performance in Parkinson's disease. *Biol.Psychol.* **24**: 23-37.

Pang, S., Borod, J.C., Hernandez, A., Bodis-Wollner, I., Raskin, S., Mylin, L., Coscia, L., and Yahr, M,D., 1990, The auditory P300 correlates with specific cognitive deficits in Parkinson's disease. *J. Neural Transm.(P-D section)* **2**: 249-264.

Polich, J., 1991, P300 in evaluation of aging and dementia. In: Brunia CHM, Mulder G, Verbaten MN, eds, Event-related brain potential research. (EEG suppl. 42), Elsevier, Amsterdam, pp 304-322.

Polich, J., Eischen, S.E., and Collins, G.E., 1994, P300 from a single auditory stimulus. *Electroencephalogr. Clin Neurophysiol.* **92**: 253-261.

Polich, J., and Kok, A., 1995, Cognitive and biological determinants of P300: an integrative review. *Biol. Psychol.* **41**: 103-146.

Posner, M.I., 1980, Orienting of attention. *Q.J. Exp. Psychol.* **32**: 3-25.

Prasher, D., and Findley, L., 1991, Dopaminergic induced changes in cognitive and motor processing in Parkinson's disease. *J. Neurol. Neurosurg. Psychia.* **54**: 603-609.

Rafal, R.D., Posner, M.I., Walker, J.A., and Friedrich, F.J., 1984, Cognitive and basal ganglia: separating mental and motor components of performance in Parkinson's disease. *Brain* **107**: 1083-1094.

Raven, J., 1960, Guide to the standard progressive matrices. Lewis, London.

Riklan, M., Wheliham, W., and Cullinan, T., 1976, Levodopa and psychometric test performance in parkinsonism- 5 years later. *Neurology* **26**: 173-179.

Ritter, W., Simson, R., and Vaughan, Jr., H,G., 1972, Association cortex potentials and reaction time in auditory discrimination. *Electroencephalogr.Clin. Neurophysiol.* **33**: 547-555.

Roth, W., Ford, J., and Kopell, B., 1978, Long-latency evoked potentials and reaction time. *Psychophysiology* **15**: 17-23.

Rumbach, L., Tranchant, C., Viel, J.F., and Warter, J.M., 1993, Event-related potentials in Parkinson's disease: a 12 month follow-up study. *J. Neurol. Sci.* **116**:148-151.

Snyder, E., and Hillyard, S.A., 1979, Changes in visual event-related potential in older persons. In: Hoffmeister F, ed. Brain function in old age, Springer, Berlin, pp 112-125.

Stanzione, P., Fattaposta, F., Giunti, P. et al., 1991, P300 variations in parkinsonian patients before and during dopaminergic monotherapy: a suggested dopamine component in P300. *Electroencephalogr. Clin Neurophysiol.* **80**: 446-453.

Tachibana, H., Toda, K., and Sugita, M., 1992, Actively and passively evoked P3 latency of event-related potentials in Parkinson's disease. *J. Neurol. Sci.***11**: 134-142.

Wechsler, D., 1981, Wechsler adult intelligence scale - Revised. Psychological Corporation, New York.

Wickens, C., Kramer, A., Vanasse, L., and Donchin, E., 1983, Performance of concurrent tasks: a psychological analysis of the reciprocity of information-processing resources. *Science* **221**: 1080-1082.

Yamaguchi, S., Tsuchiya, H., and Kobayashi, S., 1995, Electrophysiologic correlates of visuo-spatial attention shift. *Electroencephalogr. clin. Neurophysiol.* **94**: 450-461.

Zimmerman, P., Sprengelmeyer, R., Fimm, B., and Wallesch, C.W., 1992, Cognitive slowing in decision tasks in early and advanced Parkinson's disease. *Brain and Cognition* **18**: 60-69.

NEURONAL INTRANUCLEAR INCLUSIONS IN NEOSTRIATAL STRIOSOMES AND MATRIX IN HUNTINGTON'S DISEASE

John C. Hedreen*

Abstract: Neuronal intranuclear inclusions have recently been described in the cerebral cortex and neostriatum in Huntington's disease brain that are demonstrable immunocytochemically with antibodies to N-terminal huntingtin and to ubiquitin. Their distribution by region and by neuron type matches the known vulnerability pattern in Huntington's disease, and it is thought that the presence of inclusions is a marker for future cell degeneration. We recently described a preferential loss of medium spiny neurons in neostriatal striosomes relative to the matrix, as defined by calbindin immunostaining, in early to middle stages of Huntington's disease. We now show that ubiquitinated neuronal intranuclear inclusions are present in a greater proportion of the medium spiny neurons of striosomes than in those of the matrix in Huntington's disease brains (Vonsattel grade 0-3). The proportion in striosomes correlates strongly with age at death, probably a reflection of CAG repeat length, whereas in the matrix the proportion with inclusions correlates with Vonsattel grade. The finding of a preferential involvement of striosomal medium spiny neurons early in the course of disease is consistent both with the hypothesis of greater vulnerability of striosomal neurons to the pathogenic process in Huntington's disease, and with the idea that the inclusions are harbingers of cell death.

1. INTRODUCTION

In postmortem brains of individuals with Huntington's disease (HD) many neurons in the two most vulnerable regions, cerebral neocortex and neostriatum, contain neuronal intranuclear inclusions (NII).[1,2] These NII are demonstrable immunocytochemically with antibodies against N-terminal huntingtin protein and against ubiquitin. We previously reported that medium spiny neurons in neostriatal striosomes are more vulnerable to cell death than those in the matrix in early to middle stages of HD.[3,4] This finding suggested that NII might

---

* John C. Hedreen, Dept. of Psychiatry, Box 1007, New England Medical Center, 750 Washington St., Boston, MA 02111

The Basal Ganglia VI
Edited by Graybiel *et al.*, Kluwer Academic/Plenum Publishers, 2002

be more frequent in striosomal than in matrix neurons early in the disease course. The NII had been suggested to be part of the cytopathogenic pathway, but may instead be bystander landmarks marking vulnerable cells.[5,6]

2. METHODS

The frequency of ubiquitinated NII was examined among medium spiny neurons located in the striosomal and matrix compartments of the neostriatum in 7 HD brains with Vonsattel pathological severity grades of 0-3.[7] Coronal paraffin sections from the neostriatum at the level of the nucleus accumbens from HD and control brains, obtained from the McLean Hospital Brain Resource Center (Belmont, MA), were immunostained for calbindin, to reveal matrix-striosome compartments, and for ubiquitin, to demonstrate the NII Sections were pretreated with antibody-specific antigen retrieval procedures.[8] A drawing tube map of the calbindin-defined striosome-matrix compartments and tissue landmarks (blood vessels and white matter bundles) was made at low power (2.5x) from the calbindin-immunostained section, in the ventral two thirds region where immunostaining was still well preserved. Selection of the paired striosomal and matrix regions to be counted was made on the calbindin immunostained section. In the ubiquitin-immunostained section, the counting regions were identified by blood vessel and white matter bundle landmarks using the 2.5x or 5x lens, at which power NII are not visible, and the selected area was centered in the viewing field. Using the 40x lens, counts were then made in each sampled region of medium spiny neurons with nuclei visible in the section, with and without NIL The statistical calculations (Spearman, Wilcoxon, comparison of proportions or z-test) were done using SigmaStat (Jandel Scientific, Corte Madera, CA).

3. RESULTS

In the HD neostriatum immunostained for calbindin, in regions retaining calbindin immunoreactivity in the matrix, striosome-matrix compartmentation was observed similar to that in the control brains. Ubiquitin-positive NII were present in neostriatal medium spiny neurons as described by others.[1,2]

In every case the proportion of medium spiny neurons containing NII was significantly higher in the striosomes than in the matrix (z test; Table 1). In addition, analysis of the paired mean striosome and matrix proportion data for all seven HD cases showed that the striosome samples had a significantly greater mean proportion of neurons with NII than did the matrix samples (Wilcoxon, p=0.016).

The proportions of medium spiny neurons with NII in the matrix in the 7 cases had a significant positive correlation with pathological grade (Spearman, r_s=0.77, p=0.025); the proportions in the striosomes did not correlate significantly with grade (Table 1). While the grade 0 case had the highest proportion of striosome neurons with NII, the next most severe case (lA) had the second lowest proportion, suggesting that pathological grade does not determine this value. As brains with longer repeat lengths are reported to have more N11,[1,2] it is likely that the higher proportion in the grade 0 case relates to repeat length rather than to pathological severity. CAG repeat lengths for these two cases were 53 and 40, respectively.

Repeat length was not available in all cases; it correlates inversely with age at death.[9] Age at death had a strong inverse correlation with the proportion of medium spiny neurons with NII in the striosome samples (Spearman, r_s= -0.83, p=0.015), but not with the proportion in the matrix samples (Table 1).

Table 1. NII in neostriatal striosomes and matrix.

Proportion of sampled neurons in striosomes and matrix with NII, Vonsattel grade, age at death, sex of patients.

Grade/ Case	Age/ Sex	Proportion with NII	
		Striosomes #	Matrix§
0	26M	.401*	.120
1A	69F	.225*	.034
1B	52M	.308*	.127
2A	57F	.314*	.126
2B	68M	.208*	.126
3A	64F	.313*	.185
3B	68M	.267*	.154

* Significantly greater than the corresponding matrix proportion in the same case (z test, p<0.0001 for all except 2B, p=0.007 for 2B).
\# Significant negative correlation between proportion of striosome neurons with NII and age at death (Spearman, r_s=-0.83, p=0.015; corresponding results for matrix are -0.08, p=0.84).
§ Significant positive correlation between proportion of matrix neurons with NII and Vonsattel grade (Spearman, r_s=0.77, p=0.025; corresponding results for striosomes are -0.24, p=0.55).

4. DISCUSSION

All cases, of pathological grades 0 through 3, had a significantly greater proportion of striosome than matrix medium spiny neurons with NII. This result is consistent with our previous demonstration that striosomal neurons are more vulnerable in early to middle stages of the disease process than those in the matrix,[3] and with the idea that NII, although not necessarily themselves toxic, form in the neurons that suffer cell injury, and can therefore be seen as markers of neurons that will later die.

From the relatively high frequency of the NII even in early cases, and the 10 to 30 year duration between clinical onset and death, one must conclude that a neuron may manifest such an inclusion for a prolonged period (months or years). This implies that the inclusions are certainly not acutely toxic.

The proportion of medium spiny neurons with NII in striosomes correlated strongly with age at death, probably a reflection of CAG repeat length, whereas in the matrix the proportion with NII correlated with Vonsattel grade; these contrasting findings suggest a possible difference in pathogenetic process in the two compartments.

5. ACKNOWLEDGEMENTS

I thank the McLean Hospital Brain Tissue Resource Center, Belmont, MA for the paraffin blocks of Huntington and control brain tissue. This work was supported by grant ROI NS29484 from the NIH.

6. REFERENCES

1. DiFiglia, M., Sapp, E., Chase, K.O., Davies, S.W., Bates, G.P., Vonsattel, J.P., and Aronin, N., 1997, Aggregation of huntingtin in neuronal intranuclear inclusions and dystrophic neurites in brain. *Science*, 277:1990-1993.
2. Becher, M.W., Kotzuk, J.A., Sharp, A.H., Davies, S.W., Price, D.L., and Ross, C.A., 1998, Intranuclear neuronal inclusions in Huntington's disease and dentatorubral and pallidoluysian atrophy: correlation between the density of inclusions and IT15 triplet repeat length. *Neurobiol. Dis.*, 4:387-397.
3. Hedreen, J.C., and Folstein, S.E., 1995, Early loss of neostriatal striosome neurons in Huntington's disease. *J. Neuropath. Exp. Neurol.*, 54:105-120.
4. Augood, S.J., Faull, R.L., Love, D.R., and Emson, P.C., 1996, Reduction in enkephalin and substance P messenger RNA in the striatum of early grade Huntington's disease: a detailed cellular *in situ* hybridization study. *Neurosci.*, 72:1023-1036.
5. Ross, C.A., 1997, Intranuclear neuronal inclusions: A common pathogenic mechanism for glutamine-repeat neurodegenerative diseases? *Neuron*, 19:1147-1150.
6. Paulson, H.L., 1999, Protein fate in neurodegenerative proteinopathies: polyglutamine diseases join the (mis)fold. *Am. J. Hum. Genet.*, 64:339-345.
7. Vonsattel, J.P., Myers, R.H., Stevens, T.J., Ferrante, R.J., Bird, E.D., and Richardson, E.P. Jr., 1985, Neuropathological classification of Huntington's disease. *J. Neuropath. Exp. Neurol.*, 44:559-577.
8. Hedreen, J.C., and Mucci, L.A., 1995, Antigen retrieval for paraffin section immunohistochemistry with antibodies commonly used in studies of degenerative diseases. *Soc. Neurosci. Abstr.*, 21:1802.
9. Gusella, J.F., and MacDonald, M.E., 1995, Huntington's disease: CAG genetics expands neurobiology. *Curr. Opin. Neurobiol.*, 5:656-662.

PROBING STRIATAL FUNCTION IN OBSESSIVE COMPULSIVE DISORDER
Neuroimaging studies of implicit sequence learning

Scott L. Rauch*

1. INTRODUCTION

Our research team has sought to develop neuroimaging probes of striatal function, to investigate the pathophysiology of obsessive compulsive disorder (OCD) and related diseases. Functional neuroimaging studies have indicated reliable striatal recruitment during a variety of implicit sequence learning and procedural learning tasks. A series of experiments involving the serial reaction time task (SRT) — an implicit sequence learning paradigm — will be reviewed. Experiments employing positron emission tomography (PET) as well as functional magnetic resonance imaging (fMRI) have demonstrated striatal activation associated with implicit learning during the SRT. Moreover, a modified version of the SRT was used to explore temporal aspects of thalamic activity associated with striatal recruitment. Finally, initial findings in OCD and Tourette syndrome will also be discussed.

2. OBSESSIVE COMPULSIVE DISORDER (OCD)

2.1 Phenomenology of OCD

Although OCD is categorized among the anxiety disorders, it is fundamentally characterized by unwanted, repetitive, intrusive thoughts that typically prompt repetitive behaviors which are performed in a ritualized fashion to neutralize the thoughts and accompanying anxiety (APA 1994). OCD is a common disorder, with onset in youth. Contemporary conceptualizations of OCD consider it as one of a group of possibly related conditions, termed the OC-spectrum; the other constituents of this group of disorders include:

---

* S. Rauch, Departments of Psychiatry & Radiology, Massachusetts General Hospital, Bldg. 149, 13th Street, Charlestown, Massachusetts, USA 02129

Tourette syndrome, body dysmorphic disorder, and trichotillomania (Hollander 1993). The notion is that these conditions not only resemble one another phenomenologically, but also may share common features in terms of treatment, etiology, and pathophysiology.

2.2 Neurobiological Models of OCD

Contemporary models of OCD have focused on the cortico-striato-thalamo-cortical (CSTC) circuit that includes orbitofrontal cortex and the caudate nucleus. The striatal topography model of OC-spectrum disorders posits that the primary pathology in OCD is located within the caudate nucleus, and that related disorders entail analogous pathology within other striatal regions (Rauch et al 1998a).

Neuroimaging data provide some of the strongest support for these models of OCD pathophysiology (Rauch & Baxter 1998). Structural neuroimaging studies of OCD have shown subtle volumetric abnormalities involving the caudate nucleus, whereas similar studies of Tourette syndrome and Trichotillomania have found analogous volumetric abnormalities involving the putamen. Measurements of N-acetyl aspartate (a purported index of healthy neuronal density) have likewise found reductions in striatum associated with OCD.

Functional neuroimaging studies of patients with OCD vs. normal comparison subjects have indicated hyperactivity within orbitofrontal cortex and caudate nucleus, as well as anterior cingulate cortex. Interestingly, this hyperactivity is accentuated during OCD symptom exacerbation, and attenuated with successful treatment. Whereas anterior cingulate activation is seen non-specifically across a range of anxiety disorders as well as normal emotional states, symptom-related hyperactivity within the orbitofrontal-caudate CSTC appears to be a signature that is somewhat specific to OCD.

Complementing neuroimaging data are findings indicating that autoimmune mechanisms, that are known to cause striatal damage, likely underlie some cases of OCD and Tourette syndrome (Swedo et al 1998). Likewise, the literature is replete with case examples in which natural lesions of this circuit present clinically as OCD.

3. CORTICO-STRIATAL CIRCUITRY

3.1 Cortico-Striatal Circuitry and Implicit Learning

In the past, theories of cortico-striatal function emphasized the role of the basal ganglia in motor control. More recently, however, there has been a growing appreciation for the role of these circuits in mediating affective and cognitive functions as well (Alexander et al 1990). In particular, within the domain of learning and memory, there is now substantial evidence indicating that cortico-striatal circuitry supports a particular form of implicit (ie, nonconscious) learning and memory, called procedural learning and memory (Schacter et al 1993; Rauch & Savage 1997). Neuropsychological studies have shown a double dissociation between implicit and explicit (ie, conscious) learning and memory capabilities, suggesting that they are mediated by distinct brain systems. Specifically, explicit learning and memory is largely mediated by frontohippocampal systems (Squire 1992; Schacter et al 1996). Thus, whereas neurologic patients with gross striatal pathology exhibit implicit learning deficits and intact explicit learning abilities, patients with early classic temporal lobe dementias exhibit impaired explicit learning abilities and intact implicit learning and memory (Heindel et al 1989; Knopman & Nissen 1991; Willingham & Koroshetz 1993).

3.2 Neuroimaging Studies of Learning & Memory

With the advent of functional imaging, numerous studies have now demonstrated activation of fronto-temporal circuitry during explicit learning and memory tasks (eg, Schacter et al, 1996), as well as activation of cortico-striatal circuits during procedural learning and memory tasks (eg, Grafton et al, 1992). We were particularly interested in developing a functional imaging probe of striatal function to test hypotheses regarding cortico-striatal dysfunction in OCD and related disorders. Consequently, we developed a version of the serial reaction time task (SRT; Nissen & Bullemer, 1987) — a classic implicit sequence learning paradigm — for use in conjunction with PET. In an initial study, we demonstrated significant right striatal activation during implicit sequence learning, whereas a different constellation of regions was recruited during an explicit version of the same task (Rauch et al, 1995). These findings were entirely consistent with similar studies carried out in other laboratories (Grafton et al 1995, Doyon et al 1996, Berns et al 1997).

4. COGNITIVE ACTIVATION STUDIES IN OCD

4.1. PET-SRT Paradigm in OCD

Equipped with a validated neuroimaging probe of striatal function, we next sought to answer three research questions about patients with OCD vs. healthy comparison subjects: 1) would patients with OCD exhibit deficits in implicit sequence learning performance; 2) would they show normal striatal activation; 3) would they manifest aberrant recruitment of ancillary brain regions not seen in normal subjects? In an initial study using the PET-SRT paradigm, we found that patients with OCD showed entirely normal implicit learning performance, but failed to exhibit normal right striatal activation, and instead manifested aberrant medial temporal activation (not seen in normal subjects) (Rauch et al, 1997a).

4.2. Functional MRI-SRT Paradigm in Healthy Subjects

We next developed a more powerful form of the SRT probe, by adapting it for use in conjunction with fMRI. This would allow us to meaningfully assess the presence or absence of significant activation in individual subjects, rather than averaged across groups of subjects. In a validation study with healthy subjects, we again showed reliable, significant recruitment of right striatum during implicit sequence learning and an absence of significant medial temporal activation (Rauch et al, 1997b).

Before applying this probe to the study of OCD and related disorders, however, we wanted to extend our analysis to investigate the role of the thalamus during SRT performance. Models of cortico-striatal function suggest that the striatum is capable of "recognizing" temporo-spatial constellations of cortical inputs, and then influencing activity at the level of the thalamus (Jackson & Houghton 1995). This striato-thalamic influence can take the form of facilitation (also called enhancement or amplification) via the direct pathway, or inhibition (also called gating or filtering) via the indirect pathway. Thus, it would be most illuminating to determine whether changes in activity at the level of the thalamus during implicit sequence learning took the form of activation (most consistent with amplification) or deactivation (most consistent with gating). Since it was plausible that both of these processes were at work, we looked at different time domains to see if we could isolate one or another of these profiles. In a pair of experiments, we showed and then replicated a finding of reliable thalamic deactivation during an early phase of implicit sequence learning (Rauch et al 1998b). Thus, the fMRI-SRT paradigm promised a capacity to test hypotheses regarding: striatal recruitment, aberrant medial temporal recruitment, and thalamic gating, in OCD.

4.3. Functional MRI-SRT Paradigm in OCD

Employing the fMRI-SRT paradigm, we completed a preliminary study of patients with OCD, patients with Tourette syndrome, and healthy comparisons subjects (Rauch et al, 2001). Briefly, the findings of that experiment replicated those of our earlier PET-SRT study: the OCD group again exhibited normal SRT performance, but failed to recruit right striatum and instead showed aberrant medial temporal recruitment. In contrast, patients with Tourette syndrome exhibited a trend toward impaired SRT performance, and while they also failed to recruit right striatum, they showed no significant aberrant medial temporal recruitment. This is consistent with the notion that the medial temporal recruitment in OCD serves some compensatory function. Further extending our previous findings, both the patients with OCD and those with Tourette syndrome exhibited a failure to significantly deactivate the thalamus. This is consistent with current theories of OCD and related disorders as diseases of cortico-striatal dysfunction that entail deficient thalamic gating.

5. COGNITIVE NEUROSCIENCE OF OCD

Neuropsychological studies have consistently indicated that patients with OCD have normal general intelligence (Savage 1998). Similar to other disorders of purported fronto-striatal dysfunction, however, patients with OCD exhibit learning and memory deficits attributable to an impairment in the ability to spontaneously recognize and exploit organizational elements in their environment (Savage et al, 1999).

It is challenging to reconcile these various aspects of OCD, from a cognitive neuroscience perspective. However, we speculate that CSTC dysfunction is responsible for this neuropsychological profile, as well as the intrusive cognitions of OCD (as a consequence of ineffective thalamic gating). Moreover, we submit that compulsions represent repetitive behaviors that serve to recruit healthy elements of CSTC circuitry, and thereby increase thalamic gating. Aberrant medial temporal activation appears to minimize implicit learning deficits in a compensatory fashion; however, one cannot help but wonder whether adaptive use of this system somehow contributes to the affective or cognitive manifestations of the disease. The facts that frontohippocampal circuitry is normally engaged in the service of explicit (ie, conscious) information processing and that the hippocampus is a central constituent of the limbic system, raise the possibility that aberrant medial temporal recruitment mediates the cognitive intrusions and/or their affective accompaniments (ie, anxiety) in OCD.

6. FUTURE RESEARCH

These emerging findings in OCD prompt a host of follow-up questions, and hence ideas for subsequent experiments:

First, it must be determined whether the observed findings in OCD and Tourette syndrome are in any way specific to those disorders versus generalizable across categories of disease. For instance, failed striatal recruitment may simply reflect cortico-striatal dysfunction that is generalizable across basal ganglia diseases (eg, Parkinson and Huntington diseases) as well as OC-spectrum disorders. In fact, analogous findings of both failed striatal recruitment and aberrant medial temporal recruitment have recently been reported by Dagher and colleagues (1998).

Second, it will be important to determine whether these abnormalities represent "state" or "trait" markers of disease. In the case of OCD, further experiments will be necessary, for instance, to determine whether these brain imaging profiles return to normal following successful treatment with medication or behavior therapy.

Third, it will be important to explore the neuropsychological consequences of this abnormal recruitment pattern in OCD. We are currently employing cognitive neuroscience methods to investigate the neuropsychological cost of the apparent compensatory use of medial temporal (rather than striatal) circuitry in support of implicit sequence learning. We have hypothesized that, whereas healthy individuals possess a capacity for parallel processing (via the fronto-temporal and cortico-striatal systems), patients with OCD suffer a parallel processing deficiency as a consequence of cortico-striatal dysfunction. While simple tests that normally recruit the cortico-striatal system can be performed by individuals with OCD (via use of the fronto-hippocampal system) without evidence of overt impairment, grossly deficient performance will be revealed during tasks that require dual processing capabilities. An ongoing study of OCD using a dual task version of the SRT is currently underway in our laboratory.

Fourth, it will be important to explore alternative research methods for more fully characterizing the role of CSTC circuitry in implicit learning. In particular, it may prove valuable to examine brain activity during a modified SRT paradigm using methods that provide superior spatial and/or temporal resolution. For instance, use of fMRI with a higher field strength magnet, thinner slices, and an event-related design may prove fruitful in this regard.

7. CONCLUSIONS

Several lines of evidence point to CSTC circuitry in the pathophysiology of OCD. Neuroimaging as well as neuropsychological studies also implicate this system in normal implicit learning functions. Initial experiments employing the SRT (a classic implicit sequence learning paradigm) in conjunction with functional imaging have revealed a reproducible abnormal brain activity profile in OCD, and an overlapping pattern of abnormality in Tourette syndrome. Future studies using this approach promise to elaborate upon these initial findings. Such research has the potential to advance our understanding of OCD, as well as normal implicit learning and CSTC function.

8. ACKNOWLEDGMENTS

The research described has been supported in part by grants from NIMH (MH01215), the Tourette Syndrome Association, Inc., and the National Alliance for Research on Schizophrenia and Depression. Support has also been received from the David Judah research fund. I would also like to thank my numerous collaborators who helped conduct the primary research cited, especially Cary Savage, Tim Curran, Michael Jenike, Lee Baer, Lisa Shin, and Paul Whalen.

9. REFERENCES

Alexander, G.E., Crutcher, M.D., and DeLong, M.R., 1990, Basal ganglia-thalamocortical circuits: parallel substrates for motor, oculomotor, "prefrontal" and "limbic" functions. *Progress in Brain Research* **85**:119-146.

American Psychiatric Association, 1994, *Diagnostic and Statistical Manual of Mental Disorder,* Fourth Edition. Washington, DC: American Psychiatric Association.

Berns, G.S., Cohen, J.D., and Mintun, M.A., 1997, Brain regions responsive to novelty in the absence of awareness. *Science* **276**:1272-1275.

Dagher, A., Doyon, J., Owen, A.M., et al., 1998, Medial temporal lobe activation in Parkinson's disease during fronto-striatal tasks revealed by PET: evidence for cortical reorganization? [abstract]. Fifth International Congress of Parkinson's Disease and Movement Disorders, New York, P4.093.

Doyon, J., Owen, A.M., Petrides, M., et al., 1996, Functional anatomy of visuomotor skill learning in human subjects examined with positron emission tomography. *Eur. J. Neurosci.* **8**:637-648.

Grafton, S.T., Mazziotta, J.C., Presty, S., et al., 1992, Functional anatomy of human procedural learning determined with regional cerebral blood flow and PET. *J. Neurosci.* **12**:2542-2548.

Grafton,, S.T., Hazeltine, E., and Ivry, R., 1995, Functional mapping of sequence learning in normal humans. *J. Cog. Neurosci.* **7**: 497-510.

Heindel, W.C., Salmon, D.P., Shults, C.W., Walicke, P.A., and Butters, N., 1989, Neuropsychological evidence for multiple implicit memory systems: A comparison of Alzheimer's, Huntington's, and Parkinson's disease patients. *J. Neurosci.* **9**:582-587.

Hollander E (ed), 1993, Obsessive-compulsive spectrum disorders. *Psychiatric Annals* **23**:355-407.

Jackson, S., and Houghton, G., 1995, Sensorimotor selection and the basal ganglia: A neural network model. In: Houk, J.C., Davis, J.L., Beiser, D.G., (eds). *Models of Information Processing in the Basal Ganglia.* Cambridge: MIT Press, pp 337-368.

Knopman, D., and Nissen, M.J., 1991, Procedural learning is impaired in Huntington's disease: evidence from the serial reaction time task. *Neuropsychologia* **29**:245-254.

Nissen, M.J., and Bullemer, P., 1987, Attentional requirements of learning: Evidence from performance measures. *Cognitive Psychology* **19**:1-32.

Rauch, S.L., and Baxter, L.R., 1998, Neuroimaging of OCD and related disorders. In: Jenike, M.A., Baer, L., Minichiello, W.E., eds. *Obsessive-Compulsive Disorders: Practical Management.* Boston: Mosby, pp 289-317.

Rauch, S.L., and Savage, C.R., 1997, Neuroimaging and neuropsychology of the striatum. In: Miguel, E.C., Rauch, S.L., and Leckman, J.F., (eds). *Neuropsvchiatry of the Basal Ganglia.* Psychiatric Clinics of North America. Philadelphia: W.B. Saunders, 741-768.

Rauch, S.L., Savage, C.R., Brown, H.D., et al., 1995, A PET investigation of implicit and explicit sequence learning. *Hum. Brain Mapping* **3**:271-286.

Rauch, S.L., Savage, C.R., Alpert, N.M., et al., 1997a, Probing striatal function in obsessive compulsive disorder: a PET study of implicit sequence learning. *J. Neuropsychiatry* **9**:568-573.

Rauch, S.L., Whalen, P.J., Savage, C.R., et al., 1997b, Striatal recruitment during an implicit sequence learning task as measured by functional magnetic resonance imaging. *Hum. Brain Mapping* **5**:124-132.

Rauch, S.L., Whalen, P.J., Curran, T., McInerney, S., and Savage, C.R., 1998b, Thalamic deactivation during early implicit sequence learning: a functional MRI study. *NeuroReport* **9**:865-870.

Rauch, S.L., Whalen, P.J., Dougherty, D.D., and Jenike, M.A., 1998a, Neurobiological models of obsessive compulsive disorders. In: Jenike, M.A., Baer, L., Minichiello, W.E., eds. *Obsessive-Compulsive Disorders: Practical Management.* Boston: Mosby, pp 222-253.

Rauch, S.L., Whalen, P.J., Curran, T., et al., Probing striato-thalamic function in OCD and TS using neuroimaging methods. In: *Tourette Syndrome.* Cohen, D.J., Goetz, C., and Jankovic, J., (eds). Philadelphia: Lippincott, Williams & Wilkins. Advances in Neurology 2001:85; 207-224.

Savage, C.R., 1998, Neuropsychology of OCD: research findings and treatment implications. In: Jenike, M.A., Baer, L., and Minichiello, W.E., eds. *Obsessive-Compulsive Disorders: Practical Management.* Boston: Mosby, pp 254-275.

Savage, C.R., Baer, L., Keuthen, N.J., et al., 1999, Organizational strategies mediate nonverbal memory impairment in obsessive-compulsive disorder. *Biol Psychiatry* **45**:905-916.

Schacter, D.L., Chiu, P., and Ochsner, K.N., 1993, Implicit memory: A selective review. *Ann. Rev. Neurosci.* **16**:159-182.

Schacter, D.L., Alpert, N.M., Savage, C.R., Rauch, S.L., Albert, M.S., 1996, Conscious recollection and the human hippocampal formation: evidence from positron emission tomography. *Proc Nat Acad Sci USA* **93**:321-325.

Squire, L.R., 1992, Memory and the hippocampus: a synthesis from findings with rats, monkeys, and humans. *Psychol. Rev.* **99**:195-231.

Swedo, S.E., Leonard, H.L., Garvey, M., et al., 1998, Pediatric autoimmune neuropsychiatric disorders associated with streptococcal infections: clinical description of the first 50 cases. *Am. J. Psychiatry* B:264-271.

Willingham, D.B., and Koroshetz, W.J., 1993, Evidence for dissociable motor skills in Huntington's disease patients. *Psychobiology* **21**:173-182.

Section II

ANIMAL MODELS

EVOLUTION OF THE MULTIUNIT ACTIVITY OF THE BASAL GANGLIA IN THE COURSE OF DYNAMIC EXPERIMENTAL PARKINSONISM

Erwan Bezard, Thomas Boraud, Bernard Bioulac, and Christian E. Gross*

Abstract: Electrodes were implanted into the brains of two monkeys in order to allow chronic recording of the multiunit electrophysiological activity of the globus pallidus pars externalis (GPe), the globus pallidus pars internalis(GPi) and the subthalamic nucleus (STN) both before and after treatment with 1-methyl-4- phenyl-1,2,3,6-tetrahydropyridine protocol designed to mimic the gradual evolution of human Parkinson's disease. GPi and STN activity augmented significantly in the course of treatment, even before the first appearance of clinical signs ($p < 0.01$). GPe activity, on the other hand, remained stable throughout the protocol. Once symptoms appeared, their progression paralleled that of GPi and STN hyperactivity. Both symptoms. And hyperactivity stabilized once parkinsonism was fully established. These results would indicate (1) that the functional existence of the so-called indirect pathway is open to doubt, since GPe activity undergoes no modification at any stage of this experimental protocol, (2) that the structure principally responsible for the instigation of glutamatergic compensatory mechanisms is the STN, which increases its level of activity even before the end of the presymptomatic period. The new insight these data afford on the functional organization of the basal ganglia in a dynamic model of experimental parkinsonism should improve our understanding of the pathogenesis of these mechanisms.

1. INTRODUCTION

The principal pathological characteristic of Parkinson's disease (PD) is the progressive death of the pigmented neurons of the Substantia Nigra pars compacta (SNc), the nigrostriatal dopamine neurons.[1] Even with a relatively short presymptomatic period, PD is a long term evolutive disease and the numerous animal models at our disposal only offer stable models of nigral lesion. The myriad studies that have explored the physiopathology of PD and allowed the elaboration of a postulated schema for the functional organization of the motor

---

* E. Bezard, T. Boraud, B. Bioulac, and C. Gross, Basal Gang, Laboratoire de Neurophysiologie, CNRS UMR 5543, Universite de Bordeaux 11, 146 rue Leo Saignat, 33076 Bordeaux Cedex, France

circuit, both for the normal and for the parkinsonian state, [2-5] have been of capital importance but the fact remains that the models used do not reproduce the long term degenerative process that is characteristic of human PD. This problem has been raised in certain reports which have highlighted discrepancies between results obtained in classic animal models and those described in human PD.[6-8] The authors of these reports have attributed these conflicting results[7] to the fact that PD is a chronic degenerative disease, whereas animal models are produced either by acute lesion or by semi-chronic intoxication with specific neurotoxins, i.e. 6-hydroxydopamine or 1-methyl-4-phenyl-1,2,3,6-tetrahydropyridine (MPTP), and do not present the progressive and continuous nigral neurodegeneration that is observed in human PD.

In order to study the physiopathological changes occurring in the course of parkinsonism, we have begun to work on the design of a dynamic model both for the monkey[9] and for the mouse.[10] Monkeys are treated daily with a low dose of MPTP until they reach a clinical score corresponding to 30-35% of the maximum disability score on a parkinsonian monkey rating scale.[11-14] All the monkeys so far treated have presented the principal characteristics of PD, akinesia, rigidity, balance disturbances, gait disorders and not only postural but also resting tremor. In addition, histological lesions are similar to those observed in human PD, i.e. partial degeneration of the SNc, the VTA and the retrorubral area which encompasses the nucleus parabrachialis pigmentosus and the A-8 area.[15]

The present study has used this monkey model to trace the gradual modification of the electrophysiological activity of the globus pallidus pars externalis (GPe), the globus pallidus pars internalis (GPi) and the subthalamic nucleus (STN) induced by chronic MPTP intoxication in conditions which come closer than in previous models to the progressive evolution of idiopathic PD.

2. EXPERIMENTAL PROCEDURES

2.1 Animals

Experiments were carried out on two cynomolgus monkeys (*Macaca fascicularis*) weighing 4-5 kg. Animals were housed in approved individual primate cages under standard conditions of humidity (50 ± 5%), temperature (24 ± 1°C) and light (12h light/dark cycles) and they had free access to food and water. Their care was supervised by veterinarians skilled in the healthcare and maintenance of non-human primates. Our laboratory operates under the guidelines laid down by the National Institute of Health and is recognized by the French Ministry of the Environment. Surgical procedures were performed under general anesthesia (ketamine-hydrochloride 40 mg/kg i.m., Panpharma, France, and xylazine 5 mg/kg i.m., Sigma, USA).

2.2 Stereotaxic Surgery

As individual monkeys differ greatly with regard to specific intracerebral sites, the standard Horsley-Clarke stereotaxic technique was improved by using sagittal and frontal ventriculography.[16] A stereotaxic atlas[17] was used for precise adjustment before insertion into the skull of three stainless steel bipolar concentric electrodes (Plastic One, USA, 22-

gauge) to allow access to the GPe (A: 16.5 mm; L: 8 mm; 40° to the right of the vertical; length of electrode = 19 mm), the GPi (A: 16.5 mm; L: 5 mm; 40° to the left of the vertical; length of electrode = 33 mm) and the STN (A: 12 mm; L: 5 mm; vertically; length of electrode = 29 mm). Before the plastic pedestals of these electrodes were secured to the skull with stainless steel screws and dental acrylic cement, we checked by X-ray that the tip of each electrode was correctly positioned.

2.3 Experimental Protocol

Once animals had recovered from implantation of stainless steel bipolar concentric electrodes, and before MPTP treatment was started, multiunit activity recordings of each structure were performed from D-5 to D0, at 4 p.m. each day, in order to investigate the changes induced by passive limb movements and to define the basal activity of each structure. We then began a series of daily low-dose injections of MPTP, a protocol which has already been described[9,12,18] designed to reflect the slow evolution of nigral degeneration in parkinsonism.[10] Animals were treated daily (0.2 mg/kg i.v. at 9 a.m. dissolved in saline, Sigma, St. Louis, USA) until they reached a score over eight on the clinical rating scale. Clinical symptoms then continued to develop after the treatment was stopped since dopaminergic neuronal death continues for five to six days after the last injection.[19] Multiunit recordings were performed daily from D0 to D25 at 4 p.m. in order to follow the evolution of GPe, GPi and STN activity as parkinsonism progressed.

2.4 Behavioural Assessments

Animals' behaviour was evaluated daily at 2 p.m. on a parkinsonian monkey rating scale[9,11-13] using videotape recordings of monkeys in their cages as well as clinical neurological assessment. During each 30 min session two examiners evaluated animals' levels of motor performance, coaxing them to effect various tasks by offering appetizing fruits. Minimum score was 0 and maximum disability score was 25. A score of 15 corresponds to full parkinsonism and is equivalent to stage IV on the Hoen and Yahr rating scale.

2.5 Electrophysiological Recordings

Extracellular multiunit recordings were carried out in calm awake monkeys. Electrophysiological activity was amplified (EG&G, M113, bandwidth 300 - 3 KHz), monitored with the aid of an audiomonitor, and displayed and stored on a computer (PowerPc 6400, Apple) via a MacLab interface (MacLab/4S V1.0.6/1.7.2/0, AD Instruments, USA) using Chart software (Chart V3.5.6, AD Instruments, USA) allowing both on-line and off-line analysis (i.e. discrimination and integration of the multiunit activity). Each limb joint was extended and flexed and the amplitude of these passive limb movements recorded through a potentiometer (PK 16S6-MCB, Radiospares, France) linked to two articulated arms, and connected to a second input to the MacLab interface. The mean $\pm$ SD firing rate for a day was calculated as the mean frequency of the 48 samples each lasting 5s recorded each day. Statistical analysis of results was performed on these frequency data using Student's t-test. A probability level of 5% ($p < 0.05$) was considered significant. For visual representation, multiunit activity was expressed as a percentage of the mean ± SD firing rate of the control activity recorded at D0 (Fig. 1).

2.6 Immunohistochemistry and Histology

Frontal sections of monkeys' brains including the SNc were processed for visualization of TH-IR as previously described.[9] Macroscopic and microscopic examination showed that TH-immunoreactive cell body density was dramatically reduced in the SNc of the two MPTP-treated monkeys in comparison with a control monkey. TH-immunoreactive cell counts revealed a loss of 90% ($p < 0.0001$). Frontal sections including the pallidum and the STN were stained with Cresyl violet for precise location of the tips of the electrodes.

3. RESULTS

3.1 Evolution of Clinical Signs

The first clinical signs appeared after 16 days of MPTP treatment for monkey N°1 and 15 days for monkey N°2 (Fig. 1A). Two or three days later scores reached maximum values on the clinical rating scale for both monkeys (Fig. 1A). Monkeys became bradykinetic and adopted a flexed posture, with rigidity of the limbs and a decrease in vocalization. Movements became less accurate, for example when reaching for fruit, and occasional episodes of freezing were observed. Both monkeys presented postural tremor and some resting tremor.

3.2 Evolution of Multiunit Activity

3.2.1 GPe Activity

The electrophysiological activity of the GPe remained stable throughout the protocol for both monkeys from D0 (N°1 : 105 ± 16%, N°2 : 103 ± 30%) to D25 (N°1 : 101 ± 11%; N°2 : 89 ± 31%) (Fig. 1B). No significant variation was found at any stage.

3.2.2 GPi Activity

The electrophysiological activity of the GPi remained stable for both from D0 (N°1 : 98 ± 7%, N°2 : 105 ± 13%) to D12 (N°1 : 99 ± 11%; N°2 : F = 109 ± 14%). From D13 onwards, however, for both monkeys, i.e. before the first appearance of clinical signs, the mean frequency of the electrophysiological activity of GPi (N°1: 128 ± 19%, N°2 : 133 ± 21%) increased and became significantly different from basal activity ($p < 0.01$) (Fig. 1D). From this point up to D18 for monkey N°1 and D17 for monkey N°2, mean frequency increased and was sometimes significantly different from that of the preceding day (Fig. 1D) ($p < 0.05$). From D19 for monkey N°1 (208 ± 29%) and D18 for monkey N°2 (204 ± 57%), mean frequency of GPi stabilized at hyperactive level (Fig. 1D).

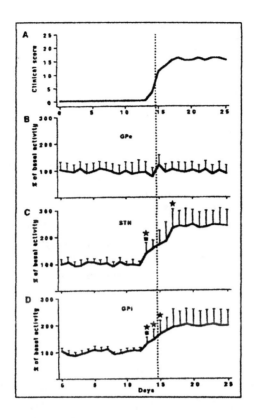

Figure 1. Evolution of clinical score (A) and frequency (mean ± SD) of electrophysiological activity of the GPe (B), STN (C) and GPi (D) of monkey N°2 in the course of experimental parkinsonism. Dotted line: monkey became parkinsonian

3.2.3 STN activity

The electrophysiological activity of the STN remained stable from D0 (N°1 :F = 102 ± 29%, N°2 :99 ± 19%) to D12 (N°1 : 109 ± 33%, N°2 : F = 95 ± 14%). From D13 onwards, however, for both monkeys, i.e. before the first appearance of clinical signs, the mean frequency of the electrophysiological activity of STN (N°1 : 162 ± 40%, N°2 : 140 ± 38%) increased and became significantly different from basal activity ($p < 0.01$) (Fig. 1 C). From this point up to D18 for monkey N°1 and D17 for monkey N°2, mean frequency increased and was sometimes significantly different from that of the preceding day (Fig. 1C) ($p < 0.05$). From D19 for monkey N°1 (281 ± 59%) and D18 for monkey N°2 (242 ± 66%), mean frequency of STN stabilized at hyperactive level (Fig. 1 C).

4. DISCUSSION

Our results show that (1) GPe electrophysiological activity remains stable throughout the protocol, (2) both STN and GPi become hyperactive as the syndrome develops, (3) this augmentation of the electrophysiological activity of both the STN and the GPi precedes the first appearance of parkinsonian motor abnormalities.

(1) The fact that the electrophysiological activity of the GPe remains stable throughout the protocol would argue against its implication in the functional indirect pathway described by Delong[4] and confirms the need for a re-evaluation of the functional anatomy of the basal ganglia in the normal and parkinsonian states.[20] Authors using *in situ* hybridization methods to measure GAD_{67} or COI mRNA expression levels[20-22] have also produced results showing no variation in GPe neuronal activity between monkeys in normal conditions and after acute intoxication by MPTP. This concordance of results strongly suggests that there is no quantitative modulation of GPe functional activity at any stage of the parkinsonian syndrome (even at full parkinsonism). In single cell electrophysiological studies of GPe activity in normal and parkinsonian monkeys already published, we have reported a slight decrease in mean frequency but this was accompanied by a modification of the firing pattern, in the form of high bursting activity,[23,24] which no doubt constitutes a compensatory phenomenom to optimize neurotransmitter release. This permits the maintenance of a stable level of activity throughout PD.

(2) Whilst the electrophysiological activity of the GPe remains stable, that of the STN and the GPi increases. The hyperactivity of both these structures has now been well documented, not only in MPTP-treated monkeys[23,25-27] but also in human PD, but virtually nothing is known of the evolution of this activity in the course of experimental parkinsonism. This is the first study to show that both the STN and the GPI, the main output structure of the basal ganglia, present an increase in electrophysiological activity *before* the first appearance of clinical signs. This increase appears to be tolerated by the motor thalamic nuclei and/or the cortical neurons. It would seem that both or one of these structures manage to integrate slightly damaged messages from the basal ganglia without there being any behavioural consequences.

It is well known that the level of GPi neuronal activity is linked to the level of STN neuronal activity, since experimental lesions[28-30] or high frequency stimulation of the STN[31,32], regularize the activity of pallidal neurons. That the electrophysiological activity of the GPi is affected by modifications in STN electrophysiological activity is therefore not surprising. What is not yet clear is the pathogenesis of STN hyperactivity. Our results show that the GPe is unlikely to be responsible since its level of activity remains stable. Levy et al., in their discussion on the different structures which can influence the STN[20], have pinpointed three possible sources of influence : glutamatergic excitatory projections from the cerebral cortex[33,34], glutamatergic excitatory projections from the parafascicular nucleus *(Pf)*[33,34] and dopaminergic inputs from the mesencephalon.[34,37] The Pf-STN and SNc-STN pathways, despite they are relatively sparse as compared to the afferents from the cerebral cortex and the GPe[38], could well be the structures responsible for this augmentation.

(3) Electrophysiological activity of both the STN and GPi begins to increase even before parkinsonian motor abnormalities appear and this augmentation continues in parallel with the aggravation of the clinical score up to the point when the syndrome stabilizes. The augmentation of frequency between D12 and D13, i.e. the first day when this phenomenom

occurs, is, however, higher in the STN (+48%) than in the GPi (+29%). Since the electrophysiological activity of the GPi does not follow exactly the increase of STN activity, one can so postulated that there is a concomittant inhibitory influence on the GPi, certainly from the striatum, which allow to counterbalance the direct excitatory influence of the STN. We have already demonstrated the existence of glutamatergic compensatory mechanisms responsible (1) for masking PD at the end of the presymptomatic period[13,18,39] and (2) for slowing down the progression of motor abnormalities during the evolutive symptomatic period[12,18,39], by blocking glutamatergic afferents from the STN and the pedunculopontine nucleus (PPN) to SNc in monkeys chronically intoxicated with MPTP along the lines of the dynamic model we have developed. The structure responsible for this compensation remained, however, unidentified. The knowledge that symptoms are masked in the presyrnptomatic period of experimental parkinsonism and alleviated in the symptomatic period implies a modification of the activity of glutamatergic inputs to the SNc in the course of the experimental disease sufficient to increase the activity of dopaminergic neurons. Since PPN neurons have recently been reported to be hypoactive in experimental parkinsonism[40], and our own results show that STN hyperactivity precedes the first appearance of clinical signs, it is highly probable that the STN is the key nucleus responsible for glutamatergic compensatory mechanisms.

Presymptomatic compensation implies that surviving SNc dopaminergic neurons increase striatal dopamine release in order to maintain striatal dopaminergic homeostasis.[41] The fact that dopaminergic neurons increase their bursting activity would constitute a compensatory mechanism designed to achieve this homeostasis.[39] It is well known that dopaminergic burst firing is triggered by glutamatergic inputs to the SNc. Excitatory amino acids have been shown to generate this type of firing in dopamine neurons[42,43] and lesion of the STN has been reported to sharply reduce their level of burst firing without affecting their firing rate.[44] It has also been shown that MPTP treatment in monkey induces a significant increase in the percentage of STN neurons discharging in burst.[25] Lisman has recently proposed that bursts should be considered the main units of neural information since they are capable of (1) making reliable synapses unreliable and (2) inducing burst firing in postsynaptic neuron.[45] It is probable that STN neurons increase their influence on SNc neurons in the period preceding the first appearance of clinical signs not only by augmenting their mean firing rate but also by modifying their firing pattern to fire preferentially in bursts. If we follow Lisman's reasoning[45], this phenomenom would therefore induce an increase of burst firing in SNc neurons and augment the number of reliable synapses between the STN and the SNc. In this way, dopaminergic homeostasis in the striatum is maintained at optimum level. Simultaneous recordings of the extracellular unit activity of both the STN and the SNc at the end of the presymptomatic period in an MPTP monkey model of this type is now necessary before we can advance further on the hypothesis raised by the results of this present study.

5. CONCLUSIONS

The progressive character of the dynamic model we have used for this study has allowed a more faithful simulation of the evolution of human PD physiopathology than was possible until now and, at the same time, produced results that lead us to reconsider the functional organization of the basal ganglia. It appears clear that the GPe plays a more minor role in this organization than that predicted by the "direct/indirect" hypothesis of basal ganglia function. It would also seem that the STN and the SNc are reciprocally linked in a closed loop from the anatomical, functional and electrophysiological aspects. Such loops are characteristic of basal ganglia circuitry where ancillary loops are frequent. This STN-SNc circuit appears, however, to be of considerable functional importance.

6. ACKNOWLEDGEMENTS

We wish to thank Stephane Guitraud, Christelle Imbert and Sandra Dovero for technical assistance. This study was supported partly by the CNRS arid the IFR of Neuroscience (INSERM N°8; *CNRS* N°13), partly by the MESR grant-N°95523629 and the University Hospital of Bordeaux.

7. REFERENCES

1. Ehringer, H. and Hornkiewicz, O. 1960. Verteilung von Noradrenalin and Dopamin (3-Hydroxytyramin) im Gehim des Menschen und ihr Verhalten bei Erkrankungen des extrapyramidalen Systems. *Klin. Wochenschr.* **38**, 1236-1239.
2. Albin R.L., Young, A.B.,and Penney J.B. 1989. The functional anatomy of basal ganglia disorders. *Trends Neurosci.* **12**, 366-375.
3. Alexander, G.E.and Crutcher, M.D. 1990. Functional architecture of basal ganglia circuits: neural substrates of parallel processing. *Trends Neurosci.* **13**, 266-271.
4. DeLong, M.R. 1990. Primate models of movement disorders of basal ganglia origin. *Trends Neurosci.* **13**, 281-285.
5. Wichmann, T. and DeLong, M.R. 1996. Functional and pathophysiological models of the basal ganglia. *Curr. Opin. Neurobiol.* **6**, 751-758.
6. Levy, R., Herrero, M.-T., Ruberg, M., Villares, J., Faucheux, B., Guridi, J., Guillen, J., Luquin, M.R., Javoy-Agid, F., Obeso, J.A., Agid ,Y., and Hirsch, E.C. 1995. Effects of nigrostriatal denervation and L-Dopa therapy on the GABAergic neurons of the striatum in MPTP-treated monkeys and Parkinson's disease: an in situ hybridization study of GAD67mRNA. *Eur. J. Neurosci.* **7**, 1199-1209.
7. Levy, R., Vila, M., Herrero, M.-T., Faucheux, B., Agid, Y., and Hirsch, E.C. 1995. Striatal expression of substance P and methionine-enkephalin genes in patients with Parkinson's disease. *Neurosci. Lett.* **199**, 220-224.
8. Stoessl, A.J. and Rajakumar, N. 1996. Effects of subthalamic nucleus lesions in a putative model of tardive dyskinesia in the rat. *Synapse* **24**, 256-261.
9. Bezard, E., Imbert, C., Deloire, X., Bioulac, B., and Gross, C. 1997. A chronic MPTP model reproducing the slow evolution of Parkinson's disease: evolution of motor symptoms in the monkey. *Brain Res.* **766**, 107-112.
10. Bezard, E., Dovero, S., Bioulac, B., and Gross, C. 1997. Kinetics of nigral degeneration in a chronic model of MPTP-treated mice. *Neurosci. Lett.* **234**, 43-46.
11. Benazzouz, A., Boraud, T., Dubedat, P., Boireau, A., Stutzmann, J.M.,and Gross C. 1995. Riluzole prevents MPTP-induced parkinsonism in the rhesus monkey: a preliminary study. *Eur. J. Pharmacol.* **284**, 299-307.
12. Bezard, E., Boraud, T., Bioulac, B., and Gross C. 1997. Compensatory effects of glutamatergic inputs to the substantia nigra pars compacta in experimental parkinsonism. *Neuroscience* **81**, 399-404.
13. Bezard, E., Boraud, T., Bioulac, B.and Gross, C.E. 1997. Presymptomatic revelation of experimental parkinsonism. *Neuroreport* **8**, 435-438.
14. Kurlan, R, Kim ,M.H., and Gash, D.M. 1991. Oral levodopa dose-response study in MPTP-induced hemiparkinsonian monkeys: assessment with a new rating scale for monkey parkinsonism. *Mov. Disord.* **6**, 111-118.

15. Hirsch, E.C. 1994. Biochemistry of Parkinson's disease with special reference to the dopaminergic systems. *Mol. Neurobiol.* **9**, 135-142.
16. Feger, J., Ohye, C., Gallouin, F., and Albe-Fessard, D. 1975. Stereotaxic technique for stimulation and recording in nonanaesthetized monkeys: application to the determination of connections between caudate nucleus and substantia nigra. In Advances in Neurology (eds. Meldrum, B.S. & Marsden, C.D.) Vol. 10 pp. 35-45. Raven Press, New York.
17. Szabo, J. and Cowan, W.M. 1984. A stereotaxic atlas of the brain of the cynomolgus monkey (macaca fascicularis). *J. Comp. Neurol.* **222**, 265-300.
18. Bezard, E., Bioulac, B., and Gross C. 1998. Glutamatergic compensatory mechanisms in experimental parkinsonism. *Prog. Neuropsychopharmacol. Biol. Psychiatry* **22**, 609-623.
19. Jackson-Lewis, V., Jakowec, M., Burke, R.E., and Przedborski, S. 1995. Time course and morphology of dopaminergic neuronal death caused by the neurotoxin 1-methyl-4-phenyl-1,2,3,6-tetrahydropyridine. *Neurodegeneration* **4**, 257-269.
20. Levy, R., Hazrati, L.-N., Herrero, M.-T., Vila, M., Hassani, O.-K., Mouroux, M., Ruberg, M., Asensi, H., Agid, Y., Feger, J., Obeso, J.A., Parent, A., and Hirsch, E.C. 1997. Re-evaluation of the functional anatomy of the basal ganglia in normal and parkinsonian states. *Neuroscience* **76**,335-343.
21. Herrero, M.T., Levy, R., Ruberg, M., Luquin, MR., Villares, J., Guillen, J., Faucheux, B., Javoy-Agid, F., Guridi, J., Agid ,Y., Obeso, J.A., and Hirsch, E.C. 1996. Consequence of nigrostriatal denervation and L-dopa therapy on the expression of glutamic acid decarboxylase messenger RNA in the pallidum. *Neurology* **47**, 219-224.
22. Vila, M., Levy, R., Herrero, M.-T., Ruberg, M., Faucheux, B., Obeso, J.A., Agid, Y., and Hirsch, E.C. 1997. Consequences of nigrostriatal denervation on the functioning of the basal ganglia in human and nonhuman primates: an in situ hybridization study of cytochrome oxidase subunit I mRNA. *J. Neurosci.* **17**, 765-773.
23. Boraud, T., Bezard, E., Bioulac, B., and Gross C. 1996. High frequency stimulation of the internal globus pallidus reduces the higher firing frequency of its neurons in the MPTP-treated monkey model of Parkinson's disease. *Neurosci. Lett.* **215**, 17-20.
24. Boraud, T., Bezard, E., Guehl, D., Bioulac, B., and Gross, C. 1998. Effects of L-DOPA on neuronal activity of the Globus Pallidus externalis (GPe) and Globus Pallidus internalis (GPi) in the MPTP-treated monkey. *Brain Res.* **787**, 157-160.
25. Bergman, H., Wichmann, T., Karmon, B.and Delong, M.R. 1994. The primate subthalamic nucleus. II. Neuronal activity in the MPTP model of parkinsonism. *J. Neurophysiol.* **72**, 507-520.
26. Filion, M. and Tremblay, L. 1991. Abnormal spontaneous activity of globes pallidus neurons in monkeys with MPTP-induced parkinsonism. *Brain Res.* **547**, 142-151.
27. Miller, W.C. and Long, M.R.D. 1987. Altered tonic activity of neurons in the globus pallidus and subthalamic nucleus in the primate MPTP model of parkinsonism. In Basal Ganglia II (eds. Carpenter, M.B. & Jayaraman, A.) Vol. 32 pp. 415-427. Plenum Press, New York.
28. Hamada, I. and DeLong, M.R. 1992. Excitotoxic acid lesions of the primate subthalamic nucleus result in reduced pallidal neuronal activity during active holding. *J. Neurophysiol.* **68**, 1859-1866.
29. Robledo, R., and Feger, J. 1990. Excitatory influence of rat subthalamic nucleus to substantia nigra pars reticulata and the pallidal complex: electrophysiological data. *Brain Res.* **518**, 47-54.
30. Ryan, L.J. and Sanders, D.J. 1993. Subthalamic nucleus lesion regularizes firing patterns in globus pallidus and substantia nigra pars reticulata neurons in rats. *Brain Res.* **626**, 327331.
31. Benazzouz, A., Piallat, B., Pollak, P., and Benabid, A.L. 1995. Responses of substantia nigra pars reticulata and globus pallidus complex to high frequency stimulation of the subthalamic nucleus in rats: electrophysiological data. *Neurosci. Lett.* **189**, 77-80.
32. Burbaud, P., Gross, C., and Bioulac, B. 1994. Effects of subthalamic high frequency stimulation on substantia nigra pars reticulata and globus pallidus neurons in normal rats. *J. Physiol.* **88**, 3-4.
33. Afsharpour, S. 1985. Topographical projections of the cerebral cortex to the subthalamic nucleus. *J. Comp. Neurol.* **236**, 14-28.
34. Canteras, N.S., Shammah-Lagnado, S.J., Silva, B.A., and Ricardo, J.A. 1990. Afferent connections of the subthalamic nucleus: a combined retrograde and anterograde horseradish peroxidase study in the rat. *Brain Res.* **513**, 43-59.
35. Mouroux, M.and Feger, J. 1993. Evidence that the parafascicular projection to the subthalamic nucleus is glutamatergic. *Neuroreport* **4**, 613-615.

36. Mouroux, M., Hassani, O.K., and Feger, J. 1995. Electrophysiological study of the excitatory parafascicular projection to the subthalamic nucleus and evidence for ipsi- and contralateral controls. *Neuroscience* **67**, 399-407.
37. Brown, L.L., Makman, M.H., Wolfson, L.I., Dvorkin, B., Warner, C., and Katzman, R. 1979. A direct role of dopamine in the rat subthalamic nucleus and an adjacent intrapeduncular area. *Science* **206**, 1416-1418.
38. Parent, A. and Hazrati, L.N. 1995. Functional anatomy of the basal ganglia. II. The place of subthalamic nucleus and external pallidum in basal ganglia circuitry. *Brain Res. Rev.* **20**, 128-154.
39. Bezard, E. and Gross, C.E. 1998. Compensatory mechanisms in experimental and human parkinsonism: towards a dynamic approach. *Prog. Neurobiol.* **55**, 93-116.
40. Ogura, M., Nakao, N., Nakai, E., Nakai, K., and Itakura, T. 1997. The firing activity of the pedunculopontine nucleus and basal ganglia in 6-OHDA lesioned rats. *Soc. Neurosci. Abstr.* **23**, 192 .
41. Zigmond, M.I., Abercrombie, E.D., Berger, T.W., Grace, A.A., and Stricker, E.M. 1990. Compensations after lesions of central dopaminergic neurons : some clinical and basic implications. *Trends Neurosci.* **13**, 290-296.
42. Grace, A.A. and Bunney, B.S. 1984. The control of firing pattern in nigral dopamine neurons: burst firing. *J. Neurosci.* **4**, 2877-2890.
43. Overton, P. and Clark, D. 1997. Burst firing in midbrain dopaminergic neurons. *Brain Res. Rev.* **25**, 312-334.
44. Smith, I.D. and Grace, A.A. 1992. Role of the subthalamic nucleus in the regulation of nigral dopamine neuron activity. *Synapse* **12**, 287-303.
45. Lisman, J.E. 1997. Bursts as a unit of neural information: making unreliable synapses reliable. *Trends Neurosci.* **20**, 38-43.

EVIDENCE FOR NEURONAL DYSFUNCTION IN A MOUSE MODEL OF EARLY STAGES OF HUNTINGTON'S DISEASE

M.-F. Chesselet, M. S. Levine, C. Cepeda, L. Menalled, and H. Zanjani*

1. INTRODUCTION

Huntington's disease is a progressive neurodegenerative disease due to the presence of an expanded CAG repeat in the gene encoding huntingtin, a widely distributed protein of unknown function (Huntington Disease Collaborative Research Group, 1993). As a result of the mutation, huntingtin contains an expanded stretch of glutamines in its N-terminal region (Aronin et al. 1995; Trottier et al. 1995). It is now considered that expansions greater than 40 repeats always lead to disease, whereas expansions between 36 and 39 have incomplete penetrance. In the most common form of the disease, patients begin to exhibit symptoms in the third or fourth decade of life (Harper, 1991). The disease is dominated by the presence of irrepressible dance-like movements of the extremities, evolving towards dystonic contractions in the course of the disease, including severe swallowing problems. Psychiatric and cognitive symptoms are frequent. Patients experience severe weight loss and the disease is usually fatal in 10-15 years. Juvenile forms of the disease are due to extreme CAG repeat expansions (usually greater than 60) and are characterized by onset before age 20. The symptoms are predominantly dystonic and include seizures.

The main neuropathological feature of Huntington's disease is the massive loss of efferent neurons in the caudate nucleus and putamen (Vonsattel et al. 1985). Various degrees of neuronal loss also occur in the cerebral cortex. Typically, the disease spares striatal interneurons (Kowall et al. 1987), as well as the hippocampus and cerebellum, except in the juvenile forms that have more extensive pathology than adult onset forms (de la Monte et al. 1988; Myers et al. 1988). There is no known muscular or peripheral pathology. This restricted pattern of cell loss contrasts with the widespread expression of huntingtin, not only in most brain regions, but also in muscles and peripheral organs such as lymphoblasts, liver and testis (Sharp et al. 1995). Since the discovery of the mutation causing Huntington's disease

---

* M.F. Chesselet, M.S. Leveine, C. Cepeda, L. Menalled, and H. Zanjani, Department of Neurology, Brain Research Institute, Neuropsychiatric Institute and Mental Retardation Research Center, UCLA School of Medicine, 710 Westwood Plaza, Los Angeles CA 90095. USA

in 1993, much effort has been devoted to resolve this paradox, and to understand how the presence of an expanded polyglutamine repeat in huntingtin leads to delayed and regionally selective neuronal death.

This paradox is not specific to Huntington's disease. At least eight other hereditary neurodegenerative diseases have now been related to an expanded glutamine stretch in widely distributed proteins (Reddy and Housman, 1997). In two cases, spino-bulbar degeneration (Kennedy disease) and spino-cerebellar ataxia type 6, the function of the affected protein is known: the androgen receptor in the first case (La Spada et al. 1991) and a calcium channel in the other (Zhuchenko et al. 1997). In all other diseases, the mutation occurs in a protein of unknown function. However, the similarities between all the "CAG repeat" diseases have led most investigators to consider that the mutation confers a common gain of function, independent of the specific role of each protein affected (Ordway et al. 1997).

The recent development of mouse models for several CAG repeat diseases represents a turning point in the field. It is now possible, for the first time, to examine the effects of well-characterized mutations in an intact, living mammalian organism. However, all transgenic models are not equal, and the constraints related to modeling a human disease in a mouse have led to some contradictory findings. In this review, we will describe findings from our laboratories in two mouse models of Huntington's disease and show how comparing different models is essential to evaluate the significance of findings.

2. INCREASED NEURONAL SENSITIVITY TO STIMULATION OF NMDA RECEPTORS IN MOUSE MODELS OF HUNTINGTON'S DISEASE

Even before the mutation causing Huntington's disease was identified, numerous studies had shown that excess stimulation of the NMDA subtype of glutamate receptor in rat striatum induces a pattern of cell loss similar to that observed in Huntington's disease (Schwarcz et al. 1983; Beal et al. 1986; Qin et al. 1992; Roberts et al. 1993). Similarly, local ischemia (Chesselet et al. 1990; Gonzales et al. 1992) and injection of mitochondrial poisons (Beal et al. 1993; Brouillet et al. 1995), which are believed to increase the stimulation of NMDA receptors by endogenous glutamate, also produce this pattern of neuronal death in rat striatum. This led to the hypothesis that NMDA-mediated excitotoxicity plays a critical role in cell death in Huntington's disease.

We tested the hypothesis that expression of the Huntington's disease mutation in two different mouse models altered cellular responses to the stimulation of NMDA receptors. The first model is a well-characterized transgenic animal that expresses a truncated huntingtin with very long CAG repeats (Mangiarini et al. 1996). In this animal, the transgene is limited to exon 1 of the huntingtin gene and contains 144 CAG repeats. These mice develop a complex phenotype including behavioral anomalies, motor symptoms and seizures, severe weight loss and diabetes in the first months of life (Mangiarini et al. 1996; Carter et al. 1999; Hurlbert et al. 1999). They usually die by 3-4 months. Previous studies have detected early anomalies in dopaminergic and non-NMDA receptors in the striatum of these mice (Cha et al. 1998), but neuronal loss was not detected until much later in the life of the animal (Turmaine et al. 2000).

In the second model, mice expressed expanded CAG repeats in their own huntingtin molecules as a result of a "knock in" insertion of a portion of the human gene with either 71 or 94 CAG repeats, interrupted by an arginine in position 42. To date, these mice have been observed up to 12 months of age: they do not display any gross neurological phenotypes and reproduce normally (Levine et al. 1999 and in preparation).

In a first series of experiments, we examined the response of striatal and cortical neurons from these two types of mice to glutamate receptor activation in an *in vitro* slice preparation by visualizing neurons in brain slices with infrared videomicroscopy and differential interference contrast optics to determine changes in somatic area (cell swelling). In cell cultures, it is well established that continuous activation of excitatory amino acid receptors induces toxicity that leads to cell death (Choi, 1988; Rothman, 1992). A first step in this toxic cascade is the induction of rapid cell swelling. Studies in cortical and striatal slices have shown that cell swelling occurs when neurons are exposed to glutamate receptor agonists and that this change may be a sign of neurotoxicity (Dodt and Zieglgänsberger, 1990; 1994; Dodt et al., 1993; Colwell et al., 1996; 1998; Colwell and Levine, 1996). Striatal and cortical neurons in both models displayed more rapid and increased swelling to N-methyl-D-aspartate (NMDA) than those in controls. Table 1A shows average percent change in cross-sectional area for striatal neurons in response to two concentrations of NMDA (25 and 50 µM) at two time points (10 and 15 min after exposure to NMDA). At both concentrations, cell swelling in the transgenics was significantly greater ($p<0.01$) than cell swelling in wild-type littermates indicating altered responses to NMDA receptor activation. A similar increase in cell swelling to activation of NMDA receptors occurred in the cortex (Levine et al. 1999)

Medium-sized striatal cells in mice with CAG knock-ins and their wild-type controls also showed concentration- and time-dependent swelling in response to NMDA. At both 10 and 25 µM concentrations, cell swelling in CAG94 knock-in mice was significantly greater ($p<0.01$) than cell swelling in wild-type controls or in CAG71 knock-in mice, indicating a CAG repeat length effect (Table 1B). Swelling in cortical neurons in CAG94 knock-in mice was significantly greater than swelling in CAG71 knock-ins or wild-type controls again suggesting effects were not specific to striatum (Levine et al., 1999).

In both mouse models these effects were specific to NMDA receptor activation as there were no consistent group differences after exposure to alpha-amino-3-hydroxy-5-methyl-4-isoxazole propionic acid (AMPA) or kainate (KA) either in striatum or cortex.

In other experiments we found using intracellular recordings that resting membrane potentials (RMPs) in the transgenic model were significantly more depolarized than those in their respective controls (Klapstein et al., 2001). RMPs in CAG94 mice also were more depolarized than those in CAG71 mice or their controls in a subset of striatal neurons (Levine et al., 1999).

These results suggest that striatal and cortical neurons expressing the Huntingon's disease mutation may run a higher risk of excitotoxic insult from exposure to endogenous glutamate. This raises two questions: 1) how does the expanded polyglutamine repeat in huntingtin lead to an increased response to NMDA receptor stimulation? and 2) is this increased sensitivity to NMDA stimulation responsible for the delayed neuronal death of striatal and cortical neurons in patients with Huntington's disease? We have begun to approach these questions by examining the brains of the two types of mice used in these experiments.

Table 1. Percent change in somatic cross-sectional area after exposure to NMDA

A. Transgenic Model

Genotype	Concentration	Percent Increase in Area	
		10 min	15 min
Transgenic	25µM	45±3.6(52)*	63±5.0(36)*
	50µM	87±14.0(19)*	114±13.0(11)*
Wildtype	25µM	33±4.3(29)	47±5.6(19)
	50µM	46±8.5(15)	63±14.0(8)

B. CAG Knock-in Model

Genotype	Concentration	Percent Increase in Area	
		10 min	15 min
CAG94	10µM	37±7.0(20)*	58±11.0(20)*
	25µM	60±11.0(17)*	70±9.4(16)*
CAG71	10µM	16±2.1(28)	25±3.0(28)
	25µM	24±6.5(14)	45±10(13)
Wildtype	10µM	22±3.0(47)	28±3.0(46)
	25µM	24±4.0(24)	33±7.0(20)

Data are mean±SEM. Indicates p<0.01 compared to corresponding wild-type mice with ANOVA. Number of cells measured in parentheses.

3. COMPARATIVE HISTOLOGY OF TWO MOUSE MODELS OF HUNTINGTON'S DISEASE.

Previous studies in the first mouse model (the transgenic mice overexpressing exon 1 with an expanded CAG repeat; Mangiarini et al. 1996) have led to the discovery of nuclear inclusion in brain neurons (Davies et al. 1997). These nuclear inclusions can be detected with antibodies against the N-terminal portion of huntingtin (i.e. an epitope present in the transgene), ubiquitin and to a certain extent, heat shock protein 70 (HSP70). Similar nuclear inclusions were also found in the brains of patients with Huntington's disease (DiFiglia et al. 1997) and with other CAG repeat diseases such as spinocerebellar ataxia 1 and 3 (Paulson et al. 1997; Skinner et al. 1997). Work *in vitro* suggests that processing of the mutated huntingtin into smaller fragments favors the formation of aggregates and that these aggregates may have a fibrillary structure similar to that of amyloid deposits (Scherzinger et al. 1997; Cooper et al. 1998; Martindale et al. 1998).

We confirmed that nuclear inclusions were present in the transgenic mice used in our study. However, they were absent in the knock-in mice that also showed an increased response to NMDA receptor stimulation (Levine et al. 1999). The absence of nuclear inclusions in these mice was confirmed with a battery of antibodies against the N-terminal fragment of huntingtin, ubiquitin and HSP 70. Therefore, detectable huntingtin inclusions do not precede the appearance of a cellular phenotype in this model of Huntington's disease. This suggests that nuclear inclusions follow rather than precede neuronal dysfunction in Huntington's disease.

Recent work in both mouse and cellular models of CAG repeat diseases supports the hypothesis that the formation of aggregates are not a primary event in the pathophysiology of Huntington's disease. Indeed, experimental manipulations that prevent the formation of aggregates did not prevent the occurrence of neuronal death (Saudou et al. 1998). A case was made, however, for the critical importance of an abnormal localization of the mutated huntingtin to the nucleus in the induction of apoptosis *in vitro*. (Saudou et al. 1998) Again, in the knock-in mice, the increased response to NMDA stimulation occurred in the absence of conspicuous nuclear staining for huntingtin, suggesting that huntingtin translocation to the nucleus is not the primary cause of cellular dysfunction (Menalled and Chesselet, 2002).

In conclusion, the comparison of data in the transgenic and knock-in mice indicate that increased response to stimulation of NMDA receptors may be a common feature of striatal and cortical neurons expressing huntingtin with an expanded CAG repeat. This effect was not related to unique characteristics of the transgenic mice such as diabetes or the formation of nuclear inclusions since it was also observed in the knock in mice. The presence of this effect in neurons from the knock in mice also indicates that it is likely to occur very early in the course of the disease, prior not only to neuronal death, but also to overt behavioral symptoms. This could be highly relevant to the pathophysiology of the disease because, in most cases, symptoms do not occur in humans carrying the mutation until the third or fourth decade of life. Furthermore, there is evidence that symptoms can precede neuronal loss in Huntington's disease. Very little, if any, neuronal loss was observed in the brains of some symptomatic patients who died early in the course of the disease (Vonsattel et al. 1985; Myers et al. 1988). Therefore, neurological dysfunction, eventually leading to neurological symptoms, is likely to precede neuronal death by many years in humans with the disease.

Are the increased responses to NMDA receptor stimulation we have observed in the mouse models isolated or are they related to other functional alterations in the affected neurons? To address this question, we have examined the level of expression of messenger RNAs (mRNAs) encoding neurotransmitters or neurotransmitter synthetic enzymes in the striatum of these two mouse models of Huntington's disease.

4. MOLECULAR ALTERATIONS IN STRIATAL NEURONS OF MOUSE MODELS OF HUNTINGTON'S DISEASE

Increased stimulation of striatal NMDA receptors by quinolinic acid rapidly induces a marked decrease in the level of the mRNA encoding enkephalin, a neuropeptide present in striatal neurons projecting to the external pallidum (Bordelon and Chesselet, 1999). In view of the increased response to NMDA that we have observed in mouse models of Huntington's disease, we have examined the expression of enkephalin mRNA in the striatum of these mice. The levels of enkephalin mRNA were compared to those of the mRNA encoding substance P, a neuropeptide present in striatal neurons projecting primarily to the internal pallidum and substantia nigra (see Chesselet, 1999 for review). In addition we have examined the levels of the mRNA encoding glutamic acid decarboxylases (Mr 67,000: GAD67 and Mr 65.000: GAD65), the enzymes of GABA synthesis which are present in most striatal neurons (Mercugliano et al. 1992).

Enkephalin mRNA levels were markedly decreased in the striatum of the transgenic mice and of the knock-in mice with 94 CAG repeats. This effect was observed in mice approximately 3 months old, i.e. at the same ages as the increased response to NMDA receptor stimulation. The decrease in enkephalin was not due to the death of enkephalinergic neurons as indicated by the absence of pycnotic nuclei and of conspicuous neuronal loss in the striatum. Furthermore, enkephalinergic immunoreactivity was similar to controls in the external pallidum, the region that contains the axon terminals of striato-pallidal enkephalinergic neurons. The decrease in enkephalin mRNA occurred in the absence of increased staining for glial acidic fibrillary protein (GFAP), a marker of reactive astrocytes. Similarly, immunostaining for calbindin28, a calcium-binding protein expressed by neurons of the extrastriosomal matrix of the striatum, did not reveal any conspicuous disruption of striatal architecture. In contrast to enkephalin mRNA, the levels of mRNAs encoding substance P, GAD67 and GAD65 were unchanged in both mouse models.

The comparison between the transgenic and knock-in mice again allowed us to conclude that the decrease in enkephalin mRNA occured independently of nuclear inclusions and obvious behavioral anomalies. This suggests that this effect could occur early in the course of the disease, and that a phase of neuronal dysfunction may precede neuronal death in patients with Huntingon's disease. As indicated previously, we have observed in rats that local injection of quinolinic acid into the striatum causes a rapid decrease in enkephalin mRNA levels (Bordelon and Chesselet, 1999). This suggests that the decrease in enkephalin mRNA in the striatum of mouse models of Huntington's disease could be secondary to an increased sensitivity of striatal neurons to NMDA receptor stimulation by endogenous glutamate, as suggested by the results of the cell swelling experiments. It is of interest that substance P and GAD mRNAs were unaffected in the same animals. Similarly, dopaminergic and AMPA, but not NMDA, receptors and their mRNAs were decreased in the same transgenic model (Cha et al. 1998). Taken together the results of both studies suggest that the mutation only affects a subset of mRNAs, at least early in the course of the disease. It is unclear whether these molecular effects result from a decreased level of transcription or an alteration in the stability of the affected mRNAs.

Although a previous study from our laboratory failed to detect a decrease in enkephalin mRNA in two presymptomatic carriers of the Huntington's disease mutation (Albin et al. 1991), such an effect has been detected in symptomatic patients who died at an early stage of the disease (Richfield et al. 1995). This suggests that the mouse models we have examined, particularly the knock-in mice, could provide a good model for the early stages of the disease. There are, however, some significant differences between our observations in mice and pathological reports in post-mortem human brains. Indeed, we have found in humans that at least a subpopulation of surviving enkephalinergic neurons in advanced stages still expressed high levels of enkephalin mRNA (Albin et al. 1991). A similar observation was made in the striatum of rats with a local injection of quinolinic acid (Bordelon and Chesselet, 1999), but not in the mouse models. Another difference is that, in humans, a decrease in enkephalin immunoreactivity in the axons of striato-pallidal neurons (Reiner et al. 1988) preceded the loss of enkephalin mRNA in the striatum (Albin et al. 1991), whereas in mice the reverse was observed. These differences could be due to the difficulty of mimicking the human disease in mice, or to the fact that different stages of the disease were examined in these various studies.

5. CONCLUSIONS

The comparison of two very different mouse models of Huntington's disease revealed common features that are likely related to the expression of a CAG repeat expansion in huntingtin: an increased response to NMDA receptor stimulation and a decreased level of enkephalin mRNA in striatal neurons. Based on previous studies in the rat, the changes in enkephalin mRNA levels may be secondary to the alterations in NMDA responses observed in the same mice. This suggests that excitotoxicity may play a role in the pathophysiology of Huntington's disease, as previously suggested by indirect evidence (Beal et al. 1986; Choi, 1988; Whetsell 1966). The fact that this cellular phenotype was observed prior to the occurrence of neurological symptoms and neuronal death suggests that treatments to prevent excitotoxic insults may be of use in Huntington's disease. Both effects occurred in the absence of detectable nuclear inclusions, cytoplasmic aggregates or motor anomalies in the knock-in mice. This indicates that they may occur very early in the course of the disease. Therefore, the Huntington's disease mutation can significantly alter the normal functioning of striatal neurons even in the absence of gross motor anomalies. If the same is true in patients, then treatment may have to be started before the onset of symptoms.

6. REFERENCES

Albin R. L., Qin Y., Young A. B., Penney J. B. and Chesselet M.-F. (1991) Preproenkephalin messenger RNA-containing neurons in striatum of patients with symptomatic and presymptomatic Huntington's disease: an in situ hybridization study. *Ann. Neurol.* **30**, 542-549.

Aronin N., Chase K., Young C., Sapp E., Schwarz C., Matta N., Kornreich R., Landwehrmeyer B., Bird E., Beal M. F. and al e. (1995) CAG expansion affects the expression of mutant Huntingtin in the Huntington's disease brain. *Neuron.* **15**, 1193-201.

Beal M. F., Kowall N. W., Ellison D. W., Mazurek M. F., Swartz K. J. and Martin J. B. (1986) Replication of the neurochemical characteristics of Huntington's disease by quinolinic acid. *Nature.* **321**, 168-171.

Beal M. F., Brouillet E., Jenkins B., Henshaw R., Rosen B. and Hyman B. T. (1993) Age-dependent striatal excitotoxic lesions produced by the endogenous mitochondrial inhibitor malonate. *J. Neurochem.* **61**, 1147-50.

Bordelon Y. M. and Chesselet M.-F. (1999) Early effects of intrastriatal injections of quinolinic acid on microtubule-associated protein-2 and neuropeptides in rat basal ganglia. *Neuroscience.* **in press.**,

Brouillet E., Hantraye P., Ferrante R. J., Dolan R., Leroy-Willig A., Kowall N. W. and Beal M. F. (1995) Chronic mitochondrial energy impairment produces selective striatal degeneration and abnormal choreiform movements in primates. *Proceedings Of The National Academy Of Sciences Of The United States Of America.* **92**, 7105-9.

Carter R. J., Lione L. A., Humby T., Mangiarini L., Mahal A., Bates G. P., Dunnett S. B. and Morton A. J. (1999) Characterization of progressive motor deficits in mice transgenic for the human Huntington's disease mutation. *J. Neurosci.* **19**, 3248-57.

Cha J. H., Kosinski C. M., Kerner J. A., Alsdorf S. A., Mangiarini L., Davies S. W., Penney J. B., Bates G. P. and Young A. B. (1998) Altered brain neurotransmitter receptors in transgenic mice expressing a portion of an abnormal human huntington disease gene. *Proc. Natl. Acad. Sci.,USA.* **95**, 6480-5.

Chesselet M. F., Gonzales C., Lin C. S., Polsky K. and Jin B. K. (1990) Ischemic damage in the striatum of adult gerbils: relative sparing of somatostatinergic and cholinergic interneurons contrasts with loss of efferent neurons. *Exp. Neurol.* **110**, 209-18.

Choi D. W. (1988) Glutamate neurotoxicity and diseases of the nervous system. *Neuron.* **1**, 623-34.

Colwell C. S., Altemus K. L., Cepeda C. and Levine M. S. (1996) Regulation of N-methyl-D-aspartate-induced toxicity in the neostriatum: a role for metabotropic glutamate receptors *Proc. Natl. Acad. Sci., USA.* **93**, 1200-4.

Colwell C. S. and Levine M. S. (1996) Glutamate receptor-induced toxicity in neostriatal cells. *Brain Res.* **724**, 205-12.

Colwell C. S., Cepeda C., Crawford C. and Levine M. S. (1998) Postnatal development of glutamate receptor-mediated responses in the neostriatum. *Dev. Neurosci..* **20**, 154-63.

Cooper J. K., Schilling G., Peters M. F., Herring W. J., Sharp A. H., Kaminsky Z., Masone J., Khan F. A., Delanoy M., Borchelt D. R., Dawson V. L., Dawson T. M. and Ross C. A. (1998) Truncated N-terminal fragments of huntingtin with expanded glutamine repeats form nuclear and cytoplasmic aggregates in cell culture. *Hum. Molec. Genet.* **7**, 783-90.

Davies S. W., Turmaine M., Cozens B. A., DiFiglia M., Sharp A. H., Ross C. A., Scherzinger E., Wanker E. E., Mangiarini L. and Bates G. P. (1997) Formation of neuronal intranuclear inclusions underlies the neurological dysfunction in mice transgenic for the HD mutation. *Cell.* **90**, 537-48.

de la Monte S. M., Vonsattel J. P. and Richardson E. P. J. (1988) Morphometric demonstration of atrophic changes in the cerebral cortex, white matter, and neostriatum in Huntington's disease. *J. Neuropathology & Experim. Neurol.* **47**, 516-25.

DiFiglia M., Sapp E., Chase K. O., Davies S. W., Bates G. P., Vonsattel J. P. and Aronin N. (1997) Aggregation of huntingtin in neuronal intranuclear inclusions and dystrophic neurites in brain. *Science.* **277**, 1990-3.

Dodt H. U. and Zieglgänsberger W. (1990) Visualizing unstained neurons in living brain slices by infrared DIC-videomicroscopy. *Brain Res.* **537**, 333-6.

Dodt H. U., Hager G. and Zieglgänsberger W. (1993) Direct observation of neurotoxicity in brain slices with infrared videomicroscopy. *J. Neurosci. Meth.* **50**, 165-71.

Dodt H. U. and Zieglgänsberger W. (1994) Infrared videomicroscopy: a new look at neuronal structure and function. *TINS.* **17**, 453-8.

Gonzales C., Lin R. C. and Chesselet M. F. (1992) Relative sparing of GABAergic interneurons in the striatum of gerbils with ischemia-induced lesions. *Neurosci Lett.* **135**, 53-8.

Huntington Disease Collaborative Research Group (1993) A novel gene containing a trinucleotide repeat that is unstable on Huntington's Disease chromosomes. *Cell.* **72**, 971-983.

Harper P. S. (1991) "Huntington's Disease, Major Problems in neurology." W.B. Saunders, London.

Hurlbert M. S., Zhou W., Wasmeier C., Kaddis F. G., Hutton J. C. and Freed C. R. (1999) Mice transgenic for an expanded CAG repeat in the Huntington's disease gene develop diabetes. *Diabetes.* **48**, 649-51.

Klapstein GJ, Fisher RS, Zanjani H, Cepeda C, Jokel ES, Chesselet M-F, Levine MS. Age-related electrophysiological and morphological changes in medium-sized striatal spiny neurons in R6/2 Huntington's disease transgenic mice. *J Neurophysiology*, 86 (2001) 2667-77

Kowall N. W., Ferrante R. J. and Martin J. B. (1987) Paterns of cell loss in Huntington's Disease. *TINS..* **10**, 24-29.

La Spada A. R., Wilson E. M., Lubahn D. B., Harding A. E. and Fischbeck K. H. (1991) Androgen receptor gene mutations in X-linked spinal and bulbar muscular atrophy. *Nature.* **352**, 77-9.

Levine M. S., Klapstein G. J., Koppel A., Gruen E., Cepeda C., Vargas M. E., Jokel E. S., Carpenter E. M., Zanjani H., Hurst R. S., Efstratiadis A. and Zeitlin S. and Chesselet M-F. (1999) Enhanced sensitivity to N-Methyl-D-Aspartate Receptor Activation in Transgenic and Knockin Mouse Models of Huntington's Disease. *J. Neurosci Res.* **58**, 515-532.

Mangiarini L., Sathasivam K., Seller M., Cozens B., Harper A., Hetherington C., Lawton M., Trottier Y., Lehrach H., Davies S. W. and Bates G. P. (1996) Exon 1 of the HD gene with an expanded CAG repeat is sufficient to cause a progressive neurological phenotype in transgenic mice. *Cell.* **87**, 493-506.

Martindale D., Hackam A., Wieczorek A., Ellerby L., Wellington C., McCutcheon K., Singaraja R., Kazemi-Esfarjani P., Devon R., Kim S. U., Bredesen D. E., Tufaro F. and Hayden M. R. (1998) Length of huntingtin and its polyglutamine tract influences localization and frequency of intracellular aggregates. *Nature Genetics.* **18**, 150-4.

Menalled L.B. and Chesselet M-F. (2002) Mouse models of Huntington's disease. *Topics in Pharmacological Science*, **23**, 32-39.

Menalled L., Zanjani H., MacKenzie L., Koppel A., Carpenter E., Zeitlin S. and Chesselet M-F. (2000) Decrease in striatal enkephalin mRNA in mouse models of Huntington's disease. *Exp Neurol.* **162**, 328-342.

Mercugliano M., Soghomonian J. J., Qin Y., Nguyen H. Q., Feldblum S., Erlander M. G., Tobin A. J. and Chesselet M. F. (1992) Comparative distribution of messenger RNAs encoding glutamic acid decarboxylases (Mr 65,000 and Mr 67,000) in the basal ganglia of the rat. *J. Comp. Neurol.* **318**, 245-54.

Myers R. H., Vonsattel J. P., Stevens T. J., Cupples L. A., Richardson E. P., Martin J. B. and Bird E. D. (1988) Clinical and neuropathologic assessment of severity in Huntington's disease. *Neurology.* **38**, 341-7.

Ordway J. M., Tallaksen-Greene S., Gutekunst C. A., Bernstein E. M., Cearley J. A., Wiener H. W., Dure L. S. t., Lindsey R., Hersch S. M., Jope R. S., Albin R. L. and Detloff P. J. (1997) Ectopically expressed CAG repeats cause intranuclear inclusions and a progressive late onset neurological phenotype in the mouse. *Cell.* **91**, 753-63.

Paulson H. L., Perez M. K., Trottier Y., Trojanowski J. Q., Subramony S. H., Das S. S., Vig P., Mandel J. L., Fischbeck K. H. and Pittman R. N. (1997) Intranuclear inclusions of expanded polyglutamine protein in spinocerebellar ataxia type 3. *Neuron.* **19**, 333-44.

Qin Y., Soghomonian J.-J. and Chesselet M.-F. (1992) Effects of quinolinic acid on messenger RNAs encoding somatostatin and glutamic acid decarboxylases in the striatum of adult rats. *Exp. Neurol.* **115**, 200-211.

Reddy R. S. and Housman D. E. (1997) The complex pathology of trinucleotide repeats. *Curr. Opin. Cell Biol.* **9**, 364-372.

Reiner A., Albin R. L., Anderson K. D., D'Amato C. J., Penney J. B. and Young A. B. (1988) Differential loss of striatal projection neurons in Huntington disease. *Proc. Natl. Acad. Sci. USA.* **85**, 5733-5737.

Richfield E. K., Maguire-Zeiss K. A., Cox C., Gilmore J. and Voorn P. (1995) Reduced expression of preproenkephalin in striatal neurons from Huntington's disease patients. *Ann. Neurol.* **37**, 335-343.

Roberts R. C., Ahn A., Swartz K. J., Beal M. F. and DiFiglia M. (1993) Interstriatal injections of quinolinic acid or kainic acid: differential patterns of cell survival and the effects of data analysis on outcome. *Exp. Neurol.* **124**, 272-282.

Rothman S. M. (1992) Excitotoxins: possible mechanisms of action. *Ann. Of The New York Acad. Of Sci.* **648**, 132-9.

Saudou F., Finkbeiner S., Devys D. and Greenberg M. E. (1998) Huntingtin acts in the nucleus to induce apoptosis but death does not correlate with the formation of intranuclear inclusions. *Cell.* **95**, 55-66.

Scherzinger E., Lurz R., Turmaine M., Mangiarini L., Hollenbach B., Hasenbank R., Bates G. P., Davies S. W., Lehrach H. and Wanker E. E. (1997) Huntingtin-encoded polyglutamine expansions form amyloid-like protein aggregates in vitro and in vivo. *Cell.* **90**, 549-58.

Schwarcz R., Whetsell W. O. and Mangano R. M. (1983) Quinolinic acid: an endogenous metabolite that produces axon-sparing lesions in rat brain. *Science.* **219**, 316-318.

Sharp A. H., Loev S. J., Schilling G., Li S. H., Li X. J., Bao J., Wagster M. V., Kotzuk J. A., Steiner J. P., Lo A. and al e. (1995) Widespread expression of Huntington's disease gene (IT15) protein product. *Neuron.* **14**, 1065-74.

Skinner P. J., Koshy B. T., Cummings C. J., Klement I. A., Helin K., Servadio A., Zoghbi H. Y. and Orr H. T. (1997) Ataxin-1 with an expanded glutamine tract alters nuclear matrix-associated structures [published erratum appears in Nature 1998 Jan 15;391(6664):307]. *Nature.* **389**, 971-4.

Trottier Y., Lutz Y., Stevanin G., Imbert G., Devys D., Cancel G., Saudou F., Weber C., David G., Tora L. and al e. (1995) Polyglutamine expansion as a pathological epitope in Huntington's disease and four dominant cerebellar ataxias. *Nature.* **378**, 403-6.

Turmaine M., Raza A., Mahal A., Mangiarini L., Bates G. P., Davies S. W. (2000) Nonapoptotic neurodegeneration in a transgenic mouse model of Huntington's disease. *Proc Natl Acad Sci U S A.* **Jul5; 97(14)**, 8093-7.

Vonsattel J. P., Myers R. H., Stevens T. J., Ferrante R. J., Bird E. D. and Richardson E. P. J. (1985) Neuropathological classification of Huntington's disease. *J. Neuropathol. Exp. Neurol.* **44**, 559-77.

Whetsell W. O. (1996) Current concepts of excitotoxicity. *J. Neuropathol. Exp. Neurol.* **55**, 1-13.

Zhuchenko O., Bailey J., Bonnen P., Ashizawa T., Stockton D. W., Amos C., Dobyns W. B., Subramony S. H., Zoghbi H. Y. and Lee C. C. (1997) Autosomal dominant cerebellar ataxia (SCA6) associated with small polyglutamine expansions in the alpha 1A-voltage-dependent calcium channel. *Nature Genetics.* **15**, 62-9.

EFFECTS OF GDNF ON NIGROSTRIATAL DOPAMINE NEURONS
Experience from rodent models of Parkinson's Disease

Carl Rosenblad, Deniz Kirik and Anders Björklund*

Abstract: Glial cell line-derived neurotrophic factor (GDNF) have been found to be a very potent neurotrophic factor for nigrostriatal dopamine neurons in vitro and in vivo. Intracerebral administration of GDNF can produce three principal structural effects. (1) increased survival of lesioned adult DA neurons, (2) sprouting of new fibers dopamine fibers and (3) pharmacological effects related to altered synthesis, turnover and metabolism of dopamine. These effects may in turn promote restoration of function in animals models of Parkinson's disease. This chapter will summarise the different GDNF induced effects observed and discuss how they may vary depending on which lesion model have been used as well as the treatment regime applied for GDNF.

1. INTRODUCTION

Parkinson's disease is caused by a protracted loss of nigrostriatal dopamine (DA) neurons over years to decades. This leads to a loss of DA in the target area, primarily the putamen and caudate nucleus, and when the DA content is reduced by 70-80% motor symptoms (characteristically tremor, rigidity and hypokinesia) start to appear (1). The currently most used treatment is based on administration of the DA precursor L-dopa which is metabolized to DA and thereby can substitute for the loss of endogenous DA. However, the effect of this therapy often subsides after several years of treatment. Therefore the prospect of interfering with the degenerative process and thereby preserve and improve the function of the remaining system has evolved as a therapeutic strategy. In the pursuit of this strategy, neurotrophic factors have been extensively tested because of their ability to promote sur-

---

* Carl Rosenblad, Deniz Kirik and Anders Björklund, Dept. of Physiological Sciences, Sect. Neurobiology, Wallenberg Neuroscience Center, Lund University, Lund Sweden

vival and phenotypic expression in many different neuron populations including the nigrostriatal DA neurons (2,3).

In 1993 Lin and collaborators (4) detected that conditioned medium from the B49 glioma cell line contained a trophic activity on fetal DA neurons in culture. From the medium they isolated a novel factor that was called glial cell line-derived neurotrophic factor (GDNF). GDNF was found to be a distant member of the TGFß superfamily and had the capacity to specifically promote survival and DA uptake in cultured DA neurons (4). Later GDNF has been found to exert trophic effects on several other peripheral and central neuron populations (5-11). Although a large number of biomolecules have been found to exert neuroprotective and growth stimulating effects on developing dopamine neurons in vitro (2,3) GDNF stands out as the clearly most interesting factor for nigrostriatal DA neurons based on the prominent neuroprotective effects observed in a wide variety of nigrostriatal DA neuron lesion models *in vivo* including neurotoxic or mechanical axotomy of the nigrostriatal axons (12-18). The role of GDNF for the nigral DA neurons is further substantiated by in situ hybridization studies of expression of GDNF and its receptors GFRa-1 and c-ret. These show that GDNF is expressed in the ventral mesencephalon by the time the tyrosine hydroxylase-positive neurons start to appear and at later time-points in the striatum during the time when this structure is innervated by the nigral DA neurons. Further, both GFRa-1 and c-ret are also strongly expressed in the nigral DA neurons indicating that they are responsive to GDNF (19-21). Lately, three additional members of what is now called the GDNF-family of neurotrophic factors have been found namely Neurturin (22), Persephin (23) and Neublastin/Artemin (24,25). Initial studies suggest that Neurturin and Neublastin/Artemin have neuroprotective effects that are similar to those of GDNF both in vitro and in vivo (24-27), while persephin appears to be less efficient (23).

From the studies of GDNF effects in vivo, it appears that GDNF can participate in preservation and regeneration of the nigrostriatal DA neurons, on one hand, and pharmacological effects related to regulation of DA metabolism, on the other. Behavioral effects have also been reported in conjunction with both types of cellular responses. However, this behavioral response may differ qualitatively depending on which type of GDNF effect is observed and also which animal model has been used. We will in this chapter discuss the effects of GDNF on adult nigrostriatal DA neurons, in vivo in the light of available data obtained from rodent models of Parkinson's disease, particularly the intrastriatal 6-OHDA lesion model which has turned out to be suitable for studies of morphological and behavioral changes associated with trophic factor treatment (see below).

2. THE INTRASTRIATAL 6-OHDA LESION MODEL FOR INVESTIGATION OF NEUROTROPHIC EFFECTS *IN VIVO*

Clearly, the morphological and behavioral characteristics of any animal lesion model determine which naturally occurring (or treatment-induced) phenomenon can be appropriately studied. Yet, this is important to point out since a number of different lesion models have been applied in studies of neurotrophic effects on the nigrostriatal DA system, including those of GDNF. Therefore, observations made with different doses, administration sites and time-points of administration have to be viewed in light of the model in which they are obtained. This can be exemplified by a comparison of two of the models most widely used to simulate degeneration of nigrostriatal DA neurons *in vivo*. Both models employ 6-hydroxydopamine (6-OHDA), a neurotoxin that specifically destroy DA (and other catecholamine) neurons. One is the standard unilateral 6-OHDA lesion where the toxin is administered into the medial forebrain bundle (MFB) close to cell bodies in the substantia nigra (SN), or even directly into the SN. This lesion has been extensively used in behavioral, pharmacological and transplantation studies (28). In the second model 6-OHDA is given onto the DA terminals in the striatum. This so-called intrastriatal 6-OHDA lesion model was initially used to study the effect of regional denervation of striatal DA terminals (29,30) but more recently also to obtain a protracted degeneration of nigral DA neurons (31-33). The intrastriatal 6-OHDA lesion model have been particularly useful for the characterization of the effects associated with neurotrophic factor treatment of nigral DA neurons, including protection of soma integrity and sparing/regeneration of axons or dendrites, for three major reasons. First, because of the protracted time-course of fiber degeneration and loss of nigral DA neurons. The striatal axon terminals are acutely damaged by an intrastriatal 6-OHDA lesion and during the first 24 hours the majority of tyrosine hydroxylase(TH)-immunoreactive fibers are lost from the striatum and the axonal stumps are found at the level of the caudal globus pallidus (34,35). Over the following days the retrograde degeneration continues so that the stumps are seen at the level of the SN by 7 days post lesion, i.e. approximately when a loss of neurons in the SN can first be appreciated. The initiation of cell loss is delayed up to about one week post lesion and then continues progressively with an acute phase of cell loss lasting for 2-3 weeks followed by a slow phase lasting for 1-2 months. By contrast, after an MFB lesion there is anterograde degeneration of the distal axons, and the cell death starts immediately and has a very rapid progression (within days). Thus, the degenerative process in the intrastriatal 6-OHDA lesion model provides a time window when a neuroprotective intervention can be applied. Secondly, the magnitude of the lesion. The standard MFB lesion causes a near-complete loss of cells in the SN and DA content in the striatum unilaterally (36,37). Instead, the partial lesion model spares 20-60% of the nigral DA neurons and 10-50% of the striatal DA innervation or DA levels depending on the dose of 6-OHDA and the number of deposits in the striatum (33,38-41). The residual nigrostriatal DA neurons can serve as a substrate for formation of new fibers that may provide functional reinnervation of DA depleted areas. Thirdly, the specific loss of nigrostriatal DA neurons. After an intrastriatal 6-OHDA lesion the majority of cells that are lost are nigrostriatal DA neurons although some loss is also observed in the ventral tegmental area (VTA), probably involving those VTA neurons that project to the dorsal striatum (18,39,41). This is at variance with the MFB lesion which affects large portions of the mesotelencephalic DA projections (mesostriatal, mesolimbic and mesocortical). A more specific nigrostriatal degeneration is analogous to the situation in Parkinson's disease where the DA depletion is most pronounced in the putamen and caudate nucleus (equivalent to the rodent striatum) whereas other DA innervated areas are relatively spared (42,43). Because of the morphological features described above, as well as the possibility to induce behavioral deficits of varying severity, the intrastriatal 6-OHDA lesion model has turned out to be particularly useful for assessment of GDNF effects in the nigrostriatal DA system.

Table 1. Summary of structural and behavioral effects of GDNF administration in 6-OHDA lesioned rats

Time and site of GDNF administration	Type of 6-OHDA lesion	Protection of cell bodies	Protection of striatal DA innervation	Regeneration of DA fibers	DA neurotransmission related changes	Behavioral effects	Refs
Prelesion							
SN	MFB	50-100%	Yes	--	DA ↑, SN + STR	--	a
	IS	45-85%	No	Locally over SN	--	No effect on amph, apo or f.a.	b
STR	MFB			*No reports available*			
	IS	85-100%	Yes	In the GP	--	No deficit in amph, apo, or f.a.	c
Early post lesion (<7days)							
SN	MFB			*No reports available*			
	IS	70-100%	No	No	--	No attenuation of amph, apo or f.a.	d
STR	MFB			*No reports available*			
	IS	90%	No	In the GP	--	No attenuation of amph	e
Delayed post lesion (>7 days)							
SN	MFB	10%	No	Locally near injection	DA ↑, SN TH activity ↑, SN	Complete reversal of apo	f
	IS			*No reports available*			
STR	MFB			*No reports available*			
	IS	20-40%	No	Locally in STR	--	Attenuation of amph, f.a.	g

References: a) Kearns and Gash, 1995; Kearns et al, 1997; Sullivan et al, 1998; Opacka Juffrey et al, 1996. b) Kearns and Gash, 1995; Kirik et al, unpublished observations. c) Shults et al, 1996; Kirik et al, unpublished observations. d) Winkler et al, 1996; Sauer et al, 1995. e) Rosenblad et al, 1999. f) Bowenkamp et al, 1995; Lapchak et al, 1997a,b; Hoffer et al, 1994. g) Kirik et al, IBAGS; Rosenblad et al, 1998.

Abbreviations: SN, substantia nigra; STR, striatum; MFB, medial forebrain bundle; IS, intrastriatal; GP, globus pallidus; DA, dopamine; TH, tyrosine hydroxylase; apo, apomorphine-induced turning; amph, amphetamine-induced turning; f.a., forelimb akinesia test (stepping test).

3. MORPHOLOGICAL AND BIOCHEMICAL EFFECTS OF GDNF

Following administration into either the intact or lesion nigrostriatal DA system GDNF has been found to exert three major structural effects: protection of cell bodies, axons and dendrites; regeneration of new DA fibers; and functional regulation of DA neurotransmission by actions on DA turnover and metabolism and on the expression or activity of the TH enzyme. A summary of these major effects are given in Table 1.

3.1 Protection of Cell Bodies and Axons

The most robust GDNF effect on adult nigral DA neurons, and the first one to be observed *in vivo*, is the protection of the cell bodies from lesion-induced cell death. The first studies showing a prominent protective effect of GDNF utilised administration close to the cell bodies in the SN following both partial progressive, as well as more complete lesions (12,15,16,44,45). Using retrograde pre-labelling of the nigral DA neurons with fluorogold, followed 1 week later by an intrastriatal 6-OHDA lesion, Sauer et al (15) found that the retrograde cell death and lesion-induced atrophy could be completely prevented by injections of 10μg GDNF over the SN every 2nd day for four weeks, as assessed four weeks post lesion. Similarly, complete protection was observed by Lu and Hagg, (14) after infusion of 3 μg GDNF/day for 2 weeks after a knife cut of the MFB, and 1 μg/day was sufficient to give a half-maximal response. Beck et al, (12) reported that with daily injections of 1μg GDNF during 2 weeks after an axotomy of the MFB, 85% of the nigral DA neurons remained. Thus, administration over the cell bodies in the SN can completely rescue the nigral DA neurons from progressive cell death. The nigral administration route have also been found to be effective after the more acute and complete 6-OHDA lesion of the MFB (44-47).

Recent studies using administration onto the terminals in the striatum or into the lateral ventricle (ICV) show that these routes may be equally effective in protecting the lesioned DA neurons. Rosenblad et al (27) found that 7 injections of 5 μg intrastriatally or 10μg ICV over three weeks (i.e. one injection every 3rd day), starting one day after an intrastriatal 6-OHDA lesion was able to provide near-complete (90-92%) rescue of retrogradely pre-labeled neurons. Using the same lesion paradigm, Shults et al, (48) have also reported a clear neuroprotective effect of intrastriatal GDNF administration (1μg/inj) when given repeatedly around the time of lesion (1 and 3 days before, and 1and 3 days after). It therefore appears that the ability of GDNF to rescue lesioned nigral DA neurons is not primarily dependent on the site of administration. From these studies it can also be inferred that the daily dose of exogenous protein for near complete protection in partially lesioned nigrostriatal systems seems to be in the range of 1-4 μg. However, tissue levels in the range of a few nanograms of GDNF have been found to be sufficient for prominent neuroprotection after direct intracerebral gene transfer of GDNF by midbrain injection of GDNF encoding adenoviral (49) or adeno associated viral vectors (50), which are expressed in nigral DA neurons.

Evidence from several studies point to the timing and duration of GDNF treatment as important factors in determining the efficiency of neuroprotection. In the intrastriatal 6-OHDA lesion model a single injection of 10μg GDNF given over the SN at 7 days post lesion, i.e. at the time when the axon degeneration is more or less complete but before any significant loss of cell bodies in the SN has taken place, was able to give a partial protection of the nigral DA neurons (15). A similar incomplete protection was also seen when the same dose was instead given 24 hours prior to the 6-OHDA lesion (45). By contrast, in the study by Sauer et al, (15) complete protection was seen when the GDNF treatment was sustained over the first 4 weeks. Further, Kirik and collaborators, (51) found that when the start of continuous pump infusions of GDNF was delayed 2 weeks after an intrastriatal 6-OHDA lesion, any cell loss that otherwise would have taken place later than two weeks, was completely abolished.

These studies indicate that after an intrastriatal 6-OHDA injection the nigral DA neurons reach a point where they "decide" to die and this decision may be prevented or blocked by the presence of sufficient amounts of GDNF. The time before this point is reached may vary in the individual neuron from a few days to weeks after the toxic insult and is most likely influenced by several factors such as the extent of destruction of its axonal tree and the capacity to withstand stress. On the molecular level, several different intracellular signalling pathways leading to transcriptional regulation may be responsible for the decision. For example, induction and phosphorylation of the transcription factor c-jun have been implicated to enhance cell death in several different neuron populations (52,53). In nigral DA neurons subjected to an intrastriatal 6-OHDA lesion it was recently found that there was a massive increase in c-jun and its phosphorylated form (phospho-jun) after an intrastriatal 6-OHDA lesion (54). This was fully developed by three days post lesion (i.e. before the loss of DA neurons in the SN became apparent). Moreover, supranigral injection of GDNF (10μg) at 3 days post lesion markedly reduced the number of c-jun and phospho-jun labeled neurons present in the SN especially at 7 days but also at 14 days, post lesion. This reduction was associated with a sparing of nigral DA neurons. However, there was still a drop in the number of TH-positive nigral neurons between 7 and 14 days post lesion suggesting that later phases of cell loss was not prevented by a single GDNF injection. Following a single injection of GDNF only those neurons that are at the decision point when GDNF is administered, as well as those reaching this stage while sufficient tissue levels of GDNF are still present, are prevented from dying. Thus, in situations with protracted DA neuron cell death such as in the partial lesion model or in Parkinson's disease, sustained GDNF administration is likely to give the best effects. Conversely, in the MFB lesion where a majority of the nigral DA neurons are likely to reach and pass their decision point within a short time period, a more short-lasting GDNF treatment with an early onset would be beneficial. Indeed, Kearns et al, (46) found that a single injection of GDNF given over the SN 6 hours before injection of 6-OHDA into the MFB completely prevented the loss of DA neurons in the SN; when given 24 hours before the insult the effect was only partial.

In early studies reporting neuroprotection by GDNF after intrastriatal 6-OHDA injections it was observed that the DA innervation of the striatum was not spared to the same extent as the cell bodies in the SN (15,44). Since symptoms in patients with Parkinson's disease is closely related to the extent of remaining DA innervation in the putamen and caudate nucleus (43) it became an important issue to find out if and how GDNF could preserve also the nigrostriatal axons.

With the loss of nigrostriatal axons being so rapid it is predictable that GDNF has to be applied within the first few hours after or even before the 6-OHDA lesion in order to preserve the terminals in the striatum. Yet, in studies where GDNF have been given over the SN either as repeated injections (15) or pump infusion (35) over 14-28 days, starting immediately after the lesion, there has been no signs of fiber sparing in the striatum or along the nigrostriatal pathway suggesting that GDNF given over the nigra is not able to prevent axon degeneration induced by 6-OHDA. Instead, Shults et al, (48) found that when intrastriatal injections of

GDNF was given both before and after the 6-OHDA lesion there was a sparing of both TH-positive fibers in the striatum and cells in the SN, and when intrastriatal GDNF treatment was initiated one day post lesion the fiber degeneration was partially halted and axonal stumps were seen in a position where the degenerated fibers are likely to be found by one day post lesion (27). Thus, it appears that the nigral and striatal administration sites differ with respect to their capacity to prevent nigrostriatal axon degeneration after an intrastriatal 6-OHDA lesion. Further evidence comes from a recent study where a single 25μg GDNF injection was given just dorsal the SN, or into the striatum (68). The injection was given 6 hours prior to an intrastriatal 6-OHDA lesion which has been shown to be the time interval required for induction of a maximal GDNF response in vivo (46). We found that the intrastriatal GDNF injection was able to preserve 95% of the TH-positive cells in the SN as well as 80% of the TH-positive innervation of the striatum. By contrast, when GDNF was given just above the SN no preservation of striatal TH-positive fibers was detected at any level along the nigrostriatal pathway in spite of a robust protection of the nigral DA cell bodies. Thus, in the partial lesion model it appears that sparing of nigrostriatal DA axons can be obtained by administration onto the terminals. However, at variance with the findings in the intrastriatal 6-OHDA lesion model, Kearns et al, (46) found both sparing of nigral TH-positive neurons and significantly attenuated striatal DA levels after intranigral GDNF (10μg) given 6 hours before an intranigral 6-OHDA lesion and Sullivan et al, (47) observed a significant increase in binding to striatal DA uptake sites after intranigral + ICV co-administration (75μg) given just prior to the lesion.

3.2 Regeneration

In the clinical situation where a large part of the nigrostriatal system has degenerated by the time of diagnosis (see Introduction), induction of sprouting and formation of new synaptic contacts with the target neurons could provide a basis for functional improvements. A wealth of data from studies in the intact and lesioned endogenous DA system as well as intracerebral transplants of fetal DA neurons, indicates that GDNF also can stimulate re-growth of new DA fibers *in vivo* (16,17,27,44,55-57). However, this phenomenon -just like preservation- appears to be dependent on the site of administration.

Sprouting in or close to the striatum have only been reported in studies employing intrastriatal GDNF administration. In a study where repeated GDNF injections were given from day 1-20 post lesion, a dense fiber network of thin irregular TH-positive fibers with multiple varicose swellings was observed in the globus pallidus at 4 weeks (27). The location of this sprouting matched exactly the position of the distal stumps of the degenerating axons by one day after an intrastriatal 6-OHDA injection i.e. at the time the GDNF treatment was initiated indicating that the newly formed TH-positive fibers emanated at least partly from axonal stumps residing in the globus pallidus. Yet, in this study the newly formed fibers did not extend into the denervated striatum. However, evidence from another study indicate that GDNF can induce re-growth of new fibers in the striatum (57). This study was designed to specifically investigate the capacity of remaining striatal DA fibers to reinnervate the striatum. The animals were left for four weeks after an intrastriatal 6-OHDA lesion in order for the lesion to be fully developed before GDNF was given every second day for 3 weeks. This treatment reduced the lesion area to ~50% of that in the control animals and increased the binding of [3H]-BTCP to high affinity DA uptake sites within the lesion area to 70-90% of the intact side. In this case the retrograde fiber degeneration was complete by the time GDNF

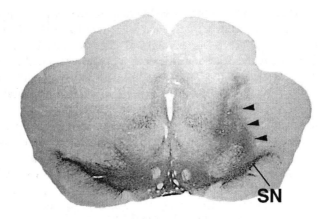

Figure 1. Coronal section throught the midbrain immuno-labelled for tyrosine hydroxylase. At 6 weeks after a single injection of GDNF dorsally to the SN, prominent sprouting can be seen around the injection site (arrowheads)(Kirik et al, unpublished observations).
3.3 Pharmacological effects on DA metabolism and TH

administration was initiated, wherefore the source of the sprouting fibers most likely were intact or partially lesioned axons of the remaining nigrostriatal projection.

Growth of TH-positive fibers have also been observed in the vicinity of the SN. Tseng et al, (17) observed prominent sprouting around a capsule containing GDNF secreting fibroblasts placed adjacent to the SN. Similar fiber outgrowth has been seen after intranigral and intrastriatal 6-OHDA lesions (18,44, 68) as well as in intact animals (55). An example is shown in Fig 1 where aberrant fibers are seen to grow from the nigral DA neurons towards the injection site. These studies thus suggests that sprouting of new DA fibers is a phenomenon that occurs locally at the site of GDNF injection and may even be influenced by the concentration gradient of GDNF. Although the potency to induce sprouting can turn out to be very useful to direct re-growth of DA fibrers into partially or completely denervated areas such aberrant sprouting may be problematic, particularly in a clinical context (see below).

The third major action of GDNF that have been observed after intracerebral administration is a pharmacological type of response with regulation of synthesis and turnover of DA levels and its metabolites DOPAC and HVA (13,16,44,55) as well as increased TH activity (58,59). Since different biochemical and histochemical techniques have been used to monitor various aspects of DA synthesis and metabolism, the exact mechanism involved may be unclear but most studies to date point to an increased activity in the DA system as the endpoint. In a series of studies using a single high bolus injection (in the range of 100-1000µg GDNF) given several weeks after a near-complete destruction of the nigrostriatal DA system by 6-OHDA lesion of the MFB, induced long-lasting increases in DA levels or turnover (as indicated by the ratio of DA/DOPAC)(13,44,59). This effect is also seen in partially lesioned or intact animals and in these cases it may suffice with somewhat lower doses. Thus, Tomac et al, (16) found that MPTP treated mice injected with 10µg of GDNF had higher striatal and nigral levels of DA, DOPAC and HVA. An interesting observation in this study was also that striatal levels was obtained by intrastriatal, but not nigral GDNF administration. Injections into intact rats also show the same type of response (55,60).

Also the activity of TH have been reported to be increased by intranigral or intrastriatal injections of GDNF in lesioned adult or intact neonatal rats (58,59,61). By contrast, two studies report that the levels of TH in neurons retrogradely pre-labeled with DiI or fluorogold prior to an axotomizing knife-cut of the MFB, (14) or an intrastriatal 6-OHDA lesion (27) were clearly reduced in a large portion of nigral neurons rescued by a nigral GDNF infusion. Moreover, when GDNF was infused for 14 days over the SN of intact animals (14), there was a dose-dependent reduction in TH-expression among the retrogradely labeled DA neurons. It is possible that these seemingly contradictory findings may simply reflect the fact that the balance between components involved in DA neurotransmission have been shifted. Decreased levels of TH (seen as a weaker intensity of TH staining) may therefore just be a compensatory mechanism for GDNF induced increases in DA turnover or TH activity due to for instance an altered phosphorylation state or increased stability of the protein (62). However, it cannot be ruled out that other components of DA neurotransmission may be affected leading to compensatory changes in TH levels.

4. FUNCTIONAL EFFECTS

Functional effects of GDNF treatment have been studied both in lesioned and intact animals. Since the response to GDNF may differ between intact and lesioned DA neurons it is important in the description of functional effects to distinguish between those that represent a protection from, or reversal of lesion-induced deficits, on one hand and those that reflect a response above normal, on the other. The ability of GDNF to prevent the development of functional deficits, or induce functional recovery, has been explored in a number of studies using different lesion models. In the lesioned system the severity of functional deficits are generally linked to the degree of striatal denervation (39,63). By contrast, the second type of response, which can be seen in both intact animals and those with mild lesions which are insufficient to produce any behavioral deficits, are based on a pharmacological over-activity in the system.

When GDNF (at a dose of 10µg/inj) was given every fourth day for 1 month, starting 5 days post lesion, and the animals were left to survive for another 4 months after the last injection (18) 72% of the nigral TH-positive neurons remained by 5 months post lesion compared to 21% in the vehicle treated animals. However, both spontaneous (forelimb akinesia) and drug-induced rotation (amphetamine) was severely and equally impaired in both treatment groups. Similarly, sequential GDNF administration above the SN for 14 days (2.5µg/day) followed by repeated intrastriatal GDNF injections from 3-5 weeks post-lesion, had no effect on the performance in the forelimb akinesia test or rotational asymmetry after apomorphine or amphetamine challenge (35). Common to both these studies were that they failed to achieve any preservation/reinstatement of the TH-positive innervation of the striatum. By contrast, studies where the nigrostriatal axons are spared or where the striatum is re-innervated by sprouting of TH-positive fibers, also report a concomitant improvement of function. For example, Kirik et al (68) found that GDNF given into the striatum prior to an intrastriatal 6-OHDA lesion completely prevented the development of functional deficits in the forelimb akinesia test, as well as in amphetamine-, or apomorphine-induced rotations. Development of amphetamine-induced rotation has also been reported to be blocked by co-administration of GDNF into the SN and ICV just prior to an intranigral 6-OHDA lesion (47). In these animals there was again a clear rescue of nigral DA neurons (70%) in combination with a marked attenuation of striatal DA levels. Further, re-growth of new DA fibers into the partially denervated striatum can also lead to functional recovery. In a study where the nigrostriatal 6-OHDA lesion was allowed to develop for 4 weeks before intrastriatal GDNF treatment was initiated (10 injections of 5 µg/inj, given over 3 weeks), the performance in the forelimb akinesia test was significantly improved and examination of the density of radioactive binding to striatal DA uptake sites (as a measure of DA innervation) showed a prominent increase in the number of striatal DA uptake sites in the striatum of GDNF treated animals. Taken together, these studies indicate that similarly to the situation in Parkinson's disease, the degree of functional deficits is well correlated to the extent of remaining DA innervation in the striatum and functional recovery is critically dependent on the extent of GDNF induced protection/re-growth of nigrostriatal axons.

Reversal of lesion-induced behavioral deficits has also been observed without changes in striatal DA levels or innervation density. A number of studies have reported that the vigorous contralateral turning upon apomorphine administration that is observed after a 6-OHDA lesion of the MFB, can be completely reversed by intranigral or ICV GDNF injections that lead to increased nigral DA levels without any measurable changes in striatal DA levels (13,44,59). This may indicate that apomorphine-induced rotation reflects a component of the behavior that is sensitive to changes in intranigral DA. The existence of an intranigrally mediated component of DA receptor-agonist induced turning has been discussed previously (64,65). However, it should be pointed out that in the intrastriatal 6-OHDA lesion model reversal of apomorphine induced rotation has not been observed after GDNF administration without simultaneous changes in striatal DA innervation (35,51). This behavioral response may therefore represent a special case that appears only with the extreme DA depletions (>99%) observed after a complete lesion of the MFB. Since GDNF has receptors on a wide range of neurons outside the nigrostriatal DA system, it can not be ruled out that pharmacological effects on other neuronal systems may influence this response.

In intact animals a different type of behavioral change has been observed after GDNF treatment. These changes reflect an increased activity above normal in the nigrostriatal DA system. Spontaneous or amphetamine-induced hyperlocomotion have been recorded in rats that have been injected with 10µg GDNF over the SN (55) or into the striatum (66), or 100µg ICV (60). The effects are seen transiently (up to 7-10 days) with a maximal response around 4-6 days after the GDNF injection (55,60,66). In a manner that resembles the reduction in apomorphine turning described above, this effect is likely to be mediated via intranigral DA release since these animals had increased tissue levels of DA in the SN but not in the striatum at 7 days post GDNF injection (55). In the study of Hudson et al, (55) it was also found that the increased levels of DA in the SN persisted and were even further increased by 3 weeks after the GDNF treatment, whereas the hyperlocomotion was no longer detectable. Thus, GDNF administration onto intact DA neurons seems to induce a rapid up-regulation in DA neurotransmission leading to a transient hyperactivity of the system. This hyperactivity is then compensated by other mechanisms than down-regulation of DA levels. Kobayashi et al, (66) reported that the increased locomotion could be blocked by administration of the D1 or D2 antaginists SCH23390 or raclopride, respectively, further supporting the view that this effect was mediated via increased DA neurotransmission.

5. CLINICAL CONSIDERATIONS

Knowing that profound effects on survival, fiber outgrowth and function that can be elicited by GDNF administration onto both intact and lesioned nigrostriatal DA neurons it is reasonable to consider the clinical prospects of neurotrophic factors. One of the most striking therapeutic effects would be to block or halt the degeneration of the nigrostriatal DA neurons and thereby slow disease progression in patients with Parkinson's disease. As we have described in this chapter, rodent studies have given insights into several potentially beneficial effects but at the same time into possible pitfalls.

First, the site of GDNF treatment. From the perspective of cell protection it is not clear which administration route is preferable but indications that sprouting of DA fibers occurs locally at the site of administration suggest that GDNF should be given into the target area where there is a reduced DA innervation. Administration in other sites may induce significant side effects from aberrant sprouting. In addition, the site of administration may be important from the point of view of specificity. There are many other neuronal systems in the brain that express the GDNF receptors which may be negatively affected by the GDNF treatment. Example of such systems would be the hippocampal formation and several cranial nerve nuclei.

Secondly, the timing of GDNF administration may be important. Parkinson's disease has a protracted cell loss over years to decades (1,67). From the observations of cell rescue in rodents it is likely that a sustained treatment is more adequate in order to obtain more efficient protection in situations with protracted cell loss. However, against this should be matched the possible effects on fiber outgrowth and DA metabolism, which may be desirable to keep within a certain physiological range.

6. REFERENCES

1. Fearnley, J. M., and Lees, A. J. 1991. Ageing and Parkinson's disease: substantia nigra regional selectivity. *Brain* **114**: 2283-2301.
2. Lindsay, R. M. 1994. Neurotrophic growth factors and neurodegenerative diseases: therapeutic potential of the neurotrophins and ciliary neurotrophic factor. *Neurobiol Aging* **15**: 249-51.
3. Unsicker, K. 1994. Growth factors in Parkinson's disease. *Prog Growth Factor Res* **5**: 73-87.
4. Lin, L.-F. H., Doherty, D. H., Lile, J. D., Bektesh, S., and Collins, F. 1993. GDNF: a glial cell line-derived neurotrophic factor for midbrain dopaminergic neurons. *Science* **260**: 1130-1132.
5. Arenas, E., Trupp, M., Akerud, P., and Ibanez, C. F. 1995. GDNF prevents degeneration and promotes the phenotype of brain noradrenergic neurons in vivo. *Neuron* **15**: 1465-73.
6. Buj-Bello, A., Buchman, V. L., Horton, A., Rosenthal, A., and Davies, A. M. 1995. GDNF is an age-secific survival factor for sensory and autonomic neurons. *Neuron* **15**: 821-828.
7. Ebendal, T., Tomac, A., Hoffer, B. J., and Olson, L. 1995. Glial cell line-derived neurotrophic factor stimulates fiber formation and survival in cultured neurons from peripheral autonomic ganglia. *J Neurosci Res* **40**: 276-84.
8. Mount, H. T., Dean, D. O., Alberch, J., Dreyfus, C. F., and Black, I. B. 1995. Glial cell line-derived neurotrophic factor promotes the survival and morphologic differentiation of Purkinje cells [published erratum appears in Proc Natl Acad Sci U S A 1995 Dec 5;92(215):11945]. *Proc Natl Acad Sci U S A* **92**: 9092-6.
9. Oppenheim, R. W., Houenou, L. J., Johnson, J. E., Lin, L. F., Li, L., Lo, A. C., Newsome, A. L., Prevette, D. M., and Wang, S. 1995. Developing motor neurons rescued from programmed and axotomy-induced cell death by GDNF [see comments]. *Nature* **373**: 344-6.

10. Trupp, M., Ryden, M., Jörnvall, H., Funakoshi, H., Timmusk, T., Arenas, E., and Ibanez, C. F. 1995. Peripheral expression and biological activities of GDNF, a new neurotrophic factor for avian and mammalian peripheral neurons. *J Cell Biol* **130**: 137-148.
11. Yan, Q., Matheson, C., and Lopez, O. T. 1995. In vivo neurotrophic effects of GDNF on neonatal and adult facial motor neurons [see comments]. *Nature* **373**: 341-4.
12. Beck, K. D., Valverde, J., Alexi, T., Poulsen, K., Moffat, B., Vandlen, R. A., Rosenthal, A., and Hefti, F. 1995. Mesencephalic dopaminergic neurons protected by GDNF from axotomy-induced degeneration in the adult rat. *Nature* **373**: 339-341.
13. Hoffer, B. J., Hoffman, A., Bowencamp, K., Huettl, P., Hudson, J., Martin, D., Lin, L.-F. H., and Gerhardt, G. 1994. Glial cell line-derived neurotrophic factor reverses toxin induced injury to midbrain dopaminergic neurons in vivo. *Neurosci Lett* **182**: 107-111.
14. Lu, X., and Hagg, T. 1997. Glial cell line-derived neurotrophic factor prevents death, but not reductions in tyrosine hydroxylase, of injured nigrostriatal neurons in adult rats. *J Comp Neurol* **388**: 484-494.
15. Sauer, H., Rosenblad, C., and Björklund, A. 1995. Glial cell line-derived neurotrophic factor but not transforming growth factor-ß prevents degeneration of nigral dopaminergic neurons following striatal 6-hydroxydopamine lesion. *Proc. Natl. Acad. Sci.* **92**: 8935-8939.
16. Tomac, A., Lindqvist, E., Lin, L.-F. H., Ögren, S. O., Young, D., Hoffer, B. J., and Olson, L. 1995. Protection and repair of the nigrostriatal dopaminergic system by GDNF in vivo. *Nature* **373**: 335-339.
17. Tseng, J. L., Baetge, E. E., Zurn, A. D., and Aebischer, P. 1997. GDNF reduces drug-induced rotational behavior after medial forebrain bundle transection by a mechanism not involving striatal dopamine. *J Neurosci* **17**: 325-33.
18. Winkler, C., Sauer, H., Lee, C. S., and Björklund, A. 1996. Short-term GDNF treatment provides long-term rescue of lesioned nigral dopaminergic neurons in a rat model of Parkinson's disease. *J. Neurosci.* **16**: 7206-7215.
19. Treanor, J. J. S., Goodman, L., de Sauvage, F., Stone, D. M., Poulsen, K. T., Beck, C. D., Gray, C., Armanini, M. P., Pollock, R. A., Moore, M. W., Buj-Bello, A., Davies, A. M., Asai, N., Takahashi, M., Vandlen, R., Henderson, C. E., and Rosenthal, A. 1996. Characterization of a multicompetent receptor for GDNF. *Nature* **382**: 80-83.
20. Trupp, M., Arenas, E., Fainzilber, M., Nilsson, A.-S., Sieber, B.-A., Grigoriou, M., Kilkenny, C., Salazar-Grueso, E., Pachnis, V., Arumäe, U., Sariola, H., Saarma, M., and Ibanez, C. F. 1996. Functional receptor for GDNF encoded by the c-ret proto-oncogene. *Nature* **381**: 785-789.
21. Trupp, M., Belluardo, N., Funakoshi, H., and Ibanez, C. F. 1997. Complmentary and overlapping expression of glial cell line-derived neurotrophic factor (GDNF), c-ret proto-oncogene, and GDNF receptor-a indicates multiple mechanisms of trophic actions in the adult rat CNS. *J Neurosci* **17**: 3554-3567.
22. Kotzbauer, P. T., Lampe, P. A., Heuckeroth, R. O., Golden, J. P., Creedon, D. J., Johnson, E. M. J., and Milbrandt, J. 1996. Neurturin, a relative of glial-cell-line-derived neurotrophic factor. *Nature* **384**: 467-470.
23. Milbrandt, J., de Sauvage, F. J., Fahrner, T. J., Baloh, R. H., Leitner, M. L., Tansey, M. G., Lampe, P. A., Heuckeroth, R. O., Kotzbauer, P. T., Simburger, K. S., Golden, J. P., Davies, J. A., Vejsada, R., Kato, A. C., Hynes, M., Sherman, D., Nishimura, M., Wang, L. C., Vandlen, R., Moffat, B., Klein, R. D., Poulsen, K., Gray, C., Garces, A., Johnson, E. M., Jr., and et al. 1998. Persephin, a novel neurotrophic factor related to GDNF and neurturin. *Neuron* **20**: 245-53.
24. Baloh, R. H., Tansey, M. G., Lampe, P. A., Fahrner, T. J., Enomoto, H., Simburger, K. S., Leitner, M. L., Araki, T., Johnson, E. M., Jr., and Milbrandt, J. 1998. Artemin, a novel member of the GDNF ligand family, supports peripheral and central neurons and signals through the GFRalpha3-RET receptor complex. *Neuron* **21**: 1291-302.
25. Rosenblad, C., Grønborg, M., Hansen, C., Blom, N., Meyer, M., Johansen, J., Dagø, L., Kirik, D., Patel, U. A., Lundberg, C., Trono, D., Björklund, A., and Johansen, T. E. 1999. In vivo protection of nigral dopamine neurons by lentiviral gene transfer of the novel GDNF family member Neublastin/Artemin. *Mol Cell Neurosci*: In press.
26. Horger, B., Nishimura, M., Armanini, M., Wang, L.-C., Poulsen, K., Rosenblad, C., Kirik, D., Moffat, B., Simmons, L., Johnson, E., Milbrandt, J., Rosenthal, A., Bjorklund, Anders, Vandlen, R., Hynes, M., and Phillips, H. 1998. Neurturin exerts potent actions on survival and function of midbrain dopaminergic neurons. *J Neurosci* **18**: 4929-4937.

27. Rosenblad, C., Kirik, D., Devaux, B., Moffat, B., Phillips, H. S., and Björklund, A. 1999. Protection and regeneration of nigral dopaminergic neurons by neurturin or GDNF in a partial lesion model of Parkinson's disease after administration into the striatum or the lateral ventricle. *Eur J Neurosci* **11:** 1554-1566.
28. Schwarting, R. K., and Huston, J. P. 1996. The unilateral 6-hydroxydopamine lesion model in behavioral brain research. Analysis of functional deficits, recovery and treatments. *Prog Neurobiol* **50:** 275-331.
29. Dunnett, S. B., and Iversen, S. D. 1982. Sensorimotor impairments following localized kainic acid and 6- hydroxydopamine lesions of the neostriatum. *Brain Res* **248:** 121-7.
30. Dunnett, S. B., and Iversen, S. D. 1982. Spontaneous and drug-induced rotation following localized 6-hydroxydopamine and kainic acid-induced lesions of the neostriatum. *Neuropharmacology* **21:** 899-908.
31. Berger, K., Przedborski, S., and Cadet, J. L. 1991. Retrograde degeneration of nigrostriatal neurons induced by intrastriatal 6-hydroxydopamine injection in rats. *Brain Res Bull* **26:** 301-7.
32. Ichitani, Y., Okamura, H., Matsumoto, Y., Nagatsu, I., and Ibata, Y. 1991. Degeneration of the nigral dopamine neurons after 6-hydroxydopamine injection into the rat striatum. *Brain Res* **549:** 350-3.
33. Sauer, H., and Oertel, W. H. 1994. Progressive degeneration of nigrostriatal dopamine neurons following axon terminal lesion by intrastriatal 6-hydroxydopamine in the rat. *Neuroscience* **59:** 401-415.
34. Jenkins, R., O'Shea, R., Thomas, K. L., and Hunt, S. P. 1993. c-jun expression in substantia nigra neurons following striatal 6- hydroxydopamine lesions in the rat. *Neuroscience* **53:** 447-55.
35. Rosenblad, C., Kirik, D., and Björklund, A. 1999. Sequential administration of GDNF into the substantia nigra and striatum promotes dopamine neuron survival and axonal sprouting but not striatal reinnervation or functional recovery in the partial 6-OHDA lesion model. *Exp Neurol*: In press.
36. Schmidt, R. H., Ingvar, M., Lindvall, O., Stenevi, U., and Bjorklund, A. 1982. Functional activity of substantia nigra grafts reinnervating the striatum: neurotransmitter metabolism and [14C]2-deoxy-D-glucose autoradiography. *J Neurochem* **38:** 737-48.
37. Ungerstedt, U., and Arbuthnott, G. 1970. Quantitative recording of rotational behavior in rats after 6-hydroxy-dopamine lesions of the nigrostriatal dopamine system. *Brain Res.* **24:** 485-493.
38. Barneoud, P., Parmentier, S., Mazadier, M., Miquet, J. M., Boireau, A., Dubedat, P., and Blanchard, J. C. 1995. Effects of complete and partial lesions of the dopaminergic mesotelencephalic system on skilled forelimb use in the rat. *Neuroscience* **67:** 837-48.
39. Kirik, D., Rosenblad, C., and Björklund, A. 1998. Characterization of behavioural and neurodegenerative changes following partial lesions of the nigrostriatal dopamine system induced by intrastriatal 6-hydroxydopamine in the rat. *Exp Neurol* **152:** 259-277.
40. Lee, C. S., Sauer, H., and Björklund, A. 1996. Dopaminergic neuronal degeneration and motor impairments following lesion by intrastriatal 6-hydroxydopamine in the rat. *Neuroscience* **72:** 641-653.
41. Przedborski, S., Levivier, M., Jiang, H., Ferreira, M., Jackson-Lewis, V., Donaldson, D., and Togaski, D. M. 1995. Dose-dependent lesions of the dopaminergic nigrostriatal pathway induced by intrastriatal injection of 6-hydroxydopamine. *Neuroscience* **67:** 631-647.
42. Kish, S. J., Shannak, K., and Hornykiewicz, O. 1988. Uneven pattern of dopamine loss in the striatum of patients with idiopathic Parkinson's disease. *N Engl J Med* **318:** 876-880.
43. Nyberg, P., Nordberg, A., Wester, P., and Winblad, B. 1983. Dopaminergic deficiency is more pronounced in putamen than in nucleus caudatus in Parkinson's disease. *Neurochemical pathology* **1:** 193-202.
44. Bowenkamp, K. E., Hoffman, A. F., Gerhardt, G. A., Henry, M. A., Biddle, P. T., Hoffer, B. J., and Granholm, A.-C. 1995. Glial cell line-deived neurotrophic factor supports survival of injured midbrain dopaminergic neurons. *J Comp Neurol* **355:** 479-489.
45. Kearns, C. M., and Gash, D. M. 1995. GDNF protects nigral dopaminergic neurons against 6-hydroxydopamine in vivo. *Brain Res* **672:** 104-111.
46. Kearns, C. M., Cass, W. A., Smoot, K., Kryscio, R., and Gash, D. M. 1997. GDNF protection against 6-OHDA: time dependence and requirement for protein synthesis. *J Neurosci* **17:** 7111-8.
47. Sullivan, A. M., Opacka-Juffrey, J., and Blunt, S. B. 1998. Long-term protection of the rat nigrostriatal dopaminergic system by glial cell line-derived neurotrophic factor against 6-hydroxydopamine in vivo. *Eur J Neurosci* **10:** 57-63.

48. Shults, C. W., Kimber, T., and Martin, D. 1996. Intrastriatal injection of GDNF attenuates the effects of 6-hydroxydopamine. *Neuroreport* **7:** 627-631.
49. Choi-Lundberg, D. L., Lin, Q., Chang, Y. N., Chiang, Y. L., Hay, C. M., Mohajeri, H., Davidson, B. L., and Bohn, M. C. 1997. Dopaminergic neurons protected from degeneration by GDNF gene therapy [see comments]. *Science* **275:** 838-41.
50. Mandel, R. J., Spratt, S. K., Snyder, R. O., and Leff, S. E. 1997. Midbrain injection of recombinant adeno-associated virus encoding rat glial cell line-derived neurotrophic factor protects nigral neurons in a progressive 6-hydroxydopamine-induced degeneration model of Parkinson's disease in rats. *Proc Natl Acad Sci U S A* **94:** 14083-8.
51. Kirik, D., Rosenblad, C., and Björklund, A. 1998. Functional recovery in the partial lesion model of PArkinson's disease after delayed infusion of GDNF. *IBAGS abstract 1.50*.
52. Dragunow, M., and Preston, K. 1995. The role of inducible transcription factors in apoptotic nerve cell death. *Brain Res Brain Res Rev* **21:** 1-28.
53. Herdegen, T., Skene, P., and Bahr, M. 1997. The c-Jun transcription factor—bipotential mediator of neuronal death, survival and regeneration. *Trends Neurosci* **20:** 227-31.
54. Vaudano, E., and Bjorklund, A. 1998. Cell death in dopaminergic nigral neurones:correlation with jun phosphorylation. *Forum of European Neuroscience abstract* **11.18**.
55. Hudson, J., Granholm, A.-C., Gerhardt, G. A., Henry, M. A., Hoffman, A., Biddle, P., Leela, N. S., Mackerlova, L., Lile, J. D., Collins, F., and Hoffer, B. J. 1995. Glial cell line-derived neurotrophic factor augments midbrain dopaminergic circuits in vivo. *Brain Res Bull* **36:** 425-432.
56. Rosenblad, C., Martinez-Serrano, A., and Bjorklund, A. 1996. Glial cell line-derived neurotrophic factor increases survival, growth and function of intrastriatal fetal nigral dopaminergic grafts. *Neuroscience* **75:** 979-985.
57. Rosenblad, C., Martinez-Serrano, A., and Björklund, A. 1998. Intrastriatal glial cell line-derived neurotrophic factor promotes sprouting of spared nigraostriatal dopaminergic afferents and induces recovery of function in a rat model of Parkinson's disease. *Neuroscience* **82:** 129-137.
58. Lapchak, P. A., Jiao, S., Collins, F., and Paul A, M. 1997. Glial cell line-derived neurotrophic factor: distribution and pharmacology in the rat following a bolus intraventricular injection. *Brain Res* **747:** 92-102.
59. Lapchak, P. A., Miller, P. J., Collins, F., and Jiao, S. 1997. Glial cell line-derived neurotrophic factor attenuates behavioural dificits and regulates nigrostriatal dopaminergic and peptidergic markers in 6-hydroxydopamine-lesioned adult rats: comparison of intraventricular and intranigral delivery. *Neuroscience* **78:** 61-72.
60. Martin, D., Miller, G., Fischer, N., Dix, D., Cullen, T., and Russel, D. 1996. Glial cell line-derived neurotrophic factor: the lateral cerebral ventricle as a site of administration for stimulation of the substantia nigra dopamine system in rats. *Eur J Neurosci* **8:** 1249-1255.
61. Beck, K. D., Irwin, I., Valverde, J., Brennan, T. J., Langston, J. W., and Hefti, F. 1996. GDNF induces a dystonia-like state in neonatal rats and stimulate dopamine and seretonin synthesis. *Neuron* **16:** 665-673.
62. Kumer, S. C., and Vrana, K. E. 1996. Intricate regulation of tyrosine hydroxylase activity and gene expression. *J Neurochem* **67:** 443-62.
63. Björklund, A., Rosenblad, C., Winkler, C., and Kirik, D. 1997. Studies on neuroprotective and regenerative effects of GDNF in a partial lesion model of Parkinson's disease. *Neurobiol Dis* **4:** 186-200.
64. Robertson, G. S., and Robertson, H. A. 1989. Evidence that L-dopa-induced rotational behavior is dependent on both striatal and nigral mechanisms. *J Neurosci* **9:** 3326-31.
65. Robertson, H. A. 1992. Dopamine receptor interactions: some implications for the treatment of Parkinson's disease. *Trends Neurosci* **15:** 201-6.
66. Kobayashi, S., Ogren, S. O., Hoffer, B. J., and Olson, L. 1998. Dopamine D1 and D2 receptor-mediated acute and long-lasting behavioral effects of glial cell line-derived neurotrophic factor administered into the striatum. *Exp Neurol* **154:** 302-14.
67. McGeer, P. L., Itagaki, S., Akiyama, H., and McGeer, E. G. 1988. Rate of cell death in parkinsonism indicate active neuropathological process. *Ann Neurol* **24:** 574-576.
68. Kirik, D., Rosenblad, C., and Björklund, A. 2000. Preservation of a functional nigrostriatal dopamine pathway by GDNF in the intrastriatal 6-OHDA lesion model depends on the site of administration of the trophic factor. *Eur. J. Neurosci* **12**:3871-3882.

A ROLE FOR THE VESICULAR MONOAMINE TRANSPORTER (VMAT2) IN PARKINSON'S DISEASE

Dwight C. German and Patricia K. Sonsalla*

1. BACKGROUND

Parkinson's disease (PD) is a neurodegenerative disorder that typically begins in the 5th to 6th decade of life. It is characterized by a resting tremor, muscular rigidity, slowness of movement and postural instability. The disease affects about 1% the population over age 50.

Monoaminergic neurons are a major target for degeneration in PD. The most affected population are the dopaminergic (DA) substantia nigra neurons[1,2]; over 50% of these neurons degenerate and these neurons are largely responsible for the motor symptoms of the disease. There is also loss of the noradrenergic neurons in the locus coeruleus (LC)[3,4], catecholaminergic neurons in the medulla[5,6], and serotonergic neurons in the median raphe nucleus.[7] However, not all monoaminergic neurons degenerate in PD. Monoaminergic neurons that do not degenerate in PD are located in the ventral tegmental area and dorsal substantia nigra pars compacta[2], hypothalamus[8] and dorsal raphe nucleus.[7] It is currently unknown why only certain monoaminergic neurons degenerate in PD. Although monoaminergic neurons are a primary target of PD, there is also degeneration of nonmonoaminergic neurons.[7,9]

2. CAUSES OF PARKINSON'S DISEASE

In a small percentage of PD patients, gene mutations have been identified that are linked to the disease. These mutations have been identified in the gene for α-synuclein on chromosome 4,[10,11] in Parkin on chromosome 6,[12] and to a locus on chromosome 2p13.[13] In cases with mutations in α–synuclein or Parkin the clinical syndrome resembles idiopathic PD, however, the age of onset is early (i.e., before age 50). In the majority of PD cases there does not appear to be a genetic basis for the disease. For example, in pairs of identical twins in which one twin has PD, the incidence of PD in the other twin is no different than the

---

* D. German, Department of Psychiatry, University of Texas Southwestern Medical School, Dallas, TX 75390-9070, P.K. Sonsalla, Department of Neurology, Robert Wood Johnson Medical School, Piscataway, NJ 08854-5635.

incidence in the general population.[14] In a recent study of twins, the same conclusion was drawn for PD cases that present after age 50.[15]

Exposure to toxins may contribute to developing PD.[16,17,18] Humans exposed to the neurotoxin 1-methyl-4-phenyl-1,2,3,6-tetrahydropyridine (MPTP) exhibit the clinical symptoms of PD and marked degeneration of DA substantia nigra neurons. The toxic metabolite of MPTP, 1-methyl-4-phenylpyridinium (MPP$^+$) inhibits Complex I of the mitochondrial respiration chain and leads to cell death.[19] Several studies have found that PD patients exhibit mitochondrial Complex I impairments.[20-22] PD patients also have MPP$^+$-like substances (e.g., (ß-carbolines) in their cerebrospinal fluid.[23] Dopamine itself may serve as an endogenous toxin since the normal metabolism of dopamine produces hydrogen peroxide as a byproduct. Dopaminergic neurons can degenerate as a result of oxidative stress caused by the formation of reactive oxygen species (ROS) like hydrogen peroxide and superoxide radicals.[24] It is possible that environmental or endogenous toxins cause impairment of mitochondrial function that leads to an increase in ROS formation in the DA neurons[25] and subsequent cell death.

3. MPTP MODEL OF PARKINSON'S DISEASE

MPTP itself is not toxic to cells and must be metabolized by the enzyme MAO-B to form the toxin MPP$^+$.[26] MPP$^+$ is accumulated selectively within monoaminergic neurons via plasma membrane monoamine transporters (e.g., [27]). Once MPP$^+$ is accumulated within DA neurons, for example, it can be sequestered into synaptic vesicles or it can inhibit mitochondrial Complex I function, increase ROS formation, and lead to neurodegeneration (see Figure 1).

MPTP administration to primates and rodents causes degeneration of the same population of midbrain DA neurons as occurs in humans with PD. In PD there is over 75% loss of DA neurons in the ventral portion of the substantia nigra pars compacta (ventral nucleus A9).[2] There is also a partial loss of DA neurons in the retrorubral field (nucleus A8) and in the ventral tegmental area (nucleus A10). In both primates[28] and mice[29], MPTP causes a similar pattern of midbrain DA neuronal degeneration. These data suggest that similar mechanisms may be responsible for the degeneration caused by MPTP and PD.

It is interesting that there is a dramatic species difference in sensitivity to MPTP toxicity. Humans are most sensitive to MPTP toxicity, followed by primates, mice and least sensitive are rats. A Parkinsonian syndrome occurs in humans and primates exposed to low mg/kg doses of MPTP. Sensitive strains of mice require one or more doses of 40 mg/kg to produce significant degeneration of midbrain DA neurons. Rats, however, are not sensitive to MPTP at doses that cause over 70% reduction in striatal dopamine concentrations in mice.[30,31] Differences in MPTP metabolism may be responsible for some of the species differences in MPTP toxicity.[32] However, studies in rats demonstrate that the MPP$^+$ concentrations in the striatum needed to produce toxicity to DA neurons far exceed those needed in mice.[30,31] This suggests that striatal content of MPP$^+$ does not necessarily predict toxicity. Differences in vesicular sequestration of MPP$^+$ in the two species may also play a role in determining toxicity (see below). The dramatic species difference in sensitivity to MPTP toxicity suggests that mechanisms have evolved that allow certain species to more effectively cope with exposure to environmental and/or endogenous toxins.

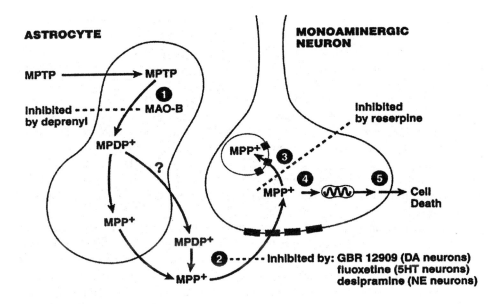

Figure 1. Hypothesis to explain the mechanism of neurotoxicity produced by MPTP. (1) MPTP is oxidized by MAO-B, presumably in astrocytes, to MPDP⁺, which undergoes further oxidation either within the astrocytes or the extracellular space to MPP⁺. (2) MPP is accumulated within monoaminergic neurons by plasma membrane transporters (not only for dopamine but also for norepinephrine [NE] and serotonin [5HT]). Inside the neurons, (3) MPP⁺ can be transported into synaptic vesicles by VMAT2 or (4) interact with Complex I mitochondrial proteins. It is hypothesized that the greater the capacity of monoamine-containing vesicles to store MPP, the less vulnerable the neurons are to the damaging effects of MPP⁺. (5) It is not known exactly how MPP⁺ causes cell death, but its ability to inhibit mitochondrial respiration is thought to play a crucial role in neurodegeneration.

4. VMAT2 AND NEUROPROTECTION

The vesicular monoamine transporter (VMAT) transports monoamines into synaptic vesicles (see [33] for review). VMAT, like other vesicular transporters, is composed of 12 transmembrane domains. VMAT2, found in the central nervous system, is localized within somata and nerve terminals in all monoaminergic neurons. At the ultrastructural level, it is incorporated in both small synaptic vesicles and primarily in the large dense cored vesicles.[34,35] VMAT2 not only has a high affinity for monoamine neurotransmitters, but also for MPP⁺.

Rat pheochromocytoma PC12 cells are relatively insensitive to MPP⁺ toxicity because they contain the peripheral nervous system form of VMAT (VMAT1).[36] Blockage of VMAT1 function, with reserpine, makes the cells sensitive to MPP⁺ toxicity.[37] Likewise, CHO cells when transfected with a VMAT2 cDNA become resistant to the toxin. These data suggest that MPP$^+$ can be sequestered within VMAT-containing vesicles and away from the mitochondria, and thus protect the cells from the toxin, as hypothesized in Figure 1.

We have found that ^{3}H-MPP$^+$ is sequestered preferentially within brain regions where VMAT2-immunoreactive monoaminergic somata and nerve terminals reside.[38] Within hours after the systemic administration of ^{3}H-MPTP to mice, ^{3}H-MPP$^+$ is found within the midbrain DA regions containing nuclei A8, A9 and A10, the noradrenergic LC, and the serotonergic dorsal raphe (DR) and median raphe nuclei. The accumulation in the LC and raphe regions are 3-fold higher than in the midbrain DA regions. MPP$^+$ was also localized within regions where VMAT2-containing nerve terminals are localized. Although little to no MPP$^+$ was found in the striatum, a region with a high density of DA nerve terminals, dense MPP$^+$ accumulation was found in the bed nucleus of the stria terminalis and in the paraventricular nucleus of the thalamus. It is interesting that the highest levels of MPP$^+$ were found in the LC yet this nucleus does not degenerate following MPTP treatment.[39,40,41] On the other hand, low levels of MPP$^+$ are accumulated in the region of the midbrain DA neurons and these neurons do degenerate.[29,39] These findings are consistent with the hypothesis that VMAT2 protects LC neurons from MPP toxicity but not nigral DA neurons because lower amounts of VMAT2 are contained within the DA neurons.

VMAT2 can protect midbrain DA neurons from degeneration in MPTP-treated animals. This has been demonstrated using transgenic mice in which VMAT2 is mutated and in some studies in which VMAT2 is inactivated by pharmacological approaches using drugs like reserpine. In such animals MPTP-toxicity to nucleus A9 neurons is enhanced.[41-46] However, there is no loss of LC neurons following either MPTP treatment or MPTP plus VMAT2 inhibition treatment. It remains to be determined whether the greater amount of VMAT2 in LC neurons (especially in the numerous dense core vesicles) as compared to the nigral DA neurons serves to protect these cells from MPP$^+$ toxicity.

We have recently investigated the species difference in MPTP sensitivity in order to determine whether the difference is related to VMAT2. Rats are significantly less sensitive to MPTP toxicity than mice[30,31], although brain levels of MPP$^+$ are similar or even higher in rats. To determine whether the species difference in toxin vulnerability is related to differences in VMAT2, we isolated synaptic vesicles from nigral DA nerve terminals in the striatum of both species.[47] We found that although vesicles from the two species have a similar affinity for ^{3}H-dopamine and ^{3}H-MPP$^+$, the rates of transport were 2-fold higher in the rat than in the mouse. Also, the number of VMAT2 binding sites was 2-fold higher in the rat. Moreover, the toxicity of an intrastriatal infusion of MPP$^+$ was significantly enhanced in rats, but not mice, in which VMAT2 was inhibited pharmacologically.[48] These data are consistent with the hypothesis that rat DA neurons can sequester more MPP$^+$ than mouse DA neurons, and are thus less vulnerable to MPTP toxicity.

5. CONCLUSIONS

The possibility that dysfunctional vesicular sequestration may increase the vulnerability of monoaminergic neurons to damage has important implications for PD. First, it is possible that vesicles might be a sequestration site for exogenous or endogenous toxins that are accumulated within the neurons (e.g., MPP$^+$-like compounds or dopamine). Second, diminished vesicular function could reduce the accumulation of monoamines into vesicles, resulting in less neurotransmitter release per impulse. This could be detrimental to the neurons, particularly the DA neurons because the resulting increase in cytosolic dopamine could lead to an increase in dopamine oxidative products and ROS. Such decreased dopamine release would also cause an increase in dopamine synthesis since both presynaptic and postsynaptic effects of the transmitter would be reduced, again potentially resulting in increased ROS. Certainly as the number of midbrain DA neurons decrease as the disease progresses, the firing rates of the remaining neurons increase (e.g., [49]), potentially contributing to the generation of more ROS. Furthermore, since vesicular transport is energy dependent, it is possible that a defect in vesicular transport coupled with lower energy stores would further exacerbate the deleterious situation. A disruption of normal vesicular function could result from reduced energy stores as a result of the impaired mitochondrial function observed in PD.[21,22] Thus, vesicular function can serve a pivotal role in monoaminergic cell survival.

6. ACKNOWLEDGEMENTS

These studies were supported in part by grants from the NIH (AG08479), the Hoblitzelle Foundation, the Carl J. and Hortense M. Thomsen Chair in Alzheimer's Disease Research, the Dallas Area Parkinsonism Foundation, the Dallas Foundation, and donations in honor of Richard D. Eiseman and George Higgenbotham.

7. REFERENCES

1. R. Hassler. Zur pathologischen anatomie des senilen and des parkinsonistischen tremor. *JPsychol Neurol.* 49:193 (1939).
2. D.C. German, K. Manaye, W.K. Smith, D.J. Woodward, et al. Midbrain dopaminergic cell loss in Parkinson's disease: computer visualization. *Ann Neurol.* 26:507 (1989).
3. V. Chan-Palay, and E. Asan. Alterations in catecholamine neurons of the locus coeruleus in senile dementia of the Alzheimer type and in Parkinson's disease with and without dementia and depression. *J Comp Neurol.* 287:373 (1989).
4. D.C. German, K.F. Manaye, C.L. White III, D.J. Woodward, et al. Disease-specific patterns of locus coeruleus cell loss. *Ann Neurol.* 32:667 (1992).
5. C.B. Saper, D.M. Sorrentino, D.C. German, and S. de Lacalla. Medullary catecholaminergic neurons in the normal human brain and in Parkinson's disease. *Ann Neurol.* 29:577 (1991).
6. W.-P. Gai,, L.B. Geffen, L. Denoroy and W.W. Blessing. Loss of C1 and C3 adrenaline-synthesizing neurons in the medulla oblongata in Parkinson's disease. *Ann Neurol.* 33:357 (1993).
7. G.M. Halliday, P.C. Blumbergs, R.G.H. Cotton. W.W. Blessing et al. Loss of brainstem serotonin- and substance P-containing neurons in Parkinson's disease. *Brain Res.* 510:104 (1990).
8. M.M. Matzuk and C.B. Saper. Preservation of hypothalamic dopaminergic neurons in Parkinson's disease. *Ann Neurol.* 18:553 (1985).
9. P.J. Whitehouse, J.C. Hedreen, C.L. White III, and D.L. Price. Basal forebrain neurons in the dementia of Parkinson disease. *Ann Neurol.* 13:243 (1983).
10. M.H. Polymeropoulos, C. Lavendan, E. Leroy , S.E. Ide, et al. Mutation in the alpha-synuclein gene identified in families with Parkinson's disease. *Science* 276:2045 (1997).
11. R. Kruger, W. Kuhn, T. Muller, D. Woitalla, et al. Ala30Pro mutation in the gene encoding alpha-synuclein in Parkinson's disease. *Nature Genet.* 18:106 (1998).
12. T. Kitada, S. Asakawa, N. Hattori, and H. Matsumine. Mutations in the parkin gene cause autosomal recessive juvenile parkinsonism. *Nature* 392:605 (1998).

13. T. Gasser, B. Muller-Myhsok, Z.K. Wszolek, R. Oehlmann, et al. A susceptibility locus for Parkinson's disease maps to chromosome 2p13. *Nature Genet.* 18:262 (1998).
14. C.D. Ward, R.C. Duvoisin, S.E. Ince, J.D. Nutt, et al. Parkinson's disease in 65 pairs of twins and in a set of quadruplets. *Neurology* 33:815 (1983).
15. C.M. Tanner., R. Ottman, S.M. Goldman, J. Ellenberg, et al. Parkinson disease in twins: an etiologic study. *JAMA* 281:341 (1999).
16. S.H. Snyder, and R.J. D'Amato. MPTP: a neurotoxin relevant to the pathophysiology of Parkinson's disease. *Neurology* 36:250 (1986).
17. P. Jenner, A.H.V. Schapira, and C.D. Marsden. New insights into the cause of Parkinson's disease. *Neurology* 42:2241 (1992).
18. J.W. Langston. The etiology of Parkinson's disease with emphasis on the MPTP story. *Neurology* 47: S153 (1996).
19. W.J. Nicklas, I. Vyas, and R.E. Heikkila. Inhibition of NADH-linked oxidation in brain mitochondria by 1-methyl-4-phenylpyridine, a metabolite of the neurotoxin, 1-methyl-4-phenyl-1,2,3,6-tetrahydropyridine. *Life Sci.* 36:2503 (1985).
20. A.H. Schapira, J.M. Cooper, D. Dexter, J.B. Clark, et al. Mitochondrial complex I deficiency in Parkinson's disease. *JNeurochem.* 54:823 (1990).
21. R.H. Haas, F. Nasirian, K. Nakano, D. Ward, et al. Low platelet mitochondrial complex I and complex II/III activity in early untreated Parkinson's disease. *Ann Neurol.* 37:714 (1995).
22. W.D.J. Parker, and R.H. Swerdlow. Mitochondrial dysfunction in idiopathic Parkinson disease. *Amer JHuman Genetics* 62:758 (1998).
23. K. S. Matsubara, S. Kobayashi, Y. Kobayashi, K. Yamashita, et al. β-Carbolinium cations, endogenous MPP⁺ analogs, in the lumbar cerebrospinal fluid of patients with Parkinson's disease. *Neurology* 45:2240 (1995).
24. C.C. Chiueh, R.M. Wu, K.P. Mohanakumar, L.M. Sternberger, G. Krishna, et al. In vivo generation of hydroxyl radicals and MPTP-induced dopaminergic toxicity in the basal ganglia. *Ann NYAcad Sci.* 738:25 (1994).
25. R.H. Swerdlow, J.K. Parks, S.W. Miller, J.B. Tuttle, et al. Origin and functional consequences of the complex I defect in Parkinson's disease. *Ann Neurol.* 40:663 (1996).
26. K. Chiba, A. Trevor, and N. Castagnoli, Jr. Metabolism of the neurotoxic tertiary amine, MPTP, by brain monoamine oxidase. *Biochem Biophy Res Comm.* 120:574 (1984).
27. J.A. Javitch, R.J. D'Amato, S.M. Strittmatter, and S.H. Snyder. Parkinsonism-inducing neurotoxin, N-methyl-4-phenyl-1,2,3,6-tetrahydropyridine: uptake of the metabolite N-methyl-4-phenylpyridine by dopamine neurons explains selective toxicity. *Proc Natl Acad Sci USA.* 82:2173 (1985).
28. D.C. German, M. Dubach, S. Askari, et al. 1-Methyl-4-phenyl-1,2,3,6-tetrahydropyridine-induced parkinsonian syndrome in *Macaca Fascicularis:* which midbrain dopaminergic neurons are lost? *Neuroscience* 24:161 (1988).
29. D.C. German, E.L. Nelson, C.L. Liang, S.G. Speciale, et al. The neurotoxin MPTP causes degeneration of specific nucleus A8, A9 and A10 dopaminergic neurons in the mouse. *Neurodegen.* 5:299 (1996).
30. Giovanni, B.A. Sieber, R.E. Heikkila, and P.K. Sonsalla. Studies on species sensitivity to the dopaminergic neurotoxin 1-methyl-4-phenyl-1,2,3,6-tetrahydropyridine. Part 1: Systemic administration. *J Pharmacol Exp Ther.* 270:1000 (1994).
31. Giovanni, P.K. Sonsalla, and R.E. Heikkila. Studies on species sensitivity to the dopaminergic neurotoxin 1-methyl-4-phenyl-1,2,3,6-tetrahydropyridine. Part 2: Central administration of 1-methyl-4-phenylpyridinium. *JPharmacol Exp Ther.* 270:1008 (1994).
32. J.N. Johannessen, C.C. Chiueh, R.S. Burns, and S.P. Markey. Differences in the metabolism of MPTP in the rodent and primate parallel differences in sensitivity to its neurotoxic effects. *Life Sci.* 36:219 (1985).
33. Y. Liu and R.H. Edwards. The role of vesicular transport proteins in synaptic transmission and neural degeneration. *Annu Rev Neurosci.* 20:125 (1997).
34. M.J. Nirenberg, Y. Liu, D. Peter, R.H. Edwards, et al. The vesicular monoamine transporter 2 is present in small synaptic vesicles and preferentially localizes to large dense core vesicles in rat solitary tract nuclei. *Proc Natl Acad Sci USA.* 92:8773 (1995).
35. M.J. Nirenberg, J. Chan, Y. Liu, R.H. Edwards, et al. Vesicular monoamine transporter-2: immunogold localization in striatal axons and terminals. *Synapse* 26:194 (1997).
36. T. Denton and B.D. Howard. A dopaminergic cell line variant resistant to the neurotoxin 1-methyl-4-phenyl-1,2,3,6-tetrahydropyridine. *JNeurochem.* 49:622 (1987).

37. Y. Liu, A. Roghani, and R.H. Edwards. Gene transfer of a reserpine-sensitive mechanism of resistance to N-methyl-4-phenylpyridinium. *Proc Natl Acad Sci* USA 89:9074 (1992).
38. S.G.Speciale, C.-L. Liang, P.K. Sonsalla, R.H. Edwards, et al. The neurotoxin 1-methyl-4-phenylpyridinium is sequestered within neurons that contain the vesicular monoamme transporter. *Neuroscience* 84:1177 (1998).
39. H. Hallman, J. Lange, L. Olson, I. Stromberg, et al. Neurochemical and histochemical characterization of neurotoxic effects of 1-methyl-4-phenyl-1,2,3,6-tetrahydropyridine on brain catecholamine neurones in the mouse. *JNeurochem.* 44:117 (1985).
40. L.S. Forno, J.W. Langston, L.E. DeLanney, 1. Irwin, et al. Locus coeruleus lesions and eosinophilic inclusions in MPTP-treated monkeys. *Ann Neurol.* 20:449 (1986).
41. C.-L. Liang, P.K. Sonsalla, K.F. Manaye, and D.C. German. Pharmacological inactivation of VMAT2 potentiates MPTP-induced neurodegeneration in mice. Soc *Neurosci Abstr.* 24:2154 (1998).
42. J.F.J. Reinhard, E.J.J. Diliberto, O.H. Viveros, and A.J. Daniels. Subcellular compartmentalization of 1-methyl-4-phenylpyridinium with catecholamines in adrenal medullary chromaffin vesicles may explain the lack of toxicity to adrenal chromaffin cells. *Proc Natl Acad Sci USA.* 84:8160 (1987).
43. J.F.J. Reinhard, A.J. Daniels, and O.H. Viveros. Potentiation by reserpine and tetrabenazine of brain catecholamine depletions by MPTP (1-methyl-4-phenyl-1,2,3,6-tetrahydropyridine) in the mouse; evidence for subcellular sequestration as basis for cellular resistance to the toxicant. *Neurosci Lett.* 90:349 (1988).
44. S.M. Russo, A.J. Daniels, O.H. Viveros, and J.F. Reinhard, Jr. Differences in the reserpine-sensitive storage *in vivo* of 1-methyl-4-phenylpyridinium in rats and mice may explain differences in catecholamine toxicity to 1-methyl-4-phenyl-1,2,3,6-tetrahydropyridine. *Neurotoxicol Teratol.* 16:277 (1994).
45. N. Takahashi, L.L. Miner, I. Sora, H. Ujike, et al. VMAT2 knockout mice: heterozygotes display reduced amphetamine-conditioned reward, enhanced amphetamine locomotion, and enhanced MPTP toxicity. *Proc Natl Acad Sci USA.* 94:9938 (1997).
46. R.R. Gainetdinov, F. Fumagalli, Y.M. Wang, S.R. Jones, et al. Increased MPTP neurotoxicity in vesicular monoamine transporter 2 heterozygote knockout mice. *J Neurochem.* 70:1973 (1998).
47. R.G.W. Staal, K.A. Hogan, C.L. Liang, D.C. German, et al. *In vitro* studies of striatal vesicles containing the vesicular monoamine transporter (VMAT2): rat vs mouse differences in sequestration of 1-methyl-4-phenylpyridinium (MPP+). *JPharmacol Exp Ther.* 293:329 (2000).
48. R.G.W. Staal and P.K. Sonsalla. Inhibition of brain vesicular monoamine transporter (VMAT2) enhances MPP+ neurotoxicity *in vivo* in striata of rats but not mice. *J Pharmacol Exp Ther.* 293:336 (2000).
49. G. Bernardini, S.G. Speciale, and D.C. German. Increased midbrain dopaminergic cell activity following 2'CH3-MPTP-induced dopaminergic cell loss: an *in vitro* electrophysiological study. *Brain Res.* 527:123 (1990).

QUANTITATIVE ANALYSIS AND BEHAVIOURAL CORRELATES OF LESIONING OF THE SUBTHALAMIC NUCLEUS IN THE HEMIPARKINSONIAN MARMOSET

J.M. Henderson, L.E. Annett, E.M. Torres and S.B. Dunnett*

1. INTRODUCTION

The basal ganglia differs in size and structure between species. For example the striatum in the rat is not segregated into a distinct caudate nucleus and putamen as it is in primates (Brodal, 1981; Paxinos, 1995). Also, the rat homologue of the internal globus pallidus, the entopeduncular nucleus, is not located in the same position as it is in primates (Brodal, 1981; Paxinos, 1995). Such anatomical differences between species may result in differential effects of basal ganglia surgery or pharmacological manipulations when results from animals are extrapolated to human movement disorders.

Parkinson's disease (PD) is a human basal ganglia disorder in which patients exhibit akinesia, rigidity, resting tremor and postural instability. The major neurochemical deficit is a loss of dopamine from the pigmented cells of the substantia nigra pars compacta. Symptoms arise after approximately 80% loss of nigrostriatal doopamine (Hornykiewicz et al., 1963). Hence, dopamine replacement therapy with levodopa or dopamine agonists constitutes the main therapy for PD, however after several years the majority of patients develop dose-related motor fluctuations, freezing episodes, dyskinesias and "on-off" phenomenon (Marsden and Parkes, 1976), necessitating the search for other therapeutic strategies.

Loss of dopaminergic activation of the striatum is thought to result in hyperactivation of the subthalamic nucleus (STN) and excessive excitatory drive to the internal globus pallidus and substantia nigra pars reticulata. These basal ganglia output nuclei have major inhibitory projections to the motor thalamus which thereby underactivates motor-related cortical areas, resulting in parkinsonian symptoms (DeLong, 1990). Consequently, much interest is focussing on the STN as a novel site for neurosurgery in the treatment of PD (for

---

* J.M. Henderson, Dept of Pharmacology, University of Syndney, Sydney, Australia; L.E. Annette, University of Hertforeshire, UK; E.M. Torres, Cantab Pharmaceuticals, Cambridge, UK; S.B. Dunnett Department of Biosciences, Cardiff University, Cardiff, Wales, UK.

The Basal Ganglia VI
Edited by Graybiel *et al.*, Kluwer Academic/Plenum Publishers, 2002

review see Henderson and Dunnett, 1998). Previous clinical case reports and experimental studies have established that infarction or haemorhage of the STN results in hemiballismus, a hyperkinetic disorder characterised by involuntary ballistic movements of the affected limb(s) contralateral to the lesion (Whittier and Mettler, 1949; Hammond et al., 1979; Hamada and DeLong, 1992; Vidakovic et al., 1994). Also hemiballismus was reported in a case report of a PD patient who suffered a unilateral haemorhage of the STN (Inzelberg and Korzyn, 1994). For such reasons animal models are being used to test the utility of STN surgery prior to more widespread clinical application. In order to do this it is also necessary to establish the equivalence of the STN and other basal ganglia structures between species used in modelling human movement disorders. We have therefore quantified the size and neuronal number of the marmoset STN and evaluated behavioural correlates of lesioning this structure in hemiparkinsonian marmosets (Henderson et al 1998), a primate species currently being used to model a variety of human movement disorders (Albanese et al., 1993; Annett et al., 1994; Nomoto et al., 1994; Mitchell et al., 1995; Dias et al., 1996; Marshall & Ridley, 1996).

2. METHODS

2.1. Surgery

Thirteen common marmosets (*Callithrix jacchus*) in good health and weighing between 300-450g were anaesthetised with Saffan anaesthesia (0.5mg i.m. followed by top-up doses of 0.3mg as necessary). All received unilateral 6-hydroxydopamine lesions to the medial forebrain bundle (MFB; as detailed by Annett et al., 1992) after completion of baseline behavioural analysis. Five sites of the MFB were injected (totalling 11µl of 6-OHDA.HBr - free base weight). A 2nd lesion was made in seven marmosets approximately 6 weeks after the 1st surgery, this time targeting the ipsilateral STN (0.25µl of 0.12M N-methyl-D-aspartate NMDA). The remaining animals received sham surgery at this time involving the ipsilateral STN. For further details of surgery refer to Henderson et al. (1998).

2.2. Behavioural analysis

Head position bias was measured by observing the direction of the head before and at weekly/fortnightly intervals after each surgery. The head position was recorded at 1sec intervals for 3 mins and the test repeated on 4 consecutive days of each test week.

Dopamine agonist-induced rotation was analysed at the baseline and at weekly/ fortnightly intervals after the MFB and STN surgeries (Henderson et al., 1998). The number and direction of 360 degree turns made during 60 min (saline or apomorphine 0.1mg/kg) or 30 min (amphetamine 0.5mg/kg) was measured from videotape recordings made immediately or 30 min after injection, respectively. At least two days separated the drug challenges.

The "Staircase" reaching task was also conducted to assess skilled use of each hand independently, with ipsilateral performance serving as a form of internal control against the hand contralateral to the lesion (Henderson et al., 1998). Briefly, marmosets were required to reach through a Perspex slot on each side to remove food rewards from each of 4 ascending steps on each side of an internal barrier. Nine x 5 min trials were conducted on each test week. This consisted of triplicate ipsilateral, contralateral and bilateral trials performed in a randomised order. The latency of the 1st reach and latency to complete the task on each side were measured.

Further details of the above tests and other variables measured including 24-hour activity and ability to reach through a tube are given in Henderson et al. (1998).

2.3. Histology

At the conclusion of the behavioural analysis all marmosets were sedated (5mg of ketamine i.m.) and terminally anaesthetised (sodium pentobarbitone 60mg i.p.). Monkeys were perfused intracardially with a 4% paraformaldehyde phosphate-buffered saline solution. Brains were removed and placed in the paraformaldehyde solution for 24 hours followed by 30% sucrose solution for 48 hours prior to being cut coronally at 60µm intervals on a Leica freezing microtome. Every 6th section was stained with cresyl violet and an identical parallel series taken for tyrosine hydroxylase immunohistochemistry (for further details see Henderson et al., 1998).

2.4. Quantitative analysis

The area of the STN was estimated by drawing the boundary observed under an Olympus® motorised-stage stereology microscope (linked via a mouse to a computer image analysis system, See Scan®, Cambridge, U.K.). The volume of the STN was calculated from the areal fractions using Cavalieri's principle. An unbiased optical disector technique was used to estimate total neuronal number from the cresyl violet stained sections containing the STN (Gunderson et al., 1988).

2.5. Statistical analysis

Groups and conditions were subjected to analysis of variance with independent multiple *a posteriori* comparisons based on Sidak's multiplicative inequality (Rohlf and Sokal, 1995).

3. RESULTS

3.1. Histology

The intact STN is shown in Figure 1. There was no significant difference in volume or total neuronal number of the STN on the non-operated side in animals receiving sham or NMDA surgery to the opposite STN. Therefore, the results on the intact side were pooled for all marmosets (Table 1). On the intact side the volume and total neuronal number of the STN was $0.718 \pm 0.120 mm^3$ ($1.44 mm^3$ bilaterally) and $24,043 \pm 5924$ (48,000 bilaterally), respectively. This enabled the neuronal density of the marmoset STN to be estimated (Table 1).

In all animals who received 6-OHDA lesions there was a profound depletion of tyrosine hydroxylase immunoreactivity in both the substantia nigra and striatum on the ipsilateral side, in contrast to the non-operated side, similar to that previously described (Annett et al., 1994).

Table 1. Total neuronal number and volume of STN on non-operated side.

Monkey #	Volume of STN mm³	Total neuronal #
1	0.579	20,712
2	0.599	14,694
3	0.703	21,694
4	0.740	26,140
5	0.763	24,257
6	0.799	25,438
7	0.759	26,750
8	0.729	30,044
9	0.884	31,780
10	0.910	32,839
11	0.504	18,232
12	0.651	16,348
Mean unilateral	0.718±0.120	24,043±5924
Bilateral estimate	1.436	48,086
Neuronal density		33,486 neurons/mm³

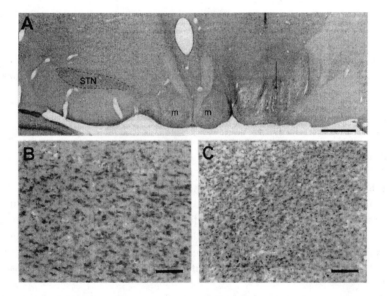

Figure 1. The intact and lesioned STN in a marmoset is observed at low power in A. At higher magnification the unlesioned STN is composed of densely packed neurons in B. Substantial cell loss and gliosis is observed in the lesioned STN in C. Scale bars in A = 1mm; B and C = 100μm. Reproduced with permission from Henderson et al. (1998).

The STN was successfully lesioned in all animals receiving NMDA injections. This produced significant atrophy, cell loss and gliosis of the STN, with little damage to surrounding structures (Figure 1). The mean atrophy and neuronal loss in these animals was 86% and 90%, respectively, and was statistically significantly different to those receiving sham STN lesions (both comparisons $P<0.01$).

3.1 Early post-operative observations

Animals awoke about 1-2 hours after 6-OHDA surgery at which time they began to exhibit a bias in head position and some rotational asymmetry toward the lesioned side. Contralateral neglect and decreased use of the contralateral hand were also observed. Approximately 1-2 hours after NMDA-STN surgery the marmosets experienced transient hemiballismus contralateral to the lesioned side. Head position and rotational asymmetry dramatically altered towards the contralateral side. This was not observed in sham STN-operated animals.

3.2. Behavioural analysis

3.2.1. Baseline

Prior to surgery the marmosets did not have any overt tendency to look in a particular direction (Figure 2). Nor did they exhibit any rotational asymmetry either spontaneously or after being injected with either amphetamine or apomorphine (Figures 3 and 4). All were able to successfully perform the staircase reaching task, generally making the 1st reach within 20 sec and clearing the rewards from each side of the staircase within a third of the maximum time allocated per trial (300 sec).

3.2.2. 6-OHDA Surgery

After 6-OHDA, all animals showed significant head position bias towards the ipsilateral side ($P<0.01$), which lasted for the entire period of observation in animals without subsequent STN NMDA lesions (Figure 2).

After 6-OHDA surgery all marmosets exhibited spontaneous rotation towards the side of the lesion ($P<0.05$). This ipsilateral rotation was more marked after administration of amphetamine (Figure 3; $P<0.01$). Marked contralateral rotation was observed after apomorphine ($P<0.01$) and lasted up to 12 weeks of observation (Figure 4).

There was approximately a ten-fold increase in the latency to make the 1st reach with the hand contralateral to the lesion after 6-OHDA lesions ($P<0.01$). On the ipsilateral side, there was also a 2-3 fold increase in latency of 1st reach ($P<0.01$). Despite some behavioural recovery over time, animals remained significantly impaired on the contralateral side. The latency to clear the rewards from each side of the staircase also significantly increased in animals after 6-OHDA, particularly on the contralateral side ($P<0.01$).

3.2.3. STN Surgery

STN lesions resulted in a contralateral bias in head position (Figure 2), reversing the directional effect of 6-OHDA lesions in five of seven marmosets. Animals with sham surgery to the STN continued to look ipsilaterally as they had done so after 6-OHDA lesions (Figure 2). The difference between STN lesioned versus sham STN animals was significant for the remainder of the study (P<0.01).

In contrast to the ipsilateral rotation observed after 6-OHDA surgery, NMDA-STN surgery resulted in a tendency for animals to rotate spontaneously away from the side of the lesion. Amphetamine injection resulted in low levels of contralateral rotation during the 1st week after this surgery and little rotational asymmetry thereafter (Figure 3). Sham STN-lesioned animals continued to rotate ipsilaterally with amphetamine as they had done so after 6-OHDA lesions (Figure 3). The difference between groups, however, failed to meet statistical significance due to the variability of the response. Whilst there was a trend towards

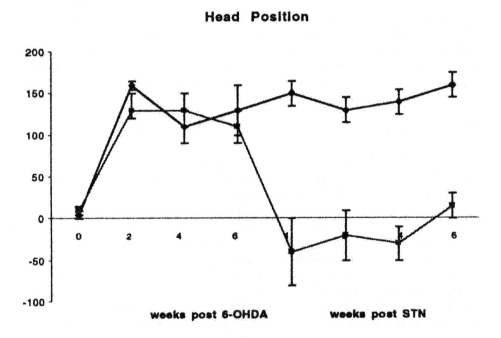

Figure 2. Net ipsilateral-contralateral head position in marmosets before and after 6-OHDA and STN surgery. Marmosets received 6-OHDA lesions and subsequent sham (diamonds) or NMDA (squares) surgery to the STN. In unoperated animals, there was no bias in head position. After 6-OHDA surgery there was a strong ipsilateral bias (P<0.01), whereas there was a reversal of head position bias for several weeks after NMDA lesions of the STN (P<0.01). Data from Henderson et al. (1998).

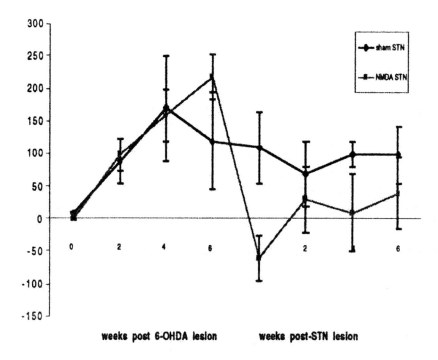

Figure 3. Net ipsilateral-contralateral rotational behaviour in response to amphetamine in unoperated marmosets and after 6-OHDA and STN surgery. Unoperated animals did not exhibit any rotational asymmetry. After 6-OHDA, there was a strong ipsilateral bias (P<0.01) which persisted in animals receiving sham STN surgery. After NMDA lesioning of the STN, there was reduced ipsilateral rotation, however this did not reach statistical significance. Data from Henderson et al. (1998).

decreased apomorphine-induced contralateral rotation over time in STN- lesioned relative to STN sham-operated animals (Figure 4), this difference also did not reach statistical significance.

STN lesions resulted in a dramatic decrease in the latency to remove the first reward using the contralateral hand and which approached baseline performance levels (P<0.05 NMDA versus sham STN groups). There was no difference between groups in this variable for the ipsilateral hand since there was evidence of similar behavioural recovery in both groups. There was, however, no significant difference in the latency to clear each side of the staircase between the STN surgery groups.

4. DISCUSSION

This study provides quantitative analysis of the volume and total neuronal number in the marmoset as well as the behavioural effects of lesioning this nucleus when using this species as a model of parkinsonism.

4.1. Characteristics of the Marmoset STN

When compared to other species, the volume of the marmoset STN (1.44mm^3 bilaterally) is approximately 14 times that of rats or 170 times smaller than that of humans (Oorshcot et al., 1996; Hardman et al., 1997). The total neuronal number (48,000 bilaterally) is approximately twice that of the rats (27,200) and one eleventh that of humans (550,000) (Oorshcot, 1996; Hardman et al., 1997). When brain size is also taken into account, there is a progressive increase in the relative volume of the STN from rats to marmosets to humans.

The neuronal density, however, decreases progressively from rats (approximately 270,000/mm^3) to marmosets (approximately 34,000/mm^3) to humans (approximately 2200/mm^3). Therefore, although the marmoset STN is closer in total neuronal number to rats, it is relatively larger. Like the human STN, this increase probably reflects greater connectivity due to increases in the neuropil and dendritic arbour, as well as the greater size of individual neurons. On this basis we propose that the marmoset may be a better animal model than the rat for therapeutic studies targeting the STN in PD.

4.2. Behavioral Correlates of STN Lesioning in Hemiparkinsonian Animals

Excitotoxic NMDA lesions of the STN resulted in a dramatic improvement in akinesia as evidenced by the marmoset's performance in the staircase task. This is similar to the findings of other studies of STN lesioning in parkinsonian monkeys and high frequency inactivation of the STN in humans (Bergman et al., 1990; Benazzouz et al., 1993; Benabid et al., 1994). This observation fits with current basal ganglia theory in that STN lesioning or high frequency stimulation reduces excessive excitation of the internal globus pallidus and substantia nigra pars reticulata, which inhibit motor thalamic nuclei, thereby facilitating the supplementary motor area and enabling increased speed of initiation of movement in the affected limb (DeLong, 1990; Wichmann and DeLong, 1996).

Other parkinsonian motoric deficits, however, persisted after STN lesioning as was supported by an absence of improvement in the latency to complete the staircase task in our monkeys. Furthermore, using a comparable staircase reaching task in the 6-OHDA rat (Montoya et al., 1990), we failed to detect a significant improvement in the number of rewards taken from the staircase during the observation period (Henderson et al., 1999).

Rotational asymmetry is commonly observed in the 6-OHDA model of parkinsonism and is thought to reflect the degree of dopaminergic imbalance (Ungerstedt, 1971). Animals given amphetamine show an ipsilateral bias whereas animals given apomorphine rotate contralaterally. It is thought that animals spontaneously rotate towards the side of containing the least dopamine, an effect which is enhanced by amphetamine which releases dopamine presynaptically, exacerbating the dopaminergic imbalance (Ungerstedt and Arbuthnott, 1970). Stimulation of residual supersensitive dopamine receptors on the lesioned side with dopamine agonists, such as apomorphine, have been proposed to cause the animal to rotate away from the direction of this increased dopaminergic activity (Ungerstedt and Arbuthnott, 1970). There was a trend towards decreased amphetamine and apomorphine rotation after STN lesioning in our hemiparkinsonian marmosets. This was so dramatic for amphetamine that one monkey rotated up to 500 times contralaterally in 30 min, however the response did not attain statistical significance due to the degree of variability between animals. In 6-OHDA lesioned rats, we have also found significant improvements in both amphetamine and apomorphine rotation after subsequent STN lesioning (Henderson et al., 1999). This supports that STN lesioning alters, but does not abolish, rotational asymmetry associated with nigrostriatal dopamine imbalance, probably by acting at a level downstream of the lesion, at the projection from the STN to the substantia nigra pars reticulata and superior colliculus. Pharmacological studies in rats demonstrate that the nigro-collicular projection is critical for the expression of rotational behaviour (Kilpatrick et al., 1982; Speller and Westby, 1996). Decreased apomorphine rotation after lesioning of the STN in 6-OHDA treated rats has been correlated with a normalisation of firing in the substantia nigra pars reticulata (Burbaud et al., 1995). Whilst PD patients do not generally exhibit high levels of rotational behaviour after dopamine agonists, some rotational asymmetry has been reported in hemiparkinsonian patients (Bracha et al., 1987). A recent comparative anatomical study found that there is a much greater representation of the substantia nigra pars reticulata of the rat, relative to humans in which the dopaminergic substantia nigra pars compacta is proportionately greater (McRitchie et al., 1996). Such anatomical differences could help to explain such interspecies differences in symptomatology.

Whilst transient hemiballismus was observed in our hemiparkinsonian marmosets in whom the STN was lesioned, more variable durations have been reported elsewhere (Bergman et al., 1990; Aziz et al., 1992; Guridi et al., 1996). Furthermore we have detected a significant alteration in head position in hemiparkinsonian marmosets after STN lesioning, similar to a study in MPTP monkeys (Butler et al., 1998). We have also investigated this pheneomena in rats with and without prior unilateral 6-OHDA lesions and have found similar effects of STN lesioning (Henderson et al., 1999). Despite benefits on some symptoms, subthalamotomy performed in PD resulted in one of several patients developing involuntary movements (Gill and Heywood, 1997; Obeso et al., 1997). The above findings would caution against unilateral subthalamotomy as a therapeutic strategy in PD patients. More suitable is high frequency bilateral STN stimulation where stimulation parameters and medication dosage can be titrated to achieve better symptomatic control with minimisation of adverse effects (Limousin et al., 1996).

5. ACKNOWLEDGEMENTS

J.M.H. was the recipient of a postdoctoral fellowship from the Swiss National Fund. The studies were supported by grants from the MRC and Wellcome Trust.

6. REFERENCES

Albanese, A., Granata, R., Gregori, B., Piccardi, M.P., Colosimo, C. and Tonali, P., 1993, Chronic administration of 1-methyl-4-phenyl-1,2,3,6-tetrahydropyridine to monkeys: behavioural, morphological and biochemical correlates, *Neuroscience*, 55: 823-832.

Annett, L.E., Rogers, D.C., Hernandez, T.D., Dunnett, S.B., 1992, Behavioural analysis of unilateral monoamine depletion in the marmoset., *Brain* 115:825-856.

Annett, L.E., Martel, F.L., Rogers, D.C., Ridley, R.M., Baker, H.F., Dunnett, S.B., 1994, Behavioural assessment of embryonic nigral grafts in the caudate nucleus and in the putamen of marmosets with unilateral 6-OHDA lesions, *Exp. Neurol.*, 125:228-246.

Aziz, T.Z., Peggs, D., Agarwa, E., Sambrook, M.A., Crossman, A.R., 1992, Subthalamic nucleotomy alleviates parkinsonism in the 1-methyl-4-phenyl-1,2,3,6-tetrahydropyridine(MPTP)-exposed primate, *Br. J. Neurosurg.* 6:575-582.
Benabid, A.L., Pollak, P., Gross, C., Hoffmann, D., Benazzouz, A., Gao, D.M., Laurent, A., Gentil, M., Perret, J., 1994, Acute and long-term effects of subthalamic nucleus stimulation in Parkinson's disease, *Stereotact. Funct. Neurosurg.* 62:76-84.
Benazzouz A, Gross C, Féger J, Boraud T, Bioulac B., 1993, Reversal of rigidity and improvement in motor performance by subthalamic high-frequency stimulation in MPTP-treated monkeys, *Eur. J. Neurosci.* 5:382-389.
Benazzouz, A., Pialat, B., Pollak, P., Benabid, A.L., 1995, Reponses of substantia nigra pars reticulata and globus pallidus complex to high frequency stimulation of the subthalamic nucleus in rats: electrophysiological data, *Neurosci. Lett.* 189:77-80.
Bergman, H., Wichman, T., DeLong, M.R., 1990, Reversal of experimental parkinsonism by lesions of the subthalamic nucleus, *Science* 249:1436-1439.
Bracha, H.S., Shults, C., Glick, 1987, Spontaneous asymmetric circling behaviour in hemi-parkinsonism: a human equivalent of the lesioned-circling rodent behaviour, *Life Sci* 40:1127-1130.
Brodal, P., 1981, *Neurological Anatomy in Relation to Clinical Medicine*, Oxford University Press, New York.
Burbaud, P., Gross, C., Benazzouz, A., Coussemacq, M., Bioulac, B., 1995, Reduction of apomorphine-induced rotational behaviour by subthalamic lesion in 6-OHDA lesioned rats is associated with a normalisation of firing rate and discharge pattern of pars reticulata neurons, *Exp. Brain Res.*, 105:48-58.
Butler, E.G., Bourke, D.W., Finkelstein, D.I., Horne, M.K., 1997, The effects of reversible inactivation of the subthalamo-pallidal pathway on the behaviour of naive and hemiparkinsonian monkeys, *J. Clin. Neurosci.* 4(2):218-227.
DeLong, M.R., 1990, Primate models of movement disorders of basal ganglia origin, *Trends Neurosci.* 13(7):281-285.
Dias, R., Robbins, T.W. and Roberts, A.C., 1996, Dissociation in prefrontal cortex of affective and attentional shifts, *Nature*, 380: 69-72.
Gill, S.S., Heywood, P., 1997, Bilateral dorsal subthalamotomy for advanced Parkinson's disease, *Lancet* 350:1224.
Gunderson, H.J.G., Bagger, P., Bendtsen, T.F., Evans, S.M., Korbo, L., Marcussen, N., Møller, A., et al., 1988, The new stereological tools: dissector, fractionator, nucleator and point sampled intercepts, and their use in pathological research and diagnosis. *Acta. Pathol. Microbiol. Immunol. Scand.*, 96:857-881.
Guridi, J., Herrero, M.T., Luquin, M.R., Guillén, J., Ruberg, M., Laguna, J., Vila, M. et al.,1996, Subthalamotomy in parkinsonian monkeys. Behavioural and biochemical analysis, *Brain* 119:1717-1727.
Hamada, I., DeLong, M.R., 1992, Excitotoxic acid lesions of the primate subthalamic nucleus result in transient dyskinesias of the contralateral limbs, *J. Neurophysiol.* 68(5):1850-1858.
Hammond, C., Féger, J., Bioulac, B., Souteyrand, J.P., 1979, Experimental hemiballism produced by unilateral kainic acid lesion in the corpus Luysii, *Brain Res.*, 171, 577-580.
Hardman, C.D., McRitchie, D.A., Halliday, G.M., Morris, J.G.L., 1997, The subthalamic nucleus in Parkinson's disease and progressive supranuclear palsy, *J. Neuropathol. Exp. Neurol.* 56: 132-42.
Henderson, J.M., Dunnett, S.B., 1998, Targeting the subthalamic nucleus in the treatment of Parkinson's disease, *Brain Res. Bull.* 46(6): 467-474.
Henderson, J.M., Annett, L.E., Torres, E.M., Dunnett, S.B., 1998, Behavioural effects of subthalamic nucleus lesions in the hemiparkinsonian marmoset (*Callithrix jacchus*), *Eur. J. Neurosci.* 10: 689-698.
Henderson, J.M., Annett, L.E., Ryan, L.J., Chaing, W., Hidaka, S., Torres, E.M., Dunnett, S.B., 1999, Subthalamic nucleus lesions induce deficits as well as benefits in the hemiparkinsonian rat, *Eur. J. Neurosci.* 11:1-9.
Hornykiewicz, O., 1963, Die topische localisation und das verhalten von noradrenalin und dopamin (3-hydroxytryptamin) in der substantia nigra des normalen und parkinsonkranken menshen, *Klin. Wochenschr.* 75:309-312.

Inzelberg, R., Korczyn, A., 1994, Persistant hemiballismus in Parkinson's disease, *J. Neurol. Neurosurg. Psychiatry* 57:1013.
Kendall, A.L., Rayment, F.D., Torres, E.M., Baker, H.F., Ridley, R.M. and Dunnett, S.B., 1998, Functional integration of striatal allografts in a primate model of Huntington's disease, *Nature Med.*, 4: 727-729.
Kilpatrick, I.C., Collingridge, G.L., Starr, M.S., 1982, Evidence for the participation of nigrotectal gamma-aminobutyrate-containing neurones in striatal and nigral-derived circling in the rat, *Neuroscience* 7:207-222.
Limousin, P., Pollak, P., Hoffman, D., Benazzouz, A., Perret, J.E., Benabid, A.L., 1996, Abnormal Involuntary Movements induced by subthalamic nucleus stimulation in parkinsonian patients, *Movement Dis.* 11(3):231-235.
McRitchie, D.A., Hardman, C.D., Halliday, G.M., 1996, Cytoarchitectural distribution of calcium binding proteins in midbrain dopaminergic regions of rat and humans, *J. Comp. Neurol.* 364(1) 121-150.
Marsden, C.D., Parkes, J.D., 1976, "On-off" effects in patients with Parkinson's disease on chronic levodopa therapy, *Lancet* 1: 292-296.
Marshall, J.W.B. and Ridley, R.M., 1996, Assessment of functional impairment following permanent middle cerebral artery occlusion in a nonhuman primate species, *Neurodegeneration*, 5: 275-286.
Mitchell, I.J., Hughes, N., Carroll, C.B. and Brotchie, J.M., 1995, Reversal of parkinsonian symptoms by intrastriatal and systemic manipulations of excitatory amino acid and dopamine transmission in the bilateral 6-OHDA lesioned marmoset, *Behav. Pharmacol.*, 6:492-507.
Montoya, C.P., Astell, S., Dunnett, S.B., 1990, Effects of nigral and striatal grafts on skilled forelimb use in the rat, *Prog. Brain Res.*, 82:459-466.
Nomoto, M., Irifune, M., Fukuzaki, K. and Fukuda, T., 1994, Effects of bifemelane on parkinsonism induced by 1-methyl- 4-phenyl-1,2,3,6-tetrahydropyridine (MPTP) in the common marmoset, *Neurosci. Lett.*, 178:95-98.
Obeso, J.A., Guridi, J., DeLong, M., 1997, Surgery for Parkinson's disease, *J. Neurol. Neurosurg. Psychiatry* 62:2-8.
Oorshcot, D.E., 1996, Total number of neurons in the neostriatal, pallidal, subthalamic, and substantial nigral nuclei of the rat basal ganglia: a stereological study using the cavalieri and optical disector methods, *J. Comp. Neurol.* 366(4):580-599.
Paxinos G., 1995, The Rat Nervous System, 2nd edition, Academic Press, San Diego.
Rohlf, F.J., Sokal, R.R., 1995, *Statistical Tables*, 3rd Ed., Freeman, New York.
Speller, J.M., Westby, G.W.M., 1996, Bicuculline-induced circling from the rat superior colliculus is blocked by GABA microinjections into the deep cerebellar nuclei, *Exp. Brain Res.*, 110,425-434.
Ungerstedt, U., 1971, Postsynaptic supersensitivity after 6-hydroxydopamine-induced degeneration of the nigrostriatal dopamine system, *Acta Physiol. Scand. Suppl.*, 367:59-73.
Ungerstedt, U., Arbuthnott, G.W., 1970, Quantitative rotational behaviour in rats after 6-hydroxydopamine lesions of the nigrostriatal dopamine system, *Brain Res.*, 24:485-493.
Vidakovic, A., Dragasevic, N., Kostic, V.S., 1994, Hemiballism: report of 25 cases, *J. Neurol. Neurosurg. Psychiatry* 57: 945-949.
Wichmann, T., DeLong, M.R., 1996, Functional and pathophysiological models of the basal ganglia. *Curr. Opinion Neurobiol.* 6:751-758.
Whittier, J.R., Mettler, F.A., 1949, Studies on the subthalamus of the rhesus monkey. *J. Comp. Neurol.* 90: 319-371.

EFFECTS OF REVERSIBLE BLOCKADE OF PEDUNCULOPONTINE TEGMENTAL NUCLEUS ON VOLUNTARY ARM MOVEMENT IN MONKEY

Masaru Matsumura and Katsushige Watanabe*

1. SUMMARY

We examined the effects of a reversible blockade of pedunculopontine tegmental nucleus (PPN) in a monkey (*macaca fuscata*) which was trained to perform a lever-pull task with an arm. Neurons of PPN presented changes in their firing rate during arm movements. The changes were either an increase or a decrease. Muscimol (a γ-aminobutyric acid (GABA) agonist) was injected into PPN area to suppress its neuronal activity while the monkey was performing the task. The blockade of the activity of PPN neurons caused slowness of movements on the both sides of arms. The peak velocities of lever movement decreased and the reaction times increased. This work suggested that PPN had an excitatory effect to the initiation and the execution of voluntary movements of the limbs. The monkey did not show any difficulty in performing the task and the success rates were kept high.

The inactivation of PPN might have resulted in a decreased activity of dopammergic neurons of the substantia nigra pars compacta. The possible pathophysiology of Parkinson's disease regarding the decreased activity of the neurons of the PPN is discussed.

2. INTRODUCTION

The pedunculopontine tegmental nucleus (PPN) has abundant connections with many motor-related areas in the brain, i. e., the primary motor cotex, the cerebellum, the internal pallidum, the subthalamic nucleus, the substantia nigra pars compacta, the thalamus and the lower brainstem nuclei.[1-4] The PPN has been studied in relation to the locomotor system and the control of muscle tones.[4-6] The role of PPN in voluntary movements of the limbs had

---

* M. Matsumura, Chuo Gunma Neurosurgery Hospital, 64-1, Nakaomachi, Takasaki, Gunma, 370-0001, JAPAN; K. Watanabe, Department of Neurosurgery, Gunma University School of Medicine, 3-39-15, Showamachi, Maebashi, Gunma, 371-8511, JAPAN

never been studied in primates. We have previously reported that neurons of PPN in monkeys presented activities related to voluntary arm movements.[7-9] The change of activities was either an increase or a decrease in the firing rate of the neurons across the arm movement.

In this work, we examined the effects of a reversible blockade of PPN by injecting muscimol in a monkey which was trained to perform a lever-pull task with an arm.

3. MATERIALS AND METHODS

A Japanese monkey *(macaca fuscata*, male, 4.5Kg) was trained on a lever-pull task. "The Guide for the Care and Use of Laboratory Animals" (National Academy Press, 1996) was followed. The monkey was given restricted fluids during periods of training and recording. Supplementary water and fruit were provided daily. The monkey was trained in the task until he performed >90% of the trials correctly.

3.1. Surgical Procedures

The monkey was prepared for microelectrode recording of the activity of single neurons by surgically anchoring to the skull a platform allowing a rigid and painless head fixation to a stereotaxic apparatus (Narishige Co., Tokyo), which was fixed to the primate chairs.[9] Under general anesthesia with ketamine hydrochloride (10 mg/kg b. wt., i. m.) followed by pentobarbital sodium (30 mg/kg b. wt, i. v.), a light metal platform with a flanged hole in its center, was fixed to the skull with acrylic screws and acrylic resin. Antibiotic therapy (piperacillin sodium, 80mg/kg, i. m.) was provided during the days following the operation.

3.2. Behavioral Paradigm

The monkey sat in a primate chair made of acrylic glass in a dimly-lit room. The animal was trained to manipulate a lever on the chair with each hand. A red light-emitting diode (LED) was illuminated in front of the animal (wait) . After a delay period of a few seconds, it was turned off and a green LED appeared (go). He was required to pull a spring-loaded lever with his hand within a short period of time (usually 800 msec.) to obtain a drop of liquid reward (Figure 1). The animal was trained to perform the task with each hand.

3.3 Recording of Neuronal Activity

We identified the location of PPN by recording the neuronal activity with microelectrodes and by a ventriculography. Microelectrode tracks were made vertically. Glass-insulated Elgiloy microelectrodes (10-30 Mohm measured at 10 Hz, exposed tips 20-40 μm) were used for the extracellular recording of single-cell activity.

The neuronal signals were amplified, filtered and displayed on a storage-oscilloscope and fed to an audiomonitor. The unitary action-potentials were passed through a window discriminator, and the times of their occurrences were stored with a resolution of one msec. A potentiometer was mounted into the lever and the position of the lever was monitored every two msec. The behavioral task as well as the storage and the display of data was

EFFECTS OF BLOCKADE OF PPN ON VOLUNTARY ARM MOVEMENT

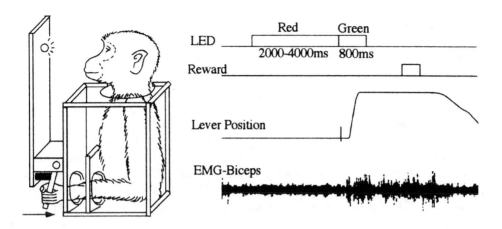

Figure 1. Behavioral Paradigm. The monkey sat on a primate chair (left). Schematic diagrams of the task (right up) and the trace of the corresponding position of the lever (right middle). A small vertical bar on the trace of the lever position indicates the time of trigger (color change of the LED). When a LED which was illuminated in front of the monkey turned from red to green, the animal was required to pull the lever to receive a drop of liquid reward. The EMG recorded from the animal's biceps muscle shows activity while the monkey pulled the lever (right down).

controlled by a laboratory computer (PC9801RA, NEC Co., Tokyo). The EMGs of limb muscles and trunk muscles were occasionally recorded using fine needle electrodes.

3.4 Microinjection of Muscimol

Muscimol (a γ-aminobutyric acid (GABA) agonist, Sigma) (5μg/μl, in PBS) was injected locally into PPN area. A volume of one to three μl of a solution was slowly (0. 2μl/min.) infused through a fine needle attached to a 10-μl Hamilton microsyringe.[10] The site of the injection was monitored with the X-ray and later confirmed with the histology. All penetrations were made vertically.

3.5. Histology

Several marking lesions were placed by passing positive currents (5 μA for 60 s) through the recording electrode. At the end of the experiments, the monkey was perfused transcardially with heparinized saline followed by 10% formalin under deep pentobarbital sodium anesthesia (60 mg/kg b. wt., i.v.). The head was then fixed to the stereotaxic device, and the brain was cut into blocks. Serial coronal sections (50 μm) were cut on a cryotome and were Nissl-stained with Cresyl violet.

4. RESULTS

4.1. Activity of PPN Neurons

The common finding of the neuronal activity of PPN was an increase of the firing rate across the lever-pull movement (Figure 2). Neurons started to change their activity usually before the movement onset and the change lasted for several hundreds msec. A decrease of the firing rate was also common (figure not shown). The change of the firing rate was found during both contralateral and ipsilateral arm movements in nearly half of the responsive neurons. This was similar to our previous study in which different type of the behavioral task was used.[9] There were some neurons that changed their activity in response to the illumination of the LED or the delivery of reward. We tried to identify the positive sharp spikes that represent the fibers of the superior cerebellar peduncle where we intended to inject muscimol to inactivate PPN efficiently (see below). These spikes often responded to the passive limb-movements.

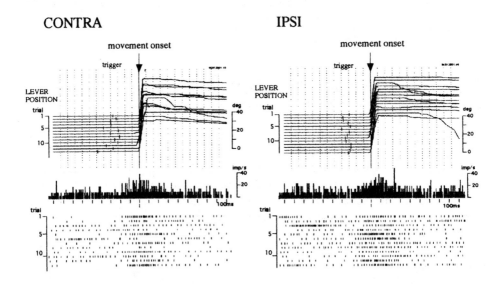

Figure 2. Neuronal activity of a PPN neuron during arm movement of the side contralateral to the recorded neuron (CONTRA) and that of during arm movement of the side ipsilateral to the recorded neuron (IPSI). Traces of lever position (top), histograms (middle) and raster displays (bottom) of activity of a single neuron are aligned with the movement onset (arrow). More than ten consecutive trials are presented A short bar on each trace of lever position represents the time of trigger presentation.

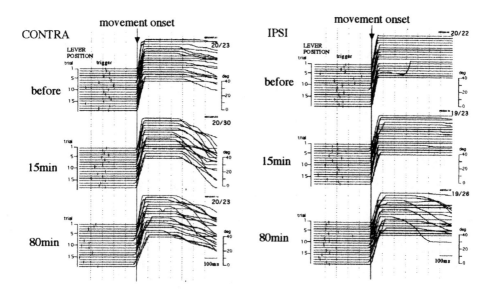

Figure 3. Traces of lever position recorded before, 15 minutes and 80 minutes after injection of muscimol into PPN area. Each trace is aligned with the movement onset (arrow). CONTRA; The monkey was performing the arm movement of the contralateral side to the muscimol injection. IPSI; The monkey was performing the arm movement of the ipsilateral side to the muscimol injection. A short bar on each trace of lever position represents the time of trigger presentation.

4.2. Microinjection of Muscicnol into PPN

Muscimol was locally injected into one side of the PPN, aiming the central part of the PPN. We chose a point at the level of the superior cerebellar peduncle, and 2.5 to 3.0 mm from the midline (see Matsumura et al.[9] for detail). After the injection, lever-pull movements became slow for the arm of the contralateral side. Soon after, the symptom appeared for both sides of the arms. Ten to fifteen minutes after the start of the injection, the velocity of the lever movement decreased and the latency increased (Figure 3 and 4). After 30 minutes, the monkey often presented a difficulty in holding the lever at the maximum position and returned the lever to the neutral position soon after the reward delivery (see the traces of the lever position at 80 minutes on Figure 3).

The monkey was alert and motivated well on the task within a couple of hours after the injection. When a large dose of muscimol (two to three μl) was injected, a hypokinesia on the contralateral limbs was apparent and the monkey could not continue the task. A circling (toward the contralateral side, 1-3 turns/min.) was observed occasionally outside the task.

The monkey returned normal on the next day. The control injection (the same amount of PBS) into PPN area did not show any significant change of the latency and the velocity of the lever movement (Figure 4). The animal's behavior was apparently normal.

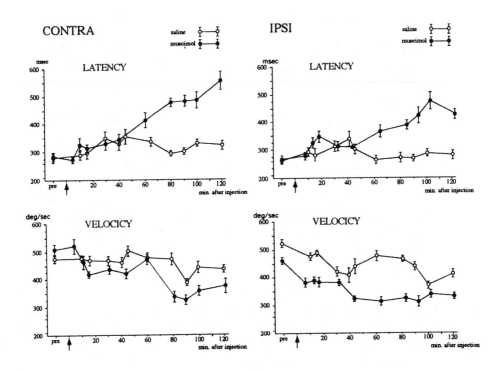

Figure 4. Mean latencies and velocities of lever-pull movements before and after injection of muscimol. open circle (control injection). closed circle (muscimol injection). arrow (time of injection). Mean values and standard deviations of 20 to 30 trials at each time were plotted. CONTRA; The monkey was performing the arm movement of the contralateral side to the recorded neuron. IPSI; The monkey was performing the aim movement of the ipsilateral side to the recorded neuron.

5. DISCUSSION

Our previous study showed that neurons of PPN presented activities related to voluntary movements of both sides of arms.[7-9] The changes of activities were either an increase or a decrease in the firing rate. In that study, we used a simple key-release task. Only the reaction time was possible to evaluate. In this study, we applied a spring-loaded manipulandum to evaluate the movement time, i. e. velocity, as well as the reaction time. A micro-injection of muscimol into one side of PPN resulted in a slowness of lever-pull movements and a prolonged reaction-time of both sides of arms.

In the PPN of primates, saccade-related neuronal activity has been reported.[11] PPN neurons changed their firing rate in relation to saccades. The changes were either an increase or a decrease. We did not record the eye movements of the animal. We could not observe any obvious abnormality of their eye movement after an injection of muscimol into PPN area.

PPN received inputs from the internal pallidum, which is known to be abnormally hyperactive in the parkinsonian state.[12] PPN sends cholinergic and glutamatergic projections to the substantia nigra pars compacta[2,13-16], whose cell loss is the major source in the Parkinson's disease. These connections are mostly bilateral, however, the ipsilateral connections are predominant. PPN also has descendent efferent projections to the lower brainstem nuclei. These were mostly bilateral. The inactivation of PPN might have produced the slowness of movement through the descending pathway. We could not conclude either the descending or the ascendin pathway was responsible for the results in this experiment.

Recently, Dormont et al.,[17,18] has reported that PPN neurons in cats showed a change of firing rate during operantly conditioned performance of a lever-release movement of a forelimb. They also reported the effects of reversible inactivation of PPN. In their study, muscimol injection induced prolonged inter-trial intervals and arrest of performance, however, the reaction times were not affected. Their results were similar to ours, however, the difference in species as well as the difference of the behavioral task should be considered for the evaluation of the results. Their work as well as our work showed that the role of PPN is to facilitate the motor system. The connections of PPN with the basal ganglia seems to be developed well in primates and the descending connections of PPN seems to be more prominent in felines than in primates, however further anatomical and physiological studies are needed.

Our group previously reported that unilateral kainic acid-induced lesion in the PPN resulted in contralateral hemiparkinsonism.[19] The monkeys presented general hypokinesia, a flexed posture and a bradykinesia on the contralateral limbs. The hemiparkinsonian symptoms observed after PPN destruction might be ascribed to a decrease in the nigrostriatal neuron activity due to excitatory input ablation. Quite similarly in this study, we observed slowness and prolonged reaction time of voluntary fast arm movements after a reversible blockade of its neuronal activity. The inactivation of PPN might have resulted in a decreased activity of dopaminergic neurons of the substantia nigra pars compacta. As stated above, in parkinsonian monkeys, as well as in parkinsonian patients, abnormally high activity of neurons in the internal pallidum have been reported.[12] This might inactivate the neurons of PPN through its GABAergic projection to the PPN. The decreased activity of the neurons in the PPN may underlie the pathophysiology of Parkinson's disease. Actually, in the postmortem study of the brains of the patients with Parkinson's disease revealed neuronal loss in the PPN.[20-22]

6. ACKNOWLEDGMENTS

The authors would like to acknowledge Prof. Emeritus Chihiro Ohye, Department of Neurosurgery, Gunma University School of Medicine, for his generous supports. This work was partly supported by the Sumitomo Foundation on the Research of Chatecholamine-related Disorders (to M. M., in 1997) and by the Life Science Foundation of Japan (to M.M., in 1999).

7. REFERENCES

1. Lavoie, B., and Parent, A., 1994, Pedunculopontine nucleus in the squirrel monkey: projections to the basal ganglia as revealed by anterograde tract-tracing methods. J. Comp. Neurol. 344:210-231.
2. Parent, A., Côté, P.-Y., and Lavoie, B., 1995, Chemical anatomy of primate basal ganglia. Prog. Neurobiol. 46:131-197.
3. Scarnati, E., and Florio, T., 1997, The pedunculopontine nucleus and related structures. Functional organization. Adv. Neurol. 74:97-110.
4. Winn, P., Brown, V. J. , and Inglis, W. L. , 1997, On the relationship between the striatum and the pedunculopontine tegmental nucleus. Crit. Rev. Neurobiol. 11:241-261.
5. Garcia-Rill, E., 1991, The pedunculopontine nucleus. Prog. Neurobiol. 36:363-389.
6. Winn, P., 1998, Frontal syndrome as a consequence of lesions in the pedunculopontine tegmental nucleus: A short theoretical review. Brain Res. Bull. 47:551-563.
7. Matsumura, M., Watanabe, K., and Ohye, C., 1996, Role of monkey pedunculopontine tegmental nucleus in lever-pull movement; A single-unit recording and effects of muscimol microinjection. Neurosci Res. Suppl 20: s 181.
8. Matsumura, M., Watanabe, K., and Ohye, C., 1996, Neuronal activity of monkey pedunculopontine tegmental nucleus area. I. Activity related to voluntary arm movements. In Basal Ganglia V (Ohye, C., Kimura, M., and McKenzie, J. eds), Plenum Press, New York, pp209-215.
9. Matsumura, M., Watanabe, K., and Ohye, C., 1997, Single-unit activity in the primate nucleus tegmenti pedunculopintinus related to voluntary arm movement. Neurosci. Res. 28:155-165.
10. Kato, M., and Kimura, M, 1992, Effects of reversible blockade of basal ganglia on a voluntary arm movement. J. Neurophysiol. 68:1516-1534.
11. Kobayashi, Y., Isa, T., Yamamoto, M., and Aizawa, H., 1998, Relationship of pedunculopontine tegmental nucleus neuron activity to saccade initiation in monkeys. Soc Neurosci Abstr 24:146.
12. Filion, M., and Tremblay, L., 1991, Abnormal spontaneous activity of globus pallidus neurons in monkeys with MPTP-induced parkinsonism. Brain Res. 547:142-151.
13. Charara, A., Smith, Y., and Parent, A., 1996, Glutamatergic inputs from the pedunculopontine nucleus to midbrain dopaminergic neurons in primates; Phaseolus vulgaris-leucoagglutinin anterograde labeling combined with postembedding glutamate and GABA immunohistochemistry. J. Comp. Neurol. 364:254-266.
14. Lavoie, B., and Parent, A., 1994, Pedunculopontine nucleus in the squirrel monkey: cholinergic and glutamatergic projections to the substantia nigra. J. Comp. Neurol. 344:232-241.
15. Takakusaki, K., Shiroyama, T., Yamamoto, T., and Kitai, S.T., 1996, Cholinergic and noncholinergic tegmental pedunculopontine projection neurons in rats revealed by intracellular labeling. J. Comp. Neurol. 371:345-361.
16. Tokuno, H., Moriizumi, T., Kudo, M. , and Nakamura, Y., 1988, A morphological evidence for monosynaptic projections from the nucleus tegmenti pedunculopontinus pars compacta (TPC) to nigrostriatal projection neurons. Neurosci. Lett. 85:1-4.
17. Condé, H., Dormont, J.F., and Farin, D., 1998, The role of pedunculopontine tegmental nucleus in relation to conditioned motor performance in the cat. II. Effects of reversible inactivation by intracerebral microinjections. Exp. Brain Res. 121:411-418.
18. Dormont , J. F., Condé, H., and Farin, D., 1998, The role of pedunculopontine tegmental nucleus in relation to conditioned motor performance in the cat. I. Context-dependent and reinforcement-related single unit activity. Exp. Brain Res. 121:401-410.
19. Kojima, J., Yamaji, Y., Matsumura, M., Nambu, A., Inase, M., Tokuno, H., Takada, M., and Imai, H., 1997, Excitotoxic lesions of the pedunculopontine tegmental nucleus produce contralateral hemiparkinsonism in the monkey. Neurosci. Lett 226:111-114.
20. Hirsch, E. C., Graybiel, A.M., Duyckaerts, C., and Javoy-Agid, F., 1987, Neuronal loss in the pedunculopontine tegmental nucleus in Parkinson disease and in progressive supranuclear palsy. Proc. Natl. Acad. Sci. U.S.A. 84:5976-5980.
21. Jellinger, K., 1988, The pedunculopontine nucleus in Parkinson's disease, progressive supranuclear palsy and Alzheimer's disease. J Neurol. Neurosurg. Psychiatry 51:540-543.
22. Zweig, R. M., Jankel, W. R., Hedreen, J. C., Mayeux, R., and Price, D. L. , 1989, The pedunculopontine nucleus in Parkinson's disease. Ann. Neurol. 26:41-46.

DIFFERENTIAL REDUCTIONS IN DOPAMINERGIC INNERVATION OF THE MOTOR-RELATED AREAS OF THE FRONTAL LOBE IN MPTP-TREATED MONKEYS

Yoshio Yamaji, Masaru Matsumura, Jun Kojima, Hironobu Tokuno, Atsushi Nambu, Masahiko Inase, Hisamasa Imai, and Masahiko Takada*

1. INTRODUCTION

Unexpectedly widespread dopaminergic innervation has been recognized in the primate cerebral cortex. The cortical dopaminergic fibers are distributed throughout the frontal-to-occipital extent, with the highest density in the prefrontal and motor-related areas of the frontal lobe.[1-10] Previous anatomical studies have shown that multiple motor-related areas of the frontal lobe, such as the primary motor cortex (MI), the premotor cortex (PM), and the supplementary motor area (SMA), contain dopaminergic fibers.[1-3,7,8,10] The dopaminergic innervation of the motor-related areas is diffuse, involving all cortical layers.[3,7,8,10] On the other hand, the cortical dopaminergic fibers as a whole arise from the three catecholaminergic cell groups—A8, A9, and A10—that are located in the ventral mesencephalon.[3] It has been reported that the cells of origin of the mesocortical dopamine projections to the dorsal frontal cortex, including the motor-related areas, are distributed predominantly in the dorsal aspects of the A8-A10 complex.[11,12] In primates, the dorsal components of these catecholaminergic cell groups largely correspond to the dorsal part of the retrorubral area, pars gamma (i.e., the dorsal tier) of the substantia nigra pars compacta, and the parabrachial pigmented nucleus of the ventral tegmental area, respectively. Given that subpopulations of dopaminergic neurons in such dorsal components may issue axon collaterals to both the frontal cortex and the striatum (for data in the rat, see refs. [13-15]), it is most likely that at least part of the dopaminergic

---

* H. Tokouno, A. Nambu, M. Takada, Tokyo Metropolitan Institute for Neuroscience, Fuchu, Tokyo 183-8526, Japan and CREST, JST (Japan Science and Technology); Y. Yamaji, Department of Public Health, Juntendo University School of Medicine, Tokyo 113-8421, Japan; M. Matsumura, Department of Neurosurgery, Gunma University School of Medicine, Maebashi, Gunma 371-8513, Japan; J. Kojima, Department of Neurology, Sayama Neurology Hospital, Sayama, Saitama 350-1302, Japan; M. Inase, Department of Physiology, Kinki University School of Medicine, Osaka-Sayama 589-8511, Japan; H. Imai, Department of Neurology, Juntendo University School of Medicine, Tokyo 113-8421, Japan.

fibers in the motor-related areas are affected in Parkinson's disease which is well known to be caused by the extensive loss of dopaminergic nigrostriatal neurons. Significant levels of depletion of dopaminergic fibers have, in fact, been shown to occur in the cortical motor-related areas of parkinsonian human and monkey brains.[8,16] Moreover, the dorsal components of the A8-A10 complex giving rise to the mesocortical dopamine system are characterized by the richness of calbindin-containing dopaminergic neurons that have been considered to spare parkinsonian insults.[11,17-21] These imply that in Parkinson's disease, the extents of reductions in cortical dopaminergic innervation may vary among the motor-related areas. Using immunohistochemistry for tyrosine hydroxylase (TH; the synthesizing enzyme for dopamine), we thus investigated the possible differential changes in dopaminergic fiber density in the motor-related areas of monkeys rendered parkinsonian by treatment with the parkinsonism-inducing drug, 1-methyl-4-phenyl-1,2,3,6-tetrahydropyridine (MPTP). As compared to a normal control, the distribution patterns of TH-immunoreactive fibers were analyzed in the MI, the dorsal and ventral divisions of the PM (PMd and PMv), the SMA, the presupplementary motor area (pre-SMA), and the cingulate motor areas (CMAs) in MPTP-treated monkeys (Fig. 1; see also refs. [22,23]).

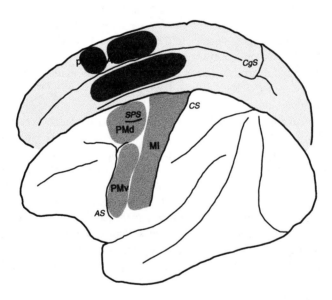

Figure 1. Motor-related areas of the frontal lobe. The primary motor cortex (MI) and the dorsal and ventral divisions of the premotor cortex (PMd and PMv) are located on the lateral surface. The supplementary and presupplementary motor areas (SMA and pre-SMA) and the rostral, dorsal, and ventral cingulate motor areas (CMAr, CMAd, and CMAv) on the medial wall. AS, arcuate sulcus; CgS, cingulate sulcus; CS, central sulcus; SPS, superior precentral sulcus.

2. MATERIALS AND METHODS

Three Japanese monkeys *(Maraca fuscata)* of either sex, weighing 4-6 kg, were used for the present study. All experimental procedures were conducted under aseptic conditions in line with NIH's *Guide for the Care and Use of Laboratory Animals*. Prior to MPTP administration, the medial wall of the frontal lobe was mapped electrophysiologically by means of intracortical microstimulation to identify the SMA and pre-SMA (for detailed procedures, see refs. [24-26]). Under general anesthesia with ketamine hydrochloride (10 mg/kg b.wt., i.m.) and sodium pentobarbital (25 mg/kg b.wt., i.v.), intracarotid infusion of MPTP was performed in two of the three monkeys. The remaining monkey was used as a control. A cannula (24 gauge) was inserted unilaterally into the common carotid artery after cramping of the external carotid artery. In each monkey, 30 ml of a solution of MPTP in the form of hydrochloride salt (Research Biochemicals International) dissolved in physiological saline was slowly infused at the dose of 0.5 mg/kg body weight over 30 min. Clinical and behavioral signs (such as hypokinesia, muscular rigidity, and postural instability) relevant to hemiparkinsonism were assessed daily.

After a survival of two weeks, the monkeys were anesthetized deeply with an overdose of sodium pentobarbital (50 mg/kg b.wt., i.v.) and perfused transcardially with physiological saline, followed by 10% formalin dissolved in 0.1 M phosphate buffer (pH 7.3) and, finally, the same fresh buffer containing 10% and then 30% sucrose. The fixed brains were removed from the skull, saturated with 30% sucrose at 4°C, and cut serially into coronal sections 60 μm thick on a freezing microtome. Every sixth section of the frontal lobe, as well as of the substantia nigra and the striatum, was immunostained for TH according to the avidin-biotin-peroxidase complex (ABC) method. The sections were first incubated with rabbit anti-TH antibody (kindly donated by Dr. I. Nagatsu) overnight, followed by biotinylated goat anti-rabbit IgG (Vector) for 2 h and, finally, ABC Elite (Vector) for 2 h. Subsequently, the sections were reacted for 10-20 min in 0.05 M Tris-HCl buffer (pH 7.6) containing 0.04% diaminobenzidine tetrahydrochloride, 0.04% $NiCl_2$, and 0.003% H_2O_2. All steps were done at room temperature. To verify the laminar organization of the cortical motor-related areas, some sections of the frontal lobe were Nissl-stained with Cresyl Violet.

3. RESULTS

First, the morphological changes in the mesostriatal dopamine system were examined in MPTP-treated monkeys. In comparison with the normal control monkey, the monkeys rendered behaviorally hemiparkinsonian exhibited large reductions in the density of TH-immunoreactive fibers in the striatum. Such a depletion of striatal TH immunoreactivity was observed bilaterally with a predominance of the side ipsilateral to MPTP infusion. In these MPTP-treated monkeys, however, only slight to moderate levels of decreases in the number of TH-immunoreactive cells were detected in the A8-A10 catecholaminergic cell groups on each side, especially in the substantia nigra pars compacta (A9). The discrepancy between MPTP neurotoxicity to striatal fibers and nigral cells may probably be ascribed to a relatively short-term post-MPTP survival (see ref. [27]).

In the normal control monkey, TH-immunoreactive fibers in the motor-related areas of the frontal lobe were distributed more densely in the areas on the medial wall, including the

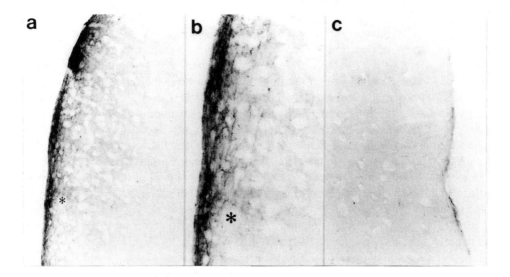

Figure 2. TH-immunoreactive fibers in the superficial layers of the SMA. a: Normal control monkey. b: Higher-power magnification of a. Asterisks in a and b indicate the same blood vessel. c: MPTP-treated monkey. a, X20; b, c, X50.

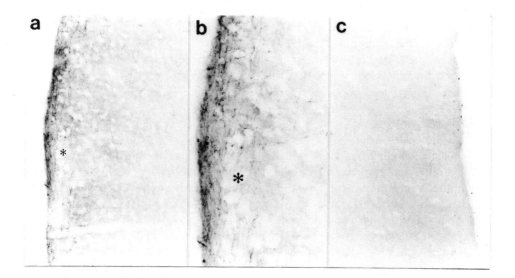

Figure 3. TH-immunoreactive fibers in the superficial layers of the pre-SMA. a: Normal control monkey. b: Higher-power magnification of a. Asterisks in a and b indicate the same blood vessel. c: MPTP-treated monkey. a, X20; b, c, X50.

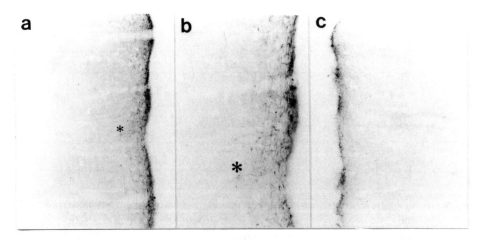

Figure 4. TH-immunoreactive fibers in the superficial layers of the MI. a: Normal control monkey. b: Higher-power magnification of a. Asterisks in a and b indicate the same blood vessel. c: MPTP-treated monkey. a, X20; b, c, X50.

SMA, pre-SMA, and CMAs, than in the areas on the lateral surface, such as the MI, PMd, PMv (Table 1). Particularly, TH immunoreactivity in the lateral surface areas was markedly seen in their convexities exposed to the cortical surface rather than in their banks facing the central, superior precentral, or arcuate sulcus. Regardless of the motor-related area, these TH-immunoreactive fibers were located mainly in the superficial layers (I and II) and, additionally, in the deep layers (III-VI). In the MPTP-treated monkeys, the density of TH-immunoreactive fibers was decreased in each of the motor-related areas of the hemisphere ipsilateral to MPTP infusion. Throughout the motor-related areas, the depletion of TH-immunoreactive fibers was prominent in layers I and II. However, the extents of such decreases in TH-immunoreactive fiber density varied among the areas (Table 1). In the medial-wall motor-related areas corresponding to the SMA and pre-SMA, TH-immunoreactive fibers were almost completely depleted (Figs. 2 and 3). Considerable reductions in TH immunoreactivity were also found in the CMAs, involving all of the rostral, dorsal, and ventral cingulate motor areas. On the other hand, TH-immunoreactive fibers were only moderately decreased in the lateral-surface motor-related areas corresponding to the MI, PMd, and PMv, as compared to those in the medial-wall motor-related areas (Fig. 4). In the hemisphere contralateral to MPTP infusion, the depletion of TH immunoreactivity in the motor-related areas was slightly less marked than in the ipsilateral hemisphere.

4. DISCUSSION

A previous immunohistochemical study has demonstrated that the density of cortical dopaminergic fibers is significantly reduced in regions corresponding presumedly to the MI and PM inpatients with Parkinson's disease.[8] A biochemical work has also shown the

Table 1. Differential reductions in TH-immunoreactive fiber density in cortical motor-related areas in MPTP-treated monkey

	medial wall			lateral surface		
	SMA	pre-SMA	CMAs	MI	PMd	PMv
control	●●●	●●●	●●●/●●	●●/●	●●/●	●●/●
MPTP	●/–	●/–	●/–	●	●	●

The number (1-3) of filled circles represents the relative density of TH-immunoreactive fibers.

decreases in dopamine levels in the SMA and the cingulate cortex of MPTP-treated monkeys.[16] The present study has confirmed and extended these data with the demonstration that the dopaminergic innervation of the motor-related areas of the frontal lobe is differentially affected among the areas in parkinsonian monkeys. Our results revealed that in MPTP-treated monkeys, TH-immunoreactive fibers in the medial-wall cortical areas, including the SMA, pre-SMA, and CMAs, were depleted to a greater extent than those in the lateral-surface cortical areas, including the MI, PMd, and PMv. Consistent with the previous report,[8] reductions in the density of TH-immunoreactive fibers in the motor-related areas were more prominent in the superficial layers (I and II) than in the deep layers (III-VI). In general, TH is expressed not only in the dopaminergic neuron system, but also in the noradrenergic one. Therefore, TH immunoreactivity alone does not enable to distinguish these two neuron systems until it turns clear whether the enzyme that converts dopamine to noradrenaline, dopamine-beta-hydroxylase (DBH), is expressed or not. However, DBH-immunoreactive fibers in the cortical motor-related areas have been shown to be much less densely distributed—especially in the superficial layers—than TH-immunoreactive fibers.[7,8,10] Further, it is surprising that only part of these DBH-immunoreactive fibers display TH immunoreactivity.[7,10] In this context, the TH-immunoreactive fibers in the motor-related areas can be considered mostly dopaminergic. Thus, the present results indicate that the dopaminergic fibers in the superficial layers of the medial-wall motor-related areas (i.e., the SMA, pre-SMA, and CMAs) are most drastically depleted in parkinsonian monkeys.

Combined retrograde labeling and TH immunohistochemical studies have reported that the mesocortical dopamine projections to the motor-related areas, such as the MI and SMA, originate mainly from the dorsal aspects of the A8-A10 complex.[11,12] Some neurons in these regions have been shown to project to more than one motor-related area by sending axon collaterals.[11] Of particular interest is that calbindin is frequently colocalized in the dorsal populations of dopaminergic neurons in the ventral mesencephalon.[11,17-21] In view of the fact that the dopaminergic neurons expressing calbindin are less vulnerable in Parkinson's disease[18-21] although the loss of dopaminergic neurons extensively occurs over the A8-A10 complex,[18,19,28-31] it is likely that these calbindin-containing neurons may preferentially project to the lateral-surface motor-related areas (i.e., the MI, PMd, and PMv) where the dopaminergic fibers have less severely been affected in parkinsonian monkeys.

Increased levels of glucose utilization have been reported in the MI and PM of MPTP-treated monkeys.[32] However, such alterations in glucose metabolism have not as yet been analyzed in the medial-wall motor-related areas. It has been well documented that dysfunctions of the SMA produce motor impairments relevant to Parkinson's disease (for example, impaired bilateral movements),[5,6,17,18,28,33-35] or that motor behaviors involved in the SMA (for example, internally instructed movements) are impaired in Parkinson's disease due to underactivation of the SMA.[34,36-38] Thus, the differential reductions in dopaminergic innervation of the medial-wall versus lateral-surface motor-related areas may reflect the specificity of motor deficits in Parkinson's disease.

5. REFERENCES

1. Berger, B., Trottier, S., Gaspar, P., Verney, C., and Alvarez, C., 1986, Major dopamine innervation of the cortical motor areas in the *Cynomolgus* monkey. A radioautographic study with comparative assessment of serotonergic afferents, *Neurosci. Lett.* 72:121-127.
2. Berger, B., Trottier, S., Verney, C., Gaspar, P., and Alvarez, C., 1988, Regional and laminar distribution of the dopamine and serotonin innervation in the macaque cerebral cortex: A radioautographic study, *J. Comp. Neurol.* 273:99-119.
3. Berger, B., Gaspar, P., and Verney, C., 1991, Dopaminergic innervation of the cerebral cortex: Unexpected differences between rodents and primates, *Trends Neurosci.* 14:21-27.
4. Björklund, A., Divac, I., and Lindvall, O., 1978, Regional distribution of catecholamines in monkey cerebral cortex, evidence for a dopaminergic innervation of the primate prefrontal cortex, *Neurosci. Lett.* 7:115-119.
5. Brown, R.M., Crane, A.M., and Goldman, P.S., 1979, Regional distribution of monoamines in the cerebral cortex and subcortical structures of the rhesus monkey: Concentrations and in vivo synthesis rates, *Brain Res.* 168:133-150.
6. De Keyser, J., Ebinger, G., and Vauquelin, G., 1989, Evidence for a widespread dopaminergic innervation of the human cerebral neocortex, *Neurosci. Lett.* 104:281-285.
7. Gaspar, P., Berger, B., Febvret, A., Vigny, A., and Henry, J.P., 1989, Catecholamine innervation of the human cerebral cortex as revealed by comparative immunohistochemistry of tyrosine hydroxylase and dopamine-beta-hydroxylase, *J. Comp. Neurol.* 279:249-271.
8. Gaspar, P., Duyckaerts, C., Alvarez, C., Javoy-Agid, F., and Berger, B., 1991, Alterations of dopaminergic and noradrenergic innervations in motor cortex in Parkinson's disease, *Ann. Neurol.* 30:365-374.
9. Levitt, P., Rakic, P., and Goldman-Rakic, P., *1984,* Region-specific distribution of catecholamine afferents in primate cerebral cortex: A fluorescence histochemical analysis, *J. Comp. Neurol.* 227:23-36.
10. Lewis, D.A., Campbell, M.J., Foote, S.L., Goldstein, M., and Morrison, J.H., *1987,* The distribution of tyrosine hydroxylase-immunoreactive fibers in primate neocortex is widespread but regionally specific, *J. Neurosci.* 7:279-290.
11. Gaspar, P., Stepniewska, I., and Kaas, J.H., 1992, Topography and collateralization of the dopaminergic projections to motor and lateral prefrontal cortex in owl monkeys, J. *Comp. Neurol.* 325:1-21.
12. Williams, S.M., and Goldman-Rakic, P.S., *1998,* Widespread origin of the primate mesofrontal dopamine system, *Cereb. Cortex* 8:321-345.
13. Fallon, J.H., and Loughlin, S.E., 1982, Monoamine innervation of the forebrain: Collateralization, *Brain Res. Bull.* 9:295-307.
14. Loughlin, S.E., and Fallon, J.H., *1984,* Substantia nigra and ventral tegmental area projections to cortex: Topography and collateralization, *Neuroscience* 11:425-435.
15. Takada, M., and Hattori, T., *1986,* Collateral projections from the substantia nigra to the cingulate cortex and striatum in the rat, *Brain Res.* 380:331-335.
16. Elsworth, J.D., Deutch, A.Y., Redmond, D.E., Jr., Sladek, J.R., Jr., and Roth, R.H., 1990, MPTP reduces dopamine and norepinephrine concentrations in the supplementary motor area and cingulate cortex of the primate, *Neuroscd. Lett.* 114:316-322.

17. Damier, P., Hirsch, E.C., Agid, Y., and Graybiel, A.M., 1999, The substantia nigra of the human brain. I. Nigrosomes and the nigral matrix, a compartmental organization based on calbindin D28K immunohistochemistry, *Brain* 122:1421-1436.
18. Damier, P., Hirsch, E.C., Agid, Y., and Graybiel, A.M., 1999, The substantia nigra of the human brain. II. Patterns of loss of dopamine-containing neurons in Parkinson's disease, *Brain* 122:1437-1448.
19. German, D.C., Manaye, K.F., Sonsalla, P.K., and Brooks, B.A., 1992, Midbrain dopaminergic cell loss in Parkinson's disease and MPTP-induced parkinsonism: Sparing of calbindin-D_{28K}-containing cells, *Ann. N. Y. Acad. Sci.* 648:42-62.
20. Lavoie, B., and Parent, A., 1991, Dopaminergic neurons expressing calbindin in normal and parkinsonian monkeys, *Neuroreport* 2:601-604.
21. Yamada, T., McGeer, P.L., Baimbridge, K.G., and McGeer, E.G., *1990*, Relative sparing in Parkinson's disease of substanfia nigra dopamine neurons containing calbindin-D_{28K}, *Brain Res.* 526:303-307.
22. Picard, N., and Strick, P.L., 1996, Motor areas of the medial wall: A review of their location and functional activation, *Cereb. Cortex* 6:342-353.
23. Tanji, J., *1994*, The supplementary motor area in the cerebral cortex, *Neurosci. Res.* 19:251-268.
24. Inase, M., Tokuno, H., Nambu, A., Akazawa, T., and Takada, M., *1999*, Corticostriatal and corticosubthalamic input zones from the presupplementary motor area in the macaque monkey: Comparison with the input zones from the supplementary motor area, *Brain Res.* 833:191-201.
25. Nambu, A., Takada, M., Inase, M., and Tokuno, H., 1996, Dual somatotopical representations in the primate subthalamic nucleus: Evidence for ordered but reversed body-map transformations from the primary motor cortex and the supplementary motor area, *J. Neurosci.* 16:2671-2683.
26. Takada, M., Tokuno, H., Nambu, A., and Inase, M., 1998, Corticostriatal projections from the somatic motor areas of the frontal cortex in the macaque monkey: Segregation versus overlap of input zones from the primary motor cortex, the supplementary motor area, and the premotor cortex, *Exp. Brain Res.* 120:114-128.
27. Herkenham, M., Little, M.D., Bankiewicz, K., Yang, S.-C., Markey, S.P., and Johannessen, J.N., 1991, Selective retention of MPP+ within the monoaminergic systems of the primate brain following MPTP administration: An *in vivo* autoradiographic study, *Neuroscience* **40**:133-158.
28. Deutch, A.Y., Elsworth, J.D., Goldstein, M., Fuxe, K., Redmond, D.E., Jr., Sladek, J.R., Jr., and Roth, R.H., 1986, Preferential vulnerability of A8 dopamine neurons in the primate to the neurotoxin 1-methyl-4-phenyl-1,2,3,6-tetrahydropyridine, *Neurosci. Lett.* 68:51-56.
29. Elsworth, J.D., Deutch, A.Y., Redmond, D.E., Jr., Sladek, J.R., Jr., and Roth, R.H., 1990, MPTP-induced parkinsonism: relative changes in dopamine concentration in subregions of substantia nigra, ventral tegmental area and retrorubral field of symptomatic and asymptomatic vervet monkeys, *Brain Res.* 513:320-324.
30. German, D.C., Dubach, M., Askari, S., Speciale, S.G., and Bowden, D.M., 1988, 1-methyl-4-phenyl-1,2,3,6-tetrahydropyridine-induced parkinsonian syndrome in *Macaca fascicularis:* Which midbrain dopaminergic neurons are lost?, *Neuroscience* 24:161-174.
31. Kastner, A., Hirsch, E.C., Herrero, M.T., Javoy-Agid, F., and Agid, Y., *1993*, Immunocytochemical quantification of tyrosine hydroxylase at a cellular level in the mesencephalon of control subjects and patients with Parkinson's and Alzheimer's disease, *J. Neurochem.* 61:1024-1034.
32. Palombo, E., Porrino, L.J., Bankiewicz, K.S., Crane, A.M., Sokoloff, L., and Kopin, I.J., *1990*, Local cerebral glucose utilization in monkeys with hemiparkinsonism induced by intracarotid infusion of the neurotoxin MPTP, *J. Neurosci.* 10:860-869.
33. Brinkman, C., 1984, Supplementary motor area of the monkey's cerebral cortex: Short- and long-term deficits after unilateral ablation and the effects of subsequent callosal section, *J. Neurosci.* 4:918-929.
34. Cunnington, R., Iansek, R., Bradshaw, J.L., and Phillips, J.G., 1995, Movement-related potentials in Parkinson's disease. Presence and predictability of temporal and spatial cues, *Brain* 118:935-950.
35. Dick, J.P., Benecke, R., Rothwell, J.C., Day, B.L., and Marsden, C.D., 1986, Simple and complex movements in a patient with infarction of the right supplementary motor area, *Mov. Dis.* 1:255-266.
36. Jahanshahi, M., Jenkins, I.H., Brown, R.G., Marsden, C.D., Passingham, R.E., Brooks, D.J., *1995*, Self-initiated versus externally triggered movements. I. An investigation using measurement of regional cerebral blood flow with PET and movement-related potentials in normal and Parkinson's disease subjects, *Brain* 118:913-933.
37. Jones, D.L., Phillips, J.G., Bradshaw, J.L., Iansek, R., and Bradshaw, J.A., *1992*, Impairment in bilateral alternating movements in Parkinson's disease?, *J. Neurol. Neurosurg. Psychiat.* 55:503-506.

38. Praamstra, P., Cools, A.R., Stegeman, D.F., and Horstink, M.W.I.M., *1996*, Movement-related potential measures of different modes of movement selection in Parkinson's disease, *J. Neurol. Sci.* 140:67-74.

CONTROL OF EPILEPTIC SEIZURES
Another function for the basal ganglia?

Colin Deransart, Véronique Riban, Laurent Vercueil, Karine Nail-Boucherie, Christian Marescaux and Antoine Depaulis*

Abstract. During the last two decades, evidence has accumulated to demonstrate the existence, in the central nervous system, of an endogenous mechanism that exerts an inhibitory control over different forms of epileptic seizures. The substantia nigra and the superior colliculus have been described as key structures in this control circuit: inhibition of GABAergic neurons of the substantia nigra pars reticulata results in suppression of seizures in various animal models of epilepsy. The role in this control mechanism of the direct GABAergic projection from the striatum to the substantia nigra and of the indirect pathway from the striatum through the globus pallidus and the subthalamic nucleus, was examined in a genetic model of absence seizures in the rat. In this model, pharmacological manipulations of both the direct and indirect pathways resulted in modulation of absence seizures. Activation of the direct pathway or inhibition of the indirect pathway suppressed absence seizures through disinhibition of neurons in the deep and intermediate layers of the superior colliculus. Dopamine D1 and D2 receptors in the nucleus accumbens, appear to be critical in these suppressive effects. Along with data from the literature, our results suggest that basal ganglia circuits play a major role in the modulation of seizures and provide a framework to understand the role of these circuits in the modulation of generalized seizures.

1. INTRODUCTION

Epilepsy refers to a collection of neurological disorders affecting approximately 50 million people worldwide. The epilepsy is characterized by the spontaneous recurrence of seizures, either convulsive or non-convulsive. Seizures result from the sudden transient oversynchronization and overactivation of a neuronal population within the brain. They are currently divided into two groups: partial (focal) and generalized (Commission on Classifica-

---

* C. Deransart, Klinikum der Albert-Ludwigs-Universitat Neurozentrum, Sektion Klinische Neuropharmakologie, Breisacherstr. 64 D-79106 Freiburg im Breisgau, Germany; V. Riban, L. Vercueil, K. Nail-Boucherie, C. Marescaux and A. Depaulis, Neurobiologie et neuropharmacologie des epilepsies generalisees INSERM U. 398, Faculte de Medecine, 11 rue Humann F-67085 Strasbourg Cedex, France.

tion and Terminology of the International League Against Epilepsy, 1981). Partial seizures initially involve a limited part of neurones within one hemisphere. They are characterized by the focal aspect of their symptoms and electroencephalographic changes. Consciousness may be impaired and partial seizures may spread to other parts of the brain, thus progressing to a secondarily generalized seizure. Primary generalized seizures simultaneously involve both cerebral hemispheres from the beginning, displaying no evidence of local onset. Electroencephalographic changes are bilateral and synchronous. Consciousness is generally impaired and motor manifestations occur, either convulsive (tonic and/or clonic movements) or non-convulsive (staring, motor inhibition with behavioral arrest).

Epileptic syndromes are built on several single factors: seizure type, severity and recurrence of seizures, aetiology, age of onset, anatomical data, precipitating factors, evolution. Thus, in addition to a distinction between partial and generalized epilepsies, the epileptic syndromes are classified as idiopathic or symptomatic disorders according to the relative contribution of genetic and acquired pathologic factors. Determination of the epileptic syndrome determine to a great extent the type of evaluation and therapy the patient will receive (Holmes, 1997). Each kind of seizure responds differently to pharmacological treatments. However, these treatments are mainly symptomatic and it is assumed that up to 30% of seizures are resistant to antiepileptic drugs, despite accurate diagnosis and carefully monitored treatments. Although surgical treatment may be an alternative, it is mainly restricted to patients suffering from partial seizures in which an epileptic focus is clearly delineated outside "eloquent" brain regions.

Several animal models of seizures and epilepsy have been developed to understand the pathophysiological mechanisms underlying the different types of epileptic seizures (for review see Danober et al., 1998; Fischer, 1989; Jobe et al., 1991; Loscher and Schmidt, 1988; Sperk, 1994). Recently, preference has been given to models of recurrent seizures whether they are genetic or induced by an initial manipulation, because they better reproduce the chronicity of this disease and the sudden occurrence and interruption of the seizures. In this respect, we have developed a genetic model of generalized non-convulsive seizures in the rat which is reminiscent of human absence-epilepsy (Danober et al., 1998). In this model developed in our laboratory, the seizures are characterized by generalized spike-and-wave discharges (SWD) recorded on the cortical electroencephalogram, concomitant with behavioural arrest (Marescaux et al., 1992; Danober et al., 1998). These paroxysmal activities are concomitant with behavioral arrests and rhythmic twitching of the vibrissae. In general, this model shows many of the features of absence epilepsy in human and is suppressed by all anti-absence drug. The cortex, the reticular nucleus and the ventrobasal relay nuclei of the thalamus play a predominant role in the generation of SWD, whereas limbic structures (e.g.. hippocampus. amygdala) are not involved. As seizures occur spontaneously every minute and last about 20 sec. this model allows to test the effectiveness and time-course of pharmacological treatments (Loscher and Schmidt, 1988; Danober et al., 1998).

2. INITIATION AND CONTROL CIRCUITS

Both clinical and experimental studies have led to the hypothesis that each kind of seizure arises from a specific neuronal circuit (Browning, 1986; Danober et al, 1998; Loscher and Ebert, 1996). Within these "initiation circuits," several dysfunctions have been described which may facilitate the occurrence of seizures. Alterations of GABAergic and/or glutamatergic neurotransmission, as well as possible mutations of the ionic channels appear to play a critical role in these circuits (Meldrum, 1995; Noebels, 1996). However, little is known about the mechanisms which allow seizures to occur and/or preclude them from evolving towards a status epilepticus and become deleterious. Interruption of a seizure can be considered as a passive phenomenon, secondary to an energetic failure of the exhausted neurons within the epileptic circuit. It can also be an active phenomenon underlied by control mechanisms, the failure of which could facilitate the reccurrence of seizures (Engel, 1995; Moshe et al., 1992). Similarly, mechanisms responsible for long lasting episodes devoid of any seizures (interictal phases) warrants further investigations. Data from animal studies have accumulated evidence over the years suggesting the existence of mechanisms controlling seizures within the central nervous system (Dragunow, 1986). More especially, studies performed by several research groups have pointed to the basal ganglia as one of the main systems involved in the control of epileptic seizures (Deransart et al., 1998). The present paper will review evidence collected in our group, as well as data from the literature, suggesting the involvement of the basal ganglia circuits in the control of seizures.

3. THE SUBSTANTIA NIGRA AS A "COMMON DENOMINATOR" IN THE CONTROL OF SEIZURES

At the beginning of the 80's, several groups reported that inhibition of the substantia nigra by local bilateral infusion of GABA agonists could suppress seizures in a wide variety of animal models (see Gale, 1985; Depaulis et al., 1994). This suppressive effect was observed both on the behavioral and electroencephalographic components of the seizures, suggesting that it was not solely due to blockade of the motor output. In addition, using the different models of absence-epilepsy, we were further able to show that such inhibition of the substantia nigra could also block non convulsive seizures (Depaulis et al., 1988; 1989). This demonstrated that inhibition of the substantia nigra could in fact interfere with very different types of epileptic circuits. Although modification of activity and/or metabolism of the substantia nigra has been described during or after some forms of seizures (Nehlig et al.. 1991; Bonhaus et al., 1991), this structure does not appear critical in the initiation of seizures. Electroencephalographic recording within the substantia nigra generally reveals the existence of delayed and/or small amplitude paroxysmal activities during seizures which are different to what is generally observed in structures involved in the genesis of seizures (e.g., cortex, hippocampus, thalamus) (Vergnes et al., 1990). Furthermore, its lesion does not preclude the occurrence of seizures (Wahnschaffe and Loscher, 1990; Depaulis et al., 1990c). For these reasons, it was suggested that neurons of the substantia nigra, and more especially the pars reticulata, belong to a distal mechanism that controls different kinds of epileptic circuits (Gale, 1985; Depaulis et al., 1994). According to this hypothesis, several studies have examined the role of both the inputs and outputs of the substantia nigra pars reticulata in the control of epileptic seizures.

4. EVIDENCE FOR THE INVOLVEMENT OF THE BASAL GANGLIA IN THE CONTORL OF SEIZURES

The possibility that a direct GABAergic striato-nigral pathway could modulate absence seizures through inhibition of the substantia nigra pars reticulata (SNpr) was first suggested by experiments showing that intranigral injections of GABAA agonists have anti-epileptic effects in most models of seizures in the rat (for review see Depaulis et al., 1994). We have recently confirmed and extended these findings in the GAERS. In this model, activation of striatal neurons by a bilateral injection of NMDA as well as intrastriatal injections of SKF-39393, an agonist of the Dl dopamine receptors, significantly suppressed absence seizures. These antiepileptic effects were obtained at doses deprived of any behavioral effects and were more pronounced when injection sites were located in the ventral part of the striatum, i.e., in the nucleus accumbens. Activation of cells in the striatum by either glutamate or D1 agonists has been shown to increase the release of GABA in the SNpr as measured by microdialysis (Morari et al., 1996) and to decrease SNpr activity (Chevalier and Deniau, 1990; Akkal et al., 1996). The involvement of the direct striato-nigral pathway in the control of absence seizures was further confirmed by the fact that blockade of the $GABA_A$ receptors in the SNpr by local injection of picrotoxin, as well as blockade of the D1 receptors in the striatum by the selective antagonist SCH23390, increased the occurrence of absence seizures in GAERS, without modifying the general behavior of the animals (Deransart et al., 2000). These data are in agreement with other studies showing that injection of either an NMDA agonist (Cavalheiro and Turski, 1986) or a GABA antagonist (Turski et al., 1989) in the striatum block seizures induced by pilocarpine. In addition, this suppression of seizures could be reversed by the subsequent injection of a GABA antagonist in the SNpr (Turski et al., 1987). However, intrastriatal injections of the Dl agonist SKF-38393 were shown to have no protective effects against pilocarpine-induced clonic seizures and even aggravated the seizures (Turski et al., 1988). Similarly, intrastriatal injections of the Dl antagonist SCH23390 did not lower the threshold for pilocarpine-induced seizures. This apparent discrepancy may account for the involvement of different control circuits in this model and warrants further studies in other models of convulsive seizures.

Because the glutamatergic input of the SNpr from the subthalamic nucleus appears critical in the maintenance of the nigral cells' activity, it was suggested that this projection — as a part of the so-called "indirect" striato-nigral pathway — was also involved in the control of epileptic seizures. In most models of convulsive seizures, intranigral injection of NMDA antagonists was shown to be antiepileptic (for review see Depaulis et al., 1994). In GAERS, injections of different NMDA antagonists in the SNpr suppressed absence seizures whereas injections of non-NMDA antagonists were without effects (Deransart et al., 1996). The possibility that inhibition of the subthalamic nucleus also leads to suppression of seizures was first established in this model. Bilateral injection of a GABA agonist in this structure, known to decrease the activity of SNpr neurons (Féger and Robledo, 1992), resulted in marked suppression of SWD (Deransart et al., 1996). Furthermore, absence seizures could be interrupted by high-frequency stimulations of the subthalamic nucleus (Vercueil et al., 1998). Such high-frequency electrical stimulation has been shown to suppress the activity of SNpr neurons (Benazzouz et al., 1995). In both cases, the suppression of absence seizures by inhibition of the subthalamic nucleus could be dissociated from any behavioral effects such as dystonia or stereotypies. Anti-epileptic effects obtained by inhibition of the subthalamic nucleus were also observed using local injections of a GABA agonist in models of generalized tonic-clonic seizures (Veliskova et al., 1996) or partial seizures with secondary generalization (Deransart et al., 1998).

Because the subthalamic nucleus receives an important GABAergic input from the pallidum, the involvement of this structure was investigated in GAERS. Injection of a GABA antagonist into the globus pallidus or the ventral pallidum resulted in seizure suppression in GAERS, at doses deprived of any behavioral effects. This antiepileptic effect was correlated with a decrease of glutamate levels in the SNpr measured by microdialysis, suggesting that this effect is mediated through a reduction of activity of the subthalamic nucleus (Deransart et al., 1999). In addition, antiepileptic effects were more pronounced when the injection sites were located in the ventral part of the pallidum. Conversely, an increase in the occurrence of absence seizures was observed after injections of a $GABA_A$ agonist into the ventral pallidum (Deransart et al., 1999). To date, no data have been collected concerning the involvement of the pallidum in the control of convulsive seizures.

The role of the striato-pallido-subthalamo-nigral pathway in the control of seizures was further confirmed by intrastriatal injections of D2 agonists in the striatum. Injections of bromocriptine or quinpirole significantly suppressed absence seizures in GAERS. Conversely, blockade of D2 receptors by intrastriatal injections of antagonists (haloperidol or sulpiride) resulted in a significant increase of absence seizures. The effects on seizures observed following intrastriatal injections of D2 agonists and antagonists were dissociated from any behavioral effects and were more pronounced when injections were located in the ventral part of the striatum, in the nucleus accumbens core region (Deransart et al., submitted). Consistent with the antiepileptic effects of D2 agonists, intrastriatal injections of D2 agonists have also been shown to block seizures in different models of convulsions, whereas injections of D2 antagonists in the same region have proconvulsant effects (Al-Tajir and Starr, 1991 a&b; Csernansky et al., 1988; Turski et al., 1988; Wahnschaffe and Loscher, 1991).

Recent data obtained in our group suggest that the two striatonigral pathways cooperate in the suppression of absence seizures. A marked suppression of absence-seizures was obtained after combined intranigral injections of muscimol, a GABA agonist and CGP 40116, an NMDA antagonist at doses shown to have no effect by themselves. An additive suppression of absence-seizures was also obtained following combined injections of low doses of a D1 and a D2 agonist in the nucleus accumbens (Deransart et al., 2000). This later effect also suggests that dopamine in the striatum plays a key role in the modulation of absence-seizures by the basal ganglia. This hypothesis is in agreement with previous results showing a suppression of absence seizures in GAERS following systemic injection of dopamimetics, whereas aggravation was observed after injection of dopamine antagonists (Warter et al., 1988). The critical influence of dopamine in the control of absence seizures was also confirmed by experiments showing that lesions of dopaminergic neurons within the ventral mesencephalon (pars compacta of the substantia nigra and ventral tegmental area) temporarily aggravate absence seizures in GAERS (Lannes, personal communication). Finally, these data are also in agreement with the antiepileptic effects of dopamimetic drugs which have been reported both in human and in animals (for review see Starr, 1996).

5. THE NIGRO-COLLICULAR PROJECTION AS A CRITICAL OUTPUT OF THE CONTROL CIRCUIT

Several data from our laboratory as well as others suggest that the collicular output of the SNpr is involved in the control of epileptic seizures. This GABAergic projection exerts a tonic inhibition of neurons located in the deep and intermediate layers of the superior colliculus and a decrease of SNpr activity results in disinhibition of these neurons (Chevalier et al., 1985). In different models of epileptic seizures, lesion of the superior colliculus antagonizes the anti-epileptic effects obtained by injecting a GABA agonist into the SN, whereas lesion of other nigral outputs has no significant consequences (Depaulis et al., 1990c; Garant, 1987). Furthermore, activation of superior colliculus neurons by local injection of glutamate agonists or GABA antagonists also results in significant anti-epileptic effects (Depaulis et al., 1990a; Gale et al., 1993; Redgrave et al., 1988). More recently, several data obtained in the electroshock model have suggested that a subpopulation of neurons located in the caudal part of the superior colliculus is more specifically involved in these effects (Shehab et al, 1995). Similarly, using electrical stimulations and local infusions of a GABA antagonist, we have recently observed a preferential involvement of neurons of the posterior colliculus (Nail-Boucherie et al., in press). The projections through which these neurons interfer with epileptic circuits remains to be determined and are currently under investigation. Whether the nigro-collicular projection is the only nigral output involved in the control of seizures also remains to be confirmed. Finally, whether these neurons are under the control of the same striato-nigral projections involved in the control of seizures will be an important point to examine.

6. CONTROL OF SEIZURES AND OTHER FUNCTIONS OF THE BASAL GANGLIA

The basal ganglia are well known to be involved in the control of motor behavior and pharmacological modifications of the activity of these structures generally lead to behavioral effects such as stereotypes or modifications of locomotion in the rat. In addition, an increase of behavioral activity is known to suppress absence-seizures both in human and in rats (Drinkenburg et al., 1995; Lannes et al., 1988; Niedermeyer. 1996) and the suppression of absence seizures observed following pharmacological manipulations of the basal ganglia could have been secondary to motoric effects. However, this point was carefully controlled and, in all cases, the suppression of absence-seizures in GAERS was clearly dissociated from behavioral effects. This was made possible by the use of low doses of compounds and small volumes of intracerebral microinjections (200 nl/side). However, using this technique and because of the interconnected circuits in the basal ganglia, the possibility of a complete dissociation between the control of seizures and other functions cannot be addressed (Joel and Wiener, 1994; Mink, 1996).

Among the different circuits described in the basal ganglia (Afifi, 1994; Alexander and Crutcher, 1990; Groenewegen and Berendse, 1994), our results in the GAERS are in favor of the preferential involvement of a ventral circuit. The most effective sites to modulate absence seizures by intracerebral microinjections were mostly located ventrally, both at the striatal (nucleus accumbens) and pallidal (ventral pallidum) levels. In the rat, a projection from the nucleus accumbens to the ventral pallidum has been described which projects to the SNpr through the subthalamic nucleus (Deniau and Thierry, 1997; Groenewegen and Berendse, 1994; Maurice et al., 1997). This ventral circuit has been implicated in the control of affective and motivational aspects of behavior (Delong, 1990; Koob and Bloom. 1988; Le Moal and Simon, 1991; Mogenson et al., 1980). The possibility that such circuits are involved in the control of absence seizures is in agreement with both clinical (Guey et al., 1969; Jung, 1962) and experimental (Vergnes et al, 1991) observations indicating the importance of the motivational state in the occurrence of absence seizures. For instance, when involved in games or school activities, epileptic children rarely display absence seizures (Guey et al., 1969). Similarly, no SWD are recorded in GAERS during the performance of various task (e.g., food rewarded operant conditionning, social interactions) (Vergnes et al., 1991). Increase in dopamine in the nucleus accumbens has been reported during different kinds of motivated behaviors in animals (Fiorino et al., 1997; Salamone, 1996; Wilson et al., 1995). Although quite speculative, it is possible that an increase of dopamine lowers the probability of absence-seizure occurrence when the patient, or the rat, is involved in a motivated behavior. Although dopamine-sensitive anticonvulsant sites were primarily located within the ventral striatum, further studies are necessary to characterize the exact basal ganglia circuits involved in the control of the different forms of generalized seizures and to determine if these circuits are the same, whatever the type of seizures (Turski et al, 1988; Wahnschaffe and Loscher, 1991).

Our data provide a general framework to understand the functioning of the basal ganglia in the control of seizures and, more generally, rhythmic activities involving forebrain structures. As a matter of fact, only seizures involving structures of the forebrain appears to be modified by such a control (Depaulis et al., 1994; Deransart et al., 2001). No anticonvulsant effects were ever observed following GABAergic inhibition of the SNpr in rats prone to develop tonic seizures in response to a loud sound (Depaulis, 1990b). Such seizures, have been shown to primarily involved brainstem structures and have no cortical expression (Browning et al., 1986). Using the same model, we similarly observed that GABAergic inhibition of the subthalamic nucleus had no effect (Deransart et al., personal observation). Thus, investigating these control mechanisms appears not only necessary to understand epilepsy but could also provide an interesting approach to the understanding of the functioning of basal ganglia and how their circuits control cortical activity. For future prospects, it is noteworthy that further investigations should be aimed to clarify whether this system is endogenously activated and to address its relevance in human epilepsy.

7. ACKNOWLEDGEMENTS

This work was supported by the Institut National de la Sante et de la Recherche Medicate and a grant from the French Ministry of Education and Research (CD).

8. REFERENCES

Afifi, A., Basal ganglia: functional anatomy and physiology. Part 2. J. *Child Neurology,* 9 (1994) 352-361.
Akkal, D., Burbaud, P., Audin, J. and Bioulac. B., Responses of substantia nigra pars reticulata neurons to intrastriatal D1 and D2 dopaminergic agonist injections in the rat, *Neurosci Lett,* 213 (1996) 66-70.
Al-Tajir, G. and Starr, M.S., Anticonvulsant effect of striatal dopamine D2 receptor stimulation: Dependence on cortical circuits?, *Neuroscience. 43* (1991a) 51-57.

Al-Tajir, G. and Starr, M.S., D-2 agonists protect rodents against pilocarpine-induced convulsions by stimulating D-2 receptors in the striatum. but not in the substantia nigra, *Pharm. Biochem. Behav.*, 39 (1991b) 109-113.

Alexander, G. and Crutcher, M., Functional architecture of basal ganglia circuits neural substrates of parallel processing, *Trends Neurosci.*, 13 (1990) 266-271.

Benazzouz, A., Piallat, B., Pollak, P. and Benabid, A.L., Responses of substantia nigra pars reticulata and globus pallidus complex to high frequency stimulation of the subthalamic nucleus in rats: Electrophysiological data, *Neurosci Lett*, 189 (1995) 77-80.

Bonhaus, D., Russell, R. and McNamara, J. Activation of substantia nigra pars reticulata neurons: role in the initiation and behavioral expression of kindled seizures. *Brain Res* 545 (1991) 41-48.

Browning, R., Neuroanatomical localization of structures responsible for seizures in the GEPR: lesion studies. *Life* Sci 39 (1986) 857-867.

Cavalheiro, E. and Turski, L., Intrastriatal N-methyl-D-aspartate prevents amygdala kindled seizures in rats, *Brain Res,* 377 (1986) 173-176.

Chevalier, G. and Deniau, J., Disinhibition as a basic procss in the expression of striatal functions, *Trends Neurosci,* 13 (1990) 277-280.

Chevalier, G., Vacher, S., Deniau, J. and Desban, M., Disinhibition as a basic process in the expression of striatal functions. I. The striato-nigral influence on tecto-spinal/tecto-diencephalic neurons, *Brain Res,* 334 (1985) 215-226.

Commission on Classification and Terminology of the International League Against Epilepsy. Proposal for revised clinical and electroencephalographic classification of epileptic seizures. *Epilepsia,* 22 (1981) 489-501.

Csernansky, J., Melletin, J., Beauclair, L. and Lombrozo, L., Mesolimbic dopaminergic supersensivity following electrical kindling of the amygdala, *Biol. Psychiatry,* 23 (1988) 285-294.

Danober, L., Deransart, C., Depaulis, A., Vergnes, M. and Marescaux, C., Pathophysiological mechanisms of genetic absence epilepsy in the rat, *Prog. Neurobiol.,* 54 (1998) 1-31.

Delong, M., Primate models of movement disorders of basal ganglia origin, *Trends Neurosci.,* 13 (1990). 281-285.

Deniau, JM and Thierry, AM., Anatomical segregation of information processing in the rat substantia nigra pars reticulata. In: The basal ganglia and new surgical approaches for Parkinson's diseases, Advances in Neurology 74 (1997) 83-96, Obeso JA, DeLong MR, Ohye C, Marsden CD (eds), Lippincott-Raven Publishers, Philadelphia.

Depaulis, A., Liu, Z., Vergnes, M., Marescaux, C., Micheletti, G. and Warter, J., Suppression of spontaneous generalized non-convulsive seizures in the rat by microinjection of GABA antagonists into the superior colliculus, *Epilepsy Res,* 5 (1990a) 192-198.

Depaulis, A., Marescaux, C., Liu, Z. and Vergnes, M., The GABAergic nigro-collicular pathway is not involved in the inhibitory control of audiogenic seizures in the rat. *Neurosci Letters 111* (1990b) 269-274.

Depaulis, A., Snead, O.I., Marescaux, C. and Vergnes, M., Suppressive effects of intranigral injection of muscimol in three models of generalized non-convulsive epilepsy induced by chemical agents, *Brain Res.,* 498 (1989) 64-72.

Depaulis, A., Vergnes, M., Liu, Z., Kempf, E. and Marescaux, C. Involvement of the nigral output pathways in the inhibitory control of the substantia nigra over generakized non-convulsive seizures in the rat, Neuroscience 39 (1990c) 339-349.

Depaulis, A., Vergnes, M. and Marescaux, C., Endogenous control of epilepsy: the nigral inhibitory system, *Prog Neurobiol,* 42 (1994) 33-52.

Depaulis, A., Vergnes, M., Marescaux, C., Lannes, B. and Warter, J., Evidence that activation of GABA receptors in the substantia nigra suppresses spontaneous spike-and-wave discharges in the rat, *Brain Res,* 448 (1988) 20-29.

Deransart, C., Marescaux, C. and Depaulis, A., Involvement of nigral glutamatergic inputs in the control of seizures in a genetic model of absence epilepsy in the rat, *Neurosci,* 71 (1996) 721-728.

Deransart C., Lê-Pham, B.T., Hirsch E., Marescaux C. and Depaulis A., Inhibition of the substantia nigra suppresses absences and clonic seizures in audiogenic rats, but not tonic seizures: evidence for seizure specificity of the nigral control. *Neuroscience* 105, (2001) 203-211.

Deransart C., Lê B.T., Marescaux C. and Depaulis A., Role of the subthalamo-nigral input in the control of amygdala-kindled seizures in the rat. Brain Res. 807 (1998) 78-83.

Deransart C., Riban V., Lê B.T., Hechler V., Marescaux C. and Depaulis A., Evidence for the involvement of the pallidum in the modulation of seizures in a genetic model of absence epilepsy in the rat.

Neurosci Letters, 265 (1999) 131-134).
Deransart C., Riban V., Lê B.T., Marescaux C. and Depaulis A., Dopamine in the striatum modulates seizures in a genetic model of absence epilepsy in the rat. *Neurosci.* 100, (2000) 335-344.
Dragunow, M., Endogenous anticonvulsant substances. *Neurosci Biobehav Rev* 10 (1986) 229-244.
Drinkenburg, W.H., Coenen, A.M., Vossen, J.M. and Vanluijtelaar, E.L., Sleep deprivation and spike-wave discharges in epileptic rats, *Sleep,* 18 (1995) 252-256.
Engel, J.Jr., Inhibitory mechanisms of epileptic seizure generation, In: Fahn, S., Hallett, M., Luders, H.O. and Marsden, C.D., ed, *Negative motor phenomena,* Advances in neurology, Vol 67 (1995) 157-171.
Feger, J. and Robledo, P., The effects of activation or inhibition of the subthalamic nucleus on the metabolic and electrophysiological activities within the pallidal complex and the substantia nigra in the rat., *Eur J Neurosci,* 3 (1992) 947-95 2.
Fiorino, D.F., Coury, A. and Phillips, A.G, Dynamic changes in nucleus accumbens dopamine efflux during the Coolidge effect in male rats. *J Neurosci.* 17 (1997) 4849-4855.
Fischer, R.S., Animal models of the epilepsies. *Brain Res. Rev..* 14 (1989) 245-278.
Gale, K., Mechanisms of seizure control mediated by gamma amino butyric acid: role of the substantia nigra, *Fed Proc,* 44 (1985) 2414-2424.
Gale, K., Pazos, A., Maggio, R., Japikse, K. and Pritchard. P.. Blockade of GABA receptors in superior colliculus protects against focally evoked limbic motor seizures, *Brain Res, 603 (1993) 279-283.*
Garant, D.S. and Gale, K., Substantia nigra-mediated anticonvulsivant actions: role of nigral output pathways, Exp *Neurol,* 97 (1987) 143-159.
Groenewegen, H. and Berendse, H., Anatomical relationships between the prefrontal cortex and the basal ganglia in the rat, In: Thierry AM, Glowinski J., Goldman Rakic PS, Christen Y (eds): *Motor and cognitive functions of the prefrontal cortex.,* Berlin: Springer-Verlag, 1994: 51-77.
Guey, J., Bureau, M., Dravet, C. and Roger, J., A study of the rhythm of petit mal absences in children in relation to prevailing situations. The use of EEG telemetry during psychological examinations, school exercises and periods of inactivity, *Epilepsia,* 10 (1969) 441-451.
Holmes, G.L., Classification of seizures and the epilepsies. In: Schachter S.C. and Schomer D.L., ed, *The comprehensive evaluation and treatment of epilepsy.* Academic Press, Inc., New York:, *1997, pp. 1-36.*
Jobe, P., Mishra, P., Ludvig, N. and Dailey, J., Scope and contribution of genetic models to understanding of the epilepsies. *Critical Rev. Neurobiol.* 6 (1991) 183-220.
Joel, D. and Weiner, I., The organization of the basal ganglia-thalamocortical circuits: open interconnected rather than closed segregated, *Neuroscience.* 68 (1994) 363-379.
Jung, R., Blocking of petit-mal attacks by sensory arousal and inhibition of attacks by an active change in attention during the epileptic aura, *Epilepsia,* 3 (1962) 435-437.
Koob, G. and Bloom, F., Cellular and molecular mechanisms of drug dependence, *Science,* 242 (1988) 715-723.
Lannes, B., Micheletti, G, Vergnes, M., Marescaux, C., Depaulis. A. and Warter, J., Relationship between spike-wave discharges and vigilance levels in rats with spontaneous petit mal-like epilepsy, *Neurosci Lett,* 94 (1988) 187-191.
Le Moal, M. and Simon, H., Mesocorticolimbic dopaminergic network: functional and regulatory roles, Physiol. *Rev.,* 71 (1991) 155-234.
Loscher, W. and Ebert, U., The role of the piriform cortex in kindling. *Progress Neurobiol* 50 (1996) 427-481.
Loscher, W. and Schmidt, D., Which animal model should be used in the search for new antiepileptic drugs? A proposal based on experimental and clinical considerations, *Epilepsy Res,* 2 (1988) 145-181.
Marescaux, C., Vergnes, M. and Depaulis, A., Genetic absence epilepsy in rats from strasbourg - A review, J *Neural Transm, Suppl 35 (1992) 37-70.*
Maurice, N., Deniau, J.M., Menetrey, A., Glowinski, J. and Thierry, A.M., Position of the ventral pallidum in the rat prefrontal cortex basal ganglia circuit, *Neuroscience, 80* (1997) 523-534.
Meldrum, B.S., Neurotransmission in epilepsy. *Epilepsia, 36 (Suppl. 1)* (1995) S30-S35.
Mink, J.W., The basal ganglia: Focused selection and inhibition of competing motor programs, *Prog Neurobiol,* 50 (1996) 381-425.
Mogenson, G, Jones, D. and Yim, C., From motivation to action: functional interface between the limbic system and the motor system, *Prog Neurobiol,* 14 (1980) 69-97.

Morari, M., O'Connor, W.T., Ungerstedt, U., Bianchi, C. and Fuxe, K., Functional neuroanatomy of the nigrostriatal and striatonigral pathways as studied with dual probe microdialysis in the awake rat .2. Evidence for striatal N-methyl-D-aspartate receptor regulation of striatonigral gabaergic transmission and motor function, *Neuroscience,* 72 (1996) 89-97.

Moshe, S.L., Sperber, E.F., Brown, L.L. and Tempel A., Age-dependent changes in the substantia nigra GABA-mediated seizure suppression. In: Avanzini, A., Engel, J. Jr., Fariello, R. and Heinemann, U., eds. *Neurotransmitter in epilepsy,* Epilepsy Res Suppl 8 (1992) 97-106.

Nail-Boucherie, K., Lê-Pham, B.T., Marescaux C., Depaulis A., Suppression of absence seizures by electrical and pharmacological activation of the caudal superior colliculus in a genetic model of absence epilepsy in the rat. *Experimental Neurology*, in press.

Nehlig, A., Vergnes, M., Marescaux, C. Boyet, S. and Lannes, B., Local cerebral glucose utilization in rats with petit mal-like seizures, *Ann Neurol,* 29 (1991) 72-77.

Niedermeyer, E., Primary (idiopathic) generalized epilepsy and underlying mechanisms, *Clin Electroencephal.,* 27 (1996) 1-21.

Noebels, J.L., Targeting epilepsy genes, *Neuron,* 16 (1996) 241-244.

Phillipson, O. and Griffiths, A., The topographic order of inputs to nucleus accumbens in the rat, *Neuroscience,* 16 (1985) 275-296.

Redgrave, P., Dean, P. and Simkins, M., Intratectal glutamate suppresses pentylenetetrazole-induced spike-and-wave discharges, *Eur. J. Pharmac.,* 158 (1988) 283-287.

Salamone, J., The behavioral neurochemistry of motivation: methodological and conceptual issues in studies of dynamic activity of nucleus accumbens dopamine, *J Neurosci Meth. 64 (1996) 137-149.*

Shehab, S., Simkins, M., Dean, P. and Redgrave, P., The dorsal midbrain anticonvulsant zone-I. Effects of locally administered excitatory amino acids or bicuculline on maximal electroshock seizures. *Neuroscience* 65 (1995) 671-679

Sperk, G., Kainic acid seizures in the rat. *Prog Neurobiol* 42 (1994) 1-32.

Starr, M.S., The role of dopamine in epilepsy, *Synapse,* 22 (1996) 159-194.

Turski, L., Cavalheiro, E.A., Bortolotto, Z.A., Ikonomidou-Turski. C., Kleinrok, Z. and Turski, W.A., Dopamine-sensitive anticonvulsant site in the rat striatum, *J. Neurosci.,* 8 (1988) 3837-3847.

Turski, L., Ikonomidou, C., Turski, W.A., Bortolotto, Z.A. and Cavalheiro, E.A., Review: cholinergic mechanisms and epileptogenesis. The seizures induced by pilocarpine: a novel experimental model of intractable epilepsy, *Synapse,* 3 (1989) 154-171.

Turski, L., Meldrum, B.S., Cavalheiro, E.A., Calderazzo-Filho. L.S.. Bortolotto, Z.A., Ikonomidou-Turski, C. and Turski, W.A., Paradoxical anticonvulsant activity of the excitatory amino acid N-methyl-D-aspartate in the rat caudate-putamen, Proc *Natl Acad* Sci USA, 84 (1987) 1689-1693.

Veliskova, J., Velsek, L. and Moshe, S.L., Subthalamic nucleus: A new anticonvulsant site in the brain, *Neuroreport,* 7 (1996) 1786-1788.

Vercueil, L., Benazzouz, A., Deransart, C., Bressand, K., Marescaux. C., Depaulis, A., and Benabid, A.L., High-frequency stimulation of the sub-thalamic nucleus suppressed absence seizures in the rat: comparison with neurotoxic lesions. *Epilepsy Res.* 31 (1998) 39-46.

Vergnes, M., Marescaux, C., Boehrer, A. and Depaulis, A., Are rats with genetic absence epilepsy behaviorally impaired?, *Epilepsy Res,* 9 (1991) 97-104.

Vergnes, M., Marescaux, C. and Depaulis, A., Mapping of spontaneous spike and wave discharges in wistar rats with genetic generalized non convulsive epilepsy, *Brain Res,* 523 (1990) 87-91.

Wahnschaffe, U. and Loscher, W., Selective bilateral destruction of substantia nigra has no effect on kindled seizures induced from stimulation of amygdala or piriform cortex in rats, *Neurosci Lett,* 113 (1990) 205-210.

Wahnschaffe, U. and Loscher, W., Anticonvulsant effects of ipsilateral but not controlateral microinjections of the dopamine D2 agonist LY 171555 into the nucleus accumbens of amygdala-kindled rats, *Brain Res,* 553 (1991) 181-187.

Warter, J., Vergnes, M., Depaulis, A., Tranchant, C., Rumbach. L., Micheletti, G. and Marescaux, C., Effect of drugs affecting dopaminergic neurotransmission in rats with spontaneous petit mal-like seizures, *Neuropharrnacology.* 27 (1988) 269-274.

Wilson, C., Nomikos, G., Collu, M. and Fibiger, H., Doparninergic correlates of motivated behavior: importance of drive, *J Neurosci,* 15 (1995) 5169-5178.

Section III

INFORMATION CODING IN THE BASAL GANGLIA

SURROUND INHIBITION IN THE BASAL GANGLIA

A brief review

Michel Filion*

1. INTRODUCTION

Neuroscience textbooks introduce the notion of surround inhibition in particular relation with the work of Kuffler (1953). Neurons with a receptive field comprising a central excitatory zone within a surrounding inhibitory background, react to contrast much more than to absolute intensity. A theoretical example (Fig 1) illustrates this property.

Could neighboring afferent signals in the basal ganglia be selected similarly by their contrasting properties? However, let us first review experimental results concerning surround inhibition in the basal ganglia.

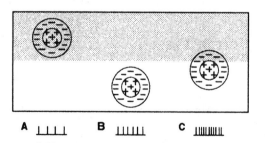

Figure 1. Stimulation excites a visual neuron exquisitely when applied in the center of its receptive field. Oppositely, it inhibits the neuron when applied in periphery of the field. Consequently, whether the receptive field is fully in the dark (A) or fully in the light (B) does not change firing rate very much. However, firing reaches its maximum when the receptive field is at the edge: mostly in the light, and partly in the dark (C).

---

* M. Filion, RC-9800 CHUL, Department of Neuroscience, 2705 Blvd. Laurier, Ste.-Foy QC G1V 4G2, Canada.

The Basal Ganglia VI
Edited by Graybiel *et al.*, Kluwer Academic/Plenum Publishers, 2002

2. SURROUND INHIBITION IN THE NORMAL

We studied the responses of globus pallidus neurons to electrical stimulation of the striatum in intact waking monkeys (Tremblay and Filion 1989). The pallidal neurons of both the internal (GPi) and external segments (GPe) responded characteristically with a decrease in firing (inhibition), followed by an increase (excitation) (Fig 2A2). A number of neurons, however, responded with excitation only. Typically, they were surrounding the firsts (Figs 2A1 and 2A3). In this center-surround organization of responses, inhibition occurs only in the center, whereas excitation occurs in both the center and the periphery. Therefore, the excitatory response appears to have two roles: the first being to curtail or control the duration of inhibition, in the temporal domain, and the second being to restrict or control the extent of inhibition, in the spatial domain.

In a double article (Chevalier et al 1985, and Deniau and Chevalier 1985), which is often cited, even recently, because it demonstrates clearly the importance of disinhibition in the basal ganglia, the authors also present results suggesting the existence of lateral inhibition in the basal ganglia. Thus, neurons of the pars reticulata of the substantia nigra in ketamine anesthetized normal rats, were inhibited following microinjection of glutamate at given sites in the striatum, but the same neurons were excited when the glutamate was injected at sites peripheral to the firsts (Fig 2B).

Lateral inhibition occurring either within the striatum or the globus pallidus or at both sites may explain the center-surround organization demonstrated in the latter two studies. The following studies provide more information on those possibilities.

Rebec and Curtis (1988) injected glutamate in the striatum of rats under local anesthesia. The striatal neurons recorded near the injection site were excited, whereas those recorded far from the injection site were inhibited. Interestingly, the reverse was true following local injection of serotonin instead of glutamate: inhibition near the injection site and excitation at a distance. The center-surround organization of responses disappeared following the intravenous injection of haloperidol, an antagonist of dopaminergic transmission. Thus, the results suggest that lateral inhibition occurs within the striatum and is dependent on dopamine.

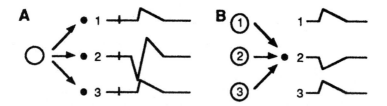

Figure 2. (A) Three neurons, located one above the other in the globus pallidus of the monkey, respond to electrical stimulation of a single site in the striatum. They show a center-surround organization of their responses. The central response (2) is an initial decrease in firing followed by an increase. The peripheral response (1 and 3) is a pure increase (from Fig 6 in Tremblay and Filion 1989, and Fig 2 in Filion et al 1994). (B) Similarly, a single neuron in the pars reticulata of the substantia nigra of the rat responds differently to microinjection of glutamate at three sites in the striatum. Injection at a central site (2) inhibits the nigral neuron. Whereas injections at surrounding sites (1 and 3) excite the same neuron (from Figs 4 and 5 in Deniau and Chevalier 1985).

Recently, Parthasarathy and Graybiel (1997) applied electrical stimulation for 1-2 hours to the sensorimotor cortex of the Nembutal anesthetized monkey. The manipulation resulted in the expression of Fos immediate early gene proteins by neurons in the putamen. A proportion of such Fos neurons contained enkephalin, which identifies them as striatal output neurons (in the indirect pathway). The other Fos neurons rather contained parvalbumin, which identifies them as striatal inhibitory interneurons using gamma-aminobutyric acid (GABA) as neurotransmitter. Interestingly, the latter neurons were sparsely distributed in both the center and the periphery of the clusters of Fos neurons. The latter results demonstrate again a center-surround organization within the striatum, with the output neurons in the center and the inhibitory interneurons in both the center and the periphery.

The last study does not test the relationship of surround inhibition with dopamine. However, Soghomonian *et al* (1992) had already shown that the level of messenger ribonucleic acid encoding glutamic acid decarboxylase (required for the synthesis of GABA) is much higher in the few parvalbumin interneurons than in the numerous output neurons of the striatum of the intact rat. However, after lesion of nigral dopamine neurons, the reverse becomes true. The output neurons appear hyperactive, whereas the GABAergic interneurons appear hypoactive. These results suggest that, when the dopamine input to the striatum is decreased, as in parkinsonism, lateral inhibition exerted by striatal interneurons is decreased, whereas that exerted by striatal output neurons, if it exists (see below), is increased.

3. SURROUND INHIBITION IN PARKINSONISM

The temporo-spatial center-surround organization of pallidal responses to electrical stimulation of the striatum, that we had observed in intact monkeys (Fig 2A), was lost when we rendered the animals parkinsonian with MPTP (1-methyl-4-phenyl-1,2,3,6-tetrahydropyridine). Moreover in the parkinsonians, more neurons responded to stimulation, and the responses were larger, longer lasting, and less selective with respect to stimulation site than in the intact monkeys (Tremblay *et al* 1989).

Such a loss of selectivity, with respect to stimulation site, was also true for pallidal responses to natural somatosensory stimulation (Filion *et al* 1988). Thus, in intact monkeys, only about 20% of GPi neurons responded to passive movement of a single contralateral joint, in only one direction, whereas in parkinsonians, 60% of GPi neurons responded to passive movement of more than one joint, of both the upper and lower limbs, bilaterally, in more than one direction, and often with excessive magnitude. Clearly the loss of center-surround organization was accompanied by a loss of selectivity of neuronal responses at the output of the basal ganglia. However, when the parkinsonians were treated with the dopamine agonist apomorphine, the center-surround organization and the selectivity of responses reappeared (Fig 3).

Treating parkinsonians with levodopa often induces dyskinesia. We reproduced such dyskinesia by injecting the GABA antagonist bicuculline in the GPe of intact monkeys (Matsumura *et al* 1995). Shortly after the injection, and just before the beginning of dyskinesia, the GPe neurons close to the injection site increased their activity, probably disinhibited by the GABA antagonist. Inversely, neurons located far from the same injection site decreased their activity, as if recurrent inhibition within the GPe had produced a center-surround organization of responses to the bicuculline injection. Simultaneously in the GPi, there was

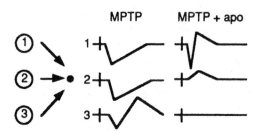

Figure 3. The 6 peristimulus histograms show the responses of the same GPi neuron to electrical stimulation of the striatum in a parkinsonian monkey, before and during the effects of apomorphine. Before apomorphine (MPTP), the neuron responds to three sites. All responses comprize large and long lasting inhibition. Following the injection of apomorphine (MPTP + apo), when the signs of parkinsonism have disappeared and the animal is even hyperactive and dyskinetic: the GPi neuron now responds to stimulation of the first site with a short inhibition followed by excitation -a central type of response (Fig 2A2)-; it responds to stimulation of the second site with a pure excitation -a peripheral type of response (Figs 2A1 and 2A3)-; and finally it does not respond any more to stimulation of the third site (from Fig 2 in Filion *et al* 1989, and Fig 1 in Filion *et al* 1994).

also a center-surround organization of activity. It was, however, the mirror-image of that observed in the GPe, with the inhibition in the center and the excitation at the periphery. Interestingly, this experimental, abnormally large center-surround organization of activity preceded and accompanied the execution of abnormal, dyskinetic movements.

4. MECHANISMS AND SITES

Let us now examine a number of mechanisms and sites likely to explain and localize surround inhibition or excitation in the basal ganglia.

Electrical and pharmacological stimulation of the striatum may have excited striatal output neurons, either directly or indirectly through afferent fibers. These output neurons are GABAergic and have recurrent collaterals, likely to produce lateral inhibition (Fig 4B). However, Jaeger et al (1994) tried hard, but did not succeed in obtaining electrophysiological results demonstrating recurrent inhibition among striatal projection neurons. The negative result suggests that the postsynaptic receptors of the recurrent axonal branches are sensitive only to neurotransmitters co-localized with GABA in the striatal output neurons, and not sensitive to GABA itself. Thus, the possibility that lateral inhibition exerted by striatal output neurons is responsible for the center-surround organization of activity, that we and others have observed in the basal ganglia, is not supported.

Stimulation of the striatum, either electrical or pharmacological, may also have excited striatal interneurons. Some of them are GABAergic and may have produced lateral inhibition (Fig 4A). They may thus have contributed to the center-surround organization of activity, even though they account for only 1 to 2% of the total population of striatal neurons.

Finally, striatal stimulation may have altered the high frequency activity of GPe neurons. Some of those pallidal GABAergic inhibitory neurons appear to have recurrent collaterals ending within the GPe itself. Indeed, Shink and Smith (1995) have shown that such local collaterals, that have been demonstrated in the rat (Kita and Kitai 1994), very probably exist

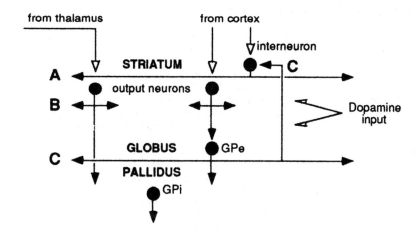

Figure 4. Diagram of the basal ganglia showing thalamic and cortical excitatory inputs (empty arrowheads) to striatal output neurons and interneurons. The latter two types of neurons are inhibitory (filled arrowheads) and have recurrent axonal collaterals that could explain surround inhibition at two levels (A and B). The striatal output neurons project to the basal ganglia output neurons in the GPi, either directly or indirectly through the GPe. GPe neurons also are inhibitory and have recurrent collaterals that could induce surround inhibition (C), not only within the GPe, but also upstream in the striatum, where the recurrent collaterals inhibit the interneurons that are very likely inducing lateral inhibition within the striatum. Finally, the proper functioning of surround inhibition appears to require dopamine.

also in the monkey. GPe neurons may therefore contribute to the center-surround organization of activity of pallidal neurons. Moreover, recent studies by Bevan et al (1998) show that GPe neurons also send recurrent axonal branches into the striatum (Fig 4), where they innervate selectively parvalbumin containing interneurons. The latter neurons are GABAergic and probably exert lateral inhibition within the striatum (see above). Therefore, GPe neurons would not only produce lateral inhibition within the pallidum, but would also decrease lateral inhibition recurrently, upstream, in the striatum.

5. CONCLUSIONS

In conclusion, we know enough about lateral inhibition in the basal ganglia to conclude (1) that it exists, in various forms and locations; (2) that it appears to be a crucial mechanism to provide selectivity, a fundamental function of the basal ganglia; (3) that it is dependent on the presence of dopamine, and therefore likely to explain abnormalities in parkinsonism and levodopa-induced dyskinesia; and finally (4) that it should be studied more extensively.

6. REFERENCES

Bevan MD, Booth PAC, Eaton SA, Bolam JP. Selective innervation of neostriatal interneurons by a subclass of neurons in the globus pallidus of the rat. J Neurosci 1998;18:9438-9452

Deniau JM, Chevalier G. Disinhibition as a basic process in the expression of striatal functions. II. The striato-nigral influence on thalamocortical cells of the ventromedial thalamic nucleus. Brain Res 1985;334:227-233

Chevalier G, Vacher S, Deniau JM, Desban M. Disinhibition as a basic process in the expression of striatal functions. I. The striato-nigral influence on tecto-spinal/tecto-diencephalic neurons. Brain Res 1985;334:215-226

Filion M, Tremblay L, Bedard PJ. Abnormal influences of passive limb movement on the activity of globus pallidus neurons in parkinsonian monkeys. Brain Res 1988;444:165-176

Filion M, Tremblay L, Bedard PJ. "Excessive and unselective responses of medial pallidal neurons to both passive movement and striatal stimulation in monkeys with MPTPinduced parkinsonism." In *Neural Mechanisms in Disorders of Movement* Allan R. Crossman and M.A. Sambrook, eds. Libbey, London, 1989, pp. 157-164

Filion M, Tremblay L, Matsumura M, Richard H. Focalisation dynamique de la convergence informationelle dans les noyaux gris centraux. Rev Neurol (Paris) 1994;150:627-633

Jaeger D, Kita H, Wilson C. Surround inhibition among projection neurons is weak or nonexistent in the rat neostriatum. J Neurophysiol 1994;72:2555-2558

Kita H, Kitai ST. The morphology of globus pallidus projection neurons in the rat: An intracellular staining study. Brain Res 1994;636:308-319

Kuffler SW. Discharge patterns and functional organization of mammalian retina. J Neurophysiol 1953;16:37-68

Parthasarathy HB, Graybiel A. Cortically driven immediate-early gene expression reflects modular influence of sensorimotor cortex on identified striatal neurons in the squirrel monkey. J Neurosci 1997;17:2477-2491

Rebec GV, Curtis SD. Reciprocal zones of excitation and inhibition in the neostriatum. Synapse 1988;2:633-635

Shink E, Smith Y. Differential synaptic innervation of neurons in the internal and external segments of the globus pallidus by the GABA- and glutamate-containing terminals in the squirrel monkey. J Comp Neurol 1995;358:119-141

Soghomonian JJ, Gonzales C, Chesselet MF. Messenger RNAs encoding glutamate-decarboxylases are differentially affected by nigrostriatal lesions in subpopulations of striatal neurons. Brain Res 1992;576:68-79

Tremblay L, Filion M. Responses of pallidal neurons to striatal stimulation in intact waking monkeys. Brain Res 1989;498:1-16

Tremblay L, Filion M, Bedard PJ. Responses of pallidal neurons to striatal stimulation in monkeys with MPTP-induced parkinsonism. Brain Res 1989;498:17-33

SURROUND INHIBITION IN THE BASAL GANGLIA

Jeffery R. Wickens*

1. INTRODUCTION

This chapter is a review of existing evidence and current thinking in relation to the concept of surround inhibition in the basal ganglia. In the following sections, the possibility of inhibitory interactions among spiny projection neurons will be reconsidered in the light of recent evidence. Statistical considerations based on quantitative neuroanatomical findings suggest a low probability of synaptic connection between neighbouring neurons. Electrophysiological studies have failed to demonstrate inhibitory interactions between spiny projection neurons. Together these findings suggest that inhibitory interactions among spiny projection neurons are relatively sparse. This suggests that the collaterals of spiny projection neurons are not the major source of surround inhibition. Instead, feedforward inhibition by a network of GABA/parvalbumin containing interneurons seems to explain surround inhibition at a macroscopic level. New models are needed to address the functional significance of powerful feedforward inhibition by GABA/parvalbumin containing interneurons, and the function of sparse inhibitory interactions between spiny projection neurons.

2. ANATOMICAL BASIS FOR SURROUND INHIBITION

The striatum is the major input structure in the basal ganglia. The vast majority of the neurons in the striatum are spiny projection neurons.[1] They receive direct inputs from the cerebral cortex[2], thalamus[3] and substantia nigra and project to output structures such as the globus pallidus, the substantia nigra and the entopeduncular nucleus.[2,3] The cellular properties of the spiny projection neurons, their extrinsic inputs, and their interactions with other striatal neurons are key determinants of the information processing operations performed by the striatum.

---

* J.R. Wickens, Department of Anatomy and Structural Biology, School of Medical Sciences, University of Otago, Dunedin, New Zealand.

The Basal Ganglia VI
Edited by Graybiel *et al.*, Kluwer Academic/Plenum Publishers, 2002

Spiny projection neurons have a relatively uniform somatodendritic architecture.[4] There are usually several primary dendrites radiating from the soma, which divide three or four times to form a dendritic tree that occupies a spherical space of radius approximately 200 μm. Each neuron gives rise to a main axon that projects to target structures. From the main axon arise multiple collateral branches. These divide repeatedly to form a network of local axon collaterals that overlaps extensively with the dendritic tree of the parent neuron.[4] The synaptic boutons of the local axon collaterals form synapses on dendritic spines and shafts of spiny projection neurons.[2] From these descriptions arose the idea of the striatum as a network of spiny projection neurons interconnected by their local axon collaterals.

The synapses formed by the local axon collaterals of the spiny projection neurons have the appearance of inhibitory synapses, with symmetrical synaptic densities and large pleomorphic vesicles.[4] The striatopallidal and striatonigral terminations of the main axon have been known for some time to use the inhibitory neurotransmitter, gamma-amino butyric acid (GABA)[5,6] and produce inhibitory effects in the target nuclei. It has, therefore, seemed probable that the local axon collaterals of spiny projection neurons should also be inhibitory.

Immunohistochemical techniques have been used to investigate the neurotransmitter used in the synapses of the local axon collaterals. Some variability has been reported in the proportion of striatal neurons that stain positively for glutamate decarboxylase (GAD), the synthesising enzyme for GABA.[7-9] When conditions are optimised for detection of GAD, however, the great majority (80%) of neurons with the morphological characteristics of spiny projection neurons stain positively for GAD.[10] It has also been shown that GAD-positive boutons form synapses with the cell body and dendritic shafts of neurones identified as projection neurones by retrograde labelling from the substantia nigra.[11] Finally, immunohistochemical staining for GABA has identified numerous synapses between GABA-positive boutons and similarly staining dendrites.[12] Thus the input to spiny projection neurons from other spiny projection neurons is GABAergic.

It is important to note that the spiny projection neurons are not the only GABAergic neurons in the striatum. In the process of determining the GABAergic nature of local circuits in the striatum, a small population of interneurons expressing intense immunoreactivity for GABA and GAD was identified.[7,9,10,12,13] These cells are also parvalbumin immunoreactive and are referred to as GABA/parvalbumin containing interneurons.

3. THE PROBABILITY OF SYNAPTIC CONTACT BETWEEN SPINY PROJECTION NEURONS

The probability of one, two or more synapses between the local axonal collaterals of one spiny neuron and the dendrites of another spiny neuron located a certain distance from the first can be estimated using statistical arguments. Such arguments have been applied to the question of connectivity between pyramidal cells in the cerebral cortex[14] and in that case have led to conclusions that are consistent with experimentally determined values.[15] These arguments assume that each receiving neuron has a number of potential sites for symmetrical synapses. These synaptic sites are intermingled with postsynaptic sites belonging to many other spiny neurons. Thus, the potential postsynaptic targets can be imagined to be a subset of a cloud-like distribution of alternative postsynaptic targets.

The proportion of the total number of postsynaptic sites that belong to a given neuron can be estimated from quantitative neuroanatomical studies of the rat striatum. These provide realistic values for the number of synaptic sites of a given neuron, and the total number of synaptic sites of all neurons in the space. Key numbers and parameters from them are summarised in Table 1. Ingham et al.,[16] give the density of asymmetrical synapses in the rat striatum as 0.91 gm^{-3} and estimate that symmetrical synapses account for about 20% of the total number of synapses. The other sources of symmetrical synapses include GABA/parvalbumin, somatostatin and cholinergic interneurons, and dopaminergic afferents which are also predominantly of the symmetrical type. Although the number of neurons giving rise to these synapses is relatively small, the extensive collateral branching of these neurons suggests they may account for the majority of symmetrical synapses in the striatum. The proportion of symmetrical synapses that come from spiny projection neurons can be estimated to be about 1 in 6.[17] Using these numbers the number of synapses of spiny projection neurons per unit volume is on the order of 0.038 mm^{-3}. The proportion of the synapses that belong to each individual neuron, on average, can be calculated from the density of medium-sized somata. This number is given by Oorschot[1] as 84,900 mm^{-3}. From these values the average number of postsynaptic sites on one spiny projection neuron is estimated as 448.

Table 1: Assumed parameter values for spiny neurons

Parameter	Value
Density of neurons[1]	84,900 mm^{-3}
Density of synapses of spiny neurons[16,17]	0.038 µm^{-3}
Number of local synapses made by each spiny neuron (estimated from the above)	448
Axonal and dendritic arborisation (diameter)	400 µm

The probability of synaptic contact as a function of distance between somata is calculated from the volume of the solid formed by the intersection of two spheres representing the dendritic and axonal arborisations of the respective neurons. The number of postsynaptic sites in the volume that belong to the neuron in question (j), and the total number of synaptic sites in the volume (n) are calculated, and from this the fraction of the postsynaptic sites belonging to the receiving neuron is calculated. This gives the probability (p) that a postsynaptic site chosen at random will belong to the postsynaptic neuron (p=j/n). The number of contacts made in the same volume by the presynaptic neuron (k) is similarly determined. Finally, the probability of the presynaptic neuron making one, two or more synapses with the postsynaptic neuron is calculated from the cumulative hypergeometric distribution with parameters j, k and n, or more simply from the cumulative binomial distribution with parameters p and k.[18]

Table 2 shows the estimated probability of at least one or more contact, and the probability of two or more contacts from one neuron to another as a function of the distance between their somata. These estimates have a number of implications for the expected connectivity in the striatum. The case in which the distance between somata is zero corresponds to the probability of a recurrent connection mediated by a synapse between the axons and dendrites of the same neuron. It can be seen that the probability of one or more such recurrent synapses is P=0.146. This means that, on average, approximately one in seven spiny projection neurons will make a synapse with itself. This leads to an experimental prediction: if a search was made of all the boutons (approximately 448 by Table 1) of a neuron that had been individually labelled by intracellular injection it would be expected that about in about one in seven neurons, or one in 3,068 boutons, a labelled bouton would be seen to terminate on the intracellularly labelled neuron of origin. Such an exhaustive search has not been reported but consistent with the prediction there does not appear to be a single instance of a recurrent autapse in the existing literature. Thus the estimate of a low probability of autaptic connection seems plausible at present.

Table 2: Probability of synaptic connections between spiny neurons

Distance between somata (μm)	Probability of one or more contacts	Probability of two or more contacts
0	0.146	0.011
50	0.119	0.007
100	0.094	0.005
150	0.070	0.002
200	0.048	0.001
250	0.028	0.000
300	0.013	0.000
350	0.003	0.000
400	0.000	0.000

Table 2 also predicts the probability of synaptic contact for neurons separated by different distances. As the distance between somata increases, the volume of the region of overlap of their dendritic and axonal arborisations decreases, and the probability of synaptic contact decreases rapidly. For two neurons with somata 250 μm apart, the probability of a synaptic contact from one neuron to the other is P=0.028. In other words, the chance that a neuron situated at this distance will receive one or more synaptic contacts is approximately one in 36.

A final observation of some significance is that the probability of two or more synaptic contacts between a pair of spiny projection neurons is exceedingly low. In the most favourable case of immediately adjacent neurons the probability of two or more contacts is just P=0.011. This low probability of two or more synaptic connections suggests that each postsynaptic target is likely to be contacted only once and therefore each spiny neuron probably makes one synaptic contact with, on average, 448 others.

The estimated connection probabilities given above are of interest in two ways. Firstly, they suggest that the network of spiny projection neurons is sparsely interconnected in the sense that the probability of synaptic contact is much less than one, decreasing with distance between somata. Thus, no synaptic contacts will exist between most pairs of neurons. Secondly, each postsynaptic neuron is most likely to be contacted only once. A corollary of this result is that functional interactions between connected neurons are likely to be mediated by a single synapse, and therefore are likely to be weak. These conclusions, however, are based on an argument which makes a number of simplifying assumptions. The validity of these assumptions will be considered below.

A core assumption of the statistical model is that the distribution of the synapses of a spiny neuron can be modelled as a uniform cloud of synaptic sites. In reality, synaptic sites are distributed along dendrites and axons. However, it turns out for these calculations that unless there is a special kind of registration of dendrites and axons it makes little difference whether synaptic sites are scattered as in a cloud, or arranged in rows along randomly oriented processes (provided these are oriented in random directions). For example, the estimates for connectivity in the cortex obtained using clouds of straight axonal and dendritic segments are very close to those obtained using spherical clouds.[19]

An assumption that might have more serious effects on the above estimates is that the sites at which the axon collaterals of spiny projection neurons terminate are uniformly distributed within the dendritic field, irrespective of distance from the soma. The synapses made by collateral branches of the axons of spiny projection neurons are widely distributed over the surface of other spiny projection neurons and not limited to the aspiny region of the proximal dendrites. The relative frequencies of the postsynaptic contacts of spiny projection neurons include: dendritic spines, 40%; dendritic shafts, 48%; and somata or initial axonal segments, 12%.[4] Consistent with this, in a different study 8% of spines along two distal portions of a dendrite from a single spiny neuron received symmetrical inputs of the type thought to be from local axon collaterals.[20] Thus, significant numbers of synapses made by collateral branches of spiny projection neurons do terminate on spiny dendrites some distance from the soma, suggesting that the uniform spherical cloud of synaptic sites is a good approximation.

The spherical dendritic and axonal arborisation assumed in the estimates is a simplification. Dendritic domains may be directionally polarised.[21,22] Spiny projection neurons can also be divided into two types on the basis of their intrastriatal axonal arborisations.[23] The more common type has local axonal arborisations within the dendritic field of the cell of origin, as assumed in the previous section. A second less common cell type has axon collaterals distributed widely in the neostriatum. Some of the implications of the more extended collateral distribution of some spiny neurons have been discussed elsewhere.[24]

Existing knowledge of the relation between subtypes of spiny projection neurons and their propensity to make synapses with one another is very limited and indirect. Immunocytochemistry combined with retrograde axonal tracing has revealed inhomogeneities among the spiny projection neurons with respect to their projections to different target areas; and their expression of different types of chemical markers such as enkephalin, substance P, and dynorphin. These markers are related to differences in the axonal projections of these cells. They might also be related to differences in the local axonal arborisations and the probability of synaptic connections between spiny projection neurons of different classes. Immunoreactivity for enkephalin and dopamine D2 receptors has been used to selectively label the neurons which give rise to the striatopallidal or "indirect" pathway. Immunoreactivity for substance P and dopamine D 1 receptors has been used to label neurons that give rise to the striatonigral/entopeduncular pathway or "direct" pathway. These labels have been combined in various ways with pathway-labelling tracers to prove the existence of synaptic contacts from direct to other direct pathway neurons[25]; from indirect to direct pathway neurons[26]; and from direct to indirect pathway neurons.[26,27] Although there has been no quantitative study of the relative number of synaptic contacts between these different classes on neurons, the existing evidence suggests no strong bias in connectivity along the lines of direct and indirectly projecting neurons.

4. ELECTROPHYSIOLOGICAL TESTS FOR SYNAPTIC INTERACTIONS

The inhibitory interactions which were initially expected on the basis of the anatomy of spiny projection neurons have proven to be very elusive in direct electrophysiological tests. Electrophysiological indications of inhibitory interactions have often turned out to have some other explanation.

Initial intracellular studies of the responses of striatal neurons to cortical stimulation revealed a characteristic response pattern consisting of a depolarising postsynaptic potential followed by a long-lasting period of hyperpolarisation. The long-lasting period of hyperpolarisation was initially interpreted as feedback inhibition from the collaterals of spiny projection neurons.[28] Subsequent investigations showed, however, that the hyperpolarising phase was not an inhibitory postsynaptic potential (IPSP) as originally suspected, but was rather due to a brief cessation of excitatory afferent activity[29] probably brought about by inhibitory interactions in the cerebral cortex.

An IPSP is seen in striatal slices following cortical stimulation[30,31] but this is not due to feedback inhibition from spiny projection neurons. This IPSP is due to feedforward inhibition from the GABA/parvalbumin containing interneurons.[10] These interneurons exert a powerful inhibitory control over projection neurons.[32] They also fire at a lower threshold than the spiny projection neurons which makes it difficult to isolate the effects of local axon collaterals of the spiny projection neurons from inhibition by the GABA/parvalbumin containing interneurons.

One way to activate a spiny projection neuron without also activating inhibitory interneurons is to cause it to fire by depolarising intracellular current injection. Park et al.[33] used this method to test for inhibition of a striatal neuron by its own collaterals. Action potentials evoked in the recorded neuron by a depolarising current pulse failed to reduce the amplitude of excitatory postsynaptic potentials (EPSPs) evoked by cortical stimulation. However, the EPSPs evoked by stimulation of the substantia nigra were attenuated if they were evoked shortly after action potential firing in the cell from which the recordings were being made. This effect was blocked by GABA antagonists. These results suggest that spiny projection neurones may be inhibited by their own collaterals, but that the inhibition is selective for a particular class of inputs activated by substantia nigra stimulation.

The EPSPs evoked by substantia nigra stimulation in the experiments just described were probably produced by collaterals of corticofugal axons passing in the nearby cerebral peduncle.[34] They thus represent a particular set of corticostriatal afferents from brainstem projecting cortical neurons. The latter innervate the striatum via collateral branches arising in the internal capsule.[35] It is possible that these represent a population of excitatory inputs that make synapses at the minority of dendritic spines which receive inputs from local axon collaterals.[4,20]

In one of the few direct studies to date, Jaeger et al.,[36] made intracellular records from 35 pairs of spiny projection neurones in slices. They found no evidence for inhibition of one spiny neuron by another: no IPSPs, no suppression of firing evoked by current injection, and no reduction of cortically-evoked EPSPs. In other intracellular experiments they found no evidence for IPSPs after antidromic activation of the substantia nigra or globus pallidus. If the local axon collaterals mediate inhibitory effects, then this test ought to have been positive. On the basis of the foregoing evidence they concluded that lateral inhibition was weak or nonexistent in the rat striatum. The failure of the antidromic inhibition test is a very strong piece of evidence against inhibitory interactions among spiny projection neurons. The effects on responses to EPSPs evoked by stimulation of corticofugal axons were not tested, however, so the possibility of the selective effect described by Park et al.,[33] was not ruled out.

The regulation of GABA release by presynaptic receptors may be an important factor determining whether IPSPs can be detected in particular experiments. Local GABAergic synapses of the spiny projection neurons may be suppressed under some conditions. For example, in striatal cultures[37] inhibitory responses are decreased by the GABA-B receptor agonist Baclofen. Similarly, one type of unitary IPSP evoked by minimal stimulation of striatal slices is regulated by GABA-B receptors.[38] The IPSP regulated by GABA-B receptors might originate from spiny projection neurons, in view of the finding that the GABA-B receptors modulate IPSPs of striatonigral projection neurons.[39] There is also some indication that IPSPs may be evident after antidromic stimulation in the presence of locally applied drugs which enhance neurotransmitter release.[10] Further work is needed to explore possible state-dependent inhibitory interactions.

5. SURROUND INHIBITION AT THE MACROSCOPIC LEVEL

Despite the lack of inhibitory interactions between pairs of individual spiny projection neurons at the cellular level, functional surround-inhibition is evident at the macroscopic level. Activation of striatal neurons by localised application of an excitatory neurotransmitter produces spatially alternating zones of excitation and inhibition in the striatum.[40] In the substantia nigra, stimulation in a single striatal area can produce simultaneous excitatory and inhibitory influences on different nigrothalamic neurons.[41] Stimulation in the striatum similarly produces a focus of inhibition in a restricted area of the globus pallidus, with a contrasting surround of excitation at the fringes.[42] These findings are compatible with functional surround-inhibition type dynamics.

The presence of surround inhibition at the macroscopic level seems paradoxical if inhibition via local axon collaterals of spiny projection neurons is weak or non-existent. However, several pieces of evidence suggest that surround inhibition is due to the small population of GABA/parvalbumin containing interneurons rather than the more numerous spiny projection neurons. Cortical afferents make asymmetric synapses on these interneurons[43,44] and they in turn make symmetrical synapses on spiny projection neurons.[45] Although these neurons account for only a small minority of the total neurons in the striatum, their axonal arborisation is very dense in the area surrounding the cell body, and the GABA/

parvalbumin containing interneurons appear to be an important source of GABAergic synaptic input to spiny projection neurons.[32] They are thus in a position to exert feed-forward inhibitory effects. This possibility is supported by the pattern of immediate early gene expression in response to focal stimulation of the cortex. Activation of spiny projection neurons occurs in focal zones surrounded by larger zones in which more widespread GABA/parvalbumin containing interneurons are activated.[46] This suggests that a localised cortical input may produce a widespread activation of GABA/parvalbumin containing interneurons and a more limited activation of a smaller number of projection neurons at the focus of the excitatory input.

The metabolic evidence for surround inhibition by GABA/parvalbumin containing interneurons is supported by a recent electrophysiological study. In dual whole-cell recordings from GABAergic interneurons and projection neurons the GABAergic interneurons were shown to exert a powerful inhibitory effect on the spiny projection neurons.[32] In contrast to the lack of inhibitory interactions seen between spiny projection neurons, IPSPs were detected in approximately 25% of spiny projection neurons recorded within 250 µm of an interneuron. Thus, feedforward inhibitory mechanisms, in which cortical afferents excite GABA/parvalbumin containing interneurons which in turn inhibit the spiny projection neurons, has emerged as a powerful source of surround inhibition in the striatum.

6. SURROUND INHIBITION: FEED-BACK VERSUS FEED-FORWARD

The evidence presented above suggests that inhibitory interactions between spiny projection neurons, if they exist, are likely to be sparse (in the sense that the probability of a connection is much less than one) and weak (typically involving only one synaptic contact). These weak interactions are overshadowed by inhibition produced by GABA/parvalbumin containing interneurons which, although relatively few, appear to be very excitable and produce strong inhibitory effects. These findings are of significance for computational models of striatal function.

Strong inhibitory interactions between spiny projection neurons have been assumed in a number of models of striatal function. In these models the role of inhibition has been to create a neural dynamic of competition in which the more active neurons suppress the less strongly excited ones. The most extreme form of this model is the inhibitory domain[47], in which every neuron inhibits every other, and a so-called "winner takes all" dynamic operates. This kind of dynamic was useful in these models to implement a number of functions of the striatum, including: reciprocal inhibition of muscular antagonists[47], suppression of competing motor programs[48] and competitive learning.[49] If the estimates of connectivity made above are correct, however, then the probability of finding sets of mutually inhibitory neurons is very small.

Feedforward inhibition by the GABA/parvalbumin containing interneurons is an alternative mechanism for producing surround inhibition in the striatum. However, the existence of this form of surround inhibition does not mean that feedback inhibition by the spiny neuron collaterals has no part to play in models of striatal function. Models in which both feedforward and feedback inhibition play an important role have been proposed.[50] The neurodynamic effects of feedforward inhibition differ from the effects of feedback inhibition. Feedback inhibition leads to a prevailing dynamic of competition in which small differences in activity become amplified. Feedforward inhibition does not produce the same dynamic of competition.

7. CONCLUSION

The idea that mutual inhibition between spiny projection neurons is a central organising principle of striatal function needs to be reconsidered in the light of recent evidence. Statistical considerations based on quantitative neuroanatomy suggest a low probability of synaptic connection between spiny projection neurons. These estimates are compatible with the failure of electrophysiological studies to demonstrate inhibitory interactions between spiny projection neurons. Together, these findings argue against the concept of the striatum as a strongly connected network of spiny projection neurons with mutually inhibitory interactions. In its place, feedforward inhibition by a powerful network of GABA/parvalbumin containing interneurons appears to play a central role.

Future computational models will need to address the functional significance of powerful feedforward inhibition by GABA/parvalbumin containing interneurons. One idea which seems compatible with recent evidence is that these interneurons function as a kind of blanking circuit which resets the spiny projection neurons to an initial value at the onset of sequences of motor activity.[50] Further work is needed to determine if the functions previously attributed to inhibitory interactions among spiny projection neurons can be carried out by feedforward inhibitory connections.

Finally, the existence of the local axon collaterals and synaptic contacts among spiny projection neurons should not be forgotten. The computational significance of weak interactions in a sparsely connected network may turn out to play a subtle but important role.

8. REFERENCES

1. Oorschot, D.E. (1996) Total number of neurons in the neostriatal, pallidal, subthalamic, and substantia nigral nuclei of the rat basal ganglia: a stereological study using the Cavalieri and optical disector methods. *J. Comp. Neurol.*, 366: 580-99.
2. Somogyi, J.P., Bolam, J.P. and Smith, A.D. (1981) Monosynaptic cortical input and local axon collaterals of identified striatonigral neurons. A light and electron microscope study using the Golgi-peroxidase transport degeneration procedure. *J. Comp. Neurol.*, 195: 567-84.
3. Dube, L., Smith, A.D. and Bolam, J.P. (1988) Identification of synaptic terminals of thalamic or cortical origin in contact with distinct medium-size spiny neurons in the rat neostriatum. *J. Comp. NeuroL*, 267:455-71.
4. Wilson, C.J. and Groves, P.M. (1980) Fine structure and synaptic connection of the common spiny neuron of the rat neostriatum: A study employing intracellular injection of horseradish peroxidase. *J. Comp. Neurol.*, 194: 599-615.
5. Precht, W. and Yoshida, M. (1971) Blockage of caudate-evoked inhibition of neurons in the substantia nigra by picrotoxin. *Brain Res.*, 32: 229-33.
6. Yoshida, M. and Precht, W. (1971) Monosynaptic inhibition of neurons of substantia nigra by caudatonigral fibres. *Brain Res.*, 32: 225-8.
7. Bolam, J.P., Powell, J.F., Wu, J.-Y. and Smith, A.D. (1985) Glutamate decarboxylase-immunoreactive structures in the rat neostriatum: A correlated light and electron microscopic study including a combination of Golgi-impregnation with immunocytochemistry. *J. Comp. NeuroL*, 237: 1-20.

8. Kita, H. and Kitai, S.T. (1988) Glutamate decarboxylase immunoreactive neurons in cat neostriatum: Their morphological types and populations. *Brain Res.,* 447: 346-52.
9. Kubota, Y., Inagaki, S., Shimada, S., Kito, S. and Wu J, Y. (1987) Glutamate decarboxylase-like immunoreactive neurons in the rat caudate putamen. *Brain Res. Bull.,* 18: 687-97.
10. Kita, H. (1993) GABAergic circuits of the striatum. *Prog. Brain Res.,* 90: 51-72.
11. Aronin, N., Chase, K. and DiFiglia, M. (1986) Glutamic acid decarboxylase and enkephalin immunoreactive axon terminals in the rat neostriatum synapse with striatonigral neurons. *Brain Res.,* 365: 151-8.
12. Pasik, P., Pasik, T., Holstein, G. and Hamori, J. (1988) GABAergic elements in the neuronal circuits of the monkey neostriatum: A light and electron microscopic immunocytochemical study. *J. Comp. Neurol.,* 270: 157-70.
13. Cowan, R.L., Wilson, C.J., Emson, P.C. and Heizmann, C.W. (1990) Parvalbumin-containing GABAergic interneurons in the rat neostriatum. *Neuroscience,* 57: 661-71.
14. Braitenberg, V. and Shüz, A. (1991) *Anatomy of the cortex: Statistics and geometry.* Berlin: Springer.
15. Deuchars, J., West, D.C. and Thomson, A.M. (1994) Relationship between morphology and physiology of pyramid-pyramid single axon connections in rat neocortex in vitro. *J. Physiol.,* 478: 423-35.
16. Ingham, C.A., Hood, S.H., Taggart, P. and Arbuthnott, G.W. (1998) Plasticity of synapses in the rat neostriatum after unilateral lesion of the nigrostriatal dopaminergic pathway. *J. Neuroscience,* 18: 4732-43.
17. Wilson, C.J. (1999) Striatal circuitry: Categorically selective, or selectively categorical?, in *Brain dynamics and the striatal complex.,* R. Miller and J.R. Wickens, Eds. Harwood Academic. p. 289-305.
18. Wickens, J.R. and Miller, R. (1997) A formalisation of the neural assembly concept 1. Constraints on neural assembly size. *Biol. Cybern.* 77, 351-8.
19. Liley, D.T.J. and Wright, J.J. (1994) Intracortical connectivity of pyramidal and stellate cells: estimates of synaptic densities and coupling symmetry. *Network,* 5: 175-89.
20. Freund, T.F., Powell, J.F. and Smith, A.D. (1984) Tyrosine hydroxylase-immunoreactive boutons in synaptic contact with identified striatonigral neurons, with particular reference to dendritic spines. *Neuroscience,* 13: 1189-215.
21. Walker, R.H., Arbuthnott, G.W., Baughman, R.W. and Graybiel, A.M. (1993) Dendritic domains of medium spiny neurons in the primate striatum: Relationships to striosomal borders. *J. Comp. Neurol.,* 337:614-28.
22. Walker, R.H. and Graybiel, A.M. (1993) Dendritic arbors of spiny neurons in the primate striatum are directionally polarized. *J. Comp. Neurol.,* 337: 629-39.
23. Kawaguchi, Y., Wilson, C.J. and Emson, P.C. (1990) Projection subtypes of rat neostriatal matrix cells revealed by intracellular injection of biocytin. *J. Neuroscience,* 10: 3421-38.
24. Oorschot, D.E. (1999) The domain hypothesis: a central organising principle for understanding neostriatal circuitry? in *Brain dynamics and the striatal complex.,* R. Miller and J.R. Wickens, Eds. Harwood Academic. p. 65-76.
25. Bolam, P. and Izzo, P.N. (1988) The postsynaptic targets of substance P-immunoreactive terminals in the rat neostriatum with particular reference to identified spiny striatonigral neurons. *Exp. Brain Res.,* 70: 361-77.
26. Yung, K.K.L., Smith, A.D., Levey, A.I. and Bolam, J.P. (1996) Synaptic connections between spiny neurons of the direct and indirect pathways in the neostriatum of the rat: Evidence from dopamine receptor and neuropeptide immunostaining. *Eur. J. Neurosci.,* 8: 861-9.
27. DiFiglia, M., Aronin, N. and Martin, J.B. (1982) Light and electron microscopic localization of immunoreactive leu-enkephalin in the monkey basal ganglia. *J. Neuroscience,* 2: 303-20.
28. Bernardi, G., Marciani, M.G., Morocutti, C. and Giacomini, P. (1975) The action of GABA on rat caudate neurones recorded intracellularly. *Brain Res.,* 92: 511-5.
29. Wilson, C.J., Chang, H.T. and Kitai, S.T. (1983) Disfacilitation and long-lasting inhibition of neostriatal neurons in the rat. *Exp. Brain Res.,* 51: 227-35.
30. Kita, T., Kita, H. and Kitai, S.T. (1985) Local stimulation induced GABAergic response in rat striatal slice preparations: intracellular recording on QX-314 injected neurons. *Brain Res.,* 360: 304-10.
31. Lighthall, J.W., Park, M.R. and Kitai, S.T. (1981) Inhibition in slices of rat neostriatum. *Brain Res.,* 212: 182-7.
32. Koos, T. and Tepper, J.M. (1999) Inhibitory control of neostriatal projection neurons by GABAergic interneurons. *Nature Neuroscience,* 2: 467-72.

33. Park, M.R., Lighthall, J.W. and Kitai, S.T. (1980) Recurrent inhibition in the rat neostriatum. *Brain Res.*, 194:359-69.
34. Wilson, C.J., Chang, H.T. and Kitai, S.T. (1982) Origins of postsynaptic potentials evoked in identified rat neostriatal neurons by stimulation in substantia nigra. *Exp. Brain Res.*, 45: 157-67.
35. Wilson, C.J. (1986) Postsynaptic potentials evoked in spiny neostriatal projection neurons by stimulation of ipsilateral and contralateral neocortex. *Brain Res.*, 367: 201-13.
36. Jaeger, D., Kita, H. and Wilson, C.J. (1994) Surround inhibition among projection neurons is weak or nonexistent in the rat neostriatum. *J. Neurophys.*, 72: 2555-8.
37. Behrends, J.C. and Bruggencate, G. (1998) Changes in quantal size distributions upon experimental variations in the probability of release at striatal inhibitory synapses. *J. Neurophys.*, 79: 2999-3011.
38. Radnikow, G., Rohrbacher, J. and Misgeld, U. (1997) Heterogeneity in use-dependent depression of inhibitory postsynaptic potentials in the rat neostriatum in vitro. *J. Neurophys.*, 77: 427-434.
39. Tepper, J.M., Paladini, C.A. and Celada, P. (1998) GABAergic control of the firing pattern of substantia nigra dopaminergic neurons. *Adv. Pharmacol.*, 42: 694-9.
40. Rebec, G.V. and Curtis, S.D. (1988) Reciprocal zones of excitation and inhibition in the neostriatum. *Synapse*, 2: 633-35.
41. Deniau,J.M. and Chevalier, G. (1985) Disinhibition as a basic process in the expression of striatal functions. II The striatonigral influence on thalamocortical cells of the ventromedial thalamic nucleus. *Brain Res.*, 334: 227-33.
42. Tremblay, L., Filion, M. and Bedard, B.J. (1989) Responses of pallidal neurons to striatal stimulation in monkeys with MPTP-induced parkinsonism. *Brain Res.*, 498: 17-33 .
43. Kita, H., Kosaka, T. and Heizmann, C.W. (1990) Parvalbumin-immunoreactive neurons in the rat neostriatum: a light and electron microscopic study. *Brain Res.*, 536: 1-15.
44. Lapper, S.R., Smith, Y., Sadikot, A.F., Parent, A. and Bolam, J.P. (1992) Cortical input to parvalbumin-immunoreactive neurones in the putamen of the squirrel monkey. *Brain Res.*, 580: 215-24.
45. Bennett, B.D. and Bolam, J.P. (1994) Synaptic input and output of parvalbumin-immunoreactive neurons in the neostriatum of the rat. *Neuroscience*, 62: 707-19.
46. Parthasarathy, H.B. and Graybiel, A.M. (1997) Cortically driven immediate-early gene expression reflects modular influence of sensorimotor cortex on identified striatal neurons in the squirrel monkey. *J. Neuroscience*, 17: 2477-91.
47. Wickens, J.R., Alexander, M.E. and Miller, R. (1991) Two dynamic modes of striatal function under dopaminergic-cholinergic control: simulation and analysis of a model. *Synapse*, 8: 1-12.
48. Mink, J.W. (1996) The basal ganglia: focused selection and inhibition of competing motor programs. *Prog. Neurobiol.*, 50: 381-425.
49. Plenz, D. and Kitai, S.T. (1999) Adaptive classification of cortical input to the striatum by competitive learning, in *Brain dynamics and the striatal complex.*, R. Miller and J.R. Wickens, Eds. Harwood Academic. p. 169-77.
50. Wickens, J.R. and Arbuthnott, G.W. (1993) The corticostriatal system on computer simulation: An intermediate mechanism for sequencing of actions. *Prog. Brain Res.*, 99: 325-39.

INFORMATION PROCESSING IN THE CORTICO-STRIATO-NIGRAL CIRCUITS OF THE RAT BASAL GANGLIA

Anatomical and neurophysiological aspects

S. Charpier, P. Mailly, S. Mahon, A. Menetrey, M. J. Besson, and J. M. Deniau*

1. INTRODUCTION

Current concepts on basal ganglia functions emphasize the role of striatum in adaptive control of behavior. As proposed by Houk[1], experimental evidence suggest that through its interrelation with the cerebral cortex, thalamus and brain stem, the striatum participates in the contextual analysis of the environment and uses this information to select and execute adapted behavioral responses. New aspects of the anatomo-functional organization of the rodent cortico-striato-nigral circuits led to clarify some of the mechanisms by which striatum may achieve its functions. Among these advances are: i) the anatomical segregation of cortical information in the striato-nigral circuits leading to the formation of specific sensory-motor associations; ii) the demonstration of a physiological long-term potentiation (LTP) at cortico-striatal synapses that provides a cellular mechanism for adaptive formation and storage of sensory-motor linkages.

2. ANATOMICAL CHANNELING OF CORTICAL INFORMATION IN THE STRIATO-NIGRAL CIRCUITS

The striatum constitutes the main input structure of the basal ganglia and receives afferent from the entire cerebral cortex. The way cortical information is processed in the striatum and in the subsequent output structures of the basal ganglia has been highly

---

* S. Charpier, P. Mailly, S. Mahon, A. Menetrey, M. J. Besson, and J. M. Deniau, Départment de Neurochimie-Anatomie, Institut des Neurosciences, UMR 7624, Université Pierre et Marie Curie, Paris, France.

debated[2,3]. Models characterized by an extreme degree of convergence (information "funnel" hypothesis) or segregation (modular parallel architecture) have been proposed (see for review Bergman et al[4]). Our studies on the rat striato-nigral circuits are consistent with a hybrid model combining both convergence and segregation.

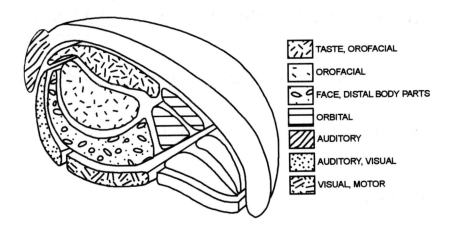

Figure 1. Lamellar representation of the striatal functional mosaic in the SNR

It is well established that the cerebral cortex is orderly mapped onto the striatum. Each cortical area projects to a defined striatal sector and within each striatal sector converge projections from functionally associated cortical areas (see for review Deniau and Thierry[5]). At the level of the substantia nigra pars reticulata (SNR), the topology of striato-nigral projections preserves a segregation between the projections of individual components of the striatal functional mosaic[6]. Using axonal tracing techniques, we have established that the various components of the striatal mosaic are mapped in the SNR under the form of longitudinal laminae organized in an onion-like manner (Fig 1).

This onion-like architecture is particularly salient in the lateral SNR where the striatal sectors related to sensory-motor cortical areas project under the form of curved laminae enveloping a dorso-lateral core. Since a similar lamellar principle of organization rules the topographical distribution of the nigral output neurons[7], we have proposed that the nigral lamination might provide an anatomical substrate to maintain in the basal ganglia output pathways the segregation of cortico-striatal inputs. However, in view of the considerable extension of nigral cell dendrites, it is clear that the SNR cannot operate as a simple relay station. To clarify the mode of processing of cortico-striatal information in the SNR, we have examined how the dendritic field of nigral neurons relates to the lamellar organization of striatal projections. Single nigral neurons were labeled using juxtacellular injection of neurobiotin[8] and three dimensional reconstruction of dendrites were performed. The results illustrated in Figs 2 to 6 demonstrate clearly that the dendritic field of SNR cells largely

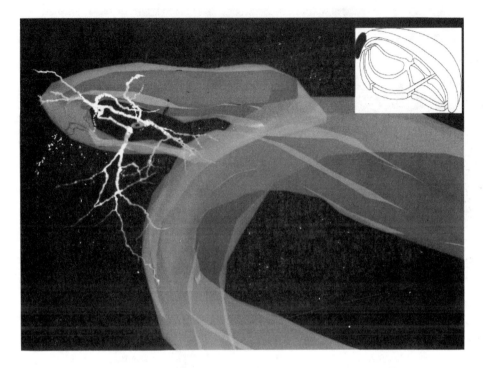

Figure 2. 3D reconstructed neuron in the auditive sector of the SNR (coronal view)

conforms to the lamination of the striato-nigral projections. As striato-nigral projections, dendrites of nigral neurons form curved laminae enveloping a dorso-lateral core.

The remarkable alignment between the projection fields of individual striatal sectors and the dendritic field of the corresponding nigral neurons suggests that the onion-like architecture of the SNR contributes to the formation of input-output registers (or processing channels) in the cortico-striato-nigral circuits.

At this point it is essential to stress that the terms input-output registers or processing channels used here to define the functional architecture of the cortico-striato-nigral circuits do not mean that these circuits are envisioned as parallel lines of transmission in which no integrative process occurs. As mentioned above, each striatal sector integrates information from multiple functionally related cortical areas and within each striatal sector a single striatal output neuron receives synaptic inputs from a considerable number of cortical cells. Similarly, in the SNR, the length of individual dendritic fields suggests that each cell integrates inputs from a large number of striatal neurons. It is particularly striking that only two or three nigral output neurons joined end to end are sufficient to cover (in its medio-lateral or rostro-caudal extent) the projection field of a given striatal sector. Therefore, the channeling mechanism of cortical information in the striato-nigral circuits has to be interpreted in term of specific integrative processes. The integrative plan of cortical information within the sensory-motor district of the nigral "onion" can be summarized as follows. The most external

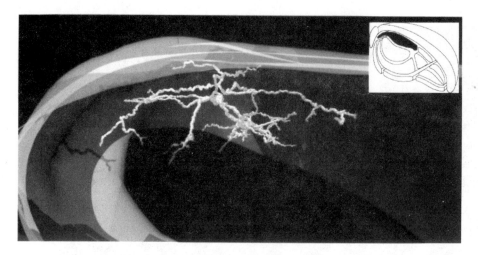

Figure 3. 3D reconstructed neuron in the gustative-orofacial sector of the SNR (coronal view)

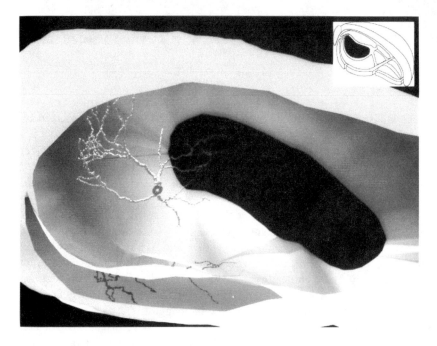

Figure 4. 3D reconstructed neuron in the orofacial sector of the SNR (coronal view)

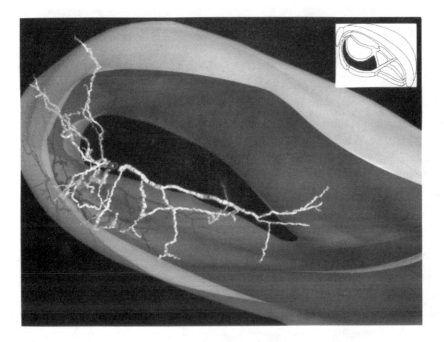

Figure 5. 3D reconstructed neuron in the facial sector of the SNR (coronal view)

nigral lamina of the SNR (lateral and ventral) receives afferent from striatal sectors related to the auditory and visual cortical areas.

Nigral neurons were found to have dendritic fields confined to this lamina. The considerable extent of dendritic fields in this lamina is consistent with the large auditory and visual receptive fields of SNR cells. Moreover, the overlap of dendritic fields suggests that some nigral cells may integrate both auditory and visual modalities.

Occupying a more internal position within the nigral "onion" is the projection field of the somatic sensory-motor district. In this region, the dorso-ventral topography of striatal projections preserves a coherent representation of the body such that the limbs and the posterior part of the face are represented ventrally and the nasal and orofacial parts of the face are represented dorsally (core of the nigral "onion"). Interestingly, the most ventral neurons of this nigral sector, which presumably process somatic information from a part of the body seen by the animal, have dendrites extending into the ventral visual lamina. These neurons probably perform visuo-somatic integration to organize movements. In contrast, neurons located in the core of the nigral "onion", where are represented parts of the body not seen by the animal, have dendrites that do not extend ventrally into the visual nigral lamina. Instead, the dendrites of these neurons extend dorsally towards the lamina receiving projections from the striatal sector related to perirhinal and gustative cortical areas.

Altogether these data support the notion that striato-nigral circuits comprise sensory-motor channels organized to integrate the various information useful for adaptation of different classes of behavior. While sensory-motor linkages are probably species specific, basal

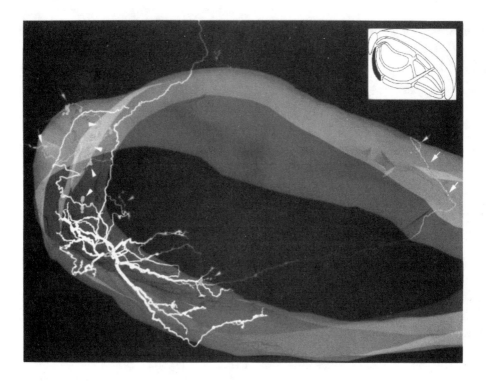

Figure 6. 3D reconstructed neuron in the auditory-visual sector of the SNR (coronal view).

ganglia channels may constitute, in a given specie, an anatomical constraint for sensory-motor learning.

3. STRIATAL OUTPUT NEURON OPERATES AS A COINCIDENCE DETECTOR OF CORTICAL INPUTS.

Among the electrophysiological features characterizing striatal output neurons is their low level of spontaneous firing is anaesthetized[9,10] (see also Fig 7A$_1$) as well as in awake animals[11]. The relative silence of these cells is explained by their non-linear intrinsic electrical membrane properties rather than a synaptic inhibition. Indeed, it is now assumed that membrane excitability of striatal output cells is mainly controlled by a set of potassium conductance activated over a wide range of potentials[12] and that account for their low resting membrane potential, weak input resistance and short time constant. As a functional consequence on synaptic integration, it has been proposed that striatal output neurons filter out small and temporally uncorrelated depolarizing synaptic inputs and require numerous synchronized excitatory afferent to produce a substantial membrane depolarization[10,12]. We have recently confirmed this hypothesis by experimental investigations in barbiturate-anaesthetized rats where intracellular activity of striatal output neurons was simultaneously recorded

PROCESSING IN CORTICO-STRIATO-NIGRAL CIRCUITS OF THE BASAL GANGLIA 205

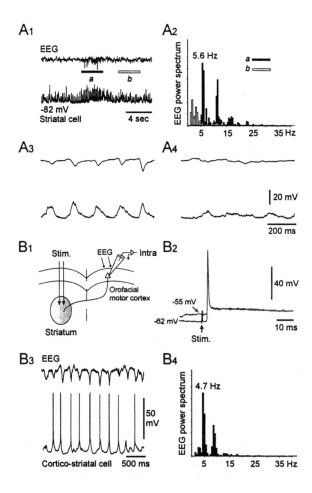

Figure 7. Relation between striatal membrane depolarizations and cortico-striatal synchronization. (A) Intracellular activity of a striatal output neuron (lower traces) and simultaneous EEG of the afferent orofacial motor cortex (upper traces). (A1) Large amplitude membrane potential oscillations in striatal neuron occurred during cortical spindle waves (a) while temporally disorganized striatal activity was associated with a desynchronized EEG (b). (A2) Fast Fourier transforms performed on the EEG periods indicated in A1. Note the clear periodicity near 5 Hz during the cortical spindling (a). (A3-4) Expanded records from the synchronized (3) and desynchronized (4) periods shown in A1. (B) Firing discharge of cortico-striatal neurons is in phase with EEG spindling. (B1) Experimental design for electrophysiological identification of crossed cortico-striatal neurons. Antidromic activation of these cells were obtained by electrical stimulation of the contralateral striatum (stim.). Intracellular recordings in the orofacial motor cortex were combined with a focal surface EEG. (B2) Example of antidromic responses. Note the constancy of the antidromic action potential latency (3.3 ms) despite a spontaneous depolarizing shift (arrow). (B3) Rhythmic synaptic depolarizations and firing activity in cortico-striatal cells coincide with EEG spindle waves. (B4) Spectral analysis of the EEG epoch shown in B3.

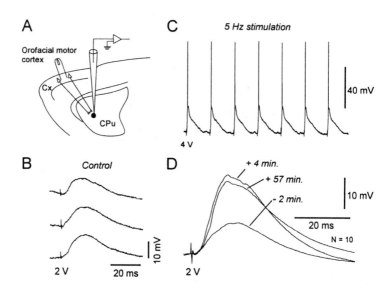

Figure 8. Physiological cortico-striatal LTP. (A) Experimental set-up. Intracellular recordings were obtained from neurons located in the striatal region related to the orofacial motor cortex. Monosynaptic EPSPs are evoked by electrical stimulation of this cortical area. Cx: cerebral cortex, Cpu: caudate putamen. (B) Three successive synaptic responses induced during the control period by cortical stimulation of weak intensity (2V). (C) Period of the conditioning procedure which consisted in iterative (N=500) cortical stimulation at 5 Hz applied with a suprathreshold intensity. (D) Superimposition of averaged (N=10) test synaptic responses recorded at the indicated times after the end of the 5-Hz conditioning.

with a focal electroencephalogram (EEG) of an identified afferent cortical region[10]. As shown in Fig 7A$_1$, striatal neurons display in vivo continuous membrane potential fluctuations which provide a sustained excitatory synaptic "noise." This spontaneous synaptic activity exhibit short periods of large amplitude oscillations (a in Fig 7A1) that are associated with 5-Hz EEG spindle waves (Fig 7A$_2$ Individual striatal oscillations are phase lock with cortical waves (Fig 7A$_3$) while small amplitude non-rhythmic synaptic potentials are related to a desynchronized EEG (Fig 7A$_4$). Since the cerebral cortex provides the main excitatory input of the striatum it is expected that the temporal link between cortical spindle waves and large amplitude depolarization in striatal cells results from a synchronization of cortico-striatal afferent converging onto the recorded cell. This is confirmed by intracellular recording of identified cortico-striatal neurons (Fig 7B$_{1-2}$), showing that synaptic depolarization and firing discharge in these neurons are in phase with the cortical waves (Fig 7B$_{3-4}$).

The sustained 5-Hz synchronization in cortico-striatal neurons that induces coherent synaptic depolarization in striatal neurons can also modulate the efficacy of cortico-striatal glutamatergic synapses. This is exemplified in Figure 8 where cortico-striatal LTP is obtained by 5-Hz cortical stimulation of the cortical afferent. This result provides an additional support for physiological LTP at connections between cerebral cortex and striatum[9].

These recent results strongly support the assumption that striatal output neurons function as coincidence detectors of converging cortical information[1]. Since firing discharge of striatal cells leads to behavioral output through disinhibition of premotor networks[13], activity-dependent facilitation in the cortico-striato-nigral channels would favor the formation of sensory-motor linkages during behavioral learning.

4. REFERENCES

1. Houk, J.C., 1995, Information processing in modular circuits linking basal ganglia and cerebral cortex. In models of information processing in the basal ganglia (J.C. Houk, J. L.Davis and G. Beiser, eds) The MIT Press, Cambridge, pp 3-9.
2. Alexander, G. E., DeLong, M; R., and Strick, P. L., 1986, Parallel organization of functionally segregated circuits linking basal ganglia and cortex. Annu. Rev. Neurosci. 9: 357-381.
3. Percheron, G., and Filion, M., 1991, Parallel processing in the basal ganglia: up to a point. Trends Neurosci. 14: 55-56.
4. Bergman, H., Feinglod, A., Nini, A., Raz, A., Slovin, H., Abeles, M., and Vaadia, E., 1998, Physiological aspects of information processing in the basal ganglia of normal and parkinsonian primates. Trends Neurosci. 21: 32-38.
5. Deniau, J. M., and Thierry, A. M., 1997, Anatomical segregation of information processing in the rat substantia nigra pars reticulata. In the basal ganglia and new surgical approaches for parkinson's disease, Advances in Neurology, vol 74, (J. A. Obeso, M. R. DeLong, C. Ohye, and C. D. Marsden, eds) Lippincott-Raven Publishers, Philadelphia, pp 83-96.
6. Deniau, J. M., Menetrey, A., and Charpier, S., 1996, The lamellar organization of the rat substantia nigra pars reticulata: segregated patterns of striatal afferents and relationship to the topography of corticostriatal projections. Neuroscience 73: 761-781.
7. Deniau, J. M., and Chevalier, G., 1992, The lamellar organization of the rat substantia nigra pars reticulata: distribution of projection neurons. Neuroscience 46: 361-377.
8. Pinault, D., 1994, Golgi-like labeling of a single neuron recorded extracellularly. Neurosci. Lett. 170: 255-260.
9. Charpier S., and Deniau J.-M., 1997, *In vivo* activity-dependent plasticity at corticostriatal connections: evidence for physiological long-term potentiation. Proc. Natl. Acad. Sci. U. S. A. 94: 7036-7040.
10. Charpier, S., Mahon, S., and Deniau, J.-M., 1999, *In vivo* induction of striatal long-term potentiation by low-frequency stimulation of the cerebral cortex. Neuroscience 91: 1209-1222.
11. Wilson, C. J., and Groves, P. M., 1981, Spontaneous firing patterns of identified spiny neurons in the rat neostriatum. Brain Res. 220:67-80.
12. Wilson, C. J., 1995, The contribution of cortical neurons to the firing pattern of striatal spiny neurons. In models of information processing in the basal ganglia (J.C. Houk, J. L.Davis and G. Beiser, eds) The MIT Press, Cambridge, pp 29-50.
13. Chevalier, G., and Deniau J. M., 1990, Disinhibition as a basic process in the expression of striatal functions. Trends Neurosci. 13: 277-280.

THE CONTROL OF SPIKING BY SYNAPTIC INPUT IN STRIATAL AND PALLIDAL NEURONS

Dieter Jaeger*

Abstract: Different patterns of simulated synaptic input were applied to striatal and pallidal neurons *in vitro* using dynamic current clamping. It was found that striatal neurons required a much larger baseline of excitatory inputs than pallidal neurons to allow spiking. Spike rates in pallidal neurons in response to applied synaptic conductances were much higher even when inhibitory input conductances dominated. Repeated applications of inputs with defined short-term correlations demonstrated that the timing of individual spikes can be controlled within 2 ms by specific input patterns both in striatum and GP. The presence of synchronization in the input led to much increased spike rates even when the mean rate of input was not changed. These results indicate specific modes of synaptic integration in striatal and GP neurons which depend on particular intrinsic voltage-gated conductances. An important role of these mechanisms in network function of the basal ganglia seems likely.

1. INTRODUCTION

Every second in a behaving animal, each striatal medium spiny neuron receives thousands of synaptic inputs from cerebral cortex, thalamus, and local interneurons. Neurons in the globus pallidus, in turn, receive a similarly high frequency of inputs from striatum and subthalamic nucleus. The main pathway of information flow is generally believed to travel from cerebral cortex to striatum and on to globus pallidus[2]. This pathway is excitatory at corticostriatal synapses and inhibitory at striatopallidal synapses. The main question underlying the present work is how output spike trains of single neurons in this pathway may be controlled by thousands of synaptic inputs per second. More specifically, our goal was to examine how the temporal precision of individual spikes and the spike rate over longer time intervals can be described as a function of the temporal pattern of excitatory and inhibitory input conductances. To analyze the control of spiking by specific synaptic input patterns, we used the technique of dynamic current clamping *in vitro*[11]. This technique allowed us to apply computer generated conductance patterns to neurons while recording their membrane

---

* Dieter Jaeger, Department of Biology, Emory University, Atlanta, GA 30322.

potential and spike pattern. We recorded from striatal and pallidal neurons to test the hypothesis that the intrinsic properties of neurons in these structures are specialized to allow different modes of spike control by synaptic inputs.

2. METHODS

Frontal or sagittal 300μm thick brain slices were prepared from 15-35 day old male Sprague-Dawley rats. Animals were perfused with a sucrose-Ringer solution under deep anesthesia before the brain was removed. Whole cell recordings were obtained at 32° C under visual guidance using a 63x water immersion lens. Electrode impedances ranged from 6 to 12 MΩ. The intracellular solution contained (in mM): K-Gluconate 140; NaCl 10; HEPES 10; MgATP 4; NaGTP 0.4; EGTA 0.2; Spermine 0.05. The extracellular solution contained (in mM): NaCl 124; KCl 3; KH_2PO_4 1.2; $MgSO_4$ 1.9; $NaHCO_3$ 26; $CaCl_2$ 2; glucose 20. Endogenous synaptic input in the slices was blocked by adding 40 μM picrotoxin, 10 μM CNQX, and 200 μM AP-5.

To add artificial synaptic inputs via the recording electrode, a computer used stored synaptic conductance patterns to calculate the current to be injected on-line at a 10 KHz refresh rate. The equation for the injected synaptic current is: $I_{syn} = g_{syn} * (V_m - V_{rev})$, where g is conductance and V_{rev} is the synaptic reversal potential. The reversal potentials for excitation and inhibition were set to 0 and -70 mV, respectively. Each excitatory input induced a conductance with a 0.5 ms rise time and a 1.2 ms decay time constant. The rise and decay time constants for inhibitory inputs were 0.93 and 20 ms, respectively. These values were chosen to match the time course of generic AMPA-type excitatory and $GABA_A$-type inhibitory inputs. Currents resulting from many individual EPSCs and IPSCs were added so that the total synaptic input pattern a neuron may receive *in vivo* is injected via the recording electrode. For details of this method please see[3].

3. RESULTS

To identify the type of recorded neurons, positive and negative current pulses were injected. As shown in previous studies[8,9], striatal medium spiny neurons expressed a hyperpolarized resting potential of -70 to -80 mV, and a strong rectification upon injection of hyperpolarizing current (Fig. 1). With depolarising current injection they showed regular spiking. The latency to the first spike was long when the current amplitude was low, which is due to a slowly inactivating K^+ current[10]. Pallidal neurons in contrast showed a more depolarised resting potential around -60 mV, a sag in the response to negative current injection, and fast regular spiking even with low amplitudes of injected positive current (Fig. 1). These properties have been found to identify pallidal projection neurons[7]. The different response properties to current injection between striatal medium spiny neurons and pallidal projection neurons point out that they express different sets of voltage-gated membrane currents. These currents are gated by voltage changes in the range caused by synaptic input and their (de)activation or (de)inactivation will thus influence the net voltage response to synaptic input. The presence of different such currents in striatal and GP neurons makes it likely that they will respond differently to identical patterns of synaptic input.

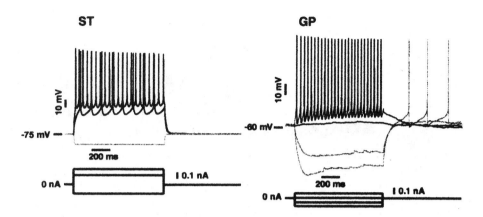

Figure 1. Current injection pulses applied to medium spiny neurons in striatum (ST) and to neurons in globus pallidus (GP)

To apply artificial synaptic conductance patterns with dynamic clamping we constructed a stimulus, which contained the activity of 100 excitatory and 100 inhibitory simulated synapses. Groups of 10 excitatory and 10 inhibitory synapses were coupled to result always in the synchronous activation of 10 synapses (see below for the significance of correlated inputs). Each of the synaptic groups was activated randomly with a mean rate of 80 Hz for excitatory and 30 Hz for inhibitory inputs. This stimulus contained on average 8000 excitatory and 3000 inhibitory inputs per second. Figure 2A shows a 2 s segment of the summed input conductances and Figure 2B illustrates a typical voltage response for striatal and pallidal neurons. The application of ongoing random excitatory and inhibitory input immediately induced an irregular spiking pattern in the recorded neurons, which is otherwise not observed *in vitro*. The total input conductance levels required to induce realistic spiking in striatum was quite different from that required in GP. Although the same pattern of synaptic input was used for both structures, the amplitude of conductance had to be scaled by different gain factors to induce realistic spike patterns. Medium spiny neurons, though smaller in size than GP neurons, required more inward synaptic current to trigger spiking. GP neurons were found to spike at a higher rate than medium spiny neurons even in the presence of a much lower level of excitatory inputs (Fig. 2A,B).

When the same synaptic input pattern was presented to a single neuron several times, we found that a majority of spikes were timed precisely within 2 ms for repeated stimulus presentations (Fig. 2C). This precise control of spike timing via synaptic conductances allows in principle that information in basal ganglia circuits is contained in the absolute timing of individual spikes.

To examine how synaptic conductance changes are related to the time of spiking we constructed spike-triggered averages of synaptic current and of excitatory and inhibitory synaptic conductances (Fig. 2D). We found that on average spikes in striatal and GP neurons were preceded by an increase in excitatory input conductance as well as a decrease in inhibitory conductance. Excitation was acting within a shorter period before each spike than

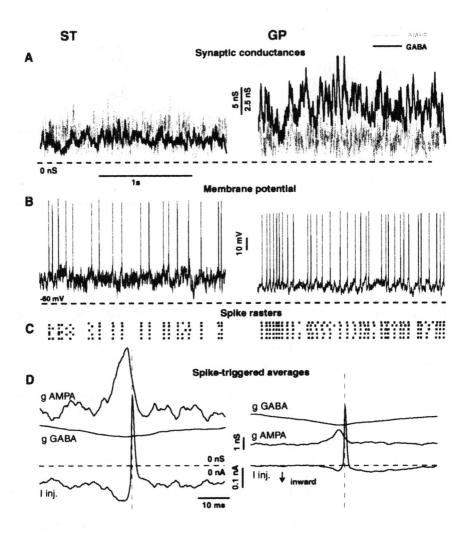

Figure 2. Dynamic current clamping of striatal and GP neurons. A. The black and grey lines represent the summed conductance traces of 100 GABA and 100 AMPA synapses, respectively. The amplitude of excitatory inputs was scaled down to 40% of the striatal input for GP stimulation, and the inhibitory input was scaled up by 20%. C. The rows of black squares denote spike times for subsequent stimulus presentations. D. Spike triggered synaptic conductance and injected current. The positive spike in injected current reflects the sharp change in driving force for synaptic input during an action potential.

inhibition, however, indicating that inhibitory inputs act with a slower time course than excitatory inputs in controlling spike timing. Striatal neurons showed a much higher peak in excitation preceding spike initiation than GP neurons, indicating that a synchronous burst of excitation is a much more important mechanism in inducing a spike in a striatal than a GP neuron. This result is in good agreement with previous evidence that striatal neurons spike preferentially in response to coincident excitatory input from many cortical neurons[13]. In contrast, GP neurons are much more sensitive to small excitatory signals, and can be active in the presence of a large inhibitory baseline. These results suggest that the intrinsic response properties of these two types of neurons may be responsible to a large degree for the different spike rates observed *in vivo*[5].

To examine how correlations in the activity of many inputs affect spiking we constructed a low-correlation and a high-correlation input pattern. The high-correlation pattern consisted of 10 groups of 10 excitatory and inhibitory inputs as described above. The low-correlation input consisted of 100 individually activated excitatory and 100 inhibitory inputs. The resulting conductance traces for the low-correlation pattern had the same mean amplitude as those resulting from high-correlation inputs, but much lower amplitude fluctuations around the mean. Neurons in striatum rarely showed any spiking in response to low-correlation inputs (Fig. 3). Pallidal neurons did spike in response to these inputs, but their spiking frequency was much increased in response to high-correlation inputs (Fig. 3). This finding substantiates the observation that striatal neurons preferentially respond to synchronous excitatory inputs[13]. This dependence of spiking on 'packets' of inputs allows for population coding schemes that use the coordinated activation of neural assemblies to contain specific information[1].

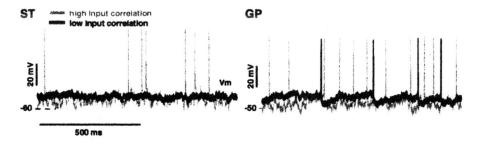

Figure 3. The effect of input correlation on spiking. Voltage traces (Vm) for low-correlation and high-correlation inputs are super-imposed. Note that subthreshold high-frequency fluctuations are more pronounced in the high-correlation condition.

The results shown above demonstrate that the temporal structure of spike trains in striatum and in GP can be precisely controlled by synaptic conductances and that correlated packets of inputs are highly effective in controlling spiking. These findings do not deny the presence of rate coding in a traditional sense, i.e. the increase of spike frequency with an increase in the mean excitatory drive. To examine how the presence of correlation and a change of mean drive may interact in the control of spiking we constructed two stimuli that both had the same high correlation in inputs, but one stimulus had an additional constant component in excitatory conductance that equaled 80% of the dynamic component. The addition of constant excitatory conductance did result in an increase of spike frequency from 2 to 8 Hz for the striatal and 10 to 17 Hz for the GP neuron shown in Figure 4. Very similar increases were seen in other neurons recorded with these stimuli. Interestingly, the precise timing of most of the spikes already present before the 80% addition of excitatory conductance was left intact (spikes noted by asterisks in Fig. 4). Instead of creating a completely new spike pattern, the addition of constant excitatory conductance inserted new spikes in the existing spike train. Therefore, the information transmitted by specific pulses of synchronous inputs can be maintained with different mean rates of spiking. The additional excitatory drive tended to push smaller synchronous events above the threshold needed to trigger a spike. It is important to note that the mean excitatory drive is given by the combined total level of inhibition and excitation. In this situation the modulation of inhibition can be seen as both changing the mean excitatory drive and thus the mean spike rate, as well as setting the threshold for the amplitude of excitatory input packets required to induce a spike.

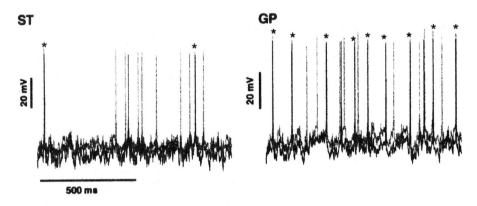

Figure 4. The effect of added constant excitatory conductance. Responses with added conductance are superimposed on the baseline response and drawn in grey.

4. DISCUSSION

An ongoing barrage of excitatory and inhibitory inputs leads to a considerable baseline of synaptic conductances. These conductances define a combined reversal potential, which sets the excitatory drive exerted on the cell. Cells that require a large depolarising current to reach spike threshold such as striatal medium spiny neurons are tuned to respond to input conditions with a large amount of excitation. In contrast, cells that require little or no depolarising input current to trigger spikes such as pallidal projection neurons are tuned to input conditions in which inhibition dominates. The rate in increase of spike rate resulting from increasing excitatory drive is further dependent on intrinsic cell properties such as depolarisation-activated K^+ currents[4]. Pallidal projection neurons show much higher spike rates *in vivo* than striatal medium spiny neurons, even though they presumably receive far greater amounts of inhibition. Our results indicate that the intrinsic properties of these neurons are well suited to account for these different spike rates.

Beyond the different requirement of pallidal and striatal neurons in the relative mean level of excitatory and input conductances, our results indicate that the correlation between synaptic inputs is important. Medium spiny neurons were particularly tuned to respond to synchronous pulses of many excitatory inputs on top of an already substantial baseline of excitation. This finding with artificial synaptic input *in vitro* is in good agreement with the subthreshold membrane trajectory and spike pattern found *in vivo*. Intracellular recordings of medium spiny neurons *in vivo* indicate the existence of a characteristic up-state, which is produced by a large baseline of excitatory input[14,15]. Spikes are triggered by additional depolarising pulses above the up-state baseline[12,13]. The present data indicate that the timing of individual spikes in medium spiny neurons may be controlled with a 2 ms accuracy by realistic synaptic inputs. Inhibitory inputs made a significant contribution to the control of spike rate as well as spike timing. This observation suggests that the activation pattern of fast spiking inhibitory interneurons in striatum may be of great significance in shaping spike responses of medium spiny neurons to cortical input. Inhibition via collaterals from other medium spiny neurons is less likely to be involved, since such inhibition has been found to be weak or absent[6].

Neurons in GP could also spike with a 2 ms accuracy in response to excitatory and inhibitory inputs, but a baseline of excitation was not required for the induction of spiking. The spike rate of GP neurons increased dramatically with small increases in excitation or decreases in inhibition, while individual spikes could still be controlled accurately by specific sets of correlated inputs. In addition, the presence of short-term correlation in the input pattern also led to a pronounced increase in the overall spike rate. Inhibitory and excitatory inputs were in general of equal importance in controlling GP spiking.

The method of dynamic current clamping *in vitro* can reveal how single neurons respond to specific patterns of simulated synaptic inputs. Which of the demonstrated features of controlling spiking are ultimately used to transmit information in basal ganglia circuits remains an exciting area of future research.

5. ACKNOWLEDGEMENTS

The author thanks Jesse Hanson and Lisa Kreiner for much assistance with the described work.

6. REFERENCES

1. Aertsen A, Diesmann M, Gewaltig MO. (1996) Propagation of synchronous spiking activity in feedforward neural networks. J.Physiol.(Paris.). 90:243-247.
2. Albin RL, Young AB, Penney JB. (1995) The functional anatomy of disorders of the basal ganglia. TINS. 18:63-64.
3. Jaeger D, Bower JM. (1999) Synaptic control of spiking in cerebellar Purkinje cells: dynamic current clamp based on model conductances. J.Neurosci. 19:6090-6101.
4. Jaeger D, De Schutter E, Bower JM. (1997) The role of synaptic and voltage-gated currents in the control of Purkinje cell spiking: a modeling study. J.Neurosci. 17:91-106.
5. Jaeger D, Gilman S, Aldridge JW. (1994) Primate basal ganglia activity in a precued reaching task: preparation for movement. Exp.Brain Res. 95:51-64.
6. Jaeger D, Kita H, Wilson CJ. (1994) Surround inhibition among projection neurons is weak or nonexistent in the rat neostriatum. J.Neurophysiol. 72:2555-2558.
7. Kita H, Kitai ST. (1991) Intracellular study of rat globus pallidus neurons: membrane properties and responses to neostriatal, subthalamic and nigral stimulation. Brain Res. 564:296-305.

8. Kita T, Kita H, Kitai ST. (1984) Passive electrical membrane properties of rat neostriatal neurons in an in vitro slice preparation. Brain Res. 300:129-139.
9. Nisenbaum ES, Wilson CJ. (1995) Potassium currents responsible for inward and outward rectification in rat neostriatal spiny projection neurons. J.Neurosci. 15:4449-4463.
10. Nisenbaum ES, Xu ZC, Wilson CJ. (1994) Contribution of a slowly inactivating potassium current to the transition to firing of neostriatal spiny projection neurons. J.Neurophysiol. 71:1174-1189.
11. Sharp AA, ONeil MB, Abbott LF, Marder E. (1993) Dynamic clamp: computer generated conductances in real neurons. J.Neurophysiol. 69:992-995.
12. Stern EA, Jaeger D, Wilson CJ. (1998) Membrane potential synchrony of simultaneously recorded striatal spiny neurons *in vivo*. Nature. 394:475-478.
13. Wickens JR, Wilson CJ. (1998) Regulation of action-potential firing in spiny neurons of the rat neostriatum *in vivo*. J.Neurophysiol. 79:2358-2364.
14. Wilson CJ, Groves PM. (1981) Spontaneous firing patterns of identified spiny neurons in the rat neostriatum. Brain Res. 220:67-80.
15. Wilson CJ, Kawaguchi Y. (1996) The origins of two-state spontaneous membrane potential fluctuations of neostriatal spiny neurons. J. Neurosci. 16:2397-2410.

EXCITATORY CORTICAL INPUTS TO PALLIDAL NEURONS THROUGH THE CORTICO-SUBTHALAMO-PALLIDAL HYPERDIRECT PATHWAY IN THE MONKEY

Atsushi Nambu, Hironobu Tokuno, Ikuma Hamada, Hitoshi Kita, Michiko Imanishi, Toshikazu Akazawa, Yoko Ikeuchi, and Naomi Hasegawa*

1. INTRODUCTION

The basal ganglia are currently considered to be composed of the two major circuits, "direct" and "indirect" pathways.[1] The striatum, i.e., the caudate nucleus and putamen (Put), is the input stage of the basal ganglia. The direct pathway arises from GABAergic striatal neurons, and projects monosynaptically to the output nuclei, i.e., the internal segment (GPi) of the globus pallidus (GP) and the substantia nigra pars reticulata (SNr). The indirect pathway also arises from GABAergic striatal neurons, and projects polysynaptically to the GPi/SNr by way of a sequence of connections involving the external segment (GPe) of the GP and the subthalamic nucleus (STN).

Recently the STN came to be regarded as another input stage of the basal ganglia.[7,10,11] The monkey STN receives somatotopically organized inputs from the primary motor cortex (MI), supplementary motor area and premotor cortex[4,12,14] and sends outputs to the GPe and GPi/SNr. Through this cortico-subthalamo-pallidal "hyperdirect" pathway[12], the motor-related cortical areas are considered to exert a direct influence on the output nuclei of the basal ganglia, bypassing the striatum. In order to understand the motor control mechanism of the forelimb movements by the basal ganglia, the present study was undertaken to characterize the influence of the MI on the activities of GP neurons through the "hyperdirect" pathway in unanaesthetized monkeys. Stimulating electrodes were chronically implanted into the forelimb region of the MI after electrophysiological mapping. Responses of GP and STN neurons to the stimulation in the MI were then recorded simultaneously. The change in the

---

* Atsushi Nambu, Hironobu Tokuno, Ikuma Hamada, Michiko Imanishi, Toshikazu Akazawa, Yoko Ikeuchi, and Naomi Hasegawa, Tokyo Metropolitan Institute for Neuroscience, Tokyo Metropolitan Organization for Medical Research, Fuchu, Tokyo 183-8526, Japan. Hitoshi Kita, Department of Anatomy and Neurobiology, College of Medicine, University of Tennessee, Memphis, TN 38163, USA.

response pattern of GP neurons to cortical stimulation after chemical inactivation of the STN was further examined by means of continuous recording of single neuronal activity.

2. METHODS

Two Japanese monkeys *(Macaca fuscata)* were used in this study. Using methods described elsewhere,[12] monkeys initially received surgery for head fixation and easy access during recording of neuronal activities. After the forelimb region of the MI was identified using intracortical microstimulation (ICMS, less than 50 μA, a train of 12 cathodal pulses of 200 μs duration at 333 Hz), pairs of bipolar stimulating electrodes (intertip distance, 2 mm) were implanted. Histological studies at the end of the experiments confirmed that tips of the stimulating electrodes were placed in the gray matter of the precentral gyrus.

During experimental sessions, the monkey was seated in the monkey chair with its head restrained. A glass-coated Elgiloy-alloy microelectrode was inserted through the dura obliquely into the GP. For recording and injection of drugs in the STN, a 10 μl-Hamilton microsyringe with a recording electrode[18] was inserted vertically into the STN. The activities

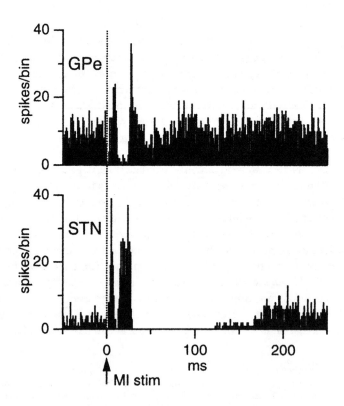

Figure 1. Peri-stimulus time histograms (PSTHs, 100 times; bin width, 1 ms) showing responses of a GPe *(upper)* and *STN (lower)* neuronal pair to the stimulation of the forelimb region of the MI at time 0 (indicated by *an arrow*).

of GP and STN neurons were simultaneously recorded, and the responses of both GP and STN neurons to the cortical electrical stimulation (0.3 ms duration single pulse, 0.3 - 0.7 mA strength at 0.4 - 0.8 Hz) were observed by constructing peri-stimulus time histograms (PSTHs, bin width, 1 ms). Through a needle in the STN, 0.5 - 2.0 µl of one of the following drugs was injected: 0.5 - 1.0 µg/µl muscimol (GABA agonist), 20 mM (±)-3-(2-(carboxypiperazin-4-yl)-propyl-1-phosphonic acid (CPP, NMDA antagonist), 10 mM 1,2,3,4-tetrahydro-6-nitro-2,3-dioxo-benzo[f]quinoxaline-7-sulfonamide disodium (NBQX, non-NMDA antagonist), each of which was dissolved in saline. The activity of GP neurons was

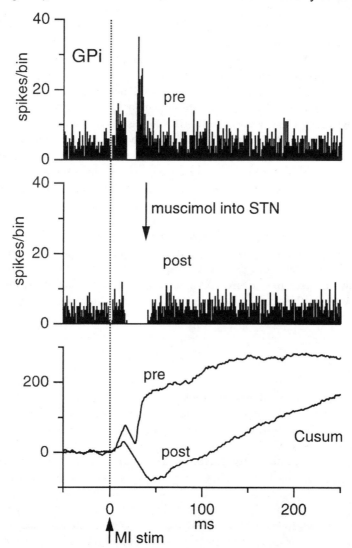

Figure 2. Effect of muscimol injection into the STN on the response of GP neurons evoked by cortical stimulation. *Upper panel,* A PSTH (100 times; bin width 1 ms) of a GPi neuron before injection. The forelimb region of the MI was stimulated at time 0 (*arrow*). *Middle panel,* A PSTH of the same neuron after muscimol injection into the STN. *Lower panel,* Cumulative sum before (*pre*) and after injection (*post*).

continuously monitored before and after injection of drugs into the STN, and responsiveness of these neurons to the cortical stimulation was examined by constructing PSTHs and calculating cumulative sum.[3]

3. RESULTS

Typical responses of a GP and STN neuronal pair evoked by MI stimulation are shown in Figure 1. Stimulation of the forelimb region of the MI induced an early, short-latency excitation, followed by an inhibition, and then a late excitation in both GPe (Fig. 1, upper panel) and GPi (Fig. 2, upper panel) neurons. The same stimulation induced simultaneously an early excitation and a late excitation, which were interrupted by a brief period of inhibition, and a succeeding long-lasting inhibition in STN neurons (Fig. 1, lower panel). The onset of the early excitation in STN neurons preceded that in GP neurons by 2 - 6 ms in most cases.

Muscimol was injected into the STN to block temporally the neuronal activities, and its effect on the cortically-evoked responses of GP neurons was examined. Stimulation of the forelimb region of the MI evoked an early, short-latency excitation, an inhibition and a late excitation in GP neurons (Fig. 2, upper panel). Muscimol injection instantly blocked the spontaneous activity of STN neurons, then reduced markedly the early and late excitations of GP neurons, and made the inhibition longer (Fig. 2, middle panel). The reduction can be well seen in the cumulative sum chart (Fig. 2, lower panel). The spontaneous discharge rate of GP neurons decreased to 0 - 78 % of the control level after muscimol injection. CPP or NBQX was injected into the STN to block the glutamatergic cortico-subthalamic transmission[2,17] with little effect on the GABAergic pallido-subthalamic transmission (indirect pathway), and its effect on the cortically-evoked responses of GP neurons was examined. CPP injection abolished most of the early excitation and part of the late excitation without a significant change in the duration of the inhibition (data not shown). On the other hand, NBQX injection had no effect on the response pattern of GP neurons.

4. DISCUSSION

The early excitation in GP neurons evoked by cortical stimulation is considered to be caused by the cortico-subthalamo-pallidal pathway based on the following findings. The early excitation in GP neurons was preceded by that in STN neurons, which is likely to be caused by the cortico-subthalamic projections. The latency difference between the two excitations was 2 - 6 ms, which is comparable to the latency of the STN-induced excitation of GP neurons in rats,[9] and the latency of the GP-induced antidromic response of STN neurons in monkeys (0.5 - 3.5 ms, Nambu et al., unpublished observation) and in rats.[8] Results obtained by injection of drugs into the STN finally support the idea that the early excitation of GP neurons is derived from the STN. Indeed, blockade of neuronal activities in the STN by muscimol injection abolished the early and late excitations in GP neurons. In addition, blockade of the cortico-subthalamic transmission by CPP injection, but not by NBQX injection, into the STN, suppressed the early excitation in GP neurons. Thus, the cortico-subthalamic glutamatergic transmission directly related to the motor control in the monkey is considered to be mediated mainly by NMDA receptors. On the other hand, the inhibition in GP neurons evoked by cortical stimulation is considered to be mediated by the cortico-striato-pallidal direct pathway.

The present study clearly shows that the cortico-subthalamo-pallidal hyperdirect pathway exerts powerful excitatory effects on the output nuclei of the basal ganglia, and is faster in signal conduction than the direct and indirect pathways connecting the cortex and the output nuclei by way of the striatum (Fig. 3). Conduction time through each projection could be estimated based on the present and previous studies examining the latency distribution of orthodromic and antidromic responses. The conduction times of the MI-STN and STN-GP projections could be 2 - 6 ms (the present study) and 0.5 - 3.5 ms (Nambu et al., unpublished observation), respectively. Thus, the total conduction time through the cortico-subthalamo-pallidal hyperdirect pathway can be estimated to be 2.5 - 9.5 ms. On the other hand, the conduction times of the MI-Put and Put-GP projections would be 6 - 12 ms[13] and 3 - 9 ms[19], respectively. Thus, the total conduction time through the cortico-striato-pallidal direct pathway is assumed to be 9 - 21 ms. These estimated values suggest that conduction time through the hyperdirect pathway is much shorter than that through the direct pathway.

Our present findings expand the "center-surround model"[11] or the simple "disinhibition model"[5,6] of basal ganglia function. When a voluntary movement is about to be initiated by cortical mechanisms, a corollary signal is sent simultaneously from the motor cortex to the GPi through the hyperdirect pathway to activate GPi neurons extensively, thereby resulting

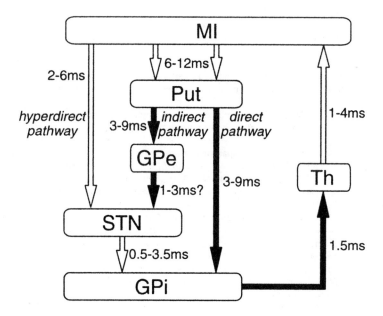

Figure 3. A schematic diagram of the hyperdirect, direct and indirect pathways. *White* and *black* arrows represent excitatory glutamatergic and inhibitory GABAergic projections, respectively. Estimated conduction time through each projection is indicated. The conduction times of the GPi-Th and Th-MI projections are based on the data by Nambu et al.[15,16] *GPe*, external segment of the globus pallidus; *GPi*, internal segment of the globus pallidus; *MI*, primary motor cortex; *Put*, putamen; *STN*, subthalamic nucleus; *Th*, thalamus.

in inhibition of their target neurons in the thalamus related to both the selected motor program and other competing programs. Then, another corollary signal through the direct pathway is sent to the GPi to inhibit a specific population of pallidal neurons. Such pallidal neurons disinhibit their target neurons in the thalamus, and provoke the selected motor program. Thus, the cortico-subthalamo-pallidal hyperdirect pathway is considered to contribute to the precise execution of the selected motor program at the appropriate time.

5. ACKNOWLEDGMENTS

This study was supported by Grants-in-Aid for Scientific Research from the Ministry of Education, Science, Sports and Culture of Japan (09680805), the Mitsubishi Foundation, and the Brain Science Foundation to AN, and NIH grants (NS-26473 and NS-36720) to HK.

6. REFERENCES

1. G.E. Alexander and M.D. Crutcher, Functional architecture of basal ganglia circuits: neural substrates of parallel processing, *Trends Neurosci.* 13:266-271 (1990).
2. M.D. Bevan, C.M. Francis, and J.P. Bolam, The glutamate-enriched cortical and thalamic input to neurons in the subthalamic nucleus of the rat: convergence with GABA-positive terminals, *J. Comp. Neurol.* 361:491-511 (1995).
3. P.H. Ellaway, Cumulative sum technique and its application to the analysis of peristimulus time histograms, *Electroenceph. clin. Neurophysiol.* 45:302-304 (1978).
4. K. Hartmann-von Monakow, K. Akert, and H. Künzle, Projections of the precentral motor cortex and other cortical areas of the frontal lobe to the subthalamic nucleus in the monkey, *Exp. Brain Res.* 33:395-403 (1978).
5. O. Hikosaka and R.H. Wurtz, Visual and oculomotor functions of monkey substantia nigra pars reticulata. I. Relation of visual and auditory responses to saccades, *J. Neurophysiol.* 49:1230-1253 (1983).
6. O. Hikosaka and R.H. Wurtz, Visual and oculomotor functions of monkey substantia nigra pars reticulata. IV. Relation of substantia nigra to superior colliculus, *J. Neurophysiol.* 49:1285-1301 (1983).
7. H. Kita, Physiology of two disynaptic pathways from the sensorimotor cortex to the basal ganglia output nuclei, in: *The Basal Ganglia IV: New Ideas and Data on Structure and Function*, G. Percheron, J.S. McKenzie, and J. Féger, eds., Plenum, New York (1994) p. 263-276.
8. H. Kita, H.T. Chang, and S.T. Kitai, Pallidal inputs to subthalamus: intracellular analysis, *Brain Res.* 264:255-265 (1983).
9. H. Kita and S.T. Kitai, Intracellular study of rat globus pallidus neurons: membrane properties and responses to neostriatal, subthalamic and nigral stimulation, *Brain Res.* 564:296-305 (1991).
10. R. Levy, L.N. Hazrati, M.T. Herrero, M. Vila, O.K. Hassani, M. Mouroux, M. Ruberg, H. Asensi, Y. Agid, J. Féger, J.A. Obeso, A. Parent, and E.C. Hirsch, Re-evaluation of the functional anatomy of the basal ganglia in normal and parkinsonian states, *Neuroscience* 76:335-343 (1997).
11. J.W. Mink and W.T. Thach, Basal ganglia intrinsic circuits and their role in behavior, *Curr. Opin. Neurobiol.* 3:950-957 (1993).
12. A. Nambu, M. Takada, M. Inase, and H. Tokuno, Dual somatotopical representations in the primate subthalamic nucleus: evidence for ordered but reversed body-map transformations from the primary motor cortex and the supplementary motor area, *J. Neurosci.* 16:2671-2683 (1996).
13. A. Nambu, K. Kaneda, H. Tokuno, and M. Takada, Organization of corticostriatal motor inputs in monkey putamen, *J. Neurophysiol.* (in press).

14. A. Nambu, H. Tokuno, M. Inase, and M. Takada, Corticosubthalamic input zones from forelimb representations of the dorsal and ventral divisions of the premotor cortex in the macaque monkey: comparison with the input zones from the primary motor cortex and the supplementary motor area, *Neurosci. Lett.* 239:13-16 (1997).
15. A. Nambu, S. Yoshida, and K. Jinnai, Projections on the motor cortex of thalamic neurons with pallidal input in the monkey, *Exp. Brain Res.* 71:658-662 (1988).
16. A. Nambu, S. Yoshida, and K. Jinnai, Discharge patterns of pallidal neurons with input from various cortical areas during movement in the monkey, *Brain Res.* 519:183-191 (1990).
17. B. Rouzaire-Dubois and E. Scarnati, Pharmacological study of the cortical-induced excitation of subthalamic nucleus neurons in the rat: evidence for amino acids as putative neurotransmitters, *Neuroscience* 21:429-440 (1987).
18. H. Tokuno, Y. Ikeuchi, A. Nambu, T. Akazawa, M. Imanishi, I. Hamada, and N. Hasegawa, A modified microsyringe for extracellular recording of neuronal activity, *Neurosci. Res.* 31:251-255 (1998).
19. S. Yoshida, A. Nambu, and K. Jinnai, The distribution of the globus pallidus neurons with input from various cortical areas in the monkeys, *Brain Res.* 611:170-174 (1993).

TANS, PANS AND STANS

Ben D. Bennett and Charles J. Wilson*

Abbreviations:
TANs, tonically active neurons
PANS, phasically active neurons
STANs, sometimes tonically active neurons

1. INTRODUCTION

During the last four decades, extracellular single unit recordings have been performed within the neostriatum to investigate the firing properties and response characteristics of the neurons contained therein. Although there are numerous morphologically distinct classes of neostriatal cells, typically only two types of neurons are reported in unit recording studies. These two classes are discriminated on the basis of firing rate, firing pattern, extracellular action potential waveform and response to different aspects of movement tasks. One class, the so-called phasically active neurons (PANS), are believed to be the spiny projection cells and exhibit increases in firing in relation to movement. The other class which spike in a tonic, irregular fashion (TANs) are thought to be cholinergic interneurons and display changes in firing in relation to sensory stimuli which trigger a rewarded movement. Data from intracellular *in vivo* and *in vitro* studies have revealed that the intrinsic membrane properties of spiny cells and cholinergic interneurons are very different and provide a cellular basis for understanding the contrasting mechanisms underlying spike generation and patterning in these two populations of cells.

2. A PANACEA FOR NEOSTRIATAL NOMENCLATURE?

In what are perhaps the first published extracellular (Albe-Fessard et al., 1960) and intracellular (Sedgwick and Williams, 1967) accounts of the spontaneous spiking activity of neostriatal neurons, phasic firing was reported and appeared to be attributable to episodic

---

* Ben D. Bennett and Charles J. Wilson, Department of Anatomy and Neurobiology, University of Tennessee, Memphis TN 38163, USA

fluctuations in the membrane potential. The earliest physiological classifications of neostriatal cells were not based on firing properties per se, but on the basis of whether stimulation of the substantia nigra or iontophoretic application of dopamine produced an increase or decrease in firing rate (Frigyesi and Purpura, 1966, 1967; Albe-Fessard et al., 1967; McLennan and York, 1967; Connor, 1968, 1970; Feltz, 1970; Hull et al., 1970; Feltz and Albe-Fessard, 1972; Richardson et al., 1977). It was also demonstrated that units which fired spontaneously were inhibited by nigral stimulation or dopamine application whereas such stimuli elicited spiking in units which were initially silent, lending support to the hypothesized existence of two neuronal populations (Connor, 1970; Feltz and Albe-Fessard, 1972). Concurrently, recordings from awake behaving monkeys were also revealing heterogenous responses of neostriatal units during learned motor tasks, in which some putamen units preferentially fired prior to the onset of *rapid* movements whereas others responded during *slower* movements (DeLong, 1972, 1973). In later studies, neostriatal units which could be differentiated on the basis of their action potential waveform, exhibited either paired pulse facilitation or inhibition in response to stimulation of the corticostriatal pathway which was suggested to reflect differential regulation by dopaminergic and GABAergic inputs (Nisenbaum et al., 1988, 1992; Nisenbaum and Berger, 1992).

The first systematic correlation of the relationship between firing rate and pattern with cellular morphology was performed by Wilson and Groves (1981). These authors utilized extracellular unit recording with subsequent intracellular impalement and filling of neurons with horseradish peroxidase in locally anaesthetized, immobilized rats to demonstrate that cells exhibiting a range of firing rates (0.24-32.7 Hz; mean = 5.54, s.d. = 7.38) and firing episode duration (0.1-3.0 sec; mean = 1.05, s.d. = 1.0) were all neostriatal spiny cells (Wilson and Groves, 1981; Wilson, 1993). Nevertheless, autocorrelograms generated from this seemingly heterogeneous sample of spiking activity exhibited a remarkably stereotyped appearance, with an early mode, corresponding to the interspike interval during firing episodes, superimposed on a lower firing probability, corresponding to the mean firing rate (Wilson and Groves, 1981; Wilson, 1993).

3. OFF ON A TANGENT

An early suggestion that neostriatal neurons might be divisible on the basis of their spontaneous firing patterns came from Anderson (1977). Interspike interval plots revealed different distributions for 'slow discharge neurons (S)' and 'low frequency burst cells (LF-B) cells' recorded in the putamen of awake monkeys. The S cells exhibited a nearly Gaussian distribution in the ISI histogram (modal ISI -200 msec), whereas the LF-B neurons presented a sharp peak at short ISI durations (modal value ~50 msec) with a long tail of longer intervals. Plots of mean vs. modal interspike intervals also revealed a clear tendency for LF-B cells to fire episodically or phasically (Anderson, 1977). Kimura and colleagues (1984) demonstrated that neurons which could be differentiated on the basis of spontaneous firing pattern and action potential width responded to different aspects of a learned and rewarded motor task. The tonically active neurons (TANs) described by Kimura et al, (1984) were similar to 'S' cells described by Anderson (1977), whereas phasically active neurons (PANs) were more similar to 'LF-B' cells. TANs did not exhibit changes in spiking during the movement portion of the task but responded instead to sensory stimuli which served to indicate an impending reward for performing the movement (Kimura et al., 1984). The response occured about 60 msec after the sensory cue and was an increased probability of firing followed by a pause in activity. These same responses were observed in TANs when no movement was required and stimuli indicated a 'free' juice reward (Kimura et al., 1984). Alexander and DeLong (1985) confirmed the presence of two classes of neurons in the putamen and caudate nucleus using similar identification procedures. These authors illustrated that the majority (92%) of units in their sample exhibited low firing rates (< 0.01 - 0.5 Hz), short duration extracellular action potential waveforms and sensorimotor-dependent phasic increases in firing rate, whereas the remaining 8% were tonically active (2-5 Hz) with longer spike duration and were unresponsive to sensorimotor stimulation (Alexander and DeLong, 1985). Subsequent studies have confirmed and extended these initial observations but estimates of the relative proportions of TANs and PANs differ greatly between investigators (Alexander and DeLong, 1985; Kimura, 1986; Aldridge et al., 1990; Kimura et al., 1990; Apicella et al., 1991). Much of this variability can doubtless be blamed upon the sampling bias against recording from PANS because a significant proportion of these cells are silent and consequently go undetected in extracellular investigations.

The observation that populations of TANs can exhibit synchronous pause responses (Aosaki et al., 1994a,b, 1995; Graybiel et al., 1994; Raz et al., 1996), suggests that there may be a common synaptic or neuromodulatory input to the TANS which might underlie such synchronization. Indeed, lesions of the dopaminergic projection to the neostriatum (Aosaki et al., 1994a) or local application of dopaminergic antagonists (Watanabe and Kimura, 1998) result in a loss of the pause response. However, inactivation of the centromedian-parafascicular complex in primates also results in a loss of the response (Matsumoto et al., 1997) suggesting that a combination of inputs might underlie the synchronization of the pause.

Studies which have utilized multivariate identification procedures to differentiate between TANs and PANs on the basis of firing rate, action potential waveform and response during the sensory or motor aspects of a learned and rewarded movement have consistently reported that many TANs exhibit bimodal interspike interval histograms (Kimura et al., 1990; Aosaki et al., 1994b; Aosaki et al., 1995). Typically a bimodal distribution of intervals is taken as evidence of bursty or phasic firing (Cocatre-Zilgien and Delcomyn, 1992). However, such distributions are very rarely exhibited by PANs owing to the irregularity of intervals between episodes of firing (see Aldridge and Gilman 1991 for a thorough quantification and discussion of the spiking patterns of PANs). It is of particular note that on occasion individual TANs switched between firing patterns, producing bimodal and unimodal interval histograms at different times during the course of a single recording (Aosaki et al., 1994b).

The morphological identity of TANs was initially (Kimura et al., 1984) suggested to be the type II large, aspiny neurons previously described in Golgi studies (DiFiglia et al., 1976). The fact that PANs but not TANs can be antidromically activated by stimulation of the globus pallidus indicates that TANs are probably interneurons (Kimura et al. 1990, 1996). Furthermore, Aosaki et al. (1995) compared the measured incidence of TANs encountered during single electrode penetrations with the incidence predicted from the known distribution density of various classes of striatal interneurons and concluded that cholinergic cells were the most likely candidate. Compelling evidence that the TANs detected in extracellular unit recordings are the cholinergic interneurons of the neostriatum comes from a study by Wilson and colleagues (1990). Utilizing intracellular recording and subsequent filling with horseradish peroxidase, Wilson et al. (1990) demonstrated that the morphological features of neurons which fire tonically are indistinguishable from choline acetyltransferase-containing neurons identified with immunocytochemical techniques (Bolam et al., 1984; Wainer et al., 1984; Phelps et al., 1985; DiFiglia, 1987).

4. PANS LOOK TONIC IF YOU CAN KEEP THEM UP

The work of Sedgwick and Williams (1967) suggested that the origin of phasic firing in PANs might be traced to episodic fluctuations in the membrane potential. In an elegant series of experiments Katayama and colleagues (1980) demonstrated that firing episodes were correlated with fluctuations in the cortical field potential and that pairs of neostriatal units tended to exhibit synchronous episodes. These data led to speculation that the cortex might be intimately involved in the production of rhythmic activity (Katayama et al., 1980). However, it was not until the study of Wilson and Groves (1981) that the neurons in which the episodic fluctuations in membrane potential occurred were unequivocally identified as neostriatal spiny cells. The two preferred levels of polarization have subsequently come to be known as the Up and Down states (see Wilson and Kawaguchi, 1996). Spikes and synaptic potentials are absent from the Down state which is very polarized (typically -80 to -90 mV; see Wilson, 1993). By contrast, the Up state exhibits noisy fluctuations in the membrane potential and spikes are triggered by rapid depolarizations (Wilson, 1993). These observations provided insight into the characteristic appearance of autocorrelograms generated from spike trains of PANs. Thus, the episodic fluctuations in membrane potential give rise to the early peak, and the seemingly random timing of spikes during Up states and lack of periodicity in state transitions are responsible for an overall low firing probability upon which the peak of the autocorrelogram is superimposed (Stern et al., 1998).

The hyperpolarized episodes or Down states, were originally considered to be due to inhibitory synaptic input (Hull et al., 1970, 1973; Buchwald et al., 1973; Bernardi et al., 1975, 1976; Katayama et al., 1980), which was thought to arise from neighboring GABAergic spiny cells providing mutual inhibition. In a series of experiments by Wilson and colleagues (Wilson et al., 1983a, 1983b, 1986) it was demonstrated that the Down state actually reflected periods of synaptic quiesence and expression of the intrinsic properties of the neuron. The Up state, by contrast, has an absolute requirement for excitatory input and is absent in preparations in which the cortical innervation of the neostriatum is interrupted (Wilson et al., 1983a). The Down state is dominated by a powerful, voltage-dependent, inwardly rectifying potassium current (I_{Kir}) which serves to clamp the membrane potential close to E_K (Kita et al., 1985b; Kawaguchi et al., 1989; Uchimura et al., 1989; Jiang and North, 1991; Nisenbaum and Wilson, 1995). Transitions to the Up state and subsequent spike generation require a massive barrage of relatively synchronous depolarizing synaptic input in order to produce sufficient depolarization to begin deactivation of I_{Kir}. Once this process begins then it leads to a very rapid transition to the Up state owing to positive feed-back i.e. depolarization causes further deactivation of I_{Kir} which causes additional depolarization owing to the increased input resistance of the neuron and collapsing electrotonic structure (see Wilson, 1993). In the absence of opposition, such a synaptic barrage would be expected to drive spiny cells towards the reversal potential of the synaptic input. However, this does not happen and instead the Up state is established, typically at a few millivolts below action potential threshold. At this depolarized membrane potential outwardly rectifying potassium currents (I_{Af}, I_{As} and a non-inactivating potassium current) are activated (Bargas et al., 1989; Surmeier et al., 1992; Nisenbaum et al., 1994, 1996; Nisenbaum and Wilson, 1995) as well as voltage-dependent inward currents (Kits et al., 1985a; Calabresi et al., 1987; Galarraga et al., 1989). Both inward and outward currents interact with the synaptic barrage to produce the Up state (Wilson and Kawaguchi, 1996). The noisy appearance of the membrane potential during the Up state and the irregularity of spike timing is accounted for by moment-to-moment fluctuations in synaptic input which are occasionally sufficient to bring the cell to threshold. Once the depolarizing drive to the cell is reduced past a critical level, the Up state is terminated by the concerted action of the outwardly rectifying potassium currents which drive the cell towards E_K, activating I_{Kir} and reestablishing the Down state. The phasic firing of spiny cells therefore reflects phasic synaptic input and when synaptic drive is prolonged, long-duration episodes of *tonic* firing can occur (see Gardiner and Nelson, 1992).

5. TANS ARE TONIC IN VITRO

In what is the only intracellular study of cholinergic cells *in vivo*, action potential generation was observed to arise from the depolarizations produced by summation of a very few, discrete depolarizing potentials (Wilson et al., 1990). The sensitivity of cholinergic cells to synaptic excitation is belived to be due to the relatively depolarized membrane potential and high input resistance of these neurons (Wilson et al., 1990; Jiang and North, 1991; Kawaguchi, 1992, 1993; Bennett and Wilson, 1998) and contrasts sharply with the behavior of spiny cells (see above). These observations led to the suggestion that the cholinergic cells behave as a sensitive coincidence detector (see Wilson, 1993) with the timing of individual spikes largely reflecting the temporal structure of the synaptic barrage. In a recent study we examined the phase-dependent ability of excitatory and inhibitory synaptic inputs to shorten or lengthen the interspike interval (Bennett and Wilson, 1998). We showed that the amplitude of the depolarization produced by EPSPs or brief depolarizing current pulses were strongly dependent upon the recent voltage history of the cell. At short post-spike intervals (~50 ms), depolarizing inputs were relatively ineffective because of a marked reduction in the apparent input resistance of the cell owing to the fact that the afterhyperpolarization is maximally activated at this time (Wilson et al., 1990; Jiang and North, 1991; Kawaguchi, 1992, 1993). The reduced input resistance diminishes the efficacy of depolarizing synaptic inputs and the afterhyperpolarization therefore places a ceiling on the firing rate by vetoing synaptic input occuring at short post-spike intervals (Bennett and Wilson, 1998).

As the membrane potential approaches threshold, depolarizing current pulses were found to produce larger and more longer lasting voltage excursions than hyperpolarizing current pulses of the same amplitude (Bennett and Wilson, 1998). This non-linearity is likely to be due to the recruitment of a subthreshold sodium current (Chao and Alzheimer, 1995; Bennett et al., 1999) and coupled with the increasing input resistance, produced as the conductance underlying the afterhyperpolarization deactivates, makes excitatory inputs more effective later in the interspike interval. In contrast to EPSPs, the efficacy of inhibitory synaptic inputs were not phase-dependent. This suggested that such inputs might act to influence the overall level of excitability rather than tightly regulating spike timing as seems to be the case for EPSPs (Bennett and Wilson, 1998). It should be noted that whilst IPSPs were effective throughout the interspike interval, they were much less readily evoked than EPSPs, indicating that inhibitory inputs to cholinergic cells are probably sparse (Bennett and Wilson, 1998).

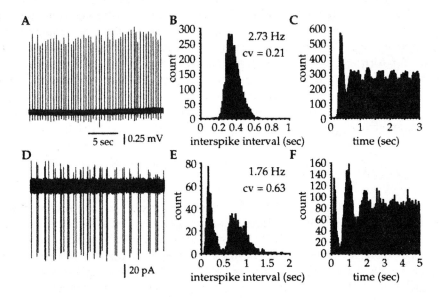

Figure 1. Extracellular single unit recordings of cholinergic cells firing spontaneously *in vitro*. A spectrum of firing rates and patterns are observed ranging from regular/irregular spiking (A-C) to rhythmic clustered or burst firing (D-F). Regular/irregular spiking gave rise to unimodal interspike interval (ISI) histograms (B). Typically spike trains with a high coefficient of variation (cv) gave rise to ISI histograms which were increasingly skewed to the right. The autocorrelogram has a large initial peak, indicating the increased likelihood of spiking at the end of the afterhyperpolarization and some additional periodic components (C). Cholinergic neurons firing bursts or clusters of spikes (D) were readily identified by their bimodal ISI histograms (E) and multipeaked autocorrelograms (F). The two modes in the ISI histogram correspond to the intraburst and interburst intervals (E). The early peak in the autocorrelogram corresponds to the increased probability of spiking within a burst and the second peak reflects the periodicity of the interburst intervals. Recordings at 35 ± 2 °C. Bins for B and E =10 ms; C and F = 20 ms. Sample for B and C = 31 min; E and F = 22 min.

The observation that tonic activity was still observed in cholinergic cells, both *in vivo* (Wilson et al., 1990) and *in vitro* (Bennett and Wilson, 1998; Calabresi et al., 1998; Lee et al., 1998), under conditions of reduced excitatory input and the presence of a regenerative, subthreshold inward current indicated that these cells might be endogenously active. Pharmacological blockade of spontaneous excitatory, inhibitory and neuromodulatory synaptic inputs to cholinergic interneurons did not influence spontaneous firing *in vitro*, demonstrating that these cells are tonically active in the absence of any input (Bennett and Wilson, 1999). Furthermore, the full range of firing rates and patterns which have been described for TANS *in vivo* are observed *in vitro* (Fig. 1) indicating that, although spike timing is clearly influenced by synaptic input (Wilson et al., 1990; Bennett and Wilson, 1998) spike generation and patterning are determined largely by the intrinsic properties of the cell (Bennett and Wilson, 1999). The fact that cholinergic cells studied *in vitro* can give rise to either unimodal or bimodal interspike interval histograms (Fig. 1) suggests that similar firing patterns observed during unit recordings *in vivo* (Aosaki et al., 1994b) do not necessarily reflect an entrainment to episodic synaptic input, but may result from a change in the intrinsic firing mode of the neuron. Whole-cell recordings (Bennett and Wilson, 1999) have revealed that when cholinergic cells spike in a regular or irregular pattern the membrane potential typically remains within 5 or 10 mV of action potential threshold during the interspike interval (Fig. 2). However, during burst firing the membrane potential traverses a much wider voltage range (Fig. 2) suggesting that ionic currents which are usually inactive during regular or irregular firing become active during bursting and vice versa. Preliminary data indicate that in the subthreshold voltage range a noninactivating tetrodotoxin-sensitive sodium current inward

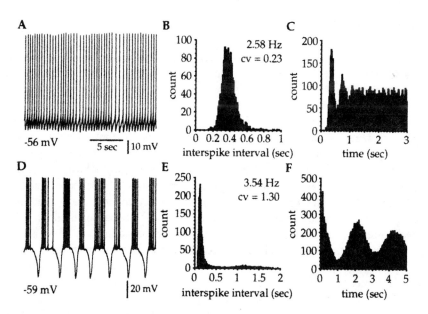

Figure 2. Whole-cell recordings illustrating the membrane potential fluctuations, interspike interval histograms and autocorrelograms associated with regular/irregular spiking (A-C) and burst firing (D-F). The membrane potential typically remains within 5-10 mV during regular or irregular spiking (A). The interspike interval (ISI) histogram is unimodal and skewed slightly to the right (B) and the corresponding autocorrelogram has a single prominent peak reflecting the increased probability of firing at the end of the afterhyperpolarization and subsequent weak periodicity (C). During burst firing the membrane potential makes very large excursions and spiking is triggered during the episodic depolarizations (D). The corresponding ISI histogram is bimodal, with the first mode corresponding to the intraburst intervals and the second to the interburst intervals (E). The autocorrelogram exhibits multiple peaks (F). The early peak is due to the increased spiking probability within a burst and the strong rhythmicity of bursting in this neuron is seen as recurring periodic components (F). The membrane potential for the initial point of each trace in A and D is indicated. Recordings at 35 ± 2 °C. Bins for B and E =10 ms; C and F = 20 ms. Sample for B and C =10 min; E and F = 12.5 min.

is available and provides a depolarizing influence (Bennett et al., 1999). Although subthreshold calcium entry is detectable, this does not appear to contribute a significant inward current. However, calcium entry is required for activation of the calcium-dependent potassium current (Kawaguchi, 1993) which underlies the afterhyperpolarization (AHP). The AHP is likely to be a pivotal determinant in the spiking pattern as pharmacological treatments which reduce the amplitude of the AHP convert regularly or irregularly spiking cells into burst firing neurons (Bennett et al., 1999). Furthermore, activation of D1 dopamine receptors has been shown to prolong the interspike interval by enhancing the AHP amplitude (Bennett and Wilson, 1998) which could explain why blockade of dopaminergic transmission results in a loss of the pause response (Aosaki et al., 1994a; Watanabe and Kimura, 1998). The currently available *in vitro* and *in vivo* data together indicate that cholinergic neurons exhibit both tonic and phasic firing and switching between these firing modes probably arises from modulation of intrinsic conductances.

6. STANS TO REASON

There are at least two major points which emerge from the preceding discussion of spiking patterns and their origins in PANs and TANs. The first is a taxonomical issue regarding the identification of TANs and PANs and the second relates to the cellular and synaptic processes which give rise to the firing characteristics peculiar to each class of neuron.

The firing rates of both TANs and PANs span a wide range and overlap to a considerable degree. Identification schemes which rely on this criteria alone are potentially inaccurate. Firing pattern is a more useful indicator of cell identity, with interspike interval histograms and autocorrelograms being especially useful. Although the firing patterns of both TANs and PANs can be diverse, the autocorrelograms generated from spike trains of spiny cells are sufficiently stereotyped as to lend considerable confidence when identifying neurons particularly when combined with characterization of action potential waveform and response properties of the same neuron. The acronymical exactitude of TANs and PANs is undermined by the fact that both classes contain neurons which can be described as tonic or phasic. We might ponder whether both TANs and PANs could best be referred to as STANs (sometimes tonically active neurons).

Examination of the mechanisms underlying spike generation and patterning in PANS and TANs reveals dramatic differences in these two populations of neurons. The spiny cells are electrically quiescent in the absence of any extrinsic perturbation and require massive, relatively synchronous excitatory input to produce state transitions and spike triggering from the Up state. The episodic nature of these inputs produces the phasic activity associated with extracellular unit recordings and the irregularity of spiking in the Up state accounts for the absence of periodicity in autocorrelograms generated from spike trains of spiny cells. If the input to spiny cells ceases to be phasic then so does their output. In contrast, cholinergic interneurons fire spontaneously in the absence of any input and produce spike trains which are remarkably similar to those reported for TANs *in* vivo. This suggests that the synaptic barrage to which these cells are ordinarily subjected acts to disrupt ongoing activity and raises the possibility that pauses in firing which occur in response to sensory stimuli might result from a reduction of excitatory input and susequent expression of the underlying firing pattern of the cell. Alternatively, such pauses could result from transient firing mode shifts produced by synaptic or neuromodulatory influences.

7. ACKNOWLEDGEMENTS

This work was supported by NIH grant NS 37760. We would like to thank Dr. James Tepper for his suggestion of the acronym, STANs.

8. REFERENCES

Albe-Fessard D, Rocha-Miranda C, Oswaldo-Cruz E (1960) Activités évoquées dans le noyau caudé du chat en réponse a des types divers d'afférences. II. Etude microphysiologique. EEG Clin Neurophysiol 12:649-661.
Albe-Fessard D, Raieva S, Santiago W (1967) Sur les relations entre la substance noire et le noyau caudé. J Physiol (Paris) 59:324-325.
Aldridge JW, Gilman S, Dauth G (1990) Spontaneous neuronal unit activity in the primate basal ganglia and the effects of precentral cerebral cortical ablations. Brain Res 516:46-56.
Aldridge JW, Gilman S (1991) The temporal structure of spike trains in the primate basal ganglia: Afferent regulation of bursting demonstrated with precentral cortical ablation. Brain Res 543:123-138.
Alexander GE, DeLong MR (1985) Microstimulation of the primate neostriatum. II. Somatotopic organization of striatal microexcitable zones and their relation to neuronal response properties. J Neurophysiol 53:1417-1430.
Anderson ME (1977) Discharge patterns of basal ganglia neurons during active maintenance of postural stability and adjustment to chair tilt. Brain Res 143:325-338.
Aosaki T, Graybiel AM, Kimura M (1994) Effect of the nigrostriatal dopamine system on acquired neural responses in the striatum of behaving monkeys. Science 265:412-415.
Aosaki T, Tsubokawa H, Ishida A, Watanabe K, Graybiel AM, Kimura M (1994b) Responses of tonically active neurons in the primate's striatum undergo systematic changes during behavioral sensorimotor conditioning. J Neurosci 14:3969-3984.
Aosaki T, Kimura M, Graybiel AM (1995) Temporal and spatial characteristics of tonically active neurons of the primate's striatum. J Neurophysiol 73:1234-1252.
Apicella P, Scarnati E, Schultz W (1991) Tonically discharging neurons of monkey striatum respond to preparatory and rewarding stimuli. Exp Brain Res 84:672-675.
Bargas J, Galarraga E, Aceves J (1989) An early outward conductance modulates the firing latency and frequency of neostriatal neurons of the rat brain. Exp Brain Res 75:146-156.
Bennett BD, Wilson CJ (1998) Synaptic regulation of action potential timing in neostriatal cholinergic interneurons. J Neurosci 18:8539-8549.
Bennett BD, Wilson CJ (1999) Spontaneous activity of neostriatal cholinergic interneurons in vitro. J Neurosci 19:5586-5596.
Bennett BD, Callaway JC, Wilson CJ (1999) Intrinsic mechanisms underlying spike generation and patterning in neostriatal cholinergic intemeurons. Soc Neurosci Abst 25:in press.
Bernardi G, Marciani MG, Morocutti C, Giacomini P (1975) The action of GABA on rat caudate neurones recorded intracellularly. Brain Res 92:511-515.
Bernardi G, Marciani MG, Morucutti C, Giacomini P (1976) The action of picrotoxin and bicuculline on rat caudate neurons inhibited by GABA. Brain Res 102:379-384.
Bolam JP, Wainer BH, Smith AD (1984) Characterization of cholinergic neurons in the rat neostriatum. A combination of choline acetyltransferase immunocytochemistry, Golgi- impregnation and electron microscopy. Neuroscience 12:711-718.
Buchwald NA, Price DD, Vernon L, Hull CD (1973) Caudate intracellular response to thalamic and cortical inputs. Exp Neurol 38:311-323.
Calabresi P, Misgeld U, Dodt HU (1987) Intrinsic membrane properties of neostriatal neurons can account for their low level of spontaneous activity. Neuroscience 20:293-303.

Calabresi P, Centonze D, Pisani A, Sancesario G, North RA, Bernardi G (1998) Muscarinic IPSPs in rat striatal cholinergic interneurones. J Physiol (Lond) 510:421-427.

Chao TI, Alzheimer C (1995) Do neurons from rat neostriatum express both a TTX-sensitive and a TTXinsensitive slow Na+ current? J Neurophysiol 74:934-941.

Cocatre-Zilgien JH, Delcomyn F (1992) Identification of bursts in spike trains. J Neurosci Methods 41:19-30.

Connor JD (1968) Caudate unit responses to nigral stimuli: Evidence for a possible nigro-neostriatal pathway. Science 160:1240-1242.

Connor JD (1970) Caudate nucleus neurones: Correlation of the effects of substantia nigra stimulation with iontophoretic dopamine. J Physiol (Lond) 208:691-703.

DeLong MR (1972) Activity of basal ganglia neurons during movement Brain Res 40:127-135.

DeLong MR (1973) Putamen: Activity of single units during slow and rapid arm movements. Science 179:1240-1242.

DiFiglia M, Pasik P, Pasik T (1976) A Golgi study of neuronal types in the neostriatum of monkeys. Brain Res 114:245-256.

DiFiglia M (1987) Synaptic organization of cholinergic neurons in the monkey neostriatum. J Comp Neurol 255:245-258.

Feltz P (1970) Relation nigro-striatale: Essai de différentiation des excitations et inhibitions par micro-iontophorese de dopamine. J Physiol (Paris) 62:151.

Feltz P, Albe-Fessard D (1972) A study of the ascending nigro-caudate pathway. EEG Clin Neurophysiol 33:179-193.

Frigyesi TL, Purpura D (1966) Electrophysiological analysis of nigro-caudate evoked activities. Trans Amer Neurol Ass 91:236-238.

Frigyesi TL, Purpura D (1967) Electrophysiological analysis of reciprocal caudato-nigral relations. Brain Res 6:440-456.

Galarraga E, Bargas J, Sierra A, Aceves J (1989) The role of calcium in the repetitive firing of neostriatal neurons. Exp Brain Res 75:157-168.

Gardiner TW, Nelson RJ (1992) Striatal neuronal activity during the initiation and execution of hand movements made in response to visual and vibratory cues. Exp Brain Res 92:15-26.

Graybiel AM, Aosaki T, Flaherty AW, Kimura M (1994) The basal ganglia and adaptive motor control. Science 265:1826-1831.

Hull CD, Bernardi G, Buchwald NA (1970) Intracellular responses of caudate neurons to brain stem stimulation. Brain Res 22:163-179.

Hull CD, Bernardi G, Price DD, Buchwald NA (1973) Intracellular responses of caudate neurons to temporally and spatially combined stimuli. Exp Neurol 38:324-336.

Jiang ZG, North RA (1991) Membrane properties and synaptic responses of rat striatal neurones in vitro. J Physiol (Lond) 443:533-553.

Katayama Y, Tsubokawa T Moriyasu N (1980) Slow rhythmic activity of caudate neurons in the cat: Statistical analysis of caudate neuronal spike trains. Exp Neurol 68:310-321.

Kawaguchi Y, Wilson CJ, Emson PC (1989) Intracellular recording of identified neostriatal patch and matrix spiny cells in a slice preparation preserving cortical inputs. J Neurophysiol 62:1052-1068.

Kawaguchi Y (1992) Large aspiny cells in the matrix of the rat neostriatum in vitro: physiological identification, relation to the compartments and excitatory postsynaptic currents. J Neurophysiol 67:1669-1682.

Kawaguchi Y (1993) Physiological, morphological, and histochemical characterization of three classes of interneurons in rat neostriatum. J Neurosci 13:4908-4923.

Kimura M, Rajkowski J, Evarts E (1984) Tonically discharging putamen neurons exhibit set-dependent responses. Proc Natl Acad Sci U S A 81:4998-5001.

Kimura M, Kato M, Shimazaki H (1990) Physiological properties of projection neurons in the monkey striatum to the globus pallidus. Exp Brain Res 82:672-676.

Kimura M, Kato M, Shimazaki H, Watanabe K, Matsumoto N (1996) Neural information transferred from the putamen to the globus pallidus during learned movement in the monkey. J Neurophysiol 76:3771-3786.

Kita H, Kita T, Kitai ST (1985a) Regenerative potentials in rat neostriatal neurons in an in vitro slice preparation. Exp Brain Res 60:63-70.

Kita T Kita H, Kitai ST (1985b) Effects of 4-aminopyridine (4-AP) on rat neostriatal neurons in an in vitro slice preparation. Brain Res 361:10-18.

Lee K, Dixon AK, Freeman TC, Richardson PJ (1998) Identification of an ATP-sensitive potassium channel current in rat striatal cholinergic interneurones. J Physiol (Lond) 510:441-453.

Liles SL (1974) Single-unit responses of caudate neurons to stimulation of frontal cortex, substantia nigra and entopeduncular nucleus in cats. J Neurophysiol 37:254-265.

Matsumoto N, Minamimoto T, Graybiel AM, Kimura M (1997) Expression of behaviorally conditioned responses of tonically active neurons depends on thalamic input from CM-Pf complex. Soc Neurosci Abst 23:464.

McLennan H, York DH (1967) The action of dopamine on neurones of the caudate nucleus. J Physiol (Lond)189:393-412.

Nisenbaum ES, Orr WB, BergerTW (1988) Evidence for two functionally distinct subpopulations of neurons within the rat striatum. J Neurosci 8:4138-4150.

Nisenbaum ES, Berger TW (1992) Functionally distinct subpopulations of striatal neurons are differentially regulated by GABAergic and dopaminergic inputs--I. In vivo analysis. Neuroscience 48:561-578.

Nisenbaum ES, Grace AA, Berger TW (1992) Functionally distinct subpopulations of striatal neurons are differentially regulated by GABAergic and dopaminergic inputs--II. In vitro analysis. Neuroscience 48:579-593.

Nisenbaum ES, Xu ZC, Wilson CJ (1994) Contribution of a slowly inactivating potassium current to the transition to firing of neostriatal spiny projection neurons. J Neurophysiol 71:1174-1189.

Nisenbaum ES, Wilson CJ (1995) Potassium currents responsible for inward and outward rectification in rat neostriatal spiny projection neurons. J Neurosci 15:4449-4463.

Nisenbaum ES, Wilson CJ, Foehring RC, Surmeier DJ (1996) Isolation and characterization of a persistent potassium current in neostriatal neurons. J Neurophysiol 76:1180-1194.

Phelps PE, Houser CR, Vaughn JE (1985) Immunocytochemical localization of choline acetyltransferase within the rat neostriatum: a correlated light and electron microscopic study of cholinergic neurons and synapses. J Comp Neurol 238:286-307.

Raz A, Feingold A, Zelanskaya V, Vaadia E, Bergman H (1996) Neuronal synchronization of tonically active neurons in the striatum of normal and parkinsonian primates. J Neurophysiol 76:2083-2088.

Richardson TL, Miller JJ, McLennan H (1977) Mechanisms of excitation and inhibition in the nigrostriatal system. Brain Res 127:219-234.

Sedgwick EM, Williams TD (1967) The response of single units in the caudate nucleus to peripheral stimulation. J Physiol (Lond)189:281-298.

Stern EA, Jaeger D, Wilson CJ (1998) Membrane potential synchrony of simultaneously recorded striatal spiny neurons in vivo.Nature 394:475-478.

Surmeier DJ, Xu ZC, Wilson CJ, Stefani A, Kitai ST (1992) Grafted neostriatal neurons express a late-developing transient potassium current. Neuroscience 48:849-856.

Uchimura N, Higashi H, Nishi S (1989) Membrane properties and synaptic responses of the guinea pig nucleus accumbens neurons in vitro. J Neurophysiol 61:769-779.

Wainer BH, Bolam JP, Freund TF, Henderson Z, Totterdell S, Smith AD (1984) Cholinergic synapses in the rat brain: a correlated light and electron microscopic immunohistochemical study employing a monoclonal antibody against choline acetyltransferase. Brain Res 308:69-76.

Watanabe K, Kimura M (1998) Dopamine receptor-mediated mechanisms involved in the expression of learned activity of primate striatal neurons. J Neurophysiol 79:2568-2580.

Wilson CJ, Groves PM (1981) Spontaneous firing patterns of identified spiny neurons in the rat neostriatum. Brain Res 220:67-80.

Wilson CJ, Chang HT, Kitai ST (1983a) Origins of post synaptic potentials evoked in spiny neostriatal projection neurons by thalamic stimulation in the rat. Exp Brain Res 51:217-226.

Wilson CJ, Chang HT, Kital ST (1983b) Disfacilitation and long-lasting inhibition of neostriatal neurons in the rat. Exp Brain Res 51:227-235.

Wilson CJ (1986) Postsynaptic potentials evoked in spiny neostriatal projection neurons by stimulation of ipsilateral and contralateral neocortex. Brain Res 367:201-213.

Wilson CJ, Chang HT, Kital ST (1990) Firing patterns and synaptic potentials of identified giant aspiny interneurons in the rat neostriatum. J Neurosci 10:508-519.

Wilson CJ (1993) The generation of natural firing patterns in neostriatal neurons. Prog Brain Res 99:277-297.

Wilson CJ, Kawaguchi Y (1996) The origins of two-state spontaneous membrane potential fluctuations of neostriatal spiny neurons. J Neurosci 16:2397-2410.

DOPAMINE AND ENSEMBLE CODING IN THE STRIATUM AND NUCLEUS ACCUMBENS
A coincidence detection mechanism

Patricio O'Donnell*, M.D., Ph.D.

Key words: accumbens, dopamine, electrophysiology, schizophrenia.

Abstract: Most medium spiny neurons in the striatum and nucleus accumbens exhibit spontaneous alternations in their membrane potential when recorded intracellularly *in vivo*. A very negative resting potential (*down state*) is periodically interrupted by plateau depolarizations that bring the membrane potential close to firing threshold (*up state*). Action potential firing can only be observed during the up state. The spatial distribution of neurons in their up state at any given moment may represent an ensemble that controls the disinhibition of a set of thalamocortical units. The modulation of transitions between states may therefore have an impact on information processing within basal ganglia circuits. The actions of dopamine in the basal ganglia have been studied mostly with a focus on whether it is an excitatory or inhibitory agent, but the diversity of receptor subtypes, behavioral responses, and electrophysiological actions has precluded a conclusive view of the overall effect of dopamine cell activation. Following M. Levine's recent work showing that D_1 receptors may enhance NMDA-mediated responses and D_2 receptors may reduce AMPA responses, we propose that DA may act by stabilizing the membrane potential state in target neurons (either *up* or *down*). Dopamine cells fire en bloc in the presence of behaviorally relevant stimuli, resulting in a massive release of dopamine in target areas that may "tag" the distribution of active and inactive neurons. Therefore, dopamine may reinforce the ongoing ensemble, in a coincidence detection mechanism that requires glutamatergic inputs and dopamine.

---

* P. O'Donnell, Department of Pharmacology & Neuroscience, Albany Medical College.

1. ENSEMBLE CODING IN THE NUCLEUS ACCUMBENS

Information processing in striatal regions is important for a variety of functions, including motor control, sensorimotor integration and cognitive processes. The dorsal striatum may subserve motor functions by virtue of its connections with motor cortical areas, whereas the nucleus accumbens (NAcc) may be involved in cognitive functions by its connections with the prefrontal cortex (PFC) and other limbic regions. Pennartz et al. (1994) have proposed that NAcc functions are based on the organization of neural ensembles. Indeed, specific connectivity patterns in different subterritories of the NAcc might determine the spatial distribution of neurons that can be linked in such an ensemble. However, neuronal ensembles would be more advantageous if they also relied on functional properties of NAcc neurons in addition to just their hardwiring. For example, an ensemble could be defined as a set of active neurons, and these sets could be differently arranged depending on contextual conditions. This would be similar to what in the hippocampus has been described as "ensemble coding" (Eichenbaum, Wiener et al. 1989). Here, a theoretical model on the modulation of information processing in striatal regions that can account for ensemble coding is presented based on previous experimental work. The spatial and temporal distribution of active and inactive medium spiny neurons (MSN), defined by their membrane potential states, can be established and reinforced by several afferent systems. This would result in dynamic neuronal ensembles participating in the striatal control of cortical motor and cognitive functions.

The anatomical organization of basal ganglia circuits in which ensemble coding may take place is rather complex. However, if we focus on circuits projecting back to cortical areas (ignoring the also important subcortical projections), a clear parallel organization emerges (Groenewegen, Wright et al. 1996; O'Donnell, Lavín et al. 1997). As dorsal striatal regions receive inputs from motor cortical areas, their projections through the globus pallidus, substantia nigra and thalamus target primarily those cortical regions. Similarly, the NAcc is the target of projections arising in the PFC, and projects back to this area via the ventral pallidum and the mediodorsal thalamic nucleus. Striatal MSN projecting to pallidal regions are GABAergic as are pallidal GABA neurons, which are normally very active providing a tonic inhibition of thalamic neurons. This organization is a disinhibitory mechanism, characteristic of the output regions of the basal ganglia (Chevalier and Deniau 1990). Basal ganglia circuits are somewhat "closed"; that is, thalamocortical neurons target primarily the cortical area projecting to the striatal region that disinhibits them (Groenewegen and Berendse 1994). These circuits, however, have also been defined as "open-interconnected" (Joel and Weiner 1994). Indeed, there have been a number of reports indicating cross-talk between these parallel circuits (Haber, Kunishio et al. 1995; Groenewegen, Wright et al. 1996; O'Donnell, Lavín et al. 1997). In addition, the NAcc receives important projections from a number of "limbic" regions. We have shown a high degree of convergence between hippocampal, prefrontal cortical and amygdaloid afferents in this region (O'Donnell and Grace 1995). The interactions among these inputs may define activity levels in the NAcc. In turn, information processing within the NAcc may influence the activity of PFC neurons, and the integration of information taking place in the NAcc may be an important element in this function.

2. UP AND DOWN STATES IN NUCLEUS ACCUMBENS NEURONS

NAcc neurons exhibit two activity states and the distribution of neurons in either state at any given moment can define a neural ensemble. NAcc MSN exhibit a very negative resting membrane potential (*down state*) interrupted by 100-1,000 ms plateau depolarizations (*up state*) (O'Donnell and Grace 1995; O'Donnell and Grace 1998). This had been previously observed in striatal MSN (Wilson and Groves 1981; Wilson and Kawaguchi 1996). Other cell types also express this pattern, including pyramidal cortical neurons (Steriade, Nuñez et al. 1993; Cowan, Sesack et al. 1994; Branchereau, van Bockstaele et al. 1996; O'Donnell and Lewis 1998).

Up events are likely to depend on synaptic inputs. This is indicated by the absence of neurons with up and down membrane potential states in the slice preparation. NAcc neurons recorded *in vitro* exhibit a very negative and stable membrane potential (Uchimura, Higashi et al. 1986; O'Donnell and Grace 1993) that resembles the down state observed *in vivo*. Furthermore, striatal MSN from decorticated animals did not exhibit up events (Wilson, Chang et al. 1982), indicating that cortical afferents may be responsible for driving these events in the caudate-putamen (CPu). In the NAcc, on the other hand, we have shown that up events may be dependent on hippocampal inputs. In animals with a transection to the fimbria-fornix, the system carrying hippocampal afferents, no up-down states could be found in the NAcc, although they could be readily detected in the CPu (O'Donnell and Grace 1995). Stimulating the fimbria-fornix typically resulted in a depolarized and prolonged response resembling spontaneous up events (O'Donnell and Grace 1995). In addition, temporarily interrupting hippocampal afferents by injection of lidocaine in the fimbria-fornix resulted in a reversible, 15-20 minutes-long suppression of up events (O'Donnell and Grace 1995). These results indicate that the excitatory input driving up events in NAcc neurons arise primarily from hippocampal regions. Indeed, the NAcc receives heavy hippocampal input (Groenewegen, Wright et al. 1996). Although both hippocampal and neocortical terminals have been typically observed contacting spines in distal dendrites of NAcc neurons, near 10% of hippocampal afferents contact proximal dendrites or cell bodies (Meredith, Wouterlood et al. 1990), a region avoided by PFC fibers. This arrangement provides an anatomical basis for hippocampal afferents having a strong influence on membrane potential state transitions. Thus, both anatomical and physiological data support the idea of hippocampal inputs as crucial in determining NAcc neuronal activity. NAcc neurons can fire action potentials in response to cortical stimulation, but only when they are in the up state (O'Donnell and Grace 1995). Since up events in the NAcc are dependent on hippocampal input, we have interpreted these findings as evidence for a hippocampal gating of prefrontal cortical throughput in this region.

The dopamine (DA) innervation of the NAcc may play an important role in the functions attributed to this brain region, as well as in schizophrenia and other mental disorders. DA has been observed to exert multiple actions within the NAcc; however, the nature of these actions is still controversial in most cases and a clear picture on how DA may contribute to the control of NAcc information processing has yet to be provided. Among the variety of DA actions is the control of neurotransmitter release. DA has been known to affect release of other neurotransmitters via presynaptic heteroreceptors; in the NAcc, such mechanism has been observed in the control of PFC (O'Donnell and Grace 1994) and hippocampal (Yang and Mogenson 1984; Pennartz, Dolleman-van der Weel et al. 1992) afferent activity. The

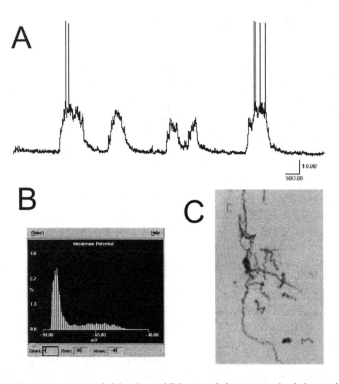

Figure 1. Most NAcc neurons recorded in vivo exhibit up and down states in their membrane potential. A: Representative tracing showing alternation between up and down states in a NAcc neuron. B: A histogram plotting the number of points recorded at any given membrane potential reveals a bimodal distribution characteristic of neurons with up and down membrane potential states (down: -81 mV; up: -63 mV). C: Neurobiotin injection revealed this neuron as a MSN located in the dorsal NAcc core.

receptor subtypes involved, as well as the direction of modulation remain controversial. For example, we have observed a D_2-mediated decrease in the responses to PFC in NAcc slices (O'Donnell and Grace 1994), whereas others have reported a D_1-mediated decrease (Harvey and Lacey 1996; Nicola, Kombian et al. 1996). The latter response may involve an atypical D_1 receptor since it required unusually high doses of D_1 antagonists to be blocked (Nicola, Kombian et al. 1996) and does not involve cAMP (Harvey and Lacey 1996). Postsynaptic actions of DA have included membrane depolarization (Akaike, Ohno et al. 1987; O'Donnell and Grace 1996), hyperpolarization (Uchimura, Higashi et al. 1986) or both, depending on the experimental conditions. Reports have also indicated either increase (Gonon and Sundstrom 1996) or decrease (O'Donnell and Grace 1996) in cell excitability. It is very difficult at present to provide a synthesis that can account for such a diverse set of findings. Some emerging ideas, however, may illuminate a path toward a unifying view on DA actions. Examples of this are some DA-glutamate interactions in the striatum, which are different depending on the receptor subtypes involved. Michael Levine has recently demonstrated in an elegant series of studies that D_1 receptor activation may enhance NMDA responses whereas D_2 DA receptors most often decrease non-NMDA responses (Levine, Altemus et al. 1996; Levine, Li et al. 1996; Cepeda and Levine 1998). These results suggest that at the negative membrane potential of the striatal neuron down states, with NMDA receptors blocked by Mg^{2+}, only the D_2 inhibition of non-NMDA receptors would be effective. On the other hand, at the depolarized levels of the up state, the Mg^{2+} inhibition would be at least partially removed, allowing for the expression of a D_1-mediated increase in NMDA responses. This would predict different actions of DA in the up vs. the down state of target neurons. Furthermore, activation of DA receptors results in the activation of a variety of second messenger cascades. Some of them have been shown to phosphorilate ion channels, an action that may have physiological consequences that depend on the membrane potential at which it occurs. For example, DA receptors may enhance the K^+ inward rectifier current ($I_{K,IR}$) (Surmeier and Kitai 1993). This current is active only at very negative membrane potentials (i.e., in the down state) and may be involved in providing stability to the down state. Only strong depolarization may overcome the action of this current, and when this occurs the channel closes (it inactivates at slightly depolarized potentials), having the effect of letting go the membrane potential and allowing a transition to the up state. On the other hand, a number of ion channels may maintain the up state. A good candidate is the slow voltage-dependent Na^+ current ($I_{Na,s}$), which is also enhanced by DA (Cepeda, Chandler et al. 1995). Calcium currents may also be involved, and L-type channels can be enhanced by D_1 receptors (Hernández-López, Bargas et al. 1997). Thus, DA may have different effects depending on the membrane potential, and these results indicate that DA receptors may reinforce the ion currents that contribute to either membrane potential state.

We are currently addressing these hypotheses using *in vivo* intracellular recordings from PFC pyramidal neurons (Lewis and O'Donnell 1998; O'Donnell and Lewis 1998). Prefrontal cortical pyramidal neurons also exhibited up and down membrane potential states (down state: -76.2 ± 7.1 mV, mean $\pm$ SD; up state: -65.8 ± 7.4 mV; n=23). Stimulation of the source of DA projections to this region, the ventral tegmental area (VTA) resulted in different effects depending on the membrane potential PFC neurons were at the time of stimulation. The most common response was a short-latency EPSP or IPSP followed by a return to the original state (Lewis and O'Donnell 1998; O'Donnell and Lewis 1998; O'Donnell, Greene et al. 1999). When the VTA was stimulated with trains imitating burst firing (i.e., 2-5 pulses at 20 Hz), long-lasting (400-3,000 ms) transitions to the up state were observed. Although administration of the D_1 antagonists SCH 23390 (0.3 mg/kg) did not prevent these transitions to the up state, their duration was significantly reduced in all neurons tested (Lewis and O'Donnell 1998). This indicates that, although a non-DA mechanism may depolarize PFC pyramidal cells to their up state, DA acting via D_1 receptors may prolong this up state for several hundred milliseconds.

3. COINCIDENCE DETECTION IN STRIATAL REGIONS

The factors triggering and maintaining the up state in striatal neurons may constitute a coincidence detection mechanism. Up events require hippocampal input activation to be triggered, and they probably require DA to be maintained. PFC inputs can then result in action potential firing only when both conditions occur. It is known that DA cells fire en bloc in the presence of behaviorally relevant stimuli (Schultz 1992). Given the state-stabilizing effect of DA, this massive firing may result in DA tagging the distribution of active and inactive units in striatal regions, reinforcing the ongoing ensemble. A coincidence detection mechanism like this one could explain findings such as those by DeFrance et al. (1981), who suggested that DA would tend to suppress NAcc cell firing unless they were activated by hippocampal input arriving within the theta range. Such a role of DA in coincidence detection within the NAcc and PFC could be important in synaptic plasticity mechanisms, setting ensembles that could be evoked later.

Furthermore, this model of ensemble coding based on membrane potential states requires synchronization of up events among sets of neurons in the NAcc. If up events are not synchronized, the NAcc output to the ventral pallidum will not be time-locked and may result in PFC units not being activated simultaneously by their MD afferents, failing to establish a functional ensemble in the PFC. NAcc neurons exhibit gap junctions, which may allow for synchronization of slow signals such as up events. Gap junctions may be the optimal means for action potential synchronization in a population of inhibitory neurons such as NAcc MSN; however, gap junctions behave as low-pass filters and therefore are not well-suited for fast signal transmission. But slower signals such as up events would be transmitted among coupled neurons without significant loss. Thus, if gap junctions are open in these structures, they would contribute to the coincidence detection function by synchronizing up events among coupled neurons, further defining the spatial domain of an ensemble. In the CPu, gap junctions are mostly closed at rest, but they can be open by activation of cortical afferents, an action that involves nitric oxide (O'Donnell and Grace 1997). DA may also contribute to up-event synchronization via its control of gap junctions in the NAcc (O'Donnell and Grace 1993) and by acting as state-stabilizer.

4. CONCLUSION

A coincidence detection mechanism like this may serve a variety of the functions in which the NAcc and other striatal regions participate. In the motor-related CPu, ensemble coding may participate in establishing presets of movement sequences as neural ensembles that can be sequentially triggered by external inputs. This may actually be a mechanism related the old and controversial concept of motor programming. In the cognitive realm, a similar preset may be established in NAcc-related circuits. Instead of movement sequences, the information encoded may be planned action sequences or even thought patterns. This may be precisely the mechanism by which the NAcc acts as a "cognitive pattern generator", as recently proposed by Ann Graybiel (1997). Indeed, the "chunking of action repertoires" also proposed by Graybiel (1998) may actually involve ensemble coding, defined as sets of neurons in the up state. Since the up state defines a condition in which a neuron is *ready* to be activated by its inputs, ensembles are defined by *activable* rather than by strictly active neurons. This provides a great flexibility in determining the distribution of active units at any given moment. Because of this, the coincidence detection is somewhat *fuzzy*, and much more plastic and adaptable that strict action potential firing coincidence. Such flexibility may underlie the very plastic characteristics of basal ganglia function and their ability to adapt routine operations (either motor or cognitive) to changing external conditions.

5. ACKNOWLEDGEMENTS

I wish to thank Ms. Barbara Lewis for her excellent assistance, and Mr. Brian Lowry (University of Pittsburgh) for providing the software used for data collection and analysis. This work was supported by a USPHS grant (MH57683).

6. REFERENCES

Akaike, A., Y. Ohno, et al. (1987). "Excitatory and inhibitory effects of dopamine on neural activity of the caudate nucleus neurons in vitro." *Brain Research* 418: 262-272.

Branchereau, P., E. J. van Bockstaele, et al. (1996). "Pyramidal neurons in rat prefrontal cortex show a complex synaptic response to single electrical stimulation of the locus coeruleus region: evidence for antidromic activation and GABAergic inhibition using in vivo intracellular recording and electron microscopy." *Synapse* 22: 313-331.

Cepeda, C., S. H. Chandler, et al. (1995). "Persistent Na$^+$ conductance in medium-sized neostriatal neurons: characterization using infrared videomicroscopy and whole cell patch clamp recordings." *Journal of Neurophysiology* 74: 1343-1348.

Cepeda, C. and M. S. Levine (1998). "Dopamine and N-methyl-D-aspartate receptor interactions in the neostriatum." *Developmental Neuroscience* 20: 1-18.

Chevalier, G. and J. M. Deniau (1990). "Disinhibition as a basic process in the expression of striatal functions." *Trends in Neuroscience* 13: 277-280.

Cowan, R. L., S. R. Sesack, et al. (1994). "Analysis of synaptic inputs and targets of physiologically characterized neurons in rat frontal cortex: combined in vivo intracellular recording and immunolabeling." *Synapse* 17: 101-114.

DeFrance, J. F., R. W. Sikes, et al. (1981). The electrophysiological effects of dopamine and histamine in the nucleus accumbens: frequency specificity. *The neurobiology of the nucleus accumbens*. R. B. Chronister and J. F. DeFrance. Brusnwick, ME, Hauer Institute: 230-252.

Eichenbaum, H., S. I. Wiener, et al. (1989). "The organization of spatial coding in the hippocampus: a study of neural ensemble activity." *Journal of Neuroscience* 9: 2764-2775.

Gonon, F. and L. Sundstrom (1996). "Excitatory effects of dopamine released by impulse flow in the rat nucleus accumbens in vivo." *Neuroscience* 75: 13-18.

Graybiel, A. M. (1997). "The basal ganglia and cognitive pattern generators." *Schizophrenia Bulletin* 23: 459-469.

Graybiel, A. M. (1998). "The basal ganglia and chunking of action repertoires." *Neurobiology of Learning and Memory* 70: 119-136.

Groenewegen, H. J. and H. W. Berendse (1994). Anatomical relationships between the prefrontal cortex and the basal ganglia in the rat. *Motor and Cognitive Functions of the Prefrontal Cortex*. A.-M. Thierry. Berlin, Springer-Verlag: 51-77.

Groenewegen, H. J., C. I. Wright, et al. (1996). "The nucleus accumbens: gateway for limbic structures to reach the motor system?" *Progress in Brain Research* 107: 485-511.

Haber, S. N., K. Kunishio, et al. (1995). "The orbital and medial prefrontal circuit through the primate basal ganglia." *Journal of Neuroscience* 15: 4851-4867.

Harvey, J. and M. G. Lacey (1996). "Endogenous and exogenous dopamine depress EPSCs in rat nucleus accumbens in vitro via D_1 receptor activation." *Journal of Physiology* 492: 143-154.

Hernández-López, S., J. Bargas, et al. (1997). "D_1 receptor activation enhances evoked discharge in neostriatal medium spiny neurons by modulating an L-type Ca^{2+} conductance." *Journal of Neuroscience* 17: 3334-3342.

Joel, D. and I. Weiner (1994). "The organization of the basal ganglia-thalamocortical circuits: open interconnected rather than closed segregated." *Neuroscience* 63: 363-379.

Levine, M. S., K. L. Altemus, et al. (1996). "Modulatory actions of dopamine on NMDA receptor-mediated responses are reduced in D_{1A}-deficient mutant mice." *Journal of Neuroscience* 16: 5870-5882.

Levine, M. S., Z. Li, et al. (1996). "Neuromodulatory actions of dopamine on synaptically-evoked neostriatal responses in slices." *Synapse* 24: 65-78.

Lewis, B. and P. O'Donnell (1998). "Effects of VTA stimulation of prefrontal cortex neurons recorded in vivo." *Society for Neuroscience Abstracts* 24: 656.
Meredith, G. E., F. G. Wouterlood, et al. (1990). "Hippocampal fibers make synaptic contact with glutamate decarboxilase-immunoreactive neurons in the rat nucleus accumbens." *Brain Research* 513: 329-334.
Nicola, S. M., S. B. Kombian, et al. (1996). "Psychostimulants depress excitatory synaptic transmission in the nucleus accumbens via presynaptic D1-like dopamine receptors." *Journal of Neuroscience* 16: 1591-1604.
O'Donnell, P. and A. A. Grace (1993). "Dopaminergic modulation of dye coupling between neurons in the core and shell regions of the nucleus accumbens." *Journal of Neuroscience* 13: 3456-3471.
O'Donnell, P. and A. A. Grace (1993). "Physiological and morphological properties of accumbens core and shell neurons recorded in vitro." *Synapse* 13: 135-160.
O'Donnell, P. and A. A. Grace (1994). "Tonic D_2-mediated attenuation of cortical excitation in nucleus accumbens neurons recorded in vitro." *Brain Research* 634: 105-112.
O'Donnell, P. and A. A. Grace (1995). "Synaptic interactions among excitatory afferents to nucleus accumbens neurons: hippocampal gating of prefrontal cortical input." *Journal of Neuroscience* 15: 3622-3639.
O'Donnell, P. and A. A. Grace (1996). "Dopaminergic reduction of excitability in nucleus accumbens neurons recorded in vitro." *Neuropsychopharmacology* 15: 87-98.
O'Donnell, P. and A. A. Grace (1997). "Cortical afferents modulate striatal gap junction permeability via nitric oxide." *Neuroscience* 76: 1-5.
O'Donnell, P. and A. A. Grace (1998). "Phencyclidine interferes with the hippocampal gating of nucleus accumbens neuronal activity in vivo." *Neuroscience* 87: 823-830.
O'Donnell, P., A. Lavín, et al. (1997). "Interconnected parallel circuits between rat nucleus accumbens and thalamus revealed by retrograde transynaptic transport of pseudorabies virus." *Journal of Neuroscience* 17: 2143-2167.
O'Donnell, P. and B. Lewis (1998). *Dopaminergic control of prefrontal cortical activity: effect of VTA stimulation on bistable membrane potential of prefrontal cortical pyramidal neurons.* Dopamine 98, Strasbourg, France.
O'Donnell, P., J. Greene et al. (1999). "Modulation of cell firing in the nucleus accumbens. " *Annals of the New York Academy of Sciences.* 877: 157-176.
Pennartz, C. M. A., M. J. Dolleman-van der Weel, et al. (1992). "Presynaptic dopamine D_1 receptors attenuate excitatory and inhibitory limbic inputs to the shell region of the rat nucleus accumbens." *Journal of Neurophysiology* 67: 1325-1334.
Pennartz, C. M. A., H. J. Groenewegen, et al. (1994). "The nucleus accumbens as a complex of functionally distinct neuronal ensembles: an integration of behavioural, electrophysiological and anatomical data." *Progress in Neurobiology* 42: 719-761.
Schultz, W. (1992). "Activity of dopamine neurons in the behaving primate." *Seminars in the Neurosciences* 4: 129-138.
Steriade, M., A. Nuñez, et al. (1993). "A novel slow (<1 Hz) oscillation of neocortical neurons in vivo: depolarizing and hyperpolarizing components." *Journal of Neuroscience* 13: 3252-3265.
Surmeier, D. J. and S. T. Kitai (1993). "D_1 and D_2 dopamine receptor modulation of sodium and potassium currents in rat neostriatal neurons." *Progress in Brain Research* 99: 309-324.
Uchimura, N., H. Higashi, et al. (1986). "Hyperpolarizing and depolarizing actions of dopamine via D-1 and D-2 receptors on nucleus accumbens neurons." *Brain Research* 375: 368-372.
Wilson, C. J., H. T. Chang, et al. (1982). "Origins of postsynaptic potentials evoked in identified rat neostriatal neurons by stimulation in substantia nigra." *Experimental Brain Research* 45: 157-167.
Wilson, C. J. and P. M. Groves (1981). "Spontaneous firing patterns of identified spiny neurons in the rat neostriatum." *Brain Research* 220: 67-80.
Wilson, C. J. and Y. Kawaguchi (1996). "The origins of two-state spontaneous membrane potential fluctuations of neostriatal spiny neurons." *Journal of Neuroscience* 16: 2397-2410.
Yang, C. R. and G. J. Mogenson (1984). "Electrophysiological responses of neurones in the accumbens nucleus to hippocampal stimulation and the attenuation of the excitatory responses by the mesolimbic dopaminergic system." *Brain Research* 324: 69-84.

NEURONAL FIRING PATTERNS IN THE SUBTHALAMIC NUCLEUS
Effects of dopamine receptor stimulation on multisecond oscillations

Kelly A. Allers, Deborah S. Kreiss, and Judith R. Walters*

1. INTRODUCTION

The subthalamic nucleus (STN) appears to play a critical role in mediating the hyper- and hypoactive behavioral states associated with basal ganglia pathophysiology[8,11,15,17,37]. Experiments designed to establish a more precise role for the STN have frequently been based on 'dual-circuit' models of basal ganglia motor circuitry[7]. These models predict that the neuronal firing rates of the STN will increase when dopamine cells of the substantia nigra pars compacta (SNc) degenerate, as occurs in Parkinson's disease (PD). This increase in firing rate has been demonstrated in animal models of PD in both the rat and monkey[8,17,23]. These models have had a profound impact on the development of effective treatment strategies for PD patients, leading to surgical procedures designed to attenuate the increase in STN neuronal firing rates. These procedures, which include high frequency stimulation and thermolytic lesion of the STN, significantly reduce the motor disturbances observed in human PD patients and in primate animal models of PD[5,15].

Increased basal STN neuronal firing rates in animal models of PD correlate well with model predictions, but firing rate changes induced by dopamine agonist administration are less consistent (see below). These discrepancies, given the well documented anatomical circuitry, have prompted this laboratory to investigate effects of dopamine receptor stimulation in the STN on modes of neural communication other than simple rate changes, i.e. changes in firing pattern. This chapter will briefly review studies reporting mean neuronal firing rate changes in the STN in response to dopamine agonist administration, previous literature on firing patterns in the STN, and current data regarding the presence of slow, multisecond oscillations in the STN firing rates which are altered by dopamine receptor

---

* K.A. Allers, D.S. Kreiss, and J.R. Walters, Experimental Therapeutics Branch, National Institute of Neurological Disorders and Stroke, National Institutes of Health, Bethesda, MD 20892. D.S. Kreiss, Department of Biology, University of Central Arkansas, 180 Lewis Science Center, Conway, AR 72035.

stimulation. Finally, the role of different patterns generated by STN neurons will be discussed in the context of basal ganglia function.

2. DOPAMINE AGONIST EFFECTS ON STN FIRING RATES

Dual circuit models of the basal ganglia predict that both D1/D2 dopamine receptor agonist and D2 selective agonist administration will decrease STN firing rates through sequential connections between striato-pallidal neurons expressing D2 receptors, and the pallido-subthalamic projection[47]. However, systemic administration of the D1/D2 agonist apomorphine to rats actually induces a robust increase in STN firing rates[22]. The D2 receptor agonist quinpirole does not induce the expected decrease in firing rates in rates, rather there is either a modest increase (0.16 mg/kg, Allers and Kreiss, unpublished observations) or no overall effect on firing rates[22].

Only in rats with 6-hydroxydopamine (6-OHDA) induced lesions of the SNc (a rat model of PD), have the predicted decreases in firing rates been observed in some STN neurons (but not all) with systemically administered apomorphine and the dopamine precursor L-DOPA[2,23]. The D2 agonist quinpirole, administered alone, does not alter firing rates in 6-OHDA rats (Aller and Kreiss, unpublished observations), while a D1 agonist (SKF38393) persists in increasing firing rates[23]. Hence, only a combination of both D1 and D2 receptor stimulation induces firing rate changes predicted by dual circuit models, only in the dopamine depleted animal and only in a subpopulation of STN neurons.

In short, studies investigating dopamine agonist-induced changes in firing rate in the basal ganglia lead one to conclude that analyzing averaged firing rates in the context of a unidirectional circuit through the basal ganglia is insufficient for explaining the results of these electrophysiological studies. Some results may be better incorporated into current models by the inclusion of additional circuitry. For instance, increases in STN neuronal firing rates may result from stimulation of D1 receptor-bearing cortical neurons projecting to the STN, or dopamine receptors located locally in the STN and other basal ganglia nuclei[13,33,45]. On the other hand, the STN and other basal ganglia nuclei may be communicating through more complex and patterned rate changes, conveying information through alterations in bursting and oscillatory activity, with varying frequencies and degrees of synchrony.

3. BURSTING, OSCILLATIONS AND SYNCHRONY IN THE STN

The idea that STN firing patterns are shaped by afferent signals, as opposed to intrinsic rhythms, is supported by a comparison of data obtained in *in vitro* preparations compared to data obtained *in vivo*. In more isolated culture preparations, STN neuronal spike trains are very regular, or pacemaker-like[10,34-36]. When the STN is co-cultured with the globus pallidus, the STN neurons fire in bursts and the coefficient of variance of STN spike trains increases[36]. Classification schemes and descriptions of *in vivo* STN spike trains assess irregularity and bursting properties, oscillatory frequencies and degree of synchrony[8,17,18,23,32,35,44]. These studies imply that as the STN is required to transmit information from an increasing number of sources, it exhibits a wide range of firing patterns. This raises questions about whether these various patterns serve different coding functions.

STN neurons are commonly described as being bursty and this burstiness may play a role in STN function. Analysis of bursting in STN neurons demonstrates an extreme example of the disparity between measuring average firing rates and measuring firing patterns. While firing rates can reach 300 Hz or more within a burst, average firing rates for these neurons are typically reported as being between 5-30 Hz. The proportion of neurons found to be bursting *in vivo* in rats and monkeys range from <10% to 93%, depending on the study[8,18,23,27,35,42]. Several investigations report changes in the proportion of neurons tonically firing in 'burst' mode after dopamine depletion in rats following 6-OHDA treatment, although both increases[17,18] and decreases in bursting[23], or burstiness, have been reported. Finally, bursting activity may occur in periodic or non-periodic episodes, with some evidence suggesting that the temporal organization of STN bursts is imparted by the globus pallidus or the cerebral cortex[27,36].

Relatively few analyses of oscillatory activity in STN neuronal spike trains have been reported. In organotypic cultures of globus pallidus and STN tissue, STN neurons exhibit oscillatory bursting activity with frequencies in a range of 0.4 – 2 Hz[36]. *In vivo*, periodic oscillatory changes in firing rates in the STN have been demonstrated in MPTP-lesioned monkeys (a primate model of PD) in the 4-5 Hz range which correlates with limb tremor induced by the MPTP lesion, and a higher 8-20 Hz range which does not correlate with movement[8,32]. In barbiturate anesthetized cats, a limb tremor with frequencies up to about 16 Hz develops which is also correlated with similar frequency oscillations in the STN[44].

Synchronous firing activity in pairs of STN neurons has been demonstrated in both rats and monkeys although reports differ as to what proportion of STN neuronal firing is correlated. In rats, nearly one-third of neuron pairs recorded exhibited correlated firing activity within 200-400 msec periods; this proportion was increased following lesions of the globus pallidus[43]. One investigation in monkeys reports correlated neuronal firing in 11-12% of neuronal pairs from both intact and MPTP lesioned monkeys, including oscillations in the 4-5 Hz range linked with limb tremors[8]. The disparity between proportions of correlated neurons may be attributable to the urethane anesthesia used in the rat study and lack of anesthe-

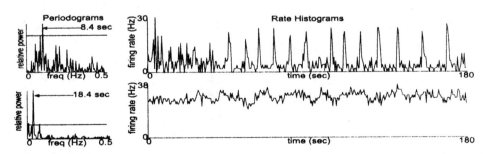

Figure 1. Lomb periodogram power spectra and rate histograms for two representative baseline STN spike trains. Rate histograms represent data from spike trains binned in 500 msec bins. The neuron in the upper trace is a bursty neuron with an oscillatory period of 8.4 sec. The neuron in the lower trace is not bursty but fluctuates with a 18.4 sec period. Periodograms show the power spectra generated with the Lomb algorithm. Arrows on periodograms indicate the main oscillatory period for these spike trains. Horizontal line on periodograms indicates significant periodicity at $p<0.05$.

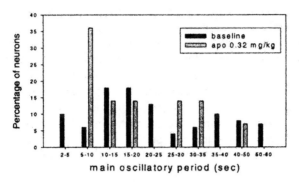

Figure 2. Distribution of main oscillatory periods in STN neurons during baseline (n = 83) and after administration of 0.32 mg/kg apomorphine (n = 14).

sia in the non-human primate study. Finally, synchronous, oscillatory bursts (0.4 – 2 Hz) have been reported between pairs of STN neurons in organotypic cultures and in ketamine/ xylazine anesthetized rats, the latter preparation exhibits oscillatory activity which is phase-locked with cortical slow-wave activity[27,36].

4. MULTISECOND OSCILLATIONS IN THE STN

Another firing pattern has recently been observed in the STN, which involves a slower time scale than those described above[2]. We have noted in single unit recording studies that an oscillatory rhythm is present in STN neuronal spike trains within the time scale of 2-60 seconds (Figure 1). These slow multisecond oscillations are present in 66% (83/126) of STN

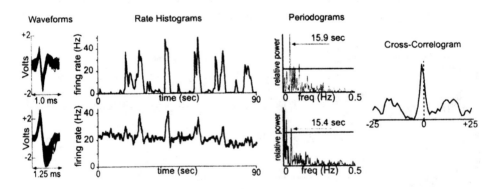

Figure 3. Waveforms, rate histograms, and periodograms from two simultaneously recorded STN neurons. Neuronal activity of two neurons are recorded from the same electrode and separated based on waveform size and shape. Rate histograms (500 msec bins) indicate similar oscillatory fluctuations between the two neurons although the average firing rates are quite different. Periodograms indicate similar oscillatory periods for the two neurons. The correlogram generated with these two spike trains indicates the neurons are 23° out of phase.

neurons recorded from animals that are awake, locally anesthetized and immobilized with gallamine. The mean main oscillatory period in these animals was 23.6 ± 1.7 sec. Figure 1 demonstrates examples of multisecond oscillations in firing rate with significant periodicity, as determined with the Lomb algorithm[2]. Multisecond oscillations in firing rate were found in both bursty and non-bursty neurons (burstiness assessed as in Kreiss, et al.[29]), and the average firing rate of a neuron did not predict the frequency of the oscillation present. As a population, STN neurons demonstrated oscillations with periods throughout the analyzed 2-60 sec range, but the disribution of intervals was skewed to the left, with the majority of periods being below 30 sec (Figure 2).

Slow oscillations in firing rate have also been found to occur in synchrony in pairs of STN neurons (Figure 3). Oscillations in the two simultaneously recorded neurons in Figure 3 have been determined by the Lomb algorithm to have periods of 15.8 sec (upper trace) and 15.3 sec (lower trace). Cross-correlation analysis confirms the coincidence of these oscillations, which are slightly out of phase. Phase relationships have been found to vary between 0° and 180°, but many are at or near 0°[3].

5. EFFECTS OF DOPAMINE RECEPTOR STIMULATION ON MULTISECOND OSCILLATIONS IN THE STN

The non-selective D1/D2 agonist apomorphine (0.32 mg/kg, i.v.) robustly shortened the mean main oscillatory period of STN neurons. Figure 4 summarizes these data. The mean main oscillatory period was 31.9 ± 5.3 sec following vehicle and 18.2 ± 3.2 sec (n = 17) following apomorphine. The shortening of mean main period was reversed by the D2 receptor antagonist haloperidol (0.2 mg/kg, i.v.) (Figure 4). As can be seen in the example in Figure 5, apomorphine induces not only the previously reported vigourous overall firing rate increase, but also, a large periodic firing rate fluctuation. The spike train in Figure 5 has an oscillation with a period of 17.5 sec in baseline. This period was shortened to 7.1 sec after apomorphine and lengthened to 27.1 sec after subsequent haloperidol treatment.

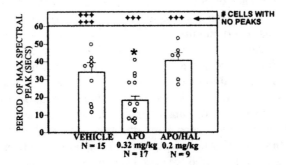

Figure 4. Periodicities of multisecond oscillations in STN neuronal firing rates after vehicle, apomorphine, and haloperidol (after apomorphine). Apomorphine significantly shortened the main oscillatory period (P<0.005), and effect that was reversed by haloperidol administration.

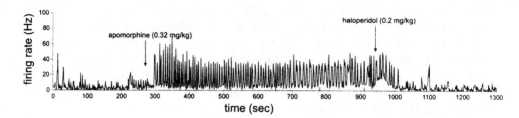

Figure 5. Rate histogram (1 sec bins) of 1300 sec of spike train from an STN neuron. The histogram demonstrates regular fluctuations in the spike train following administration of apomorphine.

6. DISCUSSION

The ability of STN neurons to exhibit a variety of distinctive firing patterns, such as bursting activity and oscillatory rate changes, raises questions about the roles these phenomena may play in basal ganglia function. One relevant issue is the possible distinction between patterns which are involved in motor control, and those which are involved in other processes, such as intrinsic learning, sequencing, and timing estimations in the seconds to minutes range which have been attributed to the basal ganglia. To the extent that the basal ganglia are involved in a range of "fast" and "slow" phasic and tonic processes, changes in spike train patterns on these different time scales may serve different functions. Another relevant issue is the possibility that changes in incidence or synchrony of these spike train patterns might play a role in pathological processes such as PD.

Bursting may be a common property of STN neurons and a means of "fast" communication, providing a brief and multi-fold increase in transmitter release. Analysis of the nature and extent of bursting in STN spike trains has been complicated by confounding influences of different anesthetics and the various methodologies for determining what constitutes a "burst" or measure of "burstiness" (see Kaneoke and Vitek[20] for discussion of burst determination). Since anesthetics appear to affect the burst properties of STN neurons[27], there are limitations to the relevance of bursting incidence from anesthetized preparations with respect to incidence in awake animals. However, two studies have been conducted in awake rats and monkeys and report that the majority of STN neurons, 62 and 69% respectively, exhibit bursting activity[8,23]. In monkeys, bursts of activity in STN neurons have been observed prior to, during, and in response to eye and limb movements[28,46]. Both excitatory and inhibitory input can lead to bursting which implicates the two major afferents to the STN, the cortex and the globus pallidus, in production of a burst response[1,24]. In fact, stimulation studies in rats have demonstrated that the typical initial response of STN neurons to cortical stimulation is a short burst of action potentials[14,29,38,41]. Research on the function of burst firing in other brain areas suggest that post-synaptic action potentials are more reliably induced by this method of neural communication than by more evenly spaced single spikes (see[26] for review). Hence, in situations where the STN must produce a fast and strong response (i.e. during movement), there are neuronal mechanisms in place to ensure trustworthy communication.

Oscillatory activity in STN neurons in the 4-7 Hz range is typically associated with limb tremor in human Parkinson's disease patients and dopamine depleted monkeys, yet it is unclear what role, if any, such oscillations have in the intact animal. To the extent that these oscillations consist of a series of bursts at regular intervals, the association with limb tremors supports the idea that the bursting phenomenon may play a role in regulation of movement. At least one study of intact primates has demonstrated that oscillations in this frequency range can occur in STN neurons in the normal state[8]. However, after MPTP-lesioning, increases were observed in the number of neurons exhibiting these oscillations and the degree of synchronization in neighboring neurons. An important question with respect to the functional role of these oscillations in the STN is whether the degree of synchronous activity fluctuates in the normal state. Increased synchrony has been postulated to facilitate the "binding" of neurons in functionally related circuits and could facilitate processing of stimuli or execution of specific movements. For example, studies in the primary motor cortex indicate that the amount of information conveyed by pairs of neurons firing in a correlated manner (over 600 msec time intervals) is greater than that conveyed by uncorrelated pairs, so that pairs or groups of neurons are better movement-parameter encoders than single neurons[30]. Enhanced synchrony could also be dysfunctional: the suggestion has been made that the observed increased synchrony in the STN and globus pallidus of MPTP lesioned monkeys represents a breakdown in circuit segregation, leading to increased interaction between neurons which normally would not be correlated[6]. Future research into these questions promises to be exciting and to have profound implications on the understanding of basal ganglia function.

The multisecond oscillations in STN neuronal activity described in the present report appear to be too slow to play a role in triggering precisely timed aspects of motor initiation or execution. However, it seems possible that the efficiency of information processing in the basal ganglia might be affected by apomorphine induced changes in oscillatory frequency in the basal ganglia network. Multisecond oscillatory activity in the STN is similar to that reported recently in other basal ganglia nuclei, specifically, the substantia nigra pars reticulata, globus pallidus and entopeduncular nucleus[39], an observation consistent with the idea that multisecond oscillations may be a property of the basal ganglia network. It should be pointed out that these results are inconsistent with current dual circuit models of basal ganglia function, which would not predict that a D1/D2 receptor agonist would have a qualitatively identical effect in multiple basal ganglia nuclei.

These slow multisecond oscillations could modulate aspects of basal ganglia function sensitive to changes in neuronal activity over relatively slow time scales. For example, some evidence suggests that the basal ganglia are involved in internal clock functions[16,19,31]. Multiple studies have implied that manipulation of dopamine receptor stimulation can alter perception of time intervals up to 60 sec in length[16,19,31]. These results have led to the suggestion that stimulation of dopamine receptors increases the speed of the internal clock involved in this temporal processing. Multisecond oscillations in the globus pallidus, similar to those observed in the STN, exhibit a shortening of oscillatory period in response to amphetamine, as well as methylphenidate, cocaine and apomorphine, all of which act to increase dopamine receptor stimulation[4,39,40]. These observations raise the possibility that multisecond oscillatory activity in the basal ganglia may play a role in temporal information processing. It has been debated whether single or multiple oscillators relating to different frequency bands may be utilized in the processing of temporal information[19]. The results described here extend the range of oscillatory periods which have been observed within the basal ganglia (0.017 – 20 Hz) and suggest that multiple oscillators exist in this network[8,27,32,36,44].

Multisecond oscillations might also be involved in cellular regulatory processes. Neuronal excitability is invariably linked to transient Ca^{2+} flux and induces periodic oscillatory increases and decreases in intracellular Ca^{2+} levels. Multisecond oscillatory activity of intracellular Ca^{2+} has been visualized *in vitro* in the pre-Botzinger complex, a nucleus that is part of a putative brainstem pattern generator[21]. Recently, cytosolic Ca^{2+} oscillations (periods tested ranged from 0.5 – several hundred seconds) have been demonstrated to be more effective in regulating gene expression than stable concentrations in non-excitable cells[12,25]. In fact, the frequency of the oscillation can determine whether gene expression is achieved, or whether a single or coordinated group of transcription factors are expressed[12,25]. Thus, the frequency of the multisecond oscillations in spiking activity may be regulating gene expression, and/or other systems that are regulated by changes in Ca^{2+} levels.

In conclusion, the subthalamic nucleus is capable of transmitting information throughout the basal ganglia using multiple modes of communication. Models incorporating STN neurons into basal ganglia function do not incorporate the fact that these neurons do not have stable baseline firing, rather some burst, some fire regularly, some show oscillations in frequencies of <1 Hz and two-thirds in the present report had large, periodic, fluctuating baseline firing rates with periods of several seconds. While it is unknown what role these properties play in determining how the STN communicates, they certainly expand the possibilities and complicate the interpretation of studies of STN function.

7. REFERENCES

1. A.K. Afifi. Basal ganglia: Functional anatomy and physiology. Part 1, *J. Child Neurol.* 9[3]:249 (1994).
2. K.A. Allers, D.S. Kreiss, and J.R. Walters. Multisecond oscillations in the subthalamic nucleus: effects of dopamine cell lesion and dopamine receptor stimulation. *Synapse* 38[1]:38 (2000).
3. K.A. Allers, D.N. Ruskin, D.A. Bergstrom, L.A. Molnar, and J.R. Walters, Correlations of multisecond oscillations in firing rate in pairs of basal ganglia neurons. *Soc. Neurosci. Abstr.* 25:1929 (1999).
4. D. Beck, D. Ruskin, D.A. Bergstrom, and J.R. Walters, D-amphetamine (AMPH) and cocaine increase the "speed" of multisecond oscillations in globus pallidus (GP) unit activity, *Soc. Neurosci. Abstr.* 25:1819 (1999).
5. A.L. Benabid, A. Benazzouz, D. Hoffmann, P. Limousin, P. Krack, and P. Pollack, Long-term electrical inhibition of deep brain targets in movement disorders, *Mov. Disord.* 13:119 (1998).
6. H. Bergman, A. Feingold, A. Nini, A. Raz, H. Slovin, M. Abeles, and E. Vaadia, Physiological aspects of information processing in the basal ganglia of normal and parkinsonian primates, *Trends Neurosci.* 21:32 (1998).
7. H. Bergman, T. Wichmann, and M.R. Delong, Reversal of experimental parkinsonism by lesions of the subthalamic nucleus. *Science* 249:1436 (1990).
8. H. Bergman, T. Wichmann, B. Karmon, and M.R. Delong, The primate subthalamic nucleus. II. Neuronal activity in the MPTP model of parkinsonism, *J. Neurophysiol.* 72:507 (1994).
9. C. Beurrier, P. Congar, B. Bioulac, and C. Hammond, Subthalamic nucleus neurons switch from single-spike activity to burst-firing mode. *J. Neurosci.* 19:599 (1999).
10. M.D. Bevan and C.J. Wilson, Mechanisms underlying spontaneous oscillation and rhythmic firing in rat subthalamic neurons, *J. Neurosci.* 19:7617 (1999).
11. F. Blandini, M. Gargia-Osuna, and J.T. Greenamyre, Subthalamic ablation reverses changes in basal ganglia oxidative metabolism and motor response to apomorphine induced by nigrostriatal lesion in rats, *Eur. J. Neurosci.* 9:1407 (1997).

12. R.E. Dolmetsch, K. Xu, and R.S. Lewis, Calcium oscillations increase the efficiency and specificity of gene expression, *Nature* 392: 933 (1999).
13. G. Flores, J.J. Liang, A. Sierra, D. Martinez-Fong, R. Quirion, J. Aceves, and L.K. Srivistava, Expression of dopamine receptors in the subthalamic nucleus of the rat: characterization using reverse transcriptase-polymerase chain reaction and autoradiography, *Neuroscience* 91:549 (1999).
14. K. Fujimoto and H. Kita, Response characteristics of subthalamic neurons to the stimulation of the sensorimotor cortex in the rat, *Brain Res.* 609: 185 (1993).
15. J. Guridi and J.A. Obeso, The role of the subthalamic nucleus in the origin of hemiballism and parkinsonism: New surgical perspectives, *Adv. Neurol.* 74:235 (1997).
16. D.L. Harrington and K.Y. Haaland, Neural underpinnings of temporal processing: A review of focal lesion, pharmacological, and functional imaging research, *Rev. in Neurosci.* 10:91 (1999).
17. O.K. Hassani, M. Mouroux, and J. Féger, Increased subthalamic neuronal activity after nigral dopaminergic lesion independent of disinhibition via the globus pallidus, *Neuroscience* 72:105 (1996).
18. J.R. Hollerman and A.A. Grace, Subthalamic nucleus cell firing in the 6-OHDA-treated rat: Basal activity and response to haloperidol, *Brian Res.* 590:291 (1992).
19. R.B. Ivry, The representation of temporal information in perception and motor control, *Curr. Opin. Neurobiol.* 6:851 (1996).
20. Y. Kaneoke and J.L. Vitek, Burst and oscillations as disparate neuronal properties, *J. Neurosci. Methods* 68:211 (1996).
21. N. Koshiya and J.C. Smith, Neuronal pacemaker for breathing visualized in vitro, *Nature* 400:360 (1999).
22. D.S. Kreiss, L.A. Anderson, and J.R. Walters, Apomorphine and dopamine D_1 receptor agonists increase the firing rates of subthalamic nucleus neurons, *Neuroscience* 72:863 (1996).
23. D.S. Kreiss, C.W. Mastropietro, S.S. Rawji, and J.R. Walters, The response of subthalamic nucleus neurons to dopamine receptor stimulation in a rodent model for Parkinson's disease, *J. Neurosci.* 17:6807 (1997).
24. R. Levy, L.N. Hazrati, M.T. Herrero, M. Vila, O.K. Hassani, M. Mouroux, M. Ruberg, H. Asensi, Y. Agid, J. Feger, J.A. Obeso, A. Parent, E.C. Hirsch, Re-evaluation of the functional anatomy of the basal ganglia in normal and parkinsonian states, *Neuroscience* 76:335 (1997).
25. W. Li, J. Llopis, M. Whitney, G. Zlokarnik, and R.Y. Tsien, Cell-permeant caged INsP3 ester shows that Ca2+ spike frequency can optimise gene expression, *Nature* 392:936 (1999).
26. J.E. Lisman, Bursts as a unit of neural information: Making unreliable synapses reliable, *Trends Neurosci* 20:38 (1997).
27. P.J. Magill, J.P. Bolam, and M.D. Bevan, Relationship of activity in the subthalamic nuclueus-globus pallidus network to cortical EEG, *J. Neurosci.* 83[5]:3169 (2000).
28. M. Matsumara, J. Kojima, W.T. Gardiner, and O. Hikosaka, Visual and oculomotor functions of the monkey subthalamic nucleus, *J Neurophysiol.* 67:1615 (1992).
29. N. Maurice, J.M. Deniau, J. Glowinski, and A.M. Thierry, Relationships between the prefrontal cortex and the basal ganglia in the rat: Physiology of the corticosubthalamic circuits, *J. Neurosci.* 18:9539 (1998).
30. E.M. Maynard, N.G. Hatsopoulos, C.L. Ojakangas, B.D. Acuna, J.N. Sanes, R.A. Normann, and J.P. Donoghue, Neuronal interactions improve cortical population coding of movement direction, *J. Neurosci.* 19:8083 (1999).
31. W.H. Meck, Neuropharmacology of timing and time perception, *Cognit. Brain Res.* 3:227 (1996).
32. M.R. Mehta and H. Bergman, Loss of frequencies in autocorrelations and a procedure to recover them, *J.Neurosci. Methods* 62:65 (1995).
33. G. Mengod, M.T. Villaro, G.B. Landwehrmeyer, M.I. Martinez-Mir, H.B. Niznik, R.K. Sunahara, P. Seeman, B.F. O'Dowd, A. Probst, and J.M. Palacios, Visualizaiton of Dopamine D1, D2, and D3 receptor mRNAs in human and rat brain, *Neurochem. Int.* 20:33S-43S (1992).
34. H. Nakanishi, H. Kita, and S.T. Kitai, Electrical membrane properties of rat subthalamic neurons in an in vitro slice preparation, *Brain Res.* 437:35 (1987).
35. P.G. Overton and S.A. Greenfield, Determinants of neuronal firing pattern in the guinea-pig subthalamic nucleus: An *in vivo* and *in vitro* comparison, *J. Neural Transm. Park. Dis. Demet. Sect.* 10:41 (1995).
36. D. Plenz and S.T. Kitai, A basal ganglia pacemaker formed by the subthalamic nucleus and external globus pallidus, *Nature* 400:677 (1999).
37. J.M. Provenzale and J.P. Glass, Hemiballismus: CT and MR findings, *J. Comput. Assist. Tomogr.* 19:537 (1995).

38. B. Rouzaire-Dubois and E. Scarnati, Pharmacological study of the cortical-induced excitation of subthalamic nucleus neurons in the rat: Evidence for amino acids as putative neurotransmitters, *Neuroscience* 21:429 (1987).
39. D.N. Ruskin, D.A. Bergstrom, Y. Kaneoke, B.N. Patel, M.J. Twery, and J.R. Walters, Multisecond oscillations in firing rate in the basal ganglia: robust modulation by dopamine receptor activation and anesthesia, *J. Neurophysiol.* 81:2046 (1999).
40. D.N. Ruskin, D.A. Bergstrom, and J.R. Walters, Multisecond oscillations in firing rate in the globus pallidus: synergistic modulation by D1 and D2 receptors, *J. Pharmacol. Exp. Ther.* 290:1493 (1999).
41. L.J. Ryan and K.B. Clark, The role of the subthalamic nucleus in the response of glubus pallidus neurons to stimulation of the prelimbic and agranular frontal cortices in rats, *Exp. Brain Res.* 86:641 (1991).
42. L.J. Ryan and K.B. Clark, Alteration of neuronal responses in the subthalamic nucleus following globus pallidus and neostriatal lesions in rats, *Brain Res.Bull.* 29:319 (1992).
43. L.J. Ryan, D.J. Sanders, and K.B. Clark, Auto- and cross-correlation analysis of subthalamic nucleus neuronal activity in neostriatal- and globus pallidal-lesioned rats, *Brain Res.* 583:253 (1992).
44. J. Sutin, T. Tsubowkawa, and R.L. McBride, Raphe neurons and barbiturate induced rhythmic activity in the subthalamic nucleus and ventral tegmental area, *Brian Res, Bull.* 1:93 (1976).
45. J.K. Wamsley, M.E. Alburges, R.D. McQuade, and M. Hunt, CNS distribution of D_1 receptors: Use of a new specific D_1 receptor agonist, [^{3}H]SCH39166, *Neurochem. Int.* 20 Suppl.: 123S (1992).
46. T. Wichmann, H. Bergman, and M.R. Delong, The primate subthalamic nucleus. I. Functional properties in intact animals, *J. Neurophysiol.* 72:494 (1994).
47. T. Wichmann and M.R. Delong, Functional and pathophysiological models of the basal ganglia, *Curr. Opin. Neurobiol.* 6:751 (1996).

Section IV

LEARNING FUNCTIONS OF THE BASAL GANGLIA AND ADAPTIVE MOTOR CONTROL

SELECTION AND THE BASAL GANGLIA
A role for dopamine

P. Redgrave, T. Prescott, and K. Gurney*

1. INTRODUCTION

Evidence from both the biological and artificial neural network literature suggests that many operations within the brain are carried out in parallel. Thus, it is now widely recognised that analyses of different qualities of the same stimulus are performed separately in specialised regions of the brain (Goodale, 1996). Representations of external and internal events which can determine or guide specific types of behaviour, action and movement are also distributed throughout the nervous system (Ewert, 1995). In both cases, however, distributed parallel processing appears to be constrained by, on the one hand, the need to think about or attend to only one thing at a time, and on the other, the need to avoid trying to do two different things with the same set of muscles. Selection mechanisms in the brain must therefore operate in both cognitive and motor domains to prevent parallel confusion.

A recurring theme throughout an extensive literature is that the basal ganglia have the capacity to select some things and suppress or reject others (Hikosaka, 1994; Marsden and Obeso, 1994; Mink, 1996; Rolls, 1994; Wickens, 1993). We have therefore recently proposed that a core function of the basal ganglia is to provide the vertebrate brain with a specialised selection mechanism to resolve conflict between systems competing at different functional levels for access to limited cognitive and behavioural resources (Prescott et al., 1999; Redgrave et al., 1999a). The current chapter summarises some of the arguments supporting this view and considers how the resulting framework provides for a reinterpretation of data describing the short latency dopamine response (Schultz, 1998), and the effect of dopamine on selection within a quantitative model of internal basal ganglia circuitry.

---

* P. Redgrave, T. Prescott, and K. Gurney, Dept. of Psychology, Univ. of Sheffield, Sheffield S10 2TP.
The Basal Ganglia VI
Edited by Graybiel *et al.*, Kluwer Academic/Plenum Publishers, 2002

2. SELECTION: SOLUTIONS TO A KEY PROBLEM

The problem of selecting between behavioural alternatives has a long history in the ethology literature (McFarland, 1989) and more recently it has emerged as a practical issue in the control of mobile (Brooks, 1994) and other artificial agents (Maes, 1995; Prescott et al., 1999). The current discussion will focus on circumstances where two or more representations, which have the capacity to guide specific but mutually exclusive acts, are in conflict over access to a limited resource that is the final common motor path (Fig 1).

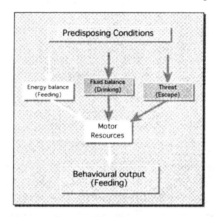

Figure 1. A mechanism is required to ensure that parallel processing behavioural systems which are mutually exclusive have orderly access to limited motor resources The density of shading indicates the level of activation (lighter = high)

In general terms, the level of support (*salience*) in each competing command system should depend on relevant causal factors which can derive from both extrinsic and intrinsic sources. Successive selections could, therefore, be made on the basis of relative levels of salience within competing systems. Over time the comparative salience values within competing systems are likely to be weighted to provide appropriate dominance relationships, and adaptable to cope with a non-stationary world. When circumstances change, current selections should be terminated and resources switched to other channels. Re-prioritisation generally occurs when a current behaviour is successful (e.g. a key opens the door) or if it proves to be ineffective (it's the wrong key). Selections are also switched if a competing system unexpectedly registers a higher level of support (the fire-alarm sounds).

A variety of architectures have been proposed to deal with the selection problem in both artificial and biological systems (Brooks, 1994; Maes, 1995; McFarland, 1989; Snaith and Holland, 1990). Two of these will now be described and considered as possible templates for interpreting patterns of connectivity that could implement selection within the vertebrate brain.

Distributed selection architectures: In architectures of this type all competitors are reciprocally connected so that each one has an inhibitory link to every other— *recurrent reciprocal inhibition* (Wickens, 1993)— and an excitatory link to the shared output resource. Such networks display a form of positive feedback since increased activity in one

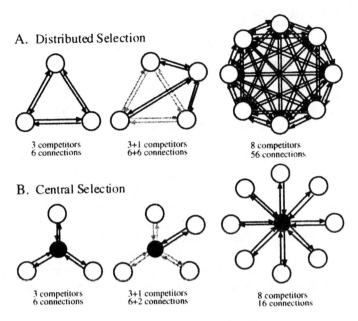

Figure 2. Numbers of additional connections required as additional competitors are added to Distributed and Central Selection devices

competitor causes increased inhibition on all others thereby reducing their reciprocal inhibitory effect. Reciprocally inhibiting networks are widespread in the central nervous system (Gallistel, 1980) (including the basal ganglia), however, connection costs are likely to preclude it from being the direct arbiter of selection between functional units distributed widely throughout the brain. Specifically, it has been noted (McFarland, 1965) that to arbitrate between n competing behaviours, a fully connected network with reciprocal inhibition requires $n(n-1)$ connections; to add a new competitor requires a further $2n$ connections (Figure 2A). Reciprocally connected architectures are therefore high cost both in terms of the density of connections between rivals and in the cost of integrating a new competitor into an existing network.

Centralised selection architectures: There are, however, good reasons why both artificial and biological control systems can benefit by exploiting a *centralised* selection mechanism for overall behavioural control. For example Snaith and Holland (1990) have pointed out that an architecture with centralised selection requires only two connections for each competitor (to and from the selection mechanism) resulting in a total of $2n$ connections (Figure 2B). This is a considerable saving over the $n(n-1)$ connections required by the distributed architecture. Moreover, to add a new competitor to the central selector only two further connections need be incorporated compared to the $2n$ required for reciprocal inhibition between all competitors. Since in central selection architectures selection and sensorimotor control processes are independent they can be modified separately. The advantages incurred by modularity in dissociating functionally distinct components of the system are probably as significant for evolved systems as they are for engineered ones.

3. SELECTION: A CORE FUNCTION

We have recently proposed (Prescott et al., 1999; Redgrave et al., 1999a) that the channelled architecture of the basal ganglia (Alexander et al., 1986; Strick et al., 1995) provide the vertebrate brain with a specialised, central selection mechanism to resolve conflict between competing systems at different functional levels. Characteristics of basal ganglia circuitry that match the requirements identified above will now be used to support this assertion. We further suggest that a series of distributed selection mechanisms (Figure 2A) are employed *within* basal ganglia circuitry in a manner that exploits their useful switching properties whilst minimising the undesirable overheads incurred by reciprocal inhibition.

Figure 3 summarises aspects of basal ganglia architecture which correspond to essential features of a central selection device (Figure 2B). In the vertebrate brain, functional systems capable of specifying action, (*command systems* (Ewert, 1995), are distributed throughout all levels of the neuraxis. A basic feature of central selection architectures is that competitors which initiate and subsequently guide actions have direct connections with the shared motor resource. This also appears to be the case in the vertebrate brain. A wealth of anatomical evidence shows that command systems at all levels communicate directly with cortical and/or hindbrain pre-motor and motor mechanisms (Holstege, 1991).

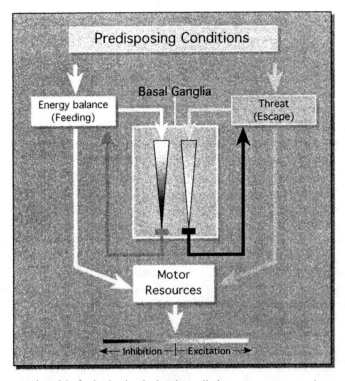

Figure 3. A conceptual model of selection by the basal ganglia between two command systems competing for access to a shared motor resource. Excitatory connections are represented by light grey/white and inhibitory ones by dark grey/black.

All competitors also have connections to a common selection mechanism (the basal ganglia) which could act to resolve conflict over access to the limited resource. The channelled input projections to the striatum from the brainstem (via the thalamus), limbic system and most regions of cerebral cortex (Gerfen and Wilson, 1996) suggest the basal ganglia may be uniquely connected with a wide range of potential command systems. It is, however, unlikely that all afferents to the basal ganglia represent the salience of competing command systems since a range of evidence indicates the presence of several qualitatively different inputs(Gerfen andWilson, 1996). One likely possibility is that a wide variety of contextual information is also made available to the striatum (e.g. Graybiel, 1991; Schultz et al., 1995) which could serve either to enhance or reduce the salience of specific 'goal-related' inputs.

Several authors have proposed that selective processes could be supported by local reciprocal inhibitory connections within the basal ganglia (Mink, 1996; Wickens, 1993). Although the anatomical and neurochemical basis of these connections has yet to be identified (Jaeger et al., 1994), functional data (Brown and Sharp, 1995; Deniau and Chevalier, 1985) provide evidence for a general excitatory centre/inhibitory surround organisation which is a characteristic of circuits with some form of reciprocal inhibitory connections (Figure 2A). It therefore seems reasonable to assume that internal circuitry of the basal ganglia can be configured to select the most salient channels and, at the level of the output nuclei, selected channels are inhibited while non-selected channels are excited (Figure 3).

Finally the selection device must enable connections between the winning command generator(s) and the motor plant whilst simultaneously denying access to the losers. In the default state tonic inhibitory output from the basal ganglia to multiple targets in the thalamus and brainstem acts to block direct connections between command systems and brainstem motor plant. Following the suggestion of others (Chevalier and Deniau, 1990; Mink, 1996), we assume that focused disinhibition of selected channels would unblock direct connections between winning command systems and the brainstem. Maintaining or increasing the tonic inhibitory output on non-selected channels would deny the losing command systems access to the motor resource. In this manner, contradictory motor commands would be suppressed and *distortion* of the selected action prevented. Concurrent disinhibition of selected channels and increased inhibition of non-selected ones would also provide the necessary positive feedback to prevent *dithering* between closely matched competitors.

This general view of basal ganglia function extends previous suggestions (Albin et al., 1989; Marsden and Obeso, 1994; Mink, 1996) that the basal ganglia act to promote desired movements and suppress unwanted ones. Our proposal maintains that the basal ganglia *is* the selector mechanism and chooses between multiple potentially conflicting actions and movements 'desired' by independent parallel processing command systems, and ensures that the most urgent or salient is given unhindered access to the motor plant.

We have recently instantiated the ideas outlined above into a quantitative computational model (Gurney et al., 1998, 2001a, 2001b; Humphries and Gurney 2002). In formulating the model we applied the computational premise of selection to the functional architecture of the basal ganglia and discovered a natural reinterpretation of the role of the external globus pallidus as a source of control signals to a principal selection mechanism. This decomposition into 'selection' and 'control' pathways contrasts with the more usual split into 'direct' and 'indirect' pathways (Albin et al., 1989). The model corroborates our selection hypothesis by demonstrating appropriate reduction of output signals under salience input to striatum. Further, a simulated increase in the levels of dopamine suppressed overall output on all channels (thereby giving more promiscuous selection). This is consistent with the work of Robbins and Sahakian (1983) whose review of available biological data led them to conclude that mild to moderate increases in dopaminergic activity tend to facilitate switching while reductions tend to suppress switching. Full details of our computational model have been published elsewhere (Gurney et al 1998, 2001a, 2001b; Humphries and Gurney 2002).

4. THE SHORT-LATENCY DOPAMINE RESPONSE: A REINTERPRETATION

An important feature of the above model is that when a highly salient event occurs in a non-selected channel current actions should be interrupted and resources switched to deal with the novel event. Unexpected stimuli are normally made salient (i.e. have special biological significance) by virtue of their novelty, their status as primary reinforcers or by an association with primary reinforcers. It may be expected, therefore, that such stimuli would have special access to a central selection mechanism. In this regard, we have noticed that ventral mesencephalic dopamine neurones exhibit a stereotyped short-latency (50-110ms), short duration (<200ms) response to all classes of salient event provided the stimulus is unexpected (Schultz, 1998). An influential interpretation is that this short-latency response signals an error in predicted reward (Schultz et al., 1997) which is used by target structures to adjust future response probabilities. In the context of the selection model proposed above, however, an important property of rewarding events is their ability to interrupt ongoing behaviour and initiate a switch of resources. We have therefore suggested that the short-latency dopamine response may participate more in the process of switching selections to significant events rather than signalling reward error (Redgrave et al., 1999b). The suggestion that dopamine may promote behavioural switching has been made previously with respect to general modulatory effects of dopamine on basal ganglia function (e.g. Cools, 1980; Robbins and Sahakian, 1983).

In support of this 'switching hypothesis' we have noted that in most experimental paradigms used to characterise the short-latency dopamine response, the presumed reward error signal is invariably confounded with a requirement for the animal to switch attention and/or behavioural strategy. For example, when an unpredicted reward is presented, the animal typically disengages from the experimental task and switches to reward acquisition. The same confound applies when unexpected conditioned stimuli predicting reward are presented. Presumably, if a current task is interrupted and an acquisition strategy selected when a predictor is presented, any further switching on reward delivery would be unnecessary, even counterproductive. Interestingly, in situations where a stimulus predicts reward, the dopamine response transfers from the primary reward to the predictor (Schultz, 1998).

What mechanisms might underlie the gradual transfer of dopamine sensitivity from primary reward to a predicting stimulus? Part of the answer may be linked to the observation that unexpected novel or intense stimuli also elicit a robust dopamine response; although it has been pointed out that such responses are only temporary and quickly habituate (Schultz, 1998). It may also be relevant that evoked responses in primary sensory areas of the brain can be influenced by reinforcement outcome (Hernandez-Peon, 1961). For example, Wurtz and Goldberg (1972) showed that non-reinforced presentation of light spots to a monkey quickly led to habituation of the neuronal responses within sensory receptive fields in the midbrain superior colliculus. However, by associating a stimulus with reward, a previously habituated sensory response was re-instated. Thus, if the magnitude of stimulus-related activity in primary sensory networks can be influenced by association with reward, and this parameter was made available to dopamine neurones, it would explain how reward-predicting stimuli can elicit the observed response. In other words, short-latency dopaminergic activity may simply reflect a reaction to habituation and reinforcement-related enhancement of stimulus-evoked activity in primary sensory networks.

The precise timing of the short latency dopamine response relative to other events in the brain is also consistent with its involvement in the process of switching behavioural selections. In most circumstances, the dopamine signal occurs prior to, or at best, during the saccadic response designed to bring the initiating event on to the fovea for analysis. This point can be appreciated by recalling that unexpected visual events normally elicit two distinct responses in neurones of the intermediate and deep layers of the superior colliculus (Figure 4A). An initial short latency (~50ms) response is followed by a longer latency (>150ms) pre-saccadic motor burst (Jay and Sparks, 1987). The saccadic response brings an unexpected event onto the fovea for more detailed analyses involving feature extraction and object recognition. It is therefore interesting to note that dopamine responses seem to fit neatly between the sensory and pre-saccadic motor burst recorded in the primate superior colliculus (Figures 4A & 4B). It may also be significant that they typically precede the disinhibitory output signal from substantia nigra which is instrumental in facilitating the pre-

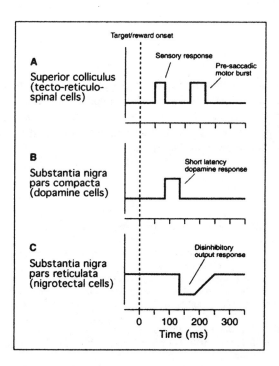

Figure 4. Short latency dopamine responses normally occur prior to or during saccadic eye movements which bring unexpected events onto the fovea.

saccadic burst recorded from tectal target neurones (Hikosaka and Wurtz, 1983)(Figure 4C). If short latency dopamine activity were to signal reward error, the computations required to recognise that an event is a reward, or predicts reward, and is unexpected, would have to be conducted *before* the animal switched its gaze to see what the stimulus was.

The idea that short-latency dopamine signals could play an essential part in switching resources to unexpected salient events suggests they may have a more general role in associative learning. The disruption of processes by which salient events are linked with specific resource selections could explain why experimental manipulations of dopamine transmission can effect both positively and negatively reinforced associative learning, and associative learning in the absence of reinforcement (Salamone et al., 1997). For example, the slowing of conditioning that occurs in the latent inhibition paradigm (Weiner, 1990) may, in part, arise from the inability of pre-habituated stimuli to evoke a short latency dopamine response. In terms of the present hypothesis, latent inhibition would occur because the habituated 'to-be-conditioned' stimulus fails to attract a diversion of resources, which would impair the processes by which it is linked both with specific cognitive and behavioural selections, and with primary reinforcement. In this context, a specific role for dopamine in the re-allocation process could be to 'bind' the representation of a significant biological event in the striatum to the selection of a particular action. The strength of this link could be enhanced, or weakened by subsequent reinforcement signals indicating outcome. Such processes could provide a basis for the assembly of successful 'chunks' in the acquisition of sequences of adaptive behaviour proposed recently by Graybiel (Graybiel, 1998).

5. CONCLUSION

Our basic proposal is that the basal ganglia provide the brain with a computational architecture which can schedule the access of competing command systems to restricted cognitive and motor resources. Experimental findings suggest that within this framework afferent dopaminergic neurotransmission has an important role in the processes which switch resources between salient command systems.

6. REFERENCES

Albin, R. L., Young, A. B. and Penney, J. B., 1989, The functional anatomy of basal ganglia disorders. *Trends Neurosci.*, 12: 366-375.
Alexander, G. E., DeLong, M. R. and Strick, P. L., 1986, Parallel organization of functionally segregated circuits linking basal ganglia and cortex. *Ann. Rev. Neurosci.*, 9: 357-381.
Brooks, R. A., 1994, Coherent behaviour from many adaptive processes. *in:* "From Animals to Animats 3: Proceedings of the Third International Conference on the Simulation of Adaptive Behaviour" D. Cliff, P. Husbands, J.-A. Meyer and S. W. Wilson Eds., MIT Press, Cambridge, MA, pp. 22-29.
Brown, L. L. and Sharp, F. R., 1995, Metabolic mapping of rat striatum: somatotopic organization of sensorimotor activity. *Brain Res.*, 686: 207-222.
Chevalier, G. and Deniau, J. M., 1990, Disinhibition as a basic process in the expression of striatal functions. *Trends Neurosci.*, 13: 277-281.
Cools, A. R., 1980, Role of the neostriatal dopaminergic activity in sequencing and selecting behavioural strategies: Facilitation of processes involved in selecting the best strategy in a stressful situation. *Behav. Brain Res.*, 1: 361-378.

Deniau, J. M. and Chevalier, G., 1985, Disinhibition as a basic process in the expression of striatal functions II. The striato-nigral influence on thalamocortical cells of the ventromedial thalamic nucleus. *Brain Res.*, 334: 227-233.

Ewert, J.-P., 1995, Command neurons and command systems. *in:* "The Handbook of Brain Theory and Neural Networks" M. A. Arbib Ed., MIT Press, Cambridge, MA, pp. 215-220.

Gallistel, C. R., 1980, "The organization of action: A new synthesis.", Lawrence Erlbaum Assoc., New Jersey.

Gerfen, C. R. and Wilson, C. J., 1996, The basal ganglia. *in:* "Handbook of chemical neuroanatomy, Vol 12: Integrated systems of the CNS, Part III." L. W. Swanson, A. Bjorklund and T. Hokfelt Eds., Elsevier, Amsterdam, pp. 371-468.

Goodale, M. A., 1996, Visuomotor modules in the vertebrate brain. *Can. J. Physiol. Pharmacol.*, 74: 390-400.

Graybiel, A., 1991, Basal ganglia - input, neural activity, and relation to the cortex. *Current Opinion Neurobiol.*, 1: 644-651.

Graybiel, A. M., 1998, The basal ganglia and chunking of action repertoires. *Neurobiol. Learn. Memory*, 70: 119-136.

Gurney, K., Prescott T.J., Redgrave P. (2001a) A computational model of action selection in the basal ganglia I: A new functional anatomy. *Biological Cybernetics* 84, 401-410.

Gurney, K., Prescott T.J., Redgrave P. (2001b) A computational model of action selection in the basal ganglia II:Analysis and simulation of behaviour. *Biological Cybernetics* 84, 411-423.

Gurney, K., Redgrave, P. and Prescott, T. J., 1998, A computational model of selective properties with the basal ganglia: a reinterpretation of functional anatomy. *Soc. Neurosci. Abstr.*, 24: 649.13.

Hernandez-Peon, R., 1961, Reticular mechanisms of sensory control. *in:* "Sensory communication." W. A. Rosenblith Ed., MIT Press, Cambridge, MA., pp. 497-520.

Hikosaka, O., 1994, Role of basal ganglia in control of innate movements, learned behaviour and cognition - A hypothesis. *in:* "The basal ganglia IV: New ideas and data on structure and function." G. Percheron, J. S. McKenzie and J. Feger Eds., Plenum Press, New York, pp. 589-596.

Hikosaka, O. and Wurtz, R. H., 1983, Visual and oculomotor function of monkey substantia nigra pars reticulata. I. Relation of visual and auditory responses to saccades. *J. Neurophysiol.*, 49: 1230-1253.

Holstege, G., 1991, Descending motor pathways and the spinal motor system: limbic and non limbic components. *Prog. Brain Res.*, 87: 307-421.

Humphries M.D., Gurney K. (2002) The role of intra-thalamic and thalamocortical circuits in action selection. *Network: Computation in Neural Systems* 13, 131-156.

Jaeger, D., Kita, H. and Wilson, C. J., 1994, Surround inhibition among projection neurones is weak or nonexistent in the rat neostriatum. *J. Neurophysiol.*, 72: 2555-2558.

Jay, M. F. and Sparks, D. L., 1987, Sensorimotor integration in the primate superior colliculus. I. Motor convergence. *J. Neurophysiol.*, 57: 22-34.

Maes, P., 1995, Modelling adaptive autonomous agents. *in:* "Artificial Life: An Overview" C. G. Langton Ed., MIT Press, Cambridge, MA, pp. 135-162.

Marsden, C. D. and Obeso, J. A., 1994, The function of the basal ganglia and the paradox of stereotaxic surgery in Parkinson's disease. *Brain*, 117: 877-897.

McFarland, D., 1989, "Problems of animal behaviour.", Longman Scientific and Technical, London.

McFarland, D. J., 1965, Flow graph representation of motivational systems. *Brit. J. Math. Stat. Psychol.*, 18: 25-43.

Mink, J. W., 1996, The basal ganglia: Focused selection and inhibition of competing motor programs. *Prog. Neurobiol.*, 50: 381-425.

Prescott, T. J., Redgrave, P. and Gurney, K. N., 1999, Layered control architectures in robots and vertebrates. *Adaptive Behavior*, 7: 99-127.

Redgrave, P., Prescott, T. and Gurney, K. N., 1999a, The basal ganglia: A vertebrate solution to the selection problem ? *Neuroscience*, 89: 1009-1023.

Redgrave, P., Prescott, T. J. and Gurney, K., 1999b, Is the short latency dopamine burst too short to signal reward error ? *Trends Neurosci.*, 22: 146-151.

Robbins, T. W. and Sahakian, B. J., 1983, Behavioural effects of psychomotor stimulant drugs: Clinical and neuropsychological implications. *in:* "Stimulants: Neurochemical, behavioural and clinical perspectives." I. Creese Ed., Raven Press, New York, pp. 301-338.

Rolls, E. T., 1994, Neurophysiology and cognitive functions of the striatum. *Revue Neurologique*, 150: 648-660.

Salamone, J. D., Cousins, M. S. and Synder, B. J., 1997, Behavioural functions of nucleus accumbens dopamine: empirical and conceptual problems with the anhedonia hypothesis. *Neurosci. Biobehav. Rev.*, 21: 341-359.

Schultz, W., 1998, Predictive reward signal of dopamine neurons. *J. Neurophysiol.*, 80: 1-27.

Schultz, W., Apicella, P., Romo, R. and Scarnati, E., 1995, Context-dependent activity in primate striatum reflecting past and future behavioural events. *in:* "Models of information processing in the basal ganglia." J. C. Houk, J. L. Davis and D. G. Beiser Eds., MIT Press, Cambridge, MA., pp. 11-27.

Schultz, W., Dayan, P. and Montague, P. R., 1997, A neural substrate of prediction and reward. *Science*, 275: 1593-1599.

Snaith, S. and Holland, O., 1990, An investigation of two mediation strategies suitable for behavioural control in animals and animats. *in:* "From Animals to Animats: Proceedings of the First International Conference on the Simulation of Adaptive Behaviour" J.-A. Meyer and S. Wilson Eds., MIT Press, Cambridge, MA, pp. 255-262.

Strick, P. L., Dum, R. P. and Picard, N., 1995, Macro-organization of the circuits connecting the basal ganglia with the cortical motor areas. *in:* "Models of information processing in the basal ganglia." J. C. Houk, J. L. Davis and D. G. Beiser Eds., MIT Press, Cambridge, MA., pp. 117-130.

Weiner, I., 1990, Neural substrates of latent inhibition - the switching model. *Psychological Bulletin*, 108: 442-461.

Wickens, J., 1993, "A theory of the striatum.", Pergamon, Oxford.

Wurtz, R. H. and Goldberg, M. E., 1972, The primate superior colliculus and the shift of visual attention. *Invest. Ophthalmol.*, 11: 441-50.

MOVEMENT INHIBITION AND NEXT SENSORY STATE PREDICTION IN THE BASAL GANGLIA

Amanda Bischoff-Grethe, Michael G. Crowley, and Michael A. Arbib*

1. INTRODUCTION

The basal ganglia (BG) have received increasing attention over the last decade from both experimentalists and computational modelers in an effort to more fully understand their role in motor control. Their suggested role has ranged from motor preparation and facilitation[2,18] to initiation[21] and program selection[7,22] to motor inhibition[51]. All these models have the BG more or less directly involved in the control of movements, either by selecting the motor command to be executed, or through the facilitation of a motor command presumably selected by cortical mechanisms. However, researchers have found that patients with diseases of the BG, particularly Huntington's disease and Parkinson's disease (PD), do not have significant motor control difficulties when visual input is available[12,16,53], but do have problems with specific forms of internally driven *sequences* of movements[20] as well as certain forms of motor memory tasks[47,52]. This implies that the basal ganglia are less involved in the selection of a single, sensorially shaped motor command, but may instead be involved in assisting cortical planning centers in some fashion as well as provide sequencing information for cross-modal movements, e.g., simultaneous arm reach, hand grasp, head movement and eye movement.

We propose that the basal ganglia have at least two primary tasks represented by some complex interplay of the direct and indirect pathways[3,46] (Figure 1). We suggest that the *direct path*, involving projections directly from the striatum onto the BG output nuclei, is primarily responsible for providing an estimate of the next sensory state, based upon the current sensory state and the planned motor command, to cortical planning centers. The primary role of the *indirect path,* where the striatum projects onto the external globus pallidus (GPe), in turn projecting to the subthalamic nucleus (STN), and finally to the BG output structures, is to inhibit motor activity while cortical systems are either determining which motor command to execute, or are waiting for a signal indicating the end of a delay period.

---

* A. Bischoff-Grethe, Psychology Service, San Diego VA Healthcare System, and Department of Psychiatry, University of California, San Diego, La Jolla, CA 92093. M.G. Crowley and M.A. Arbib, Department of Computer Science, University of Southern California, Los Angeles, CA 90089

2. CONCEPTUAL MODEL OF THE BASAL GANGLIA

2.1. The Inhibition of Movement

The idea of basal ganglia involvement in movement inhibition has been well supported in human and in monkey studies. In a series of well-known experiments, Hikosaka and Wurtz[28-30] examined the role of substantia nigra pars reticulata (SNr), a basal ganglia output region, in oculomotor function. They showed that SNr tonically inhibited the superior colliculus (SC), a region responsible for the generation of saccades. In ventral STN there is a cluster of visuo-oculomotor neurons which, when stimulated, suppresses saccades via excitation of SNr's tonic inhibition of SC[42]. While SC receives a number of cortical and subcortical inputs, only the basal ganglia input is inhibitory. Without basal ganglia inhibition, SC would be barraged by saccadic commands. Hikosaka and Wurtz demonstrated this effect by injecting a GABA antagonist into SC or SNr; the monkeys were unable to maintain fixation and made saccades continually[31,32]. They also demonstrated a retinotopically appropriate release of SNr inhibition prior to each saccade.

Studies of limb movement also demonstrate a role for the indirect pathway in movement inhibition. Stimulation of STN does not evoke limb movements in monkey[56]. When African green monkeys were treated with MPTP, STN increased its spontaneous firing rate; this induced the parkinsonian signs of tremor, akinesia, and rigidity[6]. Application of muscimol, a GABA agonist, to STN reduced neuronal activity, leading to reductions in tremor, akinesia, and rigidity[57]. Alexander and Crutcher[2] recorded from supplementary motor area (SMA), motor cortex (MC), and putamen in monkeys performing an elbow flexion-extension task. They reported activity during the premovement phases of the task in the arm region of all three areas. This activity began before target presentation and decreased just prior to movement. Jaeger et al.[34] reported premovement behavior in putamen, caudate and both segments of globus pallidus in a precued reaching task. Activity within internal globus pallidus (GPi) was likely influenced by input from the indirect pathway. Kimura et al.[37] also reported premovement activity in putamen medium spiny neurons.

Cortical segregation of projections to the direct and indirect pathways also exists. Movement-related regions, such as MC and the arcuate premotor area (caudal to the arcuate sulcus) project to different portions of globus pallidus than do the preparatory-related regions of SMA-cingulate, prefrontal cortex, and rostral premotor cortex[45,58]. The movement-related neurons were located ventrolaterally in the caudal part of globus pallidus (primarily GPi), while the preparatory-related regions were located dorsomedially in the rostral part of globus pallidus (primarily GPe). Since matrisomes within the striatum tend to project to either GPe or to GPi but rarely to both, we were able to segregate both the cortical projections to putamen and the putamen projections to the direct and indirect pathways within our model.

2.2. Next Sensory State Information

It is highly unlikely that movement-related cells within basal ganglia are directly responsible for the initiation of movement. Crutcher and Alexander[18] found that movement-related putamen neurons tended to fire an average of 33 ms after the onset of movement, as well as after activation of motor cortex (about 56 ms later) and SMA (about 80 ms later). Mink and

Thach[43] found that movement-related activity within both GPi and GPe was also late. This has been supported in more recent studies by Turner and Anderson[54], who showed that although pallidal discharge is related to the kinematics of reaching, GP neurons rarely change their discharge before the initial activity of agonist muscles. Kimura[36] found putamen neurons for reaching/grasping that are active prior to the muscles, as well as after the movement begins. He suggests the premovement neurons have visuomotor sensitivity. In saccadic behavior, SNr releases its inhibition of SC prior to movement[30]; this may due to the influence of visuomotor premovement activity within putamen upon the output nuclei of the BG.

While there is evidence to correlate the direction of eye movement[13,44] and arm movement[18,19] with neuronal activity, relating neuronal activity to other movement parameters is less clear. There is evidence in reaching that activity within both segments of globus pallidus may be correlated to amplitude and peak velocity[24,54], while putamen neurons are at least

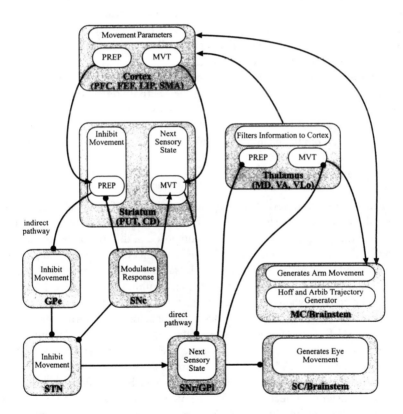

Figure 1. Model diagram illustrating the proposed functions for each component of the oculomotor and skeletomotor circuit. Although we have only shown the brainstem as divided into the movement-specific regions, we also follow the "loop hypothesis" that all other regions are subdivided into relatively distinct regions forming the separate oculomotor and skeletomotor circuits. Our key hypothesis is the shared functionality of the corresponding regions in the two circuits. **GPe**: external globus pallidus; **GPi**: internal globus pallidus; **MC**: motor cortex; **MVT**: movement neurons; **PREP**: premovement neurons; **SC**: superior colliculus; **SNc**: substantia nigra pars compacta; **SNr**: substantia nigra pars reticulata; **STN**: subthalamic nucleus.

correlated to movement direction and load application[18,19]. Since the basal ganglia activation occurs after cortical activation, it is unlikely the basal ganglia are providing these parameters directly to cortex, but they may have a modulatory influence upon the cortical areas responsible for movement. Further, these parameters can assist in determining the next sensory state. While Brotchie et al.[13,14] found only directional activity, they did suggest the basal ganglia provide the signal to SMA as to the completion of a portion of a sequence. We propose that the basal ganglia are not signaling the end of a sequence per se, but are informing the cortex of the expected next state, thus allowing these cortical regions to prepare the next movement within the sequence.

Since that microstimulation of thalamic pallidal receiving areas tends not to elicit arm movement[15,55] the basal ganglia may not in actuality control the precise dynamics of movement but instead may be permitting a movement to be made. That is, the disinhibition of the thalamus may increase the thalamic firing rate, which in turn may provide the cortical regions enough input to reach threshold and fire. Without this disinhibition, which appears to be lacking in Parkinson's disease, cortical cells have a more difficult time achieving threshold, leading to difficulties in movement performance. In Huntington's disease, therefore, one might suggest that too much thalamic disinhibition allows for indiscriminate motor activity, since cortical cells more easily reach threshold.

Our hypothesis of the direct pathway is based upon evidence that the basal ganglia do not command or initiate movements, but rather their activity reflects cortical activation and calculates the next state of the system, as well as has some influence over such movement kinematics as velocity and goal position. We have designed our model representation of the direct pathway with both the neuroanatomical and neurophysiological data in mind.

2.3. The Functional Pathway

Based upon our hypothesis, we suggest the following functional breakdown of the cortex and the basal ganglia (Figure 1). Although our diagram shows a generic cortical projection to basal ganglia, this projection is composed of different areas of cortex. Within the oculomotor system, the frontal eye fields (FEF) project the current goal position to the movement-related areas of caudate, the mediodorsal and ventral anterior thalamus, and SC. FEF also contains neurons related to visual activity within the eye's fovea. These neurons inhibit saccadic activity when active; they project to the next sensory state area of caudate. The lateral intraparietal cortex (LIP) provides retinotopic location information regarding saccadic targets to SC, FEF, and the next sensory state areas of caudate[4,41]. The prefrontal cortex (PFC) is involved with the planning of saccadic eye movements and the maintaining of a working memory of saccade targets[25]. It proves the "go" signal to FEF and the movement-related areas of caudate to begin a saccade. SC transmits the saccade motor signal to the brainstem to effect a saccade. The movement areas of caudate are responsible for releasing SNr's tonic inhibition of SC, allowing a saccade to occur. The indirect pathway therefore provides information regarding the upcoming saccade. This is important when multiple saccades are to be performed, and becomes disrupted in multiple memory saccades when dopamine is reduced.

Within the skeletomotor pathway, PFC is responsible for working memory and for sequence learning, as seen in learning studies in monkey[5] and in human[35]. Within the skeletomotor pathway, PFC projects to presupplementary motor area (pre-SMA)[40]. Since pre-SMA is responsible for preparing a movement sequence, particularly if visually guided[27,49], PFC may be a source for the sequences pre-SMA "knows." Pre-SMA projects sequential information to both SMA-proper and to the basal ganglia's indirect pathway. SMA-proper is involved in the internal generation of sequences and repetitive movements[48,50]. SMA-proper neurons have been shown to perform several functions: they contain information on the overall sequence to be performed; they keep track of which movement is next according to a sequence and which movement is currently being performed; they project the current movement to be performed to MC and to the direct pathway of the basal ganglia; and they project the preparations of the next movement of the sequence to another population of premovement neurons within MC and to the indirect (inhibitory) pathway of basal ganglia. SMA-proper acts as the engine that drives the motor sequence. The basal ganglia may be viewed as containing the brake pedal (indirect pathway) and the gas pedal (direct pathway), providing the "engine" with performance information. The motor cortex, then, carries out the motor command dictated by SMA-proper, and handles the fine tuning of the movement (e.g., target position, amplitude, force) partly based upon information provided by the basal ganglia. The motor cortex projects the motor parameters to both the brainstem and the basal ganglia's direct pathway. The basal ganglia's two pathways thus perform two different roles: the indirect pathway inhibits upcoming motor commands from being performed while the current movement is in progress, while the direct pathway projects the next sensory state back to cortex. This informs SMA-proper and MC of the expected next state and allows these regions to switch to the next movement of the sequence, typically the one that is being inhibited by the basal ganglia's indirect pathway. Thus, the simple box marked "Cortex" in Figure 1 becomes a complex structure in our actual computer modeling: it combines models of PFC, FEF and LIP in our saccade model, and MC, pre-SMA, and SMA-proper in our model of limb movements.

3. BRIEF SUMMARY OF MODEL RESULTS

3.1. Saccade Generation

We used the double saccade paradigm, for both visually-guided and memory-guided saccades, to test sequential movements in our model[17]. This allowed us to simulate the impact of reduced dopamine in substantia nigra pars compacta (SNc) upon both cases. In the visually-guided double saccade experiment, the maximum velocity of the saccade and the time to complete the saccade are essentially unaffected by the level of dopamine. In the memory-guided cases, the second saccade is not performed when dopamine is reduced to half or below normal levels (Figure 2a-b). The increased tonic inhibition from the BG onto the cortico-thalamic memory loops causes a significant reduction in the level of LIP activation for the remapped saccade target causing it to decay before the second saccade can be generated. This is similar to a symptom that PD patients sometimes display, in that they are unable to switch from the first task to the second task in a sequence without external cueing. However, our model developed this "akinesia" with only a 50% decrease in dopamine. Parkinson's patients may not develop significant symptoms until the level of dopamine loss is more than 80%. It is believed that striatal dopamine receptors become sensitized to the presence of dopamine as an adaptation to the decreased levels of dopamine in the striatum.

We found that the model adaptation required to allow for the second saccade was not linear as the level of dopamine was decreased, but required a substantially larger increase in weights for progressively less dopamine. When dopamine was reduced by a factor of 6 (16% of normal), our "adaptive" weights had to be increased by a factor of 20.

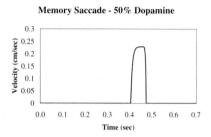

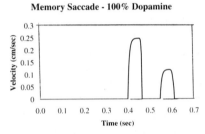

Figure 2. Saccade velocities for the oculomotor model. **a)** The memory-guided saccade is performed correctly under normal conditions. When dopamine is reduced to 50% of normal within the model, **b)** the memory-guided saccade neglects to perform the second saccade.

In analyzing the firing rates of the memory neurons in both LIP and PFC, we found that LIP neurons begin to fire first, in our model, and they must continue to fire until the prefrontal memory neurons become active so that the reciprocal connections between them can "override" the increased inhibition from the BG. Thus, the inability of the model to generate the second saccade in a double memory saccade is a timing issue. The remapped memory signal must "survive" until other cortical regions receiving input from LIP are also able to increase their activity. It may be that the akinesia in Parkinson's disease has a similar mechanism as our model. This could also explain why PD patients perform much better when they have visual feedback. With visual information, it is not necessary to maintain remapped memories that may decay.

3.2. Sequential Arm Movements

We also used the skeletomotor model to perform sequential movements[8,10]. The model was provided with the location of three tasks and the order in which contact should be made (Figure 3). SMA-proper contained neurons similar to those seen by Tanji and Shima[50]; these neurons represented an overall task sequence and the subsequences within the task. A trajectory generator[33] performed the movements as dictated by MC. Under normal conditions the model successfully moved from target to target (Figure 3a), and exhibited a velocity similar to normal subjects (Figure 3b). When we depleted dopamine, we encountered several behavioral results. As SNc reduced its ability to affect the direct and indirect pathways, there was an increase in the firing rates of STN and both sections of globus pallidus. The increase in inhibitory output led to a decrease in MC firing rate, as MC was unable to completely overcome GPi's inhibition via ventrolateral pars oralis thalamus (VLo) projections. This

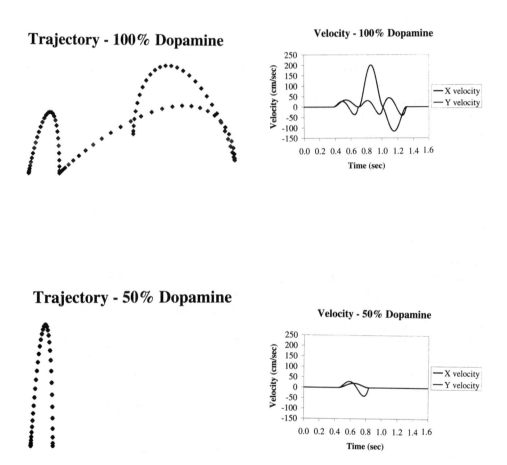

Figure 3. The arm trajectories and velocities of the skeletomotor sequencing model. **a)** Under normal conditions, the model successfully reaches to all three targets and **b)** produces velocity curves similar to normal subjects. **c)** With a 50% reduction of dopamine, the model fails to reach for the second and third targets, and **d)** the velocities are reduced.

reduced firing rate translated to a slower movement time. Thus, the natural balance between the two pathways (and the two hypotheses of how these pathways function) is disturbed when dopamine is lost. In Parkinson's disease, this leads to an increase in activation of the indirect pathway (movement inhibition) and a decrease in effectiveness of the direct pathway (next sensory state). We also saw a slowdown in overall neural activity; the basal ganglia and the motor cortex took longer to reach peak firing rates. The slowdown in neural activity, coupled with the increased inhibition of movement, was responsible for pauses between movements within the sequence. We therefore began to see pauses between movements from one target to the next. However, we also found that for each subsequent movement, the velocity was less. This is similar to the bradykinesia seen in PD patients when asked to trace the edges of a polygon[1]. The model also exhibited a reduction in cortical firing rates; this is consistent with positron emission tomography (PET) studies of PD patients, which show a reduction in activation of cortical association areas[26]. When dopamine was further depleted, the model was capable of performing only the movement towards the first target (Figure 3c). This is not dissimilar to the difficulties Parkinson's disease patients have in generating saccades to remembered targets[16]. The arm movement to the first target was also slow, with a velocity similar to that seen in PD patients (Figure 3d). This model has also been used to reproduce results seen in normal and PD subjects in a reciprocal aiming task[8,11] and normal behavior during conditional elbow movements[8,9].

4. DISCUSSION

Why should the basal ganglia have responsibility for two seemingly disparate functions? We propose that in order for the BG to pass their next sensory state estimate to the cortex at the proper time, they must know when these same regions have completed their planning for the next motor command to be executed. Thus, the cortical go signal which initiates the release of the BG inhibition on the currently selected motor command also provides the go signal for the BG to pass their next sensory state prediction to cortical planning centers. By linking the inhibition of motor commands while planning is in process with a mechanism providing advance information on a future sensory state in the BG, the brain has ensured that the timing of the two activities is correlated. Thus, planning for the second movement in a sequence does not begin before the motor program for the first movement has been passed to the circuitry that carries out the movement.

If the two tasks are interrelated, as we believe, then it may be difficult to define experiments that can test the two tasks independently. For instance, lesioning the direct cortical projections to STN should impact the ability of the basal ganglia to inhibit motor commands prior to receiving a cortical go signal. However, the increased distractibility that should result would also affect the remapping task in BG; either the wrong motor command would be executed prior to the correct motor command, e.g., antisaccade tasks in Huntington's disease[38,39], or the timing of motor command execution and next sensory state prediction would be disrupted. Another experiment would be to make a small lesion between SNr and thalamus or between thalamus and LIP. The lesioned projections should not be able to convey the next sensory state signal, where nonlesioned projections should function normally. Thus, the growth in remapped target location activity in parietal cortex seen by Duhamel et al.[23] for the second eye target location, which occurred prior to the onset of eye movement to the first target, would not be seen due to the lesioned pathway between BG and cortex.

5. REFERENCES

1. R. Agostino, A. Berardelli, A. Formica, N. Accornero, and M. Manfredi, Sequential arm movements in patients with Parkinson's disease, Huntington's disease and dystonia, *Brain.* 115:1481-1495 (1992).
2. G.E. Alexander and M.D. Crutcher, Preparation for movement: Neural representations of intended direction in three motor areas of the monkey, *J Neurophysiol.* 64:133-150 (1990).
3. G.E. Alexander, M.D. Crutcher, and M.R. DeLong, Basal ganglia-thalamocortical circuits: Parallel substrates for motor, oculomotor, "prefrontal" and "limbic" functions, in *Progress in Brain Research*, H.B.M. Uylings, C.G.V. Eden, J.P.C.D. Bruin, M.A. Corner, and M.G.P. Feenstra, eds., Elsevier Science Publishers B. V., New York (1990).
4. R.A. Andersen, C. Asanuma, and W.M. Cowan, Callosal and prefrontal associational projecting cell populations in area 7A of the macaque monkey: A study using retrogradely transported fluorescent dyes, *J Comp Neurol.* 232:443-455 (1985).
5. P. Barone and J.-P. Joseph, Prefrontal cortex and spatial sequencing in macaque monkey, *Exp Brain Res.* 78:447-464 (1989).
6. H. Bergman, T. Wichmann, B. Karmon, and M.R. DeLong, The primate subthalamic nucleus. II. Neuronal activity in the MPTP model of parkinsonism, *J Neurophysiol.* 72:507-520 (1994).
7. G.S. Berns and T.J. Sejnowski, A model of basal ganglia function unifying reinforcement learning and action selection, *Joint Symposium on Neural Computation:* 129-148 (1995).
8. A. Bischoff, Modeling the Basal Ganglia in the Control of Arm Movements. Ph.D. Thesis, University of Southern California (1998).
9. A. Bischoff-Grethe and M.A. Arbib, Modeling the basal ganglia in a conditional elbow flexion-extension task, (in preparation).
10. A. Bischoff-Grethe and M.A. Arbib, Sequential movements: A computational model of the roles of the basal ganglia and the supplementary motor area, (in preparation).
11. A. Bischoff-Grethe, M.A. Arbib, and C.J. Winstein, A computational model of the basal ganglia and its performance in a reciprocal aiming task, (submitted).
12. A.M. Bronstein and C. Kennard, Predictive ocular motor control in Parkinson's disease, *Brain.* 108:925-940 (1985).
13. P. Brotchie, R. Iansek, and M. K. Horne, Motor function of the monkey globus pallidus. 1. Neuronal discharge and parameters of movement, *Brain.* 114:1667-1683 (1991).
14. P. Brotchie, R. Iansek, and M.K. Horne, Motor function of the monkey globus pallidus. 2. Cognitive aspects of movement and phasic neuronal activity, *Brain.* 114:1685-1702 (1991).
15. J.A. Buford, M. Inase, and M.E. Anderson, Contrasting locations of pallidal-receiving neurons and microexcitable zones in primate thalamus, *J Neurophysiol.* 75:1105-1116 (1996).
16. T.J. Crawford, L. Henderson, and C. Kennard, Abnormalities of nonvisually-guided eye movements in Parkinson's disease, *Brain.* 112:1573-1586 (1989).
17. M.G. Crowley, Modeling Saccadic Motor Control: Normal Function, Sensory Remapping, and Basal Ganglia Dysfunction. Ph.D. Thesis, University of Southern California (1997).
18. M.D. Crutcher and G.E. Alexander, Movement-related neuronal activity coding direction or muscle pattern in three motor areas of the monkey, *J Neurophysiol.* 64:151-163 (1990).
19. M.D. Crutcher and M.R. DeLong, Single cell studies of the primate putamen. II. Relations to direction of movement and pattern of muscular activity, *Exp Brain Res.* 53:244-258 (1984).
20. A. Curra, A. Berardelli, R. Agostino, N. Modugno, C.C. Puorger, N. Accornero, and M. Manfredi, Performance of sequential arm movements with and without advance knowledge of motor pathways in Parkinson's disease, *Mov Disord.* 12:646-654 (1997).
21. P.F. Dominey and M.A. Arbib, A cortico-subcortical model for generation of spatially accurate sequential saccades, *Cereb Cortex.* 2:153-175 (1992).
22. P.F. Dominey, M.A. Arbib, and J.-P. Joseph, A model of corticostriatal plasticity for learning oculomotor associations and sequences, *J Cogn Neurosci.* 7:311-336 (1995).
23. J.-R. Duhamel, C.L. Colby, and M.E. Goldberg, The updating of the representation of visual space in parietal cortex by intended eye movements, *Science.* 255:90-92 (1992).

24. A.P. Georgopoulos, M.R. DeLong, and M.D. Crutcher, Relations between parameters of step-tracking movements and single cell discharge in the globus pallidus and subthalamic nucleus of the behaving monkey, *J Neurosci.* 3:1586-1598 (1983).
25. P.S. Goldman-Rakic, Circuitry of primate prefrontal cortex and regulation of behavior by representational memory, in *Handbook of Physiology, The Nervous System, Higher Functions of the Brain,* American Physiological Society, Bethesda, MD (1987).
26. S.T. Grafton, C. Waters, J. Sutton, M.F. Lew, and W. Couldwell, Pallidotomy increases activity of motor association cortex in Parkinson's disease: A positron emission tomographic study, *Ann Neurol.* 37:776-783 (1995).
27. U. Halsband, Y. Matsuzaka, and J. Tanji, Neuronal activity in the primate supplementary, pre-supplementary and premotor cortex during externally and internally instructed sequential movements, *Neurosci Res.* 20:149-155 (1994).
28. O. Hikosaka and R. Wurtz, Visual and oculomotor functions of monkey substantia nigra pars reticulata. I. Relation of visual and auditory responses to saccades, *J Neurophysiol.* 49:1230-1253 (1983).
29. O. Hikosaka and R. Wurtz, Visual and oculomotor functions of monkey substantia nigra pars reticulata. II. Visual responses related to fixation of gaze, *J Neurophysiol.* 49:1254-1267 (1983).
30. O. Hikosaka and R.H. Wurtz, Visual and oculomotor functions of monkey substantia nigra pars reticulata. III. Memory continent visual and saccade responses, *J Neurophysiol.* 49:1268-1284 (1983).
31. O. Hikosaka and R.H. Wurtz, Modification of saccadic eye movements by GABA-related substances. I. Effect of muscimol and bicuculline in monkey superior colliculus, *J Neurophysiol.* 53:266-291 (1985).
32. O. Hikosaka and R.H. Wurtz, Modification of saccadic eye movements by GABA-related substances. II. Effects of muscimol in monkey substantia nigra pars reticulata, *J Neurophysiol.* 53:292-308 (1985).
33. B. Hoff and M.A. Arbib, A model of the effects of speed, accuracy, and perturbation on visually guided reaching, in *Control of Arm Movement in Space: Neurophysiological and Computational Approaches,* R. Caminiti, P.B. Johnson, and Y. Burnod, eds., Springer-Verlag, Berlin (1992).
34. D. Jaeger, S. Gilman, and J.W. Aldridge, Primate basal ganglia in a precued reaching task: Preparation for movement, *Exp Brain Res.* 95:51-64 (1993).
35. I.H. Jenkins, D.J. Brooks, P.D. Nixon, R.S.J. Frackowiak, and R.E. Passingham, Motor sequence learning: A study with positron emission tomography, *J Neurosci.* 14:3775-3790 (1994).
36. M. Kimura, Behaviorally contingent property of movement-related activity of the primate putamen, *J Neurophysiol.* 63:1277-1296 (1990).
37. M. Kimura, M. Kato, H. Shimazaki, K. Watanabe, and N. Matsumoto, Neural information transferred from the putamen to the globus pallidus during learned movement in the monkey, *J Neurophysiol.* 76:3771-3786 (1996).
38. A.G. Lasker, D.S. Zee, T.C. Hain, S.E. Folstein, and H.S. Singer, Saccades in Huntington's disease: Initiation defects and distractibility, *Neurol.* 37:364-270 (1987).
39. C.J. Lueck, S. Tanyeri, T.J. Crawford, L. Henderson, and C. Kennard, Antisaccades and remembered saccades in Parkinson's disease, *J Neurol.* 53:284-288 (1990).
40. G. Luppino, M. Matelli, R. Camarda, and G. Rizzolatti, Corticocortical connections of area F3 (SMA-proper) and area F6 (pre-SMA) in the macaque monkey, *J Comp Neurol.* 338:114-140 (1993).
41. J.C. Lynch, A.M. Graybiel, and L.J. Lobeck, The differential projection of two cytoarchitectonic subregions of the inferior parietal lobule of macaque upon the deep layers of the superior colliculus, *J Comp Neurol.* 235:241-254 (1985).
42. M. Matsumura, J. Kojima, T.W. Gardiner, and O. Hikosaka, Visual and oculomotor functions of monkey subthalamic nucleus, *J Neurophysiol.* 67:1615-1632 (1992).
43. J.W. Mink and W.T. Thach, Basal ganglia motor control. II. Late pallidal timing relative to movement onset and inconsistent pallidal coding of movement parameters, *J Neurophysiol.* 65:301-329 (1991).
44. S.J. Mitchell, R.T. Richardson, F.H. Baker, and M.R. DeLong, The primate globus pallidus: Neuronal activity in direction of movement, *Exp Brain Res.* 68:491-505 (1987).
45. A. Nambu, S. Yoshida, and K. Jinnai, Discharge patterns of pallidal neurons with input from various cortical areas during movement in the monkey, *Brain Res.* 519:183-191 (1990).
46. A. Parent and L.-N. Hazrati, Anatomical aspects of information-processing in primate basal ganglia, *TINS.* 16:111-116 (1993).
47. A. Pascual-Leone, J. Grafman, K. Clark, M. Stewart, S. Massaquoi, J.-S. Lou, and M. Hallett, Procedural learning in Parkinson's disease and cerebellar degeneration, *Ann Neurol.* 34:594-602 (1993).
48. R.E. Passingham, D.E. Thaler, and Y. Chen, Supplementary motor cortex and self-initiated movement, in *Neural Programming,* M. Ito, ed., Karger, Basel (1989).

49. J. Tanji and H. Mushiake, Comparison of neuronal activity in the supplementary motor area and primary motor cortex, *Cogn Brain Res.* 3:143-150 (1996).
50. J. Tanji and K. Shima, Role for supplementary motor area cells in planning several movements ahead, *Nature.* 371:413-416 (1994).
51. W.T. Thach, J.W. Mink, H.P. Goodkin, and J.G. Keating, Combining versus gating motor programs: Differential roles for cerebellum and basal ganglia?, in *Role of the Cerebellum and Basal Ganglia in Voluntary Movement,* N. Mano, 1. Hamada, and M.R. DeLong, eds., Elsevier Science Publishers, Amsterdam (1993).
52. V. Thomas-Ollivier, J.M. Reymann, S. Le Moal, S. Schuck, A. Lieury, and H. Allain, Procedural memory in recent-onset Parkinson's disease, *Dement Geriatr Cogn Disord.* 10:172-180 (1999).
53. J.R. Tian, D.S. Zee, A.G. Lasker, and S.E. Folstein, Saccades in Huntington's disease: Predictive tracking and interaction between release of fixation and initiation of saccades, *Neurol.* 41:875-881 (1991).
54. R.S. Turner and M.E. Anderson, Pallidal discharge related to the kinematics of reaching movements in two dimensions, *J Neurophysiol.* 77:1051-1074 (1997).
55. J.L. Vitek, J. Ashe, M.R. DeLong, and G.E. Alexander, Physiological properties and somatotopic organization of the primate motor thalamus, *J Neurophysiol.* 71:1498-1513 (1994).
56. T. Wichmann, H. Bergman, and MR. DeLong, The primate subthalamic nucleus. I. Functional properties in intact animals, *J Neurophysiol.* 72:494-506 (1994).
57. T. Wichmann, H. Bergman, and M.R. DeLong, The-primate subthalamic nucleus. III. Changes in motor behavior and neuronal activity in the internal pallidum induced by subthalamic inactivation in the MPTP model of parkinsonism, *J Neurophysiol.* 72:521-530 (1994).
58. S. Yoshida, A. Nambu, and K. Jinnai, The distribution of the globus pallidus neurons with input from various cortical areas in the monkeys, *Brain Res.* 611:170-174 (1993).

BASAL GANGLIA NEURAL CODING OF NATURAL ACTION SEQUENCES

J. Wayne Aldridge and Kent C. Berridge*

Abstract: In this study we present evidence that neurones in the basal ganglia code the serial order of syntactic (rule-driven) sequences of natural motor behaviour (rodent grooming). Neuronal activity was recorded from the striatum in freely behaving rats while they spontaneously groomed themselves. Offline, we analysed sequential patterns of movement in a frame-by-frame scan of video taped behaviour and evaluated the correlation of neuronal activity to syntactic and nonsyntactic grooming movements. We found that sequential patterns of grooming movements activated striatal neurones. Neurones were activated preferentially during particular serial patterns (syntactic chains) of grooming and were inactive, weaker or different during the same movements in other sequences. Syntactic chains of grooming were preferentially coded by neurones in a dorsolateral striatum site where lesions disrupt syntactic grooming patterns[1,13]. The timing of neuronal activation, which was generally synchronised with or followed movement onset, suggested a role in the execution of the behavioural sequence rather than one of related to the initiation of phasic elements. We conclude that that neuronal activity in rodent neostriatum codes the serial order of natural actions and not the simple motor properties of constituent actions within a sequence.

1. INTRODUCTION

The idea that the basal ganglia have a role in motor control, first proposed by Magendie[26] more than 150 years ago, is now well accepted. But what exactly do the basal ganglia do for movement? One clue can be obtained from studies of Huntington's and Parkinson's diseases. The devastating impact on movement caused by these degenerative disorders of the basal ganglia strongly supports a motor function. However, close scrutiny suggests that the elemental properties of motor control are less effected by basal ganglia pathology than the organisational aspects of motor control. Parkinson's patients can perform motor tasks that require them to control kinematic and dynamic features of movement such as force and

---

* J.W. Aldridge, Department of Neurology, Department of Psychology, University of Michigan. K.C. Berridge, Department of Psychology, University of Michigan.

direction, however, their difficulty in performing *sequences of movements*[20] suggests that a higher, organisational aspect of motor control is disturbed by this disorder. Huntington's patients have been shown to have special "ideomotor" deficits when asked to make movements of the type, for example that would be involved in using a particular tool[34]. Some have suggested that the neostriatum may even be involved in sequential disorders of human language[11,36]. Marsden[27] suggested that "The *sequencing* of motor action and the *sequencing* of thought could be a uniform function carried out by the basal ganglia." Other evidence supports the idea that injury to the basal ganglia may produce an inability to control behavioural sequences in general. The pathological repetitions of spoken words in Tourette's syndrome[14] and the tormenting habits and thoughts of obsessive compulsive disorder[32], both of which are associated with pathology of the basal ganglia. These disorders suggest that the basal ganglia might even participate in the organisation of the sequential aspects of "cognitive" behaviour. Four decades ago Karl Lashley[25] noted the continuity between serial order at different levels of psychological complexity and the continuity this may imply for the underlying neural substrates of syntax. Several have suggested that behavioural sequencing functions of the basal ganglia might originally have evolved to co-ordinate sequences of instinctive behaviour and later have been modified to control learned behaviour[2,31]. We believe that Marsden may be right; circuitry within the neostriatum may provide a common link for sequencing phenomena as diverse as actions, words, or thoughts and that Lashley's idea of syntax can provide a profitable *formulation of function* for these structures[12].

2. SYNTACTICAL GROOMING SEQUENCES IN RODENTS

All purposeful behaviour is sequential, so what do we mean by *syntactical sequence?* In the simplest terms, a syntactic sequence is one that follows rules that determine the temporal progression of its elements. These rules impart a mathematical predictability to the sequence. Language has syntax. Given an arbitrary word, it is possible to predict with some level of probability what the next word in a language sequence will be. Other behaviour can be described as having properties of syntax if one can demonstrate lawful sequential dependencies. For example, rodent grooming behaviour has distinct syntactical properties[7,15,16,33]. The most stereotyped serial pattern that occurs in rodent grooming is the 'syntactic chain' pattern shown in the choreograph *(Figure 1)*.

Syntactic grooming chains have approximately 25 movements that are linked in a sequence that follows a fixed serial order of four phases[7,9]. Phase 1 consists of 5-9 rapid elliptical strokes over the nose and mystacial vibrissae lasting for about one second. Phase 2 is short (0.25 s) and consists of small asymmetrical strokes of increasing amplitude. Phase 3 consists of large bilateral strokes that take 2-3 s for the animal to complete. The chain concludes with Phase 4, which consists of a postural turn followed by a period (1-3 s) of body licking directed to the flank. Once the pattern begins, each remaining action can be predicted with over 90% accuracy. The four grooming actions that contribute to this sequence also occur in unpredictable order and combination outside of the syntactical chain sequence. The entire syntactical chain occurs with a frequency that is over 13,000 times greater than could be expected by chance (based upon the relative probabilities of the component 25 actions obtained from grooming outside of this syntactic chain[9]).

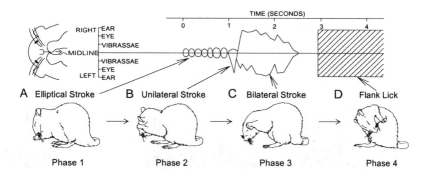

Figure 1. Syntactic grooming chains. The 4 syntactic phases, A) elliptical strokes, B) unilateral strokes, C) bilateral strokes and D) body licking are schematised in the drawings. The top graph, expresses choreographed forepaw movement as distance from the midline (y-axis) as a function of time (x-axis, tics=1 sec) for a typical syntactic chain (left paw represented by line below the axis, right paw represented by line above the axis).

We have exploited this natural sequence as a window into the role of neural systems in sequential co-ordination. The more common approach is to train animals to perform responses in an arbitrary order. That approach allows one to study very complex sequences, but it has the pitfall of confounding neural mechanisms of learning and memory with mechanisms of action syntax or sequential co-ordination per se. Failure on learned tasks can be due either to disruption of memory processes or to disruption of sequencing per se: the two cannot be discriminated. Natural behavioural sequences, on the other hand, do not depend upon explicit training, and provide a way to study neuronal mechanisms of behavioural sequencing independent of memory and explicit training.

3. NEURONAL CORRELATIONS OF GROOMING MOVEMENTS

3.1 Methods

These studies were based on neuronal recordings from freely moving rats. Briefly, Sprague-Dawley rats (~250g) were anaesthetised with ketamine-xylazine and implanted with a permanent multisite recording electrode in dorsolateral or ventromedial neostriatum. The lightweight implant did not interfere with normal behaviour and caused no discomfort. The electrodes were connected to a preamplifier and a computer through a commutator, which permitted free movement in a circular recording chamber. Spontaneous behaviour was videotaped and neuronal discharge activity was recorded for one or more hours while the animals groomed and moved about freely. A frame-by-frame analysis of the videotaped grooming sequences was conducted off-line[7,8] to find the onset and offset times of movements. Neuronal activity was analysed in relation to grooming actions by the construction of perievent time histograms. At the completion of recording the animals were killed by an overdose of anaesthetic, brains were removed and prepared histologically for verification of recording sites.

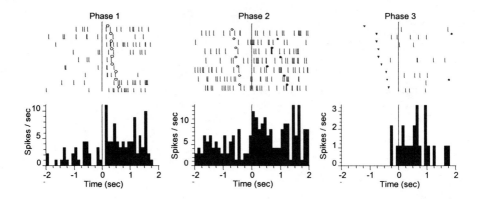

Figure 2. Three neurones responsive to syntactic grooming from dorsolateral (left, middle) and ventromedial (right) neostriatum. The phase onset is at time = 0. The histogram represents the average firing rate (y-axes) in bins 50 ms wide. Rasters of spike trains indicate neuronal activity (one spike train per chain) and the marks in each spike train indicate the time at which the preceding or following Phase began. Neuronal activity generally occurs at about the same time as movement onset.

3.2 Results

Syntactic chain grooming was a potent activator of striatal neurones (41% of tested neurones) in both dorsolateral and ventromedial striatum (Figure 2). Nonsyntactic grooming, in contrast, activated a much smaller proportion of neurones (14%) even though these less stereotyped grooming bouts incorporate the same movements and they occur much more frequently than syntactic chain sequences. Excitatory responses were more common (99%) than inhibitions (20%). All but one instance of inhibitory activity was accompanied by an excitation as well. Each phase of the grooming chain was associated with activity changes. Although the proportion of phasic responses dorsolateral striatum was greater (20%, 14%, 18%, 18%; Phases 1 to 4 respectively) than ventromedial neurones (16%, 11%, 8%, 5%; Phases 1 to 4), the differences were not significant.

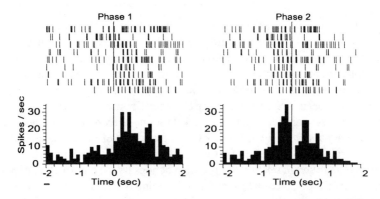

Figure 3. Multiple chain responses from the same cell. This striatal neurone had increases in activity after both Phases 1 and 2.

The dorsolateral region, whose integrity is crucial for chain grooming, had other similarities and differences to the ventromedial striatum. Most neurones were responsive to a single action in the grooming chain (dorsolateral 28% vs. ventromedial 27%), but in the dorsolateral region the proportion responding to 2 or more actions (Figure 3, 18%) was more than 3 times larger than the proportion (5%) in ventromedial striatum. Multiphase responses such as this are consistent with encoding of the sequential pattern as a whole. The predominance of this kind of response in the dorsolateral region suggests a tighter link to syntax properties of the stereotyped pattern.

Other factors also point to the much more potent influence of syntactic chain grooming. Most chain-related neurones were not activated during nonchain grooming. Even highly similar movements evoked different responses from most neurones depending on whether they occurred inside or outside a syntactic chain sequence. For some of these neurones, the pattern of neuronal activation related to equivalent strokes was radically different (Figure 4). In a few neurones there were some similarities between chain and nonchain responses, however, even in these cases the equivalent nonchain grooming actions were associated with much weaker activation than their chain counterparts. Overall, our findings suggest that information processing in the striatum is concerned preferentially with syntactic grooming sequences. Activation by nonchain grooming is not only less common, but the patterns of responses also differ and they are weaker than activation related to syntactic chain sequences. The ability of sequential context to gate neuronal firing to equivalent movements in an all-or-none fashion provides one of the strongest indications that these neurones coded *action syntax* or sequential properties of behaviour rather than the mere component movements.

Does striatal activity *initiate* grooming movements in a chain? The answer may generally be no. In our sample of neurones responsive to Phase 1 onset, which was the most reliable activator of striatal activity, not a single neurone in dorsolateral striatum had a change in neuronal activity and only one ventromedial neurone had activity before Phase 1. This timing relationship also appeared to be true for other movements in the sequence with a general pattern of movement bout onset contiguous with or followed by neuronal activa-

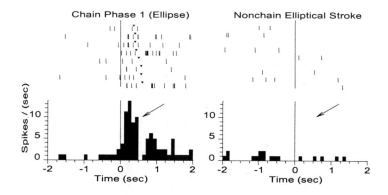

Figure 4. Chain versus nonchain activity. This dorsolateral striatal neurone had an increase of activity associated with the onset of Phase 1 (elliptical strokes) during chain grooming (left panel, arrow). The same neurone had no change associated with the onset of elliptical strokes during nonchain grooming (right panel, arrow).

tion. Neuronal activity following the onset of a movement could be initiating subsequent movements in the grooming chain, however, the variability between the onset of neuronal activity and the onset of movement was usually less to the preceding than the following phase. This timing of neuronal activation suggests that these neurones may in fact be contributing in some manner to the initiation of the next action in the sequence, or else to coding of the sequential pattern as a whole. Such a role in movement control is in keeping with our idea that the basal ganglia are facilitating the execution of syntactic action sequences. In general it seems that striatal neurones do not play a direct role in movement *initiation so* much as in movement pattern. This functional relationship to the movement sequence is important since previous work has shown that dorsolateral striatal damage does not impair the animals' ability to initiate syntactic chains, but the ability to complete the sequential pattern is severely disrupted.

3.3 Rate Coding

Besides the phasic changes in neuronal activity noted above, the neuronal code may be distributed over assemblies of neurons in a manner not easily detectable as phasic changes in perievent time histograms. Target structures, in the pallidum for example, could retrieve this information by combining input from many striatal neurons. We searched for possible rate codes[3] across periods of chain grooming, nonchain grooming (i.e., grooming movements in other patterns of serial order) and quiescent behaviour (i.e., quiet resting). Median rates differed significantly with these 3 categories behaviour (Figure 5, ANOVA, $p = 0.032$) in dorsolateral striatum but not in ventromedial striatum ($p > 0.05$). Syntactic grooming chains were associated with higher spike rates than either resting or nonchain grooming (Figure 5). Faster firing during syntactic grooming reinforces the unique relation of this sequentially stereotyped behavioural sequence to activity in the dorsolateral neostriatum.

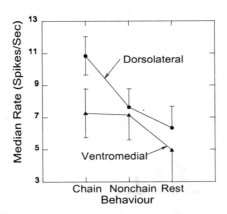

Figure 5. Rate coding. Median firing rates (y-axis) were greater for chain than nonchain or rest in dorsolateral striatum (top line). There were no significant differences in the sample of ventromedial neurones.

4. CONCLUSIONS

We have shown that neuronal activity in the neostriatum of rodents is preferentially correlated to specific syntactic sequences of grooming movements. It appears that the serial order of the movements and not the grooming movements per se was the critical factor in determining neuronal activation. In other words, it was the action syntax that was crucial for the activation of these neurones.

Our findings also demonstrate that the dorsolateral region of the neostriatum may be especially important in coding the serial order of grooming movements, in comparison to ventromedial neostriatum. Both regions had neurones that were sensitive to syntactic grooming chains, but dorsolateral neurones were more vigorously activated than ventromedial neurones during syntactic chains. Dorsolateral neurones were also more likely to respond during multiple phases of a syntactic grooming chain. This suggests that dorsolateral neurones may code syntactic patterns of movement serial order as a higher-order property, distributed over the duration of the chain. By contrast, activity of neurones in the ventromedial region actually declined during some phases of syntactic grooming chains, and ventromedial neurones were less likely to code either multiple phases or terminal phases of the chain pattern. These findings suggest that the dorsolateral region may be concerned with syntactic phase to phase transitions, or overall sequential structure, while ventromedial activity is concerned more simply with the onset of the chain pattern.

The timing of striatal neuronal activity with respect to movement suggests that it likely does not *initiate* or *generate* the syntactic sequence. Instead, we believe it is more likely to have a role in the *implementation* of the sequence that is initiated and programmed elsewhere in the brain. This role is consistent with the results of lesion and transection studies of the neural basis of behavioural grooming syntax[6,10]. Elementary generation of the basic 4-phase syntactic pattern in rats can be carried out by the isolated pontine brainstem. Even when the brain has been transected above the superior colliculus or pons and cerebellum the rat can still generate occasional syntactic chain patterns of grooming more often than chance even though they have marked deficits in sequence implementation[6]. However, the neostriatum is needed for the implementation of the pattern into normal streams of behaviour. Rather than a role in generating the syntactic sequence, these findings support an executive function for the striatum regarding serial co-ordination in which brainstem-generated syntactic patterns are incorporated into behaviour. This is consistent with suggestions that the basal ganglia may serve a role to focus selection and inhibit competing motor programs[28] and a specific role in sequence control 5. This role is also compatible with recent models of the neostriatum[17-19,21]. A recent review points out that several computational models of basal ganglia "have emphasised pattern recognition or mutual competition, or a combination of the two, to form pattern classification networks"[4].

How does this proposed role for the basal ganglia in facilitating stereotyped sequences in rats relate to basal ganglia function in humans? One possibility is that striatal circuitry preadapted for coordinating innate sequences of movements may now also participate in coordinating sequences of learned movement, and even of language and thought. The basal ganglia are activated by learned including sequential patterns of movements[22], and deficits in humans with basal ganglia disease have been suggested to include linguistic grammar errors as well as habit learning[24,29,35,23,30]. Thus serial coordination of action syntax by basal ganglia may provide a window into normal human behavior and into pathologies related to complex behavioral sequences.

5. REFERENCES

1. Aldridge J.W. and Berridge K.C. (1998) Coding of serial order by neostriatal neurons: a 'natural action' approach to movement sequence. *J. Neurosci.* 18, 2777-2787.
2. Aldridge J.W., Berridge K.C., Herman M., and Zimmer L. (1993) Neuronal coding of serial order: syntax of grooming in the neostriatum. *Psychol. Sci.* 4, 391-395.
3. Aldridge J.W. and Gilman S. (1991) The temporal structure of spike trains in the primate basal ganglia – afferent regulation of bursting demonstrated with precentral cerebral cortical ablation. *Brain Res.* 543, 123-138.
4. Beiser D.G., Hua S.E., and Houk J.C. (1997) Network models of the basal ganglia. *Curr. Opin. Neurobiol.* 7, 185-190.
5. Berns G.S. and Sejnowski T.J. (1998) A computational model of how the basal ganglia produce sequences. *J. Cogn Neurosci.* 10, 108-121.
6. Berridge K.C. (1989) Progressive degradation of serial grooming chains by descending decerebration. *Behav. Brain Res.* 33, 241-253.
7. Berridge K.C. (1990) Comparative fine structure of action: rules of form and sequence in the grooming patterns of six rodent species. *Behav.* 113, 21-56.
8. Berridge K.C. and Fentress J.C. (1986) Contextual control of trigeminal sensorimotor function. *J. Neurosci.* 9, 325-330.
9. Berridge K.C., Fentress J.C., and Parr H. (1987) Natural syntax rules control action sequence of rats. *Behav. Brain Res.* 23, 59-68.
10. Berridge K.C. and Whishaw I.Q. (1992) Cortex, striatum, and cerebellum: control of serial order in a grooming sequence. *Exp. Brain. Res.* 90, 275-290.
11. Brunner R.J., Kornhuber H.H., Seemuller E., Suger G., and Wallesch C.W. (1982) Basal ganglia participation in language pathology. *Brain. Lang.* 16, 281-299.
12. Bullock T.H. (1993) Integrative systems research on the brain: resurgence and new opportunities. *Annu. Rev. Neurosci.* 16, 1-15.
13. Cromwell H.C. and Berridge K.C. (1996) Implementation of action sequences by a neostriatal site: A lesion mapping study of grooming syntax. *J. Neurosci.* 16, 3444-3458.
14. Cummings J.L. and Frankel M. (1985) Gilles de la Tourette syndrome and the neurological basis of obsessions and compulsions. *Biol. Psychiatry.* 20, 117-126.
15. Fentress J.C. (1972) Development and patterning of movement sequences in inbred mice. In *The Biology of Behavior* (ed. Kiger K.), pp. 83-132. Oregon State University, Corvallis.
16. Fentress J.C. and Stilwell F.P. (1973) Letter: Grammar of a movement sequence in inbred mice. *Nature* 244, 52-53.
17. Gabrieli J. (1996) Contribution of the basal ganglia to skill learning and working memory in humans. In *Models of Information Processing in the Basal Ganglia* (ed. Houk J.A., Davis J.L., and Beiser D.B.), pp. 277-294. MIT Press, Cambridge.
18. Graybiel A.M. (1995) Building action repertoires: memory and learning functions of the basal ganglia. *Current. Opinion. in Neurobiology.* 5, 733-741.
19. Graybiel A.M. and Kimura M. (1995) Adaptive neural networks in the basal ganglia. In *Models of Information Processing in the Basal Ganglia* (ed. Houk J.A., Davis J.L., and Beiser D.B.), pp. 103-116. The MIT Press, Cambridge.
20. Harrington D.L. and Haaland K.Y. (1991) Sequencing in Parkinson's disease - abnormalities in programming and controlling movement. *Brain* 114, 99-115.
21. Jackson S. and Houghton G. (1995) Sensiormotor selection and the basal ganglia. In *Models of Information Processing in the Basal Ganglia* (ed. Houk J.A., Davis J.L., and Beiser D.B.), pp. 337-369. The MIT Press, Cambridge.
22. Kermadi I. and Joseph J.P. (1995) Activity in the caudate nucleus of monkey during spatial sequencing. *J. Neurophysiol.* 74, 911-933.
23. Kermadi L, Jurquet Y., Arzi M., and Joseph J.P. (1993) Neural activity in the caudate nucleus of monkeys during spatial sequencing. *Exp. Brain. Res.* 94, 352-356.

24. Knowlton B.J., Mangels J.A., and Squire L.R. (1996) A neostriatal habit learning system in humans. *Science* 273, 1399-1402.
25. Lashley K.S. (1951) The problem of serial order in behavior. In *Cerebral Mechanisms in Behavior* (ed. Jeffress L.A.), pp. 112-146. Wiley, New York.
26. Magendie F. (1975) Lecons sur les fonctions et les maladus du systeme Nerveur. *Proc. roy. Soc. Med.* 68, 203-210.
27. Marsden C.D. (1984) Which motor disorder in Parkinson's disease indicates the true motor function of the basal ganglia? In *Functions of the Basal Ganglia*. Ciba Foundation Symposium 107, pp. 225-241. Pitman, London.
28. Mink J.W. (1996) The basal ganglia: focused selection and inhibition of competing motor programs. *Prog. Neurobiol.* 50, 381-425.
29. Mishkin M., Malamut B., and Bachevalier J. (1984) Memories and habits: Two neural systems. In *Neurobiology of Human Learning and Memory* (ed. Lynch G., McGaugh J.L., and Wienberger N.M.), pp. 65-77. Guilford Press.
30. Mushiake H. and Strick P.L. (1995) Pallidal neuron activity during sequential arm movements. *J. Neurophysiol.* 74, 2754-2758.
31. Rapoport J.L. (1989) *The Boy Who Couldn't Stop Washing*. Penquin Books, New York.
32. Rapoport J.L. and Wise S.P. (1988) Obsessive-compulsive disorder: evidence for basal ganglia dysfunction. *Psychopharmacol. Bull.* 24, 380-384.
33. Richmond G. and Sachs B.D. (1978) Grooming in Norway rats: the development and adult expression of a complex motor pattern. *Behaviour* 75, 82-96.
34. Shelton P.A. and Knopman D.S. (1991) Idcomotor apraxia in Huntington's disease. *Arch. Neurol.* 48, 35-41.
35. Ullman M.T., Corkin S., Coppola M., Hickok G., Growdon J.H., Koroshetz W.J., and Pinker S. (1997) A neural dissociation within language: Evidence that the mental dictionary is part of declarative memory, and that grammatical rules are processed by the procedural system. *J. Cog. Neurosc.* 9, 266-276.
36. Volkmann J., Hefter H., Lange H.W., and Freund H.J. (1992) Impairment of temporal organization of speech in basal ganglia diseases. *Brain. Lang.* 43, 386-399.

ACTIVITY OF THE PUTAMENAL NEURONS IN THE MONKEY DURING THE SEQUENTIAL STAGES OF THE BEHAVIOURAL TASK

B.F. Tolkunov, A.A. Orlov., S.V. Afanas'ev and E.V. Selezneva*

1. INTRODUCTION

Our knowledge of the role of the striatal neurons in the organization of various cognitive and motor functions is basing mainly on the results of the experiments of different kinds. Therefore, these data are difficult to compare, even when the investigators use one and the same model of experiment. Thus, using go/no go task some authors write, that activations are absent in trials in which the movement was withheld[2,12], others, indicate that similar number of neurons (33%, 29% and 25%) respond to both signals and to each of them separately[8]. In other experiment striatal neurons increased their activity after the trigger stimulus[13].

The aim of the present work was to study the putamenal neurons during the performance by the animal of the multistage behavioral task and thus to obtain the comparable data on the participation of the putamenal neurons in the formation of functionally different actions.

2. METHODS

The monkey *(Macaca nemestrina)* was taught to perform sequentially the number of functionally different actions. The starting signal was given by the experimenter. In response to the starting signal (opening of the screen), the animal pressed two levers. The holding of the levers during 300 ms initiated the conditioned signal (one of the two small bulbs), indicating by its position through which of the two feeders (either right or left) a food reward will be given. In order to obtain the reward, the monkey had to complete one of two sequential movements: either to keep the right-sided lever in the pressed position transferring the left

---

* B.F. Tolkunov, A.A. Orlov., S.V. Afanas'ev and E.V. Selezneva, Sechenov Institute of Evolutionary Physiology and Biochemistry. 44 Torez Street, St.-Petersburg, 194223, Russia.

hand to a special manipulator and using the fingers to switch the left feeder on, or to carry out the mirror-symmetrical sequence of movements toward the feeder on the right.

Nine fragments of unit activities were distinguished using event markers on actogram. 1. The response to the start-signal. 2. The movement of both hands to the levers. 3. Pause (holding of levers during 300ms). 4. The response to the conditioned signal. 5. Premotor period. 6. The movement of one hand, taken off the lever to the manipulator. 7. The movement of the fingers for the switching of the feeder. 8. The response on the click of the feeder (the first 200ms after the click). 9. Getting of the food reward. Fragments 1 - 2 and 4 - 5 were defined by halving the times between signal starts and the resulting motor response, as consideration that the beginnings and the ends of these periods were different.

Unit activity was recorded in the central part of left putamen, at atlas coordinates A - 20, L - 8, H - 9-4[16]. The block of six platinum-iridium microelectrodes was used. Every electrode was moved individually with the use of miniature driver located on the platform, installed on the head of the animal in advanced. The whole construction was so light and stable that allowed to record the neuronal activity without commonly used rigid fixation of the animal's head.

The purpose of the study required obtaining of the averaged data on the dynamics of the neuronal activity during the whole experimental task continuously. Peristimulus histograms traditionally used are created for each action of the animal separately. They cannot be incorporated into the common time axis and do not correspond to the continuity of behavior. Therefore, non-conventional methods of data processing were used: creation of the relative

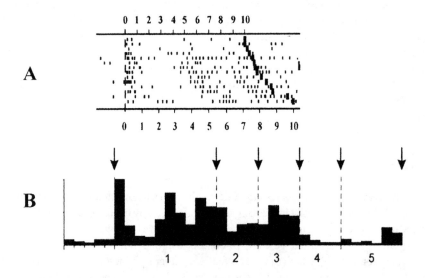

Figure 1. The method of relative histograms construction.
A - Raster presentation of neuron discharges at the first stage of behavioral task.. Each row corresponds to a separate trial. The rows are sorted by the duration of the first stage performance. 1-10 – the parts onto which the activity of the given stage is divided in each trial independently of it duration. Large dots – the beginning of the next stage. B – Relative histogram constructed using performance duration as the time scale. The arrows designate moments of the beginning of each stage of the behavioral task. The performance duration at the first stage is divided into 10 parts, at the stages 2-4 – into 4 parts, at the stage 5 – into 6 parts.

time histograms (RTH). The principle of their construction is shown in Fig. 1, it was described in more details earlier[20].

3. RESULTS

The experimental material includes RTH made on the records of activity of 63 neurons during nine sequential animal's actions, performed in two variants: in right-sided and left-sided task. Thus, altogether 63 x 9 x 2 = 1134 RTH of different fragments of neuronal activity have been considered.

The neuron was considered to be reactive to the given stage of behavior when in the RTH of its activity statistically significant (P<5%) changes in comparison with the activity before the performance of behavioral task had been discovered. Event related activity was found in all neurons. The distribution of such task-related neurons (TRN) by the stages of the behavioral task is shown in Fig.2.

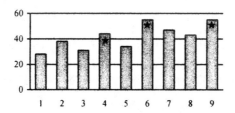

Figure 2. The number of task related neurons at each stage of the behavioral program. Asterisks indicate statistically significant difference with the previous stage (p<0.05).
1. The response to the start-signal. 2. The movement of both hands to the levers. 3. Pause (holding of levers during 300ms). 4. The response to the conditioned signal. 5. Premotor period. 6. The movement of one hand, taken off the lever to the manipulator. 7. The movement of the fingers for the switching of the feeder. 8. The response on the click of the feeder (the first 200ms after the click). 9. Getting of the food reward.

It can be seen that TRN are found at all stages of the task. The number of TRN (if compared with the number of TRN at the previous stage), increases significantly at three important stages. These are the stages of switching on the conditioned signal, beginning of the movement of the chosen hand and the obtaining of the food reward. The character of involvement of the putamenal neurons in the behavior becomes more complicated if TRN reacting by excitement are considered separately from the TRN, reacting by inhibition (Fig.3).

In such distribution the stage of the starting signal is distinct. At this stage 92% of TRN were with reactions in the form of excitement. The TRN with the activation also prevailed during the movement of both hands, in reaction to the conditioned signal, before the choosing of the working hand, in the reaction on the signal of the correct solving of the problem and during the obtaining of the food reward. However, during the movement of the chosen hand the number of TRN with the inhibition of activity was one and a half times greater than

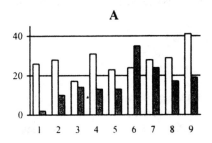
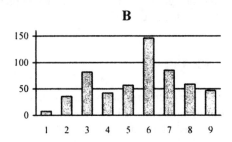

Figure 3. The number of neurons with excitatory and inhibitory reactions (A) and their percent ratio (B). A - abscissa - the number of neurons, light columns -excitatory cells, filled - inhibitory. B - abscissa - % of inhibitory cell from the number of excitatory. For further details see caption to Fig.2.

that of activated cells. Thus, the ratio of inhibited and activated neurons during the movement of one chosen hand was directly opposite to the one that was observed with stereotypically repeating in every realization of the movement of both hands.

The used behavioral task was alternative. The monkey in response to the right or left conditioned signal had to perform all further actions either by the right or left hand respectively. The other hand, therewith, retained its position on the lever. All neurons were recorded only in the left putamen. Nevertheless no difference in the distribution of the TRN by the program stages during the left and right-sided task have not been discovered. There was also no difference in the distribution of TRN with the excitement and inhibition (Fig.4).

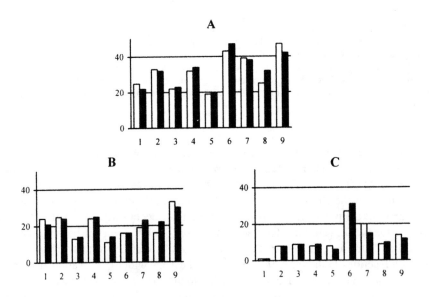

Figure 4. Neurons reactive to left and right-sided task.
A - The total number of neurons. B - the number of neurons with excitatory, C - with inhibitory reactions. Light columns -left-sided, filled - right-sided task. For further details see caption to Fig.2.

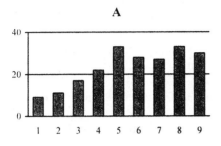
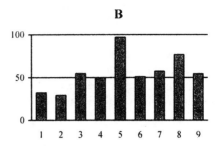

Figure 5. The number of neurons with differentiating left- and right-sided task reactions. A - Total number of cells. B - % of differentiating cells from the number of task related neurons at the given stage. For further details see caption to Fig.2.

The different attitude of TRN towards the behavioral task was discovered when out of the total number of studied cells at every stage of the program the differentiating TRN that reacted either during left or right-sided task were distinguished. Such neurons in the various quantities were at all stages of the behavioral task, both before and after the conditioned signal. It means, that the connection between the side of task performance and responses of TRN in 25-50% of cases may be accidental. The statistically significant increase in the number of differentiating TRN occurred in the premotor period, before the start of the movement by the hand, chosen on conditioned signal. The activity of the vast majority (97%) of putamenal cells varied significantly at this stage when different stimuli instructed different action (Fig.5). It is interesting to note, that the percent of differentiating TRN, after the obtaining of conditioned signal and during the movement by one hand, was the same as during the pause, when both hands of the monkey were placed symmetrically on the levers and the meaning of the conditioned signal was still unknown. The click of the solenoid, evidencing the correct performance of the task, caused the increased percent of differentiating TRN. This increase, however, was not statistically significant in comparison with the number of differentiating TRN at the stages of the task, preceding the release of the conditioned signal.

4. DISCUSSION

Comparison of activity of the same neurons with different actions of the animal had shown, that the putamenal neurons are involved in the control of all stages of behavioral task. Most of the TRN (70-87%) was found after the release of the conditioned signal, during the movement of the chosen hand and receiving of the reward.

The involvement of the putamenal neurons into sensory reactions is well known[10]. The number of neurons, responsive to sensory stimuli increased after conditioning[1,17]. There are data that the putamenal neurons can distinguish the quality of food reward, shown to the animal[3]. In our experiments 70% of the studied neurons reacted to the conditioned signal. The percent of TRN, differentiating the task, however, was the same as before the release of conditioned signal, i.e. was not beyond the limits of accidental variations of neurons' responses. Apparently, the activation of the putamenal neurons, observed at this stage is not directly connected with the differentiation of the conditioned signal. Probably, it reflects the general increase of afferent flow from the different fields of the cortex. The premotor period represents just another case. At this stage of behavior the number of TRN was smaller, than with release of conditioned signal, but significant changes of activity, connected with the meaning of the latter and the choice of the hand, were discovered in 97% of TRN. Thus, continuous analysis of the unit activity from the moment of release of conditioned signal up to the beginning of motor response to it shows, that the reaction to the release of the conditioned signal is not connected with its meaning. This connection appears later, directly before the beginning of the movement.

The important role of premotor activity of putamenal neurons in the formation of behavior was emphasized by many researchers. It is supposed, that this activity is connected with the initiation of movement, preparing of motor programs[4,9,11,15]. Alongside with this, in the character of activity before and after onset of movement some difference is found, which suggests that the neuronal mechanisms, mediating the premotor activity, may differ from those that occur during movement execution[4]. In our experiments these differences were displayed in the significant increase of the number of TRN with inhibition of activity after the onset of the movement. The other variance was that in the premotor period 97% of TRN responded either during the left or right-sided behavior. Involvement of different neurons during the performance of the left- or right-sided task indicates that in this period the influence of the putamen on the neuronal mechanisms of behavior formation is particularly essential.

The correct solution of the problem in the given experimental task is in the right choice of the hand for the further actions. The monkey performs the necessary movements by right or left hand in response to different conditioned signals, i.e. different motor centers are working. The percentage of neurons differentiating the task during the movement, however, falls to the accidental level that is observed during the pause before the release of conditioned signal. Probably, the changes in the activity of TRN observed during one hand movement execution are not directly connected with the movement. This is confirmed by the fact, that the correlation of TRN in the left putamen during the left- and right-sided movements is similar, as before the release of conditioned signal, when the actions of the animal were not lateralized.

According to the literature there are changes in the activity of neurons in the striatum, connected with receiving of the reward, its quality and even with the expectation of reward[6,7,14,21]. In the given experimental task the activity, connected with the reward is divided into two fragments. One is the reaction to the click of the feeder, i.e. to the signal that the task is solved correctly and there will be the reward. At this stage of the task the number of the TRN decreased and the percentage of differentiating TRN increased. It means that the changes of the same character as in the premotor period were observed. These changes do not achieve the level of P<0,05 but at the next stage, during receiving of the reward they disappear. Probably, this is the evidence, suggesting, that at the moment of the release of the signal of successful performing of the task, the neurons of the putamen are taking part in the fixation of correctness of the taken solution.

5. CONCLUSION

The data obtained indicate that the participation of putamenal neurons in the preparing of a certain motor program is unlikely. Apparently, the selective responding of neurons in the premotor period is connected with neuronal mechanisms of mainly general character, related to adoption of the solution of the appeared problematic situation and to the choice of the field of possible actions. The convergence of signals from the functionally specialized parts of the cortex in the striatum allows to suppose that corticofugal influences create in the striatum the generalized neural model of cortex activity, necessary for the formation of the behavior[18,19] or execute the compression of the cortical signals[5]. This process has a constant character and reveals at all stages of the behavioral task. The study of activity of one and the same neurons during the performance by the animal of functionally different actions gave the complicated picture of involvement of the putamenal neurons into the organization of behavior. Specific character of participation of the putamenal neurons in the organization of the different actions displays in the general number of TRN, in the number of TRN, differentiating different tasks and in the ratio of excitement and inhibition.

6. REFERENCES

1. Aosaki T., Tsubokawa H., Ishida A., K. Watanabe K., Graybiel A.M. and Kimura M., Responses of tonically active neurons in the primate's striatum undergo systematic changes during behavioral sensorimotor conditioning, *J. Neurosci,* 14(6):3969 (1994).
2. Apicella P., Scarnati E., Ljungberg T. and Schultz W.J., Neuronal activity in monkey striatum related to the expectation of predictable environmental events, *Neurophysiol.* 68(3):945 (1992).
3. Fukuda M, Ono T., Nishijo H. and Tabuchi E., Neuronal responses in monkey anterior putamen during operant bar-press behavior, *Brain Res. Bull.* 32(3): 227 (1993).
4. Gardiner T.W. and Nelson R.J., Striatal neuronal activity during the initiation and execution of hand movements made in response to visual and vibratory cues, *Exp. Brain Res.* 92(1):15 (1992).
5. Graybiel A.M., The basal ganglia and chunking of action repertoires, *Neurobiol. Learn. Mem.* 70(1-2):119 (1998).
6. Hassani O.K. and Schultz W., Reward discrimination in primate striatum, *Soc. Neurosci. Abstr.* 24: 1652 (1998).
7. Hollerman JR., Tremblay L. and Schultz W., Influence of reward expectation on behavior-related neuronal activity in primate striatum, *J. Neurophysiol.* 80(24947 (1998).
8. Inase M., Li B.M. and Tanji J., Dopaminergic modulation of neuronal activity in the monkey putamen through D 1 and D2 receptors during a delayed go/nogo task, *Exp. Brain Res.* 117(2):207 (1997)
9. Jaeger D.. Gilman S., and Aldridge JW., Primate basal ganglia activity in a precued reaching task: preparation for movement, *Exp. Brain Res.* 95(1):51 (1993).
10.1 O.Lidsky T.I. and Schneider J.S., Effects of experience on striatal sensory responses. *Neurosci. Lett.* 174(2): 141 (1994).
11.Montgomery E.B. and Buchholz S.R., The striatum and motor cortex in motor initiation and execution, *Brain Res.* 549(2).22.2 (1991),
12.Richardson R.T. and DeLong M.R., Context-dependent responses of primate nucleus basalis neurons in a go/no-go task, *J. Neurosci.* 10(8):2528 (1990).
13.Romo R., Scarnati E. and Schultz W., Role of primate basal ganglia and frontal cortex in the internal generation of movements. II. Movement-related activity in the anterior striatum, *Exp. Brain Res.* 91(3):385 (1992).
14.Schultz W., Apicella P., Scarnati E. and Ljungberg T. Neuronal activity in monkey ventral striatum related to the expectation of reward, *J. Neurosci.* 12(12):4595 (1992).
15.Schultz W. and Romo R., Neuronal activity in the monkey striatum during the initiation of movements, *Exp. Brain Res.* 71(2):431 (1988).

16. Sneider S.A. and Lie J.C. *A Stereotaxic Atlas of Monkey Brain (Macaca mulatta)*, University of Chicago Press, Chicago (1961).
17. Thomas T.M., Hanson J.E. and Crutcher M.D., Tonically active neurons in primate striatum respond to stimuli before, during and after learning a joystick task, *Soc. Neurosci. Abstr.* 24:1651 (1998).
18. Tolkunov B.F. Organization of the specialized image of significant signals in the neuronal network of the nonspecific brain structures, in: *Neuronal Plasticity and Memory Formation*. International Brain Research Organization monograph series. Raven press. N.Y. v.9. (1982).
19. Tolkunov B.F., Orlov A.A. and Afanas'ev S.V., Model of the function of the neostriatum and its neuronal activity during behavior in monkeys, *Neurosci. Behav. Physiol.* 27(1):68 (1997)
20. Tolkunov B.F., Question of time in studies of the neuronal correlates of behavior, *Neurosci. Behav. Physiol.* 28(4):447 (1998)
21. Tremblay L., Hollerman J.R. and Schultz W.J., Modifications of reward expectation related neuronal activity during learning in primate striatum, *J. Neurophysiol.* 80(2):964 (1998).

LEARNING - SELECTIVE CHANGES IN ACTIVITY OF THE BASAL GANGLIA

Masahiko Inase, Bao-Ming Li, Ichiro Takashima, and Toshio Iijima*

1. INTRODUCTION

The basal ganglia have been thought to be primarily involved in movement control[3,5]. However, recent evidence suggests that the basal ganglia are also essential for procedure learning[7,10], and even for association learning. For example, patients with Parkinson's disease showed impairments in the probabilistic classification task, in which they required the gradual, incremental learning of associations[8].

Conditional motor learning, or sensorimotor association learning, is the ability to develop multifarious and rapidly changeable stimulus-response relationships. In an experimental condition it can be defined as the acquisition of a motor response to an arbitrarily linked visual stimulus[12]. Electrophysiological studies have demonstrated that the premotor cortex[9] and supplementary eye field[2] are involved in the conditional motor learning. Since the basal ganglia receive inputs from the inferior temporal cortex, a cortical area for higher visual processing, and send outputs to the prefrontal and premotor cortical areas through the thalamus, they might also play an important role in the learning.

In order to examine whether the basal ganglia are involved in the conditional motor learning, we recorded neuronal activity in the two output nuclei of the basal ganglia, the internal segment of the globus pallidus (GPi) and the substantia nigra pars reticulata (SNr), while monkeys were acquiring a new stimulus-response association.

2. METHODS

Two male Japanese monkeys (*Macaca fuscata*), weighing 8.2 kg and 6.8 kg, were used in the present study. All experimental procedures complied with the National Institutes of

---

* Masahiko Inase, Bao-Ming Li, Ichiro Takashima, and Toshio Iijima, Electrotechnical Laboratory, Tsukuba 305-8568, Japan. Masahiko Inase, Kinki University School of Medicine, Osaka-Sayama 589-8511, Japan. Bao-Ming Li, Shanghai Institute of Physiology, Shanghai 200031, China.

Health Guide for the Care and Use of Laboratory Animals, and were approved by the institutional animal experiment committee.

The monkeys were trained to perform a stimulus-response association learning task using their right arm. Each monkey sat in a primate chair with his right arm attached to a manipulator, and faced a computer monitor on which visual signals were presented. When the monkey moved the manipulator to a center hold position, a blue spot appeared on the center of the monitor. After an initial hold period (0.8s), a simple geometric pattern (cue), which instructed a proper response to the monkey, was displayed for 0.5 s. After the blue spot presented again for 1.0-2.0s (delay period), it was replaced with a red spot, which triggered a response. The monkey was required to push, pull, or turn the manipulator from the center hold position within 0.8s after the trigger according to the cue presented shortly before. After the correct response, a green spot appeared on the monitor for 0.5 s, and then a drop of juice was given to the monkey.

The monkeys performed this task under two conditions in which a group of given cues were different. Under the control condition, we repeatedly used three cues, circle, triangle, and square. Circle instructed the monkey to push the manipulator, triangle did it to pull the manipulator, and square cued the turn movement. The three cues were randomly presented in a block of trials. The animals were familiar to these cues, and always performed the task correctly. In the learning condition, we also presented three different cues in a block, but replaced one of the familiar cues with a novel visual pattern. Therefore, one novel cue and two familiar ones were used in a learning block. The new cue instructed the same response as the replaced one did. At first the monkey did not know what the new cue instructed, and had to learn it by trial and error. The three cues were also presented randomly in a block of trials although the previous cue was repeated following an error trial. After a few to tens of trials he could correctly respond to the new cue.

After the monkeys achieved a constant correct performance rate, they received a surgery for head fixation and access to the brain. The animals were anesthetized with ketamine hydrochloride (10 mg/kg, i.m.) followed by sodium pentobarbital (30 mg/kg i.p.). Under aseptic conditions, we widely exposed the skull, opened the skull portions, and put on recording chambers so that we approached the SNr vertically and the GPi at an angle of 45 degree laterally from the vertical. We put several small screws on the skull for anchoring points and completely covered the exposed skull and the screws with dental acrylic. Then two stainless steel tubes for head fixation were mounted in parallel at the frontal and occipital regions with dental acrylic.

Following a recovery period from the surgery, we started neuronal recording experiments. Using a glass-coated Elgiroy electrode, single unit activity was extracellularly recorded from the GPi and SNr contralateral to the performing arm. The activity was shown on line with peri-event raster display on a computer monitor and analyzed off line. Horizontal eye movements were monitored with EOG during the neuronal recording, and EMGs were recorded in several muscles separately from the recording experiment.

3. RESULTS

3.1. Neuronal Activity in the GPi

We recorded neuronal activity from 157 GPi neurons during both the control and learning blocks of the stimulus-response association learning task. Of 157 GPi neurons, 49 neurons changed their activity during the delay period in the control and/or learning conditions. Furthermore, in 27 of these neurons the activity change during the delay period was enhanced in the learning condition compared to that in the control condition. Figure 1 shows an example of GPi neuron exhibiting an enhanced change in activity during the learning

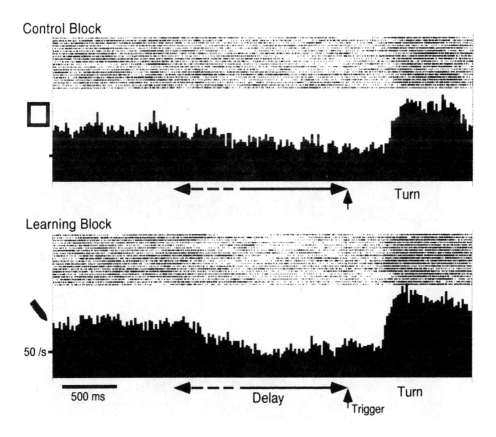

Figure 1. An example of GPi neuron presenting enhanced changes in activity during the delay period in the learning condition compared to that in the control condition. In the control block *(top)* the activity slightly decreased during the delay period (1.0-1.5s; shown as an arrow bar) in Turn. This decrease in activity was more obvious in the learning condition *(down)*. Raster displays and perievent time histograms are aligned at the time when the trigger was presented. Only correct trials were demonstrated. In the raster displays dots represent individual discharges of a single neuron in turn trails under each condition, and heavy dots denote the time of the trigger. Perievent time histograms demonstrate discharges over trials shown in raster displays. Visual patterns for the cue signal in each condition are shown at the left.

condition. In this neuron the activity slightly decreased during the delay period of turn trials in the control condition. This decrease in activity was more obvious in the learning condition than that in the control condition.

The changes in activity during the delay period were selective to one of the three responses. For example, in the neuron shown in Figure 1 the activity decreased during the delay period only in turn trials, but not in push or pull trials in the control condition. In the learning condition, when the cue signal for Turn was switched to a new one, the activity change was enhanced. On the other hand, when the cue signal for Push or Pull was switched to a new one, the neuron did not change its activity during the delay period even though the monkey tried to learn new association between the visual pattern and the response. Moreover, the activity changes during the delay period were observed only in correct trials, but not in error ones. For example, in the learning condition the neuron shown in Figure 1

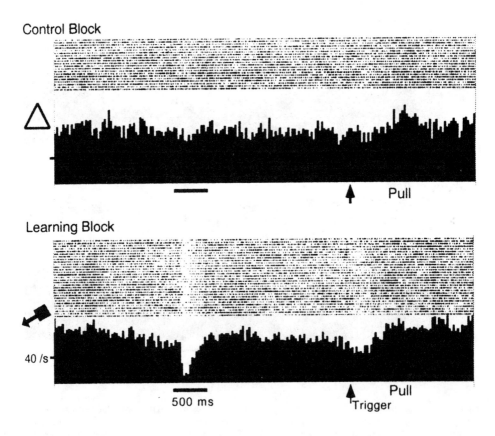

Figure 2. An example of SNr neuron responding to a new cue in the learning condition, but not to familiar cues in the control condition. In the control block *(top)* the neuron did not change its activity during the cue period (500ms; shown as a bar below the histogram). The activity clearly decreased during the cue period in the learning block *(down)*. Raster displays and perievent time histograms are aligned at the time when the trigger was presented. The delay period was constantly 2.0s in this example. Only correct trials were demonstrated. Visual patterns used for the cue signal in each condition are presented at the left.

changed its activity during the delay period when the new cue for Turn was presented and the monkey turned the manipulator. But the neuron did not change its activity when the same cue was presented and the animal erroneously push or pulled it.

Histological examinations revealed that the neurons exhibiting the changes in activity during the delay period were mainly distributed in the rostral and dorsal part of the GPi.

3.2. Neuronal Activity in the SNr

We recorded neuronal activity 35 SNr neurons during both the control and learning blocks of the stimulus-response association learning task. Of 35 SNr neurons, 12 neurons, which did not respond to familiar cues during the control condition, did respond to a new cue during the learning condition. Figure 2 shows an example of SNr neuron responding to the cue signal only in the learning condition. In this neuron the activity did not change during the cue period in pull trials in the control block. On the other hand, in the learning block the activity did decrease when the new cue was presented.

The change in activity during the cue period in the learning condition was not selective to a visual pattern, nor selective to the response following the cue. For example, the neuron shown in Figure 2, which responded to the new cue in pull trials in the learning condition, responded to another new visual pattern which also instructed the pull movement. Furthermore, in this neuron the activity decreased during the cue period when the cue for the push or turn movement was switched to a new one.

Histological examinations revealed that the recorded neurons were mainly located in the rostral and lateral part of the SNr.

4. DISCUSSION

The present study demonstrates that changes in activity were enhanced during the conditional visuomotor learning in the output nuclei of the basal ganglia, the GPi and SNr. A group of GPi neurons, which changed their activity during the delay period between presentation of the learned stimulus and execution of the response, exhibited more changes in the delay activity when the monkey tried to learn new association between a novel stimulus and the response. On the other hand, a group of SNr neurons exhibited clear responses to a new cue in the learning condition while they showed no or subtle responses to learned cues in the control condition. These results suggest that the basal ganglia play important roles in the learning of new stimulus-response association.

It has been demonstrated that the dorsal part of the premotor cortex (PMd) is necessary for the conditional motor learning[4,9,11]. Lesions of this area severely impaired learning of conditional motor associations[11] and relearning of the associations[4]. A neurophysiological study also support that the PMd has the specific role in the learning. Mitz et al.[9] reported that neurons in the PMd systematically change their set-related activity during the learning of new stimulus-response associations. In addition to the PMd, lesions of the thalamic nuclei, especially the rostral part of the ventrolateral nucleus (VLo) and ventroanterior nucleus (VA) affected the conditional motor behavior[1]. These thalamic nuclei relay outputs from the basal ganglia to the PMd[6]. Therefore, the conditional motor learning appears to depend on the integrity of two brain structures, the PMd and the basal ganglia[12].

The present results support the idea that the basal ganglia are a part of the neuronal system necessary for this learning. The enhanced changes in GPi activity demonstrated in this study appear a counterpart of the learning-dependent changes in PMd activity shown by Mitz et al.[9] Both changes were observed during the delay period between cue presentation and response execution, and selective to the response following the cue. However, the changes in GPi activity were larger early in the learning and getting smaller as the learning progressed, while the PMd activity progressively increased as the animal learned a visuomotor association. Therefore, one can speculate that the GPi may be involved in the early stage of the learning, such as encoding and short-term storage of the association, and that the PMd may be necessary for consolidation and retrieval process. The enhanced changes in SNr activity were observed during the cue presentation, and were also larger early in the learning and getting smaller as the learning progressed. These changes were not selective to a following response and this non-selectivity of the cue-related SNr activity contrasts with the selectivity of the GPi activity during the delay period. The SNr, which sends output to the prefrontal cortex through the thalamus, might be related to focusing attention to the novel cue, but not to leaning per se. Obviously, further studies are needed to clarify cellular mechanisms for establishing a new stimulus-response association.

5. REFERENCES

1. A.G.M. Canavan, P.D. Nixon, and R.E. Passingham, Motor learning in monkeys *(Macaca fascicularis)* with lesions in motor thalamus, *Exp. Brain Res.*, 77:113-126 (1989).
2. L.L. Chen and S.P. Wise, Neuronal activity in the supplementary eye field during acquisition of conditional oculomotor associations, *J. Neurophysiol.*, 73:1101-1121 (1995).
3. M.R. DeLong, G.E. Alexander, M.D. Crutcher, S.J. Mitchell, and R.T. Richardson, Role of basal ganglia in limb movements, *Human Neurobiol.*, 2:235-244 (1984).
4. U. Halsband and R.E. Passingham, Premotor cortex and the conditions for movement in monkeys *(Macaca fascicularis)*, *Behav. Brain Res.*, 18:269-277 (1985).
5. M. Inase, J.A. Buford, and M.E. Anderson, Changes in the control of arm position, movement, and thalamic discharge during inactivation in the globus pallidus of the monkey, *J. Neurophysiol.*, 75:1087-1104 (1996).
6. M. Inase and J. Tanji, Projections from the globus pallidus to the thalamic areas projecting to the dorsal area 6 of the macaque monkey: a multiple tracing study, *Neurosci. Lett.*, 180:135-137 (1994).
7. M. Kimura, Role of basal ganglia in behavioral learning, *Neurosci. Res.*, 22:353-358 (1995).
8. B.J. Knowlton, J.A. Mangels, and L.R. Squire, A neostriatal learning system in humans, *Science*, 273:1399-1402 (1996).
9. A.R. Mitz, M. Godschalk, and S.P. Wise, Learning-dependent neuronal activity in the premotor cortex: activity during the acquisition of conditional motor associations, *J. Neurosci.*, 11:1855-1872 (1991).
10. S. Miyachi, O. Hikosaka, K. Miyashita, Z. Karadi, and M.K. Rand, Differential roles of monkey striatum in learning of sequential hand movement, *Exp. Brain Res.*, 115:1-5 (1997).
11. M. Petrides, Motor conditional associative-learning after selective prefrontal lesions in the monkey, *Behav. Brain Res.*, 5:407-413 (1982).
12. S. P. Wise, Evolution of neuronal activity during conditional motor learning, in: *The Acquisition of Motor Behavior in Vertebrates*, J. R. Bloedel, T. J. Ebner, and S. P. Wise, eds, The MIT Press, Cambridge (1996).

COGNITIVE DECISION PROCESSES AND FUNCTIONAL CHARACTERISTICS OF THE BASAL GANGLIA REWARD SYSTEM

Moti Shatner, Gali Havazelet-Heimer, Aeyal Raz and Hagai Bergman[*]

1. INTRODUCTION

Cognitive behavior of individuals is generally described in terms of reactions to rewards or to predictions concerning future rewards. Physiological systems that are involved in reward mechanism might thus be correlated with various cognitive effects. Such basal ganglia systems include the midbrain dopaminergic system (Schultz 1998), and the striatal tonically active neurons - TANs (Aosaki et al. 1995, Raz et al. 1996). In this chapter we propose that both midbrain dopaminergic and striatal cholinergic interneurons (TANs) continuously emit a complex tri-phasic neural message (neural signature of reward) which is modulated by the fitness of the environment to the animal predictions.

A major field of cognitive psychology research is Decision Theory, which describes decision processes and anomalies. A key finding in Decision Theory (Kahneman and Tversky, 1979) is that the behavior of an individual is shifting from risk-aversion (when possible gains are predicted) to risk seeking (when possible losses are predicted). The second section of the current chapter presents an analysis of this effect from the basal ganglia point of view, and offers insights for the origin of the behavioral asymmetry.

It was found (e.g., Schultz, 1998), that dopamine neurons tend not to respond to stimuli which predict future aversive rewards. The third section of this chapter proposes an evolutionary explanation to the asymmetrical nature of the basal ganglia reward system. We propose that responses to aversive stimuli are not handled by the basal ganglia system, since this system is devoted for the more complex control of sequential behavior. Other more primitive systems, based on pattern detection algorithms, are called into action following aversive stimuli.

---

[*] M. Shatner, G. Havazelet-Heimer, A. Raz, and Hagai Bergman, Department of Physiology and the Center for Neural Computations, The Hebrew University - Hadassah Medical School, Jerusalem, Israel

2. DECISION THEORY AND PHYSIOLOGICAL REWARD SYSTEMS

2.1 Review of Major Cognitive Psychology Findings

In classical decision theory (Von Neuman and Morgenstern, 1953) it was an established axiom that decision-makers are risk averse, i.e. when facing two alternatives having the same expectancy, they will choose the one with the lower variance. Nevertheless, the pioneering work of Kahneman and Tversky in the field of cognitive psychology (1979, 1982) have severely undermined such classical paradigms.

In particular, Kahneman and Tversky (1982) have conducted several experiments to test decision making under uncertainty (see fig. 1). They showed that when potential profits are concerned, decision-makers are indeed risk averse, but when potential losses are concerned, subjects become risk seeking. This dichotomy in the attitude towards risk is in contradiction with classical paradigms, which assume that decision makers should always be risk averse, both when a potential profit and when a possible loss are predicted.

In Experiment #1 subjects are asked to choose between two alternatives which contained a (hypothetical) potential to gain money. Most subjects prefer alternative (A) to (B), thus showing risk aversion (preferring low variance to a slightly higher expectancy), in accordance with the "rationality paradigm". Nevertheless, when asked to choose between alternatives which refer to a possible loss (Experiment #2), most subjects prefer alternative (B*), which contains higher variance and higher risk. The subjects become risk seeking when losses were concerned.

Tversky and Kahneman concluded that the attitude towards risk is determined according to a "reference point", which is the basis for evaluating possible outcomes of the decision: When future rewards are perceived as profits compared to the current reference point, a risk aversion behavior is observed. When future rewards are perceived as losses, a risk-seeking behavior emerges. These observations were termed "Prospect Theory."

2.2 Main Functional Characteristics of the Basal Ganglia Reward System

The main functional characteristics of the basal ganglia reward systems can be summarised as follows (based on Schultz 1998):
i) The dopaminergic system codes the error between the prediction about the reward and the actual reward. That is - if the reward is stronger or sooner than expected, the dopamine neurons' activity rises. If the reward is weaker than expected, the dopamine neurons'

>   Experiment #1
>   The subject has to choose between:
>   (A) A sure gain of $80.
>   (B) 85% chance of winning $100 and 15% of winning nothing.
>
>   Experiment #2
>   The subject has to choose between:
>   (A*) A sure loss of $80
>   (B*) 85% chance of losing $100 and 15% chance of losing nothing.

Figure 1. Typical Risk Aversion / Risk Seeking experiments (After Kahneman & Tversky, 1982)

activity declines. Finally, the continuos background dopaminergic activity signals that the environment is as good as predicted.

ii) The dopaminergic system hardly responds to predictions about aversive rewards.

A simple heuristics that connects physiological signals with decision making would therefore be (Schultz et al., 1997): An organism should take actions correlated with increased reward-signal activity and avoid actions correlated with decreases in the rate of the neuronal reward-signal.

The responses of striatal cholinergic tonically active neurons (TANs) are very similar to the dopaminergic responses, yet with an opposite polarity, e.g., the main response of TANs to an unpredicted rewarding event is a depression (pause) of their background tonic activity (Aosaki et al. 1995, Raz et al. 1996). An exciting observation is that in both cases, the main response (the dopaminergic burst or the pause of the TANs) is flanked with activity of opposite polarity. Thus, excitatory bursts flank the pause of the TANs (figure 2) and short pauses can be seen preceding and following the burst of dopaminergic neurons (see for example, fig. 3, in Schultz et al., 1993 and figs. 1,2 in Mirenowicz and Schultz, 1994).

A recent review (Redgrave et al., 1999) indicated that the short-latency dopamine response might be too short (50-110 ms) to signal reward error, and suggested that the dopaminergic burst participates in the process of switching attentional and behavioral resources to unexpected, behaviorally important stimuli. Previous studies of TAN activity also reported very short (68±21 ms) latency responses in line with the early reports of the dopaminergic activity. Still, more recent studies indicated much longer latency for both the dopaminergic neurons (151±3 ms, mean±SEM, Mirenowicz and Schultz, 1994) and TANs (119±33 ms, 150±78 ms depending on the behavioral mode, Apicella 1997). Similarly, our studies of TAN responses revealed a stereotypical response with mean latency to reward around 200 ms from the reward-cue onset (figure 2).

Another support for the role of dopaminergic system in switching behavior (Redgrave et al., 1999) is the contradiction between the studies of the electrical activity of dopaminergic neurons and the neuro-chemical studies of striatal dopamine level in response to aversive events. While the electro-physiological studies indicate that the dopaminergic neurons are insensitive to aversive events, many neurochemical studies (see review in Redgrave et al., 1999) have indicated that aversive stimuli can increase the release of dopamine in the stria-

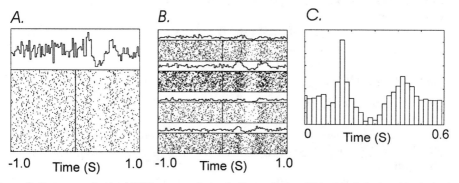

Figure 2. Responses of striatal TANs to reward-cue. A. Raster display (126 trials, below) and Peri-stimulus histogram (above) of a single TAN (HN07, unit 8). B. Raster displays (135 trials) and PSTHs of four simultaneously recorded TANs (HN06, units 2,8,15&13). C. Average PSTH of the reward responses of 23 TANs recorded on different recording sessions. Bin size =20 ms in all PSTHs.

tum, and that dopamine depletion impair the behavior that is elicited by these stressful events. Our recent preliminary studies of TANs' activity might resolve this contradiction. Using a multi-neuronal rate template we found that the characteristic response of TANs to reward (burst, pause, burst) is a collaborative phenomenon, that can be spotted in all behavioral epochs, and not only in time locking with the reward cue. The rate of this "neuronal reward signal" increases following the reward cue in correct trials, and drops below the background level following behavioral errors. We therefore suggest that the actual signal emitted by the striatal cholinergic or mid-brain dopaminergic neurons is not simply the number of spikes or the amount of released neurotransmitters. Rather, both basal ganglia reward systems elicit a continuos background level of complex tri-phasic neuronal signal. The rate of this neuronal reward signal is (up or down) modulated by the fitness between the animal prediction for future reward and the actual environmental events.

2.3 Physiological Basis of Decision Theory

Tversky, Kahneman and their successors concentrated primarily on the cognitive aspects of decision making. A basal-ganglia model can explain their findings in a broader perspective. We assume that although the basal ganglia reward systems were mainly studied with basic stimuli such as food or drink, it also respond to money stimuli, even if they are of a less "physical" nature. We shall therefore refer to money gains and losses as appetitive and aversive stimuli respectively, and analyze the cognitive experiments in the physiological framework.

Experiment 1: Profits
A = { (\$80, p=1.) }
B = { (\$100 , p=0.85) , (\$0, p=0.15) }

Reference point (P=\$80) is positive, therefore the dopaminergic system (D) is active.

Choose (A) → R=P → D unchanged (No error in prediction)
Choose (B) →
        If R=\$100 → R>P → D increased
        If R=\$0 → R<P → D decreased

Experiment 2: Losses
A* = { (-\$80, p=1) }
B* = { (-\$100 , p=0.85) , (\$0, p=0.15) }

Reference point is negative (P*= -\$80), therefore the dopaminergic system is not active.

Choose (A*) → R=P* → D still not active
Choose (B*) →
        If R= - \$100 → R<P* → D unchanged (still not active)
        If R=\$0   → R>P* → D increased (Unexpected reward)

Figure 3. Physiological analysis of Prospect Theory

2.3.1 Analysis of Experiment (1) - Risk Aversion in Profits

To recall, the subjects have to choose between a sure gain of $80 (alternative A) and a gamble (alternative B) between a gain of $100 (probability 0.85) and a gain of $0 (probability 0.15). The basal ganglia reward systems code errors in predictions (characteristic (i) above). Therefore, a reference point P (i.e. a prediction) is needed so that the reward systems could properly respond to the actual reward R. Since the two alternatives are both circa $80, which is the assured gain, this value makes the best candidate for becoming a reference point in this case.

When a subject chooses alternative (A), the actual reward will be $80. In this case the actual reward and the prediction (reference point) are identical, so there is no change in the neuronal activity level (see fig. 3). If a subject chooses alternative (B), the reward could either be $100 (which implies an increase in the neural activity, since Reward > Prediction) or $0 (which implies a decrease in the neural activity level). Kahneman and Tversky (1982) have shown that most subjects prefer alternative (A), where there is no change in the neural activity to alternative (B), where there could be an increase or a decrease in the activity level.

If one assumes a ternary model of the dopaminergic neurons, the above physiological analysis of Tversky and Kahneman's results could imply the following amendment to Schultz et al 's (1997) heuristics cited above:
1. Take actions correlated with increased reward signal activity and avoid actions that are correlated with decreases in basal ganglionic reward activity.
2. It is more important to avoid actions that lead to decreases in reward activity than to take actions that tend to increase basal ganglionic reward activity.

If we assume a continuous model of the dopaminergic and striatal cholinergic neurones, the cognitive results can be analyzed as stemming from a weighted average index for the possible increases and decreases in the reward system's activity for alternative (B). Under this assumption, the index calculated for alternative (B) must be lower than the index calculated for (A) in order for the subjects to prefer alternative (A).

2.3.2 Analysis of Experiment (2) - Risk Seeking in Losses

In this experiment, the two alternatives involve possible losses: either a sure loss of $80 (alternative (A*)), or a gamble between a loss of $100 (with probability 0.85) and $0 (with probability 0.15). The reference point in this case could be P= -$80, since this is the "sure loss", based on which the gamble is measured. The key element in the analysis of experiment #2 is property (ii) above, which asserts that the dopaminergic system does not respond to aversive stimuli. A loss of money is considered as an aversive stimulus. When a subject chooses alternative (A*), he receives a reward R= -$80. Analyzing the case similarly to Experiment 1 above (see figure 3), we would have concluded that since the reward R is equal to the predictions or since there is no coding of aversive cases (losses), there is no change in the dopaminergic level. An analysis of the choice of alternative (B*) is a little different. If the reward R is -$100, there is no response of the basal ganglia reward system, since the reward system does not code the aversive stimuli. However, if R=$0, the reward is referred to as an "appetitive surprise", since the reference point is P= -$80. In this case the reward systems will increase their activity.

The analysis for both a ternary and a continuous model will be as follows: When choosing (A*), no increase in the reinforcement activity will have a chance to occur. Choosing (B*) may lead to such an increase, but it will not lead to any decrease in the activity level. Therefore, a subject who is driven by basal ganglionic reinforcement impulses will be inclined to choose (B*), similar to Tversky and Kahneman's findings.

The shift from risk aversion to risk seeking can thus be explained using the asymmetrical nature of the basal ganglia systems (property (ii)). Moreover, the characteristics of basal ganglia neurons can also give insights for the emergence of a reference point in cognitive decision processes, as is described above.

3. THE ASYMETRICAL NATURE OF THE BASAL GANGLIA REWARD SYSTEM

It has been shown that the dopaminergic neurons respond preferably to appetitive stimuli (Schultz et al., 1997, Schultz, 1998). When considering the physiological reward systems in the context of cognitive decision making, the underrating of the information carried by aversive stimuli may have negative implications on the organism's survival, since aversive stimuli can help the organism avoid future hazards.

The underreaction of the reward system to aversive stimuli can be given a physiological explanation. A dopaminergic neuron's base firing level is 4-10 spikes per second (Schultz 1998). When an unpredicted appetitive stimulus occurs, the neuron's activity can easily go up to a higher level. Nevertheless, when an unpredicted aversive or disappointing event occurs, the firing rate can only go a few spikes per second down (since the base level is very close to 0). This means that even if dopaminergic neurons were responding equally to aversive and appetitive stimuli, they could not present a significant decrease in their activity. The average frequency of the tri-phasic TAN's signature of reward is even lower than the frequency of the background firing of the dopaminergic and cholinergic neurons. Therefore the possible modulations of this signal are also truncated by the zero level, and are naturally asymmetric.

The asymmetrical nature of the reward systems can also be given a functional explanation. While lower organisms primarily exhibit pattern recognition behavior (e.g., when receiving an air puff from the left, the organism will move to the right), higher organisms in the phylogenetic tree also apply sequential decision processes. A pattern recognition process will be most suitable for a "fight or flight" situation, where a quick decision is required. Sequential decision processes may produce better decisions, but they are harder to realise using neural networks, considering the parallel nature of such processes and the amount of calculations needed (see Beiser, & Houk, 1998 for the role of basal ganglia networks in sequential behavior).

Sequential decision processes may be thus preferable to pattern recognition when speed of decision making is of less importance. The plausibility of an immediate hazard could therefore serve as a rough indicator for the shifts between sequential and pattern-recognition decision making. When considering a possible source for this dichotomy, one may notice the growth of the basal ganglia over the phylogenetic tree (Marin et al. 1998). Basal ganglia play an important role in sequential processes (Aldridge and Berridge, 1998, Graybiel 1998) and are controlled by the midbrain dopaminergic and striatal cholinergic system.

It may thus be that the basal ganglia reward system is used to maintain sequential decision processes, while pattern-recognition faster, simpler neural networks maintain behavior. We can therefore hypothesise that the basal ganglia reward system does not have to respond to aversive stimuli, since they are already handled by other sub-systems that exhibit a pattern-recognition behavior. A possible sub-system, which performs this task, may be located in the cholinergic Nucleus Basalis (NB). The NB receives inputs from the limbic and the paralimbic structures, and sends projections to the entire cortex. Kilgard and Merzenich (1998) have shown that the NB can signify the behavioral importance of different stimuli. They propose that the NB may enable the cortex to ignore irrelevant stimuli, while focusing on other stimuli. The Nucleus Basalis might therefore constitute a good candidate for a complementary system that resolves the asymmetry nature of the basal ganglia reward system.

4. CONCLUSIONS

Cognitive behavior of individuals is generally described in terms of reactions to rewards or to predictions concerning future rewards. The neural activity of the dopaminergic and cholinergic systems of the basal ganglia may thus be correlated with various cognitive effects.

We presented some striking similarities between cognitive decision processes and functional characteristics of the basal ganglia reward system. In particular, Tversky and Kahneman's Prospect Theory (describing cognitive decision processes) was analyzed in terms of the functional characteristics of midbrain dopaminergic and striatal cholinergic interneurons. We propose that the dichotomy between risk aversion and risk seeking modes of behavior is dependent upon the asymmetrical nature of these neuronal systems. Moreover, we propose that the emergence of sequential behavior (as opposed to pattern-detection behavior) comparatively late in the process of evolution has led to the asymmetrical nature of the basal ganglia system.

The correlations between cognitive descriptions of decision making and physiological features of the basal ganglia are still mostly obscure. We do not claim to have considered the whole picture in this chapter, but we do think we presented a promising track for future research. The comparison of the two levels of research, physiological and psychological, may give insights for further experiments in both disciplines. We believe that further study of the basal ganglia using gradient of aversive and appetite stimuli will help us shed light on the critical role of these neuronal networks in normal and pathological behavior.

5. ACKNOWLEDGEMENT

This study was supported by grants of the Israeli Academy of Science and the US-Israel binational scientific foundation. We thank Naftali Tishbi and Itay Gat (Computer Science, The Hebrew University) for help with data analysis.

6. REFERENCES

Aldridge, J.W. and Berridge, K.C. Coding of serial order by neostriatal neurons: a "natural action" approach to movement sequence. J. Neurosci. 18: 2777-2787, (1998).

Aosaki, T., Kimura, M., and Graybiel, A.M., Temporal and spatial characteristics of tonically active neurons of the primate's striatum, J. Neurophysiol. 73: 1234-1252 (1995).

Apicella, P., Legallet, E., and Trouche, E., Responses of tonically discharging neurons in the monkey striatum to primary rewards delivered during different behavior states, Exp. Brain Res. 116: 456-466 (1997).

Beiser, D.G. and Houk, J.C. Model of cortical-basal ganglionic processing: encoding the serial order of sensory events. J. Neurophysiol. 79: 3168-3188 (1998).

Graybiel, A.M. The basal ganglia and chunking of action repertoires. Neurobiol. Learn. Mem. 70: 119-136, (1998).

Hollerman, J.R., Schultz, W., Dopamine neurones report an error in the temporal prediction of reward during learning, Nature Neurosci. 1: 304-309 (1998).

Kahneman, D. & Tversky, A., Prospect theory - an analysis of decision under risk, Econometrica 47:263-291, (1979).

Kahneman, D., & Tversky, A., The psychology of preferences, Sci. Am. 246:136-142 (1982).

Kilgard, M.P., & Merzenich, M.M., Cortical map reorganization enabled by Nucleus Basalis activity, Science 279: 1714-1718 (1998).

Marin, O., Smeets, W. J. A. J. & Gonzalez, A., Evolution of the basal ganglia in tetrapods: A New Perspective Based on Recent Studies in Amphibians, TINS, 21, 487-495 (1998).

Mirenowicz J. and Schultz W., Importance of unpredictability for reward response in primate dopamine neurons. J. Neurophysiol. 72: 1024-1027 (1994)

Raz, A., Feingold, A., Zelanskaya, V., and Bergman, H., Neuronal synchronization of tonically active neurons in the striatum of normal and Parkinsonian primates. J. Neurophysiol. 76: 2083-2088 (1996).

Redgrave P., Prescott, T.J, and Gurney K. Is the short latency dopamine response too short to signal reward error? TINS, 22: 146-151 (1999).

Robbins, T.W., Everitt, B.J., Neurobehavioral mechanisms of reward and motivation, Current opinion in Neurobiology 6, 228-236 (1996).

Schultz, W., Apicella P., and Ljungberg T. Responses of monkey dopamine neurons to reward and conditioned stimuli during successive steps of learning a delayed response task. J. Neurosci. 13: 900-913 (1993).

Schultz, W., Dayan, P., and Montague, P.R. , A Neural substrate of prediction and reward, Science 275, 1593-1599 (1997).

Schultz, W., Predictive reward signal of Dopamine neurons,. J. Neurophysiol. 80:1-27 (1998).

Von Neuman, J. and Morgenstern, O., Theory of Games and Economic Behavior, Princeton University Press, 3rd edition (1953).

A PARTIAL DOPAMINE LESION IMPAIRS PERFORMANCE ON A PROCEDURAL LEARNING TASK:

Implications for Parkinson's Disease

Julie L. Fudge, David D. Song, Suzanne N. Haber*

Abstract: Parkinson's disease (PD) is a prevalent neurodegenerative disease which affects the dopamine (DA) cells of the ventral midbrain. Although PD has primarily been thought of as a disorder of the motor system, clinical evidence now shows that subtle neuropsychological abnormalities are present relatively early in the disease process. These deficits are dissociated from Alzheimer-type impairments of short-term memory. Instead PD patients have difficulties learning routines in new circumstances, known as procedural learning. Frontostriatal networks are believed to mediate procedural learning, but little is known about the specific circuitry involved. We used low, chronic doses of 1-methyl-4-phenyl-1, 2, 3, 6-tetrahydropyridine (MPTP) to create a model of early PD. We then tested normal and partially lesioned animals on a simple procedural learning task. All normal animals succeeded on the task on the first trial. Partially lesioned animals naive to the task were unable to solve it over repeated trials. A partially lesioned animal that had solved the task prior to MPTP lesioning retained the strategy, and successfully completed the task on the first trial after MPTP. Behavioral results were correlated with regional changes in staining for tyrosine hydroxylase in the midbrain and striatum.

1. INTRODUCTION

In idiopathic Parkinson's disease (PD), cognitive problems range from mild neuropsychological deficits to global dementia (Celesia and Wanamaker, 1972; Lieberman et al., 1979; Mayeux et al., 1981; Lees and Smith, 1983; Huber et al., 1989; Taylor et al., 1990a; Cooper et

---

* J.L. Fudge, Department of Psychiatry, Department of Neurobiology and Anatomy, University of Rochester Medical Center Rochester, NY 14642. D.D. Song, Department of Neurology, University of California at San Diego, San Diego, CA 92161-3064. S.N. Haber, Department of Neurobiology and Anatomy, Department of Neurology, University of Rochester Medical Center Rochester, NY 14642.

al., 1991). Patients experiencing global dementias at end stage disease are often found at autopsy to have degeneration of other neuronal populations in addition to dopaminergic cells, including superimposed Alzheimer's disease(Braak et al., 1997). Recently, interest has focused on the less dramatic, more insidious cognitive deficits present in early PD(Stern et al., 1993; Jacobs et al., 1996; Owen et al., 1997; Vakil and Herishanu-Naaman, 1998). In contrast to Alzheimer patients, PD patients have difficulties on tasks of procedural learning and set-shifting, but intact short-term memory(Saint-Cyr et al., 1988a; Massman et al., 1990; Taylor et al., 1990a; Knowlton et al., 1996). Procedural learning tasks assess the ability to learn skills and to generate strategies in novel situations(Cohen and Squire, 1980; Martone et al., 1984; Mishkin et al., 1984; Butters et al., 1985; Cohen et al., 1985; Heindel et al., 1988; Saint-Cyr and Taylor, 1992). There is evidence that in PD, procedural learning is impaired early in the disease process(Saint-Cyr et al., 1988a; Owen et al., 1992; Knowlton et al., 1996; Vakil and Herishanu-Naaman, 1998; Thomas-Ollivier et al., 1999). The neuroanatomical correlates of procedural learning deficits in PD patients are unknown since most post-mortem data reflect end-stage disease. To examine the relationship between procedural learning and early brain changes in PD, we employed a low-dose MPTP protocol in monkeys. The objective of this pilot study was to: 1. design a task sensitive to procedural learning in MPTP-lesioned monkeys that were relatively intact motorically, and 2. correlate cognitive and motor deficits with changes in dopamine innervation associated with specific basal ganglia circuits.

2. METHODS

2.1 Rotating Object Retrieval Box

We developed a Rotating Object Retrieval puzzle for monkeys to solve (Fig. 1). This task consists of a freely rotating Plexiglas cube, with a circular opening on one side. A raisin reward is placed in the center of the cube, and the opening is positioned to face directly away from the animal. The animal sees the reward, and then must turn the cube to bring the opening within reach in order to retrieve the reward. In this task, adapted from one used in previous studies(Diamond, 1990; Taylor et al., 1990b; Schneider and Roeltgen, 1993), the animal must simultaneously consider two conflicting perceptions: 1. the reward is directly in the animal's line of sight, and 2. a barrier is between the reward and the animal. These inputs must be held "on line" while the animal develops a strategy to get the reward(Diamond, 1990). Unlike previous versions, this task required that the animal manipulate the barrier to get the reward rather than reaching around it. To get the reward from the Rotating Object Retrieval box, the animal must devise a plan for how and when to remove the barrier.

2.2 MPTP Dosing

Four old world monkeys (M. *nemestrina*) were randomly selected to receive a low dose MPTP exposure. Three animals were treated to a partially affected state. One animal was treated to a state of severe parkinsonism in order to compare histochemical changes in the severe state with those in partially lesioned animals. Three times per week, animals were briefly anesthetized (ketamine 10 mg/kg intramuscular) and administered 0.15 mg/kg MPTP via saphenous vein infusion. MPTP was administered until there was evidence of emergence

Figure 1. Rotating Object Retrieval task.

of action tremor or a qualitative decline in dexterity upon presentation of a simple motor task (9 well raisin tray). MPTP dosing was stopped in the severely lesioned animal when there was significant akinesis and stooped posture.

2.3 Behavioral assessment

MPTP treated animals were administered the first trial of the Rotating Object Retrieval task within 24 hours of receiving the last MPTP dose. Three normal control animals were also tested. All animals were maintained in the home cage, and were not food deprived. They were visually isolated from one another during testing, and sessions were videotaped. One animal (case 103) was allowed to solve the task prior to MPTP lesioning. All other animals were naive to the task. The Rotating Object Retrieval box was placed in front of the home cage for two minutes. The trial was a "success" if the animal retrieved the raisin, and a "fail" if the animal did not retrieve the raisin within this time. All control animals scored a "success" on the first trial. Lesioned animals were scored twice a week during a three-week recovery period. All strategies used by the animals were noted during scoring of the videotape.

2.4 Histology

Four weeks after cessation of MPTP administration, the animals were deeply anesthetized and sacrificed by intracardiac perfusion using saline followed by a 4% paraformaldehyde solution in 0.1M phosphate buffer, pH 7.4. The brains were cryoprotected in increasing gradients of sucrose (10%, 20%, and 30%). Serial sections of 50 µm. were prepared on a freezing microtome and processed for immunocytochemistry (ICC) to tyrosine hydroxylase (TH), a marker of dopamine. Tissue was incubated in primary mouse antisera diluted

1:20,000(Eugene Tech) at 4°C for 4 nights. TH staining was visualized using the avidin biotin reaction (Vector Elite kit, Vector Labs USA). Tissue was incubated for 10-12 minutes in 3,3' diaminobenzidine tetra-hydrochloride and 0.03% H_2O_2 intensified with 1 % cobalt chloride and 1 % nickel ammonium sulfate to yield a black reaction product. Sections were then thoroughly rinsed, mounted, and coverslipped.

3. RESULTS

3.1 Behavior

3.1.1 Motor function

Severity of clinical parkinsonism was rated by a blinded trained observer on a modified version of the rating scale developed by Kurlan et al(Kurlan et al., 1991). The rating scales was expanded to have severity scores from 0 (normal) to 4 (maximally affected) on all seven parameters of the scale (tremor, posture, gait, bradykinesia, balance, gross motor skills, and defensive behavior) for a maximal score of 28. At the end of MPTP treatment, motor scores were as follows: case 101= 3, case 95= 15, case 103= 18, and case 100= 24. All animals in the partially lesioned group (cases 101, 95, and 103) were able to feed and groom independently. The severely lesioned animal (case 100) required assisted feeds.

3.1.2 Rotating Object Retrieval Task

All control animals solved the Rotating Object Retrieval task on the first trial. Control animals typically looked at the raisin, reached out and touched the box with a forelimb, and then turned the box. The two MPTP lesioned animals naive to the task (cases 101 and 95) were unable to solve the task, and repeatedly reached into the barrier. Case 101 employed several strategies to retrieve the reward, including shaking the apparatus, and standing and hitting the top of the box. Case 95 persisted in clawing at the side of the box directly in the line of sight. These animals were also unable to solve the task during the three-week recovery period. In contrast, the pre-trained lesioned animal (case 103) solved the puzzle on the first trial at the end of MPTP treatment. As expected, the severely affected animal (case 100) was unable to perform the task due to severe akinesia.

3.2 Anatomy

3.2.1 Midbrain

Normal TH staining in the ventral midbrain is shown in case 90 (Fig. 2). In partially lesioned animals (cases 101, 103, and 95), there is variable loss of TH staining in the midbrain. Compared with control, case 101 has little to no loss of TH staining. Cases 95 and 103 have intermediate levels TH loss compared with the control, with the most prominent depletion in ventrolateral areas of the SNc (cases 95 and 103, arrows). In contrast, TH staining is markedly depleted in the SNc of the severely lesioned animal (case 100). Most of the remaining TH

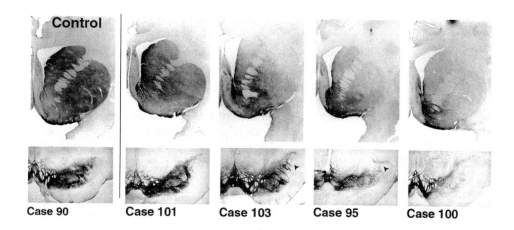

Figure 2. Coronal sections through the basal ganglia of control (case 90) and chronically lesioned MPTP monkeys (cases 101, 103, 95 and 100). Top row: Photomicrographs of TH-immunoreactivity (IR) in the striatum. TH-IR decreases most in the dorsolateral striatum, and is relatively spared in the ventral striatum. Note the remaining patches of TH-IR in the central striatum in cases 103 and 95. Bottom row: Photomicrographs of TH-IR in the ventral midbrain. TH-IR cells are lost primarily in the ventrolateral cell columns of the ventral tier (arrowheads), while those in the dorsal tier are relatively spared. However, TH is relatively spared throughout the ventral tier, even in case 100, compared to high-dose models.

positive neurons in case 100 are found in the medial SN/VTA and in a thin dorsal band spanning the dorsal SN.

3.2.2 Striatum

TH immunoreactivity in the normal striatum is characterized by a heterogeneous, patchy pattern, consistent with previous reports in primates(Holt et al., 1997)(Fig 2, case 90). TH loss in the striatum followed a dorsoventral gradient, consistent with previous reports of regional loss of striatal dopamine in Parkinson's disease and in partially lesioned MPTP animals (Fig. 2, cases 101, 103, 95, 100)(Kish et al., 1988; Hantraye et al., 1993; Schneider, 1990). Case 101 had minimal loss of TH in the dorsolateral caudate and putamen. This case, which was least affected motorically, revealed greater sparing of TH staining than cases 95 and 103, the two other partially lesioned animals. Cases 95 and 103 show greater degrees of TH loss, with a similar pattern of denervation along a dorsoventral gradient. In these animals, the dorsolateral striatum is most depleted, and there is a patchy, intermediate loss of TH immunoreactivity in the central region of the striatum. TH immunoreactivity is relatively spared in the ventral striatum. The severely symptomatic animal (case 100) showed a profound loss of TH-positive fibers in the dorsolateral and central striatum, with relative sparing in the ventromedial striatum, including the dorsomedial shell of the nucleus accumbens.

3.2.3 Dopaminergic Denervation Across Functional Domains

The striatum is divided into functional domains based on its cortical inputs. The dorsolateral striatum is considered sensorimotor-related, and the central and ventromedial striatum are considered associationand limbic-related, respectively(Haber and Fudge, 1997). Normally, TH staining throughout the striatum is made up of fine beaded fibers which are extremely dense, making it difficult to visualize individual fibers (Fig 3A-C). This pattern is altered in the MPTP-treated animals. Despite the appearance of almost complete denervation at low power, the dorsolateral striatum of a partially lesioned animal (Fig. 3D, case 95) contains some TH positive fibers. While some normal-appearing fibers remain, many TH-positive fibers are thick, and less dense, compared with control. In the central striatum of case 95, there is less fiber density compared to controls, but denser innervation by TH-positive fibers than in dorsolateral regions. In this case, as in case 103 (not shown), patches of tissue devoid of staining, interrupt regions of light staining in the central striatum (Fig. 3E). Of the surviving fibers, there are more fine, beaded axons than in the dorsolateral region.

Although relatively spared, there is also decreased TH staining in the ventral striatum compared to control (Fig. 3F).

In the severely treated animal (case 100), the dorsolateral striatum contains only scattered fibers(Fig 3G). Therefore, although the dorsolateral striatum looks equally depleted of TH in the partially and severely lesioned animals under low power, there is a much lower fiber density in severely treated animal. Remaining fibers in the dorsolateral striatum are thick, with a nonvaricose appearance(Fig 3G). As well, the central and ventromedial regions are much less densely innervated than in the partially lesioned animal(Fig 3H, I). Within the central striatum of case 100, the patch-like pattern of dopaminergic innervation is no longer seen.

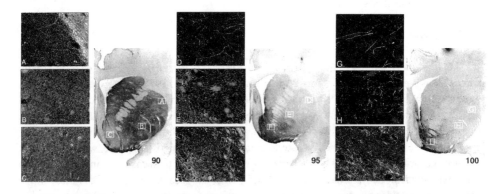

Figure 3. High power (6.3x) dark field photomicrographs of TH stained fibers across functional striatal domains in control (case 90), partially lesioned (case 95), and severely lesioned (case100) animals. Normally, TH staining across all domains consists of dense, fine beaded fibers which are difficult to visualize individually (A-C). Under high magnification, TH positive fibers in the dorsolateral striatum of case 95 are thicker and less dense (D), compared to normal (A). The central striatum has patches of lightly stained fibers interrupted by patches devoid of fibers (E). Although relatively spared, there is also decreased fiber density in the ventromedial striatum (F) compared to normal (C). Only scattered thick fibers remain in the dorsolateral striatum of the severely treated animal (G). Compared to case 95, the central (H) and ventromedial (I) regions are also less densely innervated.

4. CONCLUSIONS

Our preliminary data support the idea that MPTP treatment affects neural networks involved in learning a procedural skill. Other studies have demonstrated frontostriatal disturbances in partially lesioned MPTP animals, using classic paradigms such as delayed nonmatching to sample (DNMS)(Schneider and Kovelowski, 1990). Another category of task associated with frontostriatal function is detour-reaching tasks. Derived from developmental studies in human infants (Diamond, 1985; Piaget, 1954; Lockman, 1984), detour reaching, or "object retrieval", tasks employ a stationary transparent box in which the orientation of the open side changes according to trials set by the experimenter(SaintCyr et al., 1988b; Diamond, 1990; Taylor et al., 1990b; Schneider and Roeltgen, 1993). Human infants, monkey infants, and MPTP lesioned monkeys have all been shown to be impaired on tasks of detour reaching, presumably because of underdevelopment or lesions to the dorsolateral prefrontal cortex (Diamond, 1990). Like classic object retrieval tasks, the Rotating Object Retrieval task employs a clear Plexiglas barrier in front of a reward and gives conflicting information to the animal. To retrieve the reward, the animal must prioritize sensory information to determine that a barrier exists in front of the reward (i.e. tactile, not visual, information determines that a barrier exists in front of the reward). The Rotating Object Retrieval task changes a dimension of classic object retrieval tasks by requiring that the animal manipulate the barrier, rather than reaching around it. To do this, the animal must temporally prioritize motor plans, i.e. to manipulate the barrier before reaching for the reward. The execution of this prioritized sequence requires that the habit of reaching directly for a reward in the line of sight be temporarily inhibited(Diamond, 1990). Turning the box, either with a single push or several small pushes, is the most efficient way to get the reward. The Rotating task also allows for observation of a range of strategies.

In this study, the MPTP lesioned animals naive to the task tried to reach directly into the box. One tried this by shaking and hitting the box from different perspectives (case 101), and one (case 95) perseverated in clawing at the side of the box facing it. Neither animal attempted to turn the box. Furthermore, these two animals did not appreciate wrong moves, and failed to replace them with alternate strategies, similar to PD patients' performance on procedural learning tasks(Saint-Cyr et al., 1995;Saint-Cyr et al., 1988a; Owen et al., 1992; Knowlton et al., 1996). The fact that the pre-trained animal (case 103) retained the ability to solve the task suggests that, once learned, the successful strategy was subsumed by neural networks that are not affected by MPTP. Electrophysiologic studies in normal animals show that once a task is learned, the dopamine system no longer plays a role in cognitive processes related to getting the reward. The DA cells respond to novel stimuli during task acquisition, and cease to respond in overtrained animals(Ljungberg et al., 1992; Mirenowicz and Schultz, 1994). Case 103 was also more affected motorically than both of the naive animals who failed to solve the task, suggesting that motoric impairment did not play a role in successful manipulation of the task. Moreover, the retained ability of animal 103 to solve the task suggests that the negotiation of three-dimensional space *per se* is not affected by MPTP.

Depletion of TH in specific striatal regions indicates that subpopulations of DA neurons, and their associated nigrostriatal pathways, are affected before others, consistent with previous reports(Schneider, 1990; Hantraye et al., 1993). The degree of striatal TH loss correlated with the severity of motor symptoms across cases. Loss of TH staining in the midbrain corresponded selectively to the region of the ventral tier which provides the main input to the dorsolateral striatum(Lynd-Balta and Haber, 1994)(Figure 3b). Selective loss of TH staining in the dorsolateral striatum may be related to characteristics of ventral tier neurons which project to this area (i.e. higher levels of dopamine transporter mRNA or lower levels of calbindin-Dk28(Lavoie and Parent, 1991; Shimada et al., 1992; Uhl et al., 1994; Haber et al., 1995)).

In contrast to the direct relationship between motor symptoms and increasing TH loss in the dorsolateral striatum, our behavioral data suggest that procedural learning deficits can occur prior to marked changes in striatal TH immunoreactivity. While case 95 had a clear loss of TH staining in dorsolateral and central striatal regions, case 101 showed little loss of TH immunoreactivity. Both animals consistently failed to solve the Rotating Object Retrieval task. We conclude that behavioral deficits on the Rotating Object Retrieval task can occur prior to the loss of the majority of DA cells. Consistent with these findings our data show that cognitive deficits can occur *prior* to major changes in motor deficits. This conclusion is supported by previous studies which document cognitive deficits in motorically asymptomatic animals(Schneider and Kovelowski, 1990; Taylor et al., 1990b) and humans (Stern et al., 1990) exposed to MPTP.

The fact that cognitive changes can occur prior to significant DA cell loss raises the possibility that subtle changes in DA homeostasis, rather than actual cell death, may underlie cognitive deficits. Partial DA lesions in rats result in increased DA turnover, presumably from surviving neurons(Robinson et al., 1990; Zigmond et al., 1990; Robinson et al., 1994; Abercrombie and Zigmond, 1995). A recent study in primates using 6-[18F]fluoro-L-DOPA suggests that dysregulation of DA occurs with clinically "silent" or subtle deficits(Doudet et al., 1998). Asymptomatic MPTP monkeys had DA turnover rates that were increased relative to normals, and decreased relative to motorically affected animals. Partial lesions of DA cells may also affect regulation of the D3 receptor, which is downregulated by DA denervation, and upregulated in hyperdopaminergic conditions(Levesque et al., 1995; Bordet et al., 1997; Gurevich et al., 1997; Morissette et al., 1998; Ryoo HL, 1998; Gurevich et al., 1999; Schneider et al., 1999). D3 regulation after MPTP lesioning is complex, as emphasized in a study which showed dramatic increases in D3 receptor mRNA during an extended "recovery period" in motorically normal MPTP-lesioned animals(Todd et al., 1996). The temporal relationship between D3mRNA expression and increased in DA turnover after MPTP lesioning has yet to be explored. One question to be addresseed is whether compensatory mechanisms during the "preclinical" stages of PD and after MPTP lesions correlate with cognitive deficits. One limitation of studies to date is the use of animals with significant DA loss. Available data, including the present findings, show that cognitive impairment occurs relatively early in the disease process prior to a major loss of DA cells(Levin et al., 1989; Jacobs et al., 1996). Future studies aiming to correlate cognitive impairment with compensatory changes such DA turnover and D3 receptor regulation will benefit by using cognitively affected animals with minimal DA lesions.

5. ACKNOWLEDGEMENTS

This work was supported by the Rochester Area Pepper Center (J.F.), NS22511, the National Parkinson's Foundation, and the Lucille P. Markey Charitable Trust (S.H.)

6. REFERENCES

Abercrombie ED and Zigmond MJ (1995) Modification of central catecholaminergic systems by stress and injury. In: Psychopharmacology: The Fourth Generation of Progress (Bloom FE and Kupfer DJ, eds.), pp 355-350. New York: Raven Press.

Bordet R, Ridray S, Carboni S, Diaz J, Sokoloff P, and Schwartz JC (1997) Induction of dopamine D3 receptor expression as a mechanism of behavioral sensitization to levodopa. Proc Natl Acad Sci 94(7):3363-7.

Braak H, Braak E, Yilmazer D, de Vos RA, Jansen EN, and Bohl J (1997) Neurofibrillary tangles and neuropil threads as a cause of dementia in Parkinson's disease. J Neural Transm Supp51:49-55.

Butters N, Wolfe J, Martone M, Granholm E, and Cermak LS (1985) Memory disorders associated with Huntington's disease: Verbal recall, verbal recognition and procedural memory. Neuropsychologia 23:729-743.

Celesia G and Wanamaker WM (1972) Psychiatric disturbances in Parkinson's disease. Diseases of the Nervous System 33.

Cohen N and Squire L (1980) Preserved learning and retention of pattern-analyzing skill in amnesia: Dissociation of knowing how and knowing that. Science 210:207-210.

Cohen NJ, Eichenbaum H, Deacedo BS, and Corkin S (1985) Different memory systems underlying acquisition of procedural and declarative knowledge. Ann N Y Acad Sci 444:54-71.

Cooper JA, Sagar HJ, Jordan N, Harvey NS, and Sullivan EV (1991) Cognitive impairment in early, untreated Parkinson's disease and its relationship to motor disability. Brain 114:2095-2122.

Diamond A (1985) Development of the ability to use recall to guide action, as indicated by infants' performance on AB. Child Dev 56(4):868-83.

Diamond A (1990) Developmental time course in human infants and infant monkeys, and the neural bases of, inhibitory control in reaching. Ann N Y Acad Sci 608:637-669.

Doudet DJ, Chan GL, Holden JE, McGeer EG, Aigner TA, Wyatt RJ, and Ruth TJ (1998) 6-[18F]Fluoro-L-DOPA PET studies of the turnover of dopamine in MPTP-induced parkinsonism in monkeys. Synapse 29(3):225-32.

Gurevich EG, Joyce JN, and Ryoo H (1999) Ventral striatal circuits, dopamine D3 receptors and parkinson's disease. Soc Neurosci Astr. 25: 1600

Gurevich EV, Bordelon Y, Shapiro RM, Arnold SE, Gur RE, and Joyce JN (1997) Mesolimbic dopamine D3 receptors and use of antipsychotics in patients with schizophrenia. A postmortem study. Arch Gen Psych 54(3):225-32.

Haber SN and Fudge JL (1997) The primate substantia nigra and VTA: Integrative circuitry and function. Crit Rev Neurobiol 11(4):323-342.

Haber SN, Ryoo H, Cox C, and Lu W (1995) Subsets of midbrain dopaminergic neurons in monkeys are distinguished by different levels of mRNA for the dopamine transporter: Comparison with the mRNA for the D2 receptor, tyrosine hydroxylase and calbindin immunoreactivity. J Comp Neurol 362:400-410.

Hantraye P, Varastet M, Peschanski M, Riche D, Cesaro P, Willer JC, and Maziere M (1993) Stable parkinsonian syndrome and uneven loss of striatal dopamine fibres following chronic MPTP administration in baboons. Neuroscience 53(1):169-178.

Heindel WC, Butters N, and Salmon DP (1988) Impaired learning of a motor skill in patients with Huntington's disease. Behav Neurosci 102:141-147.

Holt DJ, Graybiel AM, and Saper CB (1997) Neurochemical architecture of the human striatum. J Comp Neurol 384:1-25.

Huber SJ, Freidenberg DL, Shuttleworth EC, Paulson GW, and Christy JA (1989) Neuropsychological impairments associated with severity of Parkinson's disease. J Neuropsychiatry Clin Neurosci 1:154-150.

Jacobs DM, Marder K, Cote LJ, Sano M, Stern Y, and Mayeux R (1996) Neuropsychological characteristics of preclinical dementia in Parkinson's disease. Neurology 45(9):1691-1696.

Kish SJ, Shannak K, and Hornykiewicz O (1988) Uneven pattern of dopamine loss in the striatum of patients with idiopathic parkinson's disease. N Engl J Med :876-880.

Knowlton BJ, Mangels JA, and Squire LR (1996) A neostriatal habit learning system in humans [see comments]. Science 273:1399-1402.

Kurlan R, Kim MH, and Gash DM (1991) Oral levodopa dose-response study in MPTPinduced hemiparkinsonian monkeys: Assessment with a new rating scale for monkey parkinsonism. Mov Disord 6:111-118.

Lavoie B and Parent A (1991) Dopaminergic neurons expressing calbindin in normal and parkinsonian monkeys. Neuroreport 2, No. 10:601-604.

Lees AJ and Smith E (1983) Cognitive deficits in the early stages of Parkinson's disease. Brain 106:257-270.

Levesque D, Martres MP, Diaz J, Griffon N, Lammers CH, Sokoloff P, and Schwartz JC (1995) A paradoxical regulation of the dopamine D3 receptor expression suggests the involvement of an anterograde factor from dopamine neurons. Proc Natl Acad Sci 92(5):1719-23.

Levin BE, Llabre MM, and Weiner WJ (1989) Cognitive impairments associated with early Parkinson's disease [see comments]. Neurology 39(4):557-61.

Lieberman A, Dziatolowski M, Kupersmith M, Serby M, Goodgold A, Korein J, and Goldstein M (1979) Dementia in Parkinson disease. Ann Neurol 6:355-359.

Ljungberg T, Apicella P, and Schultz W (1992) Responses of monkey dopamine neurons during learning of behavioral reactions. J Neurophysiol 67(1):145-163.

Lockman JJ (1984) The development of detour ability during infancy. Child Dev 55(2):482-91.

Lynd-Balta E and Haber SN (1994) The organization of midbrain projections to the striatum in the primate: Sensorimotor-related striatum versus ventral striatum. Neuroscience 59:625-640.

Martone M, Butters N, Payne M, Becker JT, and Sax DS (1984) Dissociations between skill learning and verbal recognition in amnesia and dementia. Arch Neurol 41:965-970.

Massman PJ, Delis DC, Butters N, Levin BE, and Salmon DP (1990) Are all subcortical dementias alike? Verbal learning and memory in Parkinson's and Huntington's disease patients. J Clin Exp Neuropsychol 12:729-744.

Mayeux R, Stern Y, Rosen J, and Leventhal J (1981) Depression, intellectual impairment, and Parkinson disease. Neurology 31:645-650.

Mirenowicz J and Schultz W (1994) Importance of unpredictability for reward responses in primate dopamine neurons. J Neurophysiol 72:1024-1027.

Mishkin M, Malamut B, and Bachevalier J (1984) Memories and habits: Two neural systems. In: The Neurobiology of Learning and Memory (McGaugh JL, Lynch G, and Weinberger NM, eds.), pp New York: Guilford Press.

Morissette M, Goulet M, Grondin R, Blanchet P, Bedard PJ, Di Paolo T, and Levesque D (1998) Associative and limbic regions of monkey striatum express high levels of dopamine D3 receptors: effects of MPTP and dopamine agonist replacement therapies. Eur J Neurosci 10(8):2565-73.

Owen AM, Iddon JL, Hodges JR, Summers BA, and Robbins TW (1997) Spatial and non-spatial working memory at different stages of Parkinson's disease. Neuropsychologia 35:519-532.

Owen AM, James M, Leigh PN, Summers BA, Marsden CD, Quinn NP, Lange KW, and Robbins TW (1992) Fronto-striatal cognitive deficits at different stages of Parkinson's disease. Brain 115:1727-1751.

Piaget J (1954) The Construction of Reality in the Child (Margaret Cook, Trans.). New York: Basic Books, Inc.

Robinson TE, Castaneda E, and Whishaw IQ (1990) Compensatory changes in striatal dopamine neurons following recovery from injury induced by 6-OHDA or methamphetamine: a review of evidence from microdialysis studies. Can J Psychol 44(2):253-75.

Robinson TE, Mocsary Z, Camp DM, and Whishaw IQ (1994) Time course to recovery of extracellular dopamine following partial damage to the nigrostriatal dopamine system. J Neurosci 14:2687-2696.

Ryoo HL PB, Joyce JN (1998) Dopamine D3 receptor is decreased and D2 receptor is elevated in the striatum of Parkinson's disease. Mov Disord13(5):788-797.

Saint-Cyr JA and Taylor AE (1992) The mobilization of procedural learning: The "key signature" of the basal ganglia. In: Neuropsychology of Memory (Squire LR and Butters N, eds.), pp 188-202. New York: The Guilford Press.

Saint-Cyr JA, Taylor AE, and Lang AE (1988a) Procedural learning and neostriatal dysfunction in man. Brain 111:941-959.

Saint-Cyr JA, Taylor AE, and Nicholson K (1995) Behavior and the basal ganglia. In: Behavioral Neurology of Movement Disorders (Weiner WJ and Lang AE, eds.), pp 1-28. New York: Raven Press, Ltd.

Saint-Cyr JA, Wan RQ, Aigner TG, and Doudet D (1988b) Impaired detour reaching in rhesus monkeys after MPTP lesions. Soc Neurosci Abst 14:389.

Schneider JS (1990) Chronic exposure to low doses of MPTP. II. Neurochemical and pathological consequences in cognitively-impaired, motor asymptomatic monkeys. Brain Res 534:25-36.

Schneider JS and Kovelowski CJ, II (1990) Chronic exposure to low doses of MPTP. I. Cognitive deficits in motor asymptomatic monkeys. Brain Res 519:122-128.

Schneider JS and Roeltgen DP (1993) Delayed matching-to-sample, object retrieval, and discrimination reversal deficits in chronic low dose MPTP-treated monkeys. Brain Res 615:351-354.

Schneider JS, Rothblat DS, Wade T, Joyce JN, and Ryooo H (1999) Differential modulation of dopamine D3 and D2 receptor number in symptomatic and recovered parkinsonian cats. Soc Neurosci Astr. 25: 1598.

Shimada S, Kitayama S, Walther D, and Uhl G (1992) Dopamine transporter mRNA: dense expression in ventral midbrain neurons. Mol Brain Res 13:359-362.

Stern Y, Marder K, Tang MX, and Mayeux R (1993) Antecedent clinical features associated with dementia in Parkinson's disease [see comments]. Neurology 43:1690-1692.

Stern Y, Tetrud JW, Martin WRW, Kutner SJ, and Langston JW (1990) Cognitive change following MPTP exposure. Neurology 40:261-264.

Taylor AE, Saint-Cyr JA, and Lang AE (1990a) Memory and learning in early Parkinson's disease: Evidence for a "frontal lobe syndrome". Brain Cogn 13:211-232.

Taylor JR, Elsworth JD, Roth RH, Sladek JR, Jr., and Redmond DE, Jr. (1990b) Cognitive and motor deficits in the acquisition of an object retrieval/detour task in MPTP-treated monkeys. Brain 113:617-637.

Thomas-Ollivier V, Reymann JM, Le Moal S, Schuck S, Lieury A, and Allain H (1999) Procedural memory in recent-onset Parkinson's disease. Dementia & Geriatric Cognitive Disorders 10(2):172-80.

Todd RD, Carl J, Harmon S, O'Malley KL, and Perlmutter JS (1996) Dynamic changes in striatal dopamine D2 and D3 receptor protein and mRNA in response to 1-methyl-4-phenyl-1,2,3,6-tetrahydropyridine (MPTP) denervation in baboons. J Neurosci 16:7776-7782.

Uhl GR, Walther D, Mash D, Faucheux B, and Javoy-Agid F (1994) Dopamine transporter messenger RNA in Parkinson's disease and control substantia nigra neurons. Ann Neurol 35:494-498.

Vakil E and Herishanu-Naaman S (1998) Declarative and procedural learning in Parkinson's disease patients having tremor or bradykinesia as the predominant symptom. Cortex 34(4):611-20.

Zigmond MJ, Abercrombie ED, Berger TW, Grace AA, and Stricker EM (1990) Compensations after lesions of central dopaminergic neurons: some clinical and basic implications. Trends Neurosci 13:290-296.

NEUROCHEMICAL EVIDENCE THAT MESOLIMBIC NORADRENALINE DIRECTS MESOLIMBIC DOPAMINE, IMPLYING THAT NORADRENALINE, LIKE DOPAMINE, PLAYS A KEY ROLE IN GOAL-DIRECTED AND MOTIVATIONAL BEHAVIOR

A.R. Cools and T. Tuinstra*

Key words: Nucleus accumbens, Dopamine, Noradrenaline, beta-adrenoceptors, alpha-adrenoceptors, alpha-methyl-para-tyrosine, vesicles, High responders to novelty, Low responders to novelty

Abstract: The primary goal of this chapter is to provide a short survey of the recently collected neurochemical evidence that mesolimbic noradrenaline has a dual role in the nucleus accumbens: (a) it stimulates alpha-like adrenoceptors on mesolimbic terminals of dopaminergic neurons that likely arise in the ventral tegmental area, with the result that the release of dopamine from alpha-methyl-para-tyrosine-insensitive pools in these terminals is inhibited, and (b) it stimulates beta-adrenoceptors on mesolimbic terminals of dopaminergic neurons that likely arise in the substantia nigra, pars compacta, with the result that the release of dopamine from alpha-methyl-para-tyrosine-sensitive pools in these terminals is facilitated. In addition, it is shown that exposure to a mild stressor changes the relatively low (alpha/beta) adrenergic activity that marks the non-challenged High responders to novelty, into a relatively high (alpha/beta) adrenergic activity, and that the same stressor changes the relatively high (alpha) adrenergic activity that marks the non-challenged Low responders to novelty, into a relatively low (alpha/beta) adrenergic activity. The overall impact of these findings is that mesolimbic NE fulfills all functions that are hitherto ascribed to mesolimbic DA, implying that mesolimbic NE, like mesolimbic DA, has a key role in goal-directed and motivational behavior and that it, like DA, is involved in the responses to stressful, reinforcing and aversive stimuli, including drugs of abuse.

---

* A.R. Cools and T. Tuinstra, Department of Psychoneuropharmacology, Nijmegen Institute of Neurosciences, University of Nijmegen, P.O.box 9101, 6500 HB Nijmegen
The Basal Ganglia VI
Edited by Graybiel *et al.*, Kluwer Academic/Plenum Publishers, 2002

1. INTRODUCTION

For a long time, it is known that there exists a particular relationship between dopamine (DA) and noradrenaline (NE) in the brain.[1] In this chapter attention is focused on the ability of NE to direct DA. In particular, the nature of the NE-DA interaction in the nucleus accumbens (NACC) is discussed.

Classical studies on the anatomical basis of the NE-DA interaction mainly deal with the dopaminergic, nigrostriatal system: (1) NE has been found to affect DA activity in dopaminergic, nigrostriatal fibers at the level of the substantia nigra,[2] and (2) NE has been found to affect indirectly DA activity by changing the serotonergic activity either in raphe-nigral[3] or in raphe-neostriatal fibers.[4] More recently, attention is focused on mesolimbic and mesocortical systems. Concerning the mesocortical system, there is neurochemical and behavioral evidence that stimulation of prefrontal alpha-adrenoceptors inhibits the prefrontal DA activity at the level of D1 receptors.[5]

As far as it concerns the mesolimbic system, the situation is more complicated. Pharmaco-behavioral studies have provided evidence in favor of the hypothesis that mesolimbic NE has a dual role [6,7]: it has been hypothesized to stimulate mesolimbic, alpha-like adrenoceptors that inhibit the release of mesolimbic DA from alpha-methyl-para-tyrosine-insensitive (MpT) pools, as well as to stimulate mesolimbic, beta-adrenoceptors that facilitate the release of mesolimbic DA from MpT-sensitive pools. The goal of this chapter is to provide a short survey of the recently collected neurochemical evidence that supports this view.

The impact of this finding is that mesolimbic NE fulfills all functions that are hitherto ascribed to mesolimbic DA. It implies that mesolimbic NE, like DA, plays a key role in goal-directed and motivational behavior and that it, like DA, is involved in the response to stressful, aversive and reinforcing stimuli such as drugs of abuse. This insight in turn opens new perspectives for the development of completely new classes of therapeutic drugs: drugs that interact with mesolimbic adrenoceptors, must have therapeutic effects that are hitherto ascribed to drugs that interact with mesolimbic dopaminergic receptors. In this context, it is interesting to note that olanzapine, namely a prototype of atypical antipsychotics, indeed inhibits mesolimbic, alpha-adrenoceptors, opening the perspective that this feature may also contribute to its peculiar therapeutic profile.[8]

2. RELEVANCE OF MESOLIMBIC NORADRENALINE

There are at least 7 sets of data showing that mesolimbic NE requires more attention than it has got in the past:
1. The NACC contains DA-beta-hydroxylase, namely an enzyme synthesizing NE from DA in NE-containing structures.[9]
2. The NACC receives noradrenergic fibers from the locus coeruleus and/or subcoeruleus[10] as well as from the nucleus tractus solitarius.[11]
3. The NACC contains both alpha- and beta-adrenoceptors (for review: [12]).
4. NE can be released from the NACC.[13]
5. The NACC contains specific NE-sensitive adenylate cyclases.[14]

6. At least certain behavioral responses to intra-accumbens injections of NE and NE-selective agonists differ from those elicited by such injections of DA and DA-selective agonists.[15]
7. Behavioral responses to intra-accumbens administration of NE and alpha-adrenergic agonists can be used to distinguish so-called responders and non-responders in normal outbred strains of Wistar rats.[16] In this context it is relevant to note that the originally discovered dichotomy of responders and non-responders has been found to be similar to the dichotomy of High Responders to novelty (HR) and Low Responders to novelty (LR),[17,18] respectively.
8. Priming mesolimbic alpha-adrenoceptors with a single injection of agents that directly or indirectly enhance the postsynaptic NE activity in the NACC changes the state of these receptors for a period of 24 h: responders become temporarily non-responders, and vice versa.[16]
9. Mesolimbic adrenoceptors are essential for the stress-induced sensitization to subsequent stressors –be these pharmacological or environmental.[6,12]
10. Psychostimulants and environmental challenges are interchangeable as far as it concerns their ability to sensitize mesolimbic alpha-adrenoceptors: secretion of corticosteroids that act at the level of mesolimbic mineralocorticoid receptors play a critical role in this process.[6]

3. PHARMACO-BEHAVIORAL STUDIES ON THE NE-DA INTERACTION IN THE NUCLEUS ACCUMBENS

3.1 1974-1983

Our initial analysis of the NE-DA interaction was limited to the caudate nucleus of cats. These studies have clearly revealed that alpha-like adrenoceptors exerts an inhibitory control on the function of DA at the level of a particular subclass of DA receptors, namely the so-called DAi receptors.[19] Although the DAi-DAe dichotomy has been fully overruled by the more recently introduced D1-D2 dichotomy, recent neurochemical data (see below) have provided evidence that (1) DAi receptors that are concentrated in "mesolimbic" regions and marked by so-called dotted DA fluorescence, are stimulated by DA that is derived from alpha-methyl-para-tyrosine (MpT) insensitive pools, and (2) DAe receptors that are concentrated in "striatal regions" and marked by so-called diffuse DA regions, are stimulated by DA that is derived from MpT sensitive pools. Given the regained interest in these receptors, we summarize the main features of these receptors[12,20]: (I) DAi receptors are stimulated by (3,4-dihydroxyphenyl)-2-imidazoline (DPI), but not by the mixed D1/D2 agonist apomorphine, the D2 agonist quinpirole, the D1 agonist SKF 38393 and NE; DAi receptors are inhibited by ergometrine, but not by the D2 antagonists haloperidol and l-sulpiride, the D1 antagonist SCH 23390, the alpha-adrenoceptor antagonist phentolamine and the antagonist of 5-HT2 receptors ritanserine; DAi receptors occur in terminal regions of dopaminergic fibers arising in the A8 and A10 area and mediate sedation and orofacial dyskinesia in rats and cats as well as inhibition of spontaneously firing cells in the snail Helix aspersa; (II) DAe receptors share nearly all their properties with those of D2 receptors and are concentrated in terminal regions of dopaminergic fibers arising in the A9 area: these are stimulated by apomorphine and inhibited by haloperidol and mediate typical striatal functions in rats and cats as well as excitation of spontaneously firing DA cells in the snail Helix aspersa. For additional similarities and dissimilarities between the DAi/DAe concept and the D1/D2 concept, see Cools et. (1988)[9].

3.2 1987-1991

Once it was found that stimulation of adrenoceptors in the NACC of rats elicits NE-specific and NE-selective behavioral responses,[12,16] it became interesting to investigate whether the DAi receptors that are present in the brain of rats, especially in the NACC, are also under the inhibitory control of adrenoceptors. As mentioned below, this is indeed the case. In the early nineties, we have summarized evidence in favor of the hypothesis that a high functional activity of NE in the NACC has two effects: it stimulates alpha-like adrenoceptors that in turn inhibit the mesolimbic DA activity at the level of DAi receptors, and it stimulates beta-adrenoceptors that in turn facilitate the mesolimbic DA activity at the level of DAe /D2 receptors.[12] According to this hypothesis, DAi receptors are localized on terminals of fibers arising in the basolateral amygdala (BLA), whereas the DAe/D2 receptors are localized on the terminals of fibers arising in the ventral subiculum of the hippocampus (VSH). This so-called gating theory states that mesolimbic alpha-like adrenoceptors direct the arrival of BLA-information in contrast to beta-adrenoceptors that direct the arrival of VSH-information. Indeed, Roozendaal and Cools (1994)[21] have provided pharmaco-behavioral evidence that intra-accumbens administration of both alpha- and beta-adrenergic agents directs the BLA-dependent neophobia according to this gating theory. More recently, additional evidence in favor of this theory has been collected in experiments on a VSH-dependent task: in this case too, intra-accumbens administration of alpha- and beta-adrenergic agents has been found to direct the VSH-dependent spatial Morris maze task according to the mentioned gating theory.[22]

3.3 1991-1999

As previously reported,[16] environmental and pharmacological challenges have been found to increase the functional mesolimbic NE activity in subjects with a low baseline NE activity, namely High responders to novelty (HR), but to decrease it in subjects with a high baseline NE activity, namely Low responders to novelty (LR). These findings have been incorporated in the above-mentioned gating theory with the following predictions as result (for review: [18]; see also: [7]):

HR that are at rest, are marked by a low mesolimbic NE activity with the result that (a) the low activity at the level of the mesolimbic alpha-like adrenoceptors disinhibits the release of mesolimbic DA from MpT-insensitive pools, producing thereby a high DA activity at the level of DAi receptors, and (b) the low activity at the level of mesolimbic beta-adrenoceptors is accompanied by a low release of mesolimbic DA from MpT sensitive pools, resulting thereby in a low DA activity at the level of DAe/D2 receptors. As soon as HR are confronted with mild stressors, the low baseline NE activity increases with the result that (a) the high activity at the level of alpha-like adrenoceptors inhibit the release of DA from MpT-insensitive pools, producing thereby a low DA activity at the level of DAi receptors, and that (b) the high activity at the level of beta-adrenoceptors facilitates the release of DA from MpT-sensitive pools, producing thereby a high DA activity at the level of DAe/D2 receptors.

LR that are at rest, are marked by a relatively high NE that decreases as soon these animals are confronted with mild stressors, implying that non-challenged LR can be at best compared with challenged HR, and that challenged LR can be at best compared with non-challenged HR.

4. MICRODIALYSIS STUDIES ON THE MESOLIMBIC NE-DA INTERACTION IN HR AND LR

4.1 HR and LR

Because the remainder of this chapter deals with studies on HR and LR, it is important to mention the following. When these two types of rat are selected with the help of a particular open-field procedure from the Nijmegen outbred strain of Wistar rats, these have been labeled as Nijmegen high responders to novelty (HR) and Nijmegen low responders to novelty (LR). Using a particular breeding procedure, it has been possible to breed these two types of rat; because the gnawing response to apomorphine is used in this breeding program, HR and LR have been labeled as apomorphine-susceptible and apomorphine-unsusceptible rats, respectively. As discussed elsewhere[7], the ultimate structure and function of the adult brain and body of these two types of animal (HR being similar to apomorphine-susceptible rats and LR being similar to apomorphine-unsusceptible rats) are determined, among others, by genetic and early postnatal factors. Owing to these factors and to the distinct selection procedures used in studies on individuals that differ in their response to novelty, one has to be aware of the fact that HR and LR that are studied in different research centers are not necessary fully identical. Nevertheless, it is important to note that both the HR/LR studied by Piazza et al. (1990)[23] and the HR/LR studied by Hooks et al. (1992)[24] share so many features with the Nijmegen HR/LR that these are interchangeable in our opinion. Therefore, we use the labels HR and LR when we refer to the studies of Piazza, Hooks and ourselves.

4.2 Microdialyis Studies on Mesolimbic Dopamine in HR and LR

As shown by Hooks et al. (1992),[24] who have used a quantitative microdialysis technique, HR have a significantly larger baseline value of extracellular DA than LR, when measured in the NACC. This type-specific difference is too small in order to be detected with the classic semi-quantitative method. According to the above-mentioned predictions, these finding imply that the release of DA from MpT-insensitive pools exceeds the release of DA from MpT-sensitive pools in HR.

According to the above-mentioned predictions, a mild stressor enhances the release of DA from MpT-sensitive pools in HR, but decreases it in LR, whereas this stressor decreases the release of DA from MpT-insensitive pools in HR, but increases it in LR. Saigusa et al. (1998)[7] have indeed found that a mild stressor such as novelty produces an increase in mesolimbic DA that is not only far greater in HR than in LR, but also completely prevented by MpT in HR, but not in LR, providing direct neurochemical evidence that exposure to a mild stressor increases the release of DA from MpT-sensitive pools in HR, but not in LR. The rather small release of DA in LR was not only not suppressed by MpT, but actually significantly increased. The latter finding reveals three important features. First, the stressor-induced release of DA in LR is not derived from MpT-sensitive pools. Second, the DA that is released in LR, appears to be under the inhibitory control of an intrinsic compound of which the synthesis is apparently inhibited by MpT: for, inhibition of its synthesis enhances the release of DA, opening thereby the perspective that it is indeed NE that exerts this inhibitory control on the DA that is released from MpT-insensitive pools as it has been predicted (see above). Finally, these findings imply that the stressor-induced release of DA from MpT-insensitive pools is similar, or just a little bit larger, than the stressor-induced decrease of DA from MpT-sensitive pools.

4.3 MicrodialysisSstudies on the NE-DA Interaction in the Nucleus Accumbens of HR and LR

Although our ongoing studies are not yet fully finished and, therefore, not yet published, we will summarize the data that are presented during recent congresses in the USA and The Netherlands [25,26]; see also [27-31].

4.3.1 Studies in Non-challenged HR and LR

First, intra-accumbens administration of the alpha-adrenoceptor agonist phenylephrine has been found to enhance dose-dependently the release of mesolimbic DA in HR and LR that are at rest: this increase is significantly larger in HR than in LR. In a separate set of experiments, it has been found that the phenylephrine-induced increase in DA can be antagonized by an alpha-adrenoceptor antagonist, showing that the effects under discussion are mediated via alpha-adrenoceptors. Remarkably, intra-accumbens administration of the alpha-adrenoceptor antagonist also dose-dependently increases the release of DA in HR and LR that are at rest. In this case, however, the effects are significantly smaller in HR than in LR. Given the finding that both the agonist and the antagonist produces more or less similar adrenoceptor-selective effects, it can be concluded that each of these agents acts at its own target site. The most likely explanation is that the agonist acts at the presynaptic site and the antagonist at the postsynaptic site, implying that stimulation of the presynaptic adrenoceptors inhibits the release of NE from its terminals and that stimulation of the corresponding postsynaptic alpha-adrenoceptors on the terminals of DA neurons inhibits the release of DA from these terminals. This explanation fully fits in with the above-mentioned pharmaco-behavioral data, showing that stimulation of alpha-like adrenoceptors inhibits the function of DA at the level of DAi receptors. The alternative explanation that the agonist acts at the postsynaptic site and the antagonist at the presynaptic site can be rejected. For, this action would imply that NE stimulates the release of DA, namely a finding that is fully in conflict with the previously reported pharmaco-behavioral data. Moreover, such a stimulatory action of NE upon the release of DA would imply that the release of DA in non-challenged HR is lower than that of DA in non-challenged LR, because the baseline value of NE is low in non-challenged HR, and high in non-challenged LR. As mentioned above, Hooks et al. (1992)[24] have shown that the rest release of DA is significantly larger in non-challenged HR than in non-challenged LR, providing direct support in favor of the former explanation, namely that alpha-like adrenoceptors inhibit the release of DA. Finally, the finding that the phenylephrine-induced increase in DA in non-challenged HR is larger than that in non-challenged LR shows that the alpha-adrenergic tonus in non-challenged HR is lower than that in non-challenged LR. Again, this finding nicely fits in with the outcome of pharmaco-behavioral studies (see above).

Second, intra-accumbens administration of the beta-adrenoceptor agonist isoproterenol has also been found to enhance the release of mesolimbic DA in a dose-dependent manner. Moreover, these effects are antagonized by the beta-adrenoceptor antagonist l-propranolol, showing that the effects under discussion are mediated via beta-adrenoceptors. In this case, the effects seen in non-challenged HR are identical to those seen in non-challenged LR. In contrast to stimulation of beta-adrenoceptors, inhibition of these receptors has no effect on the release of DA in non-challenged HR and LR. These data together show that the beta-adrenergic tonus is low, both in non-challenged HR and in non-challenged LR. The most likely explanation for the observed effects is that both the agonist and the antagonist act at the same site, namely postsynaptic receptors on terminals of DA neurons, implying that stimulation of these postsynaptic receptors facilitates the release of DA from the terminals of DA neurons. This explanation fully fits in not only with the above-mentioned pharmaco-behavioral data, showing that stimulation of beta-adrenoceptors enhances the function of DA at the level of DAe/D2 receptors, but also with the outcome of the neurochemical data that are mentioned below.

Third, intra-accumbens administration of MpT has been found to produce a biphasic effect on the release of DA in non-challenged HR and LR: it produces an initial increase in DA (about 15-20%) that disappears as soon as the perfusion with MpT (40 min) is stopped, and it is immediately followed by a decrease in DA (about 30%) that lasts 100-120 min. The MpT-induced decrease in DA is counteracted by intra-accumbens administration of the beta-adrenoceptor agonist isoproterenol in both HR and LR. These data provide evidence that the mesolimbic DA that is released from MpT-sensitive pools, is indeed under the excitatory control of beta-adrenoceptors, as it has been predicted on the basis of the pharmaco-behavioral studies.

4.3.2 Studies in Challenged HR and LR

First, intra-accumbens administration of the alpha-adrenoceptor agonist phenylephrine has been found to enhance the stressor-induced increase in mesolimbic DA in HR and LR to a similar extent. The same holds true for the intra-accumbens administration of the alpha-adrenoceptor antagonist phentolamine. These data again underlie the notion that the agonist and antagonist must act at different sites. Because there is no valid reason to assume that the agonist and antagonist act at sites that differ from the sites that are attacked by these agents under resting conditions, it is hypothesized that, in this case too, the agonist stimulates presynaptic adrenoceptors on NE neurons that impinge upon terminals of DA neurons, producing thereby an inhibition of the release of NE that in turn disinhibits the release of DA from MpT-insensitive pools with the result that the DA activity at the level of DAi receptors increases. Apart from this, these data reveal that the adrenergic tonus in HR does not anymore differ from that in LR: for, both the agonist-induced change in the DA activity and the antagonist-induced change in the DA activity does not anymore differ between the challenged HR and LR. These data together with the above-mentioned notion that the adrnergic tonus is far lower in non-challenged HR than in non-challenged LR, imply that the stressor has increased the alpha-NE-activity in HR, but decreased it in LR.

Second, intra-accumbens administration of the beta-adrenoceptor agonist isoproterenol has been found to produce nearly no increase in the stressor-induced release of mesolimbic DA in HR. On the other hand, the intra-accumbens administration of the beta-adrenoceptor antagonist l-propranolol has been found to produce a nearly complete inhibition of the stressor-induced increase in DA in HR. Both sets of data reveal that the stressor-induced increase in DA is under the excitatory control of beta-adrenoceptors, and that it has reached its ceiling in HR, implying that the beta-adrenergic tonus is very high in challenged HR Because the stressor-induced increase in DA that is seen in HR, is known to be derived from MpT-sensitive pools, it is also evident that the DA that is derived from MpT-sensitive pools, is under the excitatory control of beta-adrenoceptors: for, propranolol inhibits the stressor-induced release of DA from this pool. Again, this is in line with the outcome of the pharmaco-behavioral studies as well as with the outcome of the neurochemical experiments that are performed under resting conditions (see above). The fact that intra-accumbens administration of isoproterenol has been found to produce a large increase of DA (about 50%) in LR, is understandable in view of the pharmaco-behavioral data that show that isoproterenol facilitates the release of DA from the MpT-sensitive pools in LR. The finding that isoproterenol is able to produce an increase in DA in LR, but not in HR, together with the finding that the beta-adrenergic tonus is equally low in non-challenged HR and LR again reveal that the ability of the stressor to enhance NE is far larger in HR than in LR (see also above). Although the vast majority of neurochemical data fully fit in with the available pharmaco-behavioral data, there is still one discrepancy. Our microdialysis study on the effect of a mild stressor in LR show that the beta-adrenergic tonus that is low during rest, is not affected by the exposure to the stressor, whereas the pharmaco-behavioral studies imply that exposure to stress decreases the adrenergic tonus in LR. However, this apparent discrepancy can be solved, because the neurochemical studies deal with changes at the level of the beta-adrenoceptors, whereas the pharmaco-behavioral studies deal with the alpha-adrenoceptors. Indeed, it should not be forgotten that the factors that direct the NE activity at the level of alpha-adrenergic adrenoceptors, are not necessarily the factors that also direct the NE activity at the level of the beta-adrenoceptors. In fact, this is highly unlikely in view of the finding that the alpha-adrenergic tonus in non-challenged LR has an intermediate value, since the agonist and the antagonist can modulate it, whereas the beta-adrenergic tonus is minimal, since it can only be modulated by the agonist (see above).

5. CONCLUSION

The reviewed neurochemical evidence that is described elsewhere in detail[27-31], has clearly shown that mesolimbic NE directs the release of mesolimbic DA in a complicated manner (see Figure 1). Its dual role is (a) to stimulate alpha-like adrenoceptors on mesolimbic terminals of dopaminergic neurons that likely arise in the ventral tegmental area, with the result that the release of DA from MpT-insensitive pools in these terminals is inhibited, and (b) to stimulate beta-adrenoceptors on mesolimbic terminals of dopaminergic neurons that likely arise in the substantia nigra, pars compacta, with the result that the release of DA from MpT-sensitive pools in these terminals is facilitated. In addition, it is shown that so-called

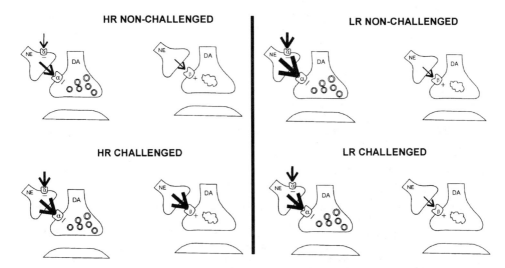

Figure 1. Model for NE-DA interaction in the Nucleus Accumbens in High responders to novelty and Low responders to novelty. Arrows indicate the adrenergic tonus at the level of the receptors. Circles: MpT-insensitive DA; cloud-shape: MpT-sensitive DA.

High responders to novelty that are at rest, have a relatively low (alpha/beta) adrenergic tonus that changes into a relatively high (alpha/beta) adrenergic activity as soon as these rats are exposed to a mild stressor, and that so-called Low responders to novelty that are at rest, have a relatively high (alpha) adrenergic tonus that changes into in a relatively low (alpha/beta) adrenergic tonus during exposure to a mild stressor. The overall impact of these findings is that mesolimbic NE fulfills all functions that are hitherto ascribed to mesolimbic DA, implying that mesolimbic NE, like mesolimbic DA, has a key role in goal-directed and motivational behavior and that it, like DA, is involved in the responses to stressful, reinforcing and aversive stimuli, including drugs of abuse.

6. REFERENCES

1. Antelman S. M. and Caggiula A. R. (1977) Norepinephrine-dopamine interactions and behavior. Science 195, 646-653.
2. Donaldson I. M., Dolphin A., Jenner P., Marsden C. D. and Pycock C. (1976) The involvement of noradrenaline in motor activity as shown by rotational behaviour after unilateral lesions of the locus coeruleus. Brain 99, 427-446.
3. Kostowski W., Samanin R., Bareggi S. R., Marc V., Garattini S. and Valzelli L. (1974) Biochemical aspects of the interaction between midbrain raphe and locus coeruleus in the rat. Brain Res. 82, 178-182.
4. Cools A. R. and Janssen H. J. (1974) The nucleus linearis intermedius raphe and behaviour evoked by direct and indirect stimulation of dopamine-sensitive sites within the caudate nucleus of cats. Eur. J. Pharmacol. 28, 266-275.
5. Tassin J. P., Studler J. M., Herve D., Blanc G. and Glowinski J. (1986) Contribution of noradrenergic neurons to the regulation of dopaminergic (D1) receptor denervation supersensitivity in rat prefrontal cortex. J. Neurochem. 46, 243-248.

6. Cools A. R. (1991) Differential role of mineralocorticoid and glucocorticoid receptors in the genesis of dexamphetamine-induced sensitization of mesolimbic, alpha 1 adrenergic receptors in the ventral striatum. Neuroscience 43, 419-428.
7. Saigusa T., Tuinstra T., Koshikawa N. and Cools A. R. (1999) High and low responders to novelty: Effects of a catecholamine synthesis inhibitor on novelty-induced changes in behaviour and release of accumbal dopamine. Neuroscience 88, 1153-1163.
8. Bymaster F. P., Calligaro D. O., Falcone J. F., Marsh R. D., Moore N. A., Tye N. C., Seeman P. and Wong D. T. (1996) Radioreceptor binding profile of the atypical antipsychotic olanzapine. Neuropsychopharmacology. 14, 87-96.
9. Hökfelt, T., Fuxe, K., Goldstein, M., Johansson, O., Park, D., Fraser, H. and Jeffcoate, S.L., Immunofluorescence mapping of central monoamine and releasing hormone (LRH) systems. In Anatomical neuroendocrinology, Karger, Basel, 1975, pp. 381-392.
10. Cedarbaum J. M. and Aghajanian G. K. (1978) Afferent projections to the rat locus coeruleus as determined by a retrograde tracing technique. J. Comp. Neurol. 178, 1-16.
11. Delfs J. M., Zhu Y., Druhan J. P. and Aston J. G. (1998) Origin of noradrenergic afferents to the shell subregion of the nucleus accumbens: anterograde and retrograde tract-tracing studies in the rat. Brain Res. 806, 127-140.
12. Cools, A. R., Van den Bos, R., Ploeger, G. and Ellenbroek, B.A., Gating function of noradrenaline in the ventral striatum: its role in behavioural responses to environmental and pharmacological challenges. In P. Willner and K.J. Scheel-Krüger (Eds.) The Mesolimbic Dopamine System: from Motivation to Action, Wiley, Toronto, 1991, pp. 141-173.
13. Li X. M., Perry K. W., Wong D. T. and Bymaster F. P. (1998) Olanzapine increases in vivo dopamine and norepinephrine release in rat prefrontal cortex, nucleus accumbens and striatum. Psychopharmacology Berl. 136, 153-161.
14. Robinson S. E., Mobley P. L., Smith H. E. and Sulser F. (1978) Structural and steric requirements for beta-phenethylamines as agonists of the noradrenergic cyclic AMP generating system in the rat limbic forebrain. Naunyn Schmiedebergs Arch. Pharmacol. 303, 175-180.
15. Cools, A. R., Spooren, W., Cuypers, E., Bezemer, R. and Jaspers, R., Heterogenous role of neostriatal and mesostriatal pathology in disorders of movement: a review and new facts. In A.R. Crossman and M.A. Sambrook (Eds.) Neural Mechanisms in Disorders of Movement, John Libbey and Company Ltd, London, 1988, pp. 111-119.
16. Cools A. R., Ellenbroek B., van-den B. R. and Gelissen M. (1987) Mesolimbic noradrenaline: specificity, stability and dose-dependency of individual-specific responses to mesolimbic injections of alpha-noradrenergic agonists. Behav. Brain Res. 25, 49-61.
17. Cools A. R., Brachten R., Heeren D., Willemen A. and Ellenbroek B. (1990) Search after neurobiological profile of individual-specific features of Wistar rats. Brain Res. Bull. 24, 49-69.
18. Cools A. R. and Gingras M. A. (1998) Nijmegen high and low responders to novelty: a new tool in the search after the neurobiology of drug abuse liability. Pharmacol. Biochem. Behav. 60, 151-159.
19. Cools A. R., van D. P., Janssen H. J. and Megens A. A. (1978) Functional antagonism between dopamine and noradrenaline within the caudate nucleus of cats: a phenomenon of rhythmically changing susceptibility. Psychopharmacology Berl. 59, 231-242.
20. Cools A. R. and Van Rossum J. (1980) Multiple receptors for brain dopamine in behavior regulation: concept of dopamine-E and dopamine-I receptors. Life Sci. 27, 1237-1253.
21. Roozendaal B. and Cools A. R. (1994) Influence of the noradrenergic state of the nucleus accumbens in basolateral amygdala mediated changes in neophobia of rats. Behav. Neurosci. 108, 1107-1118.
22. Tuinstra T., Verheij M., Willemen A., Iking J., Heeren D. J. and Cools A. R. (2000) Retrieval of spatial information in Nijmegen High and Low responders: involvement of beta-adrenergic mechanisms in the nucleus accumbens. Behavioral Neuroscience submitted.
23. Piazza P. V., Deminière J., Maccari S., Mormede P., Le Moal M. and Simon H. (1990) Individual reactivity to novelty predicts probability of amphetamine self-administration. Behav. Pharmacology 1, 339-345.
24. Hooks M. S., Colvin A. C., Juncos J. L. and Justice-JB J. (1992) Individual differences in basal and cocaine-stimulated extracellular dopamine in the nucleus accumbens using quantitative microdialysis. Brain Res. 587, 306-312.
25. Tuinstra, T. and Cools, A.R., Newly-synthesized dopamine in the nucleus accumbens is regulated by beta-adrenergic, but not alpha-adrenergic receptors, Soc.Neuro.Sci., 25 (1999) 2212(Abstract).

26. Tuinstra, T., Saigusa, T., Koshikawa, N. and Cools, A.R., Different dopamine-noradrenaline interaction in the nucleus accumbens between Nijmegen high and low responders to novelty, Soc.Neuro.Sci., 24 (1998) 955(Abstract).
27. Tuinstra T. (2000) The role of noradrenaline and dopamine in the nucleus accumbens: Individual differences. PhD-Thesis, Nijmegen University, Nijmegen, The Netherlands.
28. Tuinstra, T., & Cools, A.R. (2000) High and Low responders to novelty: 3effects of adrenergic agents on the regulation of accumbal dopamine under challenged and non-challenged condition. Neuroscience 99, 1, 55-64.
29. Tuinstra, T., & Cools a.R. (2000) Newly synthesised dopamine in the nucleus accumbens is regulated by β-adrenergic, but not α-adrenergic receptors, Neuroscience 98, 4, 743-747.
30. Tuinstra, T., Verheij, M., Willemen, A., Iking, J., Heeren, DJ., & Cools, A.R. (2000) Retrieval of spatial information in Nijmegen High and Low responders: involvement of β-adrenergic mechanisms in the nucleus accumbens. Behavioural Neuroscience 114, 6, 1088-1095.
31. Tuinstra, T., Cobelens, P., Lubbers, L., Verheij, M., & Cools, A.R. (2002) High and Low responders to novelty and mesolimbic noradrenaline: effects of noradrenergic agents on radial-maze performance. Behavioural Neuroscience (in press).

TONICALLY ACTIVE NEURONS IN THE STRIATUM OF THE MONKEY RAPIDLY SIGNAL A SWITCH IN BEHAVIORAL SET

Traci M. Thomas and Michael D. Crutcher*

1. INTRODUCTION

It is very well established that the basal ganglia play an important role in the control of movement. In the last 20 years a great deal of convincing evidence has accumulated indicating that the basal ganglia are also involved in much more than just motor functions, including cognition, motivation and reward. For example, it has been proposed that, among other things, the basal ganglia play an important role in learning stimulus-response associations[1-4], forming stimulus-reward associations[5-6], prediction of upcoming rewards[6-8], attention[9-12], and the switching of behavioral set[13-15]. There is a class of striatal interneurons, the tonically active neurons (TANs), whose activity may underlie some of these proposed functions.

The TANs are a subset of striatal interneurons, which display irregular, tonic activity. These are the putative cholinergic interneurons of the striatum[16,17]. They are broadly distributed throughout the entire neostriatum, including the caudate nucleus and putamen. The widespread distribution of these neurons and their direct synaptic contact with striatal output neurons puts them in a position to play a pivotal role in striatal function. TANs have been shown to develop dramatic responses to conditioned sensory stimuli that elicit movement or precede reward[17,18]. Thus, it has been suggested that the TANs are specifically involved in learning sensorimotor associations[18,19] or in the prediction of upcoming rewards[20,21]. We recorded the activity of TANs in two, awake rhesus monkeys while they learned and performed a series of tasks designed to investigate the relationship of TAN responsiveness to various cognitive functions. More specifically, our behavioral paradigms allowed us to explore the TAN's responsiveness to visual stimuli and rewards as well as their involvement in the formation of conditional sensorimotor associations, in global attention mechanisms, and in the switching of behavioral set.

---

* Traci M. Thomas and Michael D. Crutcher, Emory University, Department of Neurology, Atlanta, GA 30322.

2. MATERIALS AND METHODS

Two rhesus monkeys (*Macaca mulatta*) were trained to sit in a primate chair and to use a joystick to move a cursor on a video monitor situated in front them. They received rewards associated with moving the joystick. No visual targets were displayed during this pre-surgical training. After this training, two stainless steel recording chambers were surgically implanted on each monkey. One chamber was situated to permit an anterior approach to the head of the caudate and the pre-commissural putamen; the associative striatum. The other provided a lateral approach to the caudal, sensorimotor portion of the putamen. Because of similarities in connectivity and function, neurons in the precommissural putamen will be included with caudate neurons. We hypothesized that TANs in the associative and sensorimotor portions of the striatum would have different functional characteristics. We usually alternated recordings in the two chambers. Using their tonic activity and long after-hyperpolarizations as defining characteristics, extracellular single cell recordings were made from TANs as the monkeys learned and performed the following tasks.

To gather baseline information, recording was initially performed while the monkeys watched the computer screen as visual stimuli were randomly presented. The monkeys received random food rewards, which were not associated with the visual stimuli. Eye position was monitored to insure that the monkeys were watching the computer screen. The purpose of this was to determine the proportion of TANs that responded to visual stimuli and to rewards before they were associated. After gathering sufficient data in this *Baseline*

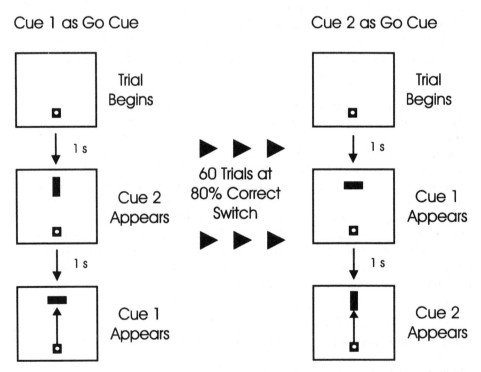

Figure 1. Sequence of events on the computer screen in front of the monkey during performance of the Go/NoGo task. The sequence of events for each version of the task is from top to bottom.

task, the monkeys were then taught to capture a visual target on the computer screen with the cursor representing joystick position to receive a reward. Single unit recording from TANs occurred during and after learning this *Simple* task.

In the final, *Go/NoGo* task, the monkeys first had to initiate a trial by capturing the initial visual target (black square) presented at the bottom of the computer screen with the cursor (white circle; Figure 1). After a delay, one of two visual stimuli (a vertical or horizontal rectangle) was randomly presented to the monkey at the top of the computer screen. One of the cues (the Go cue) instructed the monkeys to capture the cue with the cursor and the other (the NoGo cue) was to be ignored. The monkey had to determine the behavioral significance of the cues by trial-and-error. If the Go cue came up first (50% of trials), the monkeys captured the cue and then received a food reward. If the NoGo cue came up first, the monkeys ignored the cue, watched it disappear, waited for the Go cue to appear, captured this cue, and then received a food reward. Once the monkeys thoroughly learned the significance of each cue (for example: vertical rectangle = Go cue and horizontal rectangle = NoGo cue), the significance of the cues was switched so that the previous Go cue became the NoGo cue and the previous NoGo cue became the Go cue. In most cases there was a one second interval between each task event. As with the Simple task, single unit recording was performed during and after learning of this task.

3. RESULTS

3.1 Responses to Initial Visual Stimuli

For each of the three behavioral tasks a visual stimulus was presented to the monkey at the beginning of each trial. During performance of the Baseline task the visual stimulus had

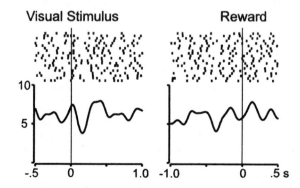

Figure 2. Example of a neuron with a modest decrease in activity following presentation of the initial visual stimulus in the trial. This pattern and magnitude of response was typical. For this and all subsequent figures each row of tics indicates the times of occurrence of action potentials in one trial. The rasters are aligned on the behavioral event indicated above each raster. The trials are presented in the order in which they were collected. The figure below each raster is the smoothed, spike density function. The vertical scale is in spikes/s.

no behavioral significance, did not require a response from the monkey, and was not paired with reward. During performance of the Simple task the visual stimulus represented the target that the monkey had to capture with the cursor in order to receive a reward. In the Go/NoGo task the initial visual stimulus was also captured with the cursor by the monkey to initiate the trial and alerted the monkey that the instruction signals were imminent. Generally, the frequency and robustness of the responses to this stimulus were fairly modest. An example of such a response is shown in Figure 2. As the monkey sequentially learned the more complex tasks and the initial visual stimulus acquired behavioral significance there was a modest increase in the proportion of TANs that responded to it. The percentages of neurons responding to this stimulus in each task are shown in Table 1.

Table 1. Responses to the initial visual stimulus in each task

Task	% Responding	Total (N)
Baseline	28%	43
Simple	52%	54
GoINoGo	51%	92
Total (N)		189

Chi Square, $p < 0.05$

3.2 Responses to Reward

A modest percentage of neurons responded to the presentation of reward in each of the tasks. An example of a typical response to the reward is shown in Figure 3. In the Baseline task the reward was randomly presented and not paired with the visual stimuli. In the Simple and Go/NoGo tasks the reward occurred at a predictable time after the monkey captured the correct target with the cursor. There was a modest increase in the proportion of neurons that responded to the reward in the Simple task, but not in the Go/NoGo task (see Table 2.) Very few TANs (approximately 2) had a response that anticipated the very predictable reward.

Table 2. Responses to reward in each task

Task	% Responding	Total (N)
Baseline	28%	43
Simple	54%	54
GoINoGo	35%	92
Total (N)		189

Chi Square, $p < 0.05$

3.3 Responses to Instruction Cues in the Go/NoGo Task

During performance of the Go/NoGo task the vast majority (98%) of recorded TANs responded to the instruction (Go or NoGo) cues. For 64% of TANs in the caudate and 17% of TANs in the putamen, the pattern of response depended upon the behavioral significance of

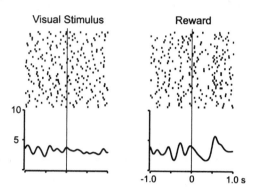

Figure 3. Typical example of a neuron that responded to the reward with a decrease in activity, but did not respond to the initial visual stimulus. For this and all subsequent figures one second of data is shown before and after the event to which the data are aligned.

the cue, i.e., whether the cue was a Go or a NoGo cue, independent of the form of the cue (vertical or horizontal rectangle). Examples of such responses are shown in Figures 4-6. This difference between the caudate and putamen was highly significant (Chi-square, $p<0.01$). The preferential involvement of the caudate is consistent with its being the associative portion of the striatum and its suggested involvement in decision-making processes, including the withholding of responses in the NoGo situation. Usually it only took the monkey one trial to respond correctly when the behavioral significance of the cues was changed. The TANs also changed their pattern of response to the instruction cues within one trial.

Other patterns of response to the instruction cues were also observed. Only two cells had a cue-specific pattern of response (not shown). For example, they responded to the vertical rectangle regardless of whether it was a Go or a NoGo cue. A modest number of cells had similar patterns of response to both the Go and NoGo cues. A substantial minority of cells were placed in the Other category. These cells had a variety of patterns of response. An example of one such cell is shown in Figure 7. A moderate number of cells in the Other category had similar patterns of response to each NoGo cue, but had a different pattern of response to one of the Go cues. Thus, these cells still contained information about the behavioral significance of the cue, but the pattern of response was highly context-dependent. The percentages of cells in the different categories in both the caudate and putamen are shown in Table 3.

For a limited number of TANs we collected data in two conditions. In the first condition there was a constant 1 s. interval between each event in the trial. In the second, the interval between events was randomized between 1-2 s. Many cells showed an enhanced response to the instruction cues when the epochs were randomized. An example is shown in Figure 8. Other TANs even showed a different overall pattern of activity in the task in the random condition (not shown). For example, a cell might respond only to the NoGo cues in the constant interval task, while in the random interval task the same cell might show the same response to the NoGo cues, but have a different response to the Go cues. This is a striking demonstration of the context-dependency of these responses.

Table 3. Responses to instruction cues in the Go/NoGo task

Pattern	Caudate	Putamen	Total (N)
Behav. Significance			
Go Cue	3%	0%	2
NoGo Cue	11%	0%	7
Both Go & NoGo	50%	17%	36
Cue Specific	2%	4%	2
All Cues Same	3%	25%	8
Other	30%	50%	31
No Response	2%	4%	2
Total (N)	64	24	88

Chi Square, p < 0.01

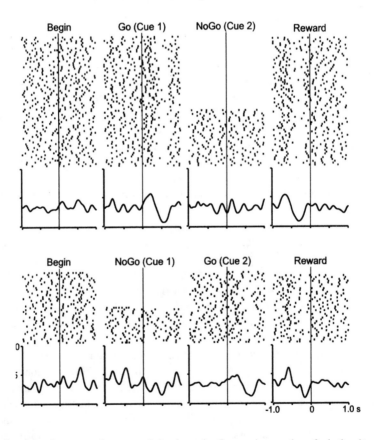

Figure 4. Example of a neuron that responded only to the Go cue, irrespective of whether it was cue 1 or cue 2. The response is a weak, early increase followed by a clear decrease that peaks at approximately 550 ms. The decrease prior to the reward is actually the earlier response to the Go cue, since there was very little variability in the time between the Go cue and the onset of the reward.

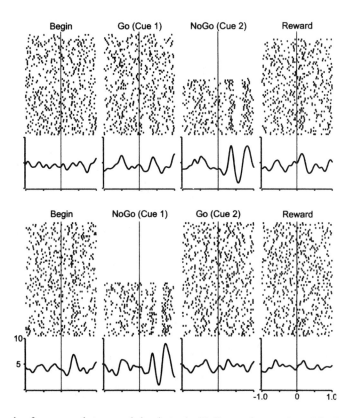

Figure 5. Example of a neuron that responded only to the NoGo cue, irrespective of the form of the cue. There is an early, weak decrease followed by a sequence of increases and decreases. Such multiphasic responses were not uncommon. Occasionally a neuron showed a multiphasic response in one condition and a monophasic response in another. Many different patterns of response were observed, e.g., increase only, decrease/increase, decrease/increase/decrease, etc. This suggests that the pattern of the response may contain information about the behavioral significance and/or context of the cue.

4. DISCUSSION

Our data demonstrate that TANs show significant responses to conditioned sensory stimuli during performance of a joystick task. Within our tasks, individual TANs responded primarily to the behavioral significance of visual stimuli rather than to the sensory characteristics of those cues. A similar phenomenon has been observed in a different type of Go/NoGo task for TANs in the putamen[19,22]. The TANs rapidly and significantly changed their pattern of activity to the instruction cues in the Go/NoGo task as soon as those cues changed their meaning, suggesting that TANs are rapidly signaling a switch in behavioral set. In addition, the fact that a subtle change in the task from constant to random epochs had a significant impact upon the responses of some TANs demonstrates that these responses are exquisitely

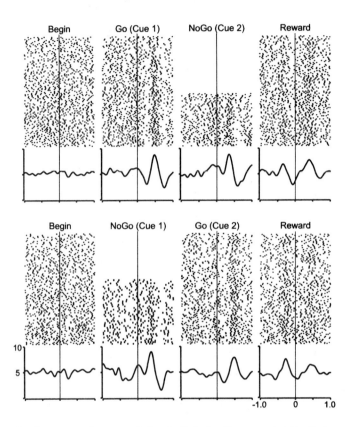

Figure 6. Example of a neuron that responded to both instruction cues, but the timing and pattern of response was different depending upon whether the cue was a Go cue or a NoGo cue. Following the Go cue there was a decrease that peaked at 200 ms, an increase that peaked at 500 ms and another decrease at 775 ms. Following the NoGo cue there was a slight, non-significant decrease at 100 ms, an increase that peaked at 400 ms and a prominent decrease at 550 ms.

context-dependent. The different patterns of response of TANs in our task were quite variable and complex. This suggests that TANs are part of a distributed network of neurons that signal the behavioral significance of stimuli depending upon task context.

As demonstrated by other researchers, TAN responsiveness to sensory stimuli that have been paired with rewards, as well as to rewards themselves, increased with training in the tasks[17,18,21]. This increase in frequency of responses to these cues was fairly modest, however. It is important to emphasize that 98% of TANs responded to the instruction cues, while much smaller percentages of TANs responded to the initial visual stimulus in the trial or to the reward.

It has been suggested that the basal ganglia play a role in attention mechanisms[9-12]. The pattern of responsiveness of TANs in these tasks suggests that they are not particularly involved in attention. Both the initial visual cue of the trial and the reward, especially in the

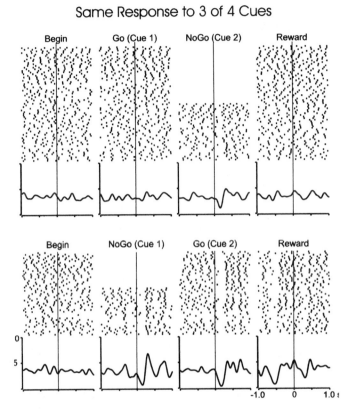

Figure 7. Example of a neuron in the Other category. This neuron responded to the instruction cue with a short-latency decrease in activity following three of the four instruction cues. There was no response when cue 1 was the Go cue.

Baseline task when the reward was unpredictable, should be very salient and alerting signals that would precipitate a TAN response if TANs were directly involved in global attention. Only 28% of cells responded to the reward in the Baseline task, while 98% responded to the instruction cues in the Go/NoGo task. In addition, the TANs did not always respond to the Go cue, which should be the most alerting signal in the Go/NoGo task.

The data in this study also do not support the hypothesis that TANs are involved in predicting upcoming rewards, which has been suggested previously[20,21]. The TANs routinely responded to the NoGo cue without responding to the Go cue. The Go cue is the most powerful predictor of an upcoming reward as it directly precedes reward delivery. Finally, the reward occurred at a predictable time in each trial in the Simple and Go/NoGo tasks. Virtually none of the TANs had a response that anticipated this very predictable reward. Furthermore, 98% of TANs responded to the instruction cues, while many fewer responded to the initial visual cue, which served as the first predictor of reward in each trial, or to the reward itself.

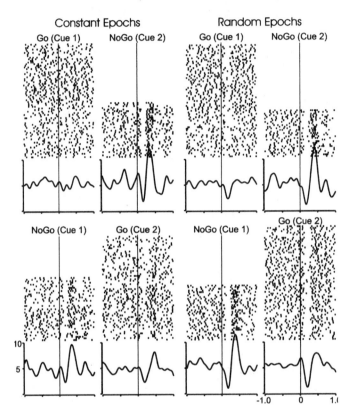

Figure 8. Example of enhanced responses to the instruction cues when the intervals between events were of random rather than constant duration. The four panels on the left of the figure show the responses to the instruction cues when there was 1 s. between each event, while those on the right show the responses when the interval between events was randomized from 1-2 s. When cue 2 was the Go cue (2nd and 4th panels on the bottom), the decrease in activity was much clearer with random expochs (4th panel). Likewise, when cue 1 was the NoGo cue (1st and 3rd panels on the bottom), the decrease and rebound increase were much clearer with random epochs (3rd panel). In the 1st panel on the top there was little, if any, decrease while in the 3rd panel (random epochs) there was a clear decrease in activity.

This suggests that TANs seem to be much better related to the behavioral significance of stimuli than to predicting reward.

It has been hypothesized that the TANs play an important role in learning sensorimotor associations[18,19]. Ours tasks were not complicated enough to clarify the role of TANs in such learning processes. Clearly the TANs developed their pattern of responsiveness as a result of learning, but this does not necessarily mean that they play a role in the process of sensorimotor learning *per se*. The monkeys learned to switch their behavioral response within one or two trials after the behavioral significance of the cues was reversed. The TANs also changed their pattern of response just as quickly. This quick change in behavior indicates that our task was not actually assessing the learning of new associations, but was assessing the switching of behavioral set. A much more difficult task in which the animal is continually presented with completely new behavioral associations must be used to truly address the issue of the role of TANs in sensorimotor learning.

The Go and NoGo cues served as the primary indicators of behavioral output in our task and the TANs showed their highest pattern of activity in relation to these cues. Furthermore, the behavioral significance of these cues was the primary determinant of a particular cell's activity pattern. The majority of TANs changed their response to a particular cue as soon as that cue changed meaning. In other words, the TANs changed their response pattern simultaneously with the animal's change in behavioral set.

These results provide strong evidence that the TANs and possibly the basal ganglia as a whole are involved in the switching of behavioral set. This hypothesis has been advanced previously in relation to basal ganglia functioning[13-15]. The activity patterns of the TANs reflect the behavioral significance of pertinent stimuli indicating that they could impart contextual significance to striatal output neurons. This information could then in turn be used by the basal ganglia to aid in the selection of appropriate behaviors. It has become clear that the basal ganglia are involved in much more than just motor processing. All aspects of behavior (movement, emotion, and cognition) require that an animal make stimulus-behavior associations and that it be able to quickly detect the context in which the stimulus is presented, so that it can perform the appropriate behavior. It is likely that the cortico-basal ganglia-frontal cortical loops play a role in these processes. Further studies must be conducted to address the role of TAN's in these processes.

5. ACKNOWLEDGEMENTS

We would like to extend our deepest appreciation to Dr. Robert Turner, who provided advice throughout these experiments. We would also like to thank Colleen Oliver for technical assistance. This research was supported by NIH grant PO NS31937.

6. REFERENCES

1. McDonald, R.J. and White, N.M. A triple dissociation of memory systems: hippocampus, amygdala, and dorsal striatum. Behav Neurosci, 1993. 107(1): 3-22.
2. Graybiel, A.M., et al. The basal ganglia and adaptive motor control. Science, 1994. 265: 1826-1831.
3. Boussaoud, D. and Kermadi, I. The primate striatum: neuronal activity in relation to spatial attention versus motor preparation. Eur J Neurosci, 1997. 9(1 0): 2152-68.
4. Passingham, R.E., et al. How do visual instructions influence the motor system? Novartis Found Symp, 1998. 218: 129-41.
5. Schultz, W., et al. Reward prediction in primate basal ganglia and frontal cortex. Neuropharmacology, 1998. 37(4-5):421-9.
6. Tremblay, L., et al. Modifications of reward expectation-related neuronal activity during learning in primate striatum. J Neurophysiol, 1998. 80(2): 964-77.
7. Ljungberg, T., et al. Responses of monkey dopamine neurons during learning of behavioral reactions. Journal of Neurophysiology, 1992. 67: 145-163.
8. Barto, A., Adaptive Critics and the Basal Ganglia, in *Models of Information Processing in the Basal Ganglia*, Houk, J., et al., Editors. 1995, The MIT Press: Cambridge. 215-232.

9. Schultz, W., et al. Responses of monkey dopamine neurons to reward and conditioned stimuli during successive steps of learning a delayed response task. J Neurosci, 1993. 13(3): 900-13.
10. Jackson, S. and Houghton, G., Sensorimotor Selection and the Basal Ganglia: A Neural Network Model, in *Models of Information Processing in the Basal Ganglia*, Houk, J., et al., Editors. 1995, The MIT Press: Cambridge. 337-367.
11. Kermadi, I. and Boussaoud, D. Role of the primate striatum in attention and sensorimotor processes: comparison with premotor cortex. Neuroreport, 1995. 6(8): 1177-81.
12. Brown, P. and Marsden, C.D. What do the basal ganglia do? [see comments]. Lancet, 1998. 351(9118): 1801-4.
13. Phillips, A.G. and Carr, G.D. Cognition and the basal ganglia: a possible substrate for procedural knowledge. Can J Neurol Sci, 1987. 14(3 Suppl): 381-5.
14. Hayes, A.E., et al. Toward a functional analysis of the basal ganglia J Cogn Neurosci, 1998. 10(2):178-98.
15. Redgrave, P., et al. Is the short-latency dopamine response too short to signal reward error? Trends Neurosci, 1999. 22(4):146-51.
16. Wilson, C.J., et al. Firing patterns and synaptic potentials of identified giant aspiny interneurons in the rat neostriatum. J.Neurosci., 1990. 10: 508-519.
17. Aosaki, T., et al. Temporal and spatial characteristics of tonically active neurons of the primate's striatum. Journal of Neurophysiology, 1995. 73: 1234-1252.
18. Aosaki, T., et al. Responses of tonically active neurons in the primate's striatum undergo systematic changes during behavioral sensory-motor conditioning. J.Neurosci, 1994. 14: 3969-3984.
19. Kimura, M. Behavioral modulation of sensory responses of primate putamen neurons. Brain Res, 1992. 578(1-2): 204-14.
20. Kimura, M. Role of basal ganglia in behavioral learning. Neurosci Res, 1995. 22(4): 353-8.
21. Apicella, P., et al. Responses of tonically discharging neurons in monkey striatum to visual stimuli presented under passive conditions and during task performance. Neurosci Lett, 1996. 203(3): 147-50.
22. Kimura, M. and Matsumoto, N. Neuronal activity in the basal ganglia. Functional implications, in *The Basal Ganglia and New Surgical Approaches for Parkinson's Diesase Advances in Neurology*, Obeso, J., et al., Editors. 1997, Lippincott-Raven: Philadelphia. 111-118.

TONICALLY ACTIVE NEURONS IN THE MONKEY STRIATUM ARE SENSITIVE TO SENSORY EVENTS IN A MANNER THAT REFLECTS THEIR PREDICTABILITY IN TIME

Paul Apicella, Sabrina Ravel, Pierangelo Sardo, and Eric Legallet*

Abstract: It is now well established that tonically active neurons (TANs) in the monkey striatum respond to motivationally relevant sensory events, such as conditioned stimuli to which the animal had to react correctly to obtain reward. Recent findings obtained in our laboratory suggested that stimulus prediction may influence the responsiveness of the TANs. In the present study we specifically investigated the effects of temporal aspects of prediction on the responses of single TANs recorded both in the caudate nucleus and putamen of two macaque monkeys. Three different behavioral situations were employed: (1) an instrumental task, in which a visual stimulus triggering a rewarded movement was preceded by an instruction stimulus presented at a fixed interval of 1.5 s before the trigger onset; (2) a classically conditioned task, in which a visual stimulus was followed after a fixed interval of 1 s by the delivery of a liquid reward without requiring the monkey to react to the stimulus; (3) a *free reward* condition, in which a liquid reward was delivered at unpredictable times (5.5-8.5 s). Both monkeys received extensive training on the two tasks having a fixed time interval between the predictive cue and the trigger stimulus or reward. To study the effect of changes in the temporal predictability of stimuli, the interval between instruction and trigger stimuli was prolonged to 4.5 s in the instrumental task and the reward was given earlier (0.3 s) or later (2 s) than its usual time of delivery in the classically conditioned task. The percentage of TANs showing responses was increased when stimuli were less predictable in time, compared to the situations in which the onset time of stimuli was highly predictable. Responses to reward given outside of any task were reduced with repeated liquid delivery at the same 2 s intervals, further suggesting that the temporal predictability of stimuli was an important factor for eliciting neuronal responses. The present results demon-

---

* P. Apicella, S. Ravel, P. Sardo, and E. Legallet, Institut de Neurosciences Physiologiques et Cognitives, CNRS, 31 chemin Joseph Aiguier, 13402 Marseille cedex 20, France. P. Sardo, Istituto di Fisiologia Umana, Università di Palermo, Corso Tukőry 129, 90134 Palermo, Italy.

The Basal Ganglia VI
Edited by Graybiel *et al.*, Kluwer Academic/Plenum Publishers, 2002

strate that the efficacy of stimuli to modulate the firing of TANs is determined both by the motivational relevance of stimuli and by predictions about stimulus timing.

1. INTRODUCTION

The tonically active neurons (TANs) in the monkey striatum constitute a particular class of neurons showing characteristic phasic responses to stimuli associated with reward in the context of a behavioral task[1-4] and to innately appetitive stimuli[5-6]. Although it has been proposed that TANs are involved in retention of learned stimulus-response associations[7], the functional significance of these stereotyped, homogeneous changes in firing remains uncertain. It is possible that the responses of TANs reflect brief changes in the animal's vigilance level. However, it was found that TANs lack responses to novel auditory stimuli that presumably generate a behavioral orienting reaction[3,6]. This argues against a simple relationship of the TAN responses to shifts in general arousal. Recently, we reported that TANS respond to a mildly aversive air puff stimulation, as well as to a liquid reward delivered passively outside of a task, suggesting that these neurons are not specialized for processing appetitive stimuli[6]. A striking feature of the TANs is that their responses are selective for specific behavioral contexts. It was soon found that the responses of TANs depend critically on the reinforcement association of the stimuli[1,3,8,9] and further studies have shown that the neuronal responses to conditioned stimuli and to primary rewards themselves also show specificities depending on the context in which they occur[2,5]. More recently, our research has been directed toward the contribution of stimulus predictability as a possible determinant of the context dependency of the TAN responses. A particularly interesting finding is that the responses of TANs to stimuli that triggered movements were diminished or abolished in the presence of an instruction cue given prior to the presentation of the trigger stimulus[10]. The objective of the present study was to examine whether the responsiveness of the TANs is dependent on variations in the temporal occurrence of motivationally relevant stimuli presented in a variety of behavioral states, thus extending previous work[5,10]. As a result of these experiments it is suggested that this particular class of striatal neurons plays an essential role in the detection of motivationally relevant sensory events according to their predictability in time.

2. METHODS

Two male macaque monkeys (*Macaca fascicularis*), weighing 5 and 6 kg, were used in the present experiments. All experimental procedures were conducted according to NIH guidelines and the French laws on animal experimentation. The monkey sat in a Plexiglas restraint box designed and built in our laboratory, facing a vertical panel placed at reaching distance of the arm. A two-colored LED (red/green) and a contact-sensitive metal knob were mounted in the middle of the panel, at eye level of the animal. The monkey was trained to keep the hand on a central bar located on the lower part of the panel. At the appearance of the red light, the animal was required to release the bar to touch the knob directly below the light. The position of the trigger remained constant thus eliciting the same reaching movement across trials. Correct responses were reinforced by a small amount of apple juice (0.3 ml) delivered through a metal tube that ended in front of the monkey's mouth. The reaction time (RT) was measured as the time between trigger onset and bar release. Each animal was at first trained on two task conditions: (1) in the uncued condition, we presented the trigger stimulus alone and at unpredictable times; (2) in the cued condition, the onset of a green light was followed after a fixed 1.5 s interval by the presentation of the trigger stimulus. Once the monkeys had been overtrained with the same 1.5 s interval, they were subjected to an additionnal cued condition in which the fixed instruction-trigger interval was prolonged to 4.5 s. This condition varied the prediction that the upcoming trigger will occur at a specific time after the instruction onset. These different task conditions were presented to the monkey in separate blocks of 30-40 trials. In a classically conditioned task, the contact of the hand with the bar was prevented by closing the sliding door at the front of the restraining box and the monkey remained motionless, holding his arms relaxed in a natural position. A red light appeared at the same location across trials and was followed after a fixed 1 s interval by the delivery of liquid reward, independent of the monkey's behavior. Since both monkeys were given extended training with the same 1 s interval, the visual signal acted as a reliable cue signaling the delivery of reward. On some occasions, the duration of the interval between the visual signal and reward was shortened to 0.3 s or lengthened to 2 s . As in the instrumental task, the temporal relationships of task stimuli were fixed in each block of trials. In a third behavioral situation, referred to here as the *free reward* condition, only the fruit juice was administered without engaging the monkey in any specific task. In this condition, liquid was delivered at irregular intervals of 5.5-8.5 s in the absence of any external predictive cues. In a variation of this condition, liquid was repeatedly delivered at constant 2 s intervals.

After the monkeys were fully trained in the uncued and cued 1.5-s conditions, they underwent aseptic surgery to implant a stainless steel recording chamber to the skull under sodium pentobarbital anesthesia (35 mg/kg, iv). Standard electrophysiological techniques for extracellular single neuron recording were used during a period of 8-10 months per animal to record from the putamen and the caudate nucleus. Striatal neurons were categorized as TANS on the basis of striking and consistent electrophysiological characteristics, such as their spontaneous firing rates and their typical spike morphology (Alexander & DeLong, 1985; Aosaki *et al.*, 1994; Apicella *et al.*, 1997). Presentation of the visual stimuli and reward and acquisition of the behavioral and neuronal data were controlled by a computer. A Wilcoxon signed-rank test was used to statistically analyze changes in neuronal activity after stimulus presentation (Apicella *et al.*, 1997). Neuronal responsiveness was evaluated in terms of proportions of neurons showing significant changes in this test ($P < 0.05$). Differences in proportions of responding neurons in the different conditions were compared with chi-square tests. To confirm our recording sites, we made small electrolytic lesions (10 µA for 10-20 s, cathodal current) at several points along selected microelectrode tracks. After completion of neuronal recording, monkeys were sacrificed with an overdose of pentobarbital and the marking lesions were histologically verified and all recording sites were reconstructed.

3. RESULTS

3.1 Performance in the Reaction Time Task

In both monkeys, RTs were significantly shorter in the cued 1.5-s condition than in the uncued condition, demonstrating that animals used the instruction as a referent for predicting the onset time of the trigger stimulus and for preparing the associated response. When the instruction-trigger interval was prolonged to 4.5 s, RTs were significantly longer than in the cued 1.5-s condition. The RTs in the cued 4.5-s condition remained essentially unchanged throughout the whole period of neuronal recording. This result indicates that monkeys did not predict the exact time of trigger onset as much as they did with the highly practiced 1.5 s interval.

3.2 Responses to the Trigger Stimulus in the Instrumental Task

A total of 100 neurons were studied while the monkey performed the instrumental task under the two basic conditions and the cued 4.5-s condition. Of these, 77 (77%) and 40 neurons (40%) responded to the trigger stimulus in the uncued and the cued 1.5-s conditions, respectively. The number of neurons responding to the trigger was increased to 68

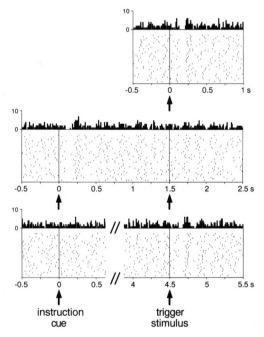

Figure 1. Response to instruction and trigger stimuli influenced by task condition. The activity of the same tonic striatal neuron was recorded while the monkey was performing the RT task under three conditions. Top and middle: uncued and cued 1.5-s conditions, respectively. Bottom: cued 4.5-s condition. The change of testing condition occurred over three successive blocks of ~30 trials. Each dot represents one neuronal impulse, and each line of dots, the neuronal activity occurring during a single trial. Rasters and perievent histograms are aligned on the presentation of the trigger stimulus in the presence and absence of the instruction cue preceding trigger onset by a fixed 1.5 s or 4.5 s interval. Vertical calibration, 10 impulses/bin; binwidth, 10 ms.

(68%) when the instruction-trigger interval was lengthened to 4.5 s. Frequencies of neurons showing trigger responses varied significantly in the cued 4.5-s condition, as compared to the cued 1.5-s condition (X2 test, P < 0.05), but not between the cued 4.5-s and the uncued conditions (P > 0.05). Interestingly, the increased responsiveness to the trigger stimulus was paralleled by a decrease in the proportion of neurons that responded to the instruction stimulus when passing from the usual 1.5 s interval (56 of 100 neurons; 56%) to the longer interval (40 of 100 neurons; 40%) (P < 0.05). Figure 1 illustrates how the trigger stimulus became effective for eliciting responses when it appeared more distant in time from the instruction. Conversely, this same neuron completely abolished its responses to the instruction when the monkey was shifted from the usual 1.5 s interval to the longer interval.

3.3 Responses to Reward in the Classically Conditioned Task

We studied the effect of delivering reward 1 s after the onset of the visual signal on neuronal activity in a total of 29 neurons. Of these, only 8 neurons (28%) showed changes in response to reward. The relationship between the neuronal responses and the time elapsing between the visual signal and reward was investigated in this condition by changing the duration of the interval. The proportion of reward responses was increased when the normally employed signal-reward interval was shortened to 0.3 s (9 of 17 neurons; 53%) or lengthened to 2 s (18 of 21 neurons; 86%). Conversely, 27 neurons (93%) showed responses to the visual signal predicting the time of reward at the end of the 1 s interval and the proportions of neurons responding to the visual signal were not affected by the manipula-

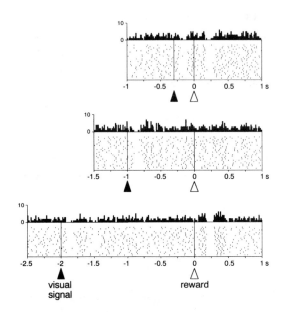

Figure 2. Influence of changing the usual time interval between the visual signal and reward. Data from three testing conditions are ordered from top to bottom in the sequence in which they were administered. Conventions are the same as in Fig. 1. A response to reward became apparent upon changing the usual 1 s interval between the visual signal and reward. This neuron remained responsive to the first stimulus in all three conditions.

tions of the signal-reward interval (0.3 s interval: 16 of 17 neurons; 94%; 2 s interval: 21 of 21 neurons; 100%). Figure 2 shows an example of a neuron tested with the visual signal presented at the three intervals before reward. The reward normally delivered 1 s after the visual signal did not elicit a response. When reward was delivered 0.3 s after the signal, the same neuron responded to reward. When reward was given 2 s after the signal, the neuron also showed a response to reward. The fact that responses to both the visual signal and reward were present with the 0.3 s interval indicated that the reduced responsiveness to reward observed with the 1 s interval did not represent inherent limitations of the neurons to process new information after the first stimulus. Although we cannot rule out a relation to the preparation for consummatory motor behavior, it seems likely that the change in responsiveness was due to differences in the temporal predictability of reward.

3.4 Responses to Reward Delivered Outside of a Task

The activity of 128 neurons was examined in relation to the delivery of free reward in the absence of any predictive cues. All neurons in this sample were also tested when the interval between liquid deliveries was held constant and shortened to 2 s. The proportions of neurons responding to reward were 70% and 55% in the irregular and regular 2-s conditions, respectively. As with the task conditions, responses largely disappeared when the monkey became able to precisely predict the time of reward. The neuron illustrated in Figure 3 gave a response when the liquid came at irregular intervals of 5.5-8.5 s, while the response disappeared during the course of repeated trials with the same 2 s intervals. In the latter condition, the response was consistent from the beginning of the testing session but gradually declined after the first 10-13 trials. Such a gradual decrease would be expected as the time of liquid delivery became fully predicted with repeated stimulus presentation.

Histological reconstructions showed that recordings were obtained from TANS distributed throughout the striatum, mostly at the levels of the dorsal and medial regions of the caudate nucleus and putamen (Apicella et al., 1998).

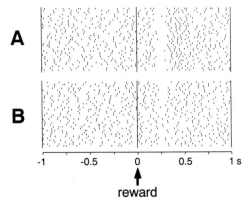

Figure 3. Disappearance of responses to reward while increasing the rate and regularity of liquid deliveries outside of a task. This tonic striatal neuron responded when the fruit juice was delivered once every 5.5-8.5 s (A), but not in trials in which the same liquid was given every 2 s (B). In this latter condition, the neuronal response gradually disappeared over the course of the testing period. The rasters are aligned on the delivery of reward and the trials are presented in the original order from top to bottom.

4. DISCUSSION

In this report, we provide evidence that sensory events modulated many TANs of the striatum if monkeys did not predict exactly when stimuli will occur, whereas the neuronal responsiveness was reduced if the onset time of stimuli was highly predictable. In the instrumental task, RT performance was used as a quantitative, objective measure of different levels of predictability about the timing of the trigger stimulus. Since the instruction-trigger interval was fixed within a block of trials, the monkey must be able to predict the moment of presentation of the trigger stimulus with either the 1.5 s or 4.5 s intervals, unless it would be more difficult to estimate a longer time interval than the usual 1.5 s interval. However, the extensive training with the same 1.5 s interval potentially had allowed the acquisition of a strong habitual temporal association between instruction and trigger stimuli which rendered the timing of the trigger highly predictable as soon the monkey detected the instruction. It was in this condition that RTs were faster than those in the other conditions and that a low proportion of TANS responded to the trigger stimulus. On the other hand, there was an increase in RT when the interval was prolonged to 4.5 s and neuronal responsiveness to the trigger stimulus increased concurrently. This might reflect the monkey's inability to take advantage of the predictability of the trigger stimulus when it was switched from the well-practiced 1.5 s interval to a longer interval. In the classically conditioned task, habitual occurrence of reward at the end of a fixed 1 s interval after the visual signal did not constitute an optimal condition to elicit reward responses, while changes in the accustomed time of reward delivery were paralleled by an increased responsiveness to reward. Finally, a majority of TANs tested in the *free reward* condition responded to liquid when it was delivered randomly in time, while a large number of neurons became unresponsive with repeated liquid delivery at the same 2 s intervals. Together, these results suggest that the sensitivity of TANs is dependent on mechanisms by which the timing of stimuli is predicted thus demonstrating that the learned or innate motivational value of stimuli is not the sole variable controlling responses of this class of striatal neurons.

Although it is clear that the temporal predictability of stimuli affects the responsiveness of TANs, other factors, such as preparation for a particular movement and nonspecific arousal also may have an influence. However, the modulatory influence of temporal context upon neuronal responsiveness was also evidenced when monkeys did not have to organize actively a motor response, therefore arguing against interpretations based on motor aspects of readiness. The question still remains whether a general level of alertness could account for the present results. Nevertheless, despite the observation that some TANs can respond to stimuli that are in themselves intense or conspicuous (Aosaki et al., 1994), a role of these neurons in regulating attention to the external environment does not appear to be supported by available experimental evidence (Ravel et al., 1999). Further studies are necessary, however, to better determine whether the sensitivity of TANS to temporal conditions depended on the animal's vigilance level and overall alertness.

Our results show that TANS respond to instruction stimuli, in agreement with data obtained before (Apicella et al., 1991; 1998). In the instrumental task, the relevance of the instruction cue as a predictor of the time of the subsequent trigger may explain these responses. Interestingly, the neuronal responsiveness to the instruction cue was reduced if the interval was too long to allow the monkey to predict precisely the moment of the trigger onset. Conversely, in the classically conditioned task, the increased neuronal responsiveness when the timing of reward was made less predictable was not paralleled by a reduction of neuronal responses to the visual signal predictive of reward. These findings indicate that the responses of TANs to predictive cues appear to be dependent on the underlying associative process for learning (classical vs. instrumental conditioning).

In conclusion, one of the novel feature introduced by our data is the interplay between the relevance of stimuli and temporal context in which they are presented that may be important for the responsiveness of TANS. The present findings suggest that TANS need to be phasically modulated in conditions in which external cues and events are not well organized and, therefore, lack an overall level of predictability. Since TANS are supposed to be the cholinergic interneurons of the striatum (Kimura et al., 1990; Wilson et al., 1990; Aosaki et al., 1995), one would expect the pause in discharge of these presumed local circuit neurons to modulate the activity of surrounding neurons of the medium spiny type projecting to the output structures of the basal ganglia (Aosaki et al., 1995). In particular, we postulate that the TAN response may promote or prevent the projection neurons from becoming active in relation to sensory events and associated behavioral reactions other than the most common or most automatic reaction for a given context. The present findings are of interest in view of hypotheses concerning the function of the basal ganglia in procedural representations underlying habit performance and certain motor skills (Marsden, 1982, Mishkin et al., 1984; Graybiel, 1995). Impairment of procedural learning has been described in patients with either Parkinson's or Huntington's disease (Saint-Cyr et al., 1988; Harrington et al., 1990; Knopman & Nissen, 1991; Knowlton et al., 1996) and functional neuroimaging studies have reported activations of the striatum during learning and automatization of sequential movements (Jenkins et al., 1994; Jueptner et al., 1997a; 1997b). Since learning about the temporal relationship among external cues and events is an important factor in the automatic execution of motor acts, TANs might be the essential constituents of striatal networks that intervene in the development of behavioral sequences that follow predictable patterns in a habit-like manner.

5. ACKNOWLEDGEMENTS

We thank R. Massarino and C. Wirig for expert technical assistance. This work was supported by grants from the European Commission (CHRX-CT94-0463, BMH4-CT95-0608).

6. REFERENCES

Alexander, G. E. & DeLong, M. R. (1985) Microstimulation of the primate neostriatum. II. Somatotopic organization of striatal microexcitable zones and their relation to neuronal response properties. *Journal of Neurophysiology* 53, 1417-1430.

Aosaki, T., Tsubokawa, H., Ishida, A., Watanabe, K., Graybiel, A. M. & Kimura, M. (1994) Responses of tonically active neurons in the primate's striatum undergo systematic changes during behavioral sensorimotor conditioning. *Journal of Neuroscience* 14, 3969-3984.

Aosaki, T., Kimura, M. & Graybiel, A. M. (1995) Temporal and spatial characteristics of tonically active neurons of the primate's striatum. *Journal of Neurophysiology* 73, 1234-1252.

Apicella, P., Schultz, W. & Scarnati, E. (1991) Tonically discharging neurons of monkey striatum respond to preparatory and rewarding stimuli. *Experimental Brain Research* 84, 672-675.

Apicella, P., Legallet, E. & Trouche, E. (1996) Responses of tonically discharging neurons in monkey striatum to visual stimuli presented under passive conditions and during task performance. *Neuroscience Letters* 203, 147-150.

Apicella, P., Legallet, E. & Trouche, E. (1997) Responses of tonically discharging neurons in the monkey striatum to primary rewards delivered during different behavioral states. *Experimental Brain Research* 116, 456-466.

Apicella, P., Ravel, S., Sardo, P. & Legallet, E. (1998) Influence of predictive information on responses of tonically active neurons in the monkey striatum. *Journal of Neurophysiology* 80, 3341-3344.

Graybiel, A. M. (1995) Building action repertoires: memory and learning functions of the basal ganglia. *Current Opinion in Neurobiology* 5,733-741.

Graybiel, A. M., Aosaki, T., Flaherty, A. W. & Kimura, M. (1994) The basal ganglia and adaptive motor control. *Science* 265, 1826-1831.

Harrington, D. L., York Haaland, K., Yeo, R. A. & Marder, E. (1990) Procedural memory in Parkinson's disease. Impaired motor but not visuoperceptual learning. *Journal of Clinical and Experimental Neuropsychology* 12, 323-339.

Jenkins, I. H., Brooks, D. J., Nixon, P. D., Frackowiak, R. S. J. & Passingham, R. E. (1994) Motor sequence learning: A study with positron emission tomography. *Journal of Neuroscience* 14, 3775-3790.

Jueptner, M., Stephan, K. M., Frith, C. D., Brooks, D. J., Frackowiak, R. S. J. & Passingham, R. E. (1997a) Anatomy of motor learning. I. Frontal cortex and attention to action. *Journal of Neurophysiology* 77, 1313-1324.

Jueptner, M., Frith, C. D., Brooks, D. J., Frackowiak, R. S. J. & Passingham, R. E. (1997b) Anatomy of motor learning. II. Subcortical structures and learning by trial and error. *Journal of Neurophysiology* 77, 1325-1337.

Kimura, M., Rajkowski, J. & Evarts, E. V. (1984) Tonically discharging putamen neurons exhibit set dependent responses. *Proceedings of the National Academy of Sciences of the United States of America* 81, 4998-5001.

Kimura, M. (1986) The role of primate putamen neurons in the association of sensory stimuli with movement. *Neuroscience Research* 3, 436-443.

Kimura, M., Kato, M. & Shimazaki, H. (1990) Physiological properties of projection neurons in the monkey striatum to the globus pallidus. *Experimental Brain Research* 82, 672-676.

Knopman, D. & Nissen, M. J. (1991) Procedural learning is impaired in Huntington's disease: Evidence from the serial reaction time task. *Neuropsychologia* 29, 245-254.

Knowlton, B. J., Mangels, J. A. & Squire, L. R. (1996) A neostriatal habit learning system in humans. *Science* 273, 1399-1402.

MARSDEN, C. D. (1982) The mysterious motor function of the basal ganglia : the Robert Wartenberg Lecture. *Neurology* 32, 514-539.

Mishkin, M., Malamut, B. & Bachevalier, J. (1984) Memories and habits: two neural systems. In *Neurobiology of Learning and Memory* (edited by Lynch, G., McGaugh, J. L. & Weinberger, N. M.) pp. 65-77. New York: Guilford Press.

Ravel, S., Legallet, E. & Apicella, P. (1999) Tonically active neurons in the monkey striatum do not preferentially respond to appetitive stimuli. *Experimental Brain Research* 128, 531-534.

Raz, A., Feingold, A., Zelanskaya, V., Vaadia, E. & Bergman, H. (1996) Neuronal synchronization of tonically active neurons in the striatum of normal and parkinsonian primates. *Journal of Neurophysiology* 76, 2083-2088.

Saint-Cyr, J. A., Taylor, A. E. & Lang, A. E. (1988) Procedural learning and neostriatal dysfunction in man. *Brain* 111, 941-959.

Wilson, C. J., Chang, H. T. & Kitai, S. T. (1990) Firing patterns and synaptic potentials of identified giant aspiny interneurons in the rat neostriatum. *Journal of Neuroscience* 10, 508-519.

Section V

NEW INSIGHTS INTO THE ANATOMY OF THE BASAL GANGLIA

DISTRIBUTION OF SUBSTANTIA NIGRA PARS COMPACTA NEURONS WITH RESPECT TO PARS RETICULATA STRIATO-NIGRAL AFFERENCES: COMPUTER-ASSISTED THREE-DIMENSIONAL RECONSTRUCTIONS

Contribution of 3D modeling to neuroanatomical studies

B. Banrezes[1], P. Andrey[1], A. Menetrey[2], P. Mailly[3], J.-M. Deniau[2], and Y. Maurin[1]

Abstract: Following ionophoretic injections of wheat germ agglutinin coupled to horseradish peroxidase into identified functional striatal territories of the rat brain, the distribution of anterogradely labelled striatonigral projections and retrogradely labelled nigrostriatal neurons was studied by three-dimensional modeling. Our models confirmed and extended the organizational scheme of striatonigral projections, already reported by our laboratory. The relationship between striatonigral projections and nigrostriatal neurons was also studied. For each striatal injection site, two subpopulations of labelled nigral neurons were distinguished by their position with respect to the striatal projection field. The first one occupied a proximal position, in register with the labelled striatal projections, while the second one was more distal. The subpopulations of proximal neurons innervating different functional striatal sectors were segregated in the mediolateral, dorsoventral and rostrocaudal directions, while the distal neurons, more scattered, showed a lesser degree of spatial segregation. These results suggest that the substantia nigra might control the flow of cortical information through the striatum via two different modalities based respectively on a closed nigro-striatal loop involving the proximal neurons, and an open loop involving the distal ones. We propose an organization scheme which may account for both the segregation and the intermingling of cell populations innervating distinct striatal sectors. Finally, the contribution of 3D modeling to neuroanatomical studies is discussed, along with the perspectives towards more quantitative and predictive 3D modeling.

[1] Analyse et Modélisation en Imagerie Biologique, INRA JE 1060/77, 78352 Jouy-en-Josas, France.
[2] INSERM U 114, Chaire de Neuropharmacologie, Collège de France, 75231 Paris Cedex 05, France.
[3] Neurobiologie des signaux Intercellulaires, CNRS UMR 7101, Université Pierre et Marie Curie, 9 Quai St Bernard, 75252 Paris Cedex 05, France.

1. INTRODUCTION

It is now generally accepted that the flow of cortical information that enters the basal ganglia at the level of the striatum is processed along the striatonigral pathway in a parallel manner[3,4,8,9,13-15,22,28]. In other words, the functional districts of the cortical mantle have their counterpart in the striatum, and this mapping is preserved at the level of the substantia nigra (SN) and of the internal segment of the globus pallidus[1,2,8,25], although this segregation does not exclude a certain degree of convergence[26]. It has recently been shown[7] that these parallel cortico-striato-nigral functional channels determine, in the substantia nigra pars reticulata (SNR) a set of curved functional layers centered around the striatal projections affiliated to the orofacial sensorimotor cortical areas. As compared to this well characterized functional nigral mosaic, the spatial organization of nigrostriatal neurons is far from clear. Early studies proposed a precise topography of nigrostriatal projections[5,11,16] while subsequent reports have shown intermingling populations of nigral neurons projecting on different striatal regions[10,20-22,29]. Topographical studies[12,19] as well as studies carried out on a functional basis[20,21] support the general idea that the transfer of information along the nigrostriatal pathway shows a lower degree of selectivity than along the cortico-striato-nigral one. Since the major role of the nigrostriatal pathway is to control the integration of cortical information within the basal ganglia, it seemed to us important to address the question as to whether a topographical organization of nigrostriatal neurons exists, which would reflect the compartmentation of the striatonigral system. This issue had not been addressed by Deniau and coworkers in their study on the compartmentation of the descending pathway[6,7], partly because of the lack of a three-dimensional (3D) representation tool. In the present report, we have taken advantage of a 3D reconstruction software developed in our laboratory[27] to address this issue and analyse the spatial distribution of retrogradely labeled nigral neurons following injections of wheat germ agglutinin coupled to horseradish peroxidase (WGA-HRP) into restricted striatal regions associated to characterized functional cortical areas.

2. THREE DIMENSIONAL ORGANIZATION OF NIGROSTRIATAL NEURONS

Serial coronal sections through the SN, obtained 48 hours after microionophoretic injection of WGA-HRP were digitized as mosaics of high resolution 2D images. These digitized mosaics were subsequently segmented to delineate the SNR and the striatal projection fields and to plot the labelled neurons. Following rigid 2D registration along the Z-axis, the stack of images was concatenated to yield 3D wire-frame or solid models such as the ones presented in Figure 1.

Figure 1A illustrates the processing of 2D mosaics before the calculation of 3D models. This mosaic corresponds to the assembly of 16 individual 512x512 pixel images resulting from the digitization of a serial section through the SN after WGA-HRP injection in the orofacial sensorimotor region of the striatum (see the schematic drawing). Digitization was made by means of a digitization board connected to a CCD camera fitted on a microscope equiped with a motorized stage, as described elsewhere[23,27]. The SN is delineated by the outer black line, the striatal projection field by the inner grey one, and the neurons marked by white squares. Subpopulations of proximal and distal neurons were identified according to the following rule: on each mosaic, the separation between the SNR and the SN pars compacta

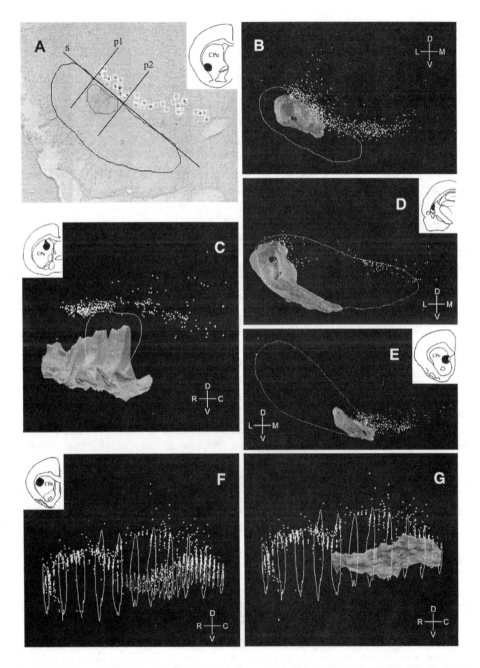

Figure 1. Spatial organization within the SN of anterogradely labelled striatal projections and retrogradely labelled neurons following WGA-HRP injections into five different functional regions of the striatum: orofacial sensorimotor (A, B), visuo-oculomotor (C), auditory-visual (D), dorsal prelimbic (E) and limb sensorimotor (F, G)

(SNC) was delimited (line S). This separation was orthogonally intersected by two tangents (p1 and p2) drawn at the lateral and medial limits of the striatal projection field. The neurons comprised between these two tangents were considered as proximal while the others were defined as distal. The segmentation procedure was carried out on all the mosaics of the stack.

The resulting 3D model is shown in B. For sake of clarity, the SNR is represented by only one of its contours, located in the middle of its antero-posterior extension (white line). The striatal projection field is displayed as the solid grey volume resulting from the triangulation of all the segmentation lines. The pointed nigral neurons are shown as white dots. The model is represented under a coronal orientation from its caudal end, with the lateral part of the SN on the left and the dorsal part on top, as indicated by the orientation cross on the top right of the figure. In this injection case, the striatal projection field occupied a latero-dorsal position within the SNR. The retrogradely labelled neurons adopted a curved lamellar shape extending, along the latero-medial axis, from the lateral border of the striatal projection field to the ventral tegmental area (VTA). The proximal neurons composed the densest subpopulation, in close apposition to the striatal projection field, while the distal neurons were more scattered, especially in the vicinity of the medial border of the striatal projection field. These distal neurons were mainly located in the dorsal part of the SNC and extended relatively far into the VTA. When presented under a horizontal orientation (not shown), the two subpopulations appeared to be present through the whole rostrocaudal extent of the SN, and were separated by large zones of discontinuity.

Figure 1C shows the 3D model corresponding to the injection of WGA-HRP into the visuo-oculomotor district of the striatum (see drawing). The SNR, the striatal projection field and the nigrostriatal neurons are represented according to the same conventions as in B. Under this quasi-sagittal orientation, the two subpopulations of neurons were clearly distinguishable, contrary to the coronal orientation (not shown) in which they appeared as a unique cell group. The proximal subpopulation, rather dense, was located over the rostral region of the striatal projection field, above its medial end. The distal one, located in the caudal part, was less numerous, and more scattered in the medial direction. Interestingly, from a horizontal point of view (not shown) the main axis of the two subpopulations of nigrostriatal neurons and of the striatal projection field were parallel, being oriented rostromedially to caudolaterally.

The 3D model related to the auditory-visual striatal district is shown in D, under a coronal orientation. The proximal neurons were almost all located in the lateral part of the SNR, in close apposition to the thickest part of the striatal projection field. The distal neurons were rather scarce over the medial part of the SNR, the two subpopulations being separated by a large gap devoid of neurons.

The 3D model calculated after injection of WGA-HRP in the dorsal prelimbic region of the striatum is shown in E under a coronal orientation. The two subpopulations of proximal and distal neurons are much less distinguishable than in the other injection cases. The striatal projection field was located in the middle of the SNR along the dorsoventral direction, beneath the ventral prelimbic/medial ventral orbital (VP/MVO) afference territory. The proximal neurons, which lied in close apposition to the striatal afferences, were localized within the SNR itself, being thus segregated from the VP/MVO-affiliated neurons, which occupied a dorsal position beyond the SNC (not shown). The distal neurons extended medially in the VTA.

The 3D model related to the limb sensorimotor striatal district is shown in F and G, under a quasi sagittal orientation. The striatal projection field was confined to the caudal half of the SNR. Contrary to other injection cases, proximal and distal neurons were segregated along the rostrocaudal axis, the distal neurons being localized above the rostral half of the SNR. These distal neurons did not extend beyond the medial limit of the SNR. The surface representation of the striatal projection field (G) which masks a large part of the proximal neurons (compare with the wire-frame representation in F) reveals that these neurons were largely located within the projection field itself.

3. FUNCTIONAL CONSIDERATIONS

Our 3D models revealed several organization rules of the nigrostriatal neurons with respect to the striatonigral projection fields.

1. along the medio-lateral axis, no neurons were found beyond the lateral limit of their corresponding striatal projection field, whatever the mediolateral position of the latter
2. for each striatal injection site, at least two populations of labelled neurons could be distinguished. The first (and major) one was situated in the immediate vicinity of the striatal projection field while the second one was located at distance, dorsomedially above the SNR or in the VTA
3. the proximal neurons also conformed to a dorso-ventral organization rule, being always located above or inside their correponding striatal projection field. A striking illustration of this rule is given by the proximal subpopulations of neurons projecting to the dorsal prelimbic and the limb sensorimotor regions of the striatum, the cell bodies of which penetrated deep inside the SNR, in close apposition with the striatal projection fields
4. the distal neurons, which also followed the first rule stated above, were always segregated from the proximal ones along at least one of the three main spatial directions. As a whole, they occupied a dorsal position within the SNC and the VTA.

These organization rules strongly suggest the existence of a spatial compartmentation of nigrostriatal proximal neurons (and to a lesser degree of distal ones), according to the functional striatal territory they innervate. In this respect, this organization is consistent with the parallel processing of cortical information within the basal ganglia. In addition, inasmuch as distal neurons affiliated to a given striatal functional territory may intermingle with the proximal neurons affiliated to another one, this organization also supports the hypothesis of the integration of informations from different functional regions. Finally, distal neurons affiliated to sensorimotor striatal regions were also found in the VTA, which is known to receive an innervation from structures associated with the limbic system[18]. This arrangement may thus participate to a functional link between the limbic/prefrontal and the sensorimotor channels of the striatum. As a whole, these conclusions are summarized in Figure 2, which proposes an organization scheme for functional anatomy relationships between the striatum, the SN-VTA and structures associated with the limbic system.

4. CONTRIBUTION OF 3D MODELING

Deniau and coworkers[7] have proposed that striatal projections in the SNR adopt an "onion-like" structure composed, in the coronal plane, of concentric curved lamellae centred around the orofacial sensorimotor projection field. The possibility to display 3D models under any orientation provide additional informations on the spatial organization of the striatonigral projections. This is particularly true for the longitudinal segregation of the visuo-oculomotor and the limb sensorimotor regions, the former being mostly rostral and the latter caudal. This is also true for the diagonal orientation (rostromedially-caudolaterally) of the visuo-oculomotor and of the ventral prelimbic projection fields. In this respect, 3D modeling, which confirmed that the segregation of functional channels along the corticostriatal pathway is maintained at the level of the SNR, also revealed that this segregation is not only organized in the coronal plane, as previously shown[7], but more generally in a three dimensional space.

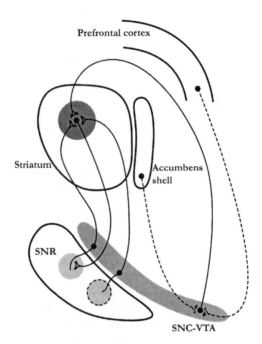

Figure 2. Hypothetical scheme for the functional anatomy of striatonigral relationships. A given functional striatal region (dark grey circle) projects to a restricted region of the SNR (light grey circle) and receives afferents from SNC-VTA neurons. These neurons may be proximal ones, located above the striatal projection field, giving then rise to a closed striatonigral regulatory loop. They may also be distal neurons, located in the SNC over projection fields associated with other functional striatal territories (dashed grey circle), then determining open regulatory loops[24] between different functional channels. Finally, they may be distal neurons situated in the VTA, thus allowing the integration with informations from the prefrontal cortex or the shell of the nucleus accumbens (dashed lines).

Regarding the nigrostriatal neurons, no clear picture emerged from the studies[5,10-12,16,19-22,29] that have been devoted to their spatial organization, possibly because most of them have been carried out irrespective of the detailed functional compartmentation which exists in the basal ganglia. In this respect, tracing experiments based upon WGA-HRP injections into functionally identified striatal regions[7] was a step towards a comprehensive view of the striatonigral relationships. However, the lack of a 3D modeling tool was a major obstacle for the visualization and the analysis of the distribution of nigrostriatal neurons. Three dimensional modeling revealed an information that was latent in the initial coronal 2D data, such as the relative 3D positions of neurons belonging to distinct serial sections. Indeed, although the reconstruction process is no more than the concatenation of a set of 2D data, it integrates the position along the Z-axis, and therefore unmasks spatial information. Furthermore, starting from sections cut along one of the 3 major directions (coronal, sagittal or horizontal), 3D models can be displayed under any of the other two. Finally, a 3D modeling tool also permits the search of the most relevant orientation (see Figure 1C).

Thus, our 3D models disclosed a spatial organization of nigrostriatal neurons that could not be perceived from the examination of 2D images. The organization rules proposed here may lead to reconsider the status of nigrostriatal neurons located within the SNR such as those affiliated to the limb sensorimotor region of the striatum. The existence of such neurons have been reported by numerous authors[10,17,20-22,29], and have sometimes been considered as a ventrally displaced population of SNC neurons[12]. Our 3D models indicate that these neurons are in fact proximal neurons which are in keeping with a logic of proximity with regard to the corresponding striatal projection fields. In this respect, 3D modeling appears to be a valuable conceptual assistant in the analysis and interpretation of biological data, which does not limit itself to being a representation tool.

5. PERSPECTIVES

All the 3D models presented here resulted from injections of WGA-HRP in the striatum of as many rats. As a consequence, the hypothesis that are put forward regarding the relationship between entities belonging to different models are based upon the subjective fusion of these models. In order to demonstrate rigorously such relationships, it is necessary to place these entities into a unique 3D model, and thus to proceed to the fusion of the individual models. This is not a trivial problem, since it cannot be solved by their simple superimposition, but requires that the models be deformed to account for the morphological interindividual variability.

Another limit of our present 3D modeling tool resides in the difficulty to evaluate the experimental and interindividual variabilities. As a matter of fact, even by building several nigral 3D models of the same entity, our modeling tool does not allow neither the quantitation of their differences nor the building of an average model. Here again, the prerequisite is the fusion of individual models, which should broaden even more the usefulness of 3D modeling for neuroanatomical investigations.

6. REFERENCES

1. Afifi A.K. (1994) Basal ganglia: functional anatomy and physiology. Part 1. *J. Child Neurol.*, **9**, 249-260.
2. Afifi A.K. (1994) Basal ganglia: functional anatomy and physiology. Part 2. *J. Child Neurol.*, **9**, 352-361.
3. Alexander G.E. and Crutcher M.D. (1990) Functional architecture of basal ganglia circuits: neural substrates of parallel processing. *Trends Neurosci.*, **13**, 266-271.
4. Alexander G.E. Crutcher M.D. and DeLong M.R. (1990) Basal ganglia-thalamocortical circuits: parallel substrates for motor, oculomotor, "prefrontal" and "limbic" functions. *Prog. Brain Res.*, **85**, 119-146.
5. Carter, D.A. and Fibiger, H.C. (1977) Ascending projections of presumed dopamine-containing neurons in the ventral tegmentum of the rat as demonstrated by horseradish peroxidase. *Neuroscience* **2**, 569-576.
6. Deniau, J.M. and Chevalier, G., (1992) The lamellar organization of the rat substantia nigra pars reticulata: distribution of projection neurons. *Neuroscience* **46**, 361-377.
7. Deniau, J.M., Menetrey, A. and Charpier, S. (1996) The lamellar organization of the rat substantia nigra pars reticulata: segregated patterns of striatal afferents and relationship to the topography of corticostriatal projections. *Neuroscience* **73**, 761-781.
8. Deniau J.M. and Thierry A.M. (1997) Anatomical segregation of information processing in the substantia nigra pars reticulata. In *The Basal Ganglia and new surgical approaches for Parkinson's disease* (eds Obeso J.A., DeLong M.R., Ohye C. and Marsden C.D.), *Advances in Neurology*, Vol. 74, pp. 83-96. Lippincott-Raven Publishers, Philadelphia.
9. Donoghue J.P. and Herkenham M. (1986) Neostriatal projections from individual cortical fields conform to histochemically distinct striatal compartments in the rat. *Brain Res.*, **365**, 397-403.
10. Druga, R. (1989) Nigro-striatal projections in the rat as demonstrated by retrograde transport of horseradish peroxidase. I. Projections to the rostral striatum. *J. Hirnforsch.* **30**, 11-21.
11. Fallon, J.H. and Moore, R.Y. (1978) Catecholamine innervation of the basal forebrain. IV. Topography of the dopamine projection to the basal forebrain and neostriatum. *J. Comp. Neurol.* **180**, 545-580.
12. Fallon J.H. and Loughlin, S.E. (1995) Substantia nigra. In *The Rat Nervous System: A Handbook for Neuroscientists* (eds. Paxinos, G. and Watson, J.), pp. 215-237. Academic Press, Sydney.
13. Goldman-Rakic P.S. and Selemon L.D. (1990) News frontiers in basal ganglia research. *Trends Neurosci.*, **13**, 241-244.
14. Graybiel A.M. and Ragsdale C.W.Jr. (1979) Fiber connections of the basal ganglia. *Prog. Brain Res.*, **51**, 239-283.
15. Groenewegen H.J., Berendse H.W., Wolters J.G. and Lohman A.H.M. (1990) The anatomical relationship of the prefrontal cortex with the striatopallidal system, the thalamus and the amygdala : evidence for a parallel organization. In *Prog. Brain Res.* (eds Uylings H.B.M., Van Eden C.G., De Bruin M.A., Corner M.A. and Feenestra M.G.P.), Vol. 85, pp. 95-118. Elsevier, Amsterdam.
16. Guyenet, P.G. and Aghajanian G.K. (1978) Antidromic identification of dopaminergic and other output neurons of the rat substantia nigra. *Brain Res.* **150**, 69-84.
17. Guyenet, P.G. and Crane, J.K. (1981) Non-dopaminergic nigrostriatal pathway. *Brain Res.* **213**, 291-305.
18. Heimer, L. and Wilson, R.D. (1975) The subcortical projections of the allocortex: similarities in the neuronal association of the hippocampus, the piriform cortex and the neocortex. In *Golgi Centennial Symposium: Perspectives in Neurobiology* (ed. Santini M.), pp. 177-193, Raven, New York.
19. Hontanilla, B., De Las Heras, S. and Gimenez-Amaya, J.M. (1996) A topographic re-evaluation of the nigrostriatal projections to the caudate nucleus in the cat with multiple retrograde tracers. *Neuroscience* **72**, 485-503.
20. Lynd-Balta, E. and Haber, S.N. (1994) The organization of midbrain projections to the striatum in the primate. *Neuroscience* **59**, 609-623.
21. Lynd-Balta, E. and Haber, S.N. (1994) The organization of midbrain projections to the striatum in the primate: sensorimotor-related striatum versus ventral striatum. *Neuroscience* **59**, 625-640.
22. Lynd-Balta, E. and Haber, S.N. (1994) Primate striato-nigral projections: a comparison of the sensorimotor-related striatum and the ventral striatum. *J. Comp. Neurol.* **345**, 562-578.

23. Maurin, Y., Banrezes, B., Menetrey, A., Mailly, P. and Deniau, J.M. (1999) Three-dimensional distribution of nigrostriatal neurons in the rat: relation to the topography of striatonigral projections. *Neuroscience*, **91**, 891-909.
24. Nauta, W.J.H. and Domesick, V.B. (1984) Afferent and efferent relationships of the basal ganglia. In *Functions of the Basal Ganglia*, Ciba Foundation Symposium 107, pp. 3-23, Pitman, London.
25. Parent A. and Hazrati L.-N. (1995) Funtional anatomy of the basal ganglia. I. The cortico-basal ganglia-thalamo-cortical loop. *Brain Res. Rev.*, **20**, 91-127.
26. Parthasarathy H.B., Schall J.D. and Graybiel A.M. (1992) Distributed but convergent ordering of corticostriatal projections: analysis of the frontal eye field and the supplementary eye field in the macaque monkey. *J. Neurosci.*, **12**, 4468-4488.
27. Roesch, S., Mailly, P., Deniau, J.M. and Maurin, Y. (1996) Computer assisted three-dimensional reconstruction of brain regions from serial section digitized images. Application to the organization of striato-nigral relationships in the rat. *J. Neurosci. Methods* **69**, 197-204.
28. Selemon L.D. and Goldman-Rakic P.S. (1985) Longitudinal topography and interdigitation of corticostriatal projections in the rhesus monkey. *J. Neurosci.*, **5**, 776-794.
29. Van der Kooy, D. (1979) The organization of the thalamic, nigral and raphe cells projecting to the medial vs lateral caudate-putamen in rat. A fluorescent retrograde double labeling study. *Brain Res.* **169**, 381-387.

THE IMMUNOCYTOCHEMICAL LOCALIZATION OF TYROSINE HYDROXYLASE IN THE HUMAN STRIATUM AND SUBSTANTIA NIGRA

A postmortem ultrastructural study

Rosalinda C. Roberts*

Abstract: The synaptic organization of tyrosine hydroxylase (TH), the synthesizing enzyme for dopamine, is reviewed for the human striatum and substantia nigra (SN). In the striatum, TH-immunoreactivity [TH-i] was present in myelinated and unmyelinated varicose axons. Both varicosities [0.75-1.5μm] and intervaricose segments [0.2-0.3μm] formed synapses with spines and dendrites. Most synapses formed by TH-i profiles were symmetric axospinous (57-62%) or symmetric axodendritic (33-35%). An occasional asymmetric synapse was observed. Synapses formed by TH-i profiles were short in length [0.226μm], and had non-perforated postsynaptic densities. In the SN, TH-i was present in SN pars compacta neurons, which had large somata and vast cytoplasm, in dendrites, and in myelinated axons. Unlabeled axon terminals with varied morphologies formed synapses with TH-i somata and dendrites. The axon terminals varied in size, vesicle packing density, vesicle morphology and the symmetry of the synapse formed with labeled profiles. The general pattern of striatal TH-i was similar, but not identical, to that of other species. The morphological characteristics of dopaminergic nigral neurons and the synapses on them were similar to those of other species.

1. INTRODUCTION

The dopaminergic nigrostriatal system is critical for the normal function of the basal ganglia, which are involved in motor, cognitive and behavioral functions. Striatal targets of dopaminergic neurons are the medium spiny projection neurons[13,20,23] as well as several types of interneurons, though the inputs to these cell types are not as extensive as to the spiny neurons[1,5,14,17,18,21,32]. Dopaminergic axons, identified by the immunocytochemical lo-

---

* Rosalinda C. Roberts, Maryland Psychiatric Research Center, Department of Psychiatry, University of Maryland School of Medicine, Baltimore MD 21228

calization of its synthesizing enzyme tyrosine hydroxylase (TH), predominantly form symmetric synapses with dendritic spines and shafts[4,9,11,13,16,23,30].

The dopamine neurons in the substantia nigra (SN) can be distinguished from other neurons, based on morphological criteria[12]. The SN receives neurochemically distinct inputs of diverse origin including GABAergic and glutamatergic neurons from several brain regions[3,24]. The morphological features of the terminals forming synapses with dopamine neurons in experimental animals are well characterized and are specific for different inputs[2,3,6,8,24,29].

While the major features of nigral dopaminergic neurons and the synaptic organization of their inputs to the striatum are relatively consistent among various species, few reports exist on the ultrastructural localization of TH in the human brain. Thus, we summarize here our previous[22] and ongoing research on the ultrastructural organization of dopaminergic neurons and synapses in the human striatum and SN. Electron microscopic observations taken from postmortem human brain will be useful for comparative purposes among species, as well as to provide normative data for future studies of the abnormal basal ganglia in human diseases which afflict this brain area.

2. METHODS

Postmortem human brain tissue was obtained from the Maryland Brain Collection. The tissue was collected within seven hours of death from adult control cases with no history of CNS or neurological disease (Table I). Coronal blocks of tissue were immersed in a cold solution of 4% paraformaldehyde and 1% glutaraldehyde in 0.1M phosphate buffer, pH 7.2-7.4 (PB) and stored at 4°C for a period of at least one week. Samples were taken from the head of the caudate and anterior putamen, or the SN.

The tissue was cut with a vibratome and free floating 40μm thick sections were processed for the immunocytochemical localization of TH [purchased from Boehringer Mannheim at a dilution of 1:1000] as described previously[22,26]. Tissue samples were flat embedded separately using standard techniques as described previously. Serial sections were collected.

Qualitative analyses was performed from the striatum in four cases. Quantitative analyses were performed using either simple profile counts (SPC) from all four cases, or unbiased stereology (from the case with the best ultrastructure and immunoreactivity); a full discussion of the technical implications of these techniques is found elsewhere[22].

Qualitative observations were obtained from the SN in four cases. Dopaminergic neurons (n=25) identified by the presence of TH-immunoreactive (-i) or by morphological criteria, were studied in serial sections. TH-i dendrites were analyzed for synaptic connections.

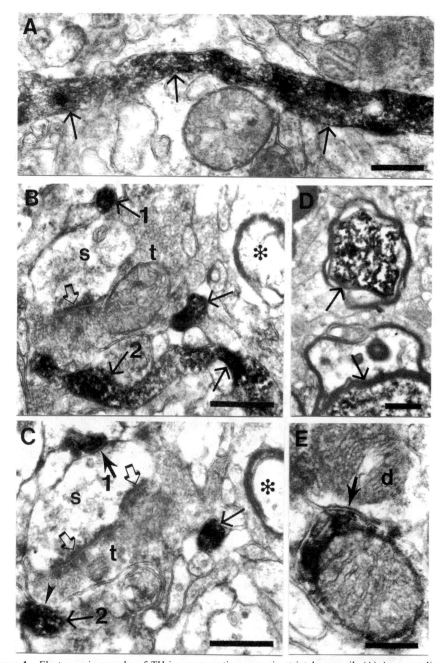

Figure 1. Electron micrographs of TH-immunoreactive axons in striatal neuropil. (A) An unmyelinated axon (arrows) coursing through the neuropil. (B,C). Two sections through a synaptic complex involving TH-i intervaricose segments (arrows), spines, and an unlabeled terminal (t) forming an asymmetric axospinous synapse (open arrow). TH-i axon #1 forms a symmetric synapse (flared arrow) with the spine. TH-i axon #2 forms an apposition (arrowhead) with the unlabeled axon terminal (t). (D) TH-i myelinated axons (arrows). (E) TH-i terminal forms a symmetric synapse (arrow) with a dendrite (d). Scale bars = 0.5 μm (AD) and 0.25 μm (B,C,E).

Table 1. Demographics

Case #	PMI (hr)	A/R/S/	Cause of Death
1	3.0	79 B M	ASCVD
2	3.0	48 B M	MVA
3	4.0	35 W M	MVA
4	4.5	47 W M	ASCVD
5	5.0	68 B M	ASCVD
6	7.0	32 W F	Cardiac arrhythmia

Control subjects used in this study were free of neurological disorders. PMI, postmortem interval. ASCVD, atherosclerotic cardiovascular disease. MVA, motor vehicle accident. Case #3 had a history of alcoholism and EtOH in blood; B, black; W, white; M, male; F, female.

3. RESULTS

At the electron microscopic level, TH-i was present in striatal myelinated (Fig. 1D) and unmyelinated axons (Fig. 1A). The majority of TH-i axons were unmyelinated, varicose and tortuous (Fig. 1A). TH-i boutons formed axospirious (Fig. 1B,C) and axodendritic synapses (Fig. 1E), which in the majority of cases were symmetric. TH-i boutons formed symmetric synapses, which were short in length, with both the head and neck of spines. Synapses were formed by TH-i varicose (usually measuring 0.75-1.5μm) and intervaricose segments (typically 0.2-0.3μm). A common arrangement of TH-i profiles was the close apposition of a TH-i

Figure 2. Photomicrograph of the SNc labeled with TH-i. Note the many labeled neurons (arrows) and processes (arrowheads). Scale bar = 250 μm

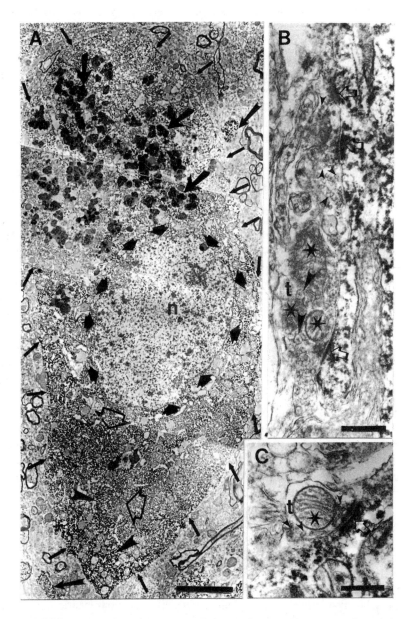

Figure 3. (A) Electron micrograph of a TH-labeled neuron in the SNc; the outline of the soma is highlighted by arrows. The nucleus (n) is highlighted by short broad arrows. The vast cytoplasm is filled with organelles, including numerous stacks of rER (open arrows), and many pigment granules (flared arrows). Reaction product is present in the cytoplasm (arrowheads). TH-i dendrites (flared arrows) are scattered throughout the neuropil. (B) A large terminal (t) forms an asymmetric perforated synapse (open arrows) with a labeled dendrite; it contains several mitochondria (stars) and is moderately packed with dense core vesicles (arrowheads) and small pleomorphic vesicles (small arrowheads). (C) A small axon terminal (t) with one mitochondrion (star) and scattered round vesicles (arrowheads) forms a synapse (open arrow) with a labeled soma. Scale bars = 5 μm (A), 0.5 μm (B,C).

bouton to both a spine and the unlabeled axon terminal synapsing with that spine. These appositions between labeled and unlabeled terminals occasionally fulfilled some, but not all, of the criteria for a synapse (Fig. 1 C).

Quantitative analyses using SPC and stereology revealed that TH-i terminals forming synapses comprised approximately 16% of all striatal synapses combined (both labeled and unlabeled). TH-i synapses comprised 35%-55% of all symmetric synapses using SPC or stereology, respectively. SPC revealed that out of 339 unmyelinated THi axons, 100 were varicosities, of which 19% formed synapses; 239 were intervaricose segments, of which 7.5% formed synapses. The results on the proportions of subtypes of TH-i synapses to each using both methods was consistent and revealed that 57-62% of TH-i synapses were symmetric axospinous, 33-35% were symmetric axodendritic and 4-8% of the synapses were asymmetric. Of the TH-i axospinous synapses, 40% were on the spine head while 60% were on the spine neck.

The substantia nigra pars compacta (SNc) contained large TH-i cells and a rich network of processes (Fig. 2). At the electron microscopic level TH-i neuronal somata were characterized by a vast cytoplasm rich in organelles, especially stacks of rough endoplasmic reticulum (rER), numerous polyribosomes and pigment granules (Fig. 3A). Somata typically received two to five axosomatic synapses in a single thin section. A moderate portion of the cell body was apposed by astrocytic processes. TH-i myelinated axons, but not terminals, were observed. Unlabeled axon terminals of various morphologies contacted TH-i dendrites, and less frequently, somata (Figs. 3&4). Unlabeled terminals, which formed synapses with TH-i profiles, varied in size, vesicle packing density, vesicle shape, amount of dense core vesicles, and in the symmetry of the synapse they formed with the labeled dendrites and somata. For example, large terminals (>3μm in length), containing several mitochondria, formed perforated asymmetric synapses with labeled dendrites.

Terminals forming asymmetric synapses were large (3B), small (3C) or medium sized (4A), contained few (3C) or several mitochondria (3B, 4A) and varied in vesicle packing density from moderately dense (3B, 4A) to sparse (3C). Terminals forming symmetric synapses were also varied in size: small (4A) medium sized (4C), and large (4B) contained 0-1 (4A,C) or several mitochondria (4C), and had variable vesicle packing density from sparse (4C) to moderate (4C).

4. DISCUSSION

In the striatum, TH-i axons were mostly unmyelinated and varicose, however a small proportion of myelinated axons were present. TH-i terminals formed predominantly symmetric synapses, but also formed a few asymmetric synapses and nonsynaptic appositions. The majority of TH-i boutons formed symmetric axospinous synapses with either the head and neck of spines. TH-i varicosities and intervaricose segments formed synapses that were short in length. Appositions were formed between TH-i boutons and unlabeled terminals forming asymmetric synapses.

Our data, that the majority of TH-i boutons formed symmetric synapses, is consistent with ultrastructural findings in other species. In rat and monkey, most studies have shown that between 94-99% of synapses formed by TH-i boutons were symmetric[9,13,23,30]. The TH-i asymmetric synapses we observed in the human material were not as thick as those formed

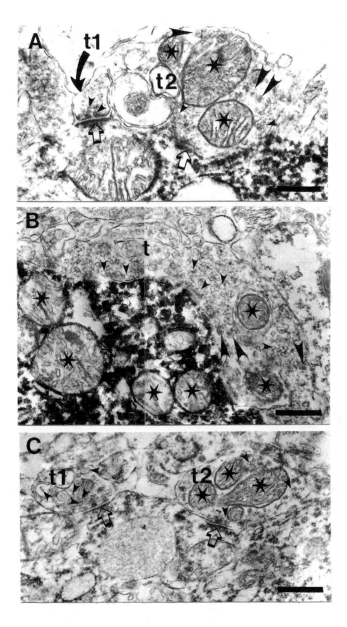

Figure 4. Axon terminals (t) of various morphologies contact TH-labeled profiles. (A) Two unlabeled terminals form synapses (open arrows) with a labeled dendrite. One terminal (t1) is very small, devoid of mitochondria and is filled with round clear vesicles (arrowheads). A larger terminal (t2) contains several mitochondria (stars) and is packed with vesicles, both clear round (small arrowheads) and dense core (longer arrowheads). (B) A large terminal filled with round vesicles (small arrowheads) an occasional dense core vesicle (large arrowhead) and two mitochondria (stars) partially encircles the dendrite. The reaction product in the dendrite obscures the postsynaptic density. (C) Two unlabeled terminals form symmetric synapses with a lightly labeled cell body. One terminal (t1) is small and contains pleomorphic vesicles. A larger terminal (t2) is loosely packed with vesicles (arrowheads) and contains several mitochondria (stars). Scale bars = 0.5μm

on spine heads by unlabeled terminals. The functional significance of these synapses is unclear, but probably not related to dopamine's differential effects on striatal neurons[7,15,19,33,34] as the asymmetric synapses represent such a minor proportion of all TH-i synapses.

The proportion of TH-i terminals forming synapses with spines versus dendrites may vary among species. Our data in human, that the majority (59-65%) of TH-i boutons formed axospinous synapses, is different from that of the monkey where scarcely one quarter of TH-i boutons formed synapses on spines[30]. The results in rodents vary, with some investigators finding that the majority of TH-i boutons formed axodendritic synapses[9], while others found that the majority of TH-i boutons formed axospinous synapses[13]. Nevertheless, in human, TH-i boutons make more axospinous synapses relative to axodendritic synapses than in monkey and possibly more than in rat. The spines of medium spiny neurons in human appear to be more heavily innervated by dopamine than in other species. This suggests that dopaminergic modulation of cortical inputs to the spines of medium spiny neurons may be more robust, and thus may exert more control over glutamatergic inputs in human striatum.

TH-i boutons in the human formed close appositions, rather than actual synapses, with postsynaptic profiles. This type of synaptic arrangement has been described in other species and is not a unique feature of the human striatum. The meaning of these contacts in the human is just as ambiguous as in other species; but, as discussed by Bouyer et al. (1984), may represent the anatomical substrate for dopamine release by glutamate[25].

In summary, the pattern of TH-i in the human is generally consistent with that of the other species. The presence of TH-i in myelinated axons and a increased proportion of TH-i axospinous vs axodendritic synapses distinguishes the ultrastructural pattern of TH-i in the human from that of other species.

Qualitatively, the pattern of TH-i in the human SN is quite similar to that of other species. The ultrastructural features of TH-i somata are consistent with those described in rat[12,27], namely a large somata with vast cytoplasm, and multiple stacks of rER. Inputs to the SNc are predominantly GABAergic (50-70%) and arise from the striatum and globus pallidus[3,24,29]. Glutamatergic inputs originate from neurons in cortex, subthalamic and pedunculopontine nuclei[2,6,28]. Terminals in the SNc also contain 5HT, acetylcholine, as well as neuropeptides[8,10]. Although in labeled dendrites and somata the reaction product sometimes obscured the thickness of postsynaptic density, the symmetry of the synapse was discernible in lighter stained profiles. The presence of terminals of variable morphology forming asymmetric synapses onto labeled profiles is expected given the glutamatergic inputs which arise from multiple sources in experimental animals. Particular terminals (ie. 3B) had similar morphological features (large asymmetric perforated synapses) to those described in animals which arise from the subthalamus and pedunculopontine nuclei[6,31]. Other terminals (ie. 4B) had features of GABAergic terminals and probably arise from the striatum or globus pallidus. The heterogeneity of unlabeled terminals in the human SNc is consistent with that of other species. Quantitative analyses in combination with immunocytochemistry will be necessary to determine how similar the synaptic morphologies and proportions of neurochemically defined subsets of terminals are in the human as compared to those of other species.

5. ACKNOWLEDGEMENTS

The author is indebted to the personnel of the Maryland Brain Collection for their special efforts in obtaining the short postmortem interval brains and would also like to thank Joyce Kelley, Lili Kung, Michelle Force and Ross McKim for excellent technical assistance and Sharon Stilling for help with the preparation of the manuscript. This work was supported in part by the Scottish Rite Benevolent Foundation's Schizophrenia Research Program, N.M.J., U.S.A., a grant from the Stanley Foundation and MH40279.

6. REFERENCES

1. Aoki, C., and Pickel, V.M., 1988, Neuropeptide Y-containing neurons in the rat striatum: Ultrastructure and cellular relations with tyrosine hydroxylase-containing terminals and with astrocytes. *Brain Res.* 459:205-225.
2. Beckstead, R.M., 1979, An autoradiographic examination of corticocortical and subcortical projections of the mediodorsal-projection (prefrontal) cortex in the rat. *J. Comp. Neurol.* 184:43-62.
3. Bolam, J.P., and Smith Y., 1990, The GABA and substance P input to dopammergic neurones in the substantia nigra of the rat. *Brain Res.* 529:57-78.
4. Bouyer, JI, Park, D.H., Joh T.H., and Pickel, V.M., 1984, Chemical and structural analysis of the relation between cortical inputs and tyrosine hydroxylase-containing terminals in rat neostriatum. *Brain Res.* 302:267-275.
5. Chang, H.T., 1988, Dopamine-acetylcholine interaction in the rat striatum: A dual-labeling immunocytochemical study. *Brain Res. Bull.* 21:295-304.
6. Chang, H.T., Kita, H., and Kitai, S.T., 1984, The ultrastructural morphology of the subthalamic-nigral axon terminals intracellularly labeled with horseradish peroxidase. *Brain Res.* 299:182-185.
7. Chiodo, L.A., and Berger, T.W., 1986, Interactions between dopamine and amino acid-induced excitation and inhibition in the striatum. *Brain Res.* 375:198-203.
8. Corvaja, N., Doucet, G., and Bolam, J.P., 1993, Ultrastructure and synaptic targets of the raphe-nigral projection in the rat. *Neuroscience* 55:417-427.
9. Descarries, L., Watkins, K.C., Garcia, S. Bosler, O., and Doucet, G., 1996, Dual character, asynaptic and synaptic, of the dopamine innervation in adult rat neostriatum: A quantitative autoradiographic and immunocytochemical analysis. *J. Comp. Neurol.* 375:167-186.
10. DiFiglia, M., Aronin, N., and Leeman, S.E., 1981, Immunoreactive substance P in the substantia nigra of the monkey: light and electron microscopic localization. *Brain Res.* 233:381-388.
11. Dimova, R., Vuillet,l., Nieoullon, A., and Kerkerian-Le Goff, L., 1993, Ultrastructural features of the choline acetyltransferase-containing neurons and relationships with nigral dopaminergic and cortical afferent pathways in the rat striatum. *Neuroscience* 53:1059-1071.
12. Domesick, V.B., Stinus, L., and Paskevich, P.A., 1983, The cytology of dopaminergic and nondopaminergic neurons in the substantia nigra and ventral tegmental area of the rat: A light- and electron-microscopic study. *Neuroscience* 37:1-9.
13. Freund, T.F., Powell, J.F., and Smith, A.D., 1984, Tyrosine hydroxylase-immunoreactive boutons in synaptic contact with identified striatonigral neurons, with particular reference to dendritic spines. *Neuroscience* 13:1189-1215.
14. Fujiyama, F., and Masuko, S., 1996, Association of dopaminergic terminals and neurons releasing nitric oxide in the rat striatum: An electron microscopic study using NADPH-Diaphorase histochemistry and tyrosine hydroxylase inununohistochemistry. *Brain Res. Bull.* 40:121-127.
15. Gerfen, C.R, McGinty, J.F., and Young III, W. S., 1991, Dopamine differentially regulates dynorphin, substance P, and enkephalin expression in striatal neurons: In situ hybridization histochemical analysis. *J. Neurosci.* 11:1016-1031.
16. Hattori, T., Takada, M., Moriizumi, T., and Van Der Kooy, D., 1991, Single dopaminergic nigrostriatal neurons form two chemically distinct synaptic types: possible transmitter segregation within neurons. *J. Comp. Neurol.* 309:391-401.

17. Karle, E.J., Anderson, K.D., and Reiner, A., 1992, Ultrastructural double-labeling demonstrates synaptic contacts between dopaminergic terminals and substance P-containing striatal neurons in pigeons. Brain Res. 572:303-309.
18. Karle, E.J., Anderson, KD., and Reiner, A., 1994, Dopaminergic terminals form synaptic contacts with enkephalinergic striatal neurons in pigeons: an electron microscopic study. Brain Res. 646:149-156.
19. Kitai, S.T., and Kocsis, J.D., 1978, An intracellular analysis of the action of the nigral-striatal pathway on the caudate spiny neuron. In *Ionophoresis and Transmitter Mechanisms in the Mammalian Central Nervous System* (R. W. Ryall and J.S. Kelly, eds), North-Holland, Amsterdam, pp. 17-23.
20. Kubota, Y., Inagaki, S., Kito, S., Takagi, H., and Smith, A.D., 1986, Ultrastructural evidence of dopaminergic input to enkephalinergic neurons in rat neostriatum. Brain Res. 367:374-378.
21. Kubota, Y., Inagaki, S., Kito, S., Shimada, S., Okayama, T., Hatanaka, H., Pelletier, G., Takagi, H., and Tohyama, M., 1988, Neuropeptide Y-immunoreactive neurons receive synaptic inputs from dopaminergic axon terminals in the rat neostriatum. Brain Res. 458:389-393.
22. Kung, L., Force, M., Chute, D.J., and Roberts, RC., 1998, Immunocytochemical localization of tyrosine hydroxylase in the human striatum: a postmortem ultrastructural study. J. Comp. Neurol. 390:52-62.
23. Pickel, V.M., Beckley, S.C., Joh, TH., and Reis, D.J., 1981, Ultrastructural immunocytochemical localization of tyrosine hydroxylase in the neostriatum. Brain Res. 225:373-385.
24. Ribak, C.E., Vaughn, J.E., Saito, K., Barber, R, and Roberts, E., 1976, Immunocytochemical localization of glutamate decarboxylase in rat substantia nigra. Brain Res. 116:287-298.
25. Roberts, P.J., and Anderson, S.D., 1979, Stimulatory effect of L-glutamate and related amino acids on (^{3}H)-dopamine from rat striatum: an in vitro model for glutamate actions. J. Neurochem. 32:1539-1545.
26. Roberts, R. C. and Kung, L., 1999, The Processing and Use of Postmortem Human Brain Tissue for Electron Microscopy. In *Using CNS Tissue in Psychiatric Research: A Practical Guide* (B. Dean, J. Kleinman and T.M. Hyde, eds.), Harwood Academic Publishers, pp.127-140.
27. Roberts, RC., McCarthy, K.E., Du, F., Okuno, E., and Schwarcz, R., 1994, Immunocytochemical localization of 3Hydroxyanthranilic acid oxygenase (3HAO) in the rat substantia nigra. Brain Res. 650:229-238.
28. Scarnati, E., Proia, A., Campana, E. and Pacitti, C., 1986, A microiontophoretic study on the nature of the putative synaptic neurotransmitter in the pedunculopontine-substantia nigra pars compacta excitatory pathway of the rat. Exp. Brain Res. 62:470-478.
29. Smith, Y., and Bolam, LP., 1990, The output neurones and the dopaminergic neurones of the substantia nigra receive a GABA-containing input from the globus pallidus in the rat. J. Comp. Neurol. 296:47-64.
30. Smith, Y., Bennett, B.D., Bolam, J.P., Parent, A., and Sadikot, A.F., 1994, Synaptic relationships between dopaminergic afferents and cortical or thalamic input in the sensorimotor territory of the striatum in monkey. J. Comp. Neurol. 344:1-19.
31. Smith, Y., Charara, A, and Parent, A., 1996, Synaptic innervation of midbrain dopaminergic neurons by glutamate-enriched terminals in the squirrel monkey. J. Comp. Neurol. 364:231-253.
32. Vuillet, J., Kerkerian, L., Bosler, O., and Nieoullon, A., 1989, Ultrastructural correlates of functional relationships between nigral dopaminergic or cortical afferent fibers and neuropeptide Y-containing neurons in the rat striatum. Neurosci. Lett. 100:99-104.
33. Williams, G.V., and Millar, J., 1990, Concentration-dependent actions of stimulated dopamine release on neuronal activity in rat striatum. Neuroscience 39:1-16.
34. Zagami, M.T., Montalbano, M.E., Sabatino, M., and LaGrutta, V., 1986, Effect of acetylcholine and dopamine iontophoretically applied on the sensory responsive caudate unit. Arch. Int. Physiol. Biochem. 94:305-316.

DENDRITIC CHANGES IN MEDIUM SPINY NEURONS OF THE WEAVER STRIATUM
A Golgi study

Diane E. Smith, Beverly Glover, and C'Lita Henry*

1. INTRODUCTION

In addition to the cerebellar granule cell deficit, for which the neurological mutant mouse weaver was first characterized[13], studies have also demonstrated that weaver has a deficit in the nigrostriatal dopamine pathway[9,14,17,22]. In the substantia nigra (SN), this deficit results in a loss of dopaminergic neurons in the pars compacta (*op. cit.*), as well as a diminution of TH (tyrosine hydroxylase)-positive dendrites in areas of the pars reticulata that contrasted sharply to the presence of thickly aligned labeled processes in the homozygous controls[16,21]. There is a concomitant decrease in the presence of dopaminergic terminals in the striatum[22].

Ingham *et al.*[7] observed a decrease in spine density in rat neostriatal medium spiny neurons after 6-hydroxydopamine injections into the nigrostriatal pathway. A subsequent study[6] showed that the loss of spines is accompanied by a loss of asymmetric synapses.

The present study was undertaken to determine the effect of the loss of dopaminergic input on the dendritic arbor of the medium spiny neurons of the neostriatum in weaver.

2. METHODS

Heterozygote breeding pairs of the neurological mutant mouse weaver were obtained from The Jackson Lab and maintained in our vivarium with a 12 hour light/dark cycle and *ad libitum* access to food and water. The pups obtained from these breeding pairs were weaned at four weeks. At eight weeks, pups homozygous for the weaver mutation and litter mate controls were anesthetized with a 0.5% sodium pentobarbital solution. Anesthesia was considered complete when there was no response to a deep toe pinch. The animals were perfused transcardially with 35 to 40 ml of a 0.15% glutaraldehyde-4% paraformaldehyde solu-

---

* D.E. Smith, B. Glover, and C. Henry, Department of Cell Biology & Anatomy, LSU Medical School - New Orleans, 1901 Perdido Street, New Orleans, LA 70112.

tion made up in a 0.1 M phosphate buffer. The brains were removed and blocked into coronal sections at the conclusion of the perfusion. Only homozygous wild-type controls were used as previous studies had indicated variations in striatal dopamine content between the hetero- and homozygous control[14]. Thionine staining of the cerebellar tissue was used to distinguish between the heterozygotes and wild type controls.

Sections through the striatum were processed for Golgi impregnation and embedded in celloidin following the procedure of Shimono and Tsuji[18]. One hundred micron coronal sections were cut on an American Optical Sliding Microtome, numbered chronologically, and dehydrated through an ascending series of alcohols to butyl alcohol, then to Cedarwood oil. The sections were next lined up in numerical order in a petri dish, placed on glass slides, trimmed of excess celloidin, further dehydrated with Xylene, and cover-slipped. The sections were viewed with a Zeiss microscope and camera lucida images of impregnated neurons within the dorsolateral striatum were projected and traced.

Dendrites were identified and marked using the centrifugal ordering method of Uyling *et al.*[23]. The camera lucida drawings were captured and analyzed using a Jandel Imaging Analysis system. The diameter of the long axis of the cell body, the size (area) of the dendritic field, the number and length of the dendritic branches, and the number of terminal dendritic branches were measured. The dendritic field was determined by drawing a straight line connecting the terminal tips of the dendrites and measuring the area within the outline. While this led to a slight exaggeration of the size of the field, we felt it provided a greater consistency between cells than if we had tried to curve the lines closer to the cell body; the

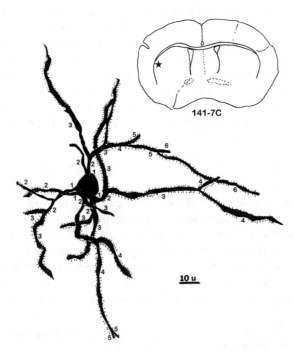

Figure 1. Medium-spiny neuron from wild type control. These neurons exhibit densely spine-laden dendrites. Location of impregnated neuron is indicated by the star in the drawing in the upper right hand corner.

size of the cell bodies did not vary between weaver and control so cell body size was not subtracted from the dendritic field measurements. These data were transferred into an ASCII file and analyzed using the nonparametric, unpaired Student t-test for cell body and dendritic field measurements and the nonparametric Mann-Whitney test for dendritic lengths and numbers. A p value of 0.05 was considered significant.

3. RESULTS

In the control animals, the majority of the cells impregnated with the Golgi solutions displayed an aspinous soma and proximal dendrites; the more distal dendrites were sparsely to densely covered with spines (Figure 1). These cells corresponded to the medium neurons with spine-laden dendrites of the mouse striatum as described by Rafols et al.[12] and to the Type I medium neurons reported in the rat striatum by Chang et al.[2]. The neurons in the weaver striatum exhibited fewer spines than the controls and appeared to have fewer dendritic processes (Figure 2).

There was no statistically significant difference between the weaver mice and their homozygous littermate controls in terms of either cell body diameter (p=0.7292) or dendritic field size (p=0.5132). The weaver neurons measured 8.980 mtm ± 0.6047 S.E.M. (Standard Error of Measurement) in diameter and the control neurons measured 9.324 mm ± 0.7473

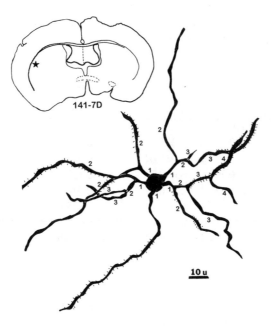

Figure 2. Medium spiny neuron from weaver striatum. Note the paucity of spines especially on the distal dendrites. The location of the neuron is indicated by the star in the drawing in the upper left hand corner.

S.E.M. The dendritic fields measured 10,731 mm² ± 1193 S.E.M. for weaver neurons and 12,192 mm² ± 1864 S.E.M. for the controls.

However, statistically significant differences were seen in the number of terminal branches, the length of the third and fourth order branches, and the number of third and fifth order branches. The medium spiny neurons in the control animals had an average of 15.85±0.90 S.E.M. terminal branches. The weaver neurons had an average of 11.6±1.13 S.E.M. terminal branches. The p value was 0.0055 which was considered very significant. The lengths of the third and fourth order dendrites were longer in weaver than in the control (Table 1), yielding a p value of 0.0435 and 0.0317 respectively. Lastly, there were significantly fewer third and fifth order dendritic branches in weaver neurons than in control neurons (Table II), p=0.0227 and 0.0109 respectively.

Table 1. Dendritic lengths

Branch Order	Control	Weaver	p value
first order	11.896±1.8	13.63±1.9	0.2264
second order	22.95±2.2	24.36±2.3	0.4699
third order	26.33±1.7	32.57±2.24	0.0435*
fourth order	30.37±2.35	35.39±2.33	0.0317*
fifth order	27.45±2.76	33.31±4.97	0.3598
sixth order	31.75±4.61	26.57±5.17	0.5835

Data are expressed as the mean (in microns) ± S.E.M. The asterisks indicate the data which are statistically significant. A $p<0.05$ is considered significant.

Table 2. Number of dendritic branches

Branch Order	Control	Weaver	p value
first order	3.8±0.22	3.3±0.3	0.1656
second order	6.1±0.37	5.3±0.5	0.2529
third order	7.75±0.51	5.6±0.64	0.0227*
fourth order	5.5±0.79	3.6±0.7	0.1288
fifth order	3.1±0.53	1.2±0.37	0.0109*
sixth order	1.3±0.3	0.6±0.21	0.1259

Data are expressed as the mean ± S.E.M. The asterisks indicate the data which are statistically significant. A $p<0.05$ is considered significant.

4. DISCUSSION

The medium spiny neurons in the striatum of the weaver mouse were found to have fewer spines, fewer terminal dendritic branches, fewer 3rd and 5th order dendritic branches, and longer 3rd and 4th order dendrites than the neurons of the homozygous controls. These dendritic alterations suggest a structural compromise in weaver as a result of the loss of the dopamine input provided by the nigrostriatal system.

Loss of afferent input has long been known to produce alterations in the target neuron[3,5,20]. The partial deafferentation that occurs in Parkinson's disease (PD), as a result of the loss of the dopamine input from the SN pars compacta, has also been found to affect the target neurons in the striatum. McNeill, et al.[10,11] reported a less extensive dendritic arbor in Golgi impregnated medium spiny neurons from the striatum of PD patients. They reported findings similar to ours in that they found the neurons from the PD brains had fewer terminal segments, a decreased total dendritic length, and fewer total number of segments/cell[11]; subsequent investigations found that the medium spiny neurons had few dendritic spines and irregular bulbous swellings[10]. They suggested that the morphological data might "...explain, in part, the declining efficacy of chronic L-DOPA replacement therapy in advanced PD".

Ingham et al.[6,7], reported dendritic changes in the medium spiny neurons of the rat neostriatum after 6-hydroxy-dopamine injections into the nigrostriatal pathway. In a Golgi study[7], they reported a lower spine density on the side ipsilateral to the injection. Spines, once thought to be static appendages on the dendrite serving only as synaptic targets, have now been recognized to be very plastic in nature and to play a significant role in the induction of information storage[1,8]. This suggests that the loss of spines we observed in weaver striatum is a contributing factor to the compromise seen in the motor function of these mutants.

In a subsequent ultrastructural investigation, Ingham et al.[6] further demonstrated that the loss of spines observed in the Golgi study is accompanied by a loss of asymmetric synapses which would have originated mainly in the cortex. Other studies have demonstrated that the input from the cortex and the midbrain have preferred sites of contact with cortical afferents synapsing on spines of proximal dendrites while dopamine afferents synapse on more distal spines[4,19]. While proximal as well as distal spines were affected in weaver, those on the distal dendrites appeared to be more severely affected. When compared to control animals there were significantly fewer distal dendritic spines because there were fewer terminal branches in weaver.

In summary, the dendritic changes observed in this Golgi study of the medium spiny neuron of weaver suggest a structural compromise in weaver that is related, in part, to the loss of the afferent dopamine input. Such a compromise would have a deleterious effect on the synaptic efficacy of the affected neurons and might help to explain some of the motor problems seen both in PD patients and in weaver.

5. REFERENCES

1. Barinaga, M.. Dendrites shed their dull image. *Science* 268:200-201 (1995).
2. Chang, H.T. Wilson, C.J. and Kitai, S.T. A Golgi study of rat neostriatal neurons: light microscopic analysis. *J. Comp. Neurol.* 208:107-126 (1982).
3. Coleman, P.D. and Riesen, A.H. Environmental effects on cortical dendritic fields. 1. Rearing in the dark. *J. Anat. (Lond)* 102:363-374 (1968).
4. Freund, T.F., Powell, J.F. and Smith, A.D. Tyrosine hydroxylase-immunoreactive boutons in synaptic contact with identified striatonigral neurons, with particular reference to dendritic spines. *Neuroscience* 13:1189-1215 (1984).

5. Garey, L.J., Fisken, R.A. and Powell, T.P. Effects of experimental deafferentation on cells in the lateral geniculate nucleus of the cat. *Brain Res.* 52:363-369 (1973).
6. Ingham, C.A., Hood, S.H., Taggart, P. and Arbuthnott, G.W. Plasticity of synapses in the rat neostriatum after unilateral lesion of the nigrostriatal dopaminergic pathway. *J. Neuroscience* 18:4732-4743 (1998).
7. Ingham, C.A., Hood, S.H., vanMaldegem, B., Weenink, A. and Arbuthnott, G.W. Morphological changes in the rat neostriatum after unilateral 6-hydroxydopamine injections into the nigrostriatal pathway. *Ex. Brain Res.* 93:17-27 (1993).
8. Koch, C., Zador, A. and Brown, T.H. Dendritic spines: Convergence of theory and experiment. *Science* 256:973-974 (1992).
9. Lane, J.D., Nadi, N.S., McBride, W.J., Aprison, M.H. and Kusano, K. Contents of serotonin, norepinephrine and dopamine in the cerebrum of the 'staggerer','weaver' and 'nervous' neurologically mutant mice. *J. Neurochem.* 29:349-350 (1977).
10. McNeill, T.H., Brown, S.A., Rafols, J.A. and Shoulson, I. Atrophy of medium spiny I striatal dendrites in advanced Parkinson's disease. *Brain Res.* 455:148-152 (1988).
11. McNeill, T.H., Brown, S.A., Shoulson, I., Lapham, L.W., Eskin, T.A. and Rafols, J.A. Regression of striatal dendrites in Parkinson's disease. In: *The Basal Ganglia II, Structure and function - Current Concepts,* edited by Carpenter, M.B. and Jayaraman, A. New York: Plenum Press, p. 475-482 (1987).
12. Rafols, J.A., Cheng, H.W. and McNeill, T.H. Golgi study of the mouse striatum: Age-related dendritic changes in different neuronal populations. *J. Comp. Neurol.* 279:212-227 (1989).
13. Rakic, P. and Sidman, R.L. Sequence of developmental abnormalities leading to granule cell deficit in cerebellar cortex of weaver mutant mice. *J. Comp. Neurol.* 152:103-132 (1973).
14. Roffler-Tarlov, S. and Graybiel, A.M. Expression of the weaver gene in dopamine-containing neural systems is dose-dependent and affects both striatal and non-striatal regions. *J. Neurosci.* 6:3319-3330 (1986).
15. Roffler-Tarlov, S. and Graybiel, A.M. The postnatal development of the dopamine-containing innervation of dorsal and ventral striatum: effects of the weaver gene. *J. Neurosci.* 7:2364-2372 (1987).
16. Roffler-Tarlov, S., Graybiel, A.M. and Martin, B. Genetic perturbation of dendrite formation. *Soc. Neurosci. Abstr.* 16:932 (1990).
17. Schmidt, M.J., Sawyer, B.D., Perry, K.W., Fuller, R.W., Foreman, M.M. and Ghetti, B. Dopamine deficiency in the weaver mutant mouse. *J. Neurosci.* 2:376-380 (1982).
18. Shimono, M. and Tsuji, N. Study of the selectivity of the impregnation of neurons by the Golgi method. *J. Comp. Neurol.* 259:122-130 (1987).
19. Smith, A.D. and Bolam, J.P. The neural network of the basal ganglia as revealed by the study of synaptic connections of identified neurones. *TINS* 13:259-265 (1990).
20. Smith, D.E. The effect of deafferentation on the development of brain and spinal nuclei. *Prog. in Neurobiol.* 8:349-367 (1977).
21. Triarhou, L.C. and Ghetti, B. The dendritic dopamine projection of the substantia nigra: Phenotypic denominator of weaver gene action in hetero- and homozygosity. *Brain Res.* 501:373-381 (1989).
22. Triarhou, L.C., Norton, J. and Ghetti, B. Synaptic connectivity of tyrosine hydroxylase immunoreactive nerve terminals in the striatum of normal, heterzygous and homozygous weaver mutant mice. *J. Neurocytol.* 17:221-232 (1988).
23. Uylings, H.B.M., Smit, G.J. and Veltman, W.A.M. Ordering Methods in Quantitative Analysis of Branching Structures of Dendritic Trees. In: *Physiology and Pathology of Dendrities,* edited by Kreutzberg, G.W. New York: Raven Press, p. 247-254 (1975).

EARLY CORTICOSTRIATAL PROJECTIONS AND DEVELOPMENT OF STRIATAL PATCH/MATRIX ORGANIZATION

Abigail Snyder-Keller, Yili Lin and David J. Graber*

1. INTRODUCTION

The patch/matrix organization of the striatum consists both of clusters of phenotypically similar striatal neurons and a preferential distribution of striatal afferents to one or the other compartment (3, 5,11). Attempts to study the role of afferent innervation in the formation of this compartmentalized pattern have been hampered by the fact that the patch/matrix organization develops prenatally in nearly all species (7,11). In the rat, nigrostriatal dopamine (DA) afferents become more heavily concentrated within the patch compartment by embryonic day (E) 20 (11,18), the earliest age patches of any markers can be consistently demonstrated. Attempts to disrupt the early-growing nigrostriatal dopaminergic innervation by 6-hydroxydopamine (11) or knife cuts severing the nigrostriatal fibers at E19 (17) failed to abolish the patch/matrix organization of striatal neurons, although postnatal lesions of the substantia nigra did alter the normal patch distribution of retrogradely-labelled striatonigral neurons (15).

The organizational influence of the corticostriatal glutamatergic innervation, the primary excitatory drive onto striatal neurons, has not been extensively studied. Glutamate and NMDA receptors are clearly involved in pattern formation in other brain regions. In the striatum, the combination of corticostriatal glutamate release, NMDA receptors concentrated in patches (13), BDNF released from corticostriatal afferents (1), and trkB receptors concentrated on patch neurons (2), makes cortex a likely candidate for influencing patch formation, stabilization, or maintenance. Because early ingrowth of corticostriatal afferents has not yet been characterized in the rat, one aim of the present set of studies was to characterize early corticostriatal innervation using both anterograde tract-tracing with biotinylated dextran-amine and transsynaptic c-fos induction after activation of corticostriatal neurons. In addition, because of the difficulties inherent in manipulating a pattern that develops prenatally, we employed organotypic cultures of prenatal as well as postnatal

---

* A. Snyder-Keller, Y. Lin and D.J. Graber, Wadsworth Center, New York State Department of Health, Empire State Plaza, Albany, NY 12201.

The Basal Ganglia VI
Edited by Graybiel *et al.*, Kluwer Academic/Plenum Publishers, 2002

striatal tissue, in order to examine the role of both corticostriatal and nigrostriatal afferents in the development of striatal patch/matrix organization.

2. EXPERIMENT 1. ANTEROGRADE TRACING OF CORTICOSTRIATAL AFFERENTS IN DEVELOPING STRIATUM

Initial tracing studies using DiI in the mouse striatum (9,10) indicated that corticostriatal arborizations are not elaborated until postnatal day (P) 2, even though many of these exist as branches of corticopontine axons that reach their target several days earlier. We have followed up on these observations using anterograde tracing of corticostriatal projections in the early postnatal period in the rat. To do this, we employed the 3000 MW form of biotinylated dextranamine (BDA; Molecular Probes), which is transported to striatum and beyond within 24 hr, thus minimizing developmental changes in the axonal arbor during the post-injection interval. Given the preferential distribution of corticostriatal afferents from prelimbic cortex to the patch compartment and afferents from somatosensory cortex to the matrix in the mature striatum (3,5), we examined the distribution of BDA-labelled fibers after injection into each of these cortical regions.

Sprague-Dawley rat pups at P1 to P8 were anesthetized on ice, and the skin overlying the anterior part of the skull was incised. A small skull flap was made just anterior to bregma, 2.5 mm lateral to the midline (for somatosensory cortex) or along the midline 2 mm anterior to bregma (for prelimbic cortex). One-half microliter of a 5% solution of the 3000 MW ("fast") form of biotinylated dextranamine (BDA; Molecular Probes) was pressure-injected through a glass micropipette, just under the pial surface (no deeper than 1 mm). Injections were made over the course of 30 seconds, and the pipette remained in place for an additional 30 seconds prior to suturing and warming of the pup. After 24 hr survival, the animals were perfused with 4% paraformaldehyde/ 0.1% glutaraldehyde in 0.1M phosphate buffer, pH 7.4. After 15-24 hr of post-fixation (depending on the age of the animal), and 8 hr in 15% sucrose, the brains were embedded in an egg yolk matrix in order to confer additional strength to the tissue. Thirty micron sections were cut frozen on a sliding microtome, and processed for BDA localization using a 2 hr incubation in ABC solution (Vector labs) and reaction in diaminobenzidine (DAB; 0.05% with 0.0015% H_2O_2) with nickel (0.25%) added to produce black fibers. Some sections were subsequently incubated in rabbit anti-DARPP-32 antibodies (1:1000; Chemicon), processed by the ABC-peroxidase technique, and reacted in DAB without nickel for a light brown stain.

Injection of BDA into cortical tissue resulted in deposits of the tracer into immediately surrounding cortex without substantial spread past the corpus callosum into the striatum. In addition to bundles of corticofugal fibers passing through the striatum, fine caliber BDA-labelled fibers were observed within the striatum. At P1 these were sparse with little axonal arbor, and increasing arborization was observed with age. Patches of BDA-labelled fibers were observed after both prelimbic and somatosensory BDA injections, from P3 through the first week of life. Double labeling with DARPP-32 to reveal the patch compartment (4) revealed that afferents from both cortical regions appeared to cluster somewhat preferentially within the patch compartment throughout the first week of life. Although somatosensory cortex projects preferentially to the matrix in adults (3,5), these early projections to the patch compartment may reflect the preponderance of corticostriatal projections arising from the deeper cortical layers in immature brain (6).

3. EXPERIMENT 2. TRANSYNAPTIC INDUCTION OF C-FOS IN STRIATAL NEURONS AFTER CORTICAL NMDA INJECTION

As an alternative means of visualizing the projections of corticostriatal neurons during development, we placed NMDA into the cortex in order to activate cortical neurons and transynaptically activate striatal neurons. We assessed this activation by examining the pattern of c-fos induction within striatal neurons using Fos immunocytochemistry. Similar studies using electrical stimulation of cortex in mature rats have shown that striatal c-fos induction is due to the release of glutamate from corticostriatal terminals (8). In order to more easily visualize the injection site after immunocytochemical processing, we injected a combination of NMDA and BDA.

Rat pups (P1-P8) were anesthetized on ice and given pressure injections of 0.5 µl of a 100 µM NMDA solution, into either prelimbic or somatosensory cortex, essentially as described above for BDA injections. The NMDA injected was made up in 5% BDA (10,000 MW "slow" form; Molecular Probes), in order to facilitate identification of the injection site. After 2-3 hr survival time, rats were perfused with 4% paraformaldehyde, and the brains prepared for immunocytochemistry as described above. Sections were immunostained using rabbit anti-Fos

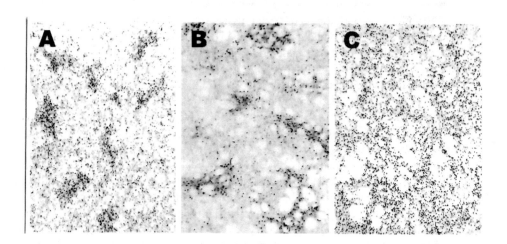

Figure 1. Fos-immunoreactive cells in striatum after cortical NMDA injections. A) Patches of Fos immunoreactivity after NMDA injection into prelimbic cortex on P2. B) Double-labeling for DARPP-32 (gray background stain of patches) and Fos (black dots) reveals overlapping patches after P6 NMDA injection into prelimbic cortex. C) Homogeneous Fos after P8 NMDA injection into somatosensory cortex.

antibodies (1:6000; Santa Cruz), and some were double-immunostained for DARPP-32 (1:1000; Chemicon).

NMDA injections into both prelimbic and somatosensory cortex induced c-fos in areas of cortex adjacent to the BDA-labeled injection site. In the striatum, numerous Fos-immunoreactive striatal neurons were evident at 2-3 hours post-injection. At the earliest postnatal ages, both prelimbic and somatosensory placements resulted in patches of Fos immunoreactive cells (Fig. 1A) that were also immunoreactive for DARPP-32 (Fig. 1B). With prelimbic NMDA placements this patch pattern was maintained through P8. NMDA placements into somatosensory cortex produced more homogeneous Fos in the striatum by the end of the first postnatal week (Fig. 1C). Postnatal day 8 was the latest age examined, due to the fact that supplementation with an anesthetic drug (tribromoethanol), necessary with older rat pups, was found to block the transynaptic c-fos induction.

4. EXPERIMENT 3. THE ROLE OF AFFERENT INNERVATION IN THE DEVELOPMENT OF STRIATAL PATCH/MATRIX ORGANIZATION IN ORGANOTYPIC CULTURES

In order to more easily examine the influence of corticostriatal afferents on prenatal as well as postnatal striatum, we employed the static organotypic culture method (14) for culturing pieces of pre- or postnatal striatum with or without cortex or substantia nigra. Again, given the known differences between corticostriatal innervation originating in prelimbic versus somatosensory cortex (3,5), in all experiments we separately examined the influence of each of these cortical regions.

Tissue was obtained from fetuses of timed-pregnant Sprague-Dawley females (day of sperm = E0). Those used as donors of striatal tissue were exposed to bromo-deoxy-uridine (BrdU; 2 i.p. injections of 75 mg/kg to the pregnant female under light ether anesthesia) on E13 and E14 in order to label cells undergoing neurogenesis at that time (specifically patch neurons) (12,16).

Because BrdU is incorporated into the DNA of dividing cells, it can be used as a permanent marker of neurons destined to populate the patches (12), which would not be altered by culture conditions. Striatal and cortical pieces were dissected from vibratome sections cut at E19-20 (300μ), E21-22 (250μ) or P1-P4 (200μ); ventral mesencephalon containing the DA neurons of the substantia nigra was dissected freehand from E12-E16 fetuses (non-BrdU-exposed). The pieces were placed 1 mm apart on the membrane of inserts (Costar "clear" inserts) in six-well culture trays containing 1.25 ml/well of 1:1 D-MEM/F12 mixture (Gibco #12500), containing either 20% horse serum or 5% fetal calf serum, and supplemented with glucose (0.3%), penicillin (100 U/ml), streptomycin (100 μg/ml), bicarbonate (1.2 mg/ml) and Hepes (4. 5 mg/ml). Cultures were incubated at 37°C in 5% CO^2 from 3 to 24 days; some were switched to serum-free medium containing N2 supplement (1% ; Gibco) at 3 days *in vitro* (div). The medium was changed every 3-4 days.

After different culture times, cultures were excised from the inserts (remaining in place on the membrane) and fixed in 4% paraformaldehyde/4% sucrose for 24 hr at 4°C. Cultures immunostained for BrdU were preincubated for 10 min in 1%H_2O_2 and 30 min in 2N HCl before neutralizing in 0.1M NaBorate for 15 min. All cultures were preincubated in 0.1M glycine (40 min), 5% normal serum and 5% bovine serum albumin (2 hr) prior to 24 hr incubations in primary antibodies: mouse anti-BrdU (1:12; Becton Dickinson), rabbit anti-tyrosine hydroxylase (TH; 1:500, Chemicon) or rabbit anti-GluRl (1:1500; Chemicon). After a 1-hr incubation in biotinylated goat anti-rabbit or horse anti-mouse secondary antibodies, sections were processed by the ABC-peroxidase technique, and reacted in 0.05% diaminobenzidine with 0.25% nickel ammonium sulfate and 0.0015% H_2O_2.

On the day of culturing, some striatal sections from the fetal or postnatal brains used in that culture prep were immediately fixed and examined for BrdU immunoreactivity. Sections from E19 fetuses exhibited a homogeneous distribution of BrdU-immunoreactive neurons throughout the striatum, with a large unlabeled subventricular zone medial to the striatum. At E20-22 the distribution of BrdU-labeled cells became progressively more patchy, and sections taken from postnatal rats were clearly patchy. Thus E19 was considered an age prior to patch formation, E20-22 was transitional, and postnatal was clearly after patch formation.

The distribution of BrdU-labeled neurons was examined at different times *in vitro*. Postnatal striatal sections maintained a distinctly patchy distribution of BrdU-labeled cells for the duration of culturing (24 div was the longest time examined). Sections from E20-22 fetuses were classified as patchy 40% of the time, and this percent was relatively constant throughout the culture period. Thus striatal patches appear to be maintained *in vitro*.

The *formation* of striatal patches was examined in sections of E19 striatum, dissected out prior to the time of *in vivo* patch formation. At short times *in vitro* (3-7 div), BrdU-labeled cells were still observed in a homogeneous distribution. At longer times (8-24 div), some of the cultures exhibited a patchy pattern of BrdU-labeled cells, representing 10% of the total number of striatal sections cultured alone that were examined. Co-culturing with substantia nigra, however, increased the percent of striatal cultures exhibiting patches of BrdU-labeled cells to 33% (Table 1). Co-culturing with cortex had no effect on promoting striatal patch formation, regardless of the source of the cortical tissue. Striatal-nigral-cortical triple co-cultures were comparable to striatal-nigral co-cultures in terms of the percent exhibiting patches. Thus co-culturing with substantia nigra appears to promote patch formation in E19 striatum grown in organotypic culture.

Other markers of the developing patch compartment were found to exhibit a patchy expression in a majority of the cultures. Immunoreactivity for the GluR1 subtype of AMPA-selective glutamate receptors, which we previously demonstrated to be expressed more by patch neurons during development (13), was patchy in over 80% of cultures derived from any age striatum. In E19 striatal cultures, GluR1 immunoreactivity was initially homogeneously distributed, and became patchy in appearance over the course of the first week *in vitro*. In co-cultures of striatum with substantia nigra, TH-immunoreactive DA fibers grew into the striatal piece in a clearly patchy fashion, even in E19 striatal cultures. Again, over 80% of striatal slices from each age group exhibited patches of TH-immunoreactive fibers by 7 div. These findings might suggest that whereas the early-forming nigrostriatal DA innervation plays a role in promoting patch formation, it may not be the only factor necessary for the complete development of the patch compartment, because not all of the E19 striatal-nigral co-cultures developed patches of BrdU-labeled cells. In addition, cortex does not appear to play a role in initial patch formation *in vitro*.

Table 1. Percent of striatal cultures exhibiting patches of BrdU-labeled cells as a function of age of striatal tissue and co-culture condition.

Age of striatal tissue	Co-culture condition[1]			
	Striatum alone	Striatum + nigra	Striatum + cortex[2]	Striatum + nigra + cortex
E19	10	33	4	29
E20-22	36	50	55	18
P1-P4	88	90	88	72

[1]Included in each group are cultures examined after 7 div. N=7-23 per group.
[2]Includes cultures with prelimbic or somatosensory cortex.

5. CONCLUSIONS

Overall, the results of these three types of experiments demonstrate that whereas the corticostriatal aferents are initially distributed to striatal patches, this is a postnatal event that appears too late to play a significant role in initial striatal patch formation. Instead, the nigrostriatal innervation that develops prenatally may function to promote initial patch formation. Corticostriatal innervation, on the other hand, may function to stabilize or maintain patches once they are formed. Pharmacological approaches will be necessary to determine whether the influences of co-cultured nigral and cortical tissue are specifically dopaminergic and glutamatergic in nature, respectively.

6. ACKNOWLEDGEMENTS

This work was supported by NIH grant MH46577.

7. REFERENCES

1. C A . Alar, N . Cai, T. Bliven, M .Juhsasz, J.M. Conner, A.L. Acheson, R.M. Lindsay and S.J. Weigand, Anterograde transport of brain-derived neurotrophic factor and its role in the brain, *Nature* 389: 856-860 (1997).
2. L.C. Costantini, S.C., Feinstein, M.J. Radeke, and A. Snyder-Keller, Compartmental expression of trkB receptor protein in developing striatum, *Neurosci.* 89: 505-513 (1999).
3. J.P. Donoghue M. Herkenham, Neostriatal projections from individual cortical fields conform to histochemically distinct striatal compartments in the rat, *Brain Res.* 365: 397-403 (1986).1986

4. G.A. Foster, M. Schultzberg, T. Hokfelt, M. Goldstein, H.C. Hemmings, Jr., C.C. Ouimet, S.I. Walaas,and P. Greengard, Development of a dopamine- and cyclic adenosine 3':5'-monophosphate-regulated phosphoprotein (DARPP-32) in the prenatal rat central nervous system, and its relationship to the arrival of presumptive dopaminergic systems, *J. Neurosci.* 7: 1994-2018 (1987).
5. C.R. Gerfen, The neostriatal mosaic: compartmentalization of corticostriatal input and striatonigral output systems, *Nature* 311: 461-463 (1984).
6. C.R Gerfen, The neostriatal mosaic: striatal patch-matrix organization is related to cortical lamination, *Science* 246: 385-388 (1989).
7. J.G. Johnston, C.R Gerfen, S.N. Haber and D. van der Kooy, Mechanisms of striatal pattern formation: conservation of mammalian compartmentalization, *Devel. Brain Res.* 57: 93-102 (1990).
8. I. Liste, G. Rozas, M.J. Guerra, and J.L. Labandeira-Garcia, Cortical stimulation induces Fos expression in striatal neurons via NMDA glutamate and dopamine receptors, *Brain Res.* 700: 1-12 (1995).
9. L.K. Nisenbaum, S.M. Webster, S.L. Chang, K.D. McQueeney and J.J. LoTurco, Early patterning of prelimbic cortical axons to the striatal patch compartment in the neonatal mouse, *Devel. Neurosci.* 20: 113-124 (1998).
10. A.N. Sheth, M.L. McKee and P.G. Bhide, The sequence of formation and development of corticostriate connections in mice, *Devel. Neurosci.* 20: 98-112 (1998).
11. A. Snyder-Keller, Development of striatal compartmentalization following pre- or post-natal dopamine depletion, *J. Neurosci.*, 11:810-821 (1991).
12. A. Snyder-Keller, The development of striatal patch/matrix organization after prenatal methylazoxymethanol (MAM): An immunocytochemical and BrdU birthdating study, *Neurosci.* 68: 751-763 (1995).
13. A. Snyder-Keller and L.C. Costantini, Glutamate receptor subtypes localize to patches in the developing striatum, *Develop. Brain Res.* 94: 246-250 (1996).
14. L. Stoppini, P.-A. Buchs, and D. Miller, A simple method for organotypic cultures of nervous tissue, *J. Neurosci. Meth.* 37: 173-182 (1991).
15. D. van der Kooy, Early postnatal lesions of the substantia nigra produce massive shrinkage of the rat striatum, disruption of patch neuron distribution, but no loss of patch neurons, *Devel. Brain Res.* 94: 242-245 (1996).
16. D. van der Kooy and G. Fishell, Neuronal birthdate underlies the development of striatal compartments, *Brain Res.* 401: 155-161 (1987).
17. D. van der Kooy and G. Fishell, Embryonic lesions of the substantia nigra prevent the patchy expression of opiate receptors, but not the segregation of patch and matrix compartment neurons, in the developing rat striatum, *Devel. Brain Res.* 66: 141-145 (1992).
18. P. Voorn, A. Kalsbeek, B. Jorritsma-Byham and H.J. Groenewegen, The pre- and postnatal development of the dopaminergic cell groups in the ventral mesencephalon and the dopaminerigc innervation of the striatum of the rat, *Neurosci.* 25: 857-887 (1988).

RE-EVALUATION OF MARKERS FOR THE PATCH-MATRIX ORGANIZATION OF THE RAT STRIATUM
Core and shell in the striatum

Richard E. Harlan, Monique Guillot, and Meredith M. Garcia*

Key words: Mu opiate receptor, somatostatin, calbindin D28k, caudate-putamen, nucleus accumbens

Abstract: To determine the degree of coincidence of patch-matrix boundaries in the rat striatum identified by several markers, we performed immunocytochemistry on adjacent 40 µm sections, using antibodies to the mu opiate receptor (MOR), somatostatin-28 and calbindin D28k. Superimposed camera lucida drawings revealed close correspondence between MOR-rich regions and regions poor in calbindin and somatostatin fibers. However, this correspondence was evident only in the ventral 2/3 of the rostral caudate-putamen (CPu) and core of the nucleus accumbens (AcbC). In the dorsal CPu at rostral levels, the shell of the accumbens (AcbSh), the olfactory tubercle, and the entire CPu caudal to the posterior limb of the anterior commissure, there was no obvious relationship among these three markers. These results suggest that the entire striatum can be divided into a core and a shell. The core is found in the rostral-ventral region, where there is close correspondence among several patch-matrix markers. The surrounding shell consists of the AcbSh, olfactory tubercle, and an extensive region of the CPu dorsal and caudal to the core, where the patch-matrix markers do not coincide.

1. INTRODUCTION

The organization of the mammalian striatum has been the subject of intense investigation. On a large scale, the striatum can be divided into dorsal and ventral regions, with the

---

* R.E. Harlan and M. Guillot, Department of Anatomy. R.E. Harlan and M.M. Garcia, Neuroscience Program. M. Guillot and M.M. Garcia, Department of Otolaryngology. Tulane University School of Medicine, 1430 Tulane Ave., New Orleans, LA 70112.

dorsal striatum consisting of the caudate and putamen, and the ventral striatum consisting of the nucleus accumbens and the olfactory tubercle[1]. The nucleus accumbens is further divided into a core region (AcbC) that is rather similar to the dorsal striatum, and a shell (AcbSh) that can be considered a transition between the striatum and the extended amygdala[2,3]. Division of the dorsal striatum into a small patch or striosome compartment and a much larger matrix compartment has been a fundamental concept since the pioneering work of Graybiel and Ragsdale[4]. These compartments differ in terms of neurochemistry, patterns of connections, and time of neurogenesis[5,6]. More recent work has suggested a non-uniformity in the matrix compartment, i.e. an organization into different matrisomes[5,7-9]. The uniformity of the patch compartment has been emphasized in several studies that have demonstrated alignment of patch-matrix boundaries identified by several markers (reviewed in 10). However, these studies have concentrated primarily in the rostral levels of the dorsal striatum, where some markers show the most abundant striosomes. To determine the correspondence among three commonly-used markers of patch-matrix organization in the rat striatum, we performed immunocytochemistry on adjacent sections throughout the entire striatum. We focused on somatostatin fibers and calbindin D28K, both of which are enriched in the matrix compartment, and on mu opiate receptors (MOR) that are enriched in the patch compartment[10]. Our results suggest a largescale division of the entire striatum into a core, where patch-matrix markers coincide, and a shell, where marked disjunction of patch-matrix markers is evident.

2. METHODS

Adult, male rats, anesthetized with an overdose of pentobarbital, were perfused transcardially with phosphate-buffered saline (PBS) followed by freshly-prepared 3% buffered paraformaldehyde, as previously described[11]. Brains were postfixed in paraformaldehyde for 2 hours, cryoprotected in 30% sucrose and frozen rapidly with crushed dry ice. Serial 40 μm sections were cut in the coronal or horizontal planes on a sliding microtome and adjacent sections were immunocytochemically stained with antibodies to somatostatin-28 (S-309, gift of R. Benoit), mu opiate receptor (Diasorin), and calbindin D28k (Sigma). Camera lucida drawings were made and superimposed to reveal correspondence among patch-matrix markers. Alignments were made by drawing myelinated fiber bundles in the caudate-putamen (CPu) or blood vessels in the AcbSh.

3. RESULTS

In the rostral striatum, a ring of MOR immunoreactivity was found surrounding the most rostral extent of the CPu and rostral pole of the Acb. The central region of both the CPu and Acb was largely devoid of MOR, but was filled with somatostatin fibers with a density inversely related to the intense MOR staining. However, there was very little calbindin, except on the lateral edge of the Acb in a region that also overlapped with intense MOR staining. Somatostatin fibers were much less frequent at the rostral pole of the CPu than in the rostral pole of the Acb, although there was a slight increase in the density of somatostatin fibers in regions devoid of MOR staining in the rostral CPu. At a more caudal level, where

AcbSh is more clearly defined (Fig. 1), good correspondence between MOR patches and absence of calbindin was seen in the ventral-medial two-thirds of the CPu. However, the region demonstrating a good alignment between MOR patches and regions relatively devoid of somatostatin fibers was considerably smaller. All three markers showed registry of patches in the ventral-medial third of the CPu and in some of AcbC. The dorsal lateral CPu had distinct MOR patches and subcallosal streak, which were not evident with either calbindin or somatostatin staining. In AcbSh, intense MOR staining was seen along the ventral medial edge. Part of this region was also stained intensely for somatostatin, but not for calbindin.

This general pattern of correspondence between the three patch-matrix markers in the ventral medial third of the CPu was evident at successively caudal sections to the level of the posterior limb of the anterior commissure (acp). Within this region of correspondence, the registry between MOR patches and striosomes defined by the absence of calbindin was somewhat more extensive than that defined by absence of somatostatin fibers. In addition, the intensity of MOR immunoreactivity was noticeably less in this region than in regions more dorsal, lateral and caudal, where there was little or no correspondence with calbindin or somatostatin.

Outside of this ventral and medial area of correspondence in the CPu, there were regions in which two markers corresponded with each other, but not with the third marker. For instance, immediately dorsal to acp, there was good correspondence between MOR patches and the absence of somatostatin fibers; however, this region was filled with intense calbindin staining. Similarly, in areas dorsal to the region of maximal correspondence of the three markers, MOR patches corresponded to regions relatively devoid of calbindin; however, it was difficult to make a distinction between patch and matrix based on density of somatostatin fibers in this region. Caudal to the acp, MOR patches appeared to form arcing rows roughly parallel to the subcallosal streak. In general, these rows did not overlap the more medial region of the CPu, which was filled fairly homogeneously with calbindin. The most medial of the MOR rows bordered the lateral edge of the globus pallidus, and overlapped extensively with a calbindin-rich zone which coincided with a small region relatively enriched in somatostatin fibers. However, calbindin immunostaining in this region was rather homogeneous, and not apparently divided into patch and matrix.

4. DISCUSSION

Throughout much of its rostral-caudal extent, the core of the nucleus accumbens contained few MOR patches. Those few patches, however, generally corresponded to regions relatively devoid of calbindin and somatostatin, although some small regions devoid of somatostatin fibers did not appear to correspond to MOR patches. The shell of the Acb was markedly heterogeneous in staining patterns. MOR patches bordered the medial edge of AcbSh, often forming a pattern similar to that of the subcallosal streak. These spots of MOR staining along the medial border of AcbSh sometimes corresponded to small regions relatively devoid of somatostatin fibers. However, because the entire AcbSh was nearly devoid of calbindin, there was no relationship between MOR and calbindin. A second line of small MOR spots or patches was seen near the lateral edge of Acb, parallel to the line along the medial edge. This line of MOR patches corresponded in general with a region relatively devoid of somatostatin and with the border between the calbindin-rich AcbC and the

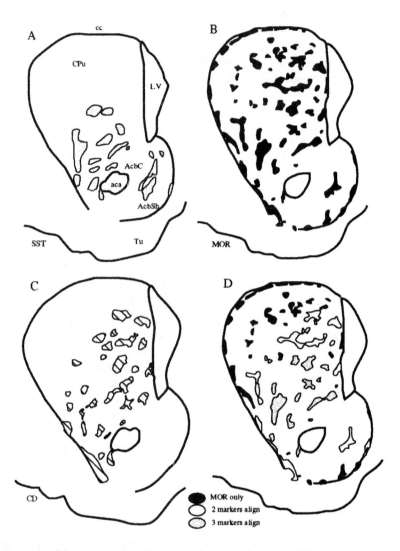

Figure 1. Drawings of three consecutive 40 μm coronal sections through the CPu and Acb, immunostained for somatostatin (SST, panel A), mu opiate receptors (MOR, panel B) and calbindin (CD, panel C). Patches were defined by intense staining for MOR, or relative absence of staining for SST fibers or CD within a region intensely stained for these two markers. Thus, large regions devoid of CD or SST fibers, such as the dorsolateral CPu, were not defined as patches. Panel D repeats the section immunostained for MOR, which was sandwiched between sections stained for SST and CD. Patches defined only by MOR are in black; patches with registry of two markers (white), or all three markers (stipple) are also indicated. The region in which all three markers are aligned is restricted to the ventral-medial CPu and part of AcbC. This core region is surrounded by a halo in which two markers coincide. Outside of this halo is the shell of the striatum, in which only MOR demarcates patches. Note the similarity between the dashed lines of MOR along the subcallosal streak, adjacent to the corpus callosum/external capsule, and a similar streak along the medial border of AcbSh. aca, anterior limb of anterior commissure; AcbC, core of nucleus accumbens; AcbSh, shell of nucleus accumbens; cc, corpus callosum; CPu, caudate-putamen; LV, lateral ventricle; Tu, olfactory tubercle.

calbindin-poor AcbSh. In the remainder of AcbSh, there was little correspondence among the three markers. In the olfactory tubercle, there was little or no MOR immunostaining, some radial stripes of calbindin, and moderate to intense somatostatin fiber staining.

An overall pattern emerges from the regions of alignment and lack of correspondence among these three markers. All three markers agree on patch-matrix boundaries in the ventral and medial region of the rostral CPu, and including the core of the Acb. In terms of the patch-matrix organization, this region can be considered the core of the striatum. Surrounding this region is a small halo where two of the three markers match, and outside of this penumbra is a much larger region with little correspondence among the markers. This larger region, which can be considered the shell of the striatum, includes the dorsal half or two-thirds of the CPu rostral to the posterior limb of the anterior commissure, essentially all of the CPu caudal to acp, most of the AcbSh, and the olfactory tubercle. This division into core and shell is most evident at a level where AcbSh is fully developed (Fig. 1). A core region, where all three markers coincide is bordered dorsolaterally with a region devoid of calbindin and containing many somatostatin cell bodies, but with a less dense concentration of somatostatin fibers. The core is bounded ventromedially by AcbSh, with some MOR patches corresponding to an absence of somatostatin fibers, but with little correspondence to calbindin. Both the dorsolateral and ventromedial shell regions contain little calbindin and are bordered by dashed lines of MOR.

Several other studies have indicated regions of alignment and nonalignment of patch-matrix markers in the striatum of human[12], monkey[13,14] cat[15] and rat[10,19]. The complexity and non-alignment of markers has been emphasized as a hallmark of the nucleus accumbens, especially in the shell (e.g. *16;* reviewed in 2, 20). In the CPu, however, most studies have concentrated on the rostral region, where the patch-matrix organization is quite prominent (e.g. 12). At this rostral level, calbindin is enriched in most of the matrix in humans [12] and rat [10]. At more caudal levels, however, calbindin is progressively restricted to the more medial region of the putamen in monkeys [17,18] or CPu in rat [18]. Nevertheless, MOR patches are clearly evident throughout the calbindin-poor zone of the dorsolateral CPu rostrally and the lateral portion caudally. It is interesting to note that the MOR patches in this calbindin-poor zone tend to line up in arcing rows parallel to the subcallosal streak. A double-migration model of striatal neurogenesis has been proposed as a potential explanation [2,6].

There may be species differences in the relative positions of the core and shell of the striatum. In the rat, we argue that the core is found in the ventral, medial and rostral extent of the CPu and AcbC. In humans, the region of greatest alignment of patch-matrix markers is in the dorsal caudate[12], although the rostral-caudal extent of this region was not explored systematically. Division of the entire striatum into core and shell may provide a new means of organizing information and patterns of connectivity and neurochemistry in the striatum.

5. ACKNOWLEDGEMENTS

Supported by grant LEQSF-RD-A-29 to MMG.

6. REFERENCES

1. Heimer, L., D.S. Zahm and G.F. Alheid. Basal Ganglia. In Paxinos, G. The Rat Nervous System. 2nd Edition, 1995 Academic Press, San Diego.
2. Heimer, L., G. Alheid, J. de Olmos, H. Groenewegen, S. Haber, R.E. Harlan, and S. Zahm *(1997)* The Accumbens: Beyond the Core-Shell Dichotomy. J. Neuropsychiatry, *9: 354-381.*
3. Heimer, L., R.E. Harlan, G.F. Alheid, M. Garcia and J. De Ohnos *(1997)* Substantia innominata: A notion which impedes clinical-anatomical correlations in neuropsychiatric disorders. Neuroscience, *76: 957-1006.*
4. Graybiel, A.M. and C.W. Ragsdale *(1978)* Histochemically distinct compartments in the striatum of human, monkey, and cat demonstrated by acetylcholinesterase staining. Proc. Natl. Acad. Sci. USA *75: 5723-5726.*
5. Graybiel, A.M., A.W. Flaherty and J.-M. Gimenez-Amaya *(1990)* Striosomes and matrisomes. In G. Bernardi (ed): The Basal Ganglia III. New York: Plenum Press, *pp. 3-12.*
6. Song, D.D. and R.E. Harlan *(1994)* Genesis and migration patterns of neurons forming the patch and matrix compartments of the rat striatum. Dev. Brain Res. *83: 233-246.*
7. Flaherty, A.W. and A.M. Graybiel *(1993)* Output architecture of the primate putamen. J. Neurosci. *13: 3222-3237.*
8. Flaherty, A.W. and A.M. Graybiel *(1994)* Input-output organization of the primate putamen. J. Neurosci. *14: 599-610.*
9. Selemom. L.D. and P.S. Goldman-Rakic *(1995)* Topographical intermingling of striatonigral and striatopallidal neurons in the rhesus monkey. J. Comp. Neurol. *297: 359-376.*
10. Gerfen, C.R. *(1992)* The neostriatal mosaic—Multiple levels of compartmental organization in the basal ganglia. Annu. Rev. Neurosci. *15: 285-320.*
11. Harlan, R.E. and M.M. Garcia. *(1995)* Charting of Jun family member proteins in the rat forebrain and midbrain: Immunocytochemical evidence for a new Jun-related antigen. Brain Res. *692: 1-22.*
12. Holt, D.J., A.M. Graybiel and C.B. Saper *(1997)* Neurochemical architecture of the human striatum. J. Comp. Neurol. *384: 1-25.*
13. Fotuhi, M., T.M. Dawson, A.H. Sharp, L.J. Martin, A.M. Graybiel and S.H. Snyder (1993) Phosphoinositide second messenger system is enriched in striosomes: Immunohistochemical demonstration of inositol 1,4,5-triphosphate receptors and phospholipase C β and γ in primate basal ganglia. J. Neurosci. 13: 3300-3308.
14. Martin, L. J., M.G. Hadfield, T.L. Dellovade, and D.L. Price (1991) The striatal mosaic in primates: Patterns of neuropeptide immunoreactivity differentiate the ventral striatum from the dorsal striatum. Neurosci. 43: 397-417.
15. Graybiel, A.M., C.W. Ragsdale, E.S. Yoneoka and R.P. Elde (1981) An immunohistochemical study of enkephalins and other neuropeptides in the striatum of the cat, with evidence that the opiate peptides are arranged to form mosaic patterns in register with the striosomal compartments visible by acetylcholinesterase staining. Neuroscience 6: 377-397.
16. Jongen-Relo, A.L., H.J. Groenewegen and P. Voorn (1993) Evidence for a multi-compartmental histochemical organization of the nucleus accumbens in the rat. J. Comp. Neurol. 337: 267-276.
17. Francois, C., J. Yelnick, G. Percheron, and D. Tande (1994) Calbindin D-28k as a marker for the associative cortical territory of the striatum in macaque. Brain Res. 633: 331-336.
18. Gerfen, C.R., K.G. Baimbridge and J.J. Miller (1985) The neostriatal mosaic: compartmental distribution of calcium binding protein and parvalbumin in the basal ganglia of the rat and monkey. Proc. Natl. Acad. Sci. USA 82: 8780-8784.
19. Mengual, E., C. Casanovas-Aquilar, J. Perez-Clausell, and J.M. Gimenez-Amaya (1995) Heterogeneous and compartmental distribution of zinc in the striatum and globus pallidus of the rat. Neurosci. 66: 523-537.
20. Zahm, D.S. and J.S. Brog (1992) On the significance of subterritories in the accumbens part of the rat ventral striatum. Neurosci. 50: 751-767.

PHYSIOLOGICAL AND MORPHOLOGICAL CLASSIFICATION OF SINGLE NEURONS IN BARREL CORTEX WITH AXONS IN NEOSTRIATUM

E.A.M. Hutton, A.K. Wright, and G.W. Arbuthnott*

Abstract: We have recorded intracellularly from the neurons of the barrel cortex and filled them with biocytin in order to identify the neurons of origin of the corticostriatal systems, which we have previously described from this area of cortex. This *in vivo* study has confirmed that whilst intermittent bursting (IB) cortical neurons and regular spiking (RS) neurons show the same characteristic firing patterns as those observed *in vitro*, the firing patterns observed *in vivo* are more complex and varied. We were able to identify morphological characteristics of cells that are the origin of the topographical corticostriatal pathway, and demonstrate the involvement of both RS and IB neuron types. So far we have been unsuccessful in morphologically identifying the cells which provide the 'diffuse' corticostriatal system although the knowledge that they occupy the septae between the barrels and can be antidromically activated from contralateral cortex has provided a distinctive physiological signature for this pathway.

1. INTRODUCTION

The mystacial vibrissae (whiskers) provide rodents with the major source of sensory information from their environment. From this information the rodent is then able to execute an appropriate motor action. This system is especially important in the assessment of stimuli immediately ahead, surrounding the snout of the animal.

The large mystacial vibrissae are arranged in 5 rows either side of the snout running dorsal to ventral (rows A-E); whiskers within a row are labeled numerically 1-8 running caudal to rostral. Whiskers that run across rows i.e. B1, C1, D1 are termed arcs[20,46]. The four most caudal vibrissae are located behind the arc of A1-E1, lie between the rows, and are labeled α-δ[46]. The remainder of the snout is covered with fine short fur known as common fur. Rodents receive their sensory information from their whiskers in two basic ways i.e. via an

---

* E.A.M. Hutton, A.K. Wright, and G.W. Arbuthnott, University of Edinburgh Centre for Neuroscience, Department of Preclinical Veterinary Sciences, Summerhall, Edinburgh, EH9 1QH, UK.

active or passive mode. The whiskers are usually used in the active mode ("whisking") where they are moved at a regular frequency of approximately 8 Hz[8,9].

The processing of vibrissal information occupies an area of approx. 20% of the primary somatosensory cortex (SI) [43]. In sections of SI cut tangentially to the pial surface and stained for the enzyme cytochrome oxidase [45], a whisker pattern can be clearly seen in posterior SI where cytochrome oxidase rich areas termed barrels [46] are related in a 1:1 manner with the facial vibrissae. The whisker representation seen in the posterior of the field is termed the Posterior Medial Barrel Subfield (PMBSF) [46]. The PMBSF appears, in coronal sections, as a cell dense, cytochrome oxidase rich, cell layer IV within which the barrels are surrounded by less cell dense and less cytochrome oxidase rich septal areas. The richness of layer IV clearly contrasts with the relatively cytochrome oxidase poor layer Va beneath. The cytochrome oxidase staining and cell density increases again within layers Vb and VI. This lamination pattern within the vibrissal region (and in SI in general) is distinct from other cortical regions. It is this detailed anatomical and functional structuring between the periphery and the ascending and descending pathways within the cortex that make SI an ideal cortical region in which to study cortical processing of peripheral information and its relay to the striatum and the basal ganglia *in vivo*.

In general terms neuronal firing patterns of neocortical neurons can be divided into 3 main classes: Fast spiking (FS), Regular spiking (RS) and Intrinsically bursting (IB). RS and IB neurons are the spiny pyramidal cells [28] and the FS neurons are the aspiny GABAergic interneurons [14].

RS neurons as a group tend to have smaller soma than IB neurons and are more spherical in shape [11]. Their apical dendrites can be clearly seen ascending from the soma towards the pial surface terminating in a relatively sparse apical tree. IB neurons tend to have more extensive basilar dendritic trees with the apical branch point occurring in layers II/III (much lower down than in RS neurons). The basilar dendrites of RS cells appear to have a horizontal extension of approximately 400 μm around the cell body whereas IB neurons tend to have more extensive (and spiny) basilar dendritic trees. RS cells have 2-8 axon collaterals that ascend vertically through the cortex terminating in either the supragranular layers or extending towards the pial matter [11,13]. The axon collaterals from IB neurons however extend horizontally through layer V for up-to 1.5-2.0 mm [13].

The most common response of RS neurons to a stimulus is to fire a single action potential [3], they do not show a tendency to burst (even with current injection) [14,18,28] and are sensitive to changes in the stimulus current and amplitude by adapting their spike frequency [14,40]. In response to a stimulus IB neurons will typically fire a "high frequency cluster of action potentials" termed a burst. Bursts typically comprise 3-5 action potentials with decreasing spike amplitude seen within a burst [1,14,15,18,27,28,33,36,41] and the number of action potentials in a burst is stimulus dependant [1]. The action potentials within cortical bursts are essentially the same as those observed within RS neurons although they do appear to have a more prominent depolarizing aspect to their afterpotentials [3]. The IB can fire single action potentials similar to those described in RS neurons and can switch between the 2 firing patterns mainly firing an initial burst followed by a train of single action potentials [2,13,15,18,27,28]. IB neurons typically fire at longer latencies than other cortical neurons [2,5] and are concentrated within lower layer V in the mouse[1,13,22,24,32,37,38,42] or in a rat sensorimotor cortex in lower layer IV and Va [5,11,12,15,28].

From the literature available for rat vibrissal primary somatosensory cortex it appears that the IB neurons are capable of receiving inputs from a wide area of sensory cortex including several barrels. From the literature it appears that the IB neurons within the middle cortical layers can amplify, initiate, co-ordinate and synchronize their activity to act as a network and can generate a cortically based rhythm [2,12,14,18]. These neurons normally fire independently but upon activation with an appropriate stimulus their firing can become coordinated even in the face of strong inhibition [11,12].

The vibrissal region of SI projects to various subcortical sites both ipsilaterally and contralaterally. Ipsilaterally SI projects to the striatum [26,29,31] thalamus [26,34,35] and brainstem [31]. Contralateral projections include thalamus[7] and striatum[6,29,30].

The corticostriatal pathway in general can be divided into two separate pathways [16]. There is a specific corticostriatal pathway arising in the superficial layer V (Va) neurons [31]. There is also a pathway arising from fine axon collaterals from deeper layer V corticofugal neurons [16,31].

The specific corticostriatal projection neurons are described as having thinner apical dendrites than the corticospinal or the corticopontine neurons and have apical dendrites that have side branches within layer V and have terminal tufts through layers III-I [21]. The input to the striatum arising from the non-columnar layer Va pyramids forms a more slowly conducting and more diffuse input to the striatum than the corticofugal input. This specific corticostriatal pathway is also independent of the brainstem and spinal inputs and is less directly influenced by the specific sensory thalamus (due to the lack of thalamic input [2]) and is more likely to be influenced by cortical interactions and processing. The striatal input from the corticofugal neurons arises from a lateral axon collateral branch innervating the striatum. The collateral leaves the parent axon as it passes through the striatal fiber bundles[17].

The layer Vb input as well as being rapid is more likely to be under thalamic influences (due to its monosynaptic input from VPm) and also provides a parallel input to the brainstem and spinal systems [17]. Similarly from barrel cortex in particular two distinct projections reach the striatum; one is topographic and seems to originate from collaterals of the corticofugal cells[47]; the other is bilateral, diffusely distributed, and seems to arise from the corticocortical cells of the septae between the barrels[4].

The layer V projection neurons of this region have been studied extensively (especially *in vitro*) and many comparable studies have been performed in both motor and visual cortex [19,24,25,44]. The aim of this study was to establish the physiological response of layer V projection neurons to whisker deflection alongside their morphology *in vivo* and to investigate their anatomical and physiological input to the striatum and the basal ganglia.

2. METHODS

Male Sprague Dawley rats (300-400 g, Harlan Olac) were anaesthetized intraperitoneally with 1.0 ml/100 g of a 10% urethane 1% chloralose mix, and additional anaesthetic (10% of original dose) administered when required. A tracheotomy was performed and the head was secured in a stereotaxic frame and the skull exposed. A CSF drain was created through the atlanto-occipital membrane. Body temperature was maintained at 37°C throughout.

An area (approximately 2 mm diameter) of the cortical surface (corresponding to a small area of SI [AP - 3.0 mm, L 5.5 mm]) was exposed and the dura deflected. A dental cement well approx. 1 cm tall was then built surrounding the exposed cortex and filled with physiological saline or low setting point paraffin wax for intracellular recording.

Single whiskers across the B1, C1 and D1 arc were isolated from each other and stimulated independently via capillary electrode glass attached to the piezoelectric bimorphs (2 mm whisker deflection, 4 mS duration). In some animals a concentric stimulating electrode was inserted into barrel cortex contralateral to the recording site.

Glass microelectrodes containing biocytin hydrochloride (5% in 1 M potassium acetate) with resistances of between 100-200 MΩ were advanced through the cortex in 3 µm steps to a maximum depth of 1200 µm or until a neuron was impaled. The cell detection stimulus used was an intracellular current pulse (100 mS long; 500 mS cycle length) which ensured rapid cell detection. Once impaled the response to whisker stimulation was established and the neuron classified as RS, IB or FS. Single whisker deflections were repeated every 2.5 S.

If the neuron was stable without holding current (i.e. low resting membrane potential, spike heights of more than 30 mV and a stable firing pattern) it was filled with biocytin. Filling was achieved using hyperpolarising current (-2 nA, 200 mS on / 200 mS off cycle). Filling continued until either the membrane became more depolarized, the firing pattern changed, and/or there was a decrease in the amplitude of the EPSPs or the action potentials. Filling was terminated after a maximum of 20 min and the electrode carefully removed from the cortex.

Animals were given a terminal dose of the chloralose urethane anaesthetic and perfused transcardially with 25 ml of heparinised (10 U/ml) saline (0.9%) followed by 300 ml of a fixative containing 2°/o paraformaldehyde and 1.25% glutaraldehyde in 0.1 M phosphate buffer pH 7.4 to ensure biocytin was fixed in the presynaptic terminals [39]. Once perfused the brains were removed and stored in 10% buffered sucrose fixative.

The brains were sectioned coronally using a freezing microtome taking 50 µm serial sections and processing the tissue as free floating sections under constant agitation. The sections were incubated for 5 hours in PBS-TX before being incubated in ABC elite (1:50 in PBS-TX) at 4°C for 60 h. They were then washed 3 times over 30 min in PBS and visualized using DAB, then washed three times in PBS, mounted onto chrome alum subbed microscope slides, dried and coverslipped with DPX. Some cases were counterstained with 5% methyl green.

3. RESULTS

Morphological classification was performed on a limited number of criteria i.e. branch point of the apical dendrite, the extension of the apical and basilar dendritic trees described *in vitro* [14]. The orientation of the axon collaterals was rarely used, as in this study they were not reliably filled or easily classifiable. As the criteria used are physical it is thought that the classification criteria (and cell class) do not differ between *in vivo* and *in vitro* recording systems. Despite limiting the number of classification criteria used compared to previous *in vitro* studies the neurons were still successfully classified as either RS or IB. In 23 cells where both morphological and physiological classification could be made there were no cells where the two methods disagreed. Cells with RS morphology all fired single spikes and all bursting cells have the morphology typical of IB cells. An important consideration when analyzing the results of intracellular tracing studies is the rarity of successful recordings

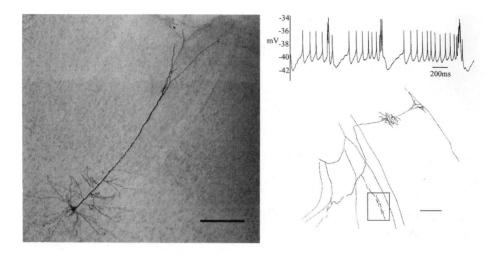

Figure 1. An IB cortical cell is shown in the photomicrograph on the left and its axonal distribution illustrated the in camera lucida drawing on the right. Above the drawing of the complete cell is an example of its normal electrophysiological activity. It clearly fires high frequency bursts of 3-5 action potentials on a depolarizing wave as well as single spikes. The calibration mark on the photomicrograph represents 100μm while the one on the drawing is 500μm.

with complete filling and tracing of the axon path and terminal fields. Of 137 cortical neurons studied only 23 were filled in layer V and of these only 4 had visible terminal fields.

Figure 1 shows a neuron for which a near complete reconstruction was possible. The neuronal cell body was located at a stereotaxic co-ordinate between AP -3.14 and AP -2.80. The neuron was classed as an IB. The axon left the cortex and had both anterior and posterior located postsynaptic targets. The axon initially ran in an anterior direction before branching within a striatal fiber bundle (AP -2.56). The branch ran within the striatal tissue at right angles to the parent axon and parallel to the corpus callosum. The striatal branch continued its path through the striatum crossing fiber bundles before ending as a hand-like arrangement of terminals within the striatal tissue (AP -2.12). As the main axon continued through the corpus callosum it appeared to branch and send an axon collateral to the reticular thalamus. The posterior projecting axon ran through the striatal fiber bundle and into the internal capsule. The axon then continued towards the brainstem.

This cell was classed as an IB neuron on the basis of its morphological characteristics. It had extensive apical and basilar dendritic trees as well as an apical dendrite that tapered as it ascended from the soma. The apical dendrite can clearly be seen to branch in layer III of cortex. In addition to the extensive basilar dendritic branching and axon collateral network within layer V the neuron also sent an axon collateral to SII. The collateral arose in layer V and terminated within layer V of SII (not shown). Unfortunately, for this neuron no physiology was recorded, so this classification is on the basis of morphology alone. The physiological trace in figure 1B is from a similar IB cell whose axon is sufficiently filled to be followed to striatal terminals.

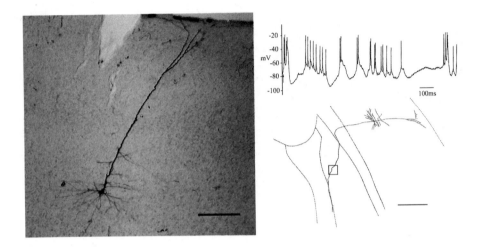

Figure 2. An RS cortical cell is shown in the photomicrograph on the left and its axonal distribution illustrated in the camera lucida drawing on the right. Above the drawing of the complete cell is an example of its electrophysiological response to whisker deflection. Although the initial doublet in response to stimulation rides on an EPSP most of the trace consists of typical single spikes. The calibration mark on the photomicrograph represents 100μm while the one on the drawing is 500μm.

In this case, and in figure 2, the photomicrographs and camera lucida reconstructions are orientated as follows: medial is always to the left of the figure, lateral is always right, dorsal to the top of the figure and ventral is always at the bottom of the figure. The position of the terminal fields in whole cell reconstructions are marked with a box.

Figure 2 shows a second neuron for which near complete reconstruction was possible. The cell has less extensive apical and basilar dendritic branching than is seen in the typical IB neurons and was classified as RS. The apical dendrite can clearly be seen to ascend from the rounded soma unlike the tapering dendritic origin that is characteristic of IB neurons. The neuron is located stereotaxically at AP -3.3. The axon leaves the cortex and enters the corpus callosum where it turns 90°, running through the corpus callosum before entering a striatal fiber bundle. There is a branch at AP -2.56 forming terminals that are orientated forwards. The terminal field is not as extensive as the IB neuron illustrated (Fig. 1A).

The physiology of this neuron appears to be similar to that described for RS neurons in that the typical response is to fire single action potentials. The cell fired doublets of action potentials but did not respond by firing bursts (classed as clusters of 3-5 action potentials associated with a depolarizing envelope).

In barrel side or septal recordings deflections of 2 whiskers give short almost identical response latencies and inhibitory periods (Fig. 3B). The slight differences between both the latency and inhibitory period values may reveal neuron position within the barrel. For example where the neuron is located almost in the center of the septum between two barrels there may be little difference in the latency and inhibitory period. Furthermore in two cases we were able to antidromically activate a neuron from a contralateral barrel field which

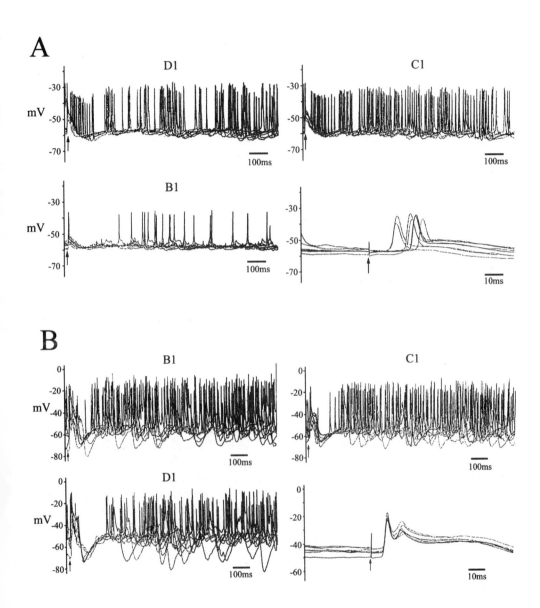

Figure 3. Electrophysiological responses (8 superimposed traces) to whiskers B1, C1, D1 are illustrated for both a barrel cell (A) and one in the septa (B). Responses to electrical stimulation of the contralateral cortex are illustrated in the lower right hand traces of each part of the figure. In (A) the responses are obviously orthodromic while those in (B) have the characteristic constant latency of antidromic activity.

identified the cell as 'septal'[4]. The whisker induced excitation in those cells (one of which is illustrated in Fig. 3B) are much less sharp than those from cells with a clear principal whisker (Fig. 3A) but whose response to contralateral cortex stimulation was orthodromic.

4. DISCUSSION

Whilst the morphology of the neurons in this study were similar to those described *in vitro* within somatosensory cortex [10,13,14] visual cortex [24,27] and motor cortex of rats [16,17,23], the physiology was more varied, in agreement with the recent report of cortical cells filled *in vivo*[48].

The physiological classification criteria used in this study with respect to burst firing were taken from *in vitro* studies [14]. In comparing results from the two experimental procedures the nature and duration of the stimulus must be taken into account. *In vitro* recordings are obtained by the injection of a current pulse into the neuron in question and the response recorded during the stimulation of the neuron. In this *in vivo* study the stimulus takes the form of a rapid (4 mS) stimulation of an individual whisker in the periphery. The response to the stimulus is then recorded after the stimulation i.e. there is no direct current pulse into the neuron and there is no continuous stimulation of the neuron. Although the stimuli in the two systems are quite different, examination of the neuronal firing patterns in the *in vivo* preparation reveals that the response patterns observed in RS and IB neurons remain basically the same. For example IB neurons fire in characteristic bursts that have been noted *in vitro* whereas RS neurons do not [14,48].

The recordings *in vivo* were never as 'clean' as those observed *in vitro*. In the *in vivo* study there was always a significant background synaptic activity associated with the resting membrane potential. It is impossible to be certain of the cause of this activity. It is possible that it may be a result either of synaptic interactions between cortical areas or of activity that is arising directly from the thalamus.

Bursting neurons appeared different to those described *in vitro*. The typical shape of a burst is described *in vitro* as 3-5 action potentials riding on a slow depolarizing envelope. In this *in vivo* preparation rather than the first action potential in the burst riding on the depolarizing envelope it appeared to precede it. The burst firing seen throughout this study was very varied and rarely were bursts the sole response to a stimulus. The most common patterns were a combination of bursts and single action potentials. In contrast the firing patterns of RS neurons in this study were similar to those described in the literature. Their only response to stimulation was to fire either single or double action potentials.

In both cell classes action potentials were followed by afterhyperpolarizations (AHP). In the case of burst firing cells, hyperpolarization followed the burst and its depth was to some extent dictated by the number of action potentials in a burst. An important observation was that after the stimulation of the neuron by a whisker the initial response was followed by a variable period where the cell was inhibited before it resumed firing.

In this study the distribution of IB or RS neurons appears to be uniform within layer V. What is not clear from this study, or the literature, is whether the distribution varies between the barrel and septal regions.

The exact location of a neuron within a specific barrel cannot be determined but there are differences in response latency, inhibitory period duration and firing pattern between IB and RS neurons when what appears to be the PW is deflected. In barrel center recordings PW deflection results in a shorter response latency and inhibitory period. This may relay precise information regarding object location to the striatum, thalamus and brainstem. For example (Fig. 3A) where D 1 is the PW for a neuron from the set of whiskers (DI, C 1 and B 1) then information from whisker C1 is likely to have a shorter latency and inhibitory period than B1.

However the latency and inhibitory period are clearly shorter for the PW. After PW deflections a clear consistent inhibitory period was observed. Deflections of a non-PW resulted in either no obvious inhibitory period or an inhibitory period that was not 'clean' i.e. was less intense and interrupted by stray action potentials.

Thus we have shown that the differentiation of the two output pathways from barrel cortex is unlikely to be along the RS - IB division in cell physiology because both of these groups of cortical cells appear to contribute to the 'topographical' input to the striatum. On the other hand from the information to date the 'diffuse' pathway originates in the septae between barrels and is defined by position and input pattern rather than by firing characteristics.

5. ACKNOWLEDGEMENTS

This work was supported by the MRC and by The Wellcome Trust (Grant No 49699)

6. REFERENCES

1. Agmon A., Connors B.W. (1989) Repetitive burst firing-neurons in the deep layers of the mouse somatosensory cortex. *Neuroscience Letters* 99, 137-141.
2. Agmon A., Connors B.W. (1992) Correlation between firing pattern and thalamocortical synaptic responses of neurons in the mouse barrel cortex. *J Neurosci 12*, 319-329.
3. Amitai Y., Connors B.W. (1995) Intrinsic physiology and morphology of single neurons in neocortex. In *The Barrel Cortex of Rodents* (eds. Jones E.G., Diamond I.T.), 7, pp. 299-232.Plenum, New York.
4. Arbuthnott G. W., Wright A. K. (1997)The cortical neurones forming the somatosensory corticostriatal projections in the rat. *Society for Neuroscience Abstracts* 23, 82.3.
5. Armstrong-James M., Fox K., Das-Gupta A. (1992) Flow of excitation within rat barrel cortex on striking a single vibrissa. *J Neurophysiol* 68, 1345-1358.
6. Berendse H.W., Galis-de Graaf Y., Groenewegen H.J. (1992) Topographical organization and relationship with ventral striatal compartments of prefrontal corticostriatal projections in the rat. *J Comp Neurol* 316, 314-347.
7. Caretta D., Sbriccoli A., Santarelli M., Pinto F., Granto A., Minciaahi D. (1996) Crossed thalamocortical and cortico-thalamic projections in adult mice. *Neuroscience Letters* 204, 69-72.
8. Carvell G.E., Simons D.J. (1990) Biometric analyses of vibrissal tactile discrimination in the rat. *J Neurosci* 10(8), 2638-2648.
9. Carvell G.E., Simons D.J. (1996) Abnormal tactile experience early in life disrupts active touch. *J Neurosci* 16(8), 2750-2757.
10. Cauller L.J., Connors B.W. (1994) Synaptic physiology of horizontal afferents to layer I in slices of rat SI neocortex. *J Neurosci* 14, 751-762.
11. Chagnac-Amitai Y., Connors B.W. (1989a) Horizontal spread of synchronised activity in neocortex and its control by GABA-mediated inhibition. *J Neurophysiol* 61, 747-758.
12. Chagnac-Amitai Y., Connors B.W. (1989b) Synchronised excitation and inhibition driven by intrinsically bursting neurons in neocortex. *J Neurophysiol* 62, 1149-1162.
13. Chagnac-Amitai Y., Luhman H.J., Prince D.A. (1990) Burst generating and regular spiking layer 5 pyramidal neurons of rat neocortex have different morphological features. *J Comp Neurol* 296, 598-613.
14. Connors B.W., Gutnick M.J. (1990) Intrinsic firing patterns of diverse neocortical neurons. *Trends Neurosci* 13, 99-104.
15. Connors BW., Gutnick M.J., Prince D.A. (1982) Electrophysiological properties of neocortical neurons in vitro. *J Neurophysiol* 48, 1302-1320.

16. Cowan R.L., Wilson C.J. (1994) Spontaneous firing patterns and axonal projections of single corticostriatal neurons in the rat medial agranular cortex. *J Neurophysiol* 71, 17-32.
17. Donoghue J. P., Kitai S.T. (1981) A collateral pathway to the neostriatum from corticofugal neurons of the rat sensory-motor cortex: An intracellular HRP study. *J Comp Neurol* 201, 1-13.
18. Franceshetti S., Guatteo E., Panzica F., Sancini G., Wanke E., Avanzini G. (1995) Ionic mechanisms underlying burst firing in pyramidal neurons: Intracellular study in rat sensorimotor cortex. *Brain Res* 696(12), 127-139.
19. Gilbert C.D., Wiesel T.N. (1979) Morphology and intracortical projections of functionally characterized neurons in the cat visual cortex. *Nature* 280, 120-125.
20. Gustafson J.W., Felbain-Keramidas S.L. (1977) Behavioural and neural approaches to the function of the mystacial vibrissae. *Psychological Bulletin* 84, 477-488.
21. Hersch S.M., White E.L. (1982) A quantitative study of the thalamocortical and other synapses in layer IV of pyramidal cells projecting from mouse SmI cortex to the caudate-putamen nucleus. *J Comp Neurol* 211, 217-225.
22. Kasper E., Larkman A.U., Blakemore C., Judge S. (1991) Physiology and morphology of identified projection neurons in the rat visual cortex studied in vitro. *Society of Neuroscience Abstracts* 17, 114-114.
23. Landry P., Wilson C.J., Kitai S.T. (1984) morphological and electrophysiological characteristics of pyramidal tract neurones in the rat. Exp *Brain Res* 57, 177-190.
24. Larkman A., Mason A. (1990) Correlations between morphology and electrophysiology of pyramidal neurones in slices of rat visual cortex.I Establishment of cells. *J Neurosci* 10(5), 1407-1414.
25. Larkman A.U. (1991) Dendritic morphology of pyramidal neurons in the visual cortex of the rat. 1. Branching patterns. *J Comp Neurol* 306, 306-319.
26. Levesque M., Charara A., Gagnon S., Parent A., Deschenes M. (1996) Corticostriatal projections from layer V cells in the rat are collaterals of long-range corticofugal axons. *Brain Res* 709(2), 311-315.
27. Mason A., Larkman A. (1990) Correlations between morphology and electrophysiology of pyramidal neurones in slices of rat visual cortex.II: Electrophysiology. *JNeurosci* 10(5), 1415-1428.
28. McCormick D.A., Connors B.W., Lighthall JW., Prince D.A. (1985) Comparative electrophysiology of pyramidal and sparsely spiny stellate neurones of the neocortex. *J Neurophysiol* 54(4), 782-806.
29. McGeorge A.J., Faull R.L.M. (1987) The organization and collateralization of corticostriate neurones in the motor and sensory cortex of the rat brain. *Brain Res* 423, 318-324.
30. McGeorge A.J., Faull R.L.M. (1989) The organization of the projection from the cerebral cortex to the striatum in the rat. *Neurosci* 29, 503-537.
31. Mercier B.E., Glickstein M., Legg C.R. (1990) Basal ganglia and the cerebellum receive different somatosensory information in rats. *Proceedings of the National Academy of Science (USA)* 87, 4388-4392.
32. Miller M.W., Chiaia N.L., Rhoades R.W. (1990) Intracellular recording and injection study of corticospinal neurons in the rat somatosensory cortex: Effect of prenatal exposure to ethanol. *J Comp Neurol* 297, 91-105.
33. Montoro R.J., Lopez-Bareno J., Jassik-Gershenfeld D. (1988) Differential bursting modes in neurons of the mammalian visual cortex in vitro. *Brain Res* 460, 168-172.
34. Pare D., Smith Y. (1996) Thalamic collaterals of corticostriatal axons: Their termination field and synaptic targets in cats. *J Comp Neurol* 372(4), 551-567.
35. Royce G.J. (1983) Cortical neurons with collateral projections to both the caudate nucleus and the centromedian-parafascicular thalamic complex: a fluorescent retrograde double labelling study in the cat. *Exp Brain Res* 50, 157-165.
36. Schwindt P., O'Brien J.A., Crill W. (1997) Quantitative analysis of firing properties of pyramidal neurons from layer V of rat sensorimotor cortex. *J Neurophysiol* 77(5), 2482-2498.
37. Silva L.R., Amitai Y., Connors B.W. (1991) Intrinsic oscillations of neocortex generated by layer 5 pyramidal neurons. *Science* 251, 432-435.
38. Silva L.R., Gumick M.J., Connors B.W. (1991) Laminar distribution of neuronal membrane properties in neocortex of normal and reeler mouse. *J Neurophysiol* 66, 2034-2040.
39. Smith Y. (1992) Antrograde tracing with PHA-L and biocytin at the electron microscopic level. In *Experimental Neuroanatomy* (ed. Bolam J.P.), 3, pp. 61-79.IRl press, Oxford.
40. Stafstrom C.E., Schwindt P.C., Crill W.E. (1984) Repetative firing in layer V neurons from cat neocortex in vitro. *J Neurophysiol* 52, 264-277.

41. Tseng G.F., Prince D.A. (1993) Heterogeneity of rat corticospinal neurons. *J Comp Neurol* 335(1), 92-108.
42. Wang Z., McCormick D.A. (1993) Control of firing mode of corticotectal and corticopontine layer V burst-generating neurons by norepinephrine, acteylcholine, and 1S,3R-ACPD. *J Neurosci* 13, 2199-2216.
43. Welker C., Woolsey T.A. (1974) Structure of layer IV in the somatosensory neocortex of the rat: Description and comparison with the mouse. *J Comp Neurol* 158, 437-454.
44. Wilson C.J. (1987) Morphology and synaptic connections of crossed corticostriatal neurons in the rat. *J Comp Neurol* 263, 567-580.
45. Wong-Riley M.T.T. (1979) Changes in the visual system of monocularly sutured or enucleated cats demonstrable with cytochrome oxidase histochemistry. *Brain Res 171*, 11-28.
46. Woolsey T.A., van der Loos H. (1970) The structural organisation of layer IV in the somatosensory region (SI) of mouse cerebral cortex. *Brain Res 17*, 205-242.
47. Wright A.K., Norrie L., Ingham C.A., Hutton E.A.M., Arbuthnott G.W. (1999) Double anterograde tracing of outputs from adjacent "barrel columns" of rat somatosensory cortex. Neostriatal projection patterns and terminal ultrastructure. *Neurosci* 88(1), 119-133.
48. Zhu J.J., Connors B. W. (1999) Intrinsic firing patterns and whisker-evoked synaptic responses of neurons in the rat barrel cortex. *JNeurophysiol* 81(3), 1171-1183.

THE ENKEPHALIN-EXPRESSING CELLS OF THE RODENT GLOBUS PALLIDUS

J.F. Marshall, B.R. Hoover, and J.J. Schuller*

Key words: Preproenkephalin, Globus pallidus, Parvalbumin, 6-OHDA, FluoroGold

Abstract: Antisense riboprobes to preproenkephalin (PPE) mRNA reveal that a population of neurons within rodent globus pallidus (GP) contains this sequence. Colocalization studies indicate that the neurons expressing PPE mRNA are almost entirely (>97%) distinct from those immunoreactive for parvalbumin (PV). These two neuron populations differ in their axonal projections. A majority of the PPE mRNA-expressing pallidal neurons are retrogradely labeled following FluoroGold (FG) iontophoresis into the striatum, whereas a small minority are labeled following FG iontophoresis into the subthalamic nucleus (STN). By contrast, our previous work showed that most PV-immunoreactive GP neurons are retrogradely labeled following STN FG iontophoresis, while a small minority are labeled by striatal FG deposits.

The integrity of the mesostriatal dopaminergic pathway contributes to regulating PPE mRNA expression within GP neurons. Rats with 6-hydroxydopamine injections along this pathway have greater PPE mRNA labeling per cell ipsilateral to the neurotoxin injection than in the unlesioned hemisphere.

Our previous work showed that the control of the immediate early gene, *c-fos*, within these GP neuron populations may differ. Whereas administration of dopamine D1/D2 receptor agonists induced Fos immunoreactivity in GP neurons irrespective of their PV content or the locus of their axonal projections, administration of the D2 receptor antagonist, eticlopride, induced Fos only in pallidostriatal neurons lacking PV immunoreactivity. These results suggest that the PV-containing and PPE-expressing pallidal neuron subpopulations are subject to differing controls by dopamine receptors.

1. INTRODUCTION

As a principal recipient of striatal efferent activity, the globus pallidus (GP) plays a critical role in the processing of movement-related information within the basal ganglia. The evidence for heterogeneous connections and functions of GP neurons suggests that the

---

* J.F. Marshall, B.R. Hoover, and J.J. Schuller, Department of Neurobiology and Behavior, University of CA, Irvine, CA 92697-4550.

The Basal Ganglia VI
Edited by Graybiel *et al.*, Kluwer Academic/Plenum Publishers, 2002

pallidal influences on information processing are likely to be achieved by several routes. Several studies have focused attention on subpopulations of GP neurons, defined on the basis of their neurochemistry, physiological properties, or axonal projections. For example, pallidal neurons can be subdivided based on their expression of the calcium binding protein parvalbumin (PV), their firing rates and patterns, or their axonal targets[1,2,3,4].

These distinctions between GP neuron populations may help to illuminate differences in the actions of dopaminergic agents on immediate early gene response within this structure. Administration to rats of dopamine D1 and D2 receptor agonists induces Fos, the protein product of the immediate early gene, *c-fos*, within GP cells, including parvalbumin-containing (PV+) and parvalbumin-lacking (PV-) neurons, as well as neurons projecting to subthalamic nucleus [STN], substantia nigra pars reticulata [SNr], entopeduncular nucleus [EP], and striatum. By contrast, administration of eticlopride, a D2-class dopamine receptor antagonist, induces Fos within a subpopulation of GP neurons that lacks PV and that projects only to striatum[4].

The apparent specificity of the GP neuron population affected by the D2 antagonist has prompted a search for more specific markers of the PV-, pallidostriatal cell population. The present study investigates the population characteristics of GP preproenkephalin mRNA-expressing (PPE+) neurons and their possible regulation by dopamine. The relationship between the PPE+ pallidal cells and those identified by their PV immunoreactivity or by their axonal projections to striatum or STN was determined. Additionally, the regulation of GP PPE mRNA expression was examined by characterizing the effects of unilateral damage to the nigrostriatal projection.

2. MATERIALS AND METHODS

To determine axonal projections of PPE+ GP cells, male Sprague-Dawley rats were anesthetized and placed in a stereotaxic instrument. The retrograde tract tracer FluoroGold (FG) was iontophoresed into the left striatum and right STN of each rat. After 12 to 14 days, the rats were perfused transcardially with buffered saline followed by 4% paraformaldehyde. Following post-fixation and cryoprotection, coronal sections (30 μm) were cut on a freezing microtome and collected.[5]

Six-hydroxdopamine (6-OHDA) lesions of the mesostriatal dopamine pathway were performed in other rats, using previously published methods.[6] Four weeks after surgery, 6-OHDA-lesioned rats and age-matched unlesioned controls were decapitated and the brains rapidly removed and frozen to -20° C. Sections (20 μm) through the striatum and GP were cut on a cryostat, slide-mounted, postfixed, and rinsed in phosphate buffer followed by Rnase-free water. Slides were air dried and stored at -20° C until used.[7]

Free-floating or slide-mounted sections through the GP were processed for the *in situ* hybridization localization of PPE mRNA. An RNA probe (970 base pairs) complementary to rat preproenkephalin mRNA was used to localize PPE mRNA. Both 'sense' and 'antisense' fragments of the riboprobe were transcribed in the presence of ^{35}S-UTP using SP6 polymerase.

To colocalize PPE mRNA and FG within GP cells, free-floating tissue was incubated sequentially in a blocking solution, an anti-FG polyclonal antibody (Chemicon; 1:20,000), biotinylated goat anti-rabbit IgG (Vector; 1:200), and avidin-biotin horseradish peroxidase complex (ABC). Following chromagenic visualization of the FG immunosignal, the sections were rinsed and mounted onto gelatin-coated slides. After drying, the slides were dipped in autoradiographic emulsion (Kodak NTB-2) and developed in Kodak D-19 after 2-3 weeks exposure.

To colocalize PPE mRNA and PV immunoreactivity in the GP, rats were perfused and brains processed as described above. To label PV, the freefloating tissue was incubated in a blocking solution, an anti-PV monoclonal antibody (Sigma; 1:1000), biotinylated horse anti-mouse IgG (Vector; 1:200), and ABC.

Fields in the GP were imaged using a Nikon Optiphot microscope interfaced to an MCID (Imaging Research Inc.; St. Catherine, Ontario, Canada) image analysis system. Two images from the same section were digitized: one of the immunoreactive cells under white light illumination and another of the ^{35}S-silver grain clusters under darkfield illumination. Grain densities were quantified as the number of pixels within an oval sampling tool whose boundary was slightly larger than the digitized image of the largest immunoreactive cell. For the PPE mRNA/PV immunocytochemistry localization experiment, the number of pixels per cell over 50 PV+ GP neurons was counted in each animal, as was the number of pixels within 50 identically sized areas over visually-identified PPE+ silver grain clusters in the GP. Frequency histograms were used to analyze the extent of PPE mRNA labeling associated with PV+ cells relative to the PPE+ clusters. Procedures for analysis of PPE mRNA/FG-labeled tissue were similar, except that the numbers of pixels per cell were measured over 50 FG-labeled pallidostriatal and 50 pallidosubthalamic neurons per animal. Background levels of grain density were determined by using the sampling tool to measure pixel densities over the corpus callosum.

3. RESULTS

Hybridization of pallidal tissue sections with ^{35}S-labeled antisense PPE riboprobes resulted in a subpopulation of GP neurons with cellular labeling, although the intensity of labeling was noticeably less than that in the striatum. The labeling was absent when sections were incubated with sense riboprobe, indicating the specificity of our hybridization (Fig. 1). To estimate the proportion of GP neurons that express PPE mRNA, the numbers of those neurons were compared to the numbers of PV-stained cells within 30 μm thick sections through GP. The PPE+ neurons (identified as ^{35}S-silver grain clusters) were found to be 60% as numerous as PV+ cells. Based on the previous estimates[2] that the PV+ GP neurons constitute two-thirds of the GP neuron population, we estimate that approximately 40% of the GP neurons express PPE mRNA.

However, the PV+ and PPE+ GP neuron populations were almost entirely distinct. The PV-containing GP neurons very rarely expressed significant levels of PPE mRNA. The densities of ^{35}S-labeled PPE mRNA associated with PV+ cells were equivalent to background levels for 341/350 of the GP PV+ cells sampled (Fig. 2). The densities of ^{35}S-labeled PPE mRNA associated with the visible PPE+ clusters were also quantified. The densities of PPE mRNA associated with these clusters, which appear to represent the PPE mRNA-containing cells of the GP (Fig. 2), were always above the established background.

The axonal projections of the PPE+ GP neurons were established by combining FG retrograde tracing with ^{35}S-PPE mRNA labeling. For all animals (N=16), above-background densities of PPE mRNA were more often associated with FG-labeled pallidostriatal than

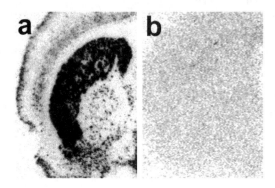

Figure 1. Film autoradiograms of coronal brain sections at the level of the GP, hybridized with ^{35}S-labeled antisense (a) or sense (b) PPE riboprobe. Spotted labeling in the GP contrasts with intense striatal signal (a). There was a lack of specific signal in sections hybridized with sense probe.

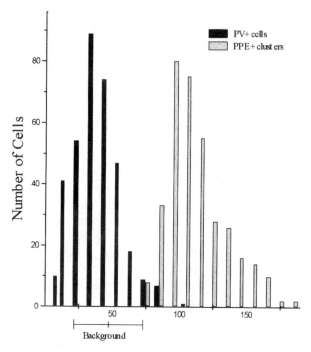

Figure 2. Histogram plotting the density of PPE mRNA (expressed as pixels per 428 µm^2) over PV+ cells and PPE+ clusters in the GP. Note that the majority of PV+ cells express PPE mRNA levels comparable with background levels, and below the level of PPE mRNA in identified clusters. Values for PPE mRNA background labeling were calculated as the mean ±2 standard deviations for the density levels obtained from the corpus callosum of all subjects (n=7). This range of values is designated by the bracketed line under the histogram.

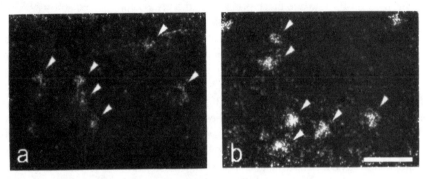

Figure 3. Darkfield photomicrographs from the same rat showing emulsion grain clusters (arrowheads) corresponding to PPE-expressing pallidal neurons contralateral (a) and ipsilateral (b) to a 6-OHDA lesion. Grain clusters are more intense ipsilateral to the lesion. Calibration bar = 100 µm.

pallido-STN neurons. For a subgroup of these rats that showed robust labeling of PPE mRNA as a result of the hybridization, we found that 57% of the FG-labeled pallidostriatal cells were PPE+. By contrast, only 9% of the pallido-STN neurons were PPE+.

To determine whether the PPE mRNA expression of GP neurons was affected by the integrity of the mesostriatal dopamine pathway, the extent of PPE mRNA labeling was determined in rats previously given unilateral 6-hydroxydopamine injections to damage this projection in one hemisphere. Four weeks after unilateral 6-OHDA lesions of the nigrostriatal pathway, labeling of PPE mRNA was elevated in the ipsilateral GP, compared to the intact hemisphere as well as to values obtained from unlesioned control animals. Analysis of individual GP neurons labeled by the ^{35}S-riboprobe revealed that the 6-OHDA lesion produced an average increase of 56% in mean pixels per cell as compared to the contralateral GP (Fig. 3).

4. DISCUSSION

The enkephalinergic neurons of the GP form a distinct neuronal population whose characteristics need to be considered in the formulation of models of basal ganglia information processing. Our estimates of the numbers of these cells suggest the GP contains approximately 60% as many PPE+ neurons as PV+ neurons. Based on previously published estimates that the PV+ cells constitute approximately two-thirds of the neuronal population of the GP[2], as well as our finding that the PPE+ and PV+ neuron populations of GP are >97% non-overlapping, we conclude that the GP neurons can be subdivided into two distinct subpopulations: (1) the PV+/PPE- neurons, which make up the majority (ca. two-thirds), and (2) the PV-/PPE+ neurons, which comprise the remainder (ca. one-third).

The PPE+ and PV+ GP cell populations also differ in the frequency with which they are retrogradely labeled from striatal versus non-striatal targets. GP neurons retrogradely labeled by FG injections into striatum are frequently (57%) PPE+ and only infrequently (16%) PV+. By contrast, GP neurons retrogradely labeled by FG injections into STN, EP, or SNr are typically (70-79%) PV+, while those projecting to STN are only infrequently (9%) PPE+ [present results[4]]. Because individual GP neurons often collateralize in their projections to these targets[3,8,9,10] a complete picture of the axonal projections of these two pallidal neuron types has yet to be obtained. However, we can conclude that the PPE+ neurons are much more likely to project axon branches to striatum than to STN, whereas the PV+ neurons are much more likely to project axon branches to STN, EP, and SNr than to striatum.

These distinctions between the PV+ and PPE+ GP neurons may have a correlate in their discharge patterns. Electrophysiological studies of awake, resting monkeys reveal two types of GP units: those with high frequency discharges interrupted by pauses (HFD), and those that fire at low frequency in bursts (LFD-B).[1] Because PV is found in a subpopulation of hippocampal and striatal neurons that can sustain high firing rates with little adaptation,[11,12] it is intriguing to consider that pallidal PV+ neurons may correspond to HFD units while GP PPE+ neurons may correspond to LFD-B units.

The present study also demonstrates that pallidal PPE mRNA expression is modulated by dopamine innervation. The upregulation of pallidal PPE mRNA following 6-OHDA lesions may reflect either the direct or indirect consequences of interrupting striatal DA neurotransmission. Considering the latter possibility first, nigrostriatal lesions profoundly alter striatal neuron function, with many lesion-induced changes occurring preferentially in striatopallidal projection neurons.[13,14] There is a corresponding decrease in pallidal firing rates[15] and GABA binding,[16] suggesting increased release of GABA from striatopallidal terminals resulting in abnormal inhibition of pallidal neurons. Another possible indirect source of pallidal PPE mRNA modulation could arise from STN glutamatergic afferents. Nigrostriatal lesions cause hyperactivity in STN neurons[17] and decrease NMDA receptor binding in the GP.[18] Delfs and colleagues have demonstrated that 6-OHDA-induced increases in pallidal GAD mRNA could be prevented by lesions of the ipsilateral STN.[19] The fact that striatal and STN afferents are predicted to have opposing influences over GP activity suggests that the effects of decreased DA neurotransmission on pallidal gene expression may be complex.

Alternatively, elevated pallidal PPE mRNA expression following 6-OHDA lesion may reflect the loss of DA receptor tone in the GP. A sparse DAergic innervation of the GP by nigral neurons has been reported,[20] and D2 receptor mRNA has been detected in GP neurons.[21] Also, iontophoretic application of DA receptor agonists onto pallidal neurons has been shown to alter their electrophysiological properties,[22] supporting a functional role for the pallidal DA innervation.

5. ACKNOWLEDGEMENTS

Research supported by Public Health Service grant NS22698 to J.F.M. and J.J.S. and NRSA grant MH14599 to B.R.H.

6. REFERENCES

1. DeLong, M.R. (1971) Activity of pallidal neurons during movement. *J Neurophysiol.* 34: 414-27.

2. Kita, H. (1994) Parvalbumin-immunopositive neurons in rat globus pallidus: a light and electron microscopic study. *Brain Res.* 657: 31-41.
3. Kita, H. and Kitai, S.T. (1991) Intracellular study of rat globus pallidus neurons: membrane properties and responses to neostriatal, subthalamic and nigral stimulation. *Brain Res.* 564: 296-305.
4. Ruskin, D.N. and Marshall, J.F. (1997) Differing influences of dopamine agonists and antagonists on Fos expression in identified populations of globus pallidus neurons. *Neuroscience.* 81:79-92.
5. Hoover, B.R. and Marshall, J.F. (1999) Population characteristics of preproenkephalin mRNA-containing neurons in the globus pallidus of the rat. *Neurosci Lett.* 265: 199-202.
6. Neve, K.A., Altar, C.A., Wong, C.A. and Marshall, J.F. (1984) Quantitative analysis of [3H]spiroperidol binding to rat forebrain sections: plasticity of neostriatal dopamine receptors after nigrostriatal injury. *Brain Res.* 302: 9-18.
7. Schuller, J.J., Billings, L.M. and Marshall, J.F. (1999) Dopaminergic modulation of pallidal preproenkephalin mRNA. *Mol Brain Res.* 69: 149-53.
8. Kita, H. and Kitai, S.T. (1994) The morphology of globus pallidus projection neurons in the rat: an intracellular staining study. *Brain Res.* 636: 308-19.
9. Schmued, L., Phermsangngam, P., Lee, H., Thio, S., Chen, E., Truong, P., Colton, E. and Fallon, J. (1989) Collateralization and GAD immunoreactivity of descending pallidal efferents. *Brain Res.* 487: 131-42.
10. Staines, W.A. and Fibiger, H.C. (1984) Collateral projections of neurons of the rat globus pallidus to the striatum and substantia nigra. *Exp Brain Res.* 56: 217-20.
11. Kawaguchi, Y. (1993) Physiological, morphological, and histochemical characterization of three classes of interneurons in rat neostriatum. *J Neurosci.* 13: 4908-23.
12. Kawaguchi, Y., Katsumaru, H., Kosaka, T., Heizmann, C.W. and Hama, K. (1987) Fast spiking cells in rat hippocampus (CAI region) contain the calcium-binding protein parvalbumin. *Brain Res.* 416: 369-74.
13. Gerfen, C.R., Engber, T.M., Mahan, L.C., Susel, Z., Chase, T.N., Monsma, F.J., Jr. and Sibley, D.R. (1990) DI and D2 dopamine receptor-regulated gene expression of striatonigral and striatopallidal neurons. *Science.* 250: 1429-32.
14. Jian, M., Staines, W.A., Iadarola, M.J. and Robertson, G.S. (1993) Destruction of the nigrostriatal pathway increases Fos-like immunoreactivity predominantly in striatopallidal neurons. *Mol Brain Res.* 19: 156-60.
15. Pan, H.S. and Walters, JR. (1988) Unilateral lesion of the nigrostriatal pathway decreases the firing rate and alters the firing pattern of globus pallidus neurons in the rat. *Synapse.* 2: 650-6.
16. Pan, H.S., Penney, J.B. and Young, A.B. (1985) Gamma-aminobutyric acid and benzodiazepine receptor changes induced by unilateral 6-hydroxydopamine lesions of the medial forebrain bundle. *J Neurochem.* 45: 1396-404.
17. Hassani, O.K., Mouroux, M. and Feger, J. (1996) Increased subthalamic neuronal activity after nigral dopaminergic lesion independent of disinhibition via the globus pallidus. *Neuroscience.* 72: 105-15.
18. Porter, R.H., Greene, J.G., Higgins, D.S., Jr. and Greenamyre, J.T. (1994) Polysynaptic regulation of glutamate receptors and mitochondrial enzyme activities in the basal ganglia of rats with unilateral dopamine depletion. *J Neurosci.* 14: 7192-9.
19. Delfs, J.M., Ciaramitaro, V.M., Parry, T.J. and Chesselet, M.F. (1995) Subthalamic nucleus lesions: widespread effects on changes in gene expression induced by nigrostriatal dopamine depletion in rats. *J Neurosci.* 15: 6562-75.
20. Lindvall, O. and Bjorklund, A. (1979) Dopaminergic innervation of the globus pallidus by collaterals from the nigrostriatal pathway. *Brain Res.* 172: 169-73.
21. Weiner, D.M., Levey, A.I., Sunahara, R.K., Niznik, H.B., BF, O.D., Seeman, P. and Brann, M.R. (1991) Dl and D2 dopamine receptor mRNA in rat brain. *Proc Natl Acad Sci USA.* 88: 1859-63.
22. Bergstrom, D.A. and Walters, J.R. (1984) Dopamine attenuates the effects of GABA on single unit activity in the globus pallidus. *Brain Res.* 310: 23-33.

CORTICOSTRIATAL PROJECTIONS FROM THE CINGULATE MOTOR AREAS IN THE MACAQUE MONKEY

Masahiko Takada,[1,2] Ikuma Hamada,[1] Hironobu Tokuno,[1,2] Masahiko Inase,[3] Yumi Ito,[1] Naomi Hasegawa,[1] Yoko Ikeuchi,[1] Michiko Imanishi,[1] Toshikazu Akazawa,[1] Nobuhiko Hatanaka,[1,2] and Atsushi Nambu[1,2]

1. INTRODUCTION

In recent years, emphasis has been placed on the importance of the functional loop linking the cerebral cortex and the basal ganglia in the execution of motor schemes based on a variety of internal and external conditions.[1,2,28,33,34] The major origin of the cortico-basal ganglia loop is located in multiple motor-related areas of the frontal lobe, such as the primary motor cortex (MI), the supplementary and presupplementary motor areas (SMA and pre-SMA), the dorsal and ventral divisions of the premotor cortex (PMd and PMv), and the cingulate motor areas (CMAs). For understanding of the mechanism underlying motor information processing in the basal ganglia, it is essential to investigate the distribution patterns of corticostriatal inputs from these motor-related areas. From such a viewpoint, a series of our recent studies have been performed to analyze the corticostriatal projections from the MI, SMA, pre-SMA, PMd, and PMv, with special reference to the organization of input zones from their forelimb representations.[12,35,36] However, almost no data have as yet been available on the distribution patterns of corticostriatal inputs from the CMAs.

The CMAs are located on the dorsal and ventral banks of the cingulate sulcus.[29] Based on the cytoarchitectonic criteria, the CMAs are divided into three parts: the rostral cingulate motor area (CMAr; area 24c), dorsal cingulate motor area (CMAd; area 6c), and ventral cingulate motor area (CMAv; area 23c).[5,19,27,39] In view of their somatotopically arranged inputs to the MI and the spinal cord, the CMAs have been considered to have a set of body part representations.[5,6,9,10,16,18,22-25,27,32,37,38] Previous electrophysiological studies with intracortical microstimulation (ICMS) also examined the somatotopy of the CMAs.[17,20] In the present study, anterograde axonal tracing was performed to analyze the corticostriatal pro-

---

* 1 Tokyo Metropolitan Institute for Neuroscience, Fuchu, Tokyo 183-8526, Japan.
2 CREST, JST (Japan Science and Technology).
3 Department of Physiology, Kinki University School of Medicine, Osaka-Sayama 589-8511, Japan.

jections from forelimb representations of the CMAr and CMAc (caudal cingulate motor area including the CMAd and CMAv). Employing extracellular unit recording and ICMS, we first mapped the medial wall of the frontal lobe of the macaque monkey to define body part representations of the CMAs, as well as of the SMA and pre-SMA. Then, the forelimb regions of the CMAr and CMAc were injected separately with the two different anterograde tracers, wheat germ agglutinin-conjugated horseradish peroxidase (WGA-HRP) and biotinylated dextran amine (BDA), to obtain terminal fields of the corticostriatal projections from both regions. Further, we compared such corticostriatal input zones with those from the forelimb regions of the SMA, pre-SMA, and MI. Thus, the present experiments were designed as shown in Table 1.

2. MATERIALS AND METHODS

Six Japanese monkeys *(Macaca fuscata)* of either sex, weighing between 4.5 and 7.5 kg, were used for this study. All experimental procedures were conducted under aseptic conditions in line with NIH's *Guide for the Care and Use of Laboratory Animals.* In each of the monkeys, a surgical operation was performed to gain easy access to electrophysiological mapping and tracer injection. Under general anesthesia with ketamine hydrochloride (10 mg/kg b.wt., i.m.) and sodium pentobarbital (25 mg/kg b.wt., i.v.), the monkeys were positioned in a stereotaxic apparatus. The skull was widely exposed; small stainless-steel screws were attached to the skull as anchors; the exposed skull and screws were completely covered with transparent acrylic resin; and two stainlesssteel pipes were mounted parallel over the frontal and occipital lobes for head fixation.

After recovery from the surgery, the monkeys were anesthetized with a combination of ketamine hydrochloride (10 mg/kg b.wt., i.m.) and xylazine hydrochloride (1-2 mg/kg b.wt., i.m.) and, then, sat quietly in a primate chair with their heads fixed in a stereotaxic frame attached to the chair. Following partial removal of the skull, a glass-insulated Elgiloy-alloy microelectrode (impedance of 0.9-1.4 MΩ at 1 kHz) was inserted perpendicular to the cortical surface through a hydraulic micromanipulator. As unit activity was recorded at 250-500-μm intervals, neuronal responses to somatosensory and visual stimuli were examined. Subsequently, the monkeys underwent ICMS with currents of less than 60 μA delivered through a

Table 1. Summary of Experiments

Monkey	Site of tracer injection*		
	WGA-HRP	BDA	RDA
CMA1	CMAc (0.1 μl x 1)	CMAr (1 μl x 1)	
CMA2	CMAr (0.1 μl x 1)	CMAc (1 μl x 2)	
CMA3	CMAr (0.1 μl x 1)	CMAc (1 μl x 2)	
CMA4	CMAc (0.1 μl x 2)	pre-SMA (1 μl x 2)	SMA (1 μl x 4)
CMA5	MI (0.1 μl x 3)	CMAr (1 μl x 1)	CMAc (1 μl x 1)
CMA6	pre-SMA (0.1 μl x 2)	CMAr (1 μl x 1)	SMA (1 μl x 4)

*The injection volume is indicated in parentheses.

constant-current stimulator (200 μs duration at 333 Hz), and evoked movements of various body parts were carefully observed. According to the microexcitability of the cortical motor-related areas, a train of cathodal pulses was switched within the range of 10-40. To confirm the exact locations of microelectrode penetration, electrolytic microlesions (anodal direct currents of 10-15 μA, 20-30 sec) were made at several loci as reference points.

Based on the electrophysiological map, injection of WGA-HRP or BDA was performed into each cortical area (Table 1). In monkeys CMA4-CMA6, tetramethylrhodamine-labeled dextran amine (RDA; Molecular Probes; 3000 M.W.) was used as the third tracer. A 4% solution of WGA-HRP (Toyobo; RZ ≥ 3.0) was deposited through a 1-μl Hamilton microsyringe, while a 20% solution of BDA (Molecular Probes; 3000 M.W.) or RDA was deposited through a 10-μl Hamilton microsyringe. The BDA and/or RDA injections were carried out 18-24 days prior to the WGA-HRP injection.

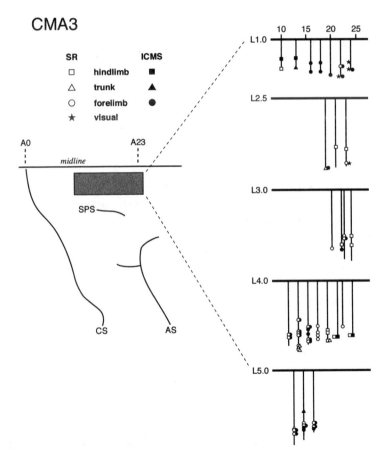

Figure 1. Results of electrophysiological mapping of the medial wall of the frontal lobe in monkey CMA3. In the stippled rectangular area viewed from the cortical surface, sensory responses (SR) and body part movements evoked by ICMS were examined. The level of L1.0 corresponds to the SMA and pre-SMA, while those of L2.5-L5.0 correspond to the CMAr and CMAc. AS, arcuate sulcus; CS, central sulcus; SPS, superior precentral sulcus.

After survival periods of 21-28 days for BDA and RDA and 3-4 days for WGA-HRP, the monkeys were anesthetized deeply with an overdose of sodium pentobarbital (50 mg/kg body weight, i.v.) and perfused transcardially with 0.1 M phosphate-buffered saline (PBS; pH 7.3), followed by 8% formalin dissolved in 0.1 M PBS (pH 7.3) and, finally, the same fresh PBS containing 10% and then 30% sucrose. The brains were removed immediately, saturated with 30% sucrose in 0.1 M PBS (pH 7.3) at 4°C, and then cut serially into 60-µm-thick coronal sections on a freezing microtome. Every sixth section was histochemically stained for WGA-HRP and BDA. The sections adjacent to WGA-HRP-stained sections were histochemically stained for BDA. Detailed histochemical procedures were as described elsewhere. [12,26,35] For light microscopic observation of RDA labeling, the sections adjacent to WGA-HRP- or BDA-stained sections were immunohistochemically stained with anti-rhodamine antibody (Molecular Probes) and ENVISION (Dako). All sections were mounted onto gelatin-coated glass slides, counterstained with 1% Neutral Red, and then coverslipped. Some sections were Nissl-stained with 1 % Cresyl Violet.

3. RESULTS

In our electrophysiological mapping of the medial wall of the frontal lobe, we initially determined the confines of the SMA and pre-SMA. Somatotopical representations of the SMA were arranged from caudal to rostral in the order of the hindlimb, trunk, forelimb, and orofacial part (Fig.l; see also refs. 11,12,26,35,36). More rostrally along the medial wall of the hemisphere, a discrete area corresponding to the pre-SMA was identified, in which only the forelimb was represented and, also, visual responses were detected (Fig. 1; see also ref. 12).

After identification of the SMA and pre-SMA, the distribution pattern of body part representations was analyzed in the dorsal and ventral banks of the cingulate sulcus, 2.5-5 mm lateral from the midline. In all of the six monkeys examined, forelimb representation was found in rostrocaudally separate regions, each of which appeared to coincide with the forelimb region of the CMAr or CMAc (Figs. 1 and 2). These regions were composed of neurons responding differently to somatosensory stimuli. The responses recorded from the CMAr were to passive joint movement, whereas those from the CMAc were not only to passive joint movement, but also to light skin touch. While ICMS in the CMAr rarely evoked movements of the forelimb, forelimb movements were readily elicited by ICMS in the CMAc.

In monkeys CMA1-CMA3, paired injections of WGA-HRP and BDA were made into the forelimb regions of the CMAr and CMAc (Table 1 and Fig. 2). Terminal labeling from the CMAr and CMAc was observed mainly in the putamen through its rostral 1/2 extent, bilaterally with an ipsilateral predominance. Accumulations of terminal label from the CMAr were widely distributed within the putamen, especially along its dorsomedial-to-ventrolateral axis (Fig. 3). The terminal labeling from the CMAr was also found in the striatal cell bridges connecting the rostral aspects of the caudate nucleus and the putamen and their surroundings (Fig. 3). Dense accumulations of terminal label from the CMAc were located in the ventrolateral portion of the putamen (Fig. 3).

Additional terminal labeling from the CMAc was seen more medially in the putamen along its dorsomedial-to-ventrolateral axis (Fig. 3). The terminal fields from the CMAr and CMAc were largely segregated, although some overlap between them was evident in the dorsomedial or ventrolateral portion of the putamen (Fig. 3). As shown in Table 1, monkeys

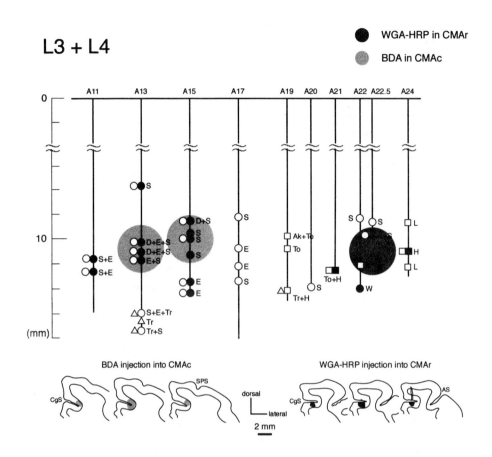

Figure 2. Details of electrophysiological mapping of the CMAr and CMAc in overlay of the levels of L3.0 and L4.0 in Figure 1. Open symbols represent the sites of somatosensory responses, and filled symbols represent the sites of body part movements evoked by ICMS. A, ankle; D, digit; E, elbow; H, hip; K, knee; L, leg; S, shoulder; To, toe; Tr, trunk; W, wrist. The sites of paired injections of WGA-HRP and BDA (specified by graded stippled areas) are depicted on the map and, also, in coronal sections. In each injection case, three equidistant sections are arranged rostrocaudally from the left to the right. AS, arcuate sulcus; CgS, cingulate sulcus; SPS, superior precentral sulcus.

CMA4-CMA6 received triple injections of WGA-HRP, BDA and RDA into the forelimb regions of three of the five target areas (i.e., the CMAr, CMAc, SMA, pre-SMA, and MI). Full investigations of terminal labeling in these three monkeys found that terminal fields from the CMAr and pre-SMA or those from the CMAc and MI at least partly overlapped in the striatal cell bridges or in the ventrolateral portion of the putamen, respectively. However, the major terminal fields from the CMAr and CMAc displayed no substantial overlap with that from the pre-SMA or MI. Moreover, the terminal fields from the CMAr and CMAc were essentially segregated from that from the SMA.

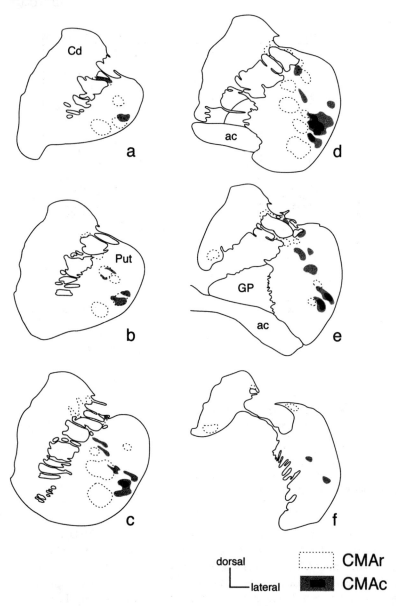

Figure 3. Distribution patterns of terminal labeling from the CMAr and CMAc, each of which was injected with WGA-HRP or BDA in monkey CMA3 (see also Fig. 2). For BDA labeling from the CMAc, blackened areas represent hot spots of stronger labeling while stippled areas represent haloes of weaker labeling. Three representative coronal sections are arranged rostrocaudally in a-f. ac, anterior commissure; Cd, caudate nucleus; GP, globus pallidus; Put, putamen.

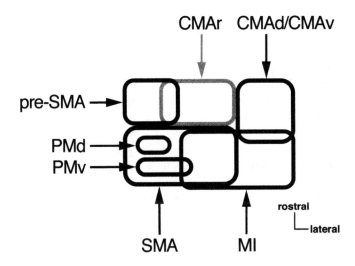

Figure 4. Schematic diagram showing the segregation-overlap relationship among the motor-related areas of the frontal lobe. "CMAd/CMAv" is equivalent to the CMAc.

4. DISCUSSION

A previous retrograde labeling study has reported that areas 24c and 23c, corresponding to the CMAr and CMAv, send projection fibers to the dorsal sensorimotor striatum (i.e., the caudal aspect of the putamen).[15] The present study has revealed that terminal fields of the corticostriatal projections from the CMAr and CMAc are located primarily in the putamen through its rostral 1/2 extent, where both terminal fields are separately distributed. Since the rostral aspect of the putamen does not receive massive afferents from the MI and SMA (see also refs. 11,35,36), the major corticostriatal input zones from the CMAr and CMAc are segregated rostrocaudally from those from the MI and SMA, except for a partial overlap of the input zones from the CMAc and MI in the ventrolateral portion of the putamen. In view of the fact that the distribution areas of corticostriatal inputs from the PMd and PMv exhibit an extensive overlap with that of the input from the SMA,[35] it is most likely that the corticostriatal input zones from the CMAr and CMAc are further segregated from those from the PMd and PMv. It has also been demonstrated in our study that the distribution areas of corticostriatal inputs from the CMAr and pre-SMA overlap to some extent in the striatal cell bridges, where the main terminal field of the corticostriatal projection from the pre-SMA is localized (see also ref. 12). However, the corticostriatal input zones from the CMAr and CMAc are still largely separate from that from the pre-SMA. These overall data indicate that corticostriatal inputs from the CMAr and CMAc occupy their own territories in the striatum, each of which is virtually segregated from corticostriatal inputs from the other motor-related areas of the frontal lobe (Fig. 4).

The present results suggest that a parallel design may probably underlie motor information processing in the cortico-basal ganglia loop arising from the CMAr and CMAc. Such a critical "parallel processing" model of basal ganglia function denotes that these cortical areas set up individual circuits coding specific motor parameters. Given the involvement of the cortico-basal ganglia loop in the cognitive aspects of motor control,[8,14] it is of interest to note that the CMAr and CMAc play differential roles in the control of voluntary movements. Accumulated evidence to date has implicated the CMAr in higherorder motor control.[29,31,32] Similarly, the pre-SMA has also been considered to have hierarchically higher motor functions.[12,29] With respect to the corticostriatal projections, both the CMAr and pre-SMA send inputs to the rostral striatum, including the striatal cell bridges, which has been reported to be responsible for motor learning rather than motor execution.[21] In addition, both motor-related areas are characterized by the strong interconnections with the prefrontal cortex.[3,4,16,18,22,23] Several anatomical studies have shown that corticostriatal inputs from the prefrontal cortex terminate rostrally within the striatum.[7,13,30,40] Thus, corticostriatal inputs from the CMAr as well as from the preSMA might at least partly be integrated with the prefrontal inputs, to exert control actions on cognitive motor behaviors.

5. REFERENCES

1. Alexander, G.E., and Crutcher, M.D., 1990, Functional architecture of basal ganglia circuits: Neural substrates of parallel processing, *Trends Neurosci.* 13:266-271.
2. Alexander, G.E., DeLong, M.R., and Strick, P.L., 1986, Parallel organization of functionally segregated circuits linking basal ganglia and cortex, *Annu. Rev. Neurosci.* 9:357-381.
3. Barbas, H., and Mesulam, M.-M., 1985, Cortical afferent input to the principalis region of the rhesus monkey, *Neuroscience* 15:619-637.
4. Bates, J.F., and Goldman-Rakic, P.S., 1993, Prefrontal connections of medial motor areas in the rhesus monkey, *J. Comp. Neurol.* 336:211-228.
5. Dum, R.P., and Strick, P.L., 1991, The origin of corticospinal projections from the premotor areas in the frontal lobe, *J. Neurosci.* 11:667-689.
6. Dum, R.P., and Strick, P.L., 1996, Spinal cord terminations of the medial wall motor areas in macaque monkeys, *J. Neurosci.* 16:6513-6525.
7. Eblen, F., and Graybiel, A.M., 1995, Highly restricted origin of prefrontal cortical inputs to striosomes in the macaque monkey, *J. Neurosci.* 15:5999-6013.
8. Graybiel, A.M., 1995, Building action repertoires: Memory and learning functions of the basal ganglia, *Curr. Opin. Neurobiol.* 5:733-741.
9. He, S.-Q., Dum, R.P., and Strick, P.L., 1995, Topographic organization of corticospinal projections from the frontal lobe: Motor areas on the medial surface of the hemisphere, *J. Neurosci.* 15:3284-3306.
10. Hutchins, K.D., Martino, A.M., and Strick, P.L., 1988, Corticospinal projections from the medial wall of the hemisphere, *Exp. Brain Res.* 71:667-672.
11. Inase, M., Sakai, S.T., and Tanji, J., 1996, Overlapping corticostriatal projections from the supplementary motor area and the primary motor cortex in the macaque monkey: An anterograde double labeling study, *J. Comp. Neurol.* 373:283-296.
12. Inase, M., Tokuno, H., Nambu, A., Akazawa, T., and Takada, M., 1999, Corticostriatal and corticosubthalamic input zones from the presupplementary motor area in the macaque monkey: Comparison with the input zones from the supplementary motor area, *Brain Res.* 833:191-201.
13. Kemp, J.M., and Powell, T.P.S., 1970, The cortico-striate projection in the monkey, *Brain* 93:525-546.
14. Kimura, M., and Graybiel, A.M., 1995, Role of basal ganglia in sensory motor association learning, in: *Functions of the Cortico-Basal Ganglia Loop,* M. Kimura, and A.M. Graybiel, eds., Springer, Tokyo, pp. 2-17.
15. Kunishio, K., and Haber, S.N., 1994, Primate cingulostriatal projection: Limbic striatal versus sensorimotor striatal input, *J. Comp. Neurol.* 350:337-356.

16. Lu, M.-T., Preston, J.B., and Strick, P.L., 1994, Interconnections between the prefrontal cortex and the premotor areas in the frontal lobe, *J. Comp. Neurol.* 341:375-392.
17. Luppino, G., Matelli, M., Camarda, R.M., Gallese, V., and Rizzolatti, G., 1991, Multiple representations of body movements in mesial area 6 and the adjacent cingulate cortex: An intracortical microstimulation study in the macaque monkey, *J. Comp. Neurol.* 311:463-482.
18. Luppino, G., Matelli, M., Camarda, R., and Rizzolatti, G., 1993, Corticocortical connections of area F3 (SMA-proper) and area F6 (pre-SMA) in the macaque monkey, *J. Comp. Neurol.* 338:114-140.
19. Matelli, M., Luppino, G., and Rizzolatti, G., 1991, Architecture of superior and mesial area 6 and the adjacent cingulate cortex in the macaque monkey, *J. Comp. Neurol.* 311:445-462.
20. Mitz, A.R., and Wise, S.P., 1987, The somatotopic organization of the supplementary motor area: Intracortical microstimulation mapping, *J. Neurosci.* 7:1010-1021.
21. Miyachi, S., Hikosaka, O., Miyashita, K., Karadi, Z., and Rand, M.K., 1997, Differential roles of monkey striatum in learning of sequential hand movement, *Exp. Brain Res.* 115:1-5.
22. Morecraft, R.J., and Van Hoesen, G.W., 1992, Cingulate input to the primary and supplementary motor cortices in the rhesus monkey: Evidence for somatotopy in areas 24c and 23c, *J. Comp. Neurol.* 322:471-489.
23. Morecraft, R.J., and Van Hoesen, G.W., 1993, Frontal granular cortex input to the cingulate (M3), supplementary (M2) and primary (MI) motor cortices in the rhesus monkey, *J. Comp. Neurol.* 337:669-689.
24. Morecraft, R.J., Schroeder, C.M., and Keifer, J., 1996, Organization of face representation in the cingulate cortex of the rhesus monkey, *Neuroreport* 7:1343-1348.
25. Muakkassa, K.F., and Strick, P.L., 1979, Frontal lobe inputs to primate motor cortex: Evidence for four somatotopically organized 'premotor' areas, *Brain Res.* 177:176-182.
26. Nambu, A., Takada, M., Inase, M., and Tokuno, H., 1996, Dual somatotopical representations in the primate subthalamic nucleus: Evidence for ordered but reversed body-map transformations from the primary motor cortex and the supplementary motor area, *J. Neurosci.* 16:2671-2683.
27. Nimchinsky, E.A., Hof, P.R., Young, W.G., and Morrison, J.H., 1996, Neurochemical, morphologic, and laminar characterization of cortical projection neurons in the cingulate motor areas of the macaque monkey, *J. Comp. Neurol.* 374:136-160.
28. Parent, A., and Hazrati, L.-N., 1995, Functional anatomy of the basal ganglia. I. The cortico-basal ganglia-thalamo-cortical loop, *Brain Res. Rev.* 20:91-127.
29. Picard, N., and Strick, P.L., 1996, Motor areas of the medial wall: A review of their location and functional activation, *Cereb. Cortex* 6:342-353.
30. Selemon, L.D., and Goldman-Rakic, P.S., 1985, Longitudinal topography and interdigitation of corticostriatal projections in the rhesus monkey, *J. Neurosci.* 5:776-794.
31. Shima, K., and Tanji, J., 1998, Role for cingulate motor area cells in voluntary movement selection based on reward, *Science* 282:1335-1338.
32. Shima, K., Aya, K., Mushiake, H., Inase, M., Aizawa, H., and Tanji, J., 1991, Two movement-related foci in the primate cingulate cortex observed in signal-triggered and self-paced forelimb movements, *J. Neurophysiol.* 65:188-202.
33. Strick, P.L., Dum, R.P., and Mushiake, H., 1995, Basal ganglia 'loops' with the cerebral cortex, in: *Functions of the Cortico-Basal Ganglia Loop*, M. Kimura, and A.M. Graybiel, eds., Springer, Tokyo, pp. 106-124.
34. Strick, P.L., Dum, R.P., and Picard, N., 1995, Macro-organization of the circuits connecting the basal ganglia with the cortical motor areas, in: *Models of Information Processing in the Basal Ganglia*, J.C. Houk, J.L. Davis, and D.G. Beiser, eds., MIT Press, Cambridge, MA, pp. 117-130.
35. Takada, M., Tokuno, H., Nambu, A., and Inase, M., 1998, Corticostriatal projections from the somatic motor areas of the frontal cortex in the macaque monkey: Segregation versus overlap of input zones from the primary motor cortex, the supplementary motor area, and the premotor cortex, *Exp. Brain Res. 120:114-128.*
36. Takada, M., Tokuno, H., Nambu, A., and Inase, M., 1998, Corticostriatal input zones from the supplementary motor area overlap those from the contra- rather than ipsilateral primary motor cortex, *Brain Res.* 791:335-340.
37. Tokuno, H., and Tanji, J., 1993, Input organization of distal and proximal forelimb areas in the monkey primary motor cortex: A retrograde double labeling study, *J. Comp. Neurol.* 333:199-209.

38. Tokuno, H., Takada, M., Nambu, A., and Inase, M., 1997, Reevaluation of ipsilateral corticocortical inputs to the orofacial region of the primary motor cortex in the macaque monkey, J. Comp. Neurol. 389:34-48.
39. Vogt, B.A., Pandya, D.N., and Rosene, D.L., 1987, Cingulate cortex of the rhesus monkey: 1. Cytoarchitecture and thalamic afferents, J. Comp. Neurol. 262:256-270.
40. Yeterian, E.H., and Pandya, D.N., 1991, Prefrontostriatal connections in relation to cortical architectonic organization in rhesus monkeys, J. Comp. Neurol. 312:43-67.

SUPERFICIAL AND DEEP THALAMO-CORTICAL PROJECTIONS FROM THE ORAL PART OF RHE VENTRAL LATERAL THALAMIC NUCLEUS (VLo) RECEIVING INPUTS TO THE INTERNAL PALLIDAL SEGMENT (GPi) AND CEREBELLAR DENTATE NUCLEUS IN THE MACAQUE MONKEY

Katsuma Nakano, Tetsuro Kayahara, Eiji Nagaoka, Hiroshi Ushiro, and Tomorari Tsutsumi*

1. INTRODUCTION

Superficial thalamo-cortical (T-C) responses were recorded in the lateral motor area and dorsal premotor area (PMd) following stimulation of the cerebellar dentate nucleus in monkeys (Sasaki, 1979). These responses were also recorded in the motor area following pallidal stimulation (Nambu et al., 1988; 1991). An anatomical study was also carried out on the T-C projections from the motor thalamic subdivisions in monkeys using autoradiography (Nakano et al., 1992). The latter findings demonstrated that the basal ganglia territories of the motor thalamic subdivisions give rise to superficial T-C projections, whereas the cerebellar territories give rise only to deep T-C projections. Recent studies, using modern axonal tracing techniques, have reported a possible overlap of basal ganglia and cerebellar inputs in the thalamic subdivisions (Rouiller et al., 1994; Sakai et al., 1996). We have reported that the lateral parts of the nucleus ventralis lateralis pars oralis (VLo) receive cerebellar afferents from the ventral part of the dentate nucleus (Nakano et al., 1996).

Further studies should address how various thalamic subdivisions link basal ganglia and cerebellar nuclei with motor related cortical areas. The present study was focused on the T-C projections from VLo which relay basal ganglia and cerebellar information.

---

* Katsuma Nakano, Tetsuro Kayahara, Eiji Nagaoka, Hiroshi Ushiro, and Tomorari Tsutsumi, Department of Anatomy, Faculty of Medicine, Me University, Tsu, Mie 514-8507, Japan.

2. MATERIALS AND METHODS

Twenty-one Japanese monkeys (*Macaca fuscata*) weighing 3.2-7.5 kg were used in the present experiments. The animals were anesthetized by an intramuscular injection of ketamine hydrochloride (7-8 mg/kg) followed by intraperitoneal injection of sodium pentobarbital (25 mg/kg). Then a total of 0.4-0.6 il of 5% biotinylated dextran amine (BDA) was stereotaxically pressure injected with a pneumatic picopump into the VLo thalamic nucleus at two or three points, from deep to superficial, or into distinct cerebellar nuclei at one or two points. After a survival time of 21 days, the monkeys were deeply reanesthetized with high dose of ketamine hydrochloride i.m., and then perfused transcardially with 1000 ml of saline, followed with 3000-7000 ml of 8% formalin and 0.15% glutaraldehyde in 0.1 M phosphate buffer (pH 7.4), and then 500 ml of the same buffer containing 10% sucrose. Sections with a thickness of 50 im were incubated with streptavidine solution and processed for diaminobenzidine tetrahydrochloride (DAB) or DAB-nickel ammonium reaction.

In another group, single injections (0.08-0.09 il) of 2% wheat germ agglutinin-conjugated horseradish peroxidase (WGA-HRP) dissolved in Tris buffer were made stereotaxi-

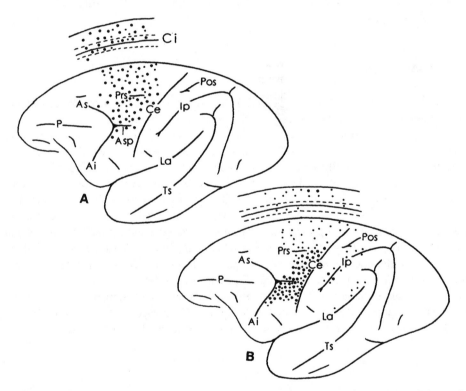

Figure 1. Diagrammatic representation of the lateral and medial (mirror image) surfaces of the cerebral hemisphere of the macaque monkey, showing superficial thalamo-cortical (T-C) projections (large dots), and deep T-C projection (small dots) from the rostral part of the nucleus ventralis lateralis pars oralis (VLo) (**A**), and caudomedial VLo (**B**). Ai, inferior arcuate sulcus; As, superior arcuate sulcus; Asp, arcuate spur, Ce, central sulcus; Ci, cingulate sulcus; 1P, intraparietal sulcus; La, lateral sulcus; P, principal sulcus; Pos, postcentral sulcus; Prs, superior precentral sulcus; Ts, superior temporal sulcus.

cally into the the various parts of the motor thalamic nuclei or cerebellar nuclei for retrograde and anterograde axonal tracing experiments. Sections for HRP experiments were processed for HRP histochemistry using the tetramethyl benzidine method.

3. RESULTS

Following injections centered in the rostral part of VLo (Fig. 1A), dense superficial T-C projections (superficial part of lamina I) and less dense deep T-C projections (laminae III-VI) were observed in the caudal part of PMd, the rostral but not ventral part of the primary motor area (MI) (Fig. 2A), and in the caudal part of the ventral premotor area (PMv) adjacent to the arcuate genu and spur as well as the supplementary motor area (SMA), mainly in the SMA-proper. In the mesial cortex, superficial T-C projections were densest in MI, moderate in the SMA-proper and sparse in the pre-SMA, while the deep T-C projections were observed mainly in MI. There were also superficial T-C projections in the banks of the cingulate sulcus, but mainly in the dorsal bank at the caudal levels. Deep T-C projections were detected also in the dorsal bank of the cingulate sulcus at the levels of rostral MI. Retrogradely BDA-labeled cells were found in the intermediate to dorsomedial parts of the medial segment of the globus pallidus (GPi).

In a case in which the injections were centered in the medial part of caudal VLo, (Fig. 1B), dense superficial and deep T-C projections were observed in the lateral 2/3 of MI (Fig. 2B) and in the dorsal part of PMv, mainly in the region ventrocaudal to the arcuate spur and genu. Terminal labelings in MI were found not only in the rostral part but also in the caudal part of the MI. In the medial part of MI, terminal labelings were seen mainly in the deep layers. In the parietal lobules, labeled fibers with varicosities and terminal specializations were sparse. These labelings were detected in the banks and fundus of the intraparietal sulcus, and only rarely in the convexity adjacent to the sulcus. Retrogradely BDA-labeled cells were detected in the ventrolateral part of caudal 2/3 GPi.

In the following set of experiments, the cerebellar projections were studied in relation to the pallidal territories in the motor thalamus. BDA or WGA-HRP injections were made into various parts of the deep cerebellar nuclei in one group, and WGA-HRP injections were attempted into the VLo, area X or nucleus ventralis posterolateralis pars oralis (VPLo) in the other group. After injections of BDA in the ventromedial parts of the nucleus dentatus (NL), anterogradely BDA-labeled fibers and terminals were observed in the rostal to lateral parts of VLo, the dorsal part of the nucleus ventralis anterior pars parvicellularis (VApc) and the dorsomedial part of the nucleus ventralis anterior pars magnocellularis (VAmc) as well as area X, whereas with BDA injections in the dorsal NL produced BDA-labeled fibers and preterminals mainly in the VPLo. These findings were confirmed with retrograde axonal tracing of WGA-HRP. Following a WGA-HRP injection in the lateral part of VLo, retrogradely labeled neurons were detected in the ventral or ventromedial part of NL, mainly at the levels of the rostral 2/3. After an injection of WGA-HRP in the lateral half of VPLo, labeled neurons in NL were seen in its dorsal half at the level of the rostral 2/3.

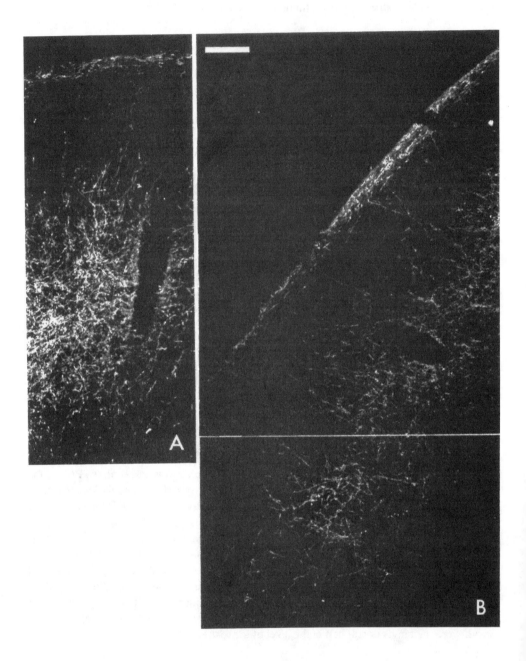

Figure 2. Darkfield photomicrographs showing superficial and deep T-C projections in the rostral part of the primary motor area (**A**), and the ventrolateral part of the primary motor area (**B**) following biotinylated dextran amine injections in the rostral VLo and caudomedial VLo, respectively. Bar = 200 ìm.

4. DISCUSSION

The most important findings in the present study were the dense superficial T-C projections from the VLo to the regions corresponding to the arm/hand representation of MI and the adjoining arcuate premotor area (APA). We also found different cortical projections from the rostral and caudal VLo. The rostromedial VLo projects mainly to the dorsomedial part of the rostral MI, and the caudal part of the PMd including the periarcuate spur region as well as the SMA-proper. The remaining major VLo projected to the entire MI including the anterior bank of the central sulcus, and the APA adjacent to the arm/hand motor area. We demonstrated superficial T-C projection to SMA from the rostromedial VLo, in addition to the VApc-VLo transitional zone. With injections of the B subunit of cholera toxin (CTB) in the periarcuate spur region, retrogradely labeled cells were seen in the transitional zone between the VApc and VLo, and lateral area X as well as the medial part of VPLo (unpublished data). This pattern of neuronal distribution is similar to that of SMA projection neurons (Nakano et al., 1993). The APA receives major thalamic input from the area X and VApc, and the MI receives its from the VLo and VPLo.

Since Schell and Strick's report (1984), segregated transthalamic projections from the GPi to SMA and from the cerebellar nuclei to the motor cortex have been accepted. However, VLo connections with MI have been demonstrated by many authors (for references see Kayahara and Nakano, 1996). Wiesendanger and Wiesendanger (1985) suggested the dual innervation of transthalamic input channels with origins in the cerebellum and basal ganglia to SMA. Accumulated data indicate that the MI and SMA receive mixed thalamic inputs originated from the GPi and cerebellar nuclei (Holsapple et al., 1991; Hoover and Strick, 1993; Jinnai et al., 1993; Nambu et al., 1988; 1991; Rouiller et al., 1994; Sakai et al., 1996).

The GPi-VLo-MI connections were demonstrated using the technique of retrograde transneuronal transport of the herpes simplex virus (Hoover and Strick, 1993). We have confirmed monosynaptic connections of GPi afferents to VLo neurons projecting to MI on the basis of electron microscopic observation (Kayahara and Nakano, 1996). In contrast to the strong VLo-SMA projection (Schell and Strick, 1984), SMA-projecting neurons occupied the lateral part of VApc or transitional zone between VApc and VLo after WGA-HRP injections in SMA (Nakano et al., 1993). Other researchers also demonstrated SMA-projecting neurons mainly in VApc (Jinnai et al., 1993) and the rostral VLo (Ilinsky et al., 1985; Jinnai et al., 1993). Overlap of MI-projecting neurons with pallidothalamic terminal fields was more prominent in the VLo following CTB injections in MI, and BDA injections in GPi (in preparation). SMA appears to receive major thalamic input from the lateral VApc, VA-VLo transitional zone, and the rostromedial VLo in addition to minor input from area X, the nucleus ventralis lateralis pars caudalis (VLc), nucleus ventralis lateralis pars postrema (VLps), nucleus paracentralis (Pc) and nucleus medialis dorsalis (MD) (Nakano et al., 1993; 1992).

According to Darian-Smith et al. (1990), each motor cortical subdivision receives differentially weighted thalamic inputs originating from the GPi, substantia nigra pars reticulata (SNr) and cerebellum, and these findings have been supported by some researchers (Rouiller et al.., 1999; Rouiller et al., 1994; Sakai et al., 1996). Thalamic neurons with double-labeling of MI and SMA injections are rare (Darian-Smith et al., 1990; Rouiller et al., 1994). Interdigitation or overlap of patches of the pallidothalamic and cerebellothalamic territories were found in the transitional border regions between each motor thalamic nuclei (Sakai et al., 1996). Rouiller at al. (1994) reported that the terminal fields of cerebellothalamic and pallidothalamic projections and the zones of origin of the thalamocortical projections to MI and SMA form a complex mosaic with interdigitating and overlapping patches in the motor thalamic nuclei and other nuclei. In our cases, the terminals of the cerebellar input in the GPi thalamic territories were seen in the lateral portion of VLo, dorsal VApc and the dorsomedial VAmc.

Using the retrograde transneuronal tracing technique of WGA-HRP, the cerebellar projections to the arm areas of MI and APA were indirectly demonstrated; the rostrodorsal part of the middle third of NL projects to the MI arm area via VPLo, and the ventrocaudal part projects to the APA arm area via area X (Orioli and Strick, 1989). Our data showed that the rostral to lateral parts of VLo, the dorsal VApc, and the dorsomedial VAmc receive cerebellar input mainly from the ventromedial part of the rostral 2/3 NL, and area X receives input from the ventral NL and the whole extent of the cadual NL. The arm/hand representations of MI and APA as well as SMA are considered to receive pallidal and cerebellar input integrated at the thalamic level; it remains to be clarified whether or not single thalamic neurons projecting to MI or APA directly receive the duel pallidal and cerebellar inputs.

Many neurons in the APA respond to visual presentations of objects around the mouth, or fire during arm movement toward the mouth (Rizzolatti et al., 1981). The PMv is involved in using visual information to guide movements in extrapersonal space (see Godschalk et al. (1981). The caudal part of PMv (F4) receives strong input from the inferior parietal lobule, and, in particular, from the ventral bank of intraparietal sulcus (VIP) (Andersen et al., 1990; Cavada and Goldman-Rakic, 1989; Chavis and Pandya, 1976). Area 7ip seems to be an important source of visual inputs to PMv (Cavada and Goldman-Rakic, 1989). The PMv is more responsive to moving than to stationary visual stimuli (Rizzolatti et al., 1981a). F4 codes space in somatocentered coordinates (Fogassi et al., 1996). It has a possible role in providing a spatial framework for the organization of head and arm movements (Fogassi et al., 1996). F4 has direct connections to the arm and mouth fields of MI (Muakkassa and Strick, 1979). Intracortical stimulation in rostral PMv (F5) evokes movements of the fingers and wrist (Gentilucci et al., 1988; Gentilucci et al., 1989). Distal movements are represented rostrally in F5, and proximal movements caudally in F4 (Gentilucci et al., 1988). In the monkey, neurons in F5 discharge during goal-directed hand movements such as grasping, holding, and tearing (Pellegrino et al., 1992).

The prefrontal cortex (PFC) is interconnected with the arm representations of multiple premotor areas (Lu et al., 1994), and has multiple routes for access to the motor system. The PFC is the highest cortical level of the motor hierarchy, and is involved in the initiation and control of motor responses (Fuster, 1989). The main output from the dorsolateral PFC to motor areas appears to be relayed in F6 (pre-SMA) (Luppino et al., 1993). F6 is connected with the other premotor areas, especially F5 and PFC, and with cingulate areas. F6 integrates visual, somatomotor, and oculomotor information and uses it to initiate arm reaching movements (Luppino et al., 1993).

The PMv connects mainly with PFC, and sends corticospinal fibers from the posterior bank of the arcuate sulcus near the spur (Dum and Strick, 1991; He et al., 1993), having a role in the initiation and control of limb movements based on visual cues and other sensory information (Gentilucci et al., 1988; Godschalk et al., 1981; Rizzolatti et al., 1988; Rizzolatti et al., 1983).

Our data demonstrated dense superficial T-C projections to the arm/hand MI and adjoining APA from the GPi territory of the VLo. The apical dendrite of the slow pyramidal cells branches into 6 shafts at the upper layer III and further branched into many finer processes extending horizontally in layers II and I (Hamada et al., 1981). The superficial T-C projections seem to contact mainly with dendritic spines of these processes. The localized reception of input by a narrower field of apical dendrites of fast pyramidal cells was considered to lead to specific movements such as skillful manipulation of the fingers (Hamada et al., 1981). Although the functional organization of the superficial T-C projections is unknown, the lateral inhibition on the local neuronal circuits in the cerebral cortex is thought to be related to the control of arm/hand movements.

5. CONCLUSION

The areas of arm/hand representation in the MI and adjacent APA receive dense superficial T-C projections from the ventrolateral part of the caudal GPi via the VLo. These cortical areas are considered to receive integrated information from the GPi and ventral NL at the thalamic in addition to the cerebral level.

6. ACKNOWLEDGMENTS

This study was supported in part by Grant-in-Aid for Scientific Research 08680814 from the Ministry of Education, Science, Sport and Culture of Japan.

7. REFERENCES

Andersen, R.A., Asanuma, C., Essick, G., and Siegel, R.M., 1990, Corticocortical connections of anatomically and physiologically defined subdivisions within the inferior parietal lobule, *J Comp Neurol,* 296:65.
Cavada, C. and Goldman-Rakic, P. S., 1989, Posterior parietal cortex in rhesus monkey: II. evidence for segregated corticocortical networks linking sensory and limbic areas with the frontal lobe, *J Comp Neurol,* 287:422.
Chavis, D. A. and Pandya, D. N., 1976, Further observations on corticofrontal connections in the rhesus monkey, *Brain Res,* 117:369.
Darian-Smith, C., Darian-Smith, I., and Cheema, S.S., 1990, Thalamic projections tosensorimotor cortex in the macaque monkey: Use of multiple retrograde fluorescent tracers, *J Comp Neurol,* 299:17.
Dum, P. R. and Strick, L. P., 1991, The origin of corticospinal projections from the premotor areas in the frontal lobe, *J Neurosci,* 11:667.
Fogassi, L., Gallese, V., Fadiga, L., Luppino, G., Matelli, M., and Rizzolatti, G., 1996, Coding of peripersonal space in inferior premotor cortex (area F4),. *J Neurophysiol,* 76:141.
Fuster, J. M., 1989, The Prefrontal Cortex (Anatomy, Physiology, and Neuropsychology of the Frontal Lobe), 2nd ed., Raven Press, New York.
Gentilucci, M., Fogassi, L., Luppino, G., Matelli, M., Camarda, R., and Rizzolatti, G., 1988, Functional organization of inferior area 6 in the macaque monkey. I. Somatotopy and the control of proximal movements, *Exp Brain Res,* 71:475.
Gentilucci, M., Fogassi, L., Luppino, G., Matelli, M., Camarda, R., and Rizzolatti, G., 1989, Somatotopic representation in inferior area 6 of the macaque monkey, *Brain Behav Evol, 33:118.*
Godschalk, M., Lemon, R. N., Nijs, H. G., and Kuypers, H. G., 1981, Behaviour of neurons in monkey peri-arcuate and precentral cortex before and during visually guided arm and hand movements, *Exp Brain Res,* 44:113.
Hamada, I., Sakai, M., and Kubota, K., 1981, Morphological differences between fast and slow pyramidal tract neurons in the mokey motor cortex as revealed by intracellular injection of horseradish peroxidase by pressure, *Neurosci Lett,* 22:233.

He, S.-Q., Dum, R. P., and Strick, P. L., 1993, Topographic organization of corticospinal projections from the frontal lobe: Motor areas on the lateral surface of the hemisphere, *J Neurosci*, 13:952.

Holsapple, J. W., Preston, J. B., and Strick, P. L., 1991, The origin of thalamic inputs to the hand representation in the primary motor cortex., *J Neurosci*, 11:2644.

Hoover, J.E. and Strick, P.L., 1993, Multiple output channels in the basal ganglia, *Science*, 259:819.

Ilinsky, I.A., Jouandet, M.L., and Goldman-Rakic, P.S., 1985, Organization of the nigrothalamocortical system in the rhesus monkey, *J Comp Neurol*, 236:315.

Jinnai, K., Nambu, A., Tanibuchi, I., and Yoshida, S:, 1993, Cerebello-and pallidothalamic pathways to areas 6 and 4 in the monkey, *Stereotact Funct Neurosurg*, 60:70.

Kayahara, T. and Nakano, K., 1996, Pallido-thalamo-motor cortical connections: an electron microscopic study in the macaque monkey, *Brain Res*, 706:337.

Lu, M.-T., Preston, J.B., and Strick, P.L., 1994, Interconnections between the prefrontal cortex and the premotor areas in the frontal lobe, *J Comp Neurol*, 341:375.

Luppino, G., Matelli, M., Camarda, R., and Rizzolatti, G., 1993, Corticocortical connections of area F3 (SMA-proper) and area F6 (pre-SMA) in the macaque monkey, *J Comp Neurol,*, 338:114.

Muakkassa, K. F. and Strick, P. L., 1979, Frontal lobe inputs to primate motor cortex: evidence for four somatotopically organized 'premotor' areas, *Brain Res*, 177:176.

Nakano, K., Hasegawa, Y., Kayahara, T., Tokushige, A., and Kuga, Y., 1993, Cortical connections of the motor thalamic nuclei in the Japanese monkey, Macaca fuscata, *Stereotact Funct Neurosurg*, 60:42.

Nakano, K., Kayahara, T., Ushiro, H., and Tsutsumi, T., 1996, Cerebello-thalamic projections in monkeys with special reference to the rostral regions of motor thalamic nuclei, *Neuroci Res, Suppl* 20, S174.

Nakano, K., Tokushige, A., Kohno, M., Hasegawa, Y., Kayahara, T., and Sasaki, K., 1992, An autoradiographic study of cortical projections from motor thalamic nuclei in the macaque monkey, *Neuroscience Res*, 13:119.

Nambu, A., Yoshida, S., and Jinnai, K., 1988, Projection on the motor cortex of thalamic neurons with pallidal input in the monkey, *Expl Brain Res*, 71:658.

Nambu, A., Yoshida, S., and Jinnai, K., 1991, Movement-related activity of thalamic neurons with input from the globus pallidus and projection to the motor cortex in the monkey, *Exp Brain Res*, 84:279.

Orioli, P. J. and Strick, L. P., 1989, Cerebellar connections with the motor cortex and the arcuate premotor area: an analysis employing retrograde transneuronal transport of WGA-HRP, *J Comp Neurol*, 288:612.

Pellegrino, G., Fadiga, L., Fogassi, L., Gallese, V., and Rizzolatti, G., 1992, Understanding motor events: a neurophysiological study, *Exp Brain Res*, 91:176.

Rizzolatti, G., Camarda, R., Fogassi, L., Gentilucci, M., Luppino, G., and Matelli, M., 1988, Functional organization of inferior area 6 in the macaque monkey. II. Area F5 and the control of distal movements, *Exp Brain Res*, 71:491.

Rizzolatti, G., Matelli, M., and Pavesi, G., 1983, Deficits in attention and movement following the removal of postarcuate (area 6) and prearcuate (area 8) cortex in macaque monkeys, *Brain*, 106:655.

Rizzolatti, G., Scandolara, C., Matelli, M., and Gentilucci, M., 1981, Afferent properties of periarcuate neurons in macaque monkeys. II. Visual responses, *Behav Brain Res*, 2:147.

Rizzolatti, G., Scandolara, C., Matelli, M., and Gentilucci, M., 1981, Afferent properties of periarcuate neurons in macaque monkeys. I. Somatosensory responses, *Behav Brain Res*, 2:125.

Rouiller, E., Tanne, J., Moret, V., and Boussaoud, D., 1999, Origin of thalamic inputs to the primary, premotor, and supplementary motor cortical areas and to area 46 in macaque monkeys: a multiple retrograde tracing study, *J Comp Neurol*, 409:131.

Rouiller, E.M., Liang, F., Babalian, A., Moret, V., and Wiesendanger, M., 1994, Cerebellothalamocortical and pallidothalamocortical projections to the primary and supplementary motor cortical areas: a multiple tracing study in macaque monkeys, *J Comp Neurol*, 345:185.

Sakai, S.T., Inase, M., and Tanji, J., 1996, Comparison of cerebellothalamic and pallidothalamic projections in the monkey (Macaca fuscata): A double anterograde labeling study, *J Comp Neurol*, 368:215.

Sasaki, K., 1979, Cerebro-cerebellar interactions in the cats and monkeys, in: *Cerebro-cerebellar interactions*, J. Massion, and K. Sasaki, eds. Elsevier, Amsterdam.

Schell, G. R. and Strick, P. L., 1984, The origin of thalamic inputs to the arcuate premotor and supplementary motor areas, *J Neurosci*, 4:539.

Wiesendanger, R. and Wiesendanger, M., 1985, Cerebello-cortical linkage in the monkey as revealed by transcellular labeling with the lectin wheat germ agglutinin conjugated to the marker horseradish peroxidase, *Exp Brain Res*, 59:105.

MORPHOLOGICAL AND ELECTROPHYSIOLOGICAL STUDIES OF SUBSTANTIA NIGRA, TEGMENTAL PEDUNCULOPONTINE NUCLEUS, AND SUBTHALAMUS IN ORGANOTYPIC CO-CULTURE

S.T. Kitai, N. Ichinohe, J. Rohrbacher and B. Teng*

1. INTRODUCTION

Recently we incorporated an organotypic culture method in our basal ganglia research. In our preparation, we were successful in co-culturing more than two structures of interest [1-3]. This *in vitro* organotypic preparation combines the advantage of *in vitro* slice preparation for ease of intracellular sharp or patch recording under improved controlled experimental chemical environment with the *in vivo* preparation in which the structure of interest is not isolated from the source of their major afferents. In this report, we would like to present a triple culture preparation consisting of the tegmental pedunculopontine nucleus (PPN), the subthalamic nucleus (STN) and the substantia nigra (SN).

Briefly, the SN is located in the ventral midbrain (VM) and consists of the pars compacta (SNc) and the pars reticulata (SNr). The principal cell type in the SNc is dopaminergic (DA) which projects mainly to the striatum [4-6] with fewer to the subthalamic nucleus (STN) [7,8] and tegmental pedunculopontine nucleus (PPN) [9]. The SNc receives cholinergic and glutamatergic inputs from the PPN/MEA, STN and cerebral cortex [10-16], and GABAergic inputs from the striatum and globus pallidus [17]. The majority of the neurons in the SNr are GABAergic [18,19] and send their axons mainly to the thalamus, superior colliculus and tegmental pedunculopontine nucleus (PPN) [4,20]. The SNr receives excitatory inputs from the STN and PPN/MEA [21,22] and inhibitory GABAergic inputs from the striatum and globus pallidus [23,24]. Electrophysiological features of DA neurons are well characterized in *in vivo* and *in vitro* slice preparation [25-37]. The PPN is situated in the mesopontine tegmentum (MPT). Wainer and co-workers [38] defined the PPN in the rat as a collection of large multipolar cholinergic neurons spatially segregated from the glutamatergic cells in the midbrain extrapyramidal area (MEA). They receive afferent inputs from the globus pallidus, entopeduncular nucleus and

---

* S.T. Kitai, N. Ichinohe, J. Rohrbacher and B. Teng, Department of Anatomy and Neurobiology The University of Tennessee, College of Medicine 855 Monroe Avenue Memphis, Tennessee 38163 U.S.A.

The Basal Ganglia VI
Edited by Graybiel *et al.*, Kluwer Academic/Plenum Publishers, 2002

SN, and are a major source of afferents to the basal ganglia. However, subsequent investigations questioned a clear distinction between the PPN and MEA in terms of their connections with the basal ganglia [16,39]. The STN is located between the zona incerta (ZI) dorsally and the internal capsule ventrally, and consists of mainly glutamatergic neurons [40-42]. The STN receives glutamatergic projection from the cerebral cortex, thalamus and PPN/MEA [43-45], cholinergic projection from the PPN [46,47] and GABAergic projection from the globus pallidus [24,48]. The STN projects, in turn, to the globus pallidus, entopeduncular nucleus, SNc/SNr and PPN/MEA [14,49-53].

We developed a PPN-SN-STN organotypic triple culture preparation to investigate a hypothesis that inputs from the PPN and STH may be responsible for bursting activities of SNDA neurons. The functional role of these inputs is poorly understood, but previous studies suggest that excitatory acetylcholine (ACh) inputs from the PPN [12,22,54,55], combined with activation of glutamatergic receptors by STH [41,56] and PPN afferents, may be a major force in modulation of DA neuronal firing. However, before we test this hypothesis, we must first establish the fact that the structures co-cultured, such as SN, PPN, and STN can retain morphological and physiological integrity found in the *in vivo* situation.

Findings on the morphology, cytoarchitechtonics and cytochemical nature of the PPN, STN and SN neurons are quite similar to the *in vivo* observations. One of the striking findings is that spontaneous firing patterns of DA neurons, unlike *in vitro* preparation, more closely resemble those of *in vivo* situation. We considered this is because two major STN and PPN afferents are intact in the organotypic culture preparation while they are severed in *in vitro* preparation. These findings indicate that the organotypic culture preparation really is a viable preparation in the study of functional circuits and adds another step in further understanding of the operation of basal ganglia circuits.

2. METHODS OF PROCEDURES

A detailed description of the methods of procedures is published elsewhere [1,57].

2.1 Preparation of Organotypic Cultures

Coronal sections (500 µm) were obtained from rat brains (Sprague Dawley, Harlan, Indianapolis, IN, U.S.A.) at postnatal day 1-2. The STN explant (STNe) (STN and the surrounding structures such as the lateral hypothalamus and the internal capsule), ventral midbrain explant (VMe) (pars compacta and pars reticulata of the SN) and mesopontine tegmentum explant (MPTe) (caudal mesopontine tegmentum including the PPN) were dissected out. The explants were placed on a small rectangular piece of Millicell-CM membrane (Millipore, Bedford, MA, U.S.A.) with 25 µl of chicken plasma (Sigma, St. Louis, MO, U.S.A.) on a coverslip. After plasma coagulation with bovine thrombin (20 µl), the cultures were put into culture tubes (Nunc, Naperville, IL, U.S.A.), with the medium (50% basal medium Eagle, 25% HBSS, and 25% horse serum, with 0.5% glucose and 0.5 mM L-glutamine) (all Gibco, Grand Island, NY, U.S.A.). The cultures were rotated in a roller tube incubator at 35°C. After 3 and 28 days *in vitro*, 10 µl of mitosis inhibitor was added for 24-48 hr (4.4 µM cytosine â-D-arabinofuranoside, 4.4 µM uridine, and 4.4 µM 5-fluoro-5'-deoxyuridine; calculated to final concentration; all Sigma, St. Louis, MO, U.S.A.). Medium was changed every 3-4 days.

2.2 Electrophysiology

Recordings were obtained from the cultures (3 to 8 weeks *in vitro*) submerged in a perfusion medium (in mM, NaCl 126, KCl 2.5, CaCl2 1.2, MgCl2 1.0, MgSO4 0.4, NaH2PO4 0.3, KH2PO4 0.3, NaHCO3 26 and glucose 11, saturated with 95%O2/5% CO2 mixture) at 35-36°C. Drugs used were 6,7-dinitroquinoxaline-2,3-dione (DNQX) (RBI Natick, MA, U.S.A.), kynurenic acid, DL-2-amino 4-methyl-5-phosphono-3-pentenoic acid (4-Methyl APPA), bicuculline, mecamylamine and atropine sulfate(Sigma, St. Louis, MO, U.S.A.). All drugs were diluted to the final concentration and applied in the perfusion medium. Whole-cell recording was obtained with electrodes (resistance: 3-5 MW) filled with (in mM): K-gluconate 120, NaCl 10, CaCl2 0.25, HEPES 10, EGTA 5, Mg-ATP 2, GTP 0.4, glucose 10 and 0.2% neurobiotin (Vector Laboratories, Burlingame, CA, U.S.A.), pH 7.3. For sharp electrode recording (resistance: 100-150 MW), electrodes were filled with 2 M K-acetate and 2% neurobiotin. An Axopatch-1A (Axon Instruments, Foster City, CA, U.S.A.) was used for whole cell recording and an Axoclamp-2A (Axon Instruments, Foster City, U.S.A.) for sharp microelectrode recording. Electrical stimulation (maximal 25 µA, 0.2 ms duration, 0.1 Hz) was delivered through a glass electrode filled with 4 mM NaCl (2-3 MW), which was placed in the PPN pars dissipata (PPNpd) or PPN pars compacta (PPNpc). The stimulation site was identified by a small lesion in the PPN area in MPTe defined by NADPH diaphorase histochemistry. All the recorded neurons were located in the SN and labeled by intracellular injection of neurobiotin with positive current pulses (250 ms, 2 Hz, 0.3-1.2 nA), double labeled with tyrosine hydroxylase (TH)-immunohistochemistry and stained with fluorescein method. Recorded neurons were reconstructed using a camera lucida drawing.

2.3 Morphological and Immunohistochemical Analysis

The choice of anatomical methods involved standard immunohistochemistry for choline acetyltransferase (ChAT), tyrosine hydroxylase (TH), parvalbumin (PV) and glutamate (Glu), and NADPH-diaphorase (NADPH-d) histochemistry and a double staining method (NADPH-d histochemistry and immunohistochemistry for TH; double immunohistochemistry for PV and TH; NADPH-d histochemistry and immunohistochemistry for Glu). For a detailed description, readers are referred to an article by Ichinohe et al, (1999) [57].

3. RESULTS

Figure 1A shows a typical triple culture consisting of the MPTe, VMe and STNe seen at 1 DIV. The STN was identified as an area surrounded by the transparent internal capsule ventrolaterally and Forel's field H_2 mediodorsally. The VMe, which contained the SN, was cut out ventrolaterally from the midbrain at the level of the rostral superior colliculus. Attention was paid to minimize the involvement of the ventral tegmental area in the VMe. The MPTe, which contained the PPN/MEA, was cut out from the mesopontine tissue at the level of the mid-inferior colliculus, excluding the raphe nucleus and the central gray. Double staining combined with ChAT immunohistochemistry and NADPH-d histochemistry showed that all the ChAT-ir neurons were also diaphorase positive (Fig. 4C,D). Therefore, we consid-

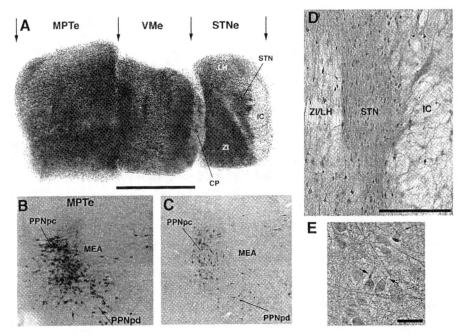

Figure 1. A: Triple culture consisting of MPTe, VMe and STNe at 1 DIV. The CP denotes the ventral aspect of the VMe. B: Distribution of cholinergic neurons in the culture (25 DIV). C: MPT in coronal sections in *in vivo* preparation. D: STNe. E: STN neurons. STN: subthalamic nucleus. IC: internal capsule. CP: cerebral peduncle. LH: lateral hypothalamus. PPNpc: PPN pars compacts. PPNpd: PPN pars dissiparta. MEA: midbrain extrapyramidal area. ZI: Zona Inserta. Scale bar: 1 mm in A, 500 imin D and 50 im in E.

ered that NADPH-d staining can be used as a reliable marker for cholinergic neurons as in the *in vivo* studies [58].

3.1. STN-explant

Figure 1D demonstrates densely packed Glu-ir neurons within the STN area which was clearly delineated from the surrounding tissue by the moderately stained neuropil. The neurons were medium in size (long axis, 18-35 μm, 24.0 ± 0.3 μm; short axis, 10-19 μm, 13.3 ± 0.1 μm: n = 280) and gave rise to 3-5 primary dendrites (Figs. lE). The internal capsule region contained only a few Glu-ir neurons. The LH/ZI area also contained Glu-ir neurons, but their density was much lower than in the STN area. In the STNe, no ChAT-ir positive neurons were found.

3.2. MPT-explant

When the distribution pattern of the NADPH-d stained neurons in the MPTe (Fig. 1B) was compared to that in the *in vivo* preparation at the level of the mid-inferior colliculus (Fig. 1C), remarkable similarities were found between these preparations. That is, dense clusters of cholinergic neurons were labeled in an area which constitutes the PPNpc (Figs. 1B,C). The

MEA, medial to the PPNpc, contained few cholinergic neurons (Figs. 1B,C). The soma size of cholinergic neurons varied from medium (22.4 ± 0.2 μm: n = 278) to large (32.2 ± 0.3 μm: n = 278). The primary and secondary dendrites of cholinergic neurons were well defined by NADPH-d staining (Figs. 2A, B). Each primary dendrite was divided into two or three secondary dendrites within 100 im from the soma. These secondary dendrites tapered gradually and sometimes had irregularly spaced varicosities (Fig. 2A) and they could be traced up to 500 μm from their soma. Within the MPTe, axons of the NADPH-d neurons emitted many thinner branches with varicosities, and these fibers were frequently in close apposition to NADPH-d positive soma and dendrites. NADPH-d positive axons originating from the labeled neurons in the MPTe were followed to the VMe and STNe (Fig. 2C). The terminal fibers were very densely distributed throughout the VMe and double labeling with NADPH-d histochemistry and TH-immunohistochemistry demonstrated that NADPH-d positive fibers with varicosities were frequently in close apposition to TH-ir somata and dendrites (Fig. not shown). The density of terminating fibers in STNe was rather moderate compared to that in the VMe.

3.3. VM-explant

Numerous TH-ir neurons (considered dopaminergic) were distributed mainly in the dorsal portion constituting the SNc of the VMe (Figs. 3A). Few TH-ir neurons were also found in the ventral portion (SNr) of the VMe. However, as reported previously by Østergaard et al. (1990), the border between the SNc and SNr in the culture preparation was not as well demarcated as in the *in vivo* preparation.

DA neurons were characterized by a medium-sized soma (long axis, 27.1 ± 0.1 μm; short axis, 18.8 ± 0.1 μm: n = 367) from which 3-6 primary dendrites emerged (Fig. 3B). The soma shape was oval, fusiform, triangular or polygonal (Fig. 3B). Primary dendrites rarely branched more than once or twice and varicosities were frequently observed on the distal portion of primary and higher order dendrites which could be traced up to several hundred micrometers from the soma. Axons originated from the soma or the proximal dendrites, and rarely gave off local collateral branches. These morphological features are similar to those described in *in vivo* [59] and organotypic culture studies [2,60,61]. TH-ir fibers originating from the VMe were observed at the border between the VMe and MPTe (Fig. 3A) and became progressively less as the distance from the border increased. Double labeling demonstrated that TH-ir fibers were in close apposition to NADPH-d positive neurons in MPTe (Fig. 3C), indicating SN dopaminergic innervation to PPN cholinergic neurons. TH-ir varicose fibers originating from VMe were also observed throughout the STN and LH/ZI. TH-ir fibers profusely ramified as they entered in the STN area and terminated as thin and highly varicose fibers reminiscent of boutons 'en passant' (Fig. not shown). PV-ir (GABAergic) neurons were also found in the VMe (Fig. 4 A, B). Double immunohistochemistry for TH and PV showed that GABAergic neurons were located generally ventral to densely packed DA neurons.

GABAergic neurons were characterized by a medium-sized soma (long axis, 23.3 ± 0.3 μm; short axis, 16.2 ± 0.3 μm: n = 250) with 3-6 thick primary dendrites. The shape of the soma was oval, fusiform, triangular or polygonal (Fig. 4B). Primary and secondary dendrites, which could be followed up to several hundred micrometers, were smooth while some terminal dendrites had varicosities. Axons emerged from the soma or the primary dendrites and axon

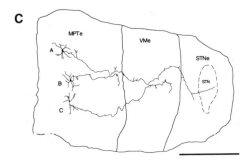

Figure 2. A: Medium-sized NADPH-d positive neurons. Note varicosities on secondary dendrite (arrows) and an axon (arrowhead) emerging from a cell body (24 DIV). B: A small-sized NADPH-d positive neuron (24 DIV). C: A drawing of projection pattern of three representative NADPH-d positive neurons (25 DIV). Axons projecting to the STNe (neuron A), the VMe (neuron C) and remain within the MPTe (neuron B). Scale bar = 50 μm in A & B, 1 mm in C.

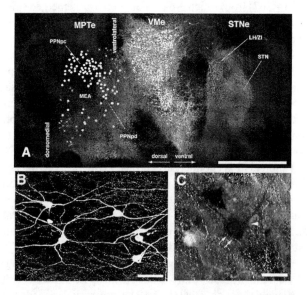

Figure 3. A: Triple cultures with TH immuno-staining (25 DIV). In the MPTe, filled circles represent NADPH-d positive neurons, and crosses TH-ir neurons. B: TH-ir neurons in the SNc. C: TH-ir axons with varicosities (arrows) are in close apposition with NADPH-d positive soma (arrowhead) in the MPTe. Scale bars = 1 mm in A; 100 μm in B; 25 μm in C.

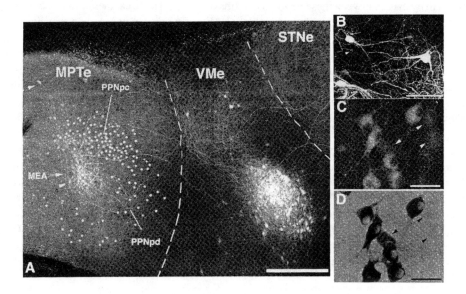

Figure 4. A: PV-ir projection from the VMe. Filled circles represent NADPH-d positive somata. Arrowheads point to PV-ir neurons in the MPTe (26 DIV). B: PV-ir neurons in the VMe. Note an axon (large arrow) with local axon collaterals (arrowheads) (27 DIV). C: ChAT-ir neurons in the MPTe. D: Same area stained for NADPH-d. Note all the ChAT-ir neurons display NADPH-d activity. Arrowheads point to neurons which are weakly positive for ChAT in C and NADPH-d negative or weak in D. Scale bars = 500 µm in A; 100 µm µm in B; 50 µm in C and D.

collaterals ramified near the soma. PV-ir neurons were also located in the STN and PPN area, but their numbers were quite limited compared to those in VMe.

Axons originated from GABAergic neurons in the VMe were traced into the MPTe and STNe (Fig. 4A). Those axons projecting to the MPTe coursed through the PPNpc and PPNpd and branched profusely within the MEA and terminated as bouton 'en passant'. In the MEA and PPN area, NADPH-d negative neurons received extensive GABAergic innervations on their soma and dendrites (Fig. not shown). Few GABAergic fibers were traced directly to the STNe (Fig. 4A), but even fewer fibers entered the STN. They rarely branched and had no sign of varicosities, indicating that the GABAergic projection from the VMe to the STN is not extensive.

3.4. Electrophysiology

3.4.1. Membrane and Action Potential Properties of Dopaminergic and Non-dopaminergic Neuron

Electrophysiological characteristics of DA and non-DA neurons in the organotypic culture preparation were quite similar to those reported from *in vivo* and *in vitro* preparations. Even though there was no difference in the mean input resistance between DA neurons (range: 80-210 MW) and non-DA neurons (range: 55-260 MW) (Table 1), a clear distinction can be made between these neurons in terms of several electrical membrane characteristics. DA neurons had broader action potentials (Fig. 5A) than non-DA neurons (Fig. 5B). Single action potentials of DA neurons were usually followed by a large AHP (Fig. 5A) while the action potentials of non-DA neurons had very little AHP (Fig. 5B). The threshold for action potentials were much more depolarized in DA neurons in comparison to non-DA neurons. Also, a marked difference between DA and non-DA neurons was found in their frequency of firing to the intensity of applied current (Fig. 5A, B), indicating that DA neurons show a significantly stronger spike frequency adaptation than non-DA neurons. In other words, non-DA neurons can fire at much higher frequencies at given current pulses.

Figure 5C and D show membrane potential changes in SN neurons to hyperpolarizing current pulses. As the intensity of the hyperpolarizing current was increased, a time-dependent inward rectification was observed in all the DA neurons tested (n=8) (Fig. 5C). On the other hand, only seven out of thirty non-DA neurons demonstrated this time-dependent inward rectification and the amount of rectification was much less compared to the DA neurons. The rest of non-DA neurons showed no rectification to hyperpolarizing current injection (Fig. 5D). DA neurons also displayed a transient outward rectification (Fig. SC, arrow). In contrast, only one out of thirty non-DA neurons showed this outward rectification.

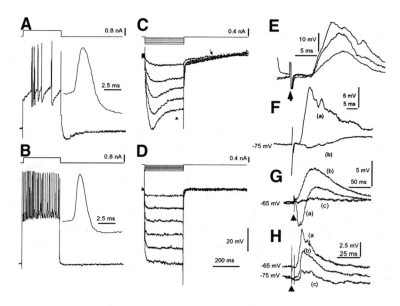

Figure 5. The response of a DA neuron (A) and a non-DA neuron (B) to the injection of a depolarizing current pulse (300 ms duration) and single action potentials with higher time resolution. Responses of DA neuron (C) and non-DA neuron (D) to injected hyperpolarizing currents. Arrow indicates a transient outward rectification and arrowhead the time-dependent inward rectification in C. Postsynaptic responses in DA neurons (E) and non-DA neurons (F-G) to PPN stimulation and their pharmacology. E: EPSPs to stimulation intensities of 15, 20, 25 µA (duration 0.2 ms applied at 0.1 Hz). Arrowheads point to the stimulation artifacts. F: Effect of DNQX (10 µM) G: (a) Depolarization of the membrane increased the amplitude of IPSPs followed by depolarizing potentials. (b) Blockage of IPSPs by bicuculline (50 µM) leaving slow depolarizing potentials which were blocked by 4-Methyl APPA (1 µm) (c). H: In another DA neuron, application of atropine sulfate (50 im) and mecamylamine (10 µm) reduced PPN-induced EPSPs (control (a), effect (b) and recovery(c)).

3.4.2. Spontaneous Firing Patterns of Dopaminergic and Non-Dopaminergic Neurons

Both DA and non-DA neurons showed three distinctively different firing patterns. Three firing patterns for DA neurons ranged from rhythmic, irregular and burst-like firing (Fig. 6). The rhythmic firing is characterized by relatively constant interspike intervals. An irregular firing is characterized by a relatively flat distribution of interspike intervals. The burst-firing pattern was defined by a train of two or more spikes occurring within a relatively short interval [27] interrupting a spontaneous irregular firing pattern. Three firing patterns for non-DA neurons consisted of irregular, burst and mixed firing (Fig. 6). In the burst firing pattern, an irregular firing pattern was interrupted completely and sudden burst of spikes were triggered from a long depolarizing potential. The mixed firing pattern is characterized by an irregular firing interrupted by a short burst firing. Unlike DA neurons, we have never observed a regular rhythmic firing in non-DA neurons.

3.4.3. Synaptic Inputs

Stimulation of the PPNpc or PPNpd induced in DA and non-DA neurons depolarizing postsynaptic potentials (DPSPs). There was no difference in the response pattern in DA and non-DA neurons to PPN stimulation. We only report the data obtained in the experiment where the stimulation site was morphologically identified to be in the PPN. Figure 5E-H show samples of PPN-induced EPSPs. Figure 5E demonstrates monosynaptically induced EPSPs following PPN stimulation since an increase in the stimulation intensity evoked an increase

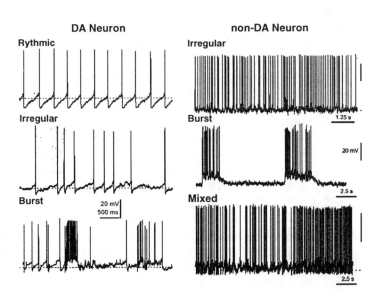

Figure 6. DA neurons with rythmic, irregular and burst firing pattern and non-DA neurons with irregular, burst and mixed firing patten.

in the amplitude of the EPSPs without changes in their latencies. The conduction velocities estimated were between 0.14-0.74 m/s (distance between stimulation and recording sites: 1-2.5mm, mean latency; 5.5 ± 0.5 ms, n=22).

These PPN-induced EPSPs were blocked by a glutamate receptor antagonist kynurenic acid (1 mM, n=4). A further analysis indicated that these glutamatergic EPSPs were composed of a fast non-NMDA receptor mediated (n=6) and slow NMDA receptor mediated component (n=3). Figures 5 F and G show sample responses obtained from a non-DA neuron. PPN stimulation induced EPSPs which were blocked by non-NMDA receptor antagonist DNQX (10 μM) applied in the medium (n=4), leaving only small IPSPs followed by slow depolarizing potentials (Fig. 5Fb). When the transmembrane potential was depolarized to -65 mV by current injection, these IPSPs were clearly seen (Fig. 5Ga). IPSPs had latencies in the range of 2.1 to 7.4 ms (mean latency: 5.3 ± 0.8 ms). When these IPSPs, most likely mediated by GABAA, were blocked by bicuculline (50 μM), large slow EPSPs were seen (Fig. 5Gb, n=3). An application of the NMDA receptor antagonist 4-Methyl APPA (1 μM) blocked these slow EPSPs (Fig. 5Gc). It was also found that the cholinergic antagonists (10 μM mecamylamine, 50 μM atropine sulfate) could reduce the amplitude of PPN-induced EPSPs by 48-59% (Fig. 5H).

4. DISCUSSION

4.1. Cellular Organization

4.1.1. MPTe

The present study demonstrated that cholinergic neurons could survive well over 21-28 DIV and retain their cytoarchitectonics, immunocytochemical nature and their projection patterns of the PPN as observed in *in vivo* studies [38]. PPN were densely packed with cholinergic neurons while the MEA was almost void of cholinergic neurons. Generally, the morphology of cholinergic neurons (soma morphology and dendritic number and branching pattern) was found to be similar to that of the *in vivo* specimen [16,38,62]. As has been described previously by others [58], we identified cholinergic neurons by NADPH-d histochemistry in this study, since a double staining combined with ChAT immunohistochemistry and NADPH-d histochemistry showed that every ChAT-ir neuron visualized by FITC was also, subsequently, visualized by NBT as NADPH-d positive. As a control, TH-ir or PV-ir neurons visualized by FITC were also stained for NADPH-d using NBT. It was found that there were no FITC positive neurons reactive to NBT. Glu-ir neurons were located mostly in the MEA, but much less in PPNpc and PPNpd demonstrated by double labeling for Glu and NADPH-d. This is similar to the *in vivo* findings which described Glu-ir neurons to be another main cell type of the PPN, and especially in the MEA in rats [63] and monkeys [64]. Even though Glu-ir neurons can be either glutamatergic or GABAergic, most of the Glu-ir neurons observed in the MPTe are likely to be glutamatergic based on their soma size (the size of most of the neurons ranged from 20-40 mm in long axis). It has been reported that single labeled Glu-ir neurons were medium-sized and larger than double labeled Glu-ir and GAD-ir neurons [65].

Double labeling studies with a combined NADPH-d histochemistry and Glu immunohistochemistry in the *in vivo* preparation have shown that acetylcholine and glutamate were contained within the same neurons in the MPT of the rat [66], and the monkey [64]. In the MPTe, approximately 10% of NADPH-d positive neurons were also found to be Glu-ir. However, this percentage may be an underestimate since we counted only those neurons labeled with NADPH-d reaction (as cholinergic) in the soma and DAB-nickel product (as Glu-ir) in the dendrites as double labeled neurons. It is possible that in some instance Glu immunoreactivity was limited only to the soma, and not in the dendrites. In the MPTe, TH-ir neurons and PV-ir neurons were also found to be intermingled with cholinergic neurons in the PPN, as has been reported by others in *in vivo* situation [38].

4.1.2. VMe

In the VMe, TH-ir neurons were distributed mainly in the dorsal portion of the VMe, while the PV-it neurons predominated in the ventral portion. This is similar to the distribution pattern of SN neurons in *in vivo* preparation in which DA neurons were densely clustered in the dorsal tier of the SN (SNc) and PV-ir neurons identified as GABAergic projection neurons [67-69] in the ventral tier of the SN (SNr). Morphological features (e.g., soma size and dendrites) of DA and GABAergic neurons were also similar to those in *in vivo* preparation. Previous studies on the rat demonstrated that a majority of SNc DA and SNr GABAergic neurons are Glu-ir [70]. Similarly, both the dorsal and ventral portion of the VMe contained neurons immunoreactive to Glu whose morphological features were similar to TH-ir and PV-ir neurons, indicating that the Glu-ir neurons represent both dopaminergic and GABAergic neurons.

4.1.3. STNe

Studies using polyclonal antisera against Glu, GABA or glycine showed that virtually all STN neurons were positive for Glu, but not for GABA or glycine [40,42]. The neurons in the STNe were immunoreactive for glutamate and mostly confined in the STN area which clearly delineated the surrounding tissue. In the STNe, no ChAT-ir neurons were found and the number of PV-ir neurons was much less compared to that of Glu-ir neurons. This is in contrast to the *in vivo* situation which showed strong representation of PV-ir neurons in the STN[67]. Plenz et al, (1998) [3] also reported a scarcity of PV-ir neurons in the STN in the organotypic culture preparation involving STN co-cultured with the cerebral cortex, striatum and globus pallidus.

4.2. Projection Patterns

4.2.1. Cholinergic Projection from MPTe to VMe and STNe

NADPH-d fibers (presumably cholinergic) originating from the MPTe projected to the VMe and STNe. This conclusion was based on (1) tracing individual NADPH-d positive fibers from the soma in the MPTe to the VMe and STNe and (2) identification of varicosities in NADPH-d positive fibers in the VMe and STNe. For this analysis, we examined those triple culture preparations in which NADPH-d positive somata were found in no other areas but the MPTe in order to eliminate a possibility that these fibers might have originated in the VMe or STNe. It was frequently found that NADPH-d positive fibers were in close apposition to TH-ir neurons in the VMe. This cholinergic projection from the PPN to the SN and STN is similar to the *in vivo* situation [11,13,16,71-73].

4.2.2. GABAergic Projection from VMe to MPTe and STNe

4.2.2a. To MPTe. Previous *in vivo* projection studies in the rat, utilizing a combined anterograde tracing technique with ChAT immunohistochemistry, have shown that fibers originating from the SNr do not terminate in the areas with a cluster of cholinergic neurons (e.g., PPN), but project heavily to the adjacent MEA, which was mostly void of cholinergic neurons and established symmetrical synaptic contacts with mostly non-cholinergic profiles [9,38]. Similar to the *in vivo* preparation, PV-ir GABAergic nigral neurons in the organotypic culture preparation projected to the MEA in a highly selective manner. That is, primary axons of nigral projection neurons from the VMe coursed for a long distance through the MPTe and branched only after they entered the MEA.

4.2.2b. To STNe: As in *in vivo* situation [74] most of the PV-ir fibers from VMe in the organotypic culture preparation did not terminate in the STN. In all cases these fibers entered into the STN-explant, but in most instances they changed their course before they reached the STN. On some occasions, PV-ir fibers managed to enter into the STN, but they rarely ramified, indicating no signs of termination.

4.2.3. Dopaminergic Projection from VMe to STNe and MPTe

In this analysis, we examined only the organotypic culture preparations (14 cultures) which contained no TH-ir somata in the STNe and MPTe in order to exclude a possibility that the TH-ir fibers originated intrinsically. TH-ir fibers originating from the VMe were mainly localized in the border between the VMe and MPTe and became progressively scarce as they entered further into the MPTe, indicating no major projection of dopaminergic fibers to the PPN/MEA area. This is similar to the *in vivo* situation where little dopaminergic projection to the PPN has been reported [75]. On the other hand, as has been observed in *in vivo* condition [76] there was a clear dopaminergic projection to the STN and LH/ZI area.

4.3. Morphological and Physiological Properties of DA and Non-DA Neurons in Co-Culture

The main morphological properties of DA neurons recorded in this study are similar to those reported in *in vitro* slice or *in vivo* studies where recorded DA neurons were intracellularly labeled for morphological analysis [28,37,77,78]. That is, the somata of the DA neurons in the triple culture were also medium in size with varying shapes (ovoid, triangle or polygonal) with 3 to 8 primary dendrites of variable thickness. Also, the electrophysiological characteristics of DA neurons were similar to those observed in *in vivo* or *in vitro* slice preparations [21,28,29,31,34,37,79-81]. More specifically, DA neurons had the input resistance of 80-210 MΩ, the time-dependent inward rectification, a delayed repolarization, long duration action potential, prominent AHP and more depolarized threshold for generating action potentials. Also, strong frequency adaptation characterized the DA neurons in our organotypic culture which was also shown *in vivo* and *in vitro* slice preparation, but the DA neurons in the triple culture were found to be able to maintain higher firing frequencies than those recorded *in vitro* slice preparation [34,37]. Possibly, this may be due to some differences between the two preparations in Ca2+-activated K+ conductance, considered to be an underlying factor for spike adaptation [27,29,80]. Another most notable firing pattern of DA neurons in the organotypic cultured preparation is that, as in *in vivo* situation, they not only fire in regular pattern, but also irregular and burst-like firing pattern [27,36]. This is in contrast to a well known observation that DA neurons, in *in vitro* slice preparation, do not burst spontaneously, but fire in a very regular pacemaker-like pattern [28,29,31,34,35,37,82]. This difference between *in vitro* slice and organotypic cultured preparation may be because critical afferent inputs are severed in the former preparation. It has been reported that stimulation of afferent inputs or direct application of glutamate agonist NMDA can cause a burst of firing in DA neurons [83-85] indicating an important role in the modulation of spontaneous firing pattern of DA neurons by afferent inputs. In our organotypic co-culture preparation two of the main excitatory inputs, namely the STN and PPN, were preserved. This would suggest that the organotypic co-culture system could retain more normalcy than the slice preparation.

Electrophysiological characteristics of the non-DA neurons in this study are also quite similar to those described for non-DA neurons in the previous *in vivo* [36] and *in vitro* slice studies. They fire in irregular non-rhythmic pattern with high firing frequencies and have short action potential duration and input resistance [21,28,31,34,37]. The above electrophysiological characteristics indicate that these non-DA neurons are most likely GABAergic projecting neurons identified in other studies [34,36]. It was also interesting to note that, unlike the non-DA neurons in *in vitro* slice preparation, the non-DA neurons in the organotypic culture had the ability to change its mode of firing from irregular to a burst firing as observed in *in vivo* situation [36].

4.4. Afferent Inputs from PPN to SN

In this study we have clearly demonstrated that PPN stimulation could either excite or inhibit SN neurons. The PPN-induced postsynaptic potentials were considered monosynaptic in nature because of their relatively constant latencies in spite of changes in stimulation intensities. The estimated conduction velocities (0.14-0.74 m/s) were slower than those (0.41 to 1.29 m/s) reported *in vivo* and *in vitro* brain slice preparations [12,16]. This may be due to our underestimation of the conduction velocities. We observed that the axons of PPN neurons in our organotypic culture take tortuous courses before entering the SN.

As observed in *in vivo* single unit recordings in the SN [22,86], the excitatory inputs from PPN were mediated some by cholinergic, but mainly glutamatergic, transmitters. Glutamate-mediated EPSPs could be pharmacologically differentiated into fast (by a non-NMDA receptor antagonist) and slow components (by a selective NMDA receptor antagonist). Previous electrophysiological studies in *in vivo* [22,86] and *in vitro* [12] preparation also reported that SNc and SNr receive excitatory inputs from the PPN/MEA and involve both NMDA and non-NMDA, and cholinergic transmission [12]. A relative scarcity of cholinergic inputs, compared to glutamatergic inputs, might be due to: 1) more abundance of the glutamate-immunoreactive neurons [57], 2) selective stimulation because of lower threshold for activation for glutamatergic neurons, and 3) acetylcholine may act primarily as a neuromodulator that changes excitability of the SN neurons [32,55].

In some occasions, we have recorded IPSPs following stimulation of PPN. Based on their reversal potentials and their responses to bicuculline application, one can conclude that they are mediated by GABAn receptors. These IPSPs could be induced by (1) the glutamate-ir small neurons in the PPN, which could be GABAergic and innervate the SN and/ or (2) an antidromic activation of the axons of GABAergic projection neurons originating from the SNr [4,9,57,87]. Both *in vivo* and *in vitro* studies indicate that the SNr cells can inhibit DA and non-DA cells via their axon collaterals [20,81,88,89].

In summary, we have clearly demonstrated that the neuronal morphologies and cytochemical nature of the PPN, SN and STN, their projection patterns and electrophysiological characteristics of SN neurons in the triple culture grown at 21-28 days are, in most part, similar to those in *in vivo* preparation. Moreover, the electrophysiological profile of our triple culture preparation is more closely related to *in vivo* situation than in *in vitro* slice preparation.. Even though further studies are needed, these findings are really encouraging to pursuit of the idea that this organotypic culture preparation may be one of the most suitable models to be adapted for the studies of basic membrane physiology and functional circuitry of the basal ganglia.

5. ACKNOWLEDGEMENTS

We thank Dr. Xiaofei Wang for her expert technical assistance with the preparation of cultures and immunohistochemistry. This study was supported by NIH, NINDS grants: NS-20702 and NS-26473.

6. REFERENCES

1. Plenz, D. & Kitai, S. T. Organotypic cortex -striatum-mesencephalon cultures: the nigrostriatal pathway. *Neurosci. Ltr.* 209, 177-180 (1996b).
2. Plenz, D. & Kitai, S. T. Regulation of the nigrostriatal pathway by metabotropic glutamate receptors during development. *J. Neurosci.* 18, 4133-4144 (1998b).
3. Plenz, D., Herrera-Marschitz, M. & Kitai, S. T. Morphological organization of the globus pallidus -subthalamic nucleus system studied in organotypic cultures. *J. Comp. Neurol.* 397, 437-457 (1998c).
4. Beckstead, R. M., Domesick, V. B. & Nauta, W. J. Efferent connections of the substantia nigra and ventral tegmental area in the rat. *Brain Res.* 175, 191-217 (1979A).
5. Bentivoglio, M., Van der Kooy, D. & Kuypers, H. G. The organization of the efferent projections of the substantia nigra in the rat. A retrograde fluorescent double labeling study. *Brain Res.* 174, 1-17 (1979).
6. Faull, R. L. & Mehler, W. R. The cells of origin of nigrotectal, nigrothalamic and nigrostriatal projections in the rat. *Neurosci.* 3, 989-1002 (1978).
7. Canteras, N. S., Shammah-Lagnado, S. J. & Ricardo, J. A. Afferent connections of the subthalamic nucleus: a combined retrograde and anterograde horseradish peroxidase study in the rat. *Brain Res.* 513, 43-59 (1990).
8. Hassani, O.-K., Frangois, Yelnik, J. & Feger, J. Evidence for a dopaminergic innervation of the subthalamic nucleus in the rat. *Brain Res.* 749, 88-94 (1997).
9. Grofova, I. & Zhou, M. Nigral innervation of cholinergic and glutamatergic cells in the rat mesopontine tegmentum: light and electron microscopic anterograde tracing and immunohistochemical studies. *J. Comp. Neurol.* 395, 359-379 (1998).
10. Beninato, M. & Spencer, R. F. The cholinergic innervation of the rat substantia nigra; a light and electron microscopic immunohistochemical study. *Exp. Brain Res.* 72, 178-184 (1988).

11. Clarke, P. B., Hommer, D. W., Pert, A. & Skirboll, L. R. Innervation of substantia nigra neurons by cholinergic afferents from pedunculopontine nucleus in the rat: neuroanatomical and electrophysiological evidence. *Neurosci.* 23, 1011-1019 (1987).
12. Futami, T., Takakusaki, K. & Kitai, S. T. Glutamatergic and cholinergic inputs from the pedunculopontine tegmental nucleus to dopanune neurons in the substantia nigra pars compacta. *Neurosci. Res.* 21, 331-342 (1995).
13. Gould, E., Woolf, N. J. & Butcher, L. L. Cholinergic projections to the substantia nigra from the pedunculopontine and laterodorsal tegmental nuclei. *Neurosci.* 28, 611-623 (1989).
14. Kita, H. & Kitai, S. T. Efferent projections of the subthalamic nucleus in the rat: light and electron microscopic analysis with the PHA L method. *J. Comp. Neurol.* 260, 435-452 (1987).
15. Naito, A. & Kita, H. The cortico-nigral projection in the rat: an anterograde tracing study with biotinylated dextran amine. *Brain Res.* 637, 317-322 (1994).
16. Takakusaki, K., Shiroyama, T., Yamamoto, T. & Kitai, S. T. Cholinergic and noncholinergic tegmental pedunculopontine projection neurons in rats revealed by intracellular labeling. *J. Comp. Neurol.* 371, 345-361 (1996).
17. Bolam, J. P. & Smith, Y. The GABA and substance P input to dopaminergic neurons in the substantia nigra of the rat. *Brain Res.* 529, 57-78 (1990).
18. Oertel, W. H., Tappaz, M. L., Berod, A. & Mugnaini, E. Two-color immunohistochemistry for dopamine and GABA neurons in rat substantia nigra and zona incerta. *Brain Res. Bull.* 9, 463-474 (1982).
19. Ribak, C. E., Vaughn, J. E., Saito, K., Barber, R. & Roberts, E. Immunocytochemical localization of glutamate decarboxylase in rat substantia nigra. *Brain Res.* 116, 287-298 (1976).
20. Deniau, J. M., Kitai, S. T., Donoghue, J. P. & Grofova, I. Neuronal interactions in the substantia nigra pars reticulata through axon collaterals of the projection neurons. An electrophysiological and morphological study. *Expl. Brain Res.* 47, 105-113 (1982).
21. Nakanishi, H., Kita, H. & Kitai, S. T. Intracellular study of rat substantia nigra pars reticulata neurons in an in vitro slice preparation: electrical membrane properties and response characteristics to subthalamic stimulation. *Brain Res.* 437, 45-55 (1987).
22. Scarnati, E., Campana, E. & Pacitti, C. Pedunculopontine-evoked excitation of substantia nigra neurons in the rat. *Brain Res.* 304, 351-361 (1984).
23. Deniau, J. M., Feger, J. & LeGuyader, C. Striatal evoked inhibition of identified nigro thalamic neurons. *Brain Res.* 104, 152-156 (1976).
24. Smith, Y., Bolam, J. P. & von Krosigk, M. Topographical and synaptic organization of the GABA-containing pallidosubthalamic projection in the rat. *Eur. J. Neurosci.* 2, 500-511 (1990A).
25. Bunney, B. S., Walters, J. W., Roth, R. H. & Aghajanian, G. K. Dopaminergic neurons: effect of antipsychotic drugs and amphetamine on single cell activity. *J. Pharmac. Exp. Ther.* 185, 560-571 (1973).
26. Grace, A. A. & Bunney, B. S. Intracellular and extracellular electrophysiology of nigral dopaminergic neurons-1. Identification and characterization. *Neurosci.* 10, 301-315 (1983).
27. Grace, A. A. & Bunney, B. S. The control of firing pattern in nigral dopamine neurons: Single spike firing. *J. Neurosci.* 4, 2866-2876 (1984).
28. Grace, A. A. & Onn, S. P. Morphology and electrophysiological properties of immunocytochemically identified rat dopamine neurons recorded in vitro. *J. Neurosci.* 9, 3463-3481 (1989).
29. Harris, N. C., Webb, C. & Greenfield, S. A. A possible pacemaker mechanism in pars compacta neurons of the guinea-pig substantia nigra revealed by various ion channel blocking agents. *Neurosci.* 31, 355-362 (1989).
30. Kang, Y. & Kitai, S. T. Calcium spike underlying rhythmic firing in dopaminergic neurons of the rat substantia nigra. *Neurosci. Res.* 18, 195-207 (1993a).
31. Lacey, M. G., Mercuri, N. B. & North, R. A. Two cell types in rat substantia nigra zona compacta distinguished by membrane properties and the actions of dopamine and opioids. *J. Neurosci.* 9, 1233-1241 (1989).
32. Lacey, M. G., Calabresi, P. & North, R. A. Muscarine depolarizes rat substantia nigra zona compacta and ventral tegmental neurons in vitro through M1-like receptors. *J. Pharmac. Exp. Ther.* 253, 395-400 (1990).
33. Matsuda, Y., Fujimura, K. & Yoshida, S. Two types of neurons in the substantia nigra pars compacta studied in a slice preparation. *Neurosci. Res.* 5, 172-179 (1987).

34. Richards, C. D., Shiroyama, T. & Kitai, S. T. Electrophysiological and immunocytochemical characterization of GABA and dopamine neurons in the substantia nigra of the rat. *Neurosci.* 80, 545-557 (1997).
35. Shepard, P. D. & Bunney, B. S. Repetitive firing properties of putative dopamine-containing neurons in vitro: regulation by an apamin-sensitive CA2+-activated K+ conductance. *Exp.Brain Res.* 86, 141-150 (1991).
36. Wilson, C. J., Young, S. J. & Groves, P. M. Statistical properties of neuronal spike trains in the substantia nigra: cell types and their interactions. *Brain Res.* 136, 243-260 (1977).
37. Yung, W. H., Hausser, M. A. & Jack, J. J. B. Electrophysiology of dopaminergic and nondopaminergic neurons of the guinea pig substantia nigra pars compacta in vitro. *J. Physiol.* 436, 643-667 (1991).
38. Rye, D. B., Saper, C. B., Lee, H. J. & Wainer, B. H. Pedunculopontine tegmental nucleus of the rat: Cytoarchitecture, Cytochemistry and some extrapryamidal connections of the mesopontine tegmentum. *J. Comp. Neurol.* 259, 483-528 (1987).
39. Spann, B. M. & Grofova, I. Origin of ascending and spinal pathways from the nucleus tegmenti pedunculopontinus in the rat. *J. Comp. Neurol.* 283, 13-27 (1989).
40. Albin, R. L., Aldridge, J. W., Young, A. B. & Gilman, S. Feline subthalamic nucleus neurons contain glutamate-like but not GABA-like or glycine-like immunoreactivity. *Brain Res.* 491, 185-188 (1989).
41. Kitai, S. T. & Kita, H. *Anatomy and physiology of the subtlialamic nucleus: a driving force of the basal ganglia* (eds. Carpenter, M. B. & Jarayaman, A.) (Plenum, New York, 1987).
42. Smith, Y. & Parent, A. Neurons of the subthalamic nucleus in primates display glutamate but not GABA immunoreactivity. *Brain Res.* 453, 353-356 (1988).
43. Bevan, M. D., Francis, C. M. & Bolam, J. P. The glutamate-enriched cortical and thalamic input to neurons in the subthalamic nucleus of the rat: convergence with GABA-positive terminals. *J. Comp. Neurol.* 361, 491-511 (1995B).
44. Kitai, S. T. & Deniau, J. M. Cortical inputs to the subthalamus: Intracellular analysis. *Brain Res.* 214, 411-415 (1981).
45. Lee, H. J., Rye, D. B., Hallanger, A. E., Levey, A. I. & Wainer, B. H. Cholinergic vs noncholinergic efferents from the mesopontine tegmentum to the extrapyramidal motor system nuclei. *J. Comp. Neurol.* 275, 469-492 (1988).
46. Bevan, M. D. & Bolam, J. P. Cholinergic, GABAergic and glutamate-enriched inputs from the mesopontine tegmentum to the subthalamic nucleus in the rat. *J. Neurosci.* 15, 7105-7120 (1995A).
47. Woolf, N. J. & Butcher, L. L. Cholinergic systems in the rat brain: III. Projections from the pontomesencephalic tegmentum to the thalamus, tectum, basal ganglia, and basal forebrain. *Brain Res. Bull.* 16, 603-637. (1986).
48. Kita, H., Chang, H. T. & Kitai, S. T. Pallidal inputs to subthalamus: intracellular analysis. *Brain Res.* 264, 255-265 (1983a).
49. Chang, H. T., Kita, H. & Kitai, S. T. The ultrastructural morphology of the subthalamonigral axon terminals intracellularly labeled with horseradish peroxidase. *Brain Res.* 299, 182-185 (1984).
50. Deniau, J. M., Hammond, C., Chevalier, G. & Feger, J. Evidence for branched subthalamic nucleus projections to substantia nigra, entopeduncular nucleus and globus pallidus. *Neurosci. Lett.* 9, 117-121 (1978).
51. Jackson, A. & Crossman, A. R. Subthalamic projection to nucleus tegmenti pedunculopontinus in the rat. *Neurosci. Lett.* 22, 17-22 (1981).
52. Kita, H., Chang, H. T. & Kitai, S. T. The morphology of intracellularly labeled rat subthalamic neurons: A light microscopic analysis. *J. Comp. Neurol.* 215, 245-257 (1983b).
53. Van der Kooy, D. & Hattori, T. Single subthalamic nucleus neurons project to both the globus pallidus and substantia nigra in rat. *J. Comp. Neurol.* 192, 751-768 (1980).
54. Lokwan, S. J., Overton, P. G., Berry, M. S. & Clark, D. Stimulation of the pedunculopontine tegmental nucleus in the rat produces burst firing in A9 dopaminergic neurons. *Neurosci.* 92, 245-254 (1999).
55. Sorenson, E. M., Shiroyama, T. & Kitai, S. T. Postsynaptic nicotinic receptors on dopaminergic neurons in the substantia nigra pars compacta of the rat. *Neurosci.* 87, 659-673 (1998).
56. Hammond, C., Deniau, J. M., Rizk, A. & Feger, J. Electrophysiological demonstration of an excitatory substhalamonigral pathway in the rat. *Brain Res.* 151, 235-244 (1978).
57. Ichinohe, N., Teng, B. & Kitai, S. T. Morphological studies of tegmental pedunculopontine nucleus, substantia nigra and subthalamic nucleus, and their interrelationship among these structures in organotypic cultures. *In Press, Anat. Embryol.* (1999).

58. Vincent, S. R., Satoh, K., Armstrong, D. M. & Fibiger, H. C. ADPH-diaphorase: a selective histochemical marker for the cholinergic neurons of the pontine reticular formation. *N Neurosci. Lett.* 43, 31-36 (1983).
59. Juraska, J. M., Wilson, C. J. & Groves, P. M. The substantia nigra of the rat: a Golgi study. *J. Comp Neurol.* 172, 585-600 (1977).
60. Jaeger, C., Gonzalo-Ruiz, A. & ., L. R. Organotypic slice cultures of dopaminergic neurons of substantia nigra. *Brain Res* Bull. 22, 981-991 (1989).
61. Ostergaard, K., Schou, J. P. & Zimmer, J. Rat ventral mesencephalon grown as organotypic slice cultures and co-cultured with striatum, hippocampus, and cerebellum. *Expl. Brain Res.* 82, 547-565 (1990).
62. Spann, B. M. & Grofova, I. Cholinergic and non-cholinergic neurons in the rat pedunculopontine tegmental nucleus. *Anat. Embryol.* 186, 215-227 (1992).
63. Jones, B. E. *Reticular formation. Cytoarchitecture, transmitters and projections.* (ed. Paxinos, G.) (Academic Press Australia, New South Wales, 1994A).
64. Lavoie, B. & Parent, A. Pedunculopontine nucleus in the squirrel monkey: Distribution of cholinergic and monoaminergic neurons in the mesopontine tegmentum with evidence for the presence of glutamate in cholinergic neurons. *J. Comp. Neurol.* 344, 190-209 (1994).
65. Ford, B., Holmes, C. J., Mainville, L. & Jones, B. E. GABAergic neurons in the rat pontomesencephalic tegmentum: Codistribution with cholinergic and other tegmental neurons projecting to the posterior lateral hypothalamus. *J. Comp. Neurol.* 363, 177-196 (1995).
66. Clements, J. R. & Grant, S. J. Glutamate-like immunoreactivity in neurons of the laterodorsal tegmental and pedunculopontine nuclei in the rat. *Neurosci. Ltrs.* 120, 70-73 (1990).
67. Celio, M. R. Calbindin D-28K and parvalbumin in the rat nervous system. *Neurosci.* 35, 375-475 (1990).
68. Rajakumar, N., Elisevich, K. & Flumerfelt, B. A. Parvalbumin-containing GABAergic neurons in the basal ganglia output system of the rat. *J. Comp. Neurol.* 350, 324-336 (1994).
69. Hontanilla, B., Parent, A. & Gimenez-Amaya, J. M. Parvalbumin and calbindin D-28k in the entopeduncular nucleus, subthalamic nucleus, and substantia nigra of the rat as revealed by double immunohistochemical methods. *Synapse* 25, 359-397 (1997).
70. Kaneko, T., Akiyama, H., Nagatsu, I. & Mizuno, N. Immunohistochemical demonstration of glutaminase in catecholaminergic and serotonergic neurons of rat brain. *Brain Res.* 507, 151-154 (1990).
71. Kimura, H., McGeer, P. L., Peng, J. H. & McGeer, E. G. The central cholinergic system studied by choline acetyltransferase immunohistochemistry in the cat. *J. Comp. Neurol.* 200, 151-201 (1981).
72. Beninato, M. & Spencer, R. F. A cholinergic projection to the rat substantia nigra from the pedunculopontine tegmental nucleus. *Brain Res.* 412, 169-174 (1987).
73. Bolam, J. P., Francis, C. M. & Henderson, Z. Cholinergic input to dopaminergic neurons in the substantia nigra: a double immunocytochemical study. *Neurosci.* 41, 483-494 (1991).
74. Faull, R. L. & Carman, J. B. Ascending projections of the substantia nigra in the rat. *J. Comp. Neurol.* 132, 73-92 (1968).
75. Steininger, T. L., Rye, D. B. & Wainer, B. H. Afferent projections to the cholinergic pedunculopontine tegemental nucleus and adjacent midbrain extrapyramidal area in the albino rat. I. Retrograde tracing studies. *J. Comp. Neurol.* 312, 515-543 (1992).
76. Fallon, J. H. & Loughlin, S. E. Substantia nigra. In G. Paxinos (ed): In The Rat Nervous System. *Sydney: Academic Press,* 353-374 (1985).
77. Preston, R. J., McCrea, R. A., Chang, H. T. & Kitai, S. T. Anatomy and physiology of substantia nigra and retrorubral neurons studied by extra- and intracellular recording and by horseradish peroxidase labeling. *Neurosci.* 6, 331-344 (1981).
78. Tepper, J. M., Sawyer, S. F. & Groves, P. M. Electrophysiologically identified nigral doparninergic neurons intracellularly labeled with HRP: light-microscopic analysis. *J. Neurosci.* 7, 2794-2806 (1987).
79. Hajos, M. & Greenfield, S. A. Topographic heterogeneity of substantia nigra neurons: Diversity in intrinsic membrane properties and synaptic inputs. *Neurosci.* 55, 919-934 (1993).
80. Kita, T., Kita, H. & Kitai, S. T. Electrical membrane properties of rat substantia nigra compacta neurons in an in vitro slice preparation. *Brain Res.* 372, 21-30 (1986).
81. Hajos, M. & Greenfield, S. A. Synaptic connections between pars compacta and pars reticulata neurones: electrophysiological evidence for functional modules within the substantia nigra. *Brain Res.* 660, 216-224 (1994).

82. Kan-. Y. & Kita, S. T. A whole cell patch-clamp study on the pacemaker potential in dopaminergic neruons of rat substantia nigra compacta. *Neuroscience Research* 18, 209-221 (1993b).
83. Charlety, P. J. *et al.* Burst firing of mesencephalic dopamine neurons is inhibited by somatodendritic application of kynurenate. *Acta Physiol. Scand.* 142, 105-112 (1991).
84. Chergui, K. *et al.* Tonic activation of NMDA receptors causes spontaneous burst discharge of rat midbrain dopamine neurons in vivo. *Eur. J. Neurosci.* 5, 137-144 (1993).
85. Johnson, S. W., Seutin, V. & North, R. A. Burst firing in dopamine neurons induced by N-methyl-D-aspartate: role of electrogenic sodium pump. *Science* 258, 665-667 (1992).
86. Di Loreto, S., Florio, T. & Scarnati, E. Evidence that non-NMDA receptors are involved in the excitatory pathway from the pedunculopontine region to nigrostriatal dopaminergic neurons. *Expl. Brain Res.* 89, 79-86 (1992).
87. Kang, Y. & Kitai, S. T. Electrophysiological properties of pedunculopontine neurons and their postsynaptic responses following stimulation of substantia nigra reticulata. *Brain Res.* 535, 79-95 (1990).
88. Hausser, M. A. & Yung, W. H. Inhibitory synaptic potentials in guinea-pig substantia nigra dopamine neurones in vitro. *J. Physiol. Lond. 479, 401-422 (1994).*
89. Tepper, J. M., Martin, L. P. & Anderson, D. R. GABAA receptor-mediated inhibition of rat substantia nigra dopaminergic neurons by pars reticulata projection neurons. *J. Neurosci.* 15, 3092-3103 (1995).

DISTRIBUTION OF THE BASAL GANGLIA AND CEREBELLAR PROJECTIONS TO THE RODENT MOTOR THALAMUS

S. T. Sakai and I. Grofovà*

Abbreviations: AM: anterior medial nucleus, APT: anterior pretectal area, AV: anterior ventral nucleus, CL: central lateral nucleus, FF: Fields of Forél, fr: fasciculis retroflexus, G: gelatinosus nucleus, LD: lateral dorsal nucleus, LHb: lateral habenular nucleus, LP: lateral posterior nucleus, MD: mediodorsal nucleus, ml: medial lemniscus, mt: mammillothalamic tract, MV: medioventral nucleus, Pf: parafascicular nucleus, pl: paralaminar region, Po: posterior group, Rt: reticular thalamic nucleus, sm: stria medullaris, SNC: substantia nigra pars compacta, SNR: substantia nigra pars reticulata, VAL: ventral anterolateral nucleus, VM: ventromedial nucleus, VPM: ventral posteromedial nucleus, ZI: zona incerta.

1. INTRODUCTION

Although the rodent cerebellothalamic distribution has been extensively studied[1,2,6], the detailed relationship to the basal ganglia thalamic distribution is yet to be determined. In a preliminary report using a double anterograde labeling strategy, Deniau and others[5] reported partially segregated and partially overlapping projections from the basal ganglia output nuclei and the cerebellum to the thalamus in the rat. The idea that the rodent motor thalamus largely receives non-overlapping basal ganglia and cerebellar input is partially based on the finding that the primary nigrothalamic target is the ventromedial nucleus (VM) while the cerebellothalamic afferents project to the ventral anterior lateral nucleus (VAL). Our recent demonstration of robust nigrothalamic projections to VAL in the rat[18] raised the possibility of overlapping basal ganglia and cerebellar afferents in the rat VAL and prompted us to re-investigate the question of segregated basal ganglia and cerebellar afferents to the rodent thalamus.

---

* S.T. Sakai and I. Grofovà, Dept. of Anatomy, College of Human Medicine, Michigan State University, East Lansing, MI 48824 U.S.A.

The Basal Ganglia VI
Edited by Graybiel et al., Kluwer Academic/Plenum Publishers, 2002

2. METHODS

Experiments were conducted in adult Sprague-Dawley rats of either sex weighing between 250-350 gm. Animals were pre-treated with an intramuscular injection of atropine sulfate (0.06 mg/kg) and were surgically anesthetized with an intraperitoneal injection of sodium pentobarbital (30 mg/kg). Iontophoretic injections of 2.5% *Phaseolus vulgaris* leucoagglutinin (PHA-L) were made into either the entopeduncular nucleus or the substantia nigra pars reticulata via a glass micropipette (tip diameter: 10-20 µm) using a Midgard Constant Current Device. Positive DC current was applied for total on-times of 20-40 minutes at 5 µA. Postoperative survival times ranged from 4-7 days. The animals were deeply anesthetized using sodium pentobarbital and intracardially perfused with saline followed by 3.0-4.0% paraformaldehyde, 0.075 M lysine, 0.01 M periodate in phosphate buffer. The brains were cut at 50 µm on a vibratome in the sagittal plane. Sections were briefly washed in tris buffered saline with 0.04% Triton-X, blocked in 3% normal rabbit serum, washed and incubated in the primary antisera (goat anti-PHA-L; dilution: 1: 2000, Vector Laboratories) and then washed and incubated in avidin-biotin complex, reacted in 0.1% diaminobenzidine, ammonium chloride/ β-D glucose in Tris buffer and glucose oxidase. In the final series of experiments, multiple injections of horseradish peroxidase conjugated to wheat germ agglutinin (WGA-HRP) were made into the contralateral cerebellar nuclei. Following a 48 hour survival time, the animals were re-anesthetized and intracardially perfused with phosphate buffered saline followed by 0.5% paraformaldehyde and 2% glutaraldehyde in phosphate buffer. The brains were sectioned on a vibratome at 50 µm thick in the sagittal plane. Sections were reacted using tetramethylbenzidine and stabilized with ammonium molybdate. Finally, the sections were counterstained using either neutral red or cresyl violet.

Selected thalamic sections were analyzed on a Leitz Orthoplan or Olympus microscope under both light and darkfield illumination. The distribution of labeling was drawn using the side arm drawing tube attached to the microscope. Sections containing the injections sites were examined in the microscope to determine the placement of the tracers. The dense core and halo of the injection sites relative to relevant nuclear boundaries were drawn using a low magnification projector. The motor thalamic cytoarchitecture and nomenclature follows that described in Sakai et al.[18].

3. RESULTS

3.1. Nigrothalamic Distribution

The PHA-L injections into the SNR including its most ventral extent resulted in the dense thalamic labeling (Fig. 1). Since we have previously described the distribution of the nigrothalamic distribution at length[18], we briefly summarize our findings here. A dense contingent of labeled nigrothalamic fibers emanated from the SNR and ascended toward the thalamus by primarily coursing through the fields of Forél and the zona incerta. The bulk of nigrothalamic fibers coursed rostrally forming dense plexuses in the rostral portion of VAL. As seen in sagittal sections, the nigrothalamic labeling occupied a crescent shaped area throughout the most rostral extent of VAL. In more medial sections, the nigral fibers distributed to VM including the portions of VM split by the mammillothalamic tract. Other thalamic

NIGROTHALAMIC   PALLIDOTHALAMIC

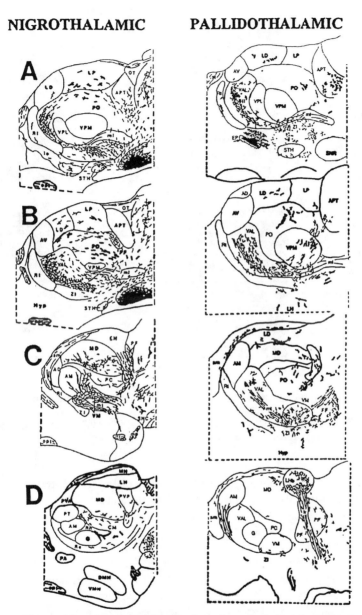

Fig. 1 Line drawings of sagittal sections depicting the PHA-L injection into the basal ganglia output nuclei and the resulting anterograde labeling. The left panel shows a SNR injections and sections A-D are cut in the sagittal plane and arranged lateral to medial. Note the density of nigrothalamic projections to VAL shown in sections A and B and to VM shown in section C. The right panel shows an EP injections and resulting thalamic labeling. Note that the pallidothalamic labeling in VAL shown in sections A and B is somewhat more lateral and caudal to the nigral labeling.

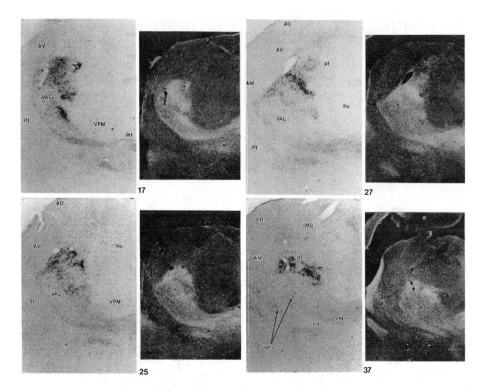

Figure 2. A sagittal series of photomicrographs showing the cerebellothalamic labeling following WGA-HRP injections into the contralateral cerebellar nuclei. Section pairs show bright and darkfield photomicrographs of a Nissl stained section arranged from lateral to medial. Note the dense black WGA-HRP labeling present throughout much of VAL.

nuclei containing nigral labeling included the mediodorsal nucleus (MD), posterior group (PO), lateral posterior (LP) and lateral dorsal (LD) nuclei and parafasicular nucleus (PF).

3.2. Pallidothalamic Distribution

The PHA-L injections into the entopeduncular nucleus (EP), the homologue of the primate internal segment of the globus pallidus, resulted in dense thalamic labeling (Fig. 1). Labeled fibers primarily coursed from EP through the ansa lenticularis traveling dorsally into VAL. The pallidothalamic fibers with numerous dense plexuses occupied a wide crescent shaped area of VAL. The bulk of the pallidothalamic labeling was located somewhat more lateral and caudal to the region occupied by the nigrothalamic label. In addition, a contingent of pallidal fibers coursed from EP dorsally through the zona incerta and the fields of Forél also traveled into VM. In VM, the location of the pallidal labeling was similar to the nigral labeling. Dense pallidal labeling was also found in the lateral habenular nucleus (LHb). Other thalamic nuclei containing pallidal labeling included PF, PO and LD.

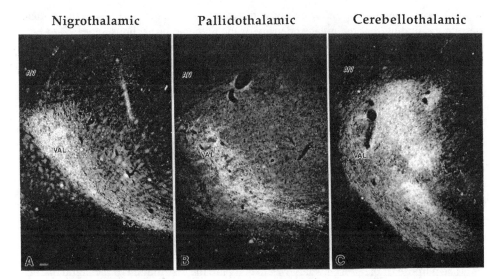

Figure 3. Darkfield photomicrographs showing the nigrothalamic (A), pallidothalamic (B) and cerebellothalamic (C) anterograde labeling in sagittal sections. Note the distribution of labeling in VAL across the cases. The greatest density of nigrothalamic labeling occurs in the rostral and medial part of VAL and to VM. The pallidothalamic fibers project primarily to rostral and lateral parts of VAL and to VM. The cerebellothalamic projections are robust to all of VAL particularly to its caudal and lateral parts and to VM. However, comparison across these cases reveals regions of overlapping nigral, pallidal and cerebellar inputs to VAL and VM. Scale bar = 100 mm.

3.3. Cerebellothalamic Distribution

Large WGA-HRP injections were made into the cerebellar nuclei involving the fastigial, interpositus and much of the lateral nuclei. The bulk of the cerebellar fibers coursed within the superior cerebellar peduncle, crossed the midline and traveled rostrally within the fields of Forél and the zona incerta (Fig. 2). A contingent of labeled fibers coursed dorsally to MD, PF, PO and the ventral posterior medial nucleus (VPM). However, the bulk of the fibers coursed through VM and traveled dorsally into VAL. The labeling in VM was extensive and included fibers with associated swellings. In VAL, the labeling occupied a wide crescent shaped region. The cerebellar labeling was quite dense filling the most central to caudal aspect of the nucleus but foci of cerebellothalamic labeling was present throughout the nucleus. The cerebellothalamic labeling generally occupied a region of VAL somewhat more caudal and lateral to the bulk of either the pallidothalamic or nigrothalamic labeling.

3.4. Comparison of the Cerebellothalamic and Basal Ganglia Distribution.

Comparison across the representative cases of the SN, EP and cerebellar injections revealed that the bulk of the thalamic labeling occupied complementary regions of VAL such that the primary target of the nigral efferents was to rostral and medial VAL while the primary target of the pallidal projections was to portions of VAL caudal and lateral to the nigral labeling. The greatest efferent projection to VAL was from the cerebellum and occupied the largest crescent shaped region primarily in the caudal and lateral portions of the nucleus. Despite the finding that the greatest density of the thalamic projections appeared to be to complementary portions of VAL, overlap between the projection systems was also noted. When similar sagittal thalamic levels from the representative cases are compared, the border zones or peripheral portions of each projection system appeared to overlap (Fig. 3). The caudal and lateral most extent of the nigral labeling appeared to coexist with the rostral and medial most extent of the pallidal labeling in VAL. Moreover, limited foci of the cerebellar labeling extended rostrally and medially in VAL and coexisted with both nigral and paladin labeling. Within VM, the pallidal, nigral and cerebellar efferents appeared to significantly overlap. Overlap from the basal ganglia sources and the cerebellum in the other thalamic nuclei was more difficult to determine due to sparser, lighter labeling. However, possible overlap from these sources in PF was also noted.

4. DISCUSSION AND CONCLUSIONS

The purpose of the present study was to compare the distribution of the nigral, pallidal and cerebellar thalamic projections in the rat. On basis of separate injections into SNR, EN and the cerebellar nuclei in the rat, we found that the basal ganglia and cerebellar efferents distributed in both a separate and overlapping manner. First, we found that the primary target of each projection system was to separate and complementary portions of the VAL. Second, the border zones of each system appeared to overlap in VAL so that the caudal and lateral nigrothalamic labeling overlapped with the rostral and medial pallidal labeling. Dense cerebellar labeling was found in caudal VAL but was present throughout the nucleus. Comparison of the nigral, pallidal and cerebellar thalamic distributions suggests that while each pathway separately targets different sectors of VAL, there are also regions of overlapping nigral, pallidal and cerebellar inputs in VAL and in VM.

A number of studies have suggested that the ascending inputs from the basal ganglia and cerebellum to motor cortical areas consist of separate and parallel circuits[9,14,22]. However, recent work directly comparing the pallidal and cerebellar distribution in the owl[20] and macaque monkey[15,19] reveal that these inputs distribute in both a segregated and possibly overlapping manner. Although the primary projections from the internal segment of the globus pallidus is to the ventral lateral nucleus pars oralis (VLo) and from the cerebellum to the ventral posterior lateral nucleus pars oralis (VPLo), there are regions of overlapping or interdigitating foci of afferent labeling, particularly in the border regions between these nuclei. Moreover, a mixture of separate as well as interdigitating and overlapping cerebellothalamic and nigrothalamic projections to subdivisions of the ventral anterior (VA) and the ventral lateral nucleus (VL) was also described in dogs[16]. Thus, the overall pattern of primarily separate projection systems with border regions receiving overlapping inputs from the basal ganglia and cerebellum is remarkably similar across these varied species. It is of interest that this overall projection pattern was also found in the rat since the rodent motor thalamus lacks the cytoarchitectonic variety and complexity characteristic of either the primate or carnivore thalamus[11]. In this regard, it is significant that Golgi analysis of the rodent VAL revealed morphologically distinct cell types which were differentially distributed to dorsal, ventral and lateral subdivisions of VAL[21]. Whether these VAL subdivisions correspond to subnuclei of the ventrolateral complex similar to that defined in other species is unclear. However, similarities in the general topography of the basal ganglia and cerebellar projection systems across these species support this possibility.

On the basis of both anatomical and physiological techniques, converging inputs from the basal ganglia and cerebellum to VM has been well documented in the rat[3,4,8,13], cat[10,12] and dog[7,16,17]. Our finding of overlapping inputs from SNR, EN and the cerebellum to VM in the rat confirms these reports. Deniau and others[5] reported an overlapping pattern of projections from the basal ganglia and cerebellum to the medial VL and VM. However, they also reported a complementary pattern of projections with little convergence in VL. In contrast, we found that the nigrothalamic projection to VAL was quite dense occupying the rostral most cap of VAL and extending caudally and laterally to regions containing pallidal and cerebellar labeling. The discrepancy between these findings may be attributed to differences in the location of the injection sites. Our injections also included the ventral most edge of the SNR, the area known to give rise to dense VAL projections[18].

The present results indicate that while the rodent motor thalamus largely receives separate projections from the basal ganglia and cerebellum, areas of overlapping inputs were also found in the border regions between each projection system. Interestingly, thalamic cells responsive to both cerebellar and basal ganglia stimulation were also detected in the border region intercalated between the cerebellar and entopeduncular projection areas in the cat[23]. However, the influence of such overlapping inputs on thalamic cell function is unclear. Although it is well known that the cerebellar input is excitatory and the basal ganglia input is inhibitory, the overall influence of such disparate inputs is complicated by the paucity of these converging inputs as well as the complexity of the synaptic arrangements in the cat motor thalamus. In this regard, the lack of thalamic interneurons in the rat motor thalamus make it a suitable model of a simple mammalian system in which to address input/output questions. Moreover, the similarities in the overall afferent and efferent organization of the motor thalamus in the rat in comparison to other species additionally affirm the use of the rodent model in studies of motor function.

5. ACKNOWLEDGEMENTS

The authors gratefully acknowledge technical assistance provided by Kathy Bruce and Michael Park. This work was supported by NS-25744 and College of Human Medicine.

6. REFERENCES

1. P. Angaut, F. Circirata and F. Serapide, Topographic organization of the cerebellothalamic projections in the rat. An autoradiographic study. *Neuroscience* 15: 389-401 (1985).
2. T.D. Aumann, J.A. Rawson, D.I. Finkelstein and M.K. Horne, Projections from the lateral and interposed cerebellar nuclei to the thalamus in the rat: a light and electron microscopic study using single and double anterograde labeling. *J. Comp. Neurol.* 349: 165-181 (1994).
3. J. Buee, J.M Deniau, and G. Chevalier, Nigral modulation of cerebellothalamo-cortical transmission in the ventromedial thalamic nucleus. *Exp. Brain Res.* 65: 241-244 (1986).
4. G. Chevalier and J.M. Deniau, Inhibitory nigral influence on cerebellar evoked responses in the rat ventromedial thalamic nucleus. *Exp. Brain Res.* 65: 241-244 (1982).

5. J.M. Deniau, H. Kita and S.T. Kitai, Patterns of termination of cerebellar and basal ganglia efferents in the rat thalamus. Strictly segregated and partly overlapping projections. *Neurosci. Letts.* 144: 202-206.
6. R.L.M. Faull and J.B. Carman, The cerebellofugal projections in the brachium conjunctivum of the rat. I. The contralateral ascending pathway. *J. Comp. Neurol.* 178: 495-518 (1978).
7. B.A.Hannah, and S.T. Sakai, The distribution of pallidothalamic and nigrothalamic projections in the dog. *Soc. Neurosci. Abstr.* 15: 1101 (1989).
8. M. Herkenham, The afferent and efferent connections of the ventromedial thalamic nucleus in the rat. J Comp. Neurol. 183: 487-58 (1979).
9. I.A. Ilinsky and K. Kultas-Ilinsky, Sagittal cytoarchitectonic maps of the *Macaca mulatta* thalamus with a revised nomenclature of the motor-related nuclei validated by observations on their connectivity. *J. Comp. Neurol.* 262: 331-364 (1987).
10. J. Jimenez-Castellanos Jr., and F. Reinoso-Suarez, Topographical organization of the afferent connections of the principal ventromedial thalamic nucleus in the cat. *J. Comp. Neurol.* 236: 297-314 (1985).
11. Jones, E.G. *The Thalamus.* New York: Plenum Press (1985).
12. K. Kultas-Ilinsky, I.A. Ilinsky, S. Warton, and K.R. Smith, Fine structure of nigral and pallidal afferents in the thalamus: An EM autoradiography study in the cat. *J. Comp. Neurol.* 26: 390-404 (1983).
13. N.K. MacLeod, and T.A. James, Regulation of cerebello-cortical transmission in the rat ventromedial thalamic nucleus. *Exp. Brain Res.* 55: 535-552 (1984).
14. G. Percheron, C. François, B. Talbi, J. Yelnik, and G. Fénelon, The primate motor thalamus. *Brain Res. Rev.* 22: 93-181 (1996).
15. Rouiller, E., F. Liang, A. Babalian, V. Moret and M. Wiesendanger, Cerebellothalamocortical and pallidothalamocortical projections to the primary and supplementary motor cortical areas: a multiple tracing study in macaque monkeys. *J. Comp. Neurol.* 345: 185-213 (1994).
16. S.T. Sakai and K. Patton, The distribution of nigrothalamic and cerebellothalamic projections in the dog: A double anterograde tracing study. *J. Comp. Neurol.* 330: 183-194 (1993).
17. S.T.Sakai, and A. Smith, Distribution of nigrothalamic projections in the dog. *J. Comp. Neurol.* 388: 83-92 (1992).
18. S.T. Sakai, I. Grofová and K. Bruce, Nigrothalamic projections and nigrothalamocortical pathway to the medial agranular cortex in the rat: single- and double-labeling light and electron microscopic studies. *J. Comp. Neurol.* 391: 506-525 (1998).
19. Sakai, S.T., M. Inase and J. Tanji, (1996) Comparison of cerebellothalamic and pallidothalamic projections in the monkey (*Macaca fuscata*): A double anterograde labeling study. *J. Comp. Neurol.* 368: 215-228.
20. S.T. Sakai, I. Stepniewska, H. Qi and J.H. Kaas, Pallidothalamic and cerebellothalamic afferents of the pre-supplementary motor area in owl monkeys: A triple labeling study. *J. Comp. Neurol.* (1999) (in press).
21. Sawyer, S.F., S.J. Young, and P.M. Groves, Quantitative Golgi study of anatomically identified subdivisions of motor thalamus in the rat. *J. Comp. Neurol.* 286:1-27 (1989).
22. G.R. Schell and P.L. Strick, The origin of thalamic inputs to the arcuate premotor supplementary motor areas. *J. Neurosci.* 5: 539-560 (1984).
23. T. Yamamoto, T. Noda, M. Miyata, and Y. Nishimura, Electrophysioological and morphological studies on thalamic neurons receiving entopedunculo- and cerebello-thalamic projections in the cat. *Brain Res.* 301: 231-242 (1984).

Section VI

THE ACTIONS OF DOPAMINE IN THE BASAL GANGLIA

EFFECTS OF DOPAMINE RECEPTOR STIMULATION ON SINGLE UNIT ACTIVITY IN THE BASAL GANGLIA

Judith R. Walters, David N. Ruskin, Kelly A. Allers, and Debra A. Bergstrom*

1. INTRODUCTION

In the past decade, research directed toward determining how dopamine receptor stimulation affects information processing in the basal ganglia has found the familiar "dual circuit" model[3,4,16] of the basal ganglia a very useful tool. This dual circuit view of the basal ganglia system emerged from investigations of the effects of dopamine cell lesion on GABA receptor changes at the sites of termination of the two types of striatal efferents: the striatonigral and striatopallidal neurons[3,35]. Also critical were studies that mapped the distribution of the two main dopamine receptor subtypes in the striatum with respect to these striatal efferents[18,32]. The conclusions drawn from these observations have been supported by additional biochemical and immunohistochemical data[18,19,36] and they led to the view that dopamine acts primarily to inhibit activity in the so-called "indirect" pathway through stimulation of the D2 receptors expressed by the striatopallidal neurons, and to enhance activity in the "direct" pathway through the D1 receptors expressed by the striatal neurons projecting to the substantia nigra pars reticulata (SNpr) and internal globus pallidus (GPi) or, in the rat, the entopeduncular nucleus (EPN).

This basic model of the basal ganglia has served as a primary frame of reference for generating hypotheses and interpreting results, providing a basis for evolving insight into how dopamine regulates movement. Its usefulness stems largely from its ability to provide an explanation for how dopamine – acting primarily in the striatum – might bring about changes in the thalamus and motor cortex which critically modulate our ability to move. The primary hypotheses derived from consideration of the basic wiring diagram of the dual circuit model are: 1) D1 and D2 receptors act separately and oppositely on the "direct" and indirect pathways, respectively, 2) the subthalamic nucleus (STN) acts primarily as a relay between the external globus pallidus (GPe) and the SNpr and GPi/EPN, 3) dopamine receptor stimulation reduces activity in the STN, GPi/EPN and SNpr, and 4) inhibition of the SNpr and

---

* J.R. Walters, D.N. Ruskin, K.A. Allers, and D.A. Bergstrom, Experimental Therapeutics Branch, National Institute of Neurological Disorders and Stroke, National Institutes of Health, Bethesda, MD 20892.

The Basal Ganglia VI
Edited by Graybiel *et al.*, Kluwer Academic/Plenum Publishers, 2002

GPi/EPN neuronal activity leads to locomotor activation by disinhibiting thalamic systems regulating cortical activity. One strategy for testing and refining these predictions is the use of extracellular single unit recording techniques to examine the changes in firing rates of different areas of the basal ganglia associated with changes in dopamine receptor stimulation. This chapter will review results from some of these studies.

2. FIRING RATE CHANGES INDUCED BY TONIC INCREASES IN DOPAMINE RECEPTOR STIMULATION

A number of studies have used extracellular single unit recording techniques to examine the effects of dopamine receptor stimulation on activity in the basal ganglia nuclei. Initial investigations in the rat led to the realization that systemic anesthesia markedly attenuates the ability of dopamine receptor stimulation to affect firing rates in the basal ganglia nuclei[10,22,29]. Presumably, this is a reflection of the fact that dopamine plays a modulatory role in the basal ganglia; the ability of dopamine agonists to alter basal ganglia output would appear to be dependent upon the level of tonic and phasic input into this system.

One of the first investigations of the effects of systemically administered dopamine agonists on firing rates of striatal neurons in awake, locally anesthesized, gallamine-immmobilized rates was conducted by Skirboll, Grace and Bunney[43]. These investigators found evidence of different cell types in the striatum: one type was typically inhibited by administration of apomorphine (which stimulates both D1 and D2 receptor subtypes) and the other showed a complex biphasic response with lower doses of apomorphine inducing neuronal excitation. Other investigators have studied effects of systemically administered indirect-acting dopamine agonists, such as amphetamine, on striatal activity in awake behaving animals with chronic indwelling electrodes. These studies have found predominantly excitatory neuronal responses to systemic d-amphetamine administration, with a relatively smaller proportion of inhibitory[27,44,51]. Overall, studies of the eletrophysiological responses of striatal neurons to systemically administered dopamine agonists in awake animals are largely consistent with one of the main hypotheses of the dual circuit model: there are two major populations of striatal neurons, the firing rates of which are differentially modulated by dopamine receptor activation. This issue has not been extensively studied *in vivo*, however, largely for technical reasons: most striatal neurons typically fire slowly and irregularly, if at all, and those giving rise to the "direct" and "indirect" pathways are not easily distinguished from one another.

On the other hand, data from single unit recording studies of the effects of various dopamine agonists on activity in the GPe, STN, EPN and SNpr, nuclei "downstream" from the striatum, do not support the model on several points or are consistent with the model only under limited conditions. Extracellular single unit recording techniques were used in awake, locally anesthetized, immobilized rats to examine the prediction that activity in the "indirect" pathway is controlled independently by D2 receptors, and that D2 receptors should increase GPe firing rates (by inhibiting the inhibitory striatopallidal neurons)[14,15,45-47,42]. The data are inconsistent with these predictions (Fig. 1). In fact, drugs, such as quinpirole, which selectively stimulate D2 receptors, induced only modest rate effects on GPe neuronal activity[15,42,45]. Similar observations, with respect to net change in rate, were also made with drugs such as SKF 38393, which selectively affect D1 receptor subtypes[14,15,42,45]. However, a com-

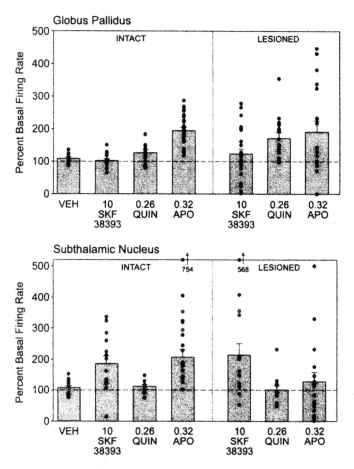

Figure 1. Effects of dopamine agonists on the single unit activity of GPe and STN neurons in intact rats or rats with 6-OHDA-induced midbrain dopamine cell lesions. The D1 agonist SKF 38393, the D2 agonist quinpirole (QUIN), and the D1/D2 agonist apomorphine (APO) were administered i.v. at the doses indicated in mg/kg. VEH indicates vehicle. Data are expressed as percent basal firing rate; dots indicate the responses of individual neurons and bar graphs indicate the mean ± SEM response of the group. Adapted from references 5, 8, 14, 15, 29, 30, 45-47.

bination of D1 and D2 agonists[14,15,42,45] or simultaneous stimulation of D1 and D2 receptors by apomorphine or amphetamine[8-11,15,25,26,37,45-47] did induce a robust increase in the activity of GPe neurons (Fig. 1). These results showed that the D1 receptors were indeed exerting an effect on the firing rates of neurons in the "indirect" pathway, in disagreement with the hypothesis that the D2 receptors are the subtype regulating activity in this pathway. Subsequent neurophysiological investigations[25] identified a second cell type in the GPe, which responded in the opposite manner to systemically administered apomorphine, supporting data from anatomical studies which indicated that the GPe has more than one type of output neuron[28,33,41].

Additional experiments showed that other predictions of the dual circuit model were also inconsistent with data from studies in the intact rat. Single unit recording data (Fig. 1) have not supported the idea that STN activity is largely controlled by the inhibitory GABAergic projection from the GPe; namely, that STN firing rates should be negatively correlated with GPe firing rates. As discussed by Allers et al. (this volume), systemic administration of D1 agonists induces a marked increase in the activity of the STN neurons, in contrast to the prediction that this receptor subtype should not exert effects on the "indirect" pathway[29] and in contrast to the prediction that dopamine agonists should inhibit STN activity. D2 agonists either had no effect on STN neuronal firing rates or induced modest increases in STN rate (instead of the predicted decreases), and apomorphine also produced substantial increases in STN unit activity[5,29,30].

Since these studies showed that the effects of dopamine agonists on the activity of the STN neurons were not consistent with the predictions of the dual circuit model, it is not surprising that it was also found that firing rates in the nuclei downstream from this area, the SNpr and EPN, were also not affected by dopamine agonist administration in the manner predicted (Ruskin et al., this volume). In intact rats, all systemically administered dopamine agonists which were tested failed to produce net decreases in neuronal firing rates in these nuclei: effects on the activity of neurons in these nuclei were typically quite variable, but the net effect was either no net change in rate in the SNpr[21,46-49], or a net increase in the EPN[37,39]. These effects were observed with doses of the agonists that produce marked increases in locomotion and stereotyped behavior. Thus, the hypothesis that an overall decrease in the activity of the basal ganglia output nuclei mediates the increase in behavioral activity induced by these drugs is also not consistent with the *in vivo* data.

It is important to point out that in the rat model of Parkinson's disease, D1 agonists and combined administration of D1 and D2 agonists induce the predicted decrease in net firing rates in the output nuclei[21,38,39,47,49,50], observations consistent with the dual circuit model. In fact, much of the biochemical and immunochemical data supporting the dual circuit model was derived from studies in rats with dopamine cell lesions, and it was generally assumed that the development of supersensitivity in the dopamine-depleted rats would provide responses quantitatively greater but qualitatively similar to those occurring in the intact rat. Interestingly, the single unit recording data argues differently: dopamine cell lesion appears to induce both quantitative and qualitative changes in the response of basal ganglia neurons to dopamine receptor stimulation. Thus, results obtained from single unit studies in the 6-OHDA-lesioned rats are more consistent with the model than those from the intact rat. However, as reviewed by Ruskin et al. (this volume), even in the dopamine depleted animals, where the predicted decreases in net activity in the EPN and SNpr nuclei are induced by D1 and D1/D2 family agonists, no clear correlation is observed between agonist doses that produce behavioral effects (i.e., rotation in unilaterally lesioned rats) and doses that produce net decreases in SNpr and EPN firing rates[38].

3. CONCLUSIONS FROM FIRING RATE STUDIES

The results described above do not support the view that drugs which increase dopamine receptor stimulation in the striatum produce increased behavioral activity in rodents by reducing the net firing rates of the basal ganglia output neurons. Firing rates in the various basal ganglia nuclei are affected by most of these drugs, but frequently not as predicted by the basic version of the dual circuit model. In fact, it is now appreciated that D1 subfamily receptors are expressed at a variety of sites capable of affecting impulse activity in the "indirect" pathway, such as in the STN nucleus, on the cell bodies and/or terminals of the corticosubthalamic projections, on the interneurons in the striatum, on afferents to the striatum and in the SNpr. Thus it appears that dopamine could be acting through dopamine receptors at a variety of sites to modulate basal ganglia activity. In addition, the "direct" and "indirect" pathways are not as cleanly anatomically segregated as suggested in the early versions of the model: there is a substantial collateral projection from the "direct" striatonigral neurons to the GPe[24]. Although it cannot be ruled out that the subpopulations of SNpr and EPN neurons which show rate decreases after dopamine agonist administration are responsible for the behavioral changes induced by the agonists, at the present time it would appear difficult to explain the general hyperactivity associated with increased dopamine receptor stimulation in rodents through mechanisms involving simple population rate decreases in the basal ganglia output to the thalamus and tectum. These studies have highlighted the fact that consideration of mean firing rate alone, in the absence of information about pattern and synchronization of spiking activity, does not adequately serve the goal of understanding how dopamine receptor stimulation is critical to the proper functioning of the basal ganglia.

4. FIRING PATTERN CHANGES INDUCED BY TONIC INCREASES IN DOPAMINE RECEPTOR STIMULATION

In view of the data described above, more efforts are being made to investigate the effects of dopamine agonists on the patterning of firing activity in the basal ganglia nuclei. In the course of such studies, we have observed that a large number of spike trains recorded from awake, locally anesthetized, immobilized rates in the SNpr, GPe, EPN and STN exhibit slow periodic changes in basal firing rate at time scales of several seconds or tens of seconds. When these oscillations were examined with the Lomb periodogram algorithm[23] for periodicities within the range of 2 to 60 s, oscillations were shown to be significantly periodic in about 50-70% of the cells recorded in the GPe, STN, SNpr and EPN[5,37,40] (Fig. 2 & 3). Power spectra showed significant periods ranging from 2 – 60 s (means: 30 – 40 s). Although the regularity and amplitude of these periodic fluctuations in firing rate varied greatly from unit to unit, the fluctuations could be quite large. Multisecond oscillatory activity in this time range was not present in recordings from rats which were systemically anesthetized with chloral hydrate or urethane[5,37] (Fig. 2).

Neurons from two areas outside the basal ganglia were also studied for oscillations in this time range. Baseline oscillations like those observed in the basal ganglia nuclei were found in spike trains from the cerebellum, but not from the septum (Fig. 3).

The multisecond oscillations observed in baseline firing rates of several basal ganglia nuclei were markedly slower than most oscillatory activity typically studied in spike trains from basal ganglia, thalamus and cortex. However, there are indications in the literature that oscillations in central activity with similar time periods can be seen in other species and under other conditions. The earliest reports in the literature are a series of studies carried out by Aladjalova[1,2], Norton and Jewett[34], and Kropotov and Gretchin[20,31]. These researchers

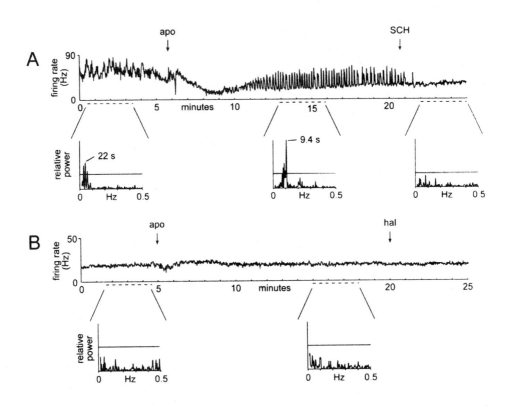

Figure 2. Spike trains from SNpr single unit recordings from an awake, immobilized rat (A) and a chloral hydrate anesthetized rat (B). Spike trains were smoothed with 1 s bins. The power spectra are from the epochs indicated. Lines across power spectra indicate the p=0.01 significance level. Apomorphine was injected i.v. at 0.32 mg/kg. In (A), a slow periodic oscillation in baseline rate of 22 s is evident; strong multisecond oscillatory activity emerges with a 9.4 s main period after apomorphine administration. This slow oscillation is eliminated by administration of the D1 antagonist SCH 23390 (0.5 mg/kg). There is no significant multisecond oscillatory activity during baseline or after apomorphine or haloperidol (0.2 mg/kg) in (B). Adapted from reference 37.

noted multisecond oscillations in pO_2 and DC-potentials in deep brain electrode recordings and EEG recordings. Notably, some of these DC-potential recordings were performed in parkinsonian patients in whom electrodes were implanted for diagnosis and therapy[20,31]. Similar low-frequency oscillations have been detected in adult humans undergoing functional magnetic resonance imaging by Hyde and coworkers, who present data supporting the view that the changes in fMRI reflect changes in neuronal firing patterns in motor cortex and associated areas[12,13]. The intriguing suggestion that these oscillations play a role in brain mechanisms underlying attention is provided by observations by Ehlers and Foote[17]. In chair-restrained squirrel monkeys, these investigators noted spontaneous periodic oscil-

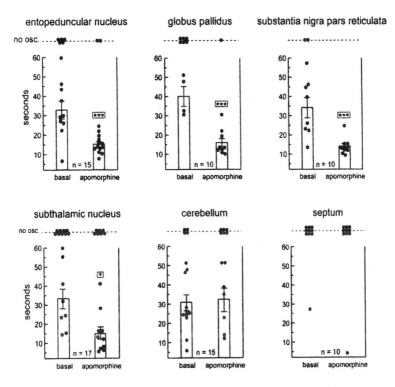

Figure 3. Main spectral peak periods for baseline and post-apomorphine epochs in four basal ganglia nuclei, cerebellum, and septum. Neurons with no significant oscillations are indicated at the top of each graph (no osc). The incidence of significant oscillations increased after apomorphine in basal ganglia nuclei (from 34/52 in baseline to 45/52). 67% of spike trains from cerebellar neurons have basal oscillations but these are not consistently altered by apomorphine. In septal neurons there was a low incidence of oscillatory activity. One neuron was recorded per rat. Data are mean ± SEM. *p<0.05, ***p<0.001 vs. basal. Unpublished observations and adapted from references 5,37.

lations in EEG and behavioral activity with a cycle length of 15-30 seconds which consisted of "alternating episodes of vigilance, characterized by visual scanning and motor movement, and inattentiveness characterized by behavioral quiescence with little eye or limb movement" (pg. 381)[17]. The periods of vigilance were associated with low amplitude desynchronized EEG and the quiescent periods with high amplitude synchronized EEG. Finally, a recent study by Wichmann and coworkers[52] reports finding slow multisecond oscillations in spike trains recorded from SNpr neurons in the monkey.

5. EFFECTS OF DOPAMINE AGONISTS ON MULTISECOND OSCILLATIONS

What has been most striking about these multisecond periodic oscillations in firing rate observed in the rodent studies is the finding that systemic dopamine agonist administration brings about a dramatic increase in the frequency and, in most cases, the regularity and

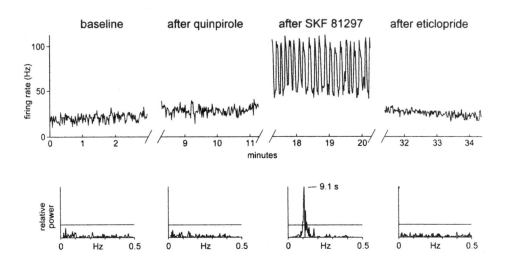

Figure 4. Spike trains and corresponding power spectra from a single unit recording of a GPe neuron. The 180 s epochs are smoothed with 1 s bins. There is no significant oscillatory activity in the multisecond range in the baseline of this neuron. After i.v. injection of the D2 agonist quinpirole, 1.0 mg/kg injected at 4.5 min, there is little change in the spectrum, although there is a moderate increase in firing rate. A subsequent i.v. injection of the D1 agonist SKF 81297, 1.0 mg/kg at 14 min, causes a large further increase in firing rate and marked oscillations which have a 9.1 s period. The D2 antagonist eticlopride, 0.2 mg/kg i.v. at 24 min, reverses the rate and pattern effects. Adapted from reference 40.

amplitude of these firing rate oscillations, and induces oscillatory activity in neurons that do not have significant baseline oscillations[5,37,39,40] (Fig. 2 - 4). As shown in the example in Fig. 2, strong, regular multisecond oscillations appear after apomorphine injection, superimposed on drug-induced changes in mean firing rate. In particular, the shift from a main spectral peak of 22 s in baseline to 9.4 s in Fig. 2A exemplifies the apomorphine-induced decrease in oscillatory period. This shift is evident in the population data from each nucleus examined in the basal ganglia but was not observed in neurons recorded in either the cerebellum or the septum (Fig. 3). In each of the four basal ganglia nuclei, baseline means for main peak period were in the 30-40 s range, and were shortened to ~ 15 s after administration of apomorphine (Fig. 2, 3, for review of STN data, see Allers et al., this volume). Apomorphine also increased the spectral power of the oscillations, causing significant increases in main spectral peak height in the GP and EPN, and a trend toward an increase in the STN and SNpr[5,37]. Apomorphine's effects on the frequency of the oscillations were reversed by administration of the D1 antagonist SCH 23390 and/or the D2 receptor antagonist haloperidol[5,37] (Fig. 2). Notably, in rats under general anesthesia, significant oscillatory activity in the multisecond range was rarely found in either baseline or after dopamine agonist administration[5,37], as shown in the example in Fig. 2B.

Further investigation with selective D1 and D2 agonists (e.g. SKF 81297 and quinpirole) and antagonists confirmed that both dopamine receptor subtypes contribute to the effects of dopamine agonists on these oscillations[5,39,40] (Fig. 4) and suggests a synergistic interaction between these receptor subtypes. Although D1 and D2 agonists each have some effect alone on multisecond oscillations in the GPe (quinpirole causes an emergence of strong oscillations in some, but not all, neurons, and SKF 81297 or SKF 82958 cause some decrease in oscillatory period[40]), the strongest modulation of multisecond oscillations is clearly caused by treatment with combined D1 and D2 agonists. In the recording of a GPe neuron shown in Fig. 4 (a cell with no significant baseline oscillation), injection of quinpirole alone does not induce significant oscillations. Later injection of SKF 81297 causes strikingly regular and high-amplitude oscillations, with the power spectrum indicating a period of 9 s. A later injection of the D2 antagonist eticlopride reverses the effect of the agonists, demonstrating that the emergence of strong oscillations after SKF 81297 was not due to that drug acting alone but was dependent on the prior injection of the D2 agonist quinpirole.

Recent studies have also shown that indirect-acting dopamine agonists such as amphetamine, cocaine and methylphenidate which block dopamine uptake and/or facilitate release exert effects on GPe oscillations similar to those induced by direct agonists[7]. In addition, preliminary observations in the rat suggest that these multisecond oscillations in firing rate may be occurring in synchrony in neurons in different regions of the basal ganglia. In pairs of basal ganglia neurons recorded simultaneously, sometimes in opposite hemispheres, highly correlated slow oscillations were observed in about 60% of cases in rats receiving drugs which bring about changes in dopamine receptor stimulation[6] (Allers et al., this volume).

6. CONCLUSIONS

Tonic changes in firing rate in basal ganglia nuclei have been documented in animal studies of dopamine agonist effects. These changes reflect a more complex integrated basal ganglia network than depicted in the useful but admittedly simplistic dual circuit model. While the manner in which these firing rate changes contribute to the therapeutic and behavioral effects of the dopaminergic drugs is clearly more complex than first hypothesized, tonic changes in firing rate are quite marked after some drug treatments and would seem likely to have a significant impact on basal ganglia function. There does not appear to be, however, a clear and simple relationship between effects of dopamine agonists on behavioral activity and on firing rate changes in basal ganglia output nuclei. The effects of dopamine receptor manipulation on additional aspects of neuronal activity, such as burstiness and oscillatory activity, may well play an important role in mediating functionally important effects of dopamine receptor stimulation on basal ganglia output. The data reviewed in the present chapter suggest that information may be carried to the postsynaptic neuron not only by mean firing rate, but also by changes in the frequency of slow oscillatory activity. Postsynaptic systems and second messenger-mediated processes with "bandpass filtering" capabilities would allow the neuron to respond to selective changes in frequency over the multisecond time scale. In addition, multisecond oscillations might act to organize information in the time domain in distributed circuits, perhaps facilitating the propagation and synchronization of faster oscillatory activity, or, alternatively, limiting information throughput in a particular circuit. The significant shift in the period of these oscillations after dopamine agonist administration suggests that abnormal organization in this time domain contributes to the hyperactivity and stereotypy caused by these drugs. If multisecond oscillations act as "temporal organizers," then the increase in power and coherence of these oscillations after dopamine agonist injection could signal that normally independent pathways are abnormally coupled. Alternatively, if these oscillations act to attenuate information flow, powerful oscillations after dopamine agonist injection would indicate abnormally low information throughput in the basal ganglia. In summary, many aspects of the generation and significance of these oscillations remain to be sorted out. However, the observation that these oscillations are markedly affected by both direct and indirect-acting dopamine agonists suggests that coordinated changes in neuronal activity at time scales longer than commonly investigated play a role in the cognitive and motor processes affected by dopamine.

7. REFERENCES

1. N.A. Aladjalova, 1957, Infra-slow rhythmic oscillations of the steady potential of the cerebral cortex, *Nature* 179:957.
2. N.A. Aladjalova, 1964, Slow electrical processes in the brain, *Prog. Brain Res.* 7:1.
3. R.L. Albin, A.B. Young, and J.B. Penney, 1989, The functional anatomy of basal ganglia disorders, *Trends Neurosci.* 12:366.
4. G.E. Alexander and M.D. Crutcher, 1990, Functional architecture of basal ganglia circuits: Neural substrates of parallel processing, *Trends Neurosci.* 13:266.
5. K.A. Allers, D.S. Kreiss, and J.R. Walters, Multisecond oscillations in the subthalamic nucleus: effects of apomorphine and dopamine cell lesion, *Synapse*, 38:38.
6. K.A. Allers, D.N. Ruskin, D.A. Bergstrom, L.R. Molnar, and J.R. Walters, 1999, Correlations of multisecond oscillations in firing rate in pairs of basal ganglia neurons, *Soc. Neurosci. Abstr.* 25:1929
7. D. Baek, D.N. Ruskin, D.A. Bergstrom, and J.R. Walters, 1999, D-amphetamine and cocaine increase the "speed" of multisecond oscillations in globus pallidus unit activity, *Soc. Neurosci. Abstr.* 25:1819.
8. D.A. Bergstrom, S.D. Bromley, and J.R. Walters, 1982, Apomorphine increases the activity of rat globus pallidus neurons, *Brain Res.* 238:266.
9. D.A. Bergstrom, S.D. Bromley, and J.R. Walters, 1982, Time schedule of apomorphine administration determines the degree of globus pallidus excitation, *Eur. J. Pharmacol.* 78:245.
10. D.A. Bergstrom, S.D. Bromley, and J.R. Walters, 1984, Dopamine agonists increase pallidal unit activity: attenuation by agonist pretreatment and anesthesia, *Eur. J. Pharmacol.* 100:3.
11. D.A. Bergstrom and J.R. Walters, 1981, Neuronal responses of the globus pallidus to systemic administration of d-amphetamine: investigation of the involvement of dopamine, norepinephrine, and serotonin, *J. Neurosci.* 1:292.
12. B. Biswal, A.G. Hudetz, F.Z. Yetkin, V.M. Haughton, and J.S. Hyde, 1997, Hypercapnia reversibly suppresses low-frequency fluctuations in the human motor cortex during rest using echo-planar MRI, *J. Cereb. Blood Flow* 17:301.
13. B. Biswal, F.Z. Yetkin, V.M. Haughton, and J.S. Hyde, 1995, Functional connectivity in the motor cortex of resting human brain using echo-planar MRI, *Magn. Reson. Med.* 34:537.
14. J.H. Carlson, D.A. Bergstrom, S.D. Demo, and J.R. Walters, 1990 Nigrostriatal lesion alters neurophysiological responses to selective and nonselective D-1 and D-2 dopamine agonists in rat globus pallidus, *Synapse* 5:83.
15. J.H. Carlson, D.A. Bergstrom, and J.R. Walters, 1987, Stimulation of both D1 and D2 dopamine receptors appears necessary for full expression of postsynaptic effects of dopamine agonists: a neurophysiological study, *Brain Res.* 400:205.
16. M.R. Delong, 1990, Primate models of movement disorders of basal ganglia origin, *Trends Neurosci.* 13:281.
17. C.L. Ehlers and S.L. Foote, 1984, Ultradian periodicities in EEG and behavior in the squirrel monkey (Saimiri sciureus), *Am. J. Primatol.* 7:381.

18. C.R. Gerfen, 1992, The neostriatal mosaic: Multiple levels of compartmental organization in the basal ganglia, *Annu. Rev. Neurosci.* 15:285.
19. C.R. Gerfen, T.M. Engber, Z. Susel, L.C. Mahan, F.J. Monsma, Jr., L.D. Mcvittie, and D.R. Sibley, 1990, Regulation by D1 and D2 dopamine receptors of striatonigral and striatopallidal peptide mRNA levels, *Science* 250:1429.
20. V.B. Gretchin and J.D. Kropotov, 1974, [On the genesis of slow oscillations of pO2 in the human brain], *Sov. J. Physiol.* 60:849.
21. K.-X. Huang and J.R. Walters, 1994, Electrophysiological effects of SKF 38393 in rats with reserpine treatment and 6-hydroxydopamine-induced nigrostriatal lesions reveal two types of plasticity in D1 dopamine receptor modulation of basal ganglia output, *J. Pharmacol. Exp. Ther.* 271:1434.
22. K.-X. Huang and J.R. Walters, 1992, D1 receptor stimulation inhibits dopamine cell activity after reserpine treatment but not after chronic SCH 23390: an effect blocked by N-methyl-D-aspartate antagonists, *J. Pharmacol. Exp. Ther.* 260:409.
23. Y. Kaneoke and J.L. Vitek, 1996, Burst and oscillation as disparate neuronal properties, *J. Neurosci. Meth.* 68:211.
24. Y. Kawaguchi, C.J. Wilson, and P.C. Emson, 1990, Projection subtypes of rat neostriatal matrix cells revealed by intracellular injection of biocytin, *J. Neurosci.* 10:3421.
25. M.D. Kelland, R.P. Soltis, L.A. Anderson, D.A. Bergstrom, and J.R. Walters, 1995, In vivo characterization of two cell types in the rat globus pallidus which have opposite responses to dopamine receptor stimulation: comparison of electrophysiological properties and responses to apomorphine, dizocilpine, and ketamine anesthesia, *Synapse* 20:338.
26. M.D. Kelland, Sr. and J.R. Walters, 1992, Apomorphine-induced changes in striatal and pallidal neuronal activity are modified by NMDA and muscarinic receptor blockade, *Life Sci.* 50:L179.
27. L.J. Kish, M.R. Palmer, and G.A. Gerhardt, 1999, Multiple single-unit recordings in the striatum of freely moving animals: effects of apomorphine and D-amphetamine in normal and unilateral 6-hydroxydopamine-lesioned rats, *Brain Res.* 833:58.
28. H. Kita, 1994, Parvalbumin-immunopositive neurons in rat globus pallidus: a light and electron microscopic study, *Brain Res.* 657:31.
29. D.S. Kreiss, L.A. Anderson, and J.R. Walters, 1996, Apomorphine and dopamine D_1 receptor agonists increase the firing rates of subthalamic nucleus neurons, *Neuroscience* 72: 863.
30. D.S. Kreiss, C.W. Mastropietro, S.S. Rawji, and J.R. Walters, 1997, The response of subthalamic nucleus neurons to dopamine receptor stimulation in a rodent model of Parkinson's disease, *J. Neurosci.* 17:6807.
31. J.D. Kropotov and V.B. Gretchin, 1975, [Correlations between slow oscillations of electrical potential and the pO2 in the human brain], *Sov. J. Physiol.* 61:331.
32. C. Le Moine, E. Normand, and B. Bloch, 1991, Phenotypical characterization of the rat striatal neurons expressing the D_1 dopamine receptor gene, *Proc. Natl. Acad. Sci. USA* 88:4205.
33. A. Nambu and R. Llinás, 1994, Electrophysiology of globus pallidus neurons in vitro, *J. Neurophysiol.* 72:1127.
34. S. Norton and R.E. Jewett, 1965, Frequencies of slow potential oscillations in the cortex of cats, *Electroencephalogr. Clin. Neurophysiol.* 19:377.
35. H.S. Pan, J.B. Penney, and A.B. Young, 1985, g-Aminobutyric acid and benzodiazepine receptor changes induced by unilateral 6-hydroxy-dopamine lesions of the medial forebrain bundle, *J. Neurochem.* 45:1396.
36. G.S. Robertson, S.R. Vincent, and H.C. Fibiger, 1992, D_1 and D_2 dopamine receptors differentially regulate *c-fos* expression in striatonigral and striatopallidal neurons, *Neuroscience* 49:285.
37. D.N. Ruskin, D.A. Bergstrom, Y. Kaneoke, B.N. Patel, M.J. Twery, and J.R. Walters, 1999, Multisecond oscillations in firing rate in the basal ganglia: robust modulation by dopamine receptor activation and anesthesia, *J. Neurophysiol.* 81:2046.
38. D.N. Ruskin, D.A. Bergstrom, C.W. Mastropietro, M.J. Twery, and J.R. Walters, 1999, Dopamine agonist-mediated rotation in rats with unilateral nigrostriatal lesions is not dependent on net inhibition in basal ganglia output nuclei, *Neuroscience* 91:935.
39. D.N. Ruskin, D.A. Bergstrom, and J.R. Walters, 1999, Firing rates and multisecond oscillations in the rodent entopeduncular nucleus: effects of dopamine agonists and nigrostriatal lesion, *Soc. Neurosci. Abstr.* 25:1929.

40. D.N. Ruskin, D.A. Bergstrom, and J.R. Walters, 1999, Multisecond oscillations in firing rate in the globus pallidus: synergistic modulation by D1 nd D2 dopamine receptors, *J. Pharmacol. Exp. Ther.* 290:1493.
41. D.N. Ruskin and J.F. Marshall, 1997, Differing influences of dopamine agonists and antagonists on Fos expression in identified populations of globus pallidus neurons, *Neuroscience* 81:79.
42. D.N. Ruskin, S.S. Rawji, and J.R. Walters, 1998, Effects of full D-1 dopamine receptor agonists on firing rates in the globus pallidus and substantia nigra pars compacta in vivo: Tests for D-1 receptor selectivity and comparisons to the partial agonist SKF 38393, *J. Pharmacol. Exp. Ther.* 286:272.
43. L.R. Skirboll, A.A. Grace, and B.S. Bunney, 1979, Dopamine auto- and postsynaptic receptors: electrophysiological evidence for differential sensitivity to dopamine agonists, *Science* 206:80.
44. J.T. Tschanz, K.E. Griffith, J.L. Haracz, and G.V. Rebec, 1994, Cortical lesions attenuate the opposing effects of amphetamine and haloperidol on neostriatal neurons in freely moving rats, *Eur. J. Pharmacol.* 257:161.
45. J.R. Walters, D.A. Bergstrom, J.H. Carlson, T.N. Chase, and A.R. Braun, 1987, D1 dopamine receptor activation required for postsynaptic expression of D2 agonist effects, *Science* 236:719.
46. J.R. Walters, D.A. Bergstrom, J.H. Carlson, B.G. Weick, and H.S. Pan, Stimulation of D-1 and D-2 dopamine receptors: synergistic effects on single unit activity in basal ganglia output nuclei, *in:* "Neurophysiology of Dopaminergic Systems: Current Status and Clinical Perspectives", L.A. Chiodo and A.S. Freeman, eds., Lakeshore Publ. Co., Detroit, Michigan (1987).
47. J.R. Walters, M.D. Kelland, K.-X. Huang, and D.A. Bergstrom, Effects of dopamine agonists on neuronal activity in the basal ganglia of normal and dopamine-denervated rats, *in:* "Advances in Parkinson's Disease Research -2.", W.J. Weiner and F. Hefti, eds., Futura Publishing Co., Mt. Kisco, New York (1992).
48. B.L. Waszczak, E.K. Lee, T. Ferraro, T.A. Hare, and J.R. Walters, 1984, Single unit responses of substantia nigra pars reticulata neurons to apomorphine: effects of striatal lesions and anesthesia, *Brain Res.* 306:307.
49. B.L. Waszczak, E.K. Lee, C.A. Tamminga, and J.R. Walters, 1984, Effect of dopamine system activation on substantia nigra pars reticulata output neurons: variable single-unit responses in normal rats and inhibition in 6-hydroxydopamine-lesioned rats, *J. Neurosci.* 4:2369.
50. B.G. Weick and J.R. Walters, 1987, Effects of D1 and D2 dopamine receptor stimulation on the activity of substantia nigra pars reticulata neurons in 6-hydroxydopamine lesioned rats: D1/D2 coactivation induces potentiated responses, *Brain Res.* 405:234.
51. M.O. West, L.L. Peoples, A.J. Michael, J.K. Chapin, and D.J. Woodward, 1997, Low-dose amphetamine elevates movement-related firing of rat striatal neurons, *Brain Res.* 745:331
52. T. Wichmann, H. Bergman, M.A. Kliem, J. Soares and M.R. DeLong, 1999, Low-frequency oscillatory discharge in the primate substantia nigra pars reticulata in the normal and parkinsonian state, *Soc. Neurosci. Abstr.* 25:1928.

IMMUNOCYTOCHEMICAL CHARACTERISATION OF CATECHOLAMINERGIC NEURONS IN THE RAT STRIATUM FOLLOWING DOPAMINE DEPLETING LESIONS

S. Totterdell and G.E. Meredith*

Abstract: In rats that receive unilateral 6-hydroxydopamine lesions, neurons immunoreactive for catecholaminergic markers appear in the striatum. If these neurons contribute to the brain's ability to convert L-DOPA to dopamine in a therapeutic context then they have obvious relevance for the treatment of Parkinson's disease. We show that the tyrosine hydroxylase immunoreactive neurons found in the dopamine-depleted striatum differ in size, morphology and location from those that are immunopositive for aromatic L-amino acid decarboxylase or dopamine. The tyrosine hydroxylase immunoreactive neurons, which lie ventrally along a rostrocaudal continuum from the core of nucleus accumbens to the fundus of the striatum, have the morphological characteristics of projection cells. Meanwhile, aromatic L-amino acid decarboxylase or dopamine-immunoreactive neurons, found in subcallosal positions in the dorsomedial caudate-putamen of both the dopamine-depleted and intact hemispheres, have the features of local circuit neurons. These catecholaminergic enzymes may be induced as a pharmacological response to the almost complete loss of dopamine from the striatum, or their appearance may reflect a general compromise in neuronal function. We present preliminary evidence for neuronal loss from the most severely depleted regions of the striatal complex suggesting that certain catecholaminergic markers may be found in neurons about to die.

1. INTRODUCTION

Injection of 6-hydroxydopamine (6-OHDA) into the rat midbrain produces a rapid but heterogeneous loss of dopaminergic axons and terminals in both the caudate-putamen (CPu) and nucleus accumbens.[45] Following such lesions, tyrosine hydroxylase (TH)- and aromatic

---

* S. Totterdell, Department of Pharmacology, University of Oxford, Mansfield Road, Oxford, OX1 3QT, UK. G.E. Meredith, Department of Anatomy, Royal College of Surgeons in Ireland, 123 St Stephen's Green, Dublin 2, Ireland.

The Basal Ganglia VI
Edited by Graybiel *et al.*, Kluwer Academic/Plenum Publishers, 2002

L-amino acid decarboxylase (AADC)- immunoreactivity appears in striatal neurons that do not normally exhibit these enzymes.[4,32,41,42,43] In rats, AADC-ir cells appear to be dopaminergic,[32] but it is not known if TH-ir cells contain dopamine; in primates, TH-ir, dopaminergic cells, present in the intact striatum, increase in numbers in dopamine-depleted animals.[4] TH-ir striatal neurons in both rodents and primates are reported either as spiny, presumably projection cells,[42,43] as small and aspiny or with an indented nucleus[32] interneurons, or a mixture of both.[4]

The presence in the dopamine-depleted striatum of neurons able to convert L-DOPA to dopamine in a therapeutic context, thus enhancing residual dopamine levels, has important implications for the pathophysiology of Parkinson's disease. These neurons may appear in response to the slow depletion of residual dopamine stores,[52] or to the reduced TH activity which accompanies dopamine loss.[16] The appearance of catecholaminergic cells only after a lesion, must be due to a fundamental change in gene expression which may signal a transient alteration in synthetic activity, as seen in the developing striatum,[22] or a more permanent change in neuronal phenotype.

Particularly interesting are the distribution and viability of these catecholaminergic neurons in nucleus accumbens, where the dopaminergic innervation is most extensively spared both after experimental lesions[49] and in parkinsonian brains[23]. Nucleus accumbens shell and core differ in the origins of their dopaminergic innervations and in their responses to manipulations of the dopamine supply,[7,11,30,34,44,49] while the rostral pole seems to be unique both in terms of its dopaminergic innervation and its function.[10,50,51]

In this report, we resolve these conflicts concerning the phenotype of striatal TH-ir neurons, including whether or not they make dopamine or contain AADC and whether they are projection or local circuit neurons. We also establish the time course for their appearance and their compartmental distribution following dopamine-depleting lesions. Finally, we provide preliminary data suggesting that there is a significant loss of neurons from the dopamine-depleted striatum/accumbens core region of the dopamine-depleted hemisphere.

2. METHODS

All procedures were carried out according to the relevant animal welfare legislation in Ireland and the United Kingdom. Injections of a 6-OHDA solution were made unilaterally into the midbrain of 39 male Sprague Dawley rats. Vehicle injected and sham operated control animals were also prepared. Following survival periods from 2 to 36 days rats were killed by transcardial perfusion with appropriate fixatives. A further seven rats received similar lesions 26 days before perfusion. Sections through the forebrain and midbrain were processed to reveal TH-ir using the ABC system with diaminobenzidine tetrahydrochloride as the chromogen. Alternate sections from 9 rats were immunoreacted for AADC (see 32) and alternate sections from 5 rats were immunoreacted for dopamine (see 48). Some TH-immunostained tissue was prepared for electron microscopic examination. In a further seven rats with 6-OHDA lesions, one set of sections from a 1:3 series was immunoreacted for TH and another processed for Nissl.

Midbrain sections were examined to localise the site of the 6-OHDA lesion and to establish the extent of loss of TH-ir neurons in the substantia nigra (SN) and ventral tegmental area (VTA). TH-, AADC-, and dopamine-ir neuronal profiles in the forebrain were identified and photographed. Profiles were classified as neurons if there was a nucleus visible and if the dendrites tapered from their origins at the cell body. Neurons were recorded as projection neurons if their dendrites were spiny and if they had smooth nuclear envelopes: neurons with smooth dendrites and indented nuclei were regarded as local circuit neurons.[5,29] Patterns of fibre loss and the immunopositive neurons in the striatal complex were plotted onto maps reproduced from the atlas of Paxinos and Watson.[36] The time course and pattern of appearance of TH-ir neurons were analysed for each of six anterior-posterior levels, in rats surviving <7 days, 7-14 days, 14-21 days and > 21 days. Data for AADC and dopamine-ir neurons were obtained in animals surviving up to 21 days post-lesion. Camera lucida drawings were made of 50 large TH-ir varicosities and of TH-ir cells in six different regions from animals that survived from 7 to 26 days post-lesion. Areas were measured using a digitising tablet linked to MacStereology, (Ranfurly Microsystems®), Renfrewshire, U.K.). All data were distributed normally. Significance level in the parametric tests was set at $p < 0.05$.

With a randomly selected first section, every third section through the nucleus accumbens and adjacent CPu was collected and stained for Nissl substance. Neurons were counted in a known fraction of the accumbal core and CPu using optical disectors at predetermined x,y positions. The height of the disector was 10μm and the area of the counting frame was 400μm^2. The optical fractionator (StereoInvestigator®, Microbrightfield, VT, USA), was used to estimate the total neuron number in the nucleus accumbens' core and in CPu. Data for both hemispheres in rats with complete unilateral lesions were compared with a two-way ANOVA. Total neuron numbers in control and 6-OHDA-treated animals were also compared using the same test.

3. RESULTS

Both coarse and fine TH-ir fibres were identified in both hemispheres of animals with unilateral lesions, as well as in control rats (Fig 1A). Some of the labelled fibres that remained on the dopamine-depleted side exhibited large, swollen structures. These were identified as varicosities, regardless of their diameter, when they had no clear nucleus, were attached to fine, immunopositive fibres of constant diameter, and exhibited an abrupt transition between swelling and fibre (Fig 1D). The mean cross-sectional area of the large varicosities (51.6 μm^2) did not differ significantly from the mean area of TH-ir perikarya (55.4 μm^2). The spared AADC- and dopamine-ir fibres and varicosities did not differ in distribution or appearance from those immunopositive for TH.

TH-ir neurons, found ventrally throughout the rostrocaudal extent of the striatal complex, were small and typically had dark nuclei with very pale cytoplasm that faded at the edges (Fig 1B&C, 2A). At high power, the envelopes of the dark nuclei appeared smooth and unindented. These neurons had either bi- or multi-polar dendrites emerging from oval cell bodies or round somata with multipolar dendrites. The extent of dendritic labelling varied but dendrites tended to be varicose with few spines. Cell sizes differed significantly between regions of the striatum and nucleus accumbens sampled ($p < 0.01$) but there was no relationship between cell size and post-lesion survival time.

TH-positive neurons with dark nuclei, identified in the light microscope, were found in the electron microscope to have nuclei enclosed in smooth envelopes (Fig. 2A&C) and cytoplasm with the normal complement of organelles. Nuclei were electron-dense and had

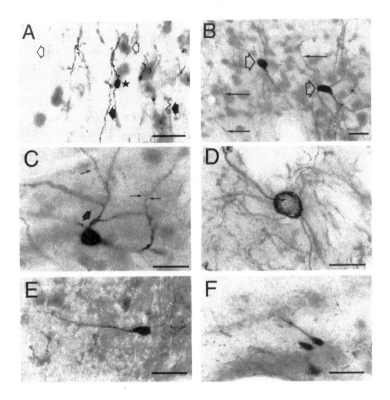

Figure 1. Catecholaminergic labelling in striatal complex after 6-OHDA lesion. (A) Residual TH-positive fibres with a swollen varicosity (star) and thin (open arrows) and thick (solid arrowheads) spared fibres. (B) Residual TH-positive fibres (thin arrows) and TH-ir neurons in the nucleus accumbens core (open arrowheads). Note their small size. (C) High power view of a TH-ir neuron in nucleus accumbens' core. Solid arrow indicates varicose primary dendrite, small arrows show spines. (D) High power view of a TH-ir varicosity. (E) Dopamine-ir neuron and (F) AADC-ir neurons in the dorsomedial CPu. Scales: A, B 50ìm; C,D 15ìm; E, F 25ìm.

clumped and dispersed chromatin. Vacuolisation of the cytoplasm was a general feature of the tissue but was most pronounced in immunolabelled neurons. TH-ir dendrites occasionally had labelled spines (Fig 2B) and also displayed some vacuolisation. Synaptic specialisations on dendritic shafts were symmetrical or possibly asymmetrical (Fig 2D) while on cell bodies they were symmetrical.

TH-ir neurons in the striatal complex were relatively uncommon, often with only one or two per section, although 58 were observed in one section. In the rostral pole of nucleus accumbens, TH-ir neurons were usually close to the anterior commissure or medially near the forceps minor of the corpus callosum: at the mid to caudal levels, they were found in the core but rarely in the shell. Caudal to accumbens, TH-ir neurons were aligned along the posterior limb of the anterior commissure and extended into the fundus striati. In CPu, TH-ir neurons were mainly found rostrally and medially, close to the lateral ventricle. Cells were occasion-

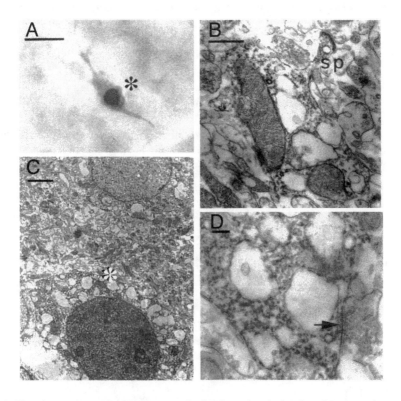

Figure 2. The ultrastructure of a TH-ir neuron in the dopamine-depleted nucleus accumbens core. An immunopositive neuron (asterisk) with dark nucleus and paler cytoplasm at light (A) and electron microscope (B) levels. Note the electron dense nucleus with a relatively intact nuclear membrane and the lighter vacuolated cytoplasm and compare with the unlabelled neuron shown in the upper part of (B). Synaptic contacts (C&D) are formed with a spine (sp) and a dendrite of the same TH-ir neuron and dendrite (arrows) of the same neuron. Scale: A, 10µm; B, 0.5µm; C, 2µm; D, 0.2µm

ally found in a dorsomedial, subcallosal position and some in the bed nucleus of the stria terminalis.

At any one survival time within the 2 to 36 day range studied, the distribution of TH-ir cells was dependent on the pattern of fibre sparing as TH-ir neurons were always located near, but not among, the spared fibres. Where destruction of the TH-positive innervation was nearly complete, cells were consistently found in the same circumscribed rostrocaudal zone. At survival times of 5-7 days, TH-ir neurons appeared only in the rostral pole and rostral core of nucleus accumbens while after 1-2 weeks, cells were found further caudally. Indeed, some neurons were seen at all six rostrocaudal levels studied, although most were in the core, 11.5mm rostral to Bregma. With survival times greater than three weeks, the absolute numbers of TH-ir neurons per section decreased and the proportion at caudal striatal sites increased.

3.1 AADC- and Dopamine-Immunoreactive Neurons: Morphology and Distribution

Neurons immunoreactive for AADC or dopamine were found in both the dopamine-depleted and intact hemispheres, although they were much less numerous in the latter. AADC-ir cells were present 2 days post-lesion, always among AADC-ir fibres, but not after longer (> 21 days) survival times, when the area was fully depleted of AADC-ir fibres. From 8 to 20 cells per animal were found at rostral and central levels of the CPu. Both AADC-ir and dopamine-ir neurons were located in the subcallosal CPu, often close to the lateral ventricle but occasionally in the dorsolateral quadrant. A few were also situated close to the forceps minor of the corpus callosum rostrally in nucleus accumbens. AADC-ir neurons (mean area, 46 µm^2) did not differ in size from dopamine-ir neurons (48µm^2) but both were significantly smaller than TH-ir neurons ($p < 0.05$). Their cytoplasm was uniformly immunoreactive and contained a pale nucleus with an indented nuclear membrane. Both AADC-ir and dopamine-ir neurons had oval to round perikarya of giving rise to one or two smooth, primary dendrites (Fig lE&F).

3.2 Tissue Volume and Total Neuronal Number

The volume of the striatum as a whole, or nucleus accumbens alone, did not differ between the dopamine-depleted and the intact hemisphere or hemispheres from control rats. However, the total number of neurons in the CPu and accumbal core regions on the side of the lesion was reduced by approximately 15% compared to the intact hemisphere ($p<0.01$).

4. DISCUSSION

The appearance and distribution of striatal neurons immunoreactive for catecholamines or their synthetic enzymes were investigated following 6-OHDA lesions of midbrain dopaminergic neurons. In the lesioned hemisphere, the distribution of the labelled cells suggests that some are capable of synthesising dopamine, but not DOPA, and others are capable of making DOPA, but not dopamine. Moreover, the greater the dopamine loss in the striatum, the fewer TH-ir neurons found, suggesting their appearance is an adaptation to the loss of the dopaminergic innervation.

The TH-ir cells described here resemble typical spiny projection neurons.[28,29,39] Although they are quite small, their perikarya fall within the size range of spiny neurons in the core of nucleus accumbens.[8,28] Their relative paucity of spines could be a direct result of dopamine deafferentation as there is significant spine loss for accumbal cells following 6-OHDA lesions.[30] Finally, the TH-ir neurons in these 6-OHDA treated rats generally have round, smooth nuclear envelopes, a feature consistently used as a marker for striatal projection cells.[6,31,39] GABA-containing striatal interneurons vary in size and have indented nuclei [5,6,17] and AADC- and dopamine-ir neurons seen in this study and by Mura et al.,[32] were aspiny, with indented nuclear envelopes. Although very small (mean cross sectional area 39µm^2/46µm^2, (32 and present results), they resemble, in both size and shape, calretinin interneurons.[2]

The presence of neurons that contain TH, or make dopamine, has implications for the recovery of striatal function after dopamine loss. However, the numbers, sizes and types of neurons seem to vary considerably and raises questions of the criteria used in other studies to identify immunoreactive profiles as neurons. It is clear that some dopaminergic fibers are spared after most lesions [14,49] and that both spared and degenerating fibres have numerous swellings along their length, some of which are greatly enlarged. These are unlikely to be the large 'C' type boutons which are reportedly not catecholaminergic[15,40]. Rather these swollen profiles are signs of degeneration, since axonal membrane functions are damaged early after 6-OHDA lesions and the ultrastructure of this tissue is similar to the changes seen with surgical axotomy.[21] The large size of these swollen varicosities makes them difficult to distinguish from catecholaminergic neurons, which have been reported to be very small.[4,32] This is particularly relevant when one considers that the swellings would be immunoreactive for TH, AADC and dopamine or its transporter.

4.1 'De Novo' Appearance of TH-it Neurons

Since the detection of catecholaminergic cells follows rather than precedes a dopamine-depleting lesion, presumably there has been a fundamental change in gene expression. It is not clear whether activation of TH or AADC genes signals a transient alteration in synthesis, as occurs in the developing striatum[22] or a more permanent change in neuronal phenotype. Results here show that a progressively greater dopaminergic denervation in the striatum is associated with a decline in the numbers of neurons expressing TH, suggesting their appearance is related to compensatory adaptations to the loss of dopaminergic innervation. Acute regulation of TH biosynthesis is related to changes in cell activity[55] and differential increases in activity may lead to TH activation in selected cells which might explain why a 6-OHDA lesion leads to raised Fos expression,[25] particularly in regions where catecholaminergic neurons are found. Nevertheless, the biosynthesis of TH is upregulated in surviving neurons following 6-OHDA lesions[53] and we know that TH enzyme activity is subject to feedback regulation, such that increased tissue concentrations of dopamine and related chemicals inhibit TH and vice versa.[55] Therefore, regional and/or lesion-induced differences in dopamine efflux may underlie the *de novo* appearance of neurons that express TH.

Why nuclei of TH-ir neurons are dark following 6-OHDA lesions remains a mystery. One possible explanation for the dense appearance of cell nuclei following such lesions could lie in the differential viability of striatal cells after dopamine deafferentation. Certainly, the vacuolisation seen in the cytoplasm of the TH-ir neurons could signal an early step in cell death.[35] There is evidence that the blood-brain barrier becomes more permeable after intracerebral injections of 6-OHDA[9] and this could result in alterations in the osmotic balance which might adversely affect striatal neurons, presumably causing cytoplasmic disruptions. However, we cannot rule out the possibility that poor fixation may itself have played a role in some cases. Nevertheless, since the number of neurons in the core and CPu decline significantly in the dopamine-depleted hemisphere three weeks following complete unilateral lesion, some form of cell death may be taking place. The fact that there appears to be no loss of overall volume accompanying the cell loss lends support the notion that there has been tissue swelling.

While 6-OHDA is widely recognised as being neurotoxic, dopamine itself has been shown to be cytotoxic, inducing apoptosis in human neuroblastoma cells.[38] It is likely that extracellular levels of dopamine are increased following 6-OHDA lesions due to the marked increase in spontaneous firing rates of striatal cells and alterations in the biochemical balance among other signal transmitters.[52] Striatal neurons are known to lose spines after such lesions and dendrites show signs of atrophy.[19,30] Accumulations of dense nuclear chromatin seem here could be early signs of cell death.[35] Other possible explanations lie with the neuronal phenotype of the TH-ir cells or in the intracellular changes in metabolism of protein production that may occur when a cell has been partially deafferented or lost its target, as occurs with nigral dopaminergic neurons when their striatal targets are destroyed at birth.[26]

4.2 Significance

The presence of neurons capable of synthesizing dopamine but not L-DOPA, or capable of making L-DOPA but not dopamine is not in itself surprising since the phenomenon has been described for other cells in a variety of species.[20,24,47] There are also numerous reports of TH being transiently expressed in forebrain neurons during development.[3,33,46] More important is whether any of these neurons actually make dopamine. The enzyme AADC is involved in the decarboxylation of both serotonin and dopamine synthetic pathways and its ability to convert L-DOPA to dopamine is the basis of levodopa treatment for Parkinson's Disease. However, it could also contribute to the fluctuations in levodopa efficacy, especially via deafferentation-induced expression of AADC by striatal cells. In contrast, cells that contain TH but are unable to decarboxylate L-DOPA are puzzling. The conversion of tyrosine to L-DOPA is dependent on the phosphorylated state of TH.[55] Although there is no direct evidence that synaptic activity regulates TH expression centrally, indirect evidence supports this possibility. Chronic stress for example, activates locus coeruleus neurons and increases TH levels.[1,54] Activity-induced elevation in TH also increase DOPA synthesis (55) and raised levels could conceivably be linked to its release. Growth factors and catecholamines trigger the *de novo* induction of TH in cultured embryonic cells, and rats infused with growth factors seem to synthesize more dopamine and show greater recovery of function after dopamine-depleting lesions.[12,13] Moreover, L-DOPA potentiates the outgrowth of neurites in cultures exposed to nerve growth factor, a response that does not require the conversion of L-DOPA to dopamine.[27] These data suggest that the novel production of TH, and presumably the synthesis of DOPA, could be important for repair and recovery in the CNS[18] and therefore play an active role in striatal recovery following 6-OHDA lesions.

5. ACKNOWLEDGEMENTS

The authors thank L. Guiver for technical assistance and the Media Services Department at the Royal College of Surgeons in Ireland for help with the photographic figures. We are also grateful to Dr. D.J. Clarke for advice on 6-OHDA lesions and Dr. B.L. Roberts for critical comments on the manuscript. The work was supported by a Health Research Board grant (140/97: GEM), and a British Council/Health Research Board Travel Grant (ST & GEM).

6. REFERENCES

1. Abercrombie, E.D., and Jacobs, B.L. 1987, Single unit response of noradrenergic neurons in the locus coeruleus of freely moving cats. II. Adaptation to chronically presented stressful stimuli. *J. Neurosci.* 7:2844.
2. Bennett, B.D. and Bolam, J.P. 1993, Characterization of calretinin-immunoreactive structures in the striatum of the rat. *Brain Res.* 609:137.
3. Berger, B., Verney, C., Gaspar, P. and Febvret, A. 1985, Transient expression of tyrosine hydroxylase immunoreactivity in some neurons of the rat neocortex during postnatal development. *Dev. Brain Res.* 23:141.
4. Betarbet, R., Turner, R., Chockkan, V., Delong M.R., Allers K.A., Walters J., Levey A.I. and Greenamyre, F.T. 1997, Dopaminergic neurons intrinsic to the primate striatum. *J. Neurosci.* 17:6761.
5. Bolam, J.P., Clarke, D.J., Smith, A.D. and Somogyi, P. 1983, A type of aspiny neuron in the rat neostriatum accumulates [3H]-gamma-aminobutyric acid: Combination of Golgi-staining, autoradiography, and electron microscopy. *J. Comp. Neurol.* 213:121.
6. Bolam, J.P., Powell, J.F., Wu, J.-Y. and Smith, A.D. 1985, Glutamate decarboxylase-immunoreactive structures in the rat neostriatum: a correlated light and electron microscopic study including a combination of Golgi impregnation with immunocytochemistry. *J. Comp. Neurol.* 237:1.
7. Broening, H.W., Cunfeng, P.U. and Vorhees, C.V. 1997, Methamphetamine selectively damages dopaminergic innervation to the nucleus accumbens core while sparing the shell. *Synapse* 27:153.
8. Chronister, R.B., Sikes, R.W., Trow, T.W. and De France, J.F. 1981, The organization of nucleus accumbens, in: *The Neurobiology of the Nucleus Accumbens,* R.B. Chronister, R.B. and J.F. De France, eds., The Haer Institute for electrophysiological research, Brunswick, Maine, pp. 147-172.
9. Cooper, P.H., Novin, D. and Butcher, L.L. 1982, Intra-cerebral 6-hydroxydopamine produced extensive damage to the blood-brain barrier in rats. *Neurosci. Lett.* 30:13.
10. De Souza, I.A.J. and Meredith, G. 1999, NMDA receptor blockade attenuates the haloperidol induction of Fos immunoreactivity in the dorsal but not the ventral striatum. *Synapse,* in press.
11. Deutch, A.Y. and Cameron, D.S. 1992, Pharmacological characterization of dopamine systems in the nucleus accumbens core and shell. *Neuroscience,* 46:49.
12. Du, X. and Iacovitti, L. 1995, Synergy between growth factors and transmitters required for catecholamine differentiation in brain neurons. *J. Neurosci.* 15:5420.
13. Du X., Stull, N.D. and Iacovitti, L. 1994, Novel expression of the tyrosine hydroxylase gene requires both acidic fibroblast growth factor and an activator. *J. Neurosci.* 14:7688.
14. Freund, T.F., Powell, J.F. and Smith, A.D. 1984, Tyrosine hydroxylase-immunoreactive boutons in synaptic contact with identified striatonigral neurons with particular reference to dendritic spines. *Neuroscience* 13:1189.
15. Gerfen, C.R., Herkenham, M. & Thibault, J. 1987, The neostriatal mosaic: 11. Patch- and matrix-directed mesostriatal dopaminergic and non-dopaminergic systems. *J. Neurosci.* 7:3915.
16. Hotchkiss, A.J. and Gibb, J.W. 1980, Long-term effects of multiple doses of methamphetamine on tryptophan hydroxylase and tyrosine hydroxylase activity in rat brain. *J. Pharmacol. Exp. Ther.* 214:257.
17. Hussain, Z., Johnson, L.R. and Totterdell, S. 1996, A light and electron microscopic study of NADPH-diaphorase-, calretinin- and parvalbumin-containing neurons in the rat nucleus accumbens. *J. Chem. Neuroanat.* 10:19.
18. Iacovitti L., Job T.H., Park I.H. and Bunge R.P. 1987, Expression of tyrosine hydroxylase in neurons of cultured cerebral cortex: evidence for phenotypic plasticity in neurons of the CNS. *J. Neurosci.* 7:1264.
19. Ingham, C.A., Hood, S.H. and Arbuthnott, G.W. 1989 Spine density on neostriatal neurones changes with 6-hydroxydopamine lesions and with age. *Brain Res.* 503:334.
20. Jaeger, C.B., Teitelman, G., Joh, T.H., Albert, V.R., Park, D.H. and Reis, D.J. 1983 Some neurons of the rat central nervous system contain aromatic L-amino acid decarboxylase but not monoamines. *Science* 219:1233.
21. Jonsson, G. 1980, Chemical neurotoxins as denervation tools in neurobiology. *Ann. Rev. Neurosci.* 3:169.
22. Kalsbeek, A., Voorn, P. and Buijs, R.M. 1991, Development of dopamine-containing systems in the CNS, in: *Handbook of Chemical Neuroanatomy: Vol 10: Ontogeny of Transmitters and Peptides in the CNS.* A. Bjorklund, T. Hokfelt and M. Tohyama, eds., Elsevier, Amsterdam, pp. 63-112.

23. Kish, S.J., Shannak, K. and Hornykiewicz, O. 1988, Uneven pattern of dopamine loss in the striatum of patients with idiopathic Parkinson's disease: Pathophysiologic and clinical implications. *New Engl. J. Med.* 318:876.
24. Kitahama, K., Denoyer, M., Raynaud, B., Borri-Voltattorni, C., Weber, M. and Jouvet, M. 1988, Immunohistochemistry of aromatic L-amino acid decarboxylase in the cat forebrain. *J. Comp. Neurol.* 270:337.
25. Labandeira-Garcia, J.L., Rozas, G., Lopez-Martin, E., Liste, I. and Guerra, M.J. 1996 Time course of striatal changes induced by 6-hydroxydopamine lesion of the nigrostriatal pathway, as studied by combined evaluation of rotational behaviour and striatal Fos expression. *Exp. Brain Res.* 108:69.
26. Marti. M.J., James, C.J., Oo, T.F., Kelly, W.J. and Burke, R.E. 1997, Early developmental destruction of terminals in the striatal target induces apoptosis in dopamine neurons of the substantia nigra. *J. Neurosci.* 17:2030.
27. Mena, M.A., Davila, V., Bogaluvsky, J. and Sulzer, D. 1998 A synergistic neurotrophic response to ldihydroxyphenylalanine and nerve growth factor. *Mol. Pharmacol.* 54:678.
28. Meredith, G.E., Agolia, R., Arts, M.P.M., Groenewegen, H.J. and Zahm, D.S. 1992, Morphological differences between projection neurons of the core and shell in the nucleus accumbens of the rat. *Neuroscience* 50:149.
29. Meredith, G.E., Wouterlood, F.G. and Pattiselanno, A. 1990, Hippocampal fibers make synaptic contacts with glutamate decarboxylase-immunoreactive neurons in the rat nucleus accumbens. *Brain Res.* 513:329.
30. Meredith, G.E., Ypma, P. and D.S. Zahm 1995 Effects of dopamine depletion on the morphology of medium spiny neurons in the shell and core of the rat nucleus accumbens. *J. Neurosci.* 15:3808.
31. Mugnaini, E. and Oertel, W.H. 1985, An atlas of the distribution of GABAergic neurons and terminals in the rat CNS as revealed by GAD immunohistochemistry, in: *Handbook of Chemical Neuroanatomy, Vol. 4, GABA and neuropeptides in the CNS*, A. Bjorklund and T. Hokfelt, eds., Elsevier, Amsterdam, pp. 75-105.
32. Mura, A., Jackson, D., Manley, M., Young, S. and Groves, P. 1995, Aromatic L-amino acid decarboxylase immunoreactive cells in the rat striatum: A possible site for the conversion of exogenous L-DOPA to dopamine. *Brain Res.* 704:51.
33. Nagatsu, I., Komori, K., Takeuchi, T., Sakai, M., Yamada, K. and Karasawa, N. 1990, Transient tyrosine hydroxylase-immunoreactive neurons in the region of the anterior olfactory nucleus of pre- and postnatal mice do not contain dopamine. *Brain Res.* 511:55.
34. O'Donnell, P. and Grace, A. 1993, Dopaminergic modulation of dye coupling between neurons in the core and shell regions of the nucleus accumbens. *J. Neurosci.* 13:257.
35. Oppenheimer, R.W. 1991, Cell death during development of the nervous system. *Ann. Rev. Neurosci.* 14:453.
36. Paxinos, G. and Watson, C. 1986, *The Rat Brain in Stereotaxic Coordinates*, Academic Press, Sydney.
37. Pickel, V.M., Johnson, E., Carson, M. & Chan, J. 1992, Ultrastructure of spared dopamine terminals in caudate-putamen nuclei of adult rats neonatally treated with intranigral 6-hydroxydopamine. *Dev. Brain Res.* 70:75.
38. Simantov, R., Blinder, E., Ratovitski, T., Tauber, M., Gabbay, M. and Porat, S. 1996, Dopamine-induced apoptosis in human neuronal cells: Inhibition by nuclei acids antisense to the dopamine transporter. *Neuroscience* 74:39.
39. Somogyi, P. and Smith, A.D. 1979 Projection of neostriatal spiny neurons to the substantia nigra. Application of combined Golgi-staining and horseradish peroxidase transport procedure at both light and electron microscopic levels. *Brain Res.* 178:3.
40. Swanson, L.W. 1982 The projections of the ventral tegmental area and adjacent regions: A combined fluorescent retrograde tracer and immunofluorescence study in the rat. *Brain Res. Bull.* 9:321.
41. Tashiro, Y., Kaneko, T., Sugimoto, T., Nagatsu, I., Kikuchi, Hand Mizuno, N. 1989a, Striatal neurons with aromatic L-amino acid decarboxylase-like immunoreactivity in the rat. *Neurosci. Letts.* 100:29.
42. Tashiro, Y., Sugimoto, T., Hattori, T., Vemura, Y., Nagatsu, I., Kikuchi, H. and Mizuno, N. 1989b, Tyrosine hydroxylase-like immunoreactive neurons in the striatum of the rat. *Neurosci. Letts.* 97:6.
43. Tashiro, Y., Kaneko, T., Nagatsu, I., Kikuchi, H. and Mizuno, N. 1990, Increase of tyrosine hydroxylase-like immunoreactive neurons in the nucleus accumbens and the olfactory bulb in the rat with the lesion in the ventral tegmental area of the midbrain. *Brain Res.* 531:159.
44. Totterdell, S. and Smith, A.D. 1989, Convergence of hippocampal and dopaminergic input onto identified neurons in the nucleus accumbens of the rat. *J. Chem. Neuroanat.* 2:285.

45. Ungerstedt, U. 1968, 6-hydroxydopamine-induced degeneration of central monoamine neurons. *Eur. J. Pharmacol.* 5:107.
46. Verney, C., Gaspar, P., Febvret, A. and Berger, B. 1988 Transient expression of tyrosine hydroxylase-like immunoreactive neurons contain somatostatin and substance P in the developing amygdala and bed nucleus of the striata terminalis of the rat. *Dev. Brain Res.* 42:45.
47. Vincent, S.R. and Hope, B.T. 1990 Tyrosine hydroxylase containing neurons lacking aromatic amino acid decarboxylase in the hamster brain. *J. Comp. Neurol.* 295:290.
48. Voorn, P., Jorritsma-Byham, B., Van Dijk, C. and Buijs, R,M. 1986, The dopaminergic innervation of the ventral striatum in the rat: a light- and electron-microscopical study with antibodies against dopamine. *J. Comp. Neurol.* 251:84.
49. Zahm, D.S. 1991, Compartments in rat dorsal and ventral striatum revealed following injection of 6hydroxydopamine into the ventral mesencephalon. *Brain Res.* 552:164.
50. Zahm, D.S. and Brog, J.S. 1992 On the significance of subterritories in the "accumbens" part of the rat ventral striatum. *Neuroscience* 50:751.
51. Zahm, D.S. & Heimer, L. 1993, Specificity in the efferent projections of the nucleus accumbens in the rat: comparison of the rostral pole projection patterns with those of the core and shell. *J. Comp. Neurol.* 327:220.
52. Zigmond, M.J., Abercrombie, E., Berger, T.W., Grace, A.A. and Stricker, E.M. 1990, Compensations after lesions of central dopaminergic neurons: Some clinical and basic implications. *Trends Neurosci.* 13:290.
53. Zigmond, M.J., Acheson, A.L., Stachowiak, M.K. and Stricker, E.M. 1984, Neurochemical compensation after nigrostriatal bundle injury in an animal model of preclinical parkinsonism. *Arch. Neurol.* 4:856.
54. Zigmond, R.E., Schon, F. and Iversen, L.L. 1974, Increased tyrosine hydroxylase activity in the locus coeruleus of rat brain stem after reserpine treatment and cold stress. *Brain Res.* 70:547.
55. Zigmond, R.E., Schwarzschild, M.A. and Rittenhouse, A.R. 1989, Acute regulation of tyrosine hydroxylase by nerve activity and by neurotransmitters via phosphorylation. *Ann Rev. Neurosci.* 12:415.

ACTIONS OF DOPAMINE ON THE RAT STRIATAL CHOLINERGIC INTERNEURONS

Toshihiko Aosaki*

1. INTRODUCTION

Based on the clinical observations it was classically assumed in the 1960s that there is a mutual antagonism between dopamine (DA) and acetylcholine (ACh) within the striatum (Barbeau, 1962; Lehmann and Langer, 1983). However, with the development of dopaminergic agents and *in vivo* microdialysis technique, it is now hypothesised that activation of D_1 receptor stimulates and activation of D_2 receptor reduces the release of ACh (Stoof et al., 1992; Consolo et al., 1993; Di Chiara et al., 1994). But, *in situ* hybridisation studies provided unexpected evidence that only a small number of the cholinergic neurons contain D_1 receptor mRNAs (Le Moine et al., 1991). To compromise the apparent controversy, several lines of evidence suggested that ACh release by D_1 receptor activation could be indirect through two alternative pathways: excitation of the striatonigral neurons containing substance P (Ajima et al., 1990; Anderson et al., 1994; Di Chiara et al., 1994) and stimulation of D_1 receptors outside the striatum (Consolo et al., 1996a, b; Damsma et al., 1991; Abercrombie and DeBoer, 1997). However, it was found recently that the majority of the neostriatal cholinergic neurons of the rat contain D_5/D_{1b} receptor mRNAs rather than D_{1a} mRNAs while all the cells express D_2 mRNAs (Bergson et al., 1995; Yan et al., 1997). Therefore, direct activation of D_5/D_{1b} DA receptors on the striatal cholinergic neurons might contribute to the evoked ACh release. This hypothesis could be tested by direct examination of the effects of D_1-like receptor activation on the membrane excitability of rat striatal giant aspiny neurons. This paper demonstrates that dopamine depolarises the striatal cholinergic cells by a D_1-mediated modulation of resting K^+ conductance and a non-selective cation channel via an adenylyl cyclase-dependent pathway.

---

* T. Aosaki, Neural Circuits Dynamics Research Group, Tokyo Metropolitan Insitute of Gerontology, 35-2, Sakae, Itabashi, Tokyo 173-0015, Japan.

2. METHODS

Whole-cell patch clamp recordings were performed in a brain slice preparation (200 μm thick) to study the actions of DA on the rat striatal giant aspiny neurons (Wistar, 15-22 days old). The solution comprised (mM): NaCl 124, KCl 3, $CaCl_2$ 2.4, $MgCl_2$ 1.2, NaH_2PO_4 1, $NaHCO_3$ 26, glucose 10, oxygenated with 95% O_2 and 5% CO_2. Whole-cell recordings were made with glass pipettes (3-4 MΩ), which contained (mM): K-methylsulfate 120, KCl 6, NaCl 6, EGTA 0.6, HEPES 121.5, $MgCl_2$ 2, ATP 4, GTP 0.3, biocytin 20 (pH 7.2). All neurons examined were filled with 0.5% biocytin and some of them were further histologically characterized after recording. Classification of the neurons obtained was based on the criteria described by Kawaguchi (1993). Briefly, the cells with a resting potential about -60 mV and those displayed long-lasting afterhyperpolarizations and strong time-dependent hyperpolarizing rectification were classified as long-lasting afterhyperpolarization cells (LA cells, cholinergic neurons).

DA and other drugs such as SCH23390, SKF38393, tetrodotoxin (TTX), and forskolin were applied by changing the solution that superfused the slice to one that contained the drug. Time taken for the drug to reach the neurons were usually about less than 1 min.

3. RESULTS

3.1 Dopamine Depolarized the LA Cells by Activation of D_1-like Receptor in a Dose-Dependent Manner

A total of 186 rat neostriatal neurons were identified as large aspiny cholinergic interneurons (long-lasting afterhyperpolarization, LA cells). As shown in Figure 1, DA depolarized the spontaneously firing cholinergic neuron and accelerated the spike frequency. Application of DA (1-100 μM) elicited a slow transient depolarization (9.4 ± 5.0 mV) in all cells

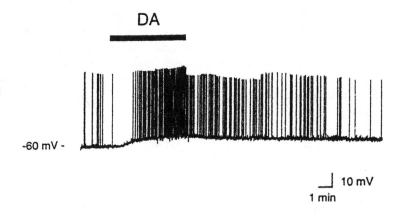

Figure 1. Effects of DA on neostriatal large aspiny neurons. DA (50 μM), bath-applied at 3 ml/min, caused a transient depolarization and a train of action potentials. Na-metabisulfite (50 μM) was present to prevent oxidative degradation of DA.

ACTIONS OF DOPAMINE ON THE RAT STRIATAL CHOLINERGIC INTERNEURONS

tested (8/8 cells). The depolarization was completely blocked by haloperidol (1 μM), a non-selective DA receptor antagonist.

The effects of DA were further analyzed under whole-cell voltage clamp (holding potential -60 mV) with TTX (0.5 μm) in the external solution. DA always induced an inward current (43 cells) in a dose-dependent manner and it was suppressed by application of SCH23390 (10 μM), an antagonist of the D_1-like DA receptor. Interestingly, application of SCH23390 itself caused an outward shift of the holding current in some cells. This indicates that they might be continuously slightly depolarized by D_1-like receptor activation, probably by spontaneously released DA from dopaminergic terminals.

3.2 Two Separate Conductances in the DA Responses

The current-voltage relationships of the DA-induced current suggested an involvement of at least two separate conductances in the DA response. In ten of 23 cells, the reversal potential of the DA response was estimated as -93.7 mV, approximately equal to the estimated K^+ equilibrium potential (E_K, -100.6 mV), with a slight reduction in membrane conductance from 7.4 to 7.0 nS. In contrast, ten of 23 cells showed an increase in membrane conductance

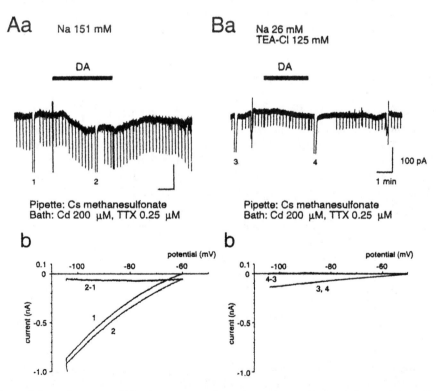

Figure 2. Lowering external Na^+ ion concentration reduces the amplitude of the DA-induced inward current. Inward currents caused by DA (100 μM) in 26 mM (Aa) and 151 mM (Ba) $[Na^+]_o$ respectively. Voltage ramps (-125 to -60 mV, 9.3 mV/sec) were applied before (1,3) and during (2,4) the inward shift of the holding current. Ab and Bb, two superimposed I-V plots, one before (1, 3) and the other during (2, 4) the peak DA response. 2-1 and 4-3, I-V plots of the net DA-induced currents.

from 5.9 to 8.3 nS. These cells showed a linear current-voltage relation with a positive slope and a reversal potential of -42.1 mV In the remaining 3 cells, the I-V relations did not cross within the voltage range tested (-120 to -40 mV) without any change in membrane conductance.

To maximize the contribution of a K^+ current component to the DA response, cells were tested in perfusion medium containing low Na^+ (27 mM) and 0.5 μM TTX. As expected, the reversal potentials of the DA-induced current were shifted to more positive values (from -88.5 to -45.0 mV) with elevated external K^+ concentrations (from 3 to 16 mM), indicating that K^+ channels were closed during flow of the DA-mediated inward current.

Next, to characterize the role of the cation conductances, 2 mM of barium (to block various K^+ channels), 200 μM of Cd^{2+} and 5 μM of nifedipine (to block voltage-sensitive Ca^{2+} channels) and 2 mM of Cs^+ (to block hyperpolarization-induced cation channel) were included in the external solution. DA (100 μM) induced an inward current with an increase in membrane conductance from 4.7 to 5.1 nS even in this condition (Fig. 2). The I-V curves obtained before and during DA exposure crossed or almost crossed at potentials around -40 mV in 10 cells. The DA-induced inward current in 26 mM Na^+ was -10.7 pA (n=5), while that under high Na^+ conditions was -34.7 pA (n=15). The Na^+ permeability was also tested with a Cs^+-containing patch pipette by lowering the external Na^+ concentration. The DA-induced inward current in 26 mM Na^+ was -10.7 pA (n=5), while that under high Na^+ conditions was -34.7 pA (n=15). These results suggest that a TTX-insensitive Na^+-permeable conductance may contribute partly to the inward current.

3.3 Effects of (±)-SKY38393 on Striatal Cholinergic Cells

(±)-SKF38393 (10 μM) depolarized and elicited action potentials in the current clamp mode. In the whole cell clamp configuration (±)-SKF38393 (1-50 μM) evoked an inward current in a dose-dependent manner (Fig. 3A). The current-voltage relations were also found similar to those with DA. The reversal potential of the net SKF38393-induced current of ten of 14 cells was about -90 mV with a negative slope conductance, while that of the remaining 4 cells did not cross the voltage axis within the voltage range tested (-120 to -40 mV), suggesting that, as in the DA case, SKF38393 evokes an inward current by two separate ionic mechanisms: blockade of K^+ conductance and opening of non-selective cation conductance.

Some cells were also analyzed for the effects of a mixture of DA (100 μM) plus sulpiride (10 μM), a D_2-like receptor antagonist, but the results were similar to those obtained with (±)-SKF38393.

3.4 Effects of SQ22536, an Adenylyl Cyclase Inhibitor, on D_5/D_1 DA Receptor-Mediated Inward Currents

D_1-class DA receptors are known to positively couple with G proteins and an adenylyl cyclase-cAMP cascade. If adenylyl cyclase mediates the effects of (±)-SKF38393, its inhibition should reduce the actions of (±)-SKF38393. This was tested by preincubation of neostriatal slices in an oxygenated saline containing 9-(tetrahydro-2-furyl) adenine (SQ22536, 300 μM) for 30 min to 6 hr, which is known to decrease the activity of adenylyl cyclase. Slices taken from the same animals were also incubated in the saline in the same manner as a control

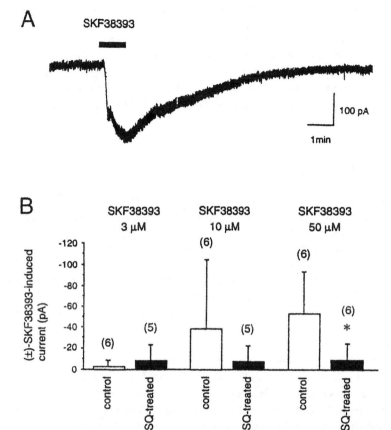

Figure 3. Activation of D1-like receptor evokes an inward current via an adenylyl cyclase pathway. A, Effect of (±)-SKF38393 (1 ìM). B, Pretreatment with SQ22536, an inhibitor of adenylyl cyclase, suppressed the (±)-SKF38393-induced currents. Open columns, the currents induced by (±)-SKF38393 in control slices. Filled columns, those in SQ22536-treated slices. SDs are shown with bars. Numbers in parentheses refer to numbers of tested cells. Comparisons were made with the Student's t test against the control group of neurons tested in the saline solution (* $p<0.05$).

and recordings were made alternately. As summarized in Figure 3B, the (±)-SKF38393-induced currents were significantly suppressed by pretreatment with SQ22536. The amplitudes in SQ22536-treated cells were -9.6 (n=5), -8.9 (n=5) and -10.9 pA (n=6) at 3, 10 and 50 μM, respectively, while those in control cells were -3.1 (n=6), -38.9 (n=6) and -53.1 pA (n=6, $p<0.05$) at 3, 10 and 50 ìM, respectively.

Forskolin, a lipophilic adenylyl cyclase activator, was also tested at 10 μM and found to evoke a large inward current in all cells tested (-98.1 pA, n=13). In contrast, dideoxyforskolin, an inactive enantiomer of the forskolin, caused no response at 10 μM (n=3). These results indicate that activation of adenylyl cyclase is required to elicit the actions of (±)-SKF38393.

Previous studies have demonstrated that a rise in intracellular Ca^{2+} mediates the effects of D_1-like receptor stimulation in oocytes from *Xenopus laevis* with messenger RNA isolated from rat striatum (Mahan et al., 1990), chromaffin cells (Artalejo et al., 1990) and in a subset of rat neostriatal neurons (Surmeier et al., 1995). Therefore, 1,2-bis(2-aminophenoxy)ethane-N,N,N',N'-tetraacetic acid (BAPTA), an agent that is known to chelate cytosolic Ca^{2+} ions, was included in patch pipettes at 10 or 20 mM and the actions of (±)-SKF38393 (10 µM) were compared with those of forskolin (10 µM). As a result, no signs of suppression of the (±)-SKF38393-induced currents evoked by (±)-SKF38393 in all cells tested, suggesting that mobilization of cytosolic Ca^{2+} ions is not associated with the observed actions of (±)-SKF38393 in striatal cholinergic neurons.

4. DISCUSSION

It has been shown that information processing in the basal ganglia is essentially in parallel in nature and that the striatal medium spiny projection neurons do not communicate with each other (Jaeger et al., 1994). Then, the activity of the medium spiny neurons should be modulated somehow by extrinsic as well as intrinsic inputs in order for the striatum to form appropriate activity patterns and relay them to subsequent basal ganglia output areas. Interneurons may act as an intrinsic modulator of the striatal projection neurons and among them cholinergic neurons have been considered particularly important since it was noticed in the early 1960s that antimuscarinic drugs ameliorate the parkinsonian symptoms.

This study demonstrates that DA induces a membrane depolarization in striatal cholinergic interneurons by activation of postsynaptic D_1-like DA receptors. This is accomplished by a D_1-mediated suppression of resting K^+ conductance and an opening of a non-selective cation channel. These results fit well with the idea that D_1-class receptors on the neostriatal cholinergic interneurons might be responsible for facilitation of ACh release.

Recent microdialysis studies have shown that DA increases net ACh release in the striatum and that this might be due to activation of D_1-like receptors (Ajima et al., 1990; Bertorelli and Consolo, 1990; Consolo et al., 1992; Zocchi and Pert, 1993). However, D_1 stimulation could take place in extrastriatal pathways such as in the parafascicular nucleus of the thalamus, cerebral cortex and substantia nigra pars reticulata, and thereby cause the release of glutamate in the striatum. This glutamate might in turn activate striatal cholinergic neurons (Consolo et al., 1996a, b; Damsma et al., 1991; Abercrombie and DeBoer, 1997). In addition, local release of substance P from the terminals of D_1-like-receptor-containing striatal medium spiny neurons would excite the cholinergic neurons (Arenas et al., 1991; Anderson et al., 1994; Aosaki and Kawaguchi, 1994; Guevara Guzman et al., 1993; Steinberg et al., 1995; Khan et al., 1996). Direct activation of the D_1-like receptors on the cholinergic neurons is also possible (Di Chiara et al., 1994). However, *in vitro* studies using striatal slice preparations have so far provided conflicting results as for the latter mechanism (Stoof and Kebabian, 1982; Gorell and Czarnecki, 1986; Scatton, 1982; Consolo et al., 1987; Dolezal et al., 1992; Tedford et al., 1992). In this regard, measurement of ACh efflux from a dissociated striatal neuronal preparation would be more advantageous to overcome these discrepancies. Using the dissociated cholinergic neuron preparation, Login and colleagues (1995, 1996) found that (±)-SKF38393 (50 µM) indeed increased the rate of release of ACh and that this release was blocked by SCH23390. Our study (Aosaki et al., 1998) therefore provided the first direct electrophysiological evidence that ACh might be secreted via direct activation of D_1-like receptors on cholinergic cells. These D_1-like receptors must be D_5, taking it into account that the striatal cholinergic neurons contain primarily D_5 DA receptors rather than D_1 receptors for the D_1-like receptors (Bergson et al., 1995; Yan et al., 1997).

Recently, it has been demonstrated that most DA receptors are located outside synaptic contacts formed by dopaminergic terminals (Hersch et al., 1995; Bergson et al., 1995; Caillé et al., 1996). DA would leak out from synaptic clefs and diffuse up to 12 μm from release sites reaching a homogeneous DA concentration of 0.2 to 1 μ

M in the extrasynaptic extracellular space before elimination by reuptake (Garris et al., 1994; Gonon, 1997). This concentration range would be high enough to depolarise striatal cholinergic neurons, which could elicit spike discharge upon small excitatory postsynaptic potentials (Wilson et al., 1990). Striatal cholinergic neurons receive a major input from the parafascicular nucleus of the thalamus. Therefore, when DA activates them, they might respond faithfully to the firing pattern of the thalamostriatal neurons.

5. REFERENCES

Abercrombie ED, DeBoer P (1997) Substantia nigra D_1 receptors and stimulation of striatal cholinergic interneurons by dopamine: a proposed circuit mechanism. J Neurosci 17: 8498-8505.

Ajima A, Yamaguchi T, Kato T (1990) Modulation of acetylcholine release by D-1, D-2 dopamine receptors in rat striatum under freely moving conditions. Brain Res 518: 193-198.

Anderson JJ, Kuo S, Chase TN, Engber TM (1994) Dopamine D_1 receptor-stimulated release of acetylcholine in rat striatum is mediated indirectly by activation of striatal neurokinin-1 receptors. J Pharmacol Exp Ther 269: 1144-1151.

Aosaki T, Kawaguchi Y (1996) Actions of substance P on rat neostriatal neurons *in vitro*. J Neurosci 16: 5141-5153.

Aosaki T, Kiuchi K, Kawaguchi Y (1998) Dopamine D_1-like receptor activation excites rat striatal large aspiny neurons *in vitro*. J Neurosci 16: 5141-5153.

Arenas E, Alberch J, Perez-Navarro E, Solsona C, Marsal J (1991) Neurokinin receptors differentially mediate endogenous acetylcholine release evoked by tachykinins in the neostriatum. J Neurosci 11: 2332-2338.

Artalejo CR, Ariano MA, Perlman RL, Fox AP (1990) Activation of facilitation calcium channels in chromaffn cells by D_1 dopamine receptors through a cAMP/protein kinase A dependent mechanism. Nature 348: 239-242.

Barbeau A (1962) The pathogenesis of Parkinson's disease: a new hypothesis. Can Med Assoc J 87: 802-807.

Bergson C, Mrzljak L, Smiley JF, Pappy M, Levenson R, Goldman-Rakic PS (1995) Regional, cellular, and subcellular variations in the distribution of D_1 and D_5 dopamine receptors in primate brain. J Neurosci 15: 7821-7836.

Bertorelli R, Consolo S (1990) D_1 and D_2 dopaminergic regulation of acetylcholine release from striata of freely moving rats. J Neurochem 54: 2145-2148.

Caillé I, Dumartin B, Bloch B (1996) Ultrastructural localization of D_1 dopamine receptor immunoreactivity in rat striatonigral neurons and its relation with dopaminergic innervation. Brain Res 730: 17-31.

Consolo S, Wu CF, Fusi R (1987) D-1 receptor-linked mechanism modulates cholinergic neurotransmission in rat striatum. J Pharmacol Exp Ther 242: 300-305.

Consolo S, Girotti P, Russi G, Di Chiara G (1992) Endogenous dopamine facilitates striatal in vivo acetylcholine release by acting on D_1 receptors localized in the striatum. J Neurochem 59: 1555-1557.

Consolo S, Girotti P, Zembelli M, Russi G, Benzi M, Bertorelli R (1993) D_1 and D_2 dopamine receptors and the regulation of striatal acetylcholine release in vivo. In: Progress in Brain Research Vol. 98 (Cuello AC, ed), pp 201-207. Amsterdam: Elsevier.

Consolo S, Baldi G, Giorgi S, Nannini L (1996a) The cerebral cortex and parafascicular thalamic nucleus facilitate *in vivo* acetylcholine release in the rat striatum through distinct glutamate receptor subtypes. Eur J Neurosci 8: 2702-2710.

Consolo S, Baronio P, Guidi G, Di Chiara G (1996b) Role of the parafascicular thalamic nucleus and N-methyl-D-aspartate transmission in the D_1-dependent control of *in vivo* acetylcholine release in rat striatum. Neuroscience 71: 157-165.

Damsma G, Robertson GS, Tham C-S, Fibiger HC (1991) Dopaminergic regulation of striatal acetylcholine release: importance of D_1 and N-methyl-D-aspartate receptors. J Pharmacol Exp Ther 259: 1064-1072.

Di Chiara G, Morelli M, Consolo S (1994) Modulatory functions of neurotransmitters in the striatum: ACh/dopamine/NMDA interactions. Trends Neurosci 17: 228-233.

Dolezal V, Jackisch R, Hertting G, Allgaier C (1992) Activation of dopamine D_1 receptors does not affect D_2 receptor-mediated inhibition of acetylcholine release in rabbit striatum. Naunyn-Schmiedeberg's Arch Pharmacol 345: 16-20.

Garris PA, Ciolkowski EL, Pastore P, Wightman RM (1994) Efflux of dopamine from the synaptic cleft in the nucleus accumbens of the rat brain. 1 Neurosci 14: 6084-6093.

Gonon PA (1997) Prolonged and extrasynaptic excitatory action of dopamine mediated by D_1 receptors in the rat striatum *in vitro*. J Neurosci 17: 5972-5978.

Gorell JM, Czarnecki B (1986) Pharmacologic evidence for direct dopaminergic regulation of striatal acetylcholine release. Life Sci 38: 2239-2246.

Guevara Guzman R, Kendrick KM, Emson PC (1993) Effect of substance P on acetylcholine and dopamine release in the rat striatum: a microdialysis study. Brain Res 622: 147-154.

Hersch SM, Ciliax BJ, Gutekunst C-A, Rees HD, Heilman CJ, Yung KKL, Bolam JP, Ince E, Yi H, Levey AI (1995) Electron microscopic analysis of D_1 and D_2 dopamine receptor proteins in the dorsal striatum and their synaptic relationships with motor corticostriatal afferents. J Neurosci 15: 5222-5237.

Jaeger D, Kita H, Wilson CJ (1994) Surround inhibition among projection neurons is weak or nonexistent in the rat neostriatum. J Neurophysiol 72: 2555-2558.

Kawaguchi (1993) Physiological, morphological, and histochemical characterization of three classes of interneurons in rat neostriatum. J Neurosci 13: 4908-4923.

Khan S, Grogan E, Whelpton R, Michael-Titus AT (1996) N- and C-terminal substance P fragments modulate striatal dopamine outflow through a cholinergic link mediated by muscarinic receptors. Neuroscience 73: 919-927.

Lehmann J, Langer SZ (1983) The striatal cholinergic interneuron: synaptic target of dopaminergic terminals? Neuroscience 10: 1105-1120.

Le Moine C, Normand E, Bloch B (1991) Phenotypical characterization of the rat striatal neurons expressing the D1 dopamine receptor gene. Proc Natl Acad Sci USA 88: 4205-4209.

Login IS, Borland K, Harrison MB, Ragozzino ME, Gold PE (1995) Acetylcholine release from dissociated striatal cells. Brain Res 697: 271-275.

Login IS, Harrison MB (1996) A D1 dopamine agonist stimulates acetylcholine release from dissociated striatal cholinergic neurons. Brain Res 727: 162-168.

Mahan LC, Burch RM, Monsma FJ, Sibley DR (1990) Expression of striatal D_1 dopamine receptors coupled to inositol phosphate production and Ca^{2+} mobilization in *Xenopus* oocytes. Proc Natl Acad Sci USA 87: 2196-2200.

Scatton B (1982) Further evidence for the involvement of D_2, but not D_1 dopamine receptors in dopaminergic control of striatal cholinergic transmission. Life Sci 31: 2883-2890.

Steinberg R, Rodie D, Souilhac J, Bougault I, Emonds-Alt X, Soubrié P, Le Fur G (1995) Pharmacological characterization of tachykinin receptors controlling acetylcholine release from rat striatum: an *in vivo* microdialysis study. J Neurochem 65: 2543-2548.

Stoof JC, Kebabian JW (1982) Independent in vitro regulation by the D-2 dopamine receptor of dopamine-stimulated efflux of cyclic AMP and K^+-stimulated release of acetylcholine from rat neostriatum. Brain Res 250: 263-270.

Stoof JC, Drukarch B, De Boer P. Westerink BHC, Groenewegen HI (1992) Regulation of the activity of striatal cholinergic neurons by dopamine. Neuroscience 47: 755-770.

Surmeier DJ, Bargas J, Hemmings HC, Nairn AC, Greengard P (1995) Modulation of calcium currents by a D_1 dopaminergic protein kinase/phosphatase cascade in rat neostriatal neurons. Neuron 14: 385-397.

Tedford CE, Crosby G, Iorio LC, Chipkin RE (1992) Effect of SCH39166, a novel dopamine D_1 receptor antagonist, on [^{3}H]acetylcholine release in rat striatal slices. Eur J Pharmacol 211: 169-176.

Yan Z, Song W-J, Surmeier DJ (1997) D2 dopamine receptors reduce N-type Ca^{2+} currents in rat neostriatal cholinergic interneurons through a membrane-delimited, protein-kinese-C-insensitive pathway. J Neurophysiol 77: 1003-1015.

Wilson CJ, Chang HT, Kitai ST (1990) Firing patterns and synaptic potentials of identified giant aspiny interneurons in the rat neostriatum. J Neurosci 10: 508-519.
Zocchi A, Pert A (1993) Increases in striatal acetylcholine by SKF-38393 are mediated through D_1 dopamine receptors in striatum and not the frontal cortex. Brain Res 627: 186-192.

EFFECT OF DIFFERENT DOPAMINERGIC AGONISTS ON THE ACTIVITY OF PALLIDAL NEURONS IN THE NORMAL MONKEY

Thomas Boraud, Erwan Bezard, Christelle Imbert, Bernard Bioulac, and Christian E. Gross*

Key words: GPe, GPi, D1, D2, electrophysiological unit recording

Abstract: The classic schema of Alexander and Crutcher (1990) proposes that cortical information is processed by basal ganglia throughout the direct and indirect pathways, which are respectively modulated by D1 and D2 dopamine receptors. The effect of dopaminergic manipulations on the electrophysiological activity of the entire pallidum, both external (GPe) and internal (GPi), remains, however, subject to debate. We tested the effects of three different dopaminergic agonists (apomorphine, piribedil, and SKF 38393) on the pallidal neuronal activity of two calm awake monkeys (Macaca fascicularis). Extracellular unit recordings gave data on firing frequency and firing pattern for the GPe and GPi neurons. The effects of drug administration were diverse. The D1/D2 agonist apomorphine and the D2 agonist piribedil decreased the firing rate of both GPe and GPi neurons but only modified the pattern of GPi neurons. The D1 agonist SKF-38393 modified the firing pattern of both GPe and GPi neurons but only decreased the firing rate of GPi neurons. These results provide considerable interest since GPe does not appeared as important that predicted by the proposal of Delong's team.

1. INTRODUCTION

According to the model of extrapyramidal motor function proposed since 1989 [1,3,9], the globus pallidus internalis (GPi), the main output structure in the extrapyramidal motor loop, receives inputs from the striatum via two pathways: one, an inhibitory gabaergic monosynaptic pathway termed the "direct pathway", the other, a stimulating trisynaptic pathway termed the "indirect pathway". Dopamine is supposed to i) induce a decrease in GPi activity

---

* T. Boraud, E. Bezard, C. Imbert, B. Bioulac, and C. E. Gross, Basal Gang, Laboratoire de Neurophysiologie, CNRS UMR 5543, Université de Bordeaux 11, 146 rue Leo Saignat, 33076 Bordeaux Cedex, France.

by activating D1 receptors in the direct pathway ; ii) at the same time activate inhibitory D2 receptors in the indirect pathway, thus decreasing the inhibitory influence of the striatum on the globus pallidus externalis (GPe). This, in turn, increases the inhibitory action of the GPe on the STN and therefore decreases the stimulating influence of the STN on the GPi.

Recently this model is seriously shaken by various immunohistochemical and in situ hybridization studies (see 13 for review). Basically, results obtained from those methods denied the role played by the GPe as a relay on the so-called indirect pathway. However it is difficult to deduce the physiology of a complex network such as basal ganglia from data obtained with techniques without any temporal discrimination.

In order to test the validity of this model, we recorded the action of three dopaminergic agonists, acting on either D1 or D2 or both of them, on the firing activity of the globus pallidus (GP) in two normal monkeys.

2. EXPERIMENTAL PROCEDURES

2.1 Animals

Experiments were carried out on two female cynomolgus monkeys *(Macaca fascicularis)*. Animals were housed in approved individual primate cages under standard conditions of humidity (50 ± 5%), temperature (24 ± 1°C) and light (12h light/dark cycles) and they had free access to food and water. Their care was supervised by veterinarians skilled in the healthcare and maintenance of non-human primates. Our laboratory operates under the guidelines laid down by the National Institute of Health and is recognised by the French Ministry of the Environment. Surgical procedures were performed under general anaesthesia (ketamine-hydrochloride 40 mg/kg i.m., Panpharma, France, and xylazine 5 mg/kg i.m., Sigma, USA).

2.2 Stereotaxic Surgery

Surgical installation of the recording chamber was performed under aseptic conditions. Monkeys were anaesthetized with ketamine (Ketalar®, Parke-Davis) and their heads fixed in a stereotaxic frame. A cylindric recording chamber (Narishige) was attached to the skull and positioned according to the intracerebral coordinates obtained by ventriculography —anterior (CA) and posterior (CP) commissures— corrected, when necessary, with the aid of a stereotaxic atlas [8]. This made it relatively easy to insert the microelectrodes with respect to the central axis of the chamber which was itself angled at 45° from the sagittal plane. We then anchored to the skull a metal platform with a hole in the center for the recording chamber, on which the head fixation device used during experimental sessions would subsequently be fixed.

2.3 Electrophysiological Recordings

Extracellular unit recordings were carried out as previously described [4] using tungsten microelectrodes (FHC : 6-8 M) in awake monkeys, once animals had gradually adapted to head restraint. Microelectrodes were lowered using a Narishige microdrive. After signal amplification (EG and G, M113), data recorded were stored online in a computer (Performa 5260, Apple) using a custom-made interface. A time interval histogram (TIH) was charted for each neuron recorded, from which we determined the mean frequency of discharge and standard deviation (F ± SD). We then built a density histogram (DH) for each cell '1. Statistical comparisons of mean frequencies were made using analysis of variance followed by a post-hoc PLSD Fisher test ($p<0.05$). The results obtained for the two monkeys were then pooled after an ANOVA test showed no significant difference between those of the two animals. Firing patterns were compared by analysis of the frequency of distribution (χ^2, ddl=2, $p<0.05$ when $\chi^2 >5,991$) of the different patterns between the two experimental situations (drug vs Normal), according to the method described by Mushiake et al. 14

2.4 Drug Administration

In order to study the effect of stimulation of the D1 and D2 dopamine receptors in the basal ganglia loop we used three different dopamine agonists at dosage known to induce dyskinesia in MPTP-treated animals apomorphine (D1/D2 agonist ; 0.1 mg/kg i.c.), SKF-38393 (partial D1 agonist ; 1.5 mg/kg i.m), piribedil (D2 agonist ; 3 mg/kg i.m.).

Once a neuron was discriminated, the firing activity was recorded three times during a 20 minute period in order to determine the basal frequency. Drug was then injected and firing activity recorded every 5 minutes until the neuron resumed basal firing activity, or was lost.

3. RESULTS

3.1 Effect on the Activity of GPi Cells (Fig. 1)

Apomorphine (D1/D2 agonist) induced a significant decrease in GPi firing frequency. Patterns are significantly modified as we observe less random and more bursting or regular cells.

SKF-38393 (D1 agonist) likewise induced a significant decrease in GPi firing frequency, but patterns are significantly modified as we observe less random and more regular cells, with fewer bursting cells.

Piribedil (D2 agonist) induced a two-phase response in GPi cells: a significant increase followed by a significant decrease. but patterns are significantly modified as we observe more random and less bursting or regular cells.

3.2 Effect on the Activity of GPe Cells (Fig. 2)

Apomorphine (D1/D2 agonist) induced a significant decrease in the firing frequency of GPe cells. There was no modification of the firing pattern.

SKF-38393 (D1 agonist) also induced a significant decrease in the firing frequency of GPe cells. Patterns are significantly modified as we observe less random and more bursting or regular cells.

Piribedil (D2 agonist) induced a two-phase response in GPe cells: a significant increase followed by a significant decrease in firing frequency. There was no modification of firing patterns.

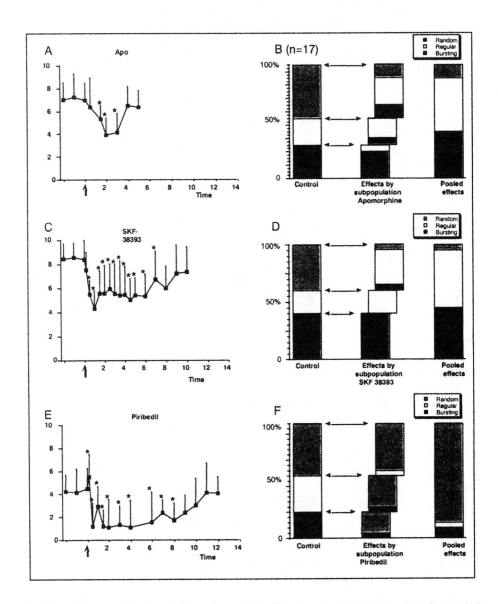

Figure 1. Influence of each drug on the activity of GPi cells. Top, Apomorphine (s.c., 0.1mg/kg) ; middle, SKF-38393 (i.m., 3mg/kg) ; bottom, Piribedil (s.c. 3mg/kg). A, C and E, effect on mean frequency. Vertical bar shows standard deviation. A star indicates a significant difference as compared to control mean frequency ($p<0.05$). B, D and F: graphic presentation of the percentage of distribution of the various patterns of discharge. First column : before drug injection. Second and third columns : after injection. The second column shows the modification induced in each subpopulation of the structure, with the first step corresponding to the pattern of cells which burst before injection, the second step to those which were regular before injection and the third step to those which were random before injection. In the third column subpopulations have been pooled.

EFFECT OF DOPAMINERGIC AGONISTS ON PRIMATE PALLIDAL NEURONS 503

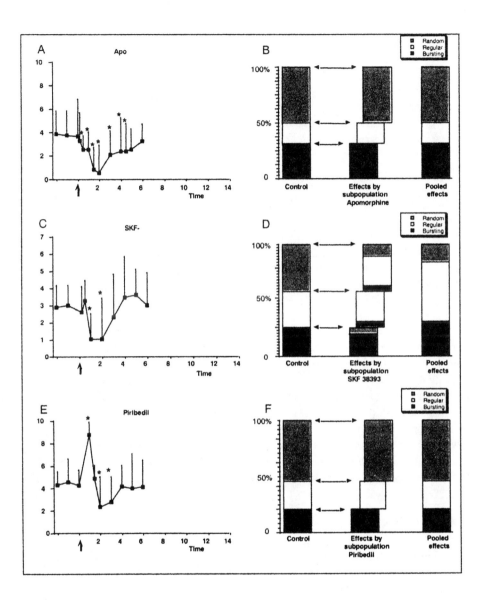

Figure 2. Influence of each drug on the activity of GPe cells. . Top, Apomorphine (s.c., 0.1mg/kg) ; middle, SKF-38393 (i.m., 3mg/kg) ; bottom, Piribedil (s.c. 3mg/kg). A, C and E, effect on mean frequency. Vertical bar shows standard deviation. A star indicates a significant difference as compared to control mean frequency ($p<0.05$). B, D and F: graphic presentation of the percentage of distribution of the different patterns of discharge. First column : before drug injection. Second and third columns : after injection. The second column shows the modification induced in each subpopulation of the structure, with the first step corresponding to the pattern of cells which burst before injection, the second step to those which were regular before injection and the third step to those which were random before injection. In the third column subpopulations have been pooled.

4. DISCUSSION

It is striking at first, that dopaminergic drugs induced no modification in the animal behaviour (data not shown) at variance with previous observations in normal rat [5,6], MPTP treated monkeys or parkinsonian patients [10].

The effect on GPi frequency of the administration of D1/D2 (apomorphine), D1 (SKF-38393) or D2 (piribedil) agonists is that which one should expect. Those dopaminergic drug induced, indeed, decrease of firing frequency.

D1/D2 or D1 stimulation induce, however, a less random pattern, whilst D2 stimulation induces a more random pattern.

This raises the question of the way basal ganglia played their role in normal situation as in GPi, their main motor control output structure, neither decrease in firing frequency nor alteration of firing patterns induced a behavioural modification. This modified output appears to be tolerated by the motor thalamic nuclei and/or the cortical neurons. It would seem that the level of activity of output structure of the basal ganglia is not the relevant feature as far as motor control is involved.

The effect on GPe activity is more surprising, in that it is similar to that of GPi activity. Both D1/D2 and D1 agonists induce a decrease in the firing rate of GPe cells, whilst the D2 agonist induces a transitory augmentation followed by a longer decrease in the firing rate. Previous studies have observed an increase in firing frequency of GPe neurons after injection of apomorphine in MPTP treated monkeys[7]. Discrepancies, between the two situations have to be explored in the same experimental conditions. But it could reflected that dopamine depletion induces not simply changes in activities of both direct and indirect pathways, but a complete reorganization of the whole network.

Only the D1 agonist modifies the firing pattern, which tends to become more regular. Those results are again at variance with the classic model of extrapyramidal function which considers that the indirect pathway is principally controlled by D2 receptors [9]. Those results should be compared to those obtained with hybridization in situ methods in rats [12]. They show that D1 agonist induced an activation of C-Fos in GP of normal rat but not in dopamine depleted rats. Once more, this tends to prove that the network of connexions between striatum and pallidum is considerably modified between both situations.

5. CONCLUSIONS

The most striking result of this study concerns the effects of dopamine agonists upon the GPe. Dopamine agonists exerted an influence upon neurons of this nucleus at variance with what we should expect from the model. However those results recovered those of previous immunuhistological and hybridization in situ studies. These results would justify a re-evaluation of the role played by the GPe in the extrapyramidal loop. The schema of dopamine transmission would appear to be more complex than hitherto imagined.

6. ACKNOWLEDGEMENTS

We wish to thank Sandra Dovero for technical assistance. This study was supported partly by the CNRS and the IFR of Neuroscience (INSERM N°8; CNRS N°13), partly by the MESR grant-N°95523629 and the University Hospital of Bordeaux.

7. REFERENCES

1. Albin R.L., Young A.B.and Penney J.B. (1989) The functional anatomy of basal ganglia disorders. Trends Neurosci. 12, 366-375.
2. Alexander G.E.and Crutcher M.D. (1990) Functional architecture of basal ganglia circuits: neural substrates of parallel processing. Trends Neurosci. 13, 266-271.
3. Alexander G.E., DeLong M.R.and Strick P.L. (1986) Parallel organization of functionally segregated circuits linking basal ganglia and cortex. Annu. Rev. Neurosci. 9, 357-381.
4. Boraud T., Bezard E., Guehl D., Bioulac B.and Gross C. (1998) Effects of L-DOPA on neuronal activity of the Globus Pallidus externalis (GPe) and Globus Pallidus internalis (GPi) in the MPTP-treated monkey. Brain Res. 787, 157-160.
5. Braun A.R.and Chase T.N. (1986) Obligatory D-1/D-2 receptor interaction in the generation of dopamine agonist related behaviors. Eur. J. Pharmacol. 131, 301-306.
6. Ernst A.M. (1967) Mode of action of apomorphine and dexamphetamine on gnawing compulsion in rats. Psychopharmacologia 10, 381-389.
7. Filion M., Tremblay L., and Bedard P.J. (1991) Effects of dopamine agonists on the spontaneous activity of globus pallidus neurons in monkeys with MPTP-induced parkinsonism. Brain Res. 547, 152-161.
8. François C., Yelnik J., and Percheron G. (1996) A stereotaxic atlas of the basal ganglia in macaques. Brain Res. Bull. 41, 151-158.
9. Gerfen C.R., Engber T.M., Mahan L.C., Susel Z., Chase T.N., Monsma F.J.and Sibley D.R. (1990) D1 and D2 dopamine receptor-regulated gene expression of striatonigral and striatopallidal neurons. Science 250, 1429-1432.
10. Jenner P. (1995) The rationale for the use of dopamine agonists in Parkinson's disease. Neurology 45 (suppl 3), S6-S12.
11. Kaneoke Y., and Vitek J.L. (1996) Burst and oscillations as disparate neuronal properties. J. Neurosci. Meth. 68, 211-223.
12. Le Moine C., Svenningsson P., Fredholm B.B.and Bloch B. (1997) Dopamine-adenosine interactions in the striatum and the globus pallidus: inhibition of striatopallidal neurons through either D2 or A2A receptors enhances D1 receptor-mediated effects on c-fos expression. J Neurosci 17, 8038-48.
13. Levy R., Hazrati L.-N., Herrero M.-T., Vila M., Hassani O.-K., Mouroux M., Ruberg M., Asensi H., Agid Y., Féger J., Obeso J.A., Parent A.and Hirsch E.C. (1997) Re-evaluation of the functional anatomy of the basal ganglia in normal and parkinsonian states. Neuroscience 76, 335-343.
14. Mushiake H., Inase M.and Tanji J. (1991) Neuronal activity in the primate premotor, supplementary and precentral motor cortex during visually guided and internally determined sequential movements. J. Neurophysiol. 66, 705-718.

D_2 DOPAMINE RECEPTOR-DEFICIENT MUTANT MICE

Tools to tease apart receptor subtype electrophysiology

M.S. Levine, C. Cepeda, R.S. Hurst, M.A. Ariano, M.J. Low and D.K. Grandy*

Key words: Dopamine Receptors, Mutant Mice, Electrophysiology, Neostriatum

Abstract: The generation of transgenic mice lacking specific dopamine (DA) receptor subtypes has allowed elucidation of receptor function in isolation. We have used DA receptor-deficient mice to test the hypothesis that D1 and D2 receptor families modulate glutamate receptor-mediated activity in opposite ways. Accordingly, D1 receptors enhance and D2 receptors reduce glutamatergic activity. Consistent with this hypothesis we found that D1 receptor-deficient mice have a reduced ability to enhance glutamate responses. In contrast, D_2 receptor-deficient mice display enhanced glutamatergic activity. This enhancement may be due to the lack of D_2 receptors on corticostriatal terminals.

1. INTRODUCTION

Molecular geneticists have been spectacularly successful in developing the technology for targeted germ-line transmission and many different types of transgenic and knockout mice have been created to aid in the elucidation of biological functions. Mouse mutants derived by targeting mutagenesis in embryonic stem cells offer many advantages to the study of the molecular and cellular mechanisms underlying receptor function, especially when pharmacological tools are unavailable. In this communication we will discuss some of our findings using dopamine (DA) receptor-deficient mutant mice to analyze the modulatory actions of DA on striatal electrophysiology.

---

* M.S. Levine, C. Cepeda, and R.S. Hurst, Mental Retardation Research Center, UCLA, Los Angeles, CA 90024; M.A. Ariano, Department of Neuroscience, Chicago Medical School, North Chicago, IL 60064; M.J. Low, Vollum Institute and D.K. Grandy, Department of Physiology and Pharmacology, Oregon Health Sciences University, Portland, OR 97201 USA

While the development of transgenic and knockout mice has great promise, it is important to note that as with any technique, there are associated pitfalls and problems in interpretation of outcomes. One problem is compensation for lack of the specific receptor subtype (Holsboer, 1997). Thus, observed effects in mutants could be due to a combination of lack of a specific receptor and changes in existing receptors. There is no way to completely control for this problem. In our studies, the mutant is used like a pharmacological antagonist, to test hypotheses about the "missing" receptor (Levine et al., 1996). If evidence is not obtained in support of the hypothesis, then it is difficult to make conclusions because of potential compensatory events. This problem will be minimized when conditional knockouts become more available. Another caveat associated with mutants concerns the genetic background of the strain. Different strains of mice have been reported to exhibit markedly different indices of neurotransmitter activity (Crawley et al., 1997). Unfortunately, this variable is difficult to control since mice are often obtained from multiple sources, and there have been few systematic attempts to standardize background strains across different laboratories. All one can do is report the background strain for each of the receptor-deficient mice and the generational data. However, for non-behavioral studies strain differences may not be a crucial variable. Our hypotheses are based on data obtained in rats. If all of the wildtype murine controls for the different receptor-deficient mutants exhibit similar responses as intact rats then it becomes difficult to argue that strain variables are confounding issues in the mutants with different genetic backgrounds.

The effects of DA are mediated by at least five identified receptor subtypes (Civelli et al., 1991; Sibley and Monsma, 1992). These subtypes have been classified into two families (D1 and D2), according to their pharmacological profiles (Creese and Fraser, 1997). This classification scheme recognizes a D1 family [D_1 and D_5 subtypes in rat or their corresponding forms in human (Sibley and Monsma, 1992)] which binds substituted benzazepines, and whose responses are transduced by elevations in target cell cAMP. The second family, D2 receptors, is composed of three subtypes, D_2, D_3 and D_4, which uses numerous transduction systems and is antagonized by neuroleptics (Sibley and Monsma, 1992). To reduce confusion, we will use D1 and D2 to refer to the families of DA receptors and subscript notation (e.g., D_1 and D_5) to refer to subtypes of each family.

Based on mRNA and protein expression using antireceptor antibodies, the D_1 receptor is heavily expressed in striatum and to a lesser degree in prefrontal cortex (Fremeau et al., 1991; Ariano and Sibley, 1994) while the D_5 receptor is more heavily expressed in cortex than striatum (Huntley et al., 1992; Bergson et al., 1995; Ariano et al., 1997b). Although there is considerable evidence that these receptor subtypes are localized postsynaptically, there is emerging electrophysiological evidence for presynaptic function, especially in the nucleus accumbens (Pennartz et al., 1992; Nicola et al., 1996; Harvey and Lacey, 1997). Pharmacologically, ligands are not available to differentiate between the functions of D_1 and D_5 subtypes. In the animal or in *in vitro* slice preparations it has been assumed that activation of D1 receptors is due to concurrent activation of both the D_1 and the D_5 subtypes, unless there is good evidence that the type of neuron examined only expresses one or the other subtype. The development of mutant mice deficient for either D_1 (Drago et al., 1994) or D_5 (Hollon et al., 1998) receptors permits characterization of these subtypes in isolation. Previously, we have used the D_1 mutant mouse to evaluate D_5 receptor function in isolation as well as to determine some of the specific functions lost following this genetic inactivation of the subtype (Levine et al., 1996).

Based on mRNA and protein expression using antireceptor antibodies in the striatum, the D_2 receptor subtype appears to be the most prevalent representative of this family, although both D_3 and D_4 receptors are expressed (Meador-Woodruff et al., 1991; Huntley et al., 1992; Ariano and Sibley, 1994; Fisher et al., 1994; Choi et al., 1995; Ariano et al., 1997a). In the nucleus accumbens the D_3 subtype is more prevalently expressed, and in the cortex the D_4 subtype is more abundant (Ariano et al., 1994; 1997a). Also in the cortex, the D_4 receptor subtype appears to be expressed on parvalbumin-positive GABAergic interneurons (Mrzljak et al., 1996). In addition to the postsynaptic D2 receptors, presynaptic receptors exist. Some of these are the autoreceptors, confined to the DA neuron by definition. Autoreceptors exhibit D2-like pharmacological profiles and provide self-regulatory function (Wolf and Roth, 1985; L'hirondel et al., 1998). It is generally believed that the D_2 subtype is the autoreceptor, although D_3 receptors have been implicated (Aretha and Galloway, 1996; Piercey et al., 1997; Tepper et al., 1997; Koeltzow et al., 1998). Other presynaptic receptors, especially in the striatum, are hypothesized to be localized to corticostriatal endings (Mercuri et al., 1985; Kornhuber and Kornhuber, 1986) or endings of medium-sized spiny neurons (Fisher et al., 1994; Sesack et al., 1994). These are generally believed to be D_2 receptors.

Unlike the D1 family in which there are no specific pharmacological agents which permit isolation of function of D_1 from D_5 receptors, there are now a number of compounds that have more selective affinities for the D_3 or D_4 receptor over the D_2 subtype. For example, there are both agonists and antagonists for the D_3 receptor and antagonists for the D_4 receptor (Hall et al., 1996; Blanchet et al., 1997; Bristow et al., 1997; Tallman et al., 1997; Mansbach et al., 1998). However, these compounds are not always selective enough for electrophysiological studies. It is now possible to examine in more detail the function of each receptor subtype within the two families because of the development of DA receptor-deficient mutant mice (Drago et al., 1994; Baik et al., 1995; Kelly et al., 1997; 1998).

The effects of DA on the electrophysiology of striatal cells are complex. Responses mediated by DA receptors are due to activation of multiple receptor subtypes and involve various second messenger systems. Both excitatory and inhibitory effects have been reported (Herrling and Hull, 1980; Akaike et al., 1987; Calabresi et al., 1987; Rutherford et al., 1988), but DA's actions are best understood if it is considered a neuromodulator (Cepeda et al., 1993; Cepeda and Levine, 1998) instead of a classical neurotransmitter. The outcome of DA modulation depends on a number of factors such as the DA receptor subtype preferentially activated, the receptor location at pre- or postsynaptic sites, the concentration of ambient DA, and the activity-state of the neuron subject to DA modulation. Although many schemes have been proposed to account for the different modulatory actions of DA, we have hypothesized a relatively simple scheme such that DA's modulatory actions are a function of the specific receptor subtype with which it interacts (either the DA receptor subtype or the neurotransmitter receptor that it will modulate). In this scheme, D1 receptor activation enhances responses due to activation of glutamate receptors, particularly those mediated by activation of NMDA receptors. In contrast, D2 receptor activation reduces responses due to activation of glutamate receptors, particularly those mediated by activation of non-NMDA receptors (at least in striatum). The enhancing effects of D1 receptor activation appear to involve postsynaptic actions, whereas the attenuating effects mediated by D2 receptors may involve both postsynaptic as well as presynaptic actions on corticostriatal terminals as we will describe below.

Two major inhibitory effects of DA have been examined, inhibition of action potentials and inhibition of synaptically-evoked excitatory postsynaptic potentials (EPSPs). Inhibition of action potentials is dependent on the ability of DA to reduce persistent and fast Na^+ currents (Calabresi et al., 1987; Surmeier et al., 1992; Surmeier and Kitai, 1993; Cepeda et al., 1995; Schiffman et al., 1995) and can be mediated by activation of either D1 or D2 receptors, although in a subset of striatal neurons activation of D2 receptors enhances fast Na^+ currents (Surmeier et al., 1992). Recent studies have demonstrated that the outcome of D1 receptor activation on striatal cell firing also depends on the initial membrane potential of the neuron. At hyperpolarized potentials activation of D1 receptors reduces the number of action potentials, but at depolarized potentials (less than −60 mV) it increases cell firing (Hernandez-Lopez et al., 1997).

Initially the reduction in the amplitude of EPSPs was believed to be dependent on activation of postsynaptic D1 receptors (Calabresi et al., 1987). D2 receptor-mediated effects were not observed unless these receptors were rendered hypersensitive by DA-depleting lesions (Calabresi et al., 1988). More recently, this result has been challenged by studies showing that activation of D2 receptors inhibits EPSPs (Jiang and North, 1991; Cepeda et al., 1994; Hsu et al., 1995). A major question concerning this role has been whether the effects of D2 receptor activation involve presynaptic and/or postsynaptic mechanisms.

There is considerable physiological and biochemical evidence favoring D2 receptor regulation of excitability and glutamate release from cortical and possibly thalamic terminals in the striatum (Mitchell and Doggett, 1980; Rowlands et al., 1980; Mercuri et al., 1985; Kornhuber and Kornhuber, 1986; Yang and Mogenson, 1986; Maura et al., 1988; Garcia-Munoz et al., 1991; Yamamoto and Davy, 1992; Flores-Hernandez et al., 1997). Recent morphological evidence using antireceptor antibodies supports localization of D_2 receptors to putative excitatory corticostriatal or thalamostriatal terminals (Fisher et al., 1994; Sesack et al., 1994), but surprisingly such detection is infrequent compared to postsynaptic localization, and localization of D_2 receptors to tyrosine hydroxylase positive terminals (DA-containing nigrostriatal inputs). These morphological studies agree with radioligand binding studies reporting difficulty in detecting appreciable D2 receptor distribution on corticostriatal terminals (Trugman et al., 1986; Joyce and Marshall, 1987). Physiological studies have also provided evidence for presynaptic modulation. There was early evidence that D2 receptor-mediated inhibition of cortical input involves both pre- and postsynaptic actions, although it was difficult to ascertain the contribution from each of these mechanisms (Brown and Arbuthnott, 1983; Mercuri et al., 1985). Recent studies have provided a more conclusive role for inhibitory actions by presynaptic D2 receptors (Hsu et al., 1995).

Clearly, conclusive determination of the role of presynaptic modulation by activation of DA in the striatum will depend upon further refinements of receptor characterization. For example, it is possible that D_4 receptors are located presynaptically on corticostriatal endings. This receptor subtype is very prevalent in pyramidal cells in the frontal cortex (Ariano et al., 1997a), an area that projects heavily to the striatum (Fonnum et al., 1981; Smith and Bolam, 1990). The presence of presynaptic D_4 receptors in the striatum could explain some of the differences between physiological and morphological outcomes since this subtype would be activated pharmacologically by D2 agonists, but not detected with morphological approaches using antireceptor antibodies to D_2 receptors.

Receptor-deficient mutant mice for each of the D1 and D2 subtypes have been generated and characterized (Drago et al., 1994; Baik et al., 1995; Kelly et al., 1997; 1998; Xu et al., 1997;

Hollon et al., 1998). In our initial studies, we characterized some of the alterations in striatal electrophysiology in the D_1 receptor-deficient mutant (Levine et al., 1996). We showed that most passive and active membrane properties of striatal medium-sized spiny neurons were similar in mutants and their wildtype littermates. We examined the hypothesis that DA-induced potentiation of NMDA-induced responses was mediated by the D_1 receptor. The results of this experiment demonstrated that D1 receptor agonists minimally potentiated responses mediated by activation of NMDA receptors in this mutant. Potentiation was not absent however, and we assumed that the remaining potentiation was mediated by activation of D_5 receptors. Evidence for this hypothesis will have to await the creation of combined D_1 and D_5 receptor-deficient mice. Only a few electrophysiological studies differentiate among the functional abilities of the D2 receptor subtypes in mutant mice. Issues surrounding synaptic plasticity in the striatum have been assessed in D_2 receptor-deficient mice (Calabresi et al., 1997) as well as how omission of the D_2 receptor alters electrophysiology of substantia nigra neurons (Mercuri et al., 1997). The consequences of deletion of the D_3 receptor have also been examined extensively (Xu et al., 1997; Koelztow et al., 1998).

In this report we describe some of our experiments using the D_2 receptor-deficient mutant to test the predictions of our hypothesis regarding the interactions between striatal DA and glutamate receptors. Our hypothesis predicts that activation of D_2 receptors should attenuate responses mediated by activation of glutamate receptors, specifically responses due to activation of non-NMDA glutamate receptors. Thus, in the D_2 receptor-deficient mouse we predict that DA would no longer be capable of attenuating or inhibiting non-NMDA receptor-mediated responses.

2. D_2 RECEPTOR-DEFICIENT MUTANT MICE DO NOT EXPRESS FUNCTIONAL D_2 PROTEIN

D_2 receptor-deficient mutant mice were obtained from Dr. David K. Grandy (Oregon Health Science Institute). The methods used for the generation of these mice have been described (Kelly et al., 1997; 1998). The animals used in the present experiments were screened for D_2 DA receptor protein expression using well-characterized antisera (McVittie et al., 1991). Briefly, antisera were applied to 10 μm thick slices of striatum derived from mutant and wildtype littermate control mice. The localization of the D_2 DA receptor protein staining was evaluated using scanning confocal microscopy. Controls for the procedure included: 1) use of multiple antisera, directed against different epitopes of the receptor protein sequence, 2) use of preimmune sera, 3) omission of primary antisera, and 4) adsorption of the primary antisera with the peptide antigen. No immunofluorescence was visible in any of the control experiments. D_2 receptor expression was examined in 4 mutant and 4 wildtype control mice. Receptor protein staining was prevalent within medium and large diameter neurons in wildtype control striata (Fig. 1A), while immunofluorescence was completely absent in the mutants (Fig. 1B). The cerebral cortex of wildtype animals also showed moderate levels of expression. Receptor expression of D_4 receptors was unaltered in the D_2 mutant, suggesting the genetic deletion did not change DA receptor staining for other closely related receptor proteins.

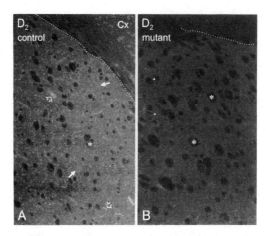

Figure 1. D_2 DA receptor protein immunofluorescent staining in coronal sections of mouse brain. A. Sections were obtained from wildtype control and showed prominent cellular expression in cells of medium diameter (arrows), within a robustly reactive neuropil. Vascular elements (open arrowhead) were also reactive for the receptor subtype, however myelinated fiber bundles of the internal capsule (asterisks) were not stained. The callosal border of the dorsal striatum is delimited by the dotted line, and the moderate receptor reaction in the overlying somatosensory cortex (Cx) is visible at the upper right of the panel. B. Sections obtained from the receptor-deficient mutant lacked immunofluorescent staining.

3. D_2 RECEPTOR-DEFICIENT MUTANT MICE HAVE ALTERED PRESYNAPTIC DA FUNCTION

To evaluate electrophysiological changes, standard brain slice techniques were used in which either current- or voltage-clamp recordings were made (Cepeda et al., 1993; 1995). Current-clamp recording was done with standard intracellular techniques using high impedance sharp electrodes. Whole-cell voltage-clamp recordings were made with patch electrodes and neurons were identified using infrared videomicroscopy and differential interference contrast optics. In general, passive and active membrane properties were similar in mutants and wildtype controls [-77.6 ± 1.5 versus -76.6 ± 0.9 mV (mean ± s.e.) for resting membrane potentials for neurons obtained from mutants and wildtypes, respectively; 75.3 ± 2.2 versus 73.3 ± 2.2 mV for action potential amplitudes for neurons obtained from mutants and wildtypes, respectively].

A marked difference between neurons obtained from mutants and wildtypes was the occurrence of spontaneous membrane depolarizations (5-20 mV in amplitude of variable duration) in about one third of the neurons from mutant animals (Fig. 2, top two traces). These depolarizations were reminiscent of the membrane oscillations observed in *in vivo* intracellular recordings (Wilson and Kawaguchi, 1996). They were not affected by addition of bicuculline (10 μM), a $GABA_A$ receptor antagonist, but could be completely blocked by the addition of 6-cyano-7-nitroquinoxaline-2,3-dione (CNQX, 5 μM), a non-NMDA receptor antagonist. In current-clamp recording, small spontaneous depolarizations of the resting membrane potential occurred with greater frequency in D_2 receptor-deficient mutants than in their wildtype littermate controls. Fig. 2C shows low amplitude spontaneous events in a wildtype mouse. In comparison, slightly more of these events occurred in the D_2 receptor-deficient mutants (Fig. 2D). In the presence of 4-aminopyridine (4-AP) (a compound that increases neurotransmitter release) and bicuculline, the frequency of low amplitude spontaneous events was increased in the wildtype mouse (Fig. 2E), but markedly increased in the mutant (Fig. 2F). In wildtype mice application of sulpiride, a D2 receptor antagonist, also increased spontaneous events, indicating that D2 family receptors were involved. We also assessed spontaneous and mini-excitatory postsynaptic currents (mEPSCs) in the D_2 receptor-deficient mutant in the presence of 4-AP and bicuculline. The data indicated that in the D_2 receptor deficient mutant, there is an increase in mEPSCs as well. We have interpreted these data to indicate that D_2 inhibition of spontaneous and evoked glutamate release is absent in these mutant mice. Together, these findings provide additional evidence for a presynaptic role for D_2 receptors in modulating corticostriatal or thalamostriatal responses (Cepeda et al., 2001). Although the majority of evidence favors D_2 inhibition of corticostriatal glutamate release, based on a recent paper, there is another interpretation that should be considered. Evidence was provided that glutamate and DA are potentially co-localized in nigrostriatal neurons (Sulzer et al., 1998). Therefore, it is also possible that DA, acting via D2

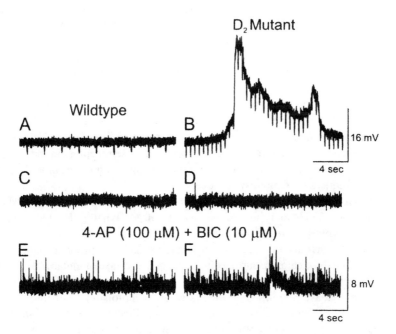

Figure 2. A and B. In the D_2 receptor-deficient mutant, in current-clamp recordings, large spontaneous depolarizations of the membrane occurred in a subset of cells (B). These could be blocked by CNQX indicating they were mediated by activation of non-NMDA receptors (data not shown). These were absent in wildtype controls (A). Membrane potentials -70 mV. C and D. In current-clamp recording, small spontaneous depolarizations of the resting membrane potential occur with greater frequency in D_2 receptor-deficient mutants. C shows low amplitude spontaneous events in a wildtype mouse. In comparison, slightly more of these events occur in D_2 receptor-deficient mutants (D). E and F. In the presence of 4-aminopyridine (4-AP) and bicuculline (BIC), the frequency of low amplitude spontaneous events is increased in the wildtype mouse (E), and markedly increased in the mutant (F). Resting potentials were -80 mV and -78 mV for mutant and wildtypes, respectively.

autoreceptors, could inhibit glutamate release from endings of nigrostriatal neurons. One would have to conclude that DA-containing cells are hyperactive after deletion of inhibitory autoreceptors and release increasing quantities of both DA and glutamate (Mercuri et al., 1997).

Although we interpret these outcomes as changes in presynaptic modulation in the striatum, we would also expect that the cortical cells that project to the striatum would be altered since they no longer express D_2 receptors. Although we have not yet examined this alternative we have found that a large population of cortical cells in the D_4 receptor deficient mutant mice are hyperactive compared to cells in wildtype controls when exposed to 4-AP and bicuculline (Rubenstein et al., 2001).

4. CONCLUSIONS

The principal novel observations we obtained from the D_2 receptor-deficient mutants were the presence of spontaneous membrane depolarizations in a subset of neurons and the marked increase in the frequency of low amplitude spontaneous activity in the presence of 4-AP and bicuculline. Studies *in vivo* have shown that the membrane potential of medium-sized spiny neurons oscillate between two preferred potentials, one hyperpolarized (down state) and one depolarized (up state) (Wilson and Kawaguchi, 1996). The depolarized state is produced by a barrage of excitatory corticostriatal inputs. Because *in vitro* most of the cortical inputs are severed by the slicing procedure, up and down states are not observed. The membrane potential is very stable and spontaneous synaptic activity is barely visible. DA plays a role in keeping the membrane potential constant, because DA-depleting lesions lead to an increase in spontaneous synaptic activity (Galarraga et al., 1987; Cepeda et al., 1989). In normal animals DA may be activating D_2 receptors on corticostriatal terminals, which in turn reduces glutamate release. The absence of these receptors in the mutant mice could explain the presence of spontaneous membrane depolarizations in some slices and the increase in the frequency of spontaneous events.

The conclusive demonstration of D2 receptors on corticostriatal terminals (Fisher et al., 1994; Sesack et al., 1994) and the relative sparsity of postsynaptic D2 receptors can be indicative of strong presynaptic effects. Indeed, D2 receptors have been shown to be able to decrease calcium currents in a number of systems. D2 receptors may be activated presynaptically to reduce calcium currents involved in the release of glutamate (Lovinger et al., 1994; Bargas et al., 1998).

If increased release of glutamate occurs in D_2 receptor-deficient mice, structural alterations in striatal neurons could be expected because of increased exposure to glutamate. Again, we have observed the occurrence of a subpopulation of medium-sized spiny neurons filled with biocytin that have reduced dendritic spines and dendritic field sizes (Cepeda et al., 2001). These data are supportive of our conclusion.

In general, the results of the present study confirm previous reports on D_2 receptor-deficient mice. For example, reduced regulatory function of DA has been demonstrated in the substantia nigra of these mice (Mercuri et al., 1997). In nucleus accumbens D_2 receptors have been shown to play an important role in prepulse inhibition (Ralph et al., 1999). In a previous study it was demonstrated that absence or blockade of D_2 receptors converts long-term depression to long-term potentiation. This finding indicates that D_2 receptors exert a negative regulation of NMDA receptor-mediated responses, since long-term potentiation in the striatum, at least in slices, only occurs when NMDA receptors are exposed (Calabresi et al., 1992; 1997). Interestingly, this conclusion agrees with our general scheme of DA actions in the striatum (Cepeda and Levine, 1998). Accordingly, D1 and D2 receptors exert opposite effects on glutamate receptor activation. In general, D1 receptor activation enhances while D2 receptor activation decreases glutamate responses. The absence of D_2 receptors will shift the direction of DA modulation towards increased responsiveness to, and probably increased release of, glutamate in striatum.

5. ACKNOWLEDGEMENTS

Supported by USPHS NS33538 and the National Association for Research in Schizophrenia and Depression.

6. REFERENCES

Akaike A., Ohno Y., Sasa M. and Takaori S. (1987) Excitatory and inhibitory effects of dopamine on neuronal activity of the caudate nucleus neurons in vitro. *Brain Res.* **418**, 262-272.

Aretha C. W. and Galloway M. P. (1996) Dopamine autoreceptor reserve in vitro: possible role of dopamine D3 receptors. *Eur J Pharmacol.* **305**, 119-122.

Ariano M. A. and Sibley D. R. (1994) Dopamine receptor distribution in the rat CNS: elucidation using antipeptide antisera directed against D1A and D3 subtypes. *Brain Res.* **649**, 95-110.

Ariano M. A., Wang J., Noblett K. L., Larson E. R. and Sibley D. R. (1997a) Cellular distribution of the rat D4 dopamine receptor protein in the CNS using anti-receptor antisera. *Brain Res.* **752**, 26-34.

Ariano M. A., Wang J., Noblett K. L., Larson E. R. and Sibley D. R. (1997b) Cellular distribution of the rat D1B receptor in central nervous system using anti-receptor antisera. *Brain Res.* **746**, 141-150.

Baik J. H., Picetti R., Saiardi A., Thiriet G., Dierich A., Depaulis A., Le Meur M. and Borrelli E. (1995) Parkinsonian-like locomotor impairment in mice lacking dopamine D2 receptors. *Nature.* **377**, 424-428.

Bargas J., Ayala G. X., Hernandez E. and Galarraga E. (1998) Ca2+-channels involved in neostriatal glutamatergic transmission. *Brain Res Bull.* **45**, 521-524.

Bergson C., Mrzljak L., Lidow M. S., Goldman-Rakic P. S. and Levenson R. (1995) Characterization of subtype-specific antibodies to the human D5 dopamine receptor: studies in primate brain and transfected mammalian cells. *Proc Natl Acad Sci U S A.* **92**, 3468-3472.

Blanchet P. J., Konitsiotis S. and Chase T. N. (1997) Motor response to a dopamine D3 receptor preferring agonist compared to apomorphine in levodopa-primed 1-methyl-4-phenyl-1,2,3,6-tetrahydropyridine monkeys. *J Pharmacol Exp Ther.* **283**, 794-799.

Bristow L. J., Collinson N., Cook G. P., Curtis N., Freedman S. B., Kulagowski J. J., Leeson P. D., Patel S., Ragan C. I., Ridgill M., Saywell K. L. and Tricklebank M. D. (1997) L-745,870, a subtype selective dopamine D4 receptor antagonist, does not exhibit a neuroleptic-like profile in rodent behavioral tests. *J Pharmacol Exp Ther.* **283**, 1256-1263.

Brown J. R. and Arbuthnott G. W. (1983) The electrophysiology of dopamine (D2) receptors: a study of the actions of dopamine on corticostriatal transmission. *Neuroscience.* **10**, 349-355.

Calabresi P., Mercuri N., Stanzione P., Stefani A. and Bernardi G. (1987) Intracellular studies on the dopamine-induced firing inhibition of neostriatal neurons in vitro: evidence for D1 receptor involvement. *Neuroscience.* **20**, 757-771.

Calabresi P., Benedetti M., Mercuri N. B. and Bernardi G. (1988) Endogenous dopamine and dopaminergic agonists modulate synaptic excitation in neostriatum: intracellular studies from naive and catecholamine-depleted rats. *Neuroscience.* **27**, 145-157.

Calabresi P., Maj R., Pisani A., Mercuri N. B. and Bernardi G. (1992) Long-term synaptic depression in the striatum: physiological and pharmacological characterization. *J Neurosci.* **12**, 4224-4233.

Calabresi P., Saiardi A., Pisani A., Baik J. H., Centonze D., Mercuri N. B., Bernardi G. and Borrelli E. (1997) Abnormal synaptic plasticity in the striatum of mice lacking dopamine D2 receptors. *J Neurosci.* **17**, 4536-4544.

Cepeda C., Walsh J. P., Hull C. D., Howard S. G., Buchwald N. A. and Levine M. S. (1989) Dye-coupling in the neostriatum of the rat: I. Modulation by dopamine-depleting lesions. *Synapse.* **4**, 229-237.

Cepeda C., Buchwald N. A. and Levine M. S. (1993) Neuromodulatory actions of dopamine in the neostriatum are dependent upon the excitatory amino acid receptor subtypes activated. *Proc Natl Acad Sci U S A.* **90**, 9576-9580.

Cepeda C., Walsh J. P., Peacock W., Buchwald N. A. and Levine M. S. (1994) Neurophysiological, pharmacological and morphological properties of human caudate neurons recorded in vitro. *Neuroscience.* **59**, 89-103.

Cepeda C., Chandler S. H., Shumate L. W. and Levine M. S. (1995) Persistent Na+ conductance in medium-sized neostriatal neurons: characterization using infrared videomicroscopy and whole cell patch-clamp recordings. *J Neurophysiol.* **74**, 1343-1348.

Cepeda C. and Levine M. S. (1998) Dopamine and N-methyl-D-aspartate receptor interactions in the neostriatum. *Dev Neurosci.* **20**, 1-18.

Cepeda C., Hurst R.S., Altemus K.L., Flores-Hernandez J., Calvert C.R., Jokel E.S., Grandy D.K., Low M.J., Rubinstein M., Ariano M. A. and Levine M. S. (2001) Facilitated glutamatergic transmission in the striatum of D_2 dopamine receptor-deficient mice. *J. Neurophysiol.* **85**, 659-670.

Choi W. S., Machida C. A. and Ronnekleiv O. K. (1995) Distribution of dopamine D1, D2, and D5 receptor mRNAs in the monkey brain: ribonuclease protection assay analysis. *Brain Res Mol Brain Res.* **31**, 86-94.

Civelli O., Bunzow J. R., Grandy D. K., Zhou Q. Y. and Van Tol H. H. (1991) Molecular biology of the dopamine receptors. *Eur J Pharmacol.* **207**, 277-286.

Crawley J. N., Belknap J. K., Collins A., Crabbe J. C., Frankel W., Henderson N., Hitzemann R. J., Maxson S. C., Miner L. L., Silva A. J., Wehner J. M., Wynshaw-Boris A. and Paylor R. (1997) Behavioral phenotypes of inbred mouse strains: implications and recommendations for molecular studies. *Psychopharmacology (Berl).* **132**, 107-124.

Creese I. and Fraser C. M. (1997). Receptor biochemistry and methodology. In: Dopamine Receptors, vol 8 (Creese I. and Fraser C. M., eds), pp 1-125. New York: Liss.

Drago J., Gerfen C. R., Lachowicz J. E., Steiner H., Hollon T. R., Love P. E., Ooi G. T., Grinberg A., Lee E. J., Huang S. P. and et al. (1994) Altered striatal function in a mutant mouse lacking D1A dopamine receptors. *Proc Natl Acad Sci U S A.* **91**, 12564-12568.

Fisher R. S., Levine M. S., Sibley D. R. and Ariano M. A. (1994) D2 dopamine receptor protein location: Golgi impregnation-gold toned and ultrastructural analysis of the rat neostriatum. *J Neurosci Res.* **38**, 551-564.

Flores-Hernandez J., Galarraga E. and Bargas J. (1997) Dopamine selects glutamatergic inputs to neostriatal neurons. *Synapse.* **25**, 185-195.

Fonnum F., Storm-Mathisen J. and Divac I. (1981) Biochemical evidence for glutamate as neurotransmitter in corticostriatal and corticothalamic fibres in rat brain. *Neuroscience.* **6**, 863-873.

Fremeau R. T., Jr., Duncan G. E., Fornaretto M. G., Dearry A., Gingrich J. A., Breese G. R. and Caron M. G. (1991) Localization of D1 dopamine receptor mRNA in brain supports a role in cognitive, affective, and neuroendocrine aspects of dopaminergic neurotransmission. *Proc Natl Acad Sci U S A.* **88**, 3772-3776.

Galarraga E., Bargas J., Martinez-Fong D. and Aceves J. (1987) Spontaneous synaptic potentials in dopamine-denervated neostriatal neurons. *Neurosci Lett.* **81**, 351-355.

Garcia-Munoz M., Young S. J. and Groves P. M. (1991) Terminal excitability of the corticostriatal pathway. I. Regulation by dopamine receptor stimulation. *Brain Res.* **551**, 195-206.

Hall H., Halldin C., Dijkstra D., Wikstrom H., Wise L. D., Pugsley T. A., Sokoloff P., Pauli S., Farde L. and Sedvall G. (1996) Autoradiographic localisation of D3-dopamine receptors in the human brain using the selective D3-dopamine receptor agonist (+)-[3H]PD 128907. *Psychopharmacology (Berl).* **128**, 240-247.

Harvey J. and Lacey M. G. (1997) A postsynaptic interaction between dopamine D1 and NMDA receptors promotes presynaptic inhibition in the rat nucleus accumbens via adenosine release. *J Neurosci.* **17**, 5271-5280.

Hernandez-Lopez S., Bargas J., Surmeier D. J., Reyes A. and Galarraga E. (1997) D1 receptor activation enhances evoked discharge in neostriatal medium spiny neurons by modulating an L-type Ca2+ conductance. *J Neurosci.* **17,** 3334-3342.

Herrling P. L. and Hull C. D. (1980) Iontophoretically applied dopamine depolarizes and hyperpolarizes the membrane of cat caudate neurons. *Brain Res.* **192,** 441-462.

Hollon T. R., Gleason T. C., Grinberg A., Ariano M. A., Huang S. P., Drago J., Crawley J. N., Westphal H. and Sibley D. R. (1998) Generation of D5 dopamine receptor-deficient mice by gene targeting. *Soc Neurosci Abstr.* in press.

Holsboer F. (1997) Transgenic mouse models: New tools for psychiatric research. *Neuroscientist.* **3,** 328-336.

Hsu K. S., Huang C. C., Yang C. H. and Gean P. W. (1995) Presynaptic D2 dopaminergic receptors mediate inhibition of excitatory synaptic transmission in rat neostriatum. *Brain Res.* **690,** 264-268.

Huntley G. W., Morrison J. H., Prikhozhan A. and Sealfon S. C. (1992) Localization of multiple dopamine receptor subtype mRNAs in human and monkey motor cortex and striatum. *Brain Res Mol Brain Res.* **15,** 181-188.

Jiang Z. G. and North R. A. (1991) Membrane properties and synaptic responses of rat striatal neurones in vitro. *J Physiol (Lond).* **443,** 533-553.

Joyce J. N. and Marshall J. F. (1987) Quantitative autoradiography of dopamine D2 sites in rat caudate-putamen: localization to intrinsic neurons and not to neocortical afferents. *Neuroscience.* **20,** 773-795.

Kelly M. A., Rubinstein M., Asa S. L., Zhang G., Saez C., Bunzow J. R., Allen R. G., Hnasko R., Ben-Jonathan N., Grandy D. K. and Low M. J. (1997) Pituitary lactotroph hyperplasia and chronic hyperprolactinemia in dopamine D2 receptor-deficient mice. *Neuron.* **19,** 103-113.

Kelly M. A., Rubinstein M., Phillips T. J., Lessov C. N., Burkhart-Kasch S., Zhang G., Bunzow J. R., Fang Y., Gerhardt G. A., Grandy D. K. and Low M. J. (1998) Locomotor activity in D2 dopamine receptor-deficient mice is determined by gene dosage, genetic background, and developmental adaptations. *J Neurosci.* **18,** 3470-3479.

Koeltzow T. E., Xu M., Cooper D. C., Hu X. T., Tonegawa S., Wolf M. E. and White F. J. (1998) Alterations in dopamine release but not dopamine autoreceptor function in dopamine D3 receptor mutant mice. *J Neurosci.* **18,** 2231-2238.

Kornhuber J. and Kornhuber M. E. (1986) Presynaptic dopaminergic modulation of cortical input to the striatum. *Life Sci.* **39,** 699-674.

Levine M. S., Altemus K. L., Cepeda C., Cromwell H. C., Crawford C., Ariano M. A., Drago J., Sibley D. R. and Westphal H. (1996) Modulatory actions of dopamine on NMDA receptor-mediated responses are reduced in D1A-deficient mutant mice. *J Neurosci.* **16,** 5870-5882.

L'hirondel, M., Cheramy A., Godeheu G., Artaud F., Saiardi A., Borrelli E. and Glowinski J. (1998) Lack of autoreceptor-mediated inhibitory control of dopamine release in striatal synaptosomes of D2 receptor-deficient mice. *Brain Res.* **792,** 253-262.

Lovinger D. M., Merritt A. and Reyes D. (1994) Involvement of N- and non-N-type calcium channels in synaptic transmission at corticostriatal synapses. *Neuroscience.* **62,** 31-40.

Mansbach R. S., Brooks E. W., Sanner M. A. and Zorn S. H. (1998) Selective dopamine D4 receptor antagonists reverse apomorphine-induced blockade of prepulse inhibition. *Psychopharmacology (Berl).* **135,** 194-200.

Maura G., Giardi A. and Raiteri M. (1988) Release-regulating D-2 dopamine receptors are located on striatal glutamatergic nerve terminals. *J Pharmacol Exp Ther.* **247,** 680-684.

McVittie L. D., Ariano M. A. and Sibley D. R. (1991) Characterization of anti-peptide antibodies for the localization of D2 dopamine receptors in rat striatum. *Proc Natl Acad Sci U S A.* **88,** 1441-1445.

Meador-Woodruff J. H., Mansour A., Healy D. J., Kuehn R., Zhou Q. Y., Bunzow J. R., Akil H., Civelli O. and Watson S. J., Jr. (1991) Comparison of the distributions of D1 and D2 dopamine receptor mRNAs in rat brain. *Neuropsychopharmacology.* **5,** 231-242.

Mercuri N., Bernardi G., Calabresi P., Cotugno A., Levi G. and Stanzione P. (1985) Dopamine decreases cell excitability in rat striatal neurons by pre- and postsynaptic mechanisms. *Brain Res.* **358,** 110-121.

Mercuri N. B., Saiardi A., Bonci A., Picetti R., Calabresi P., Bernardi G. and Borrelli E. (1997) Loss of autoreceptor function in dopaminergic neurons from dopamine D2 receptor deficient mice. *Neuroscience.* **79,** 323-327.

Mitchell P. R. and Doggett N. S. (1980) Modulation of striatal [3H]-glutamic acid release by dopaminergic drugs. *Life Sci.* **26**, 2073-2081.

Mrzljak L., Bergson C., Pappy M., Huff R., Levenson R. and Goldman-Rakic P. S. (1996) Localization of dopamine D4 receptors in GABAergic neurons of the primate brain. *Nature.* **381**, 245-248.

Nicola S. M., Kombian S. B. and Malenka R. C. (1996) Psychostimulants depress excitatory synaptic transmission in the nucleus accumbens via presynaptic D1-like dopamine receptors. *J Neurosci.* **16**, 1591-1604.

Pennartz C. M., Dolleman-Van der Weel M. J., Kitai S. T. and Lopes da Silva F. H. (1992) Presynaptic dopamine D1 receptors attenuate excitatory and inhibitory limbic inputs to the shell region of the rat nucleus accumbens studied in vitro. *J Neurophysiol.* **67**, 1325-1334.

Piercey M. F., Hyslop D. K. and Hoffmann W. E. (1997) Excitation of type II anterior caudate neurons by stimulation of dopamine D3 receptors. *Brain Res.* **762**, 19-28.

Ralph R. J., Varty G. B., Kelly M. A., Wang Y. M., Caron M. G., Rubinstein M., Grandy D. K., Low M. J. and Geyer M. A. (1999) The dopamine D2, but not D3 or D4, receptor subtype is essential for the disruption of prepulse inhibition produced by amphetamine in mice. *J Neurosci.* **19**, 4627-4633.

Rowlands G. F. and Roberts P. J. (1980) Activation of dopamine receptors inhibits calcium-dependent glutamate release from cortico—striatal terminals in vitro. *Eur J Pharmacol.* **62**, 241-242.

Rubenstein M., Cepeda C., Hurst R.S., Flores-Hernandez J., Ariano M.A., Falzone T.L., Kozell L.B., Meshul C.K., Bunzow J.R., Low M.J., Levine M.S. and Grandy D.K. (2001) Dopamine D4 receptor-deficient mice display cortical hyperexcitability. *J. Neurosci.* **21**, 3756-3763.

Rutherford A., Garcia-Munoz M. and Arbuthnott G. W. (1988) An afterhyperpolarization recorded in striatal cells 'in vitro': effect of dopamine administration. *Exp Brain Res.* **71**, 399-405.

Schiffmann S. N., Lledo P. M. and Vincent J. D. (1995) Dopamine D1 receptor modulates the voltage-gated sodium current in rat striatal neurones through a protein kinase A. *J Physiol (Lond).* **483**, 95-107.

Sesack S. R., Aoki C. and Pickel V. M. (1994) Ultrastructural localization of D2 receptor-like immunoreactivity in midbrain dopamine neurons and their striatal targets. *J Neurosci.* **14**, 88-106.

Sibley D. R. and Monsma F. J., Jr. (1992) Molecular biology of dopamine receptors. *Trends Pharmacol Sci.* **13**, 61-69.

Smith A. D. and Bolam J. P. (1990) The neural network of the basal ganglia as revealed by the study of synaptic connections of identified neurones. *Trends Neurosci.* **13**, 259-265.

Sulzer D., Joyce M. P., Lin L., Geldwert D., Haber S. N., Hattori T. and Rayport S. (1998) Dopamine neurons make glutamatergic synapses in vitro. *J Neurosci.* **18**, 4588-4602.

Surmeier D. J., Eberwine J., Wilson C. J., Cao Y., Stefani A. and Kitai S. T. (1992) Dopamine receptor subtypes colocalize in rat striatonigral neurons. *Proc Natl Acad Sci U S A.* **89**, 10178-10182.

Surmeier D. J. and Kitai S. T. (1993) D1 and D2 dopamine receptor modulation of sodium and potassium currents in rat neostriatal neurons. *Prog Brain Res.* **99**, 309-324.

Tallman J. F., Primus R. J., Brodbeck R., Cornfield L., Meade R., Woodruff K., Ross P., Thurkauf A. and Gallager D. W. (1997) I. NGD 94-1: identification of a novel, high-affinity antagonist at the human dopamine D4 receptor. *J Pharmacol Exp Ther.* **282**, 1011-1019.

Tepper J. M., Sun B. C., Martin L. P. and Creese I. (1997) Functional roles of dopamine D2 and D3 autoreceptors on nigrostriatal neurons analyzed by antisense knockdown in vivo. *J Neurosci.* **17**, 2519-2530.

Trugman J. M., Geary W. A. d. and Wooten G. F. (1986) Localization of D-2 dopamine receptors to intrinsic striatal neurones by quantitative autoradiography. *Nature.* **323**, 267-269.

Wilson C. J. and Kawaguchi Y. (1996) The origins of two-state spontaneous membrane potential fluctuations of neostriatal spiny neurons. *J Neurosci.* **16**, 2397-2410.

Wolf M. E. and Roth R. H. (1985) Dopamine autoreceptor stimulation increases protein carboxyl methylation in striatal slices. *J Neurochem.* **44**, 291-298.

Xu M., Koeltzow T. E., Santiago G. T., Moratalla R., Cooper D. C., Hu X. T., White N. M., Graybiel A. M., White F. J. and Tonegawa S. (1997) Dopamine D3 receptor mutant mice exhibit increased behavioral sensitivity to concurrent stimulation of D1 and D2 receptors. *Neuron.* **19**, 837-848.

Yamamoto B. K. and Davy S. (1992) Dopaminergic modulation of glutamate release in striatum as measured by microdialysis. *J Neurochem.* **58**, 1736-1742.

Yang C. R. and Mogenson G. J. (1986) Dopamine enhances terminal excitability of hippocampal-accumbens neurons via D2 receptor: role of dopamine in presynaptic inhibition. *J Neurosci.* **6**, 2470-2478.

ANTIPSYCHOTIC DRUG-INDUCED MUSCLE RIGIDITY AND D_2 RECEPTOR OCCUPANCY IN THE BASAL GANGLIA OF THE RAT

Kim M. Hemsley and Ann D. Crocker*

Key words: antipsychotic drugs, dopamine receptors, occupancy, muscle rigidity, rat, EMG, substantia nigra, striatum, raclopride, clozapine.

Abstract: The relationship between tonic electromyographic (EMG) activity, an index of muscle rigidity, and D_2 dopamine receptor occupancy is reported for a range of antipsychotic drugs. Raclopride, chlorpromazine and fluphenazine increased EMG when ~70% of striatal D_2 receptors were occupied, whereas clozapine failed to increase EMG and had a maximum D_2 occupancy of 54%. The findings for each drug are discussed in the context of their interactions with D_2 and other neurotransmitter receptors.

1. INTRODUCTION

Treatment with antipsychotic drugs is associated with the appearance of extrapyramidal motor side effects (EPS) which result in up to 35% noncompliance with drug treatment [8]. These Parkinson-like motor side effects include muscle rigidity, akinesia and tremor and are believed to result from dopamine D_2 receptor antagonism in the basal ganglia. Indeed, the likelihood of motor side effects developing appears to be directly related to the affinity of the antipsychotic drug for dopamine D_2 receptors [1,14].

One of the characteristics of typical (but not atypical) antipsychotic drugs is their propensity to induce motor side effects. Positron emission tomography (PET) studies in humans have established there is an association between striatal D_2 receptor occupancy and motor side effects by demonstrating an increased incidence of motor side effects when a threshold of >70% striatal D_2 receptors are occupied by clinically relevant doses of a range

---

* K.M. Hemsley and A.D. Crocker, Department of Clinical Pharmacology and Centre for Neuroscience, The Flinders University of South Australia, Adelaide, 5001, South Australia, Australia. All requests concerning this article should be made to Ann D. Crocker.

of typical antipsychotic drugs [5,12]. Further, it has been suggested that the reason the atypical drug clozapine is not associated with EPS is because it occupies <63% of striatal D_2 receptors [5].

There have been few experimental studies to support and extend the findings from PET studies, largely due to the lack of a relevant animal endpoint of EPS. We have developed a quantitative, objective measure of muscle rigidity or increased muscle tone, one of the cardinal features of EPS, which is assessed as changes in the tonic electromyographic (EMG) activity in the anterior tibialis and gastrocnemius muscles of the hindlimb of conscious, unrestrained rats. We have shown that EMG activity is increased in rats following reserpine treatment or bilateral 6-hydroxydopamine lesions of the nigrostriatal pathway, both of which reduce dopamine release in the basal ganglia [3].

In the current study we have investigated the effects of a range of commonly used antipsychotic drugs on tonic EMG activity and have related them to D_2 receptor occupancy in the striatum and substantia nigra, using *ex vivo* quantitative autoradiographic techniques [13]. The drugs studied were raclopride, haloperidol, chlorpromazine, fluphenazine and clozapine which differ in their pharmacological and clinical profiles [6]. Raclopride was used as a model compound because of its high selectivity and affinity as a dopamine D_2 receptor antagonist [10]. Haloperidol, fluphenazine and chlorpromazine are typical antipsychotic drugs whose D_2 affinities range from high to moderate and which exhibit differences in their interactions with other neurotransmitter receptors. Finally clozapine is the prototypic atypical antipsychotic drug, exhibiting low D_2 affinity and multiple interactions with other receptors.

2. METHODS AND ANIMALS

The study was approved by the Flinders University Animal Ethics Committee and was performed in accordance with the guidelines of the Australian National Health and Medical Research Council (NH&MRC).

Male Sprague-Dawley rats were implanted with in-dwelling electrodes in the right hindlimb [4] the morning before animals were injected with drug or vehicle. One hour following complete recovery from anaesthesia, animals were connected to a Grass polygraph, allowed to habituate and a baseline EMG was recorded. Rats were divided into groups and injected with either vehicle or drug. Raclopride (Astra, Sweden), haloperidol, chlorpromazine and fluphenazine (Sigma, Australia) and clozapine (gift) were dissolved in isotonic saline or DMSO and injected subcutaneously in a volume of 1 ml/kg, at doses described in the text. Control rats received an injection of isotonic saline or DMSO as appropriate. EMG activity was monitored simultaneously from up to four rats per recording period, (always including both vehicle and drug injected rats) for up to 5h.

Rats were killed by decapitation at 2h to enable *ex vivo* determination of D_2 and muscarinic (mACh) receptor occupancy [13].

2.1 Statistics

The data was analysed using analysis of variance techniques, followed by post-hoc testing of the differences between means using Dunnett's test.

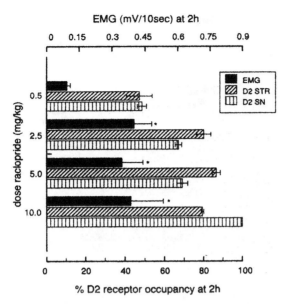

Figure 1. Effect of various doses of raclopride on EMG activity and D_2 receptor occupancy in the striatum and substantia nigra at 2h post-injection. (*$p<0.05$ c.f. control EMG)

3. RESULTS

3.1 Raclopride-Selective D_2 Antagonist

Raclopride (0.5-10 mg/kg), significantly and dose-dependently increased EMG activity over the 5 hour recording period. Significant maximal increases were obtained 2h post-injection with doses of 2.5mg/kg and greater (Figure 1).

The EMG activity was 0.39 ± 0.08mV/10sec and 0.40 ± 0.08mV/10sec in the tibialis and gastrocnemius muscles respectively, following 2.5mg/kg raclopride. EMG activity in control rats was 0.09 ± 0.02mV/10sec and 0.11 ± 0.02mV/10sec at the same time point. This increase in EMG activity was associated with D_2 receptor occupancies of 80% and 67% in the striatum and substantia nigra, respectively (Fig 1). The lowest dose tested (0.5mg/kg) did not significantly increase EMG activity at any timepoint and occupied a maximum of 47% and 48% of striatal and nigral D_2 receptors at 2h post-injection.

3.2 Haloperidol, Chlorpromazine and Fluphenazine

Haloperidol (0.025-5 mg/kg), chlorpromazine (0.1-10 mg/kg), and fluphenazine (0.05-3 mg/kg) all produced significant, dose-dependent increases in EMG activity in both muscles over time when compared with control animals. The effects on EMG activity were maximal 2-3h post-administration. A summary of the lowest dose of each drug which significantly increased EMG activity is shown in Figure 2. Striatal and nigral D2 receptor occupancy at these drug doses is also presented.

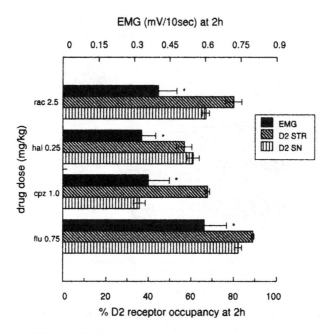

Figure 2. Summary of 'threshold' doses of antipsychotic drugs on EMG activity in the gastrocnemius muscle and dopamine D_2 receptor occupancy at 2h post-injection. (*$p<0.05$ c.f. control EMG)

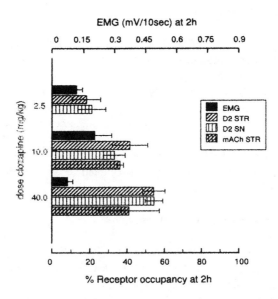

Figure 3. Effect of various doses of clozapine on EMG activity, striatal and nigral dopamine D_2 receptor occupancy and striatal muscarinic receptor occupancy at 2h post-injection.

3.3 Clozapine - Prototypic Atypical Antipsychotic Drug

Clozapine (2.5-40 mg/kg) failed to increase EMG activity at any dose administered (Fig 3). The highest occupancy of striatal (54%) and nigral (63%) D_2 receptors was observed following 40mg/kg clozapine. Interestingly, clozapine was also observed to occupy 41% and 36% of striatal muscarinic cholinergic receptors at the 40 and 10mg/kg doses, respectively (Fig 3).

4. DISCUSSION

The results of the current study show that there is an association between increases in tonic EMG activity, an objective, quantifiable measure of muscle rigidity, and the level of D_2 receptor occupancy in the rat brain induced by raclopride, haloperidol, chlorpromazine and fluphenazine. For example, large significant increases in EMG activity following the selective D_2 antagonist raclopride were observed at striatal D_2 occupancies of >80%. Chlorpromazine also produced significant increases in EMG activity when >68% of striatal D_2 receptors were occupied, whereas fluphenazine, did not significantly increase EMG activity until ~90% of striatal D_2 receptors were occupied. These occupancies are similar to those reported in PET studies to be associated with a high incidence of EPS [5].

Following haloperidol, significantly increased EMG activity was associated with striatal occupancies of >57%. The reason why this lower striatal D_2 occupancy was associated with high EMG levels is unclear, but it may relate to the significant interactions of haloperidol with other receptors. For example, it is well established that haloperidol binds strongly to sigma receptors, and drugs which bind to sigma receptors have been reported to be associated with a higher incidence of dystonias [9]. Thus antagonism of both D_2 and sigma receptors may explain haloperidol's high propensity to produce extrapyramidal side effects.

It is possible that interactions with serotonin $5HT_2$ receptors may also play a role in the production of EPS, as selective serotonin $5HT_2$ blockade has been reported to enhance some raclopride-induced behaviours [17]. In terms of the drugs used in the current study all, with the exception of raclopride, antagonise $5HT_2$ receptors with high (chlorpromazine) to moderate (haloperidol) potency [6]. Given that the relationship between EMG and D_2 occupancy is similar for raclopride and chlorpromazine, the findings do not support the view that $5HT_2$ antagonism has a significant role to play in EMG activity or determining the incidence of EPS.

Clozapine failed to produce significant increases in tonic EMG activity at any of the doses tested, which is consistent with clinical reports that its use is not associated with EPS. The explanation provided from PET studies is that clozapine does not occupy the threshold level of >70% D_2 receptors associated with motor side effects [12]. In our study we found that the largest striatal D_2 occupancy observed was 54% following a dose of 40mg/kg, approximately 4-5 times greater than doses used clinically, confirming that low D_2 occupancy may be the reason clozapine did not increase EMG. However, Seeman and Kapur [15] have argued that clozapine does occupy large numbers of dopamine D_2 receptors at clinically effective doses, but visualisation using imaging techniques is affected by the physico-chemical properties of the radioligands used. Therefore the low occupancy values obtained both in this study and in many others using PET and single photon computed tomography (SPECT), may be due to this phenomenon.

Another possible explanation for clozapine's reduced EPS potential is its potent muscarinic antagonist properties, because anti-cholinergic drugs are effective in treating EPS [11]. Significant striatal muscarinic receptor occupancies of up to 41% (Figure 3) were observed following 40mg/kg clozapine, so it is possible this receptor interaction contributed to clozapine's lack of effects on EMG activity [16]. The next highest, maximum muscarinic receptor occupancy observed was 38% with fluphenazine, then 22% with chlorpromazine, while haloperidol showed no detectable occupancy at any dose. The rank order of muscarinic occupancy for the four drugs tested, i.e. clozapine>fluphenazine>chlorpromazine>haloperidol, corresponds to their respective muscarinic receptor affinities determined in vitro [6].

Finally, in our study we also measured the occupancy of D_2 receptors in the substantia nigra because previous work from our laboratory showed that nigral D_2 receptors play a key role in the regulation of muscle tone and tonic EMG activity [2,4]. In addition we have recently reported that EMG is increased maximally when nigral D_2 receptors are inactivated by greater than 74% [7], further reinforcing their importance in muscle tone regulation. In the current study all four drugs were shown to exhibit dose dependent increases in D_2 receptor occupancy which were similar in both the striatum and the substantia nigra. Thus it is not possible to differentiate whether striatal or nigral D_2 occupancy is important in determining the magnitude of EMG activity. Similarly it can not be concluded from PET studies that D_2 occupancy in the striatum rather than in another region such as the substantia nigra, is responsible for the appearance of motor side effects. At this stage the relative importance of dopamine mechanisms in the striatum and substantia nigra in the regulation of muscle tone remains to be elucidated and is the focus of our current work.

5. ACKNOWLEDGEMENTS

KMH is a recipient of an NH&MRC 'Dora Lush' Biomedical Postgraduate Scholarship. The financial assistance of the NH&MRC and Flinders Medical Centre Foundation is gratefully acknowledged. The authors thank Dr. Ian Crosbie and Ms. Juliette Neve, Victorian College of Pharmacy, for the generous gift of clozapine and Astra (Sweden) for raclopride.

6. REFERENCES

1. Creese, I., Burt, D.R. and Snyder, S.H. (1976) Dopamine receptor binding predicts clinical and pharmacological potencies of antipsychotic drugs. Science 192, 481-483.
2. Crocker, A.D. (1997) The regulation of motor control: An evaluation of the role of dopamine receptors in the substantia nigra. Rev. Neurosci., 8, 55-76.
3. Double, K.L. and Crocker, A.D. (1993) Quantitative electromyographic changes following modifications of central dopaminergic transmission. Brain Res., 604, 342-344.
4. Double, K.L. and Crocker, A.D. (1995) Dopamine receptors in the substantia nigra are involved in the regulation of muscle tone. Proc. Nad. Acad. Sci. U.S.A. 92, 1669-1673.
5. Farde, L., Nordstrom, A.-L., Wiesel, F.-A., Pauli, S., Halldin, C. and Sedvall, G. (1992) Positron emission tomographic analysis of central D1 and D2 dopamine receptor occupancy in patients treated with classical neuroleptics and clozapine. Arch. Gen. Psychiatry 49, 538544.

6. Hacksell, U., Jackson, D.M. and Mohell, N. (1995) Does the dopamine receptor subtype selectivity of antipsychotic agents provide useful leads for the development of novel therapeutic agents? Pharmacol. & Toxicol., 76, 320-324.
7. Hemsley, K.M. and Crocker, A.D. (1998) The effects of an irreversible dopamine receptor antagonist, N-ethoxycarbonyl-2-ethoxy-1,2-dihydroquinoline (EEDQ), on the regulation of muscle tone in the rat: the role of the substantia nigra. Neurosci. Letts., 251, 77-80.
8. Hoge, S.K., Appelbaum, P., Lawlor, T., Beck, J.C., Litman, R., Greer, A., Gutheil, T.G. and Kaplan, E. (1990) A prospective multicenter study of patients' refusal of antipsychotic drugs. J. Clin. Psych. 55 29-35.
9. Jeanjean, A.P., Laterre, E.C. and Maloteaux, J.M. (1997) Neuroleptic binding to sigma receptors: possible involvement in neuroleptic-induced acute dystonia. Biol. Psychiatry, 41, 1010-1019.
10. Kohler, C., Hall, H., Ogren, S.-O. and Gawell, L. (1985) Specific in vitro and in vivo binding of ^{3}H-raclopride. Biochem. Pharmacol., 34, 2251-2259.
11. Marsden, C.D. and Jenner, P. (1980) The pathophysiology of extrapyramidal side-effects of neuroleptic drugs. Psychol. Med. 10, 55-72.
12. Nyberg, S., Nordstrom, A.-L., Halldin, C. and Farde, L.(1995) Positron emission tomography studies on D2 dopamine receptor occupancy and plasma antipsychotic drug levels in man. Int. Clin. Psychopharmacol. 10, supp 3 81-85.
13. Schotte, A., Janssen, P.F.M., Megans, A.A.H.P. and Leysen, J.E. (1993) Occupancy of central neurotransmitter receptors by risperidone, clozapine and haloperidol, measured ex vivo by quantitative autoradiography. Brain Res., 631, 191-202.
14. Seeman, P., Lee, T., Chau Wong, M. and Wong, K. (1976) Antipsychotic doses and neuroleptic/ dopamine receptors. Nature 261, 717-718.
15. Seeman, P. and Kapur, S. (1997) Clozapine occupies high levels of dopamine D2 receptors. Pharmacol. Letts., 60, 207-216.
16. Snyder, S.H., Greenberg, D. and Yamamura, HI (1974) Antischizophrenic drugs and brain cholinergic receptors. Arch. Gen. Psychiatry, 27, 169-179.
17. Wadenberg, M.-L., Salmi, P., Jimenez, P., Svensson, T. and Ahlenius, S. (1996) Enhancement of antipsychotic-like properties of the dopamine D2 receptor antagonist, raclopride by the additional treatment with the 5-HT2 receptor blocking agent, ritanserin, in the rat. Eur. Neuropsychopharmacol., 6, 305-310.

DIFFERENTIAL MODULATION OF SINGLE-UNIT ACTIVITY IN THE STRIATUM OF FREELY BEHAVING RATS BY D1 AND D2 DOPAMINE RECEPTORS

George V. Rebec and Eugene A. Kiyatkin*

Abstract: Continuous release of endogenous dopamine (DA) is believed to modulate both the activity of striatal neurons and their responsiveness to phasic fluctuations in glutamate (GLU). To determine the role of DA receptor subfamilies in these neuromodulatory effects, we combined single-unit recording with pharmacological blockade of either the D1 or D2 subfamily and assessed the effects of DA and GLU iontophoresis in rats. To ensure testing under naturally occurring physiological conditions, all data were obtained from freely behaving animals. Injection of SCH-23390 (0.2 mg/kg, sc), a D1 antagonist, elevated basal firing rate and blocked the weak inhibitory response to iontophoretic DA. This drug also enhanced the excitatory effect of GLU iontophoresis. In contrast, eticlopride (0.2 mg/kg, sc), a D2 antagonist, had a slightly inhibitory effect on basal firing rate and failed to alter the neuronal response to DA iontophoresis. Eticlopride also reduced the magnitude of GLU-induced excitations. A combination of both SCH-23390 and eticlopride increased basal firing rate and blocked the effect of DA iontophoresis, but both effects were smaller than after SCH-23390 alone. The combination attenuated, rather than enhanced, the GLU response. Collectively, these results suggest that DA acts mainly via D1 receptors to exert a tonic stabilizing influence on striatal activity, including phasic excitations induced by GLU.

1. INTRODUCTION

As the primary afferent structure of the basal ganglia, the striatum receives input from the entire cortical mantle.[16] Dorsal striatum mainly receives afferents from sensorimotor, auditory, and visual areas of neocortex, whereas input to ventral striatum or nucleus accumbens arises from allocortical and mesocortical areas as well as from the "cortical-like"

---

* G.V. Rebec and E.A. Kiyatkin, Program in Neural Science and Department of Psychology, Indiana University, Bloomington, IN 47405-7007.

basolateral amygdaloid complex.[5,8,18] In both dorsal and ventral striatum, cortical afferents terminate on the heads of the dendritic spines of medium spiny neurons, which account for roughly 95% of the neuronal population.[19]

In the absence of coordinated cortical input, these neurons are largely silent due in large part to an inwardly rectifying potassium current that keeps the membrane hyperpolarized.[22] Episodes of firing result from phasic activation of cortical afferents, which release glutamate (GLU), an excitatory amino acid.[17,21] Corticostriatal GLU release, therefore, plays a critical role in driving the activity of striatal neurons.

Another prominent set of striatal afferents arises from the ventral midbrain and releases dopamine (DA), a catecholamine modulator. DA fibers terminate en passant on the necks of spines within a few microns of GLU terminals.[19] This arrangement is interesting in light of evidence that DA may diffuse some distance beyond its release site to exert extrasynaptic effects.[3,4] It also is interesting that apart from brief fluctuations in response to certain stimuli, the level of activity in DA afferents is relatively stable.[11] Thus, mechanisms are in place to ensure continuous modulation by DA of GLU-evoked activity. Consistent with this view, a tonic increase in extracellular DA has been reported to enhance the magnitude of the neuronal response to GLU in striatum relative to the level of background activity.[9] By changing the GLU signal-to-noise ratio, DA may facilitate the phasic cortical activation of striatal circuits.

The effect of DA on the electrophysiology of the striatum is complicated by multiple DA receptors, which have been grouped into D1 and D2 subfamilies. Although both subfamilies occur throughout the striatum, their relative contributions to basal neuronal activity and neuronal responsiveness to DA and GLU remain unclear. To shed light on the role of tonic DA input in regulating striatal activity, we combined single-unit recording with DA and GLU iontopohoresis in rats treated with selective D1 (SCH-23390) or D2 antagonists (eticplopride) or their combination.[10] Data were collected from freely behaving animals to ensure a fully functioning striatal system.

2. METHODS

Male, Sprague-Dawley rats (-400 g) were anesthetized (chloropent, 0.33 ml/100g, ip) and prepared for freely moving electrophysiology and iontophoresis as previously described.[9] Animals were fitted with a skull-mounted plastic hub (1.2 mm anterior and 2.0 mm lateral to bregma), which was secured with dental cement to three stainless steel screws, one of which served as both electrical ground and attachment site for the head-mounted preamplifier. During a 3-4 day recovery period, the animals were habituated to both the recording chamber (a 1.2 m2 open-field arena) and the injection procedure in which physiological saline was injected under dorsal skin. Recording began on the following day and continued on a daily basis for the next 2-5 days.

Four-barrel, microfilament-filled, glass pipettes were pulled and broken to a diameter of 4-5 μm. The recording and balance barrels were filled with 3 M and 0.25 M solutions of NaCl, respectively. The remaining barrels contained 0.25 M solutions of L-GLU monosodium salt (pH 7.5) and DA hydrochloride (pH 4.5) dissolved in distilled water. A constant-current generator delivered the retaining (+/- 8 nA) and ejecting (+/- 0-60 nA) currents for iontophoresis. The drug-containing barrels had in vivo resistances of 20-40 MΩ (measured at constant current), and the recording barrel had an impedance of 4-10 MΩ(measured at 1 kHz). To eliminate channel cross-talk, the microfilaments were removed from the upper part of the barrels, and the opening of each barrel was separated by 2-3 mm and covered with paraffin. Freshly prepared drug solutions were injected into the barrels and the pipette was fixed in a microdrive assembly designed to mate with the skull-mounted hub. The electrode was advanced 4.0 mm below the skull surface to the starting point of unit recording.

Neuronal signals traveled from the head-mounted preamplifier to an electric swivel via shielded cable. Electrophysiological activity was amplified, filtered (bandpass: 100-3000 Hz), and stored on an audio channel of a videocassette recorder. Unit discharges were monitored on-line with a digital oscilloscope and audioamplifier, and counted in 2-sec binwidths by computer in conjunction with an amplitude-sensitive spike discriminator. A second audio channel was used to mark iontophoretic applications and behavioral events of interest. The video channel recorded behavioral activity.

After the isolation of single-unit discharges (signal-to-noise ratio of at least 3:1), data collection for each neuron usually lasted for 20-30 min. The protocol included several brief (20 sec) applications of GLU (0 to -40 nA) and DA (0 to +40 nA) at 1-2 min intervals. Silent units were detected with continuous or pulsatile ejections of GLU at low currents and then tested with DA during continuous GLU application. To eliminate movement-related changes in basal discharge activity, statistical analysis was restricted to iontophoretic applications in which the animals rested quietly with no sign of overt movement. Movement episodes that occurred during data collection were analyzed separately.

After 2-3 hr of recording, animals received a sc injection of either saline or one of the following: SCH-23390 (0.2 mg/kg), eticlopride (0.2 mg/kg), or an SCH-23390-eticlopride combination at the same dose of each. Doses were selected to ensure maximal blockade of D1 (SCH-23390) or D2 (eticlopride) receptors with minimal interaction at other binding sites.[15]

Injections were typically performed during recording of baseline unit activity to allow pre- and postdrug comparisons of baseline firing rates as well as responsiveness to DA and GLU. Further recordings were made from stochastically sampled units tested at different times after injection (up to 180 min).

When the last recording session was completed, each rat was anesthetized (chloropent), and an insulated tungsten electrode, inserted into one barrel of the pipette, was lowered into the recording area. A small current was applied to the electrode at 5.0 and 7.0 mm ventral to the skull surface to mark the electrode track. After transcardial perfusion with formosaline, the brain was removed, sectioned, and stained with cresyl violet for histological analysis. The location of each recording site was estimated from histological data and depth information noted on the microdrive assembly at the time of each recording.

Excitations or inhibitions of individual neurons during iontophoresis were considered significant when the mean rate was statistically different from the mean rate during the immediately preceeding equivalent period ($p < 0.01$, two-tailed Student's t test). Iontophoretic responses also were examined for onset and offset latencies, absolute and relative magnitudes; and dose (current)-response relationships. Group effects were analyzed with a one-way, repeated-measures ANOVA; between-group comparisons examined the number of units sensitive to DA and GLU and the excitation:inhibition ratio for each ejection current. Units tested with multiple current applications of GLU were analyzed for response threshold. Not all units could be tested equally with either DA or GLU due to shifts in firing rate caused by episodes of movement or other factors. Thus, not all recorded units were included in assessments of response thresholds and dose-response relationships. Various relationships between impulse activity and iontophoresis were evaluated with standard statistical techniques. Basal impulse activity was characterized in terms of mean rate, standard deviation, and coefficient of variation, all of which were calculated for each unit based on a 30-s period of quiet rest. Off-line analysis of videotape records was used to examine impulse activity during bouts of spontaneous movement or during the presentation of somatosensory stimuli such as a light touch of the tail with a plastic rod.

3. RESULTS

A total of 207 neurons was sampled in dorsal and ventral striatum of rats under control conditions, i.e., before injection of DA receptor antagonists or before and after saline administration. The animals rested quietly for prolonged periods with no overt movement. Consistent with previous results obtained during rest periods, [6] most neurons discharged relatively slowly; less than 30% had mean firing rates above 6 spikes/s. Although all neurons showed some variability in rate, slow-firing units had the most irregular firing pattern as evidenced by a strong negative correlation between rate and coefficient of variation ($r = -0.79$; $p < 0.001$). Interestingly, neurons in ventral striatum ($n = 66$) had a significantly higher mean rate (10.35 vs 6.39 spikes/s) and lower coefficient of variation (56.51 % vs 75.31 %) than dorsal striatal ($n = 141$) cells ($p < 0.01$ in both cases). Some neurons in both locations ($n = 35$) were very slow or silent (< 1 spike/s) and required continuous, low-current GLU application (10 - 20 nA) to maintain a stable discharge rate (2 - 41 spikes/s).

Control responses to iontophoretic DA were obtained from both spontaneously active ($n = 122$) and GLU-stimulated neurons ($n = 35$). In either case, the predominant response to DA in either dorsal or ventral striatum was a mild to moderate inhibition (baseline firing rate declined by 20 - 50%) of variable onset latency (3-12 s). The inhibition, moreover, was highly dependent on dose. A subset of spontaneously active ($n = 25$) and GLU-stimulated units ($n = 26$) was tested at 5, 10, and 20 nA DA resulting in ~25%, ~50%, and ~75% inhibitions, respectively. Although the fastest-firing units showed the greatest reduction in firing rate to DA, the relative magnitude of the DA-induced inhibition was unrelated to rate. Of the more than 650 total DA applications that occurred under control conditions, only 13 significantly increased neuronal activity.

The effects of brief applications of GLU were tested on 140 spontaneously active units. Both dorsal and ventral striatal units were highly sensitive to GLU. A dose-response analysis revealed that 87.5% of the units (14/16) tested at 5 nA GLU showed a significant increase in rate, and at 40 nA GLU all tested units (21/21) were excited. Onset latencies were relatively short (2-6 s), and the excitatory response was sustained with repeated applications. The mean threshold for excitation was 19.7 ± 1.75 nA. The magnitude of the GLU effect, moreover, depended on the level of basal firing rate in that the fastest firing cells showed the smallest relative increase. In fact, GLU rarely increased spike rates beyond 65 spikes/s even among units that had basal levels of activity near this rate. In some fastfiring units, the GLU-induced excitation was followed by a brief (< 20 s) rebound-like decrease in activity, but a complete disappearance of discharges indicative of depolarization inactivation rarely occurred.

Administration of SCH-23390 significantly altered basal firing activity. Mean firing rate after this drug was more than twice that of control (14.91 vs 6.39 spikes/s; $p < 0.0001$), but the faster rate meant a lower coefficient of variation (42.65% vs 69.31%; p < 0.001). In addition, 4 of the 5 spontaneously active units recorded both before and after SCH-23390 showed a significant increase in rate within 10-15 min after injection (p < 0.01). SCH-23390 also had an effect on behavior. Compared to the control condition when rats rested quietly with periodic displays of head and/or body movements, SCH-23390 induced prolonged periods of quiet sitting or apparent sedation. Typical pre-drug behavioral patterns did not re-emerge until 60-90 min after injection. Discharge rate, in contrast, remained elevated for ~ 180 min, and the magnitude of the increase was relatively constant across this period.

The most dramatic effect of SCH-23390 on the neuronal response to DA iontophoresis was blockade of the inhibition. Most units failed to respond to DA at any tested current. In fact, only one-third of DA applications as high as 40 nA (8/24) led to a significant inhibition of spontaneously active units, and this response was significantly weaker after SCH-23390 than in the control condition. Comparable results were obtained from five GLU-stimulated units tested at 5, 10, and 20 nA DA; in 16 of 19 applications, DA failed to have any effect on firing rate, and in three cases, firing rate actually increased. DA-induced inhibitions began to re-emerge ~120 min after SCH-23390, and DA data obtained after this time were not included in our analysis of SCH-23390 effects.

SCH-23390 enhanced several aspects of the neuronal response to GLU. For one, the absolute magnitude of the GLU-induced excitation was significantly higher at each tested current after SCH-23390 than in the control condition. Secondly, SCH-23390 significantly (p < 0.01) lowered the threshold for GLU-induced excitation to 4.17 ± 0.83 nA; many units, in fact, were excited by GLU at ejection currents < 10 nA. Finally, 5 of 17 units responded to increasing GLU doses (20-40 nA) with a partial or total cessation of firing rate after SCH-23390. This effect occurred at the peak of the GLU-induced excitation when firing rates exceeded 60 spikes/s and in some cases approached 140 spikes/s. In addition, this effect, which resembled depolarization inactivation, was short-lived (<20 s) and readily reproducible.

Like SCH-23390, eticlopride had a sedative effect on behavior,, though increases in defecation and urination occurred that were not apparent after SCH-23390. The effect of eticlopride on basal firing rate (both spontaneous and GLU-evoked) and on neuronal responses to DA and GLU iontophoresis differed substantially from those of SCH-23390. Eticlopride inhibited basal impulse activity; 32 units recorded after eticlopride had a lower mean firing rate than control (3.53 vs 6.39; p < 0.01). In fact, the fastest basal rate in the eticlopride group was < 12 spikes/s. The coefficient of variation for both groups, however, was comparable (69.31% vs 71.88%; p > 0.05). Three of five spontaneously active units recorded both before and after eticlopride showed a significant post-eticlopride inhibition; the activity of the remaining two units showed no post-eticlopride effect. Surprisingly, responses to DA in the eticlopride group were comparable to the control condition despite the generally inhibitory effect of eticlopride on basal firing rate. In fact, GLU-evoked activity, which tended to be faster than spontaneous rates, showed an even stronger DA-induced inhibition after eticlopride than GLU-evoked control units. Spontaneously active neurons responded to GLU with excitations in the eticlopride group, but the absolute magnitude of the GLU response was smaller than in the control condition. The maximum rate of the GLU-induced excitation was only one-third that of control. Another aspect of the GLU response in the eticlopride group was a post-excitatory inhibition, which in some cases resembled depolarization inactivation. Unlike the SCH-23390 group, however, this inhibitory effect occurred at rates < 25 spikes/s.

Administration of the SCH-23390-eticlopride combination, like either drug alone, suppressed behavior, but also increased defecation and urination. The behavioral effects of the combination were more prolonged and more pronounced than after either drug alone. At the neuronal level, the effects of the SCH-23390-eticlopride combination resembled those of the SCH-23390 group but were less pronounced. Thus, the combination was intermediate between the SCH-23390 and control groups in both mean firing rate and coefficient of variation. Among the five units recorded before and after injection of the combination, two showed an increase in firing rate and three failed to change. Similarly, the response to GLU was enhanced above control in the combination group, but again the enhancement was less than that observed with SCH-23390 alone. In some cases, the combination also caused a partial or complete inhibition of firing at high GLU currents (40 nA). With respect to DA iontophoresis, however, the combination closely resembled the SCH-23390 group in that DA-induced inhibitions of spontaneously active units were markedly attenuated compared to control. In fact, the attenuation was quite prolonged in the combination group, lasting for up to 130 min after

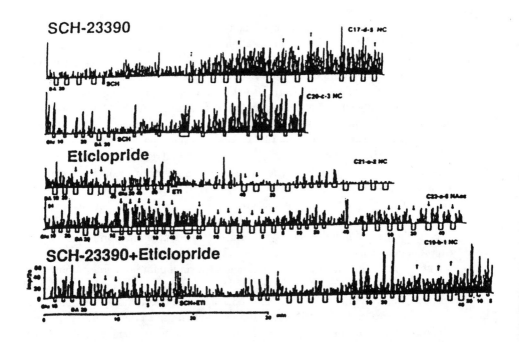

Figure 1. Rate-meter histograms of representative units in the striatum showing their responses to iontophoretic DA (long open box) and GLU (short open box) after administration of SCH-23390, eticlopride, or their combination. Numbers below iontophoretic applications indicate ejection current in nA; no numbers for subsequent applications indicate the same current. In all cases, neuronal activity is presented as impulses/2 sec. A continuous line below the histogram (e.g., C23-a-5 Nacc) indicates prolonged GLU application. Arrows above selected DA applications indicate significant decreases or increases in activity. Solid boxes indicate drug injection (SCH, ETI, or SCH + ETI). The histological location of each recording is identified in the top right corner of each histogram (striatum, NC or nucleus accumbens, NAcc) along with the number of the rat, session, and unit, respectively. Solid lines below each histogram indicate periods of spontaneous movement; at all other times the animals rested quietly.

injection. Surprisingly, the combination did not block the effects of DA applied to GLU-evoked activity.

Representative neuronal responses to DA and GLU iontophoresis before and after SCH-23390, eticlopride, or their combination are shown in Fig. 1.

5. DISCUSSION

Our use of freely behaving rats ensures that striatal circuits, including afferent fibers, are operating normally and thus are free of the complications associated with anesthetized or in vitro preparations. By using selective DA receptor antagonists, we were able to assess the receptor mechanisms through which endogenous DA regulates the basal activity level of striatal neurons and their responsiveness to GLU. Our results, which corroborate previous data from our laboratory,[9] indicate that neurons in both dorsal and ventral striatum receive tonic inhibitory input from DA terminals, which modulates both basal unit activity and the excitatory effects of GLU. We also found that these DA-mediated effects depend primarily on the D1 receptor subfamily.

A critical role for D1 receptors is supported by the increase in basal discharge rate that followed injection of SCH-23390, but not eticlopride. In fact, eticlopride had the opposite effect of slightly lowering basal activity, which may reflect an increase in endogenous DA release prompted by a blockade of D2 autoreceptors.[2] Interestingly, the SCH-23390-eticlopride combination generally had intermediate effects, further supporting a differential action of D1 and D2 receptors in the control of striatal activity by tonic release of endogenous DA.

We also found that SCH-23390 enhanced neuronal responsiveness to GLU even to the point of inducing several instances of apparent depolaization inactivation, a rare event in control conditions. Thus, D1 receptor blockade seems to remove the critical stabilizing influence of DA on striatal responsiveness to afferent stimuli. Instability in striatal firing patterns may explain the disruptive effect of SCH-23390 on performance in a wide range of learning tasks.[1,14] Eticlopride, in contrast, decreased GLU responsiveness and had an overall dampening effect on striatal activity. To the extent that eticlopride enhances DA release (see above), these effects may represent a strengthening of D1 receptor activation. Not surprisingly, some aspects of the GLU response after the SCH-23390-eticlopride combination are consistent with either D1 or D2 blockade.

Our data on DA iontophoresis, which addresses the effect of phasic DA release, also suggest a prominent role for D1 receptors. SCH-23390, for example, strongly blocked DA-induced inhibitions, while eticlopride had little or no effect. It is noteworthy, however, that combined D1-D2 receptor blockade revealed an interesting difference between spontaneously active and GLU-stimulated units. Whereas spontaneously active units showed even less response to DA after the SCH-23390-eticlopride combination than with SCH-23390 alone, the combination failed to attenuate the effects of DA on GLU-evoked activity. Possible D1 and D2 differences in regulating these two neuronal populations warrant further investigation.

Release of GLU from corticostriatal afferents is the major source of excitation of striatal neurons and appears to mediate the neuronal activation that occurs in response to sensory stimulation or during behavior. In view of evidence that DA receptors modulate GLU release, [13,20] it is conceivable that our results with SCH-23390 and eticlopride are due, at least in part, to increases or decreases in GLU release after D1 or D2 blockade, respectively. By itself, however, a change in GLU release cannot explain the altered responsiveness of striatal neurons to iontophoretic GLU. SCH-23390, for example, enhanced the sensitivity of striatal neurons to GLU, and eticlopride had the opposite effect. It appears that endogenously released DA, acting mainly via D1 receptors, exerts a tonic inhibitory influence on striatal neurons and their responsiveness to GLU. The role of D2 receptors is more difficult to characterize owing to their location on DA terminals, which allows for direct modulation of DA release. Thus, the modest inhibition of basal firing rate after eticlopride does not necessarily indicate an excitatory role for postsynaptic D2 receptors. In fact, systemic injection of quinpirole, a D2 agonist, suppresses striatal activity.[7] It may be that removal of the inhibitory influence of D 1 receptors enhances the sensitivity of striatal neurons to afferent information, and this effect persists after combined D1-D2 blockade.

The ability of DA to alter GLU responsiveness also may depend on the type of GLU receptor involved. In striatal slices, for example, DA has been reported to enhance activity at N-methyl-D-aspartate (NMDA) receptors but to attenuate non-NMDA-mediated responses.[12] Other factors are likely to add to the complexity of DA-mediated effects in behaving animals, including behavior-related changes in GLU input. Shifts in basal firing rate, which depend on corticostriatal GLU activity, strongly influence the effect of DA. Both DA and GLU inputs, moreover, are modulated by other systems. Thus, although our data support the role of DA as striatal modulator and suggest a critical function for D1 receptors, the mechanisms that regulate these effects under behaviorally relevant conditions require further study.

6. ACKNOWLEDGEMENTS

This research was supported by the National Institute on Drug Abuse (DA 02451 and DA00335). We also acknowledge the technical assistance of Paul Langley and the help of Faye Caylor in preparing the manuscript.

7. REFERENCES

1. R. J. Beninger, D-1 receptor involvement in reward-related learning, *J. Psychopharmacol.* 6: 34-42 (1992).
2. B. S. Bunney, L. A. Chiodo, and A. A. Grace, Midbrain dopamine system electrophysiological functioning: a review and new hypothesis, *Synapse* 9: 79-94 (1991).
3. K. Chergui, H. Akaoka, P. J. Charlety, C. F. Saunier, M. Buda, and G. Chouvet, Subthalamic nucleus modulates burst firing of nigral dopamine neurones via NMDA receptors, *Neuroreport* 5: 1185-1188 (1994).
4. L. Descarries, K. C. Watkins, S. Garcia, O. Bosler, and G. Doucet, Dual character, asynaptic and synaptic, of the dopamine innervation in adult rat neostriatum: A quantitative autoradiographic and immunocytochemical analysis, *J. Comp. Neurol.* 375:167-186 (1996).
5. J. P. Donoghue and M. Herkenham, Neostriatal projections from individual cortical fields conform to histochemically distinct striatal compartments in the rat, *Brain Res.* 365: 397-403 (1986).
6. J. L. Haracz, J. T. Tschanz, Z. Wang, I. M. White, and G. V. Rebec, Striatal single-unit responses to amphetamine and neuroleptics in freely moving rats, *Neurosci. Biobehav. Rev.* 17: 1-12 (1993).
7. K. C. Hooper, D. A. Banks, L. J. Stordahl, I. M. White, and G. V. Rebec, Quinpirole inhibits striatal and excites pallidal neurons in freely moving rats, *Neurosci. Lett.* 237: 69-72 (1997).

8. H. Kita and S. T. Kitai, Amygdaloid projections to the frontal cortex and the striatum in the rat, *J. Comp. Neurol.* 298: 40-49 (1990).
9. E. A. Kiyatkin and G. V. Rebec, Dopaminergic modulation of glutamate-induced excitations of neurons in the neostriatum and nucleus accumbens of awake, unrestrained rats, *J. Neurophysiol.* 75: 142-153 (1996).
10. E. A. Kiyatkin and G. V. Rebec, Striatal neuronal activity and responsiveness of striatal neurons to dopamine and glutamate after selective blockade of D1 and D2 dopamine receptors in awake rats, *J. Neurosci.* 19: 3594-3609 (1999).
11. M. Le Moal and H. Simon, Mesocorticolimbic dopaminergic network - functional and regulatory roles, *Physiol. Rev.* 71: 155-234 (1991).
12. M. S. Levine, K. L. Altemus, C. Cepeda, H. C. Cromwell, C. Crawford, M. A. Ariano, J. Drago, D. R. Sibley, and H. Westphal, Modulatory actions of dopamine on NMDA receptor-mediated responses are reduced in D-IA-deficient mutant mice, *J. Neurosci.* 16: 5870-5882 (1996).
13. G. Maura, A. Giardi, and M. Raiteri, Release-regulating D-2 dopamine receptors are located on striatal glutamatergic nerve terminals, *J. Pharnwcol. Exp. Ther.* 247: 680-684 (1988)
14. R. Miller, J. D. Wickens, and R. J. Beninger, Dopamine D-1 and D-2 receptors in relation to reward and performance: A case for the D-1 receptor as a primary site of therapeutic action of neuroleptic drugs, *Prog. Neurobiol.* 34: 143-183 (1990).
15. K. A. Neve. and R. L. Neve, Molecular biology of dopamine receptors, in: *The Dopamine Receptors* (K. A. Neve and R. L. Neve, eds.), Humana Press, Totowa, NJ, pp. 27-76 (1997).
16. A. Parent and L.-N. Hazrati, Functional anatomy of the basal ganglia. I. The corticobasal ganglia-thalamo-cortical loop, *Brain Res. Rev.* 20: 91-127 (1995).
17. A. Parent, P. Y. Cote, and B. Lavoie, Chemical anatomy of primate basal ganglia, *Prog. Neurobiol.* 46: 131-197 (1995).
18. C. M. A. Pennartz, F. H. L. Dasilva, and H. J. Groenewegen, The nucleus accumbens as a complex of functionally distinct neuronal ensembles: An integration of behavioural, electrophysiological and anatomical data, *Prog. Neurobiol.* 42: 719-761 (1994).
19. A. D. Smith and J. P. Bolam, The neural network of the basal ganglia as revealed by the study of synaptic connections of identified neurones, *Tr. Neurosci.* 13: 259-265 (1990).
20. J. Q. Wang and J. F. McGinty, Glutamatergic and cholinergic-regulation of immediate early gene and neuropeptide gene expression in the striatum, in: *Pharmacological Regulation of Gene Expression in the CNS* (K. Merchant, ed.), CRC Press, Boca Raton, FL, pp. 81-113 (1996).
21. C. J. Wilson, The generation of natural firing patterns in neostriatal neurons, in: *Chemical Signalling in the Basal Ganglia,* (G. W. Arbuthnott and P. C. Emson, eds.), Elsevier, Amsterdam, pp. 277-297. (1993).
22. C. J. Wilson, and Y. Kawaguchi, The origins of two-state spontaneous membrane potential fluctuations of neostriatal spiny neurons, *J. Neurosci.* 16: 2397-2410 (1996).

THE RELATIONSHIP OF DOPAMINE AGONIST-INDUCED ROTATION TO FIRING RATE CHANGES IN THE BASAL GANGLIA OUTPUT NUCLEI

David N. Ruskin, Debra A. Bergstrom and Judith R. Walters*

1. INTRODUCTION

Pharmacological manipulations of brain dopamine (DA) systems modify motor behavior. Current models of basal ganglia function predict that dopamine agonist-induced motor activation, such as rotation or stereotypic behavior, is mediated by decreases in activity in the output nuclei of the basal ganglia, the substantia nigra pars reticulata (SNPR) and entopeduncular nucleus (EPN), brought about by activation of striatal DA receptors. These decreases in activity are thought to lead to disinhibition in motor thalamocortical and tectofugal circuits. Electrophysiology has provided some support for these models: in animals with nigrostriatal lesions, the D_1/D_2 agonist apomorphine or high doses of the partial D_1 agonist SKF 38393 cause decreases in firing rate in most neurons of the SNPR[19,20] or the primate homolog of the EPN.[3]

Contrary to the predictions of current models, many systemic DAergic agonist treatments that induce motor activation (either rotation or stereotypy) fail to produce overall decreases in firing rate in the SNPR: A. apomorphine or amphetamine administered to normal rats produces either a range of effects on firing rates, resulting in little or no average effect[18] (Fig. 1), or a net increase in rate,[11] B. the D_2 agonist quinpirole administered to nigrostriatal-lesioned rats also produces little or no average effect on firing rate, [20] and C. SKF 38393 administered to rats that have been treated subchronically with reserpine causes consistent *increases* in firing rates[7] (Fig. 1). Few studies have examined the effect of DA agonists on firing rate in the rodent EPN.

We recently re-examined the relationship of motor behavior and SNPR/EPN firing rate in nigrostriatal-lesioned rats under conditions in which the predicted relationship (i.e. correlated rotation and decreased SNPR/EPN activity) appeared to hold true, namely following

---

* D.N. Ruskin, D.A. Bergstrom and J.R. Walters, Experimental Therapeutics Branch, National Institute of Neurological Disorders and Stroke, National Institutes of Health, Bethesda, MD, USA.

The Basal Ganglia VI
Edited by Graybiel *et al.*, Kluwer Academic/Plenum Publishers, 2002

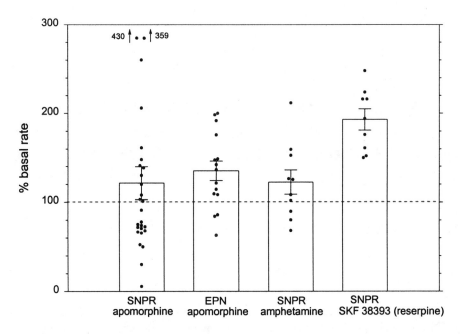

Figure 1. Several motor-activating DA agonist treatments fail to result in net decreases in firing rate in the SNPR or EPN. In neurologically-intact rats, i.v. apomorphine (0.32 mg/kg) or i.v. amphetamine (3.2 mg/kg, n=5; 6.4 mg/kg, n=5) causes a wide range of responses, including decreases, increases, or little change. Far right: In rats treated subchronically with reserpine, i.v. SKF 38393 (10 mg/kg) induces a net increase in SNPR firing rate. Adapted from refs. 6, 14, 18; amphetamine results are unpublished observations.

systemic treatment with D_1 or mixed D_1/D_2 agonist.[15] The effects of SKF 38393, chloroAPB (SKF 82958, a full D_1 agonist) and apomorphine on SNPR and/or EPN firing rates (in awake, immobilized rats) and on rotation (in freely-moving rats) were assessed. Identical drug doses and routes of administration were used for the behavioral and electrophysiological studies.

2. METHODS

Male Sprague-Dawley rats were used. Under chloral hydrate anesthesia, 6-hydroxydopamine (6µg/3µl) was infused unilaterally into the left median forebrain bundle, after i.p. desipramine pretreatment. Three to four weeks after surgery, rats were screened for lesion completeness with a s.c. injection of 0.05 mg/kg apomorphine. Non-rotators were excluded from further study.

Extracellular recordings were performed in awake, gallamine-paralyzed, artificially-respired rats held in a stereotaxic frame, as detailed in ref. 15. Briefly, surgical procedures were done under halothane anesthesia, and rats had pressure points and incisions thoroughly infiltrated with a local anesthetic. Single unit activity was recorded with glass microelectrodes. All recorded units had biphasic waveforms and were ipsilateral to the nigrostriatal lesion. At least 5 min of baseline activity was recorded before i.v. DA agonist injection, and recordings were continued for at least 10 min after injection (30 min for apomorphine). Drug effects were tested on one unit per rat. For D_1 agonists, drug-induced rate changes are given as the rate 4 - 10 min post-injection as percent of rate in the 5 min prior to injection. Recording sites were confirmed histologically. In studies of behavior, apomorphine (0.05 mg/kg) was given s.c. and rotational behavior was quantified for 30 min after injection. In other rats, SKF 38393, chloroAPB, or vehicle was given i.v. through the tail vein while rats were gently restrained. Rats were removed from restraint immediately after injection and rotational behavior was quantified for 20 min post-injection.

For immunocytochemistry, immobilized or freely-moving rats were perfused with 4% paraformaldehyde two hr after i.v. injection of SKF 38393 or vehicle. Brain sections through the striatum and subthalamic nucleus (STN) were immunostained for Fos-like antigen. The primary antibody (Oncogene Sciences, Ab-05/PC38) was used at 1:15,000, with a 24 hr room temperature incubation. Staining was visualized with diaminobenzidine. Nuclei positive for Fos-like immunoreactivity were quantified with grain counting software. Paralyzed rats used for immunocytochemistry were a separate set than those used for electrophysiology.

3. RESULTS

3.1 D_1 Agonist Effects on Rotational Behavior and Firing Rate

ChloroAPB caused significant rotation contraversive to the 6-OHDA lesioned side at all tested doses (Fig. 2A). However, this drug caused a significant net change (a decrease) in average SNPR firing rate only at the highest tested dose (Fig. 2A). While lower doses did not produce effects that were different from vehicle, inspection of individual neuronal responses demonstrates that these doses produced a range of firing rate effects in different neurons, including decreases, increases, and little change.

SKF 38393 produced significant contraversive rotation at 1.3 and 3.4 mg/kg (Fig. 2B). However, neither dose produced a net effect on SNPR firing rate, but instead produced a wide range of firing rate changes in individual neurons (Fig. 2B). Neurons of the EPN responded in a very similar manner (tested with 1.3 mg/kg; Fig. 2B). Example recordings of these varied EPN unit responses are given in Fig. 3.

3.2 Apomorphine Effects on Rotational Behavior and Firing Rate

Injection of apomorphine (0.05 mg/kg s.c.) produced contraversive rotation from 3 min post-injection onward (Fig. 4). Apomorphine's firing rate effects on SNPR neurons were delayed in comparison. A net decrease in rate did not occur until 12 - 16 min post-injection; the unit demonstrating the most rapid-onset inhibition dropped below 80% of basal rate at 9 min post-injection (Fig. 4). Also, two units had temporary rate increases before being inhibited. By 22 min, all neurons were robustly inhibited (average: 2% of basal rate). These late inhibitions were reversed by i.v. administration of haloperidol.[15]

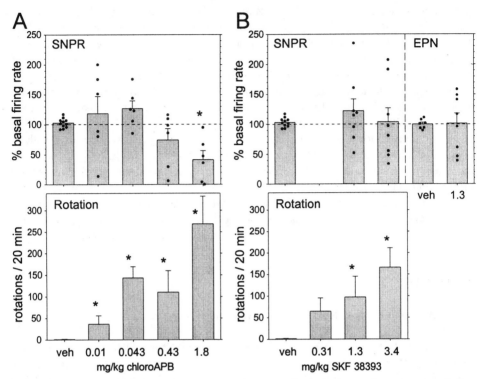

Figure 2. Effects of D_1 agonists on contraversive rotation and SNPR and EPN firing rate. A: Effects of chloroAPB. This drug induced rotation at all tested doses, but had a net effect on SNPR firing rate only at the highest dose. B: Effects of SKF 38393. This drug induced significant rotation at the two highest tested doses, yet had variable effects on SNPR and EPN firing rates. n = 5 - 6 for rotation groups. * p<.05 comparison to vehicle. Data in this figure (and subsequent figures) adapted from ref. 15.

3.3 D_1 Agonist Effects on Fos Immunoreactivity

The number of Fos-immunoreactive cell nuclei was greatly increased by SKF 38393 in both the striatum and STN in the 6-OHDA-treated hemisphere of freely-moving rats, compared to vehicle-injected rats (Fig. 5). The Fos induction in the STN supports previous findings in freely-moving rats.[16] Furthermore, SKF 38393 also increased the number of Fos-positive nuclei in the striatum and STN in the 6-OHDA-treated hemisphere of awake, immobilized rats in a manner not significantly different from that in the freely-moving rats (Fig. 5).

4. DISCUSSION

Behavioral studies using intracerebral infusion techniques have demonstrated that infusion of inhibitory substances into the SNPR or EPN produces contraversive rotation (for instance, ref. 17). Therefore, an overall inhibition of SNPR or EPN neurons seems to be

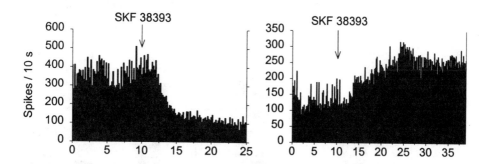

Figure 3. Rate histograms of EPN neurons recorded ipsilateral to a nigrostriatal lesion. Rate data are shown in 10 s bins. Time is shown in minutes. SKF 38393 at a rotation-inducing dose (1.3 mg/kg i.v.) has strikingly different effects on firing rate in these two neurons. The rate of the unit at left decreases by 60-70%, while the rate of the unit at right more than doubles. Later injection of D_1 antagonist (SCH 23390) completely reverses the increase at right, and partially reverses the decrease at left (data not shown).

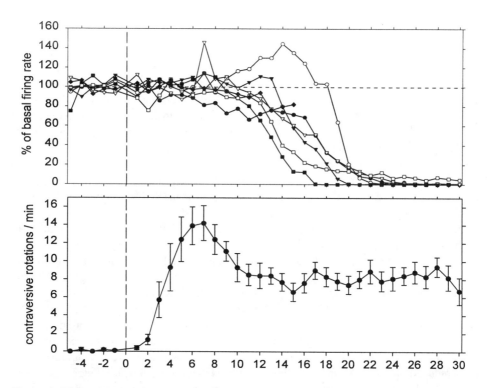

Figure 4. Effects of 0.05 mg/kg s.c. apomorphine on SNPR neuron firing rate (top) and rotation (bottom). Time is in minutes. Results from seven SNPR neurons are plotted; six are held until 30 min after injection, one is held until 15 min. Rotation, n = 10.

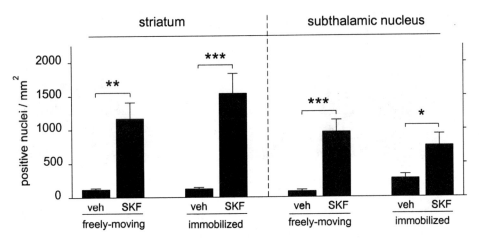

Figure 5. D_1 agonist-induced Fos-like immunoreactivity in the striatum and STN. Two-way ANOVAs demonstrate significant effects of drug treatment in both nuclei (p values <.001), but no significant effect of motor state in either structure. 1.3 mg/kg i.v. SKF 38393 ('SKF') induced Fos expression ipsilateral to the nigrostriatal lesion compared to vehicle ('veh') in both nuclei and both motor states. * p<.05, ** p<.01, *** p<.001.

sufficient to cause motor activation. However, the present data indicate that such a net reduction in activity in these nuclei is not *necessary* for DA agonist-induced motor activation, based on dose/response and time/response differences in rotational behavior and average firing rate. Such a finding is unexpected based on current models of the basal ganglia, in which striatal DA receptor activation is predicted to lead to an overall reduction in SNPR and EPN activity and thereby disinhibits motor-related structures. The present results imply that DA agonist-induced motor activation is not necessarily related to a net disinhibition of basal ganglia target structures.

Previous studies from this laboratory demonstrated that a high i.v. dose of the partial D_1 agonist SKF 38393 causes a decrease in average firing rate in the SNPR of nigrostriatal-lesioned rats.[7,20] The present results demonstrate that lower i.v. doses of this drug did not reduce average SNPR firing rate although they produced rotation. There was also a range of doses of the full D_1 agonist chloroAPB that induced rotation without causing net SNPR rate decreases. The rotation-inducing D_1 agonist treatments that failed to produce a net change in rate did cause a range of firing rate changes which varied from neuron to neuron, and included rate increases, decreases, and no change, reminiscent of responses after D_1/D_2 agonist treatment in intact rats (Fig. 1). Hence, under many of these conditions of motor activation at least some proportion of the SNPR and EPN neurons have decreased rate. It is possible that the decreased firing rate of this subpopulation is sufficient to mediate motor activation. It should be emphasized, however, that this effect is concomitant with a drug-induced rate increase in other SNPR and EPN neurons (and with little change in still others) so this state is distinct from a simple overall disinhibition of motor-related target structures. In addition, at least one motor activating treatment results in *only* firing rate increases in the SNPR (Fig. 1, right bar).

Although the morphologically-similar SNPR and EPN have been hypothesized to differ functionally on several counts,[2,8] the present results with one test dose of D_1 agonist (and also the apomorphine data presented in Fig. 1) suggest that the activity of these two structures is regulated in a similar fashion by DA receptors. However, those neurons in each structure with agonist responses of similar polarity may be functionally related. Most SNPR neurons display saccade-related or head movement-related decreases in firing rate,[5] yet the limb movement-related responses of many SNPR and EPN neurons are increases in firing rate.[2] SNPR and EPN neurons with increased activity after a motoractivating DA agonist treatment may therefore be preferentially related to limb function.

Apomorphine strongly reduces firing rate in the SNPR and EPN of animals with DA cell lesions.[3,19] In the present study, a low s.c. dose commonly used for rotational screening caused contraversive rotation as well as a robust net decrease in SNPR firing rate. The decrease in average firing rate was much delayed, however, in comparison to the commencement of rotation. Indeed, there was a notable early peak in rotation rate that occurred entirely before the beginning of the average inhibition, and largely before the inhibition of any of the individual neurons. The results support the idea that DA agonist-induced rotation is not dependent on a net decrease in SNPR rate. Also, there was no further increase in rotation rate during the epoch when average firing rate was declining. It is possible that the strong, late inhibition of SNPR units was associated with the onset of some behavior other than rotation. Stereotypical behavior (sniffing, chewing, or strong grooming) is one possibility. The continuous or intermittent expression of stereotypies might interfere with ongoing rotation, possibly explaining the reduction in rotation rate at seven to twelve minutes, when the net SNPR inhibition is commencing. This hypothesis could be tested by a more detailed behavioral analysis of the effects of this dose of apomorphine. Such a pattern of results would resemble effects noted in MPTP-treated primates, in which apomorphine-induced palliation of parkinsonism was not associated with major changes in firing rate in the primate homolog of the EPN, while apomorphine-induced dyskinesias were associated with net decreases in firing rate.[4]

The comparison of behavioral data from freely-moving rats and electrophysiological data from awake, immobilized rats is subject to the criticism that drug effects on brain function may differ in the two preparations. To address this issue, we utilized immediate-early gene protein expression as a means to assess drug influences on brain function in both freely-moving and immobilized animals. SKF 38393, at a low but rotation-inducing dose, increased the expression of Fos protein in the striatum and STN ipsilateral to the nigrostriatal lesion. Furthermore, numbers of Fos positive neurons in the striatum and STN were not significantly different in freely-moving and awake, immobilized rats following SKF 38393 treatment. D_1 agonists induce striatal Fos-like immunoreactivity primarily in striatonigral neurons.[13] These results suggest that D_1 receptor-mediated activation of striatonigral and subthalamic neurons is equivalent in freely-moving and immobilized rats, and so support the present comparison of D_1 agonist-mediated effects on behavior and basal ganglia electrophysiology in these two preparations.

The pattern of Fos expression induced by 1.3 mg/kg SKF 38393 suggests a possible mechanism for the firing rate effects in the SNPR and EPN after low doses of D_1 agonist. Although this dose of SKF 38393 increased striatal Fos expression, and therefore likely increased the level of inhibitory GABAergic input to the SNPR and EPN, it did not cause a net decrease in firing rate in these latter structures. The absence of a decrease in average firing rate may be due to increased excitatory glutamatergic input from the STN, which is activated by SKF 38393 as shown here with Fos staining and previously with single-unit recording.[9] We propose that the wide range of firing rate changes found in the SNPR and EPN after many DA agonist treatments results from an interplay of DA agonist-induced increases in both inhibitory and excitatory input to these nuclei, due to striatal and extrastriatal DA receptors, respectively. The net decrease in firing rate in the SNPR after higher doses of D_1 agonist may be due to a comparatively greater activation of the striatonigral pathway.

If DA agonist-induced rotation is not dependent on overall firing rate inhibition in the SNPR or EPN, some other feature(s) of neuronal activity in these nuclei must be involved. One possibility is the synchronization of firing activity between units. The level of DA receptor activity regulates the amount of interneuronal synchrony in the basal ganglia[10] and striatal gap junction activity which could influence synchrony.[1,12] Correlated and non-correlated activity are likely to affect target neurons in differing ways. DA agonists may change synchrony of SNPR or EPN activity in a manner that varies with behavioral state. Another possibility is periodicity. We have described multisecond periodic oscillations in firing rate in a majority of SNPR and EPN neurons.[14] In intact rats, these slow firing rate oscillations are profoundly affected by apomorphine in terms of frequency and power (both increased), in a fairly consistent manner from neuron-to-neuron (J.R. Walters et al., this volume). Apomorphine modulation of slow oscillation parameters therefore constitutes a more consistent population effect than its widely varying range of effects on mean firing rate in the SNPR and EPN of intact rats. Apomorphine also has significant effects on multisecond oscillations after nigrostriatal lesion (K.A. Allers et al., this volume). Periodic oscillations in firing rate in basal ganglia output nuclei are likely to have profound effects on the pattern of activity in the motor thalamus and tectum, thereby influencing motor output.

5. REFERENCES

1. C. Cepeda, J.P. Walsh, C.D. Hull, S.G. Howard, N.A. Buchwald, and M.S. Levine, 1989, Dye-coupling in the neostriatum of the rat: 1. Modulation by dopamine-depleting lesions, *Synapse,* 4:229-237.
2. M.R. DeLong and A.P. Georgopoulos, 1979, Motor functions of the basal ganglia as revealed by studies of single cell activity in the behaving primate, *Adv. Neurol.,* 24:131-140.
3. M. Filion, 1979, Effects of interruption of the nigrostriatal pathway and of dopaminergic agents on the spontaneous activity of globus pallidus neurons in the awake monkey, *Brain Res.,* 179:425-441.
4. M. Filion, L. Tremblay, and P.J. Bedard, 1991, Effects of dopamine agonists on the spontaneous activity of globus pallidus neurons in monkeys with MPTP-induced parkinsonism, *Brain Res.,* 547:152-161.
5. O. Hikosaka and R.H. Wurtz, 1983, Visual and oculomotor functions of monkey substantia nigra pars reticulata. 1. Relation of visual and auditory responses to saccades, *J. Neurophysiol.,* 49:1230-1253.
6. K.-X. Huang, D.A. Bergstrom, D.N. Ruskin, and JR. Walters, 1998, N-methyl-D-aspartate receptor blockade attenuates D_1 dopamine receptor modulation of neuronal activity in rat substantia nigra, *Synapse,* 30:18-29.
7. K.-X. Huang and J.R. Walters, 1994, Electrophysiological effects of SKF 38393 in rats with reserpine treatment and 6-hydroxydopamine-induced nigrostriatal lesions reveal two types of plasticity in D 1 dopamine receptor modulation of basal ganglia output, *J. Pharmacol. Exp. Then,* 271:1434-1443.
8. D. Joel and I. Weiner, 1994, The organization of the basal ganglia-thalamocortical circuits: open interconnected rather than closed segregated, *Neuroscience,* 63:363-379.
9. D.S. Kreiss, C.W. Mastropietro, S.S. Rawji, and J.R. Walters, 1997, The response of subthalamic nucleus neurons to dopamine receptor stimulation in a rodent model of Parkinson's disease, *J. Neurosci.,* 17:6807-6819.

10. A. Nini, A. Feingold, H. Slovin, and H. Bergman, 1995, Neurons in the globus pallidus do not show correlated activity in the normal monkey, but phase-locked oscillations appear in the MPTP model of Parkinsonism, *J. Neurophysiol,* 74:18001805.
11. M.E. Olds, 1988, Correlation between the discharge rate of non-dopamine neurons in substantia nigra and ventral tegmental area and the motor activity induced by apomorphine, *Neuroscience,* 24:465-476.
12. S.P. Onn and A.A. Grace, 1994, Dye coupling between striatal neurons recorded in vivo: compartmental organization and modulation by dopamine, *J. Neurophysiol.,* 71:1917-1934.
13. G.S. Robertson, S.R. Vincent, and H.C. Fibiger, 1992, D_1 and D_2 dopamine receptors differentially regulate *c-fos* expression in striatonigral and striatopallidal neurons, *Neuroscience,* 49:285-296.
14. D.N. Ruskin, D.A. Bergstrom, Y. Kaneoke, B.N. Patel, M.J. Twery, and J.R. Walters, 1999a, Multisecond oscillations in firing rate in the basal ganglia: Robust modulation by dopamine receptor activation and anesthesia, *J. Neurophysiol.,* 81:2046-2055.
15. D.N. Ruskin, D.A. Bergstrom, C.W. Mastropietro, M.J. Twery, and J.R. Walters, 1999b, Dopamine agonist-mediated rotation in rats with unilateral nigrostriatal lesions is not dependent on net inhibitions of rate in basal ganglia output nuclei, *Neuroscience,* 91:935-946.
16. D.N. Ruskin and J.F. Marshall, D_1 dopamine receptors influence Fos immunoreactivity in the globus pallidus and subthalamic nucleus of intact and nigrostriatallesioned rats, *Brain Res.,* 703:156-164.
17. J. Scheel-Krüger, J. Arnt, and G. Magelund, 1977, Behavioral stimulation induced by muscimol and other GABA agonists injected into the substantia nigra, *Neurosci. Lett.,* 4:351-356.
18. B.L. Waszczak, E.K. Lee, T. Ferraro, T.A. Hare, and J.R. Walters, 1984a, Single unit responses of substantia nigra pars reticulata neurons to apomorphine: effects of striatal lesions and anesthesia, *Brain Res.,* 306:307-318.
19. B.L. Waszczak, E.K. Lee, C.A. Tamminga, and JR. Walters, 1984b, Effect of dopamine system activation on substantia nigra pars reticulata output neurons: variable single-unit responses in normal rats and inhibition in 6-hydroxydopamine-lesioned rats, *J. Neurosci.,* 4:2369-2375.
20. B.G. Weick and J.R. Walters, 1987, Effects of D_1 and D_2 dopamine receptor stimulation on the activity of substantia nigra pars reticulata neurons in 6-hydroxydopamine lesioned rats: D_1/D_2 coactivation induces potentiated responses, *Brain Res.,* 405:234-246.

FIRING PATTERNS OF SINGLE NUCLEUS ACCUMBENS NEURONS DURING INTRAVENOUS COCAINE SELF-ADMINISTRATION SESSIONS

Laura L. Peoples*, Ph.D.

Key words: drug addiction, cocaine, rat, nucleus accumbens, neurophysiology

Abstract: Neuropharmacological studies of drug-taking behavior in laboratory animals suggest that drug actions on the nucleus accumbens (NAcc) are integrally involved in drug taking and addiction. Chronic extracellular recording techniques have been used in recent years to record the activity of single nucleus accumbens (NAcc) neurons of rats during intravenous cocaine self-administration sessions. These studies have shown that NAcc neurons exhibit a variety of firing patterns during drug-taking sessions. A large percentage of neurons show firing patterns that mirror changes in drug level and that may therefore be pharmacological. A subset of those neurons exhibit an additional change in firing that appears to be nonpharmacological and that may code various aspects of the behavior and stimulus events associated with drug seeking. The firing patterns that appear pharmacological are consistent with the conclusion that self-administered cocaine inhibits a majority of accumbal neurons. In contrast, the changes in firing most closely associated with the actual occurrence of drug self-infusion behavior consist of increases in firing rate. The latter firing patterns are similar in appearance to phasic firing patterns associated with operant behavior reinforced by natural reinforcers. Drug reinforcement and addiction are hypothesized to be mediated by drug actions on neurons that normally control instrumental behavior. The firing patterns described in this chapter are thus prototypic of what might be expected of neurons that contribute importantly to drug taking and addiction.

1. INTRODUCTION

Drug addiction is a chronic relapse disorder characterized by compelled and uncontrollable self-administration of drug.[1] Current theories of drug addiction propose that it is caused in

---

* Department of Psychology, University of Pennsylvania., Neuroscience Graduate Group, University of Pennsylvania School of Medicine, Philadelphia, PA 19106 USA. email:lpeoples@psych.upenn.edu

The Basal Ganglia VI
Edited by Graybiel *et al.*, Kluwer Academic/Plenum Publishers, 2002

part by acute and/or chronic pharmacological actions on brain areas that normally contribute to the control of instrumental behavior.[2-5] Most drugs that are abused by humans are self-administered by laboratory animals. Moreover, the patterns of drug intake exhibited by animals are similar to those exhibited by humans.[3-4] Thus, identification of the mechanisms that mediate drug taking in animals is expected to contribute to our knowledge regarding the etiology of addiction in humans.

Lesion studies have shown that both drug-induced elevations in mesoaccumbal dopamine and the impact of those dopaminergic changes on nucleus accumbens (NAcc) neurons are necessary for cocaine self-administration to be exhibited by animals.[6-7] The neurons of the nucleus accumbens (NAcc) are also critically involved in instrumental behavior maintained by natural reinforcers.[e.g., 8] This involvement of NAcc neurons in behavior maintained by non-drug reinforcers is consistent with that which is predicted for neurons that contribute to drug taking and addiction.[2-5]

The nucleus accumbens is a heterogenous brain area that consists of sub-territories as well as compartments (or zones) that differ in afferent and efferent projections, neurochemical innervation, and density of receptor subtypes.[e.g.,9-12] Consistent with these anatomical differences, neuropharmacological evidence shows that the various regions of the NAcc contribute differentially to drug self-administration, as well as to other behaviors.[12-13].
Given such heterogeneity, it is advantageous, perhaps ultimately necessary, for functional and mechanistic studies of NAcc involvement in drug taking to include investigations at the cellular level. Chronic extracellular recording techniques offer the spatial resolution necessary for such studies. The technique is additionally advantageous in that it allows for measurements of neuronal activity to be made over periods that are as short as milliseconds (msec) and as long as either hours or days. It therefore allows for simultaneous correlative investigations of: 1) the contribution of accumbal neurons to behavior, which may occur on a time base of msec or seconds (sec) and 2) drug effects on accumbal neurons, which occur on time bases of minutes (min), hours, and even days (i.e., as in the case of chronic drug effects). The recording technique is thus relatively unique in its applicability to studies of drug addiction.

Within the last decade, microwires chronically implanted within the nucleus accumbens have been used by a number of researchers to record the activity of single NAcc neurons of rats engaged in intravenous cocaine self-administration. During cocaine self-administration sessions, NAcc neurons exhibit a variety of firing patterns. These patterns can be grouped into three categories according to the time course over which they occur. This chapter will review briefly some of the most common examples of each of the three categories. The reviewed firing patterns, given their prevalence, are potentially important contributors to the NAcc efferent signal that influences drug taking. However, it is important to note that the patterns described herein are not the only firing patterns exhibited by NAcc neurons and are not the only NAcc firing patterns that should be considered as potentially influencing drug self-administration.

2. METHODS

In the studies that we have conducted thus far, male Long–Evans rats were chronically implanted with a catheter in the jugular vein and an array of 12-16 microwires in the NAcc.

Following recovery from surgery, animals were transferred to a Plexiglas chamber that was henceforth used for housing, drug self-administration sessions, and electrophysiological recordings.

Before the start of each self-administration session, a response lever was mounted on a side wall of the chamber. Onset of the session was signaled by illumination of a stimulus light above the response lever. Each lever press was followed immediately by a 0.2 ml intravenous infusion of cocaine solution, a 7.5 sec tone, and a 40 sec time out, during which the stimulus light was turned off and presses had no programmed consequence. Sessions were limited to 6 hours or 80 infusions and occurred 7 days a week.

Animals typically completed a minimum of 12-15 self-administration training sessions prior to the first recording session. At that point in time, animals exhibited a stable pattern of cocaine self-infusion behavior. Each recording experiment included three phases: 1) a predrug baseline period (20-60 min), 2) a self-administration session, and 3) and postdrug period (40-60 min). Typically, after a population of neural waveforms was isolated by post hoc discrimination procedures, that population was subjected to an interspike interval analysis to confirm that it corresponded to a discriminated single neuron. When more than a single population of neural waveforms appeared to have been recorded from a given wire, cross-correlation analyses were used to confirm that the populations corresponded to distinct neurons. All neurons were histologically verified to have been located within the NAcc. The majority of neurons that we have characterized thus far were within the core, rather than the shell. The neural sample that we have collected to date is not sufficient to allow for definitive sub-territorial comparisons. However, preliminary comparisons have been described elsewhere[16].

3. FIRING PATTERNS

3.1 Changes in tonic firing rate during the self-administration session.

In our studies, approximately 90% of neurons exhibit a change in tonic firing rate during the self-administration session, relative to the predrug baseline period. For the majority of neurons (approximately 60%), firing rate decreases rapidly during the loading period and then remains depressed, within stable limits for the duration of the self-administration session[17](Figure 1). It is possible that at least some of the changes in tonic firing are induced by pharmacological actions of cocaine and that the firing patterns reflect a component of the pharmacological transduction mechanisms that contribute to the maintenance of drug self-administration. Evidence to support this proposal includes the following. First, the decrease in tonic firing mirrors changes known to occur in drug and dopamine levels during psychomotor stimulant self-administration sessions.[e.g.,18] Additionally, the decrease in firing is consistent with evidence of anesthetized and slice recording studies which show that the primary pharmacological effect of cocaine on accumbal neurons is inhibition.[for review see 17,19] Finally, neuropharmacological data show that inhibition of certain neurons may be the primary *pharmacological* mechanism that mediates the necessary contribution of the NAcc to drug self-administration.[13]

One could alternatively argue that the change in tonic firing is not pharmacological but rather reflects the differential occurrence of drug seeking, and/or other behaviors associated

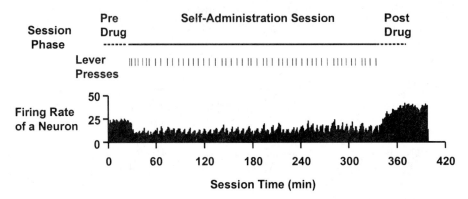

Figure 1. The change in tonic firing rate most commonly exhibited by NAcc neurons during cocaine self-administration sessions. At the top of the figure is shown the cocaine self-infusion behavior (i.e., lever presses) of a single animal during a single self-administration session. Below the display of lever presses, a stripchart displays firing rate (Hertz, Hz, calculated as a function of 0.5 min bins) (ordinate) plotted as a function of time during the recording session (abscissa). Like the neuron shown in the present figure, the majority of accumbal neurons showed a decrease in firing rate during the self-administration session, relative to the pre- and postdrug recording periods.

with it. To begin testing this hypothesis we conducted behavioral extinction tests at the end of self-administration sessions.[20] During these tests, lever presses had no programmed consequences. Animals exhibited a biphasic change in the rate of lever pressing that was characterized by an initial increase in lever pressing and a subsequent decrease and eventual cessation of responding. During the same time period, drug levels underwent a gradual decline. To the extent that the inhibition of firing during the self-administration session was related to the occurrence of drug seeking, we expected it to be enhanced during the period of extinction when drug seeking and related behaviors were increased. This was not observed to occur for a single neuron. In fact, the majority of neurons showed significant increases in firing rate. These data suggest that the inhibition of firing exhibited by accumbal neurons during the cocaine self-administration session is not related to behaviors associated with drug seeking and may instead reflect drug actions.

3.2 Phasic changes in firing during the minutes before and after self-infusion.

Approximately half of all recorded neurons exhibit a "slow" oscillation in firing rate during the minutes that elapse between successive self-infusions (Figure 2). Most typically, firing rate decreases to a relatively stable minimum during the mins after self-infusion and then gradually increases until a predictable maximum rate is attained, shortly before the occurrence of the next self-infusion (referred to as the decrease + progressive reversal pattern).[21-22] It is possible, that some of these decrease + progressive reversal firing patterns reflect behaviorally-related neural processing; although, control analyses suggest that such behavioral processing would be state- or stimulus-related rather than movement-related.[23] On the other hand, the decrease + progressive reversal firing pattern closely mirrors changes in both drug and dopamine levels that occur during cocaine self-administration sessions[24]. Moreover, a majority of neurons that exhibit the slow phasic firing pattern also show an

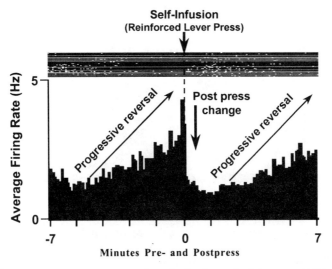

Figure 2. An example of the most common slow phasic change in firing rate exhibited by accumbal neurons during the mins before and after successive self-infusions. The figure displays average firing rate (average Hz) of a single neuron during the min (i.e., interinfusion interval) before and after all cocaine reinforced lever presses completed during a single session by an individual animal (excluding the first 8-10 presses and all lever presses bracketed by interinfusion intervals < 6.0 min). The ordinate displays average Hz calculated as a function of 0.1 min bins and the abscissa displays mins pre- and postpress. Time "0" on the abscissa represents the occurrence of the reinforced lever press. Above the histogram is a raster display that shows firing of the neuron on a trial by trial basis. The histogram demonstrates the most common phasic change in firing rate exhibited by neurons during the mins pre- and postpress (i.e., a decrease in firing rate during the min postpress followed by a progressive reversal which continued over the course of the interinfusion interval until the time of the next self-infusion).

inhibition of tonic firing rate.[17,22] Thus, the decrease + progressive reversal could reflect a cycle of drug-induced inhibition and subsequent recovery that occurs repeatedly with each successive self-infusion. To the extent that the firing pattern is pharmacological it may reflect a component of the drug actions on motivational brain circuitry hypothesized to underlie drug reinforcement.

3.3 Phasic changes in firing rate time locked to within seconds of self-infusion behavior.

Accumbal neurons exhibit a third category of firing patterns during cocaine self-administration sessions. These patterns most typically involve an increase in firing rate that occurs during the seconds (secs) pre- and or postpress. For the majority of neurons, the increase in firing begins during the 0-3 sec prepress and does not end until 0-3 secs postpress[16] (Figure 3). Similar firing patterns have been described by other researchers that have recorded neurons in the NAcc and ventral striatum of animals self-administering cocaine, heroin, or ethanol.[26-31] Although these "rapid" phasic patterns are the most frequently described patterns, they make up the smallest category of firing patterns exhibited by NAcc neurons during cocaine self-administration (i.e., approximately 25% of neurons in our studies). The close temporal association between the firing patterns and the actual drug-taking behavior is suggestive of a relationship between the two. This proposal is supported by both

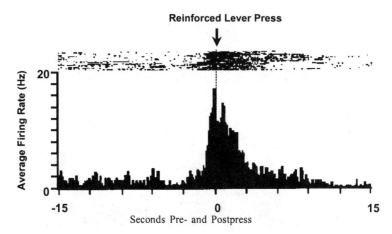

Figure 3. A "rapid" phasic change in firing rate exhibited by accumbal neurons during the secs pre- and postpress. The figure displays average firing rate (average Hz) of a single neuron during the secs before and after all cocaine reinforced lever presses completed during a single session by an individual animal (excluding the first 8-10 presses and all lever presses bracketed by interinfusion intervals < 6.0 min). The ordinate displays average Hz calculated as a function of 0.3 sec bins and the abscissa displays seconds pre- and postpress. Time "0" corresponds to the time of self-infusion. Above the histogram is a raster display that shows firing of the neuron on a trial by trial basis. The neuron represented in this figure showed an increase in firing rate that began and ended during the several secs before and after the cocaine reinforced lever press, respectively.

correlational and experimental data of several laboratories.[16,28-29,32-33] As is the case for the slow phasic decrease + progressive reversal pattern, control analyses suggest that the rapid phasic patterns are not, at least in most cases, movement-specific. The rapid patterns are similar in appearance to those associated with operant behavior maintained by natural reinforcers.[e.g., 34-35] It is thus possible that they reflect the normal contribution of accumbal neurons to instrumental behavior (i.e., are not specific to drug taking or addiction). Interestingly, the rapid and behaviorally-related phasic firing patterns are most typically exhibited by neurons that show the slow and potentially pharmacological decrease + progressive reversal firing pattern.[22] (for related findings see [29,36]) Neurons that exhibit both phasic firing related to drug-seeking behavior and a sensitivity to pharmacological actions of the self-administered drug would be prototypic of the type of neuron that one might expect to be of particular importance to mediating the accumbal contribution to drug taking (see above).

4. ACCUMBAL MECHANISMS INVOLVED IN PHARMACOLOGICAL REINFORCEMENT AND DRUG SEEKING BEHAVIOR: HYPOTHESES TO INVESTIGATE IN FUTURE STUDIES.

The electrophysiological recordings of accumbal neurons during cocaine self-administration sessions that have been conducted thus far, and that were described in the present chapter, indicate that the predominant pharmacological effect of cocaine on accumbal neurons is inhibition of firing rate.[17,22,29,34] Moreover, there is evidence (described elsewhere[e.g.,37,38]) that repeated exposure to cocaine leads to an enhancement of this effect.

Inhibition of certain neurons may thus be the primary *pharmacological* action in the NAcc that contributes to drug reinforcement and addiction.

However, the changes in firing most commonly associated with the actual occurrence of drug-seeking *behavior* involve predominantly increases, not decreases, in accumbal firing rate. This conclusion is suggested if one considers: 1) the most frequently observed phasic change in firing during the seconds before and after self-infusion (i.e., increases),[16,28,29] 2) the most common change in firing during the several mins leading up to each infusion (i.e., a progressive increase),[21] and 3) the changes in tonic firing exhibited by neurons when increases in drug-seeking behavior were induced during the behavioral extinction tests[20] (see above). It is increases in accumbal firing that are also most commonly associated with approach behaviors,[22] instrumental behaviors reinforced by natural reinforcers,[34,35] and operant behavior reinforced by electrical brain stimulation.[42]

Cumulatively, the recordings of accumbal neurons in relation to the various types of instrumental behavior are consistent with the hypothesis that the accumbens, perhaps even individual neurons within the accumbens, contributes to neural processing associated with both incentive preparatory states and consummatory or sated states. Moreover, the changes in firing associated with these antithetic states are predominantly opposed and of an excitatory and inhibitory nature, respectively. The inhibitory mechanisms associated with consummatory or sated states may modulate the throughput of the excitatory signals associated with the incentive states and additionally contribute to the neural signals critical to reinforcement of behaviors (i.e., strengthening of behaviors). These hypotheses are consistent with other interpretations regarding the neurophysiology of the NAcc.[29,40-42]

A primary therapeutic endpoint of drug addiction research is the prevention of relapse to drug taking in individuals that are attempting to maintain drug abstinence. It is also of interest to investigate the acute and chronic effects of drug on brain in the hopes of designing treatments that might actually reverse the drug-induced changes in brain that contribute to addiction. The presently discussed electrophysiological data suggest that attainment of these two therapeutic goals will necessitate both the delineation of distinct neural mechanisms and the development of distinct pharmacological treatment strategies.

5. REFERENCES

1. O'Brien, C.P., Childress, A.R., McLellan, T., and Ehrman, R. (1992) A learning model of addiction. In: Addictive States (C.P. O'Brien and J.H. Jaffe, Eds.)., Raven Press, N.Y., pp 157-177.
2. Stewart, J., deWit, H., and Eikelbloom, R. (1984) Role of unconditioned and conditioned drug effects in the self-administration of opiates and stimulants. Psychol. Rev. 91:251-268.
3. Wise, R. and Bozarth, M.A. (1987) A psychomotor stimulant theory of addiction. Psychol. Rev. 94:469-492.
4. Robinson, T.E. and Berridge, K.C. (1993) The neural basis of drug craving: An incentive-sensitization theory of addiction. Brain Res. Rev. 18:247-291.
5. Koob, G.F. and Le Moal, M. (1997) Drug Abuse: Hedonic homeostatic disregulation. Science. 278:52-58.
6. Roberts, D.C.S., Koob, G.F., Klonoff, P., and Fibiger, H.C. (1980) Extinction and recovery of cocaine self-administration following 6-hydroxydopamine lesions of the nucleus accumbens. Pharmacol. Biochem. Behav. 12:781-787.
7. Zito, K.A., Vickers, G., and Roberts, D.C.S. (1985) Disruption of cocaine and heroin self-administration following kainic acid lesions of the nucleus accumbens. Pharmacol. Biochem. Behav. 23:1029-1036.

8. Everitt, B.J., Parkinson, J.A., Olmstead, M.C., Arroyo, M., Robledo, P., and Robbins, T.W. (in press) Associative processes in addiction and reward: The role of amygdala-ventral striatal subsystems. In: Advancing from the Ventral Striatum to the Extended Amygdala: Implications for Neuropsychiatry and Drug Abuse (J.F. McGinty, Ed.), Ann. New York Acad. Sci., Vol. 877, pp 412-438.
9. Heimer, L., Zahm, D.S., Churchill, L., Kalivas, P.W., and Wohltmann, C. (1991) Specificity in the projection patterns of accumbal core and shell in the rat. Neuroscience 41(1):89-125.
10. Zahm, D.S. and Heimer, L. (1993) Specificity in the efferent projections of the nucleus accumbens in the rat: Comparison of the rostral pole projection patterns with those of the core and shell. J. Comp. Neurol. 327:220-232.
11. Jongen-Rêlo, A.L., Voorn, P., and Groenewegen, H.J. (1994) Immunohistochemical characterization of the shell and core territories of the nucleus accumbens in the rat. Euro. J. Neurosci. 6:1255-1264.
12. Groenewegen, H.D., Wright, C.I., and Beijer, A.V.J. (1996) The nucleus accumbens: Gateway for limbic structures to reach the motor system? In: Progress in Brain Research, Vol 107, (G. Holstege, B. Bandler, and C.B. Saper, Eds.), Elsevier Science, N.Y., pp 485-511.
13. Carlezon, W.A. and Wise, R.A. (1996) Rewarding actions of phencyclidine and related drugs in nucleus accumbens shell and frontal cortex. J. Neurosci. 16:3112-3122.
14. Robledo, P. and Koob, G.F. Two discrete nucleus accumbens projection areas differentially mediate cocaine self-administration in the rat. Behav. Brain. Res. 55:159-166.
15. Kelley, A.E. (in press) Functional specificity of ventral striatal compartments in appetitive behaviors. In: Advancing from the ventral striatum to the extended amygdala: Implications for neuropsychiatry and drug abuse (J.F. McGinty, Ed.)., Ann. New York Acad. Sci. Vol. 877, pp 412-438.
16. Uzwiak, A.J., Guyette, F.X., West, M.O., and Peoples, L.L. (1997) Neurons in accumbens subterritories of the rat: Phasic firing time-locked within seconds of intravenous cocaine self-infusion. Brain Res. 767:363-369.
17. Peoples, L.L., Uzwiak, A.J., Guyette, F.X., and West, M.O. (1998) Tonic inhibition of single nucleus accumbens neurons in the rat: A predominant but not exclusive firing pattern induced by cocaine self-administration sessions. Neuroscience 86:13-22.
18. Pettit, H.O. and Justice, Jr. J.B. (1991) Effect of dose on cocaine self-administration behavior and DA levels in the NAcc. Brain Res. 539:94-102.
19. White, F.J., Henry, D.J., Jeziorski, M., and Ackerman, J.M. (1992) Electrophysiologic effects of cocaine within the mesoaccumbens and mesocortical dopamine systems. In: Cocaine Pharmacology, Physiology, and Clinical Strategies (J.M. Lakoski, M.P. Galloway, and F.J. White Eds)., CRC, Boca Raton, pp 261-294.
20. Fabbricatore, A.T., Uzwiak, A.J., West, M.O., and Peoples, L.L. (1998) Comparisons of firing rates of rat nucleus accumbens neurons during cocaine self-administration and extinction. Soc. Neurosci. Abs. 24:1736.
21. Peoples, L.L. and West, M.O. (1996) Phasic firing of single neurons in the rat nucleus accumbens correlated with the timing of intravenous cocaine self-administration. J. Neurosci. 16(10):3459-3473.
22. Peoples, L.L., Uzwiak, A.J., Gee, F., Fabbricatore, A.T., Muccino, K.J., Mohta, B.D., and West, M.O. (in press) Phasic accumbal firing may contribute to the regulation of drug taking during intravenous cocaine self-administration sessions. In: Advancing from the ventral striatum to the extended amygdala: Implications for neuropsychiatry and drug abuse (J.F. McGinty, Ed.)., Ann. New York Acad. Sci., Vol 877, pp 781-787.
23. Peoples, L.L., Gee, F., Bibi, R., and West, M.O.(1998) Phasic firing time locked to cocaine self-infusion and locomotion: Dissociable firing patterns of single nucleus accumbens neurons in the rat. J. Neurosci. 18(18):7588-7598.
24. Wise, R.A., Newton, P., Leeb, K., Brunette, B., Pocock, D., and Justice Jr., J.B. (1995) Fluctuations in nucleus accumbens dopamine concentration during intravenous cocaine self-administration in rats. Psychopharmacology 120:10–20.
25. Yokel, R.A. and Pickens, R. (1974) Drug level of d- and l-amphetamine during intravenous self-administration. Psychopharmacology (Berl) 34:255-264.
26. Bowman, E.M., Aigner, T.G., and Richmond, B.J. (1996) Neural signals in the monkey ventral striatum related to motivation for juice and cocaine rewards. J. Neurophysiol. 75:1061-1973.
27. Carelli, R.M., King, V.C., Hampson, R.E., and Deadwyler, S.A. (1993) Firing patterns of nucleus accumbens neurons during cocaine self-administration in rats. Brain Res. 626:14-22.

28. Chang, J.-Y., Sawyer, S.F., Lee, R.-S., and Woodward, D.J. (1994) Electrophysiological and pharmacological evidence for the role of the nucleus accumbens in cocaine self-administration in freely moving rats. J. Neurosci. 14:1224-1244.
29. Chang, J.-Y., Paris, J.M., Sawyer, S.F., Kirillov, A.B., and Woodward, D.J. (1996) Neuronal spike activity in rat nucleus accumbens during cocaine self-administration under different fixed-ratio schedules. Neuroscience 74(2):483-497.
30. Chang, J.-Y., Janak, P. H., and Woodward, D.J. (1996) Comparison of mesocorticolimbic neuronal responses during cocaine and heroin self-administration in freely moving rats. J. Neursoci. 18(8):3098-3115.
31. Janak, P.H., Chang, J.-Y., and Woodward, D.J. (1999) Neuronal spike activity in the nucleus accumbens of behavng rats during ethanol self-administration. Brain Res. 817:172-184.
32. Carelli, R.M. and Deadwyler, S.A. (1996a) Dual factors controlling activity of nucleus accumbens cell-firing during cocaine self-administration. Synapse 24:308-311.
33. Peoples, L.L., Uzwiak A.J., Gee F., and West, M.O. (1997) Operant behavior during sessions of intravenous cocaine infusion is necessary and sufficient for phasic firing of single nucleus accumbens neurons. Brain Res. 757:280-284.
34. Carelli, R.M. and Deadwyler, S.A. (1994) A comparison of nucleus accumbens neuronal firing patterns during cocaine self-administration and water reinforcement in rats. J. Neurosci. 14(12):7735-7746.
35. Schultz, W., Apicella, P., Scarnati, E., Ljungberg, T. (1992) Neuronal activity in monkey ventral striatum related to the expectation of reward. J. Neurosci. 12:4595-4610.
36. Carelli, R.M. and Deadwyler, S.A. (1996b) Dose-dependent transitions in nucleus accumbens cell firing and behavioral responding during cocaine self-administration sessions in the rat. J. Pharmacol. Exp. Ther. 277:385-393.
37. Peoples, L.L., Uzwiak, A.J., Gee, F., and West, M.O. (1999) Tonic firing of rat nucleus accumbens neurons: Changes during the first 2 weeks of daily cocaine self-administration sessions. Brain Res. 822:321-236.
38. White, F.J., Hu, X.-T., Henry, D.J., Zhang, X.-F. (1995) Neurophysiological alterations in the mesoaccumbens dopamine system during repeated cocaine administration. In: The Neurobiology of Cocaine: Cellular and Molecular Mechanisms, (R.P. Hammer, Jr. Ed.)., CRC Press, Boca Raton. pp 95-115.
39. Wolske, M., Rompre, P.-P., Wise, R.A., and West, M.O. (1993) Activation of single neurons in the rat nucleus accumbens during self-stimulation of the ventral tegmental area. J. Neurosci 13(1):1-12.
40. Mogenson, G.J. and Yim, C.C. (1991) Neuromodulatory functions of the mesolimbic dopamine system: Electrophysiological and behavioral studies. In: The Mesolimbic Dopamine System: From Motivation to Action (P. Willner and J. Scheel-Kruger)., Wiley, N.Y., pp 105-131.
41. Kiyatkin, E.A. and Rebec, G.V. (1996) Dopaminergic modulation of glutamate-induced excitations of neurons in the neostriatum and nucleus accumbens of awake, unrestrained rats. J. Neurophysiol. 75:142-153.
42. O'Donnell, P. and Grace, A.A. (1996) Dopaminergic reduction of excitability in nucleus accumbens neurons recorded in vitro. Neuropschopharmacology 159:87-97.

Section VII

NEUROTRANSMITTER FUNCTIONS IN THE BASAL GANGLIA

IMMEDIATE EARLY GENE (IEG) INDUCTION IN THE BASAL GANGLIA UPON ELECTRICAL STIMULATION OF THE CEREBRAL CORTEX

Involvement of the MAPkinase pathway IEG induction upon cortico-striatal stimulation

M.J. Besson, V. Sgambato, P. Vanhoutte, M. Rogard, C. Pages, A.M. Thierry, N. Maurice, J.M. Deniau, and J. Caboche*

Key words: striatum, subthalamic nucleus, c-Fos, enkephalin, subtance P, *zif 268*, MKP1, CREB, Elk-1, PD 98059, phosphorylation, mRNA transcription

Abstract: The cerebral cortex is a major input to the basal ganglia through its topographically organized projections to the striatum and the subthalamic nucleus (STN). By the analysis of IEG (c-*fos* and *zif268*) mRNA expression, we found that unilateral stimulation of the cerebral cortex produced a bilateral induction of these IEG in limited striatal and subthalamic territories, which varied with the stimulated cortical area. In parallel to IEGs induction, peptide mRNAs encoding substance P and enkephalin, differentially expressed by the two efferent striatal neuronal populations, were also increased in the striatal territory activated by the cortical stimulation. The IEG induction involved the activation of ionotropic glutamate receptors (AMPA and NMDA). Among various intracellular signaling pathways that could be activated by these receptors, the Mitogen Activated Protein Kinase (MAPK) of the Extracellular Regulated-signal Kinase (ERK) subfamily, appeared to be predominant implicated in IEGs induction. In fact, the local injection of a specific inhibitor of the MAPK/ERK pathway (PD98059) totally abolished IEGmRNA upregulation. We conclude that this IEG induction could participate to adaptive mechanisms involved in long term synaptic plasticity in the basal ganglia.

---

* M.J. Besson, V. Sgambato, P. Vanhoutte, M. Rogard, C. Pages, J.M. Deniau, and J. Caboche, Dept Neurochimie-Anatomie, IDN, UPMC, 9 quai Saint Bernard, 75005 Paris, France. A.M. Thierry and N. Maurice, INSERM U114, Collège de France, 11 Pl. Marcelin Berthelot, 75005 Paris, France.

1. INTRODUCTION

The basal ganglia occupy a key position in neuronal processing for motor control. Two main structures of this complex receive a massive projection from the cerebral cortex: the striatum and the subthalamic nucleus (STN). Almost all areas of the cerebral cortex project to the striatum (Mc George & Faull, 1989) where cortical afferents impinge on the output medium spiny neurons as well as on some interneurons (Smith & Bolam, 1990). The STN receives its cortical inputs mainly from the prefrontal and motor cortices. In both structures, a potent synaptic excitatory response is produced concomitantly to the stimulation of cortical afferents, which contain glutamate as neurotransmitters. Thus, the striatum and the STN are placed in a central position to integrate cortical inputs. In turn, the efferent neurons from both structures control the output structures of basal ganglia where they exert opposite effects. A spatio-temporal integration intervenes however, since STN neurons are themselves indirectly controlled (via the globus pallidus) by striatal efferent neurons (Maurice et al., 1998).

Activation of the cerebral cortex produces activity-dependent synaptic plasticity at cortico-striatal synapses as it has been demonstrated in *in vivo* and *in vitro* models. Some aspects of synaptic plasticity can be supported by long term adaptive responses involving changes in protein synthesis and thereby gene expression. We thus examined, following the stimulation of the cerebral cortex, the induction of IEG, a class of genes rapidly and transiently activated in response to calcium influx. These genes encode for transcription factors which can control in turn the expression of other genes implicated in synaptic plasticity. We also examined the intracellular signaling pathways leading to the induction of the examined IEG since these pathways might give information about molecules which could be targeted to change synaptic plasticity.

2. PATTERN OF C-FOS INDUCTION IN THE BASAL GANGLIA FOLLOWING STIMULATION OF SELECTIVE CORTICAL AREAS

Cortical stimulations were produced by applying though pairs of wires 1.5 mm apart trains of shocks (250 Hz, 200 µA) of 50 msec duration and repeated at a frequency of 4 Hz. The expression of c-Fos protein was analyzed by immunocytochemistry one hour after the end of one-hour stimulation (Sgambato et al. 1997). Among structures of the basal ganglia, only the striatum and the STN showed a strong induction of c-Fos protein that varied, however, in their rostro-caudal and medio-lateral extension depending on the stimulated cortical area. By contrast the induction was never observed in the globus pallidus, the entopeduncular nucleus and involved only a few neurons in the substantia nigra. Besides the striatum and the STN, the homotypic contralateral cortex as well as several thalamic nuclei (reticularis, intralaminar, motor nuclei) showed c-Fos expression.

A topographical induction of c-Fos was observed in the striatum and the STN, which conformed to the distribution of cortical afferents fibers on these two structures. In this way, following the stimulation of the orofacial area of the motor cortex, c-Fos was expressed bilaterally in the ventrolateral part of the striatum in almost the complete antero-posterior axis. In the STN, labeled cells were also found, bilaterally, in the rostral and lateral quadrant of the nucleus. The bilateral expression of c-Fos observed in the striatum is consistent with the existence of a major crossed component in the cortico-striatal pathways. More striking was the bilateral induction of c-Fos in the STN, since cortical projections to this structure are exclusively ipsilateral. A possible explanation for this, is an activation of cortico-cortical connections and then activation of contralateral cortico-STN afferents, hypotheses supported by the similar cell discharges recorded in contra- and ipsi-lateral STN neurons. The stimulation of the forelimb area of the motor cortex induced c-Fos in a dorsal region of the striatum and in a caudal quadrant of the STN. Labeled neurons localized in a narrow band edging the lateral ventricle were found bilaterally in the striatum following the stimulation of the medial prefrontal cortex, which also produced c-Fos induction in the medial tiers of the STN but only ipsilaterally. The stimulation of the auditory cortex induced c-Fos in a caudo-medial region of the striatum only ipsilaterally and was ineffective to induce c-Fos in the STN, a result in agreement with the absence of cortical auditory afferent fibers in the STN (Afsharpour, 1985). In some cases, i.e. stimulation of somato-sensory, visual and orbital cortices, c-Fos was never induced in either the striatum or the STN, despite spike discharges recorded consistently in the striatum. Concordant with our patterns of c-Fos induction in the rat, a similar topographical expression of this IEG also has been reported in the striatum and STN in the monkey, subsequently to the stimulation of restricted areas of the cerebral cortex (Parthasarathy and Graybiel, 1997).

In our rat model, we also analyzed *c-fos* mRNA induction, which showed an identical spatial pattern of expression than the protein, but was already detectable after 15 min of stimulation, reaching a maximum after 30 min of stimulation and remaining sustained for at least 60 min of stimulation (Sgambato et al. 1998). A similar pattern of mRNA induction was observed with an other calcium-responsive IEG: *zif268* (Fig.1).

Glutamate released from the activated cortical afferents was at the origin of the IEG induction and both types of ionotropic glutamate receptors: NMDA and AMPA/kainate receptors were implicated. The involvement of NMDA receptors was attested by the efficacy of the NMDA receptor antagonist (MK 801) given systemically to inhibit dose-dependently the induction of c-Fos in both the striatum and the STN. In the striatum, however, few scattered neurons were resistant to the MK 801 inhibition and showed a persistent induction of Fos protein even at the higher dose utilized (4 mg/kg). AMPA receptors were also impli-

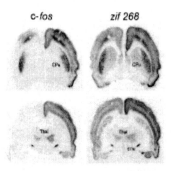

Figure 1. Effect of the electrical stimulation of the right orofacial motor cortex area on the expression of c-fos and zif268 mRNAs in the basal ganglia and some other brain structures. Note the bilateral IEG mRNA induction in the striatum (CPu) and the STN as well as in the thalamus (Thal), the ipsi-and contra-lateral cortices. Contralaterally, only the cortical area homotypic to the stimulated area was activated (see top panel, left side of the hemisphere).

cated as indicated by the abolition of c-*fos* mRNA induction in the striatum when, concomitantly to the cortical stimulation, an AMPA receptor antagonist (CNQX) was applied through a microdialysis tubing in the striatum. In this case, however, mRNA encoding IEGs (c-*fos* and *zif268*) were still induced in the contralateral striatum and, bilaterally in the STN (Sgambato et al., 1999). In the ipsilateral STN, IEG mRNA induction was as pronounced as in the contralateral side and in control animals, despite marked changes in the cortically driven discharges. Instead of having a triphasic response composed of a short and a long latency activation interrupted by a silent period (Maurice et al., 1998), the response was only composed of the short-latency excitation. Altogether these results indicate that i) the CNQX treatment was efficient to eliminate the late phase which involves the striatal and pallidal efferent neurons ii) the direct cortical activation on STN neurons is the main trigger for IEG induction.

3. STRIATAL NEURONS EXPRESSING C-FOS FOLLOWING STIMULATION OF THE CEREBRAL CORTEX

Fos immunoreactive neurons observed in the striatum after stimulation of the orofacial area of the motor cortex as well as of the prefrontal cortex showed a heterogeneous distribution, which was reminiscent of the patch/matrix compartmental organization. The comparison, on adjacent sections, of c-Fos distribution and ^{3}H-Naloxone binding sites allowed us to localize c-Fos neurons in the matrix compartment. This preferential distribution was observed in the lateral part of the striatum and at the edge of the lateral ventricle following the stimulation of the orofacial motor cortex and of the prefrontal cortex, respectively. In view of the localization of stimulation electrodes, deep as well as superficial layers, the preferential labeling of matricial neurons cannot be due to a selective activation of particular cortical layers. Thus, the induction of c-Fos in matrix neurons could reflect a difference in the density of cortico-striatal projections between matrix and patches (Cowan and Wilson, 1994). It could also be due to specific neurochemical differences since glutamate receptors and also a majority of other receptors are differentially distributed between patch and matrix compartments (Dure et al., 1992, Graybiel A. M., 1990). In the matrix compartment, the two types of efferent neurons defined by their projection sites (substantia nigra pars reticulata and globus pallidus) and their neuropeptide content (substance P versus enkephalin) were found to be Fos positive following stimulation of the orofacial motor cortex. Combining the immunocytochemical detection of Fos and the *in situ* hybridization of mRNA encoding preprotachykinin A (PPT-A) and preproenkephalin (PPE) assessed this. Among the population of Fos positive neurons an average of 72% of neurons were PPE-mRNA positive while 49% were PPT-A mRNA positive. These data are consistent with previous results obtained in monkey following microstimulation of the motor cortex showing that 75% of c-Fos positive neurons were enkephalin-immunoreactive (Parthasarathy and Graybiel, 1997). However, in this study, Fos immunoreactivity was not examined in the substance P/dynorphin neuronal population. A preferential activation of c-Fos in enkephalinergic neurons was also found in rats after cortical application of picrotoxin (Beretta et al. 1997). Altogether these results would suggest that the indirect pathway be preferentially activated by cortico-striatal afferents. However, strikingly, an equivalent percentage (around 60%) of PPE or PPT-A mRNA containing neurons were c-Fos positive (Sgambato et al., 1997), indicating that c-Fos is equally induced in these two efferent neuronal populations. This is consistent with anatomical observations indicating that both direct and indirect pathways receive afferents from the motor cortex (Hersch et al., 1995). Furthermore, in agreement with an activation of both efferent neuronal populations, we measured an increased expression of PPT-A mRNA and PPE mRNA content specifically induced in the lateral part of the striatum by stimulation of the orofacial motor cortex. The higher proportion of PPE versus PPT-A mRNA-containing neurons with c-Fos positive nuclei (see above) might reflect an upregulation of PPE mRNA in the direct striato-nigral pathway.

In interneurones characterized by their AChE mRNA content, we observed a differential expression of cFos. This marker allows defining two populations of interneurones defined, respectively, by high and moderate levels of AChE mRNA content (Bernard et al., 1995). In the lateral part of the striatum, interneurones with a moderate AChE mRNA content were frequently labeled by cFos protein whereas those with a high AChE mRNA content (corresponding to cholinergic interneurones), were never found to be c-Fos positive. This suggests that cholinergic interneurones can be hardly activated by cortical inputs or that they cannot express this IEG under a glutamate activation.

4. SIGNALING PATHWAYS INVOLVED IN THE CORTICALLY-DRIVEN IEG ACTIVATION IN THE STRIATUM

Glutamate containing cortico-striatal afferents interact on striatal efferent neurons via AMPA and NMDA receptor subtypes (Bernard & Bolam 1998). The AMPA receptor stimulation, which relieves the Mg^{2+} blockade through depolarization, allows the activation of NMDA receptor subtype, and both subtypes are involved in c-Fos induction after cortico-striatal stimulation (see above). The influx of Ca^{2+} (through the NMDA receptor associated ion-channel) acts as a second messenger activating various intracellular signaling pathways, which in turn can activate c-*fos* transcription via at least two regulatory elements located in its promoter: the cAMP/calcium responsive element (CRE) and the serum responsive element (SRE) (Badinget et al., 1993). The CRE site is regulated by the phosphorylation of the CRE binding protein (CREB) on ser133 by Ca^{2+}/calmodulin dependent kinase (CaMKs) or cAMP-dependent Protein Kinase (PKA) (Ginty 1997). The SRE regulatory element requires an assembly of a multiprotein complex, which includes a dimer of the serum response factor (SRF) and a ternary complex factor (TCF). The first TCF identified is Elk-1 (Hipskind et al., 1991), which is rapidly phosphorylated in its C-terminal region (on Ser 383 and Ser 389) consecutively to the activation the Mitogen-Activated Protein kinase (MAPK) of the Extracellular Signal-Regulated Kinase (ERK) subfamily: MAPK/ERK (Marais et al., 1993). This, in turn, leads to a strong transcriptional activation of c-*fos*.

To examine the respective role of CRE and SRE sites in c-*fos* and *zif268* mRNA induction, we analyzed, by immunocytochemistry, the activation of the regulatory factors (CREB and Elk 1) using antibodies specifically directed against their activated, phosphorylated state (phospho-Ser133 CREB and phospho-Ser383 Elk-1, respectively). The stimulation of the orofacial motor cortex, leading to the induction of c-*fos* and *zif268* mRNAs in the lateral part of the striatum (see above) was chosen for this analysis. Following a 15 min. stimulation, we found a strong increase of both phospho-CREB and phospho-Elk-1 immunoreactivity in many cells localized bilaterally in the lateral part of the striatum, in exact register with the area where c-*fos* and *zif268* mRNA were induced. We then examined the intracellular signaling cascade triggering the activation of CREB and Elk-1. Although calcium-induced CREB activation is classically attributed to the Ca/CaMK pathway, Elk-1 is known, in cell line models, as one of the main nuclear target of activated MAPK/ERK signaling pathway. Furthermore, glutamate stimulation was known to activate ERK proteins in neurons. We thus examined the ERK activation using an antibody directed against the dually phosphorylated (and thereby activated) ERK proteins. Numerous neurons showing an intense phospho-ERK immunoreactivity were found in the lateral striatum, bilaterally, in strict spatio-temporal correlation with Elk-1 and CREB activation, as well as with IEG mRNA induction. This result indicated that the cortical stimulation was able to simultaneously activate the ERK signaling cascade, to phophorylate the transcription factors Elk-1 and CREB, and to induce IEG mRNA expression.

Whereas the induction of IEG mRNA was sustained for at least one hour of stimulation, ERK activation as well as the phosphorylation of CREB and Elk-1 decreased progressively when the stimulation was prolonged (Sgambato et al., 1998). Indeed, after one hour of stimulation, phospho-CREB and phospho-Elk-1 were only found in few scattered neurons within the activated striatal area, while phospho-ERK positive neurons were reduced in number but not as markedly. The reduction of ERK activation could be due to the induction of a negative feedback mechanism linked to a specific phosphatase such as MAP Kinase Phophatase-1 (MKP-1), which was rapidly (15 min) induced by the cortical stimulation in register with ERK activation and c-*fos* and *zif268* mRNA induction. The dephosphorylation of Elk-1 and CREB could be partly due to a progressive inactivation of ERK by MKP-1, but additional mechanisms are likely to contribute to their pronounced deactivation. Indeed, calcineurin, a Ca^{2+} responsive phosphatase expressed in the striatum (Yakel, 1997) could be implicated since this phosphatase is known to dephosphorylate both Elk-1 (Sugimoto et al., 1997) and CREB (Bito et al., 1996).

Since the phophorylation of Elk-1 and CREB was in spatial register with ERK activation and IEG induction we addressed whether ERK activation was a necessary and sufficient condition to trigger IEG expression. A specific inhibitor of the ERK pathway (PD 98059) was

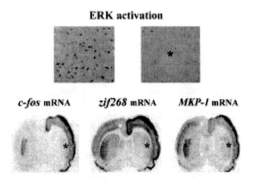

Figure 2. Effects of PD 98059 injection on ERK activation and IEG induction. PD 98059 (an inhibitor of ERK cascade) was injected unilaterally in the right striatum and the right orofacial area of the motor cortex was stimulated for 15 min. On the side of PD 98059 injection (*), ERK activation (top right panel) and IEG induction (bottom panel) were abolished whereas these activation remained unchanged on the contralateral side.

injected in the lateral part of the striatum, ipsilaterally to the stimulation side. The cortically driven ERK activation, Elk-1 and CREB hyperphosphoryla-tion as well as IEG induction were totally abolished on the PD 98059 injection side (Fig. 2). Contralaterally to the PD98059 injection side, the activation of ERK, Elk-1 and CREB as well as the strong induction of c-*fos* *zif268* and MKP-1 mRNAs attested the efficacy of the stimulation. Thus, altogether, these results stress that ERK activation is required for cortically driven IEG induction and that this signaling cascade targets Elk-1, and, more surprisingly, CREB. Usually, the Ca^{2+}-induced activation of CREB depends on CaMKs or PKA (through a Ca^{2+}/calmodulin sensitive adenylate cyclase). In our case, it appears that CREB is regulated by the ERK cascade, via a CREB kinase, for example p90RSK, a known target of ERK upon neurotrophin stimulation in PC12 cells (Xing et al., 1996) or primary cortical neurons (Finkbeiner et al., 1997) could be the link between ERK and CREB. Among the IEGs analyzed here, both c-*fos* and *zif268* promoters contain SRE and CRE sites. In contrast, the promoter for MKP-1 only contains CRE sites. Thus, the ERK signaling cascade could regulate IEGs containing either SRE or CRE sites in their promoter. How can intracellular Ca^{2+} elevation trigger ERK activation and subsequent IEG induction? This was partially addressed on a simplified model of glutamate-induced activation of c-fos on striatal slices. In this model system, reproducing in some instances a cortical stimulation, we showed that the activation of CaMKs was a prerequisite to the ERK activation, and thereby c-*fos* mRNA induction via Elk-1 and CREB activation (Vanhoutte et al., 1999). In other model systems, it seems that the Ca^{2+}-dependent non-receptor tyrosine kinases, PYK2 or FAK play a key role in activating the ras-dependent signaling pathway (ras/raf/MEK/ERK) upon intracellular Ca^{2+} elevation (Lev et al., 1995; Siciliano et al., 1996).

5. CONCLUSIONS

Cortical stimulations produce restricted induction of IEGs in the striatum and the STN, in correspondence with the activated cortical area. Our data provide evidence for a direct induction of IEGs in STN neurons by cortical stimulations. Furthermore, cortical inputs seem to control equally both direct and indirect pathways. The involvement of the MAPK/ERK signaling cascade in IEG induction is interesting in view of its critical role in long-term synaptic plasticity observed in the hippocampus (English and Sweatt 1997) as well as certain types of learning (Atkins et al., 1998). Since cortical stimulation can lead to synaptic plasticity in the striatum in striatal slices and *in vivo* (Calabresi et al 1996; Charpier and Deniau, 1997), it is tempting to speculate that MAPK/ERK activation observed in our model of cortico-striatal stimulation, participates to long term gene regulation and thereby to changes in protein expression needed for long term synaptic plasticity.

6. ACKNOWLEDGEMENTS

This work was supported by a grant from BIOMED II project N° BMH4-97-2215

7. REFERENCES

Afsharpour S., (1985) *J. Comp. Neurol.*, 236: 14-28.
Atkins C.M., Selcher J.C., Petraitis J.J., Trzaskos J.M., Sweatt J.D. (1998*) Nature Neurosci.*, 7: 602-609.
Bading H., Ginty D.D., Greenberg M.E. (1993) *Science*, 260: 181-186.
Beretta S., Parthasarathy H.B., Graybiel A.M. (1997) *J. Neurosci.*, 17: 4752-4763.
Bernard V, & Bolam JP. (1998) *Eur J Neurosci.* 10: 3721-3736
Bernard V., Legay C., Massoulie J., Bloch B. (1995) *Neurosci.*, 64: 995-1005.
Bito H., Deisseroth K., Tsien R.W. (1996) *Cell*, 87: 1203-1214.
Calabresi P., Calabresi P, Pisani A, Mercuri NB, Bernardi G. (1996) *Trends Neurosci.* 19 :19-24.
Charpier S., Deniau J.M. (1997). *Proc Natl Acad Sci* 94: 7036-7040.
Cowan R. L. and Wilson C. J., (1994) J. Neurophysiol., 71: 17-32
Dure LS 4th, Young AB, Penney JB Jr, 1992 *Proc Natl Acad Sci.*, 89: 7688-7692
English J.D., Sweatt J.D. (1997) *J. Biol. Chem.*, 272: 19103-19106.
Finkbeiner S., Tavazoie S.F., Maloratsky A., Jacobs K.M., Harris K.M., Greenberg M.E. (1997) *Neuron*, 19: 1031-1047.
Ginty D.D. (1997) *Neuron*, 18: 183-186.
Graybiel A.M., *Trends Neurosci.* (1990) 13: 244-54
Hersch SM, Ciliax BJ, Gutekunst CA, Rees HD, Heilman CJ, Yung KK, Bolam JP, Ince E, Yi H, Levey AI (1995*) J Neurosci.* 5: 5222-5237.
Hipskind R.A., Rao V.N., Mueller C.G.F., Reddy E;S.P., Nordheim A.(1991). *Nature*, 354: 531-534.
Lev, S., Moreno H., Martinez R., Canoll P., Peles E., Musacchio J.M., Plowman G.D., Rudy B., Schlessinger J.(1995) *Nature*, 376: 737-745.
Marais, R., Wynne J., Treisman R. (1993) *Cell*, 73: 381-393.
Maurice N, Deniau JM, Glowinski J, Thierry AM, (1998) *J Neurosci.*, 18: 9539-46
McGeorge AJ, Faull RL, (1989) *Neuroscience.* 29: 503-37.
Parthasarathy H.B., Graybiel A.M. (1997) *J. Neurosci.*, 17: 2477-2491.
Sgambato V, Abo V, Rogard M, Besson MJ, Deniau JM (1997) *Neuroscience*, 81: 93-112.
Sgambato V, Maurice N, Besson MJ, Thierry AM, Deniau JM., (1999) *Neuroscience*;93: 1313-21
Sgambato V., Pagès C., Rogard M., Besson M.-J., Caboche J.(1998) *J. Neurosci.*, 18: 8814-8825.
Siciliano J.C., Toutant M., Derkinderen P., Sasaki T., Girault J.A. (1996) *J. Biol. Chem.*, 271: 28942-28946.
Smith AD & Bolam JP., *Trends Neurosci.* (1990) 13: 259-65.
Sugimoto T., Stewart S., Guan K.L. (1997) *J. Biol. Chem.*, 272: 29415-29418.
Vanhoutte P., Barnier J.V.B., Guibert B., Pagès C., Besson M.J., Hipskind R.A., Caboche J. (1999) *Mol. Cell. Biol.*, 19: 136-146.
Xing J., Ginty D.D., Greenberg M.E. (1996) Science, 273: 959-963.
Yakel J.L. (1997) *Trends Pharmacol. Sci.,* 18: 124-134.

SUBSYNAPTIC LOCALIZATION OF GROUP I METABOTROPIC GLUTAMATE RECEPTORS IN THE BASAL GANGLIA

Yoland Smith, Maryse Paquet, Jesse E. Hanson, and George W. Hubert*

1. INTRODUCTION

The first evidence that glutamate may serve as a neurotransmitter in the central nervous system was published in the late 1950's. It is now well established that glutamate is the primary excitatory neurotransmitter in the vertebrate CNS. Until the middle of the 1980's, it was thought that the excitatory effects of glutamate were mediated exclusively through activation of ligand-gated cation channels. However, in 1985, Sladeczek and colleagues (1985) showed that glutamate could stimulate phospholipase C (PLC) in cultured striatal neurons via a receptor subtype that did not belong to the ionotropic receptor family. Similar findings were, then, found in hippocampal slices (Nicoletti et al., 1986a), cultured cerebellar neurons (Nicoletti et al., 1986b) and cultured astrocytes (Pearce et al., 1986). These findings opened up the possibility that glutamate, like many other neurotransmitters, activates both ligand-gated ion channels and G-protein-coupled receptors, now called metabotropic glutamate receptors (mGluRs).

Since the cloning of the first mGluR in 1991 (Houamed et al., 1991; Masu et al., 1991), our knowledge on the structure, localization and functions of this receptor family has increased substantially. Eight subtypes of mGluRs have now been cloned and subdivided into three major groups on the basis of sequence similarity, pharmacological properties and transduction pathways (Nakanishi, 1994; Pin and Duvoisin, 1995; Conn and Pin, 1997). Apart from a few exceptions (Rainnie et al., 1994; Gereau and Conn, 1995; Manzoni and Bockaert, 1995; Fiorillo and Williams, 1998), activation of group I mGluRs generally increases neuronal excitation and excitability whereas activation of group II and group III mGluRs reduces synaptic excitation.

Apart from the striatum, where the role of mGluRs in mediating long term depression in projection neurons (Calabresi et al., 1993, 1994, 1996) and presynaptic inhibition of glutamatergic transmission at corticostriatal synapses (Lovinger, 1991; Calabresi et al., 1993,

---

* Y. Smith, M. Paquet, J.E. Hanson, and G.W. Hubert, Yerkes Regional Primate Research Center and Department of Neurology, Emory University, Atlanta, GA 30329, USA.

1996; Lovinger and McCool, 1995; Cozzi et al., 1997; Manzoni et al., 1997; Pisani et al., 1997) is well established, very little is known about the functions of mGluRs in the basal ganglia. Most information on the localization of the different mGluR subtypes is limited to *in situ* hybridization and light or confocal microscope immunocytochemical findings (Martin et al., 1992; Testa et al., 1994, 1998; Kerner et al., 1997; Petralia et al., 1997; Berthele et al., 1998; Kinoshita et al., 1998; Kosinki et al., 1998; Bradley et al., 1999). As part of an ongoing series of studies on the synaptic localization of mGluRs, we present some of our recent findings on the subsynaptic distribution of group I mGluRs (mGluR1α and mGluR5) in the monkey basal ganglia.

2. MATERIALS AND METHODS

Four adult Rhesus monkeys (Macaca mulatta) and eight Sprague Dawley rats were deeply anaesthetized with an overdose of pentobarbital and perfused transcardially with cold oxygenated Ringer's solution followed by 4.0% paraformaldehyde and 0.1% glutaraldehyde in phosphate buffer (PB; 0.1 M, pH 7.4). After having been cut in 60 μm thick transverse sections with a vibrating microtome, sections were treated with sodium borohydride (1.0% in PBS), transferred to cryoprotectant and frozen at -80° C for 20 min. They were then thawed and returned to a graded series of cryoprotectant and PBS before being processed for immunocytochemistry.

2.1 Primary Antisera and Immunocytochemistry

Two commercially available affinity-purified rabbit polyclonal antibodies raised against synthetic carboxy-terminus peptides representing different amino acid sequences of mGluR1α (PNVTYASVILRDYKQSSSTL; Chemicon International, Temecula, CA) and mGluR5a/b (KSSPKYDTLIIRDYTNSSSSL; Upstate Biotech., Lake Placid, NY) were used in the present study. In immunoblot analysis of rat brain microsomes or rabbit brain extracts, both antibodies labeled a single band with an estimated molecular weight of 145 kDA which corresponds to that of mGluR1α and mGluR5 proteins (Houamed et al., 1991; Abe et al., 1992; Minakami et al., 1993). For immunocytochemistry, the mGluR1α antibodies were used at 0.5μg/ml dilution whereas the mGluR5 antiserum was used at 1.0 μg/ml. The mouse tyrosine hydroxylase (TH) antibody used for double labeling experiments in the striatum was purchased from INCSTAR (Stillwater, MN, USA) and used at 1:10 000 dilution.

2.1.1. Single Labeling Immunocytochemistry for Group 1 mGluR Localization

Both pre-embedding immunoperoxidase and silver-intensified immunogold techniques were used in the present study. Details of the immunostaining protocols for light and electron microscopy are found in a previous study (Hanson and Smith, 1999). Briefly, sections processed for immunoperoxidase were stained with diaminobenzidine (DAB) using the avidin biotin-peroxidase method (ABC, Vector Labs, Burlingame, CA, USA) whereas sections processed for immunogold were incubated with 1.4 nm gold-conjugated secondary antibodies which were silver-intensified for about 10 minutes with the HQ Silver Kit (Nanoprobes, Stonybrook, NY, USA). As controls, sections were incubated in solutions from which the primary antisera were replaced by 1% nonimmune rabbit serum while the rest of the procedure remained the same as described above. Sections processed in this way were totally devoid of gold particles or DAB staining. After immunostaining, sections were processed for electron microscopy, and ultrathin sections of various basal ganglia nuclei were prepared as described in our previous study (Hanson and Smith, 1999).

2.1.2. Double Labeling Immunocytochemistry for TH and Group I mGluRs

Some striatal sections were processed for the localization of TH and group I mGluRs. In those sections, TH was revealed first with DAB using the ABC technique whereas group I mGluRs were localized with the pre-embedding immunogold method. As controls, the TH or the two group I mGluRs antibodies were omitted in turn from the incubation solutions while the rest of the immunocytochemical procedure remained the same.

2.2 Analysis of Material

To analyse the relationships between gold particles and post-synaptic specializations, micrographs of dendrites were taken at 25 000X and 40 000X from the surface of mGluR1α- and mGluR5-immunostained sections where the labeling was optimal. The gold particles attached to the plasma membrane were then counted and pooled into three categories (extrasynaptic, perisynaptic or synaptic) based on their localization relative to post-synaptic membrane specializations visible in the plane of section (see Results for more details). Portions of dendrites where the preservation or the plane of section were not suitable to distinguish the pre- and post-synaptic membranes were omitted from the analysis. To ascertain the specificity of labeling, many immunoreactive synapses were examined in three to seven serial ultrathin sections.

To measure the surface of dendrites and length of synaptic junctions in the pallidum, micrographs of randomly selected immunolabeled dendrites taken from GPe and GPi were scanned with a digital scanner (Umax Powerlook II) and analyzed for total dendritic membrane length and total length of synaptic active zones using a Neurolucida setup and Morph software (MicroBrightField).

3. RESULTS

3.1. Striatum

Overall, the pattern of group I mGluR immunoreactivity was the same in the caudate nucleus and the putamen. Both medium-sized projection neurons and large interneurons displayed mGluR1α and mGluR5 immunoreactivities (Fig. 1C). In general, the neuropil staining was much more intense with the mGluR5 than the mGluR1α antiserum. No obvious patch/matrix pattern of distribution of immunoreactive neurons was observed with the mGluR5 antiserum, but patches of low mGluR1α immunoreactivity were occasionally seen in the caudate nucleus.

At the electron microscope level, the immunoperoxidase reaction product was mostly found in post-synaptic elements including large- and small-sized perikarya with smooth or

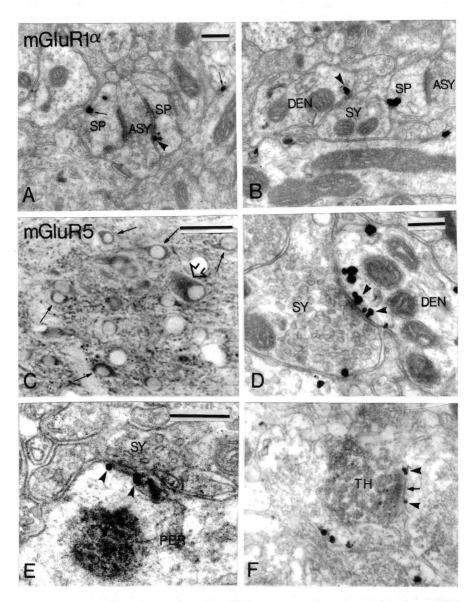

Figure 1. MGluR1α (A-B) and mGluR5 (C-F) immunoreactivity in the monkey striatum. (A) Peri- (arrowhead) and extrasynaptic (arrows) mGluR1α immunolabeling at asymmetric axo-spinous synapses. (B) Synaptic labeling at a symmetric axo-dendritic synapse (arrowhead). (C) Both medium-sized projection neurons (arrows) and large interneurons (open arrows) display mGluR5 immunoreactivity. (D) mGluR5 synaptic labeling at a symmetric axo-dendritic synapse (arrowheads). (E) Perisynaptic labeling (arrowheads) at a symmetric axo-somatic synapse. A similar pattern of labeling was found at symmetric synapses (arrow) established by TH-immunoreactive boutons (F). Abbreviations: ASY: terminals forming asymmetric synapses; DEN: dendrite, PER: perikaryon; SP: spine; SY: terminal forming a symmetric synapse; TH: tyrosine hydroxylase. Scale bars: A: 0.25 μm (valid for B); C: 50 μm; D: 0.25 μm; E: 0.25 μm (valid for F).

indented nuclei, dendritic processes of various sizes and dendritic spines. In addition, a few axon terminals which formed asymmetric axo-spinous synapses displayed light mGluR1α immunoreactivity, but presynaptic labeling was never encountered in mGluR5-immunostained sections. In sections labeled with immunogold, both mGluR1α and mGluR5 immunoreactivities were commonly found perisynaptically to asymmetric post-synaptic specializations of axo-spinous and axo-dendritic synapses (Fig. 1A). Gold particles were also seen in the main body of symmetric post-synaptic specializations established by striatal-like GABAergic boutons (Bolam et al., 1985) (Fig. 1B,D) or at the edges of "en passant" type symmetric synapses formed by vesicle-filled axonal processes that resembled dopaminergic nigrostriatal afferents (Freund et al., 1984; Smith et al., 1994) (Fig. 1E). Consistent with these observations, mGluR5 immunoreactivity was located perisynaptically to symmetric synapses formed by TH-positive terminals (Fig. 1F). A large number of extrasynaptic gold particles were also seen along the membrane of dendrites and spines (Fig. 1A,B,D).

3.2. Pallidum

Neuronal perikarya and dendritic processes displayed strong mGluR5 and mGluR1α immunoreactivity in both GPe and GPi (Fig. 2A). The pattern and intensity of staining for the two mGluR subtypes was the same throughout the extent of both pallidal segments. In general, the intensity of cytoplasmic labeling was stronger with the mGluR5 than the mGluR1α antibodies.

Overall, the pattern of distribution of the immunogold labeling was consistent with the light microscope peroxidase staining, e.g. gold particles were associated exclusively with post-synaptic elements and the density of gold particles in the cytoplasm of mGluR5-immunoreactive structures was significantly higher than was found in mGluR1α-containing elements (Hanson and Smith, 1999). There was no significant difference in the subsynaptic distribution of mGluR1α and mGluR5 immunoreactivity between the two pallidal segments. Although the majority of gold particles were extrasynaptic, a substantial proportion of immunolabeling (27-44% gold particles) was found in the main body of symmetric synapses established by striatal-like terminals in GPe and GPi (Fig. 2B,D,E). In contrast, immunogold particles associated with asymmetric synapses established by subthalamic-like terminals were always found in a perisynaptic position at the edges of synaptic junctions (Fig. 2B,C,E).

To make sure that the labeling at symmetric synapses was not merely coincidental to the fact that pallidal dendrites are tightly surrounded by striatal boutons (Smith et al., 1998), the predicted percentage of gold particles at synapses based on a random distribution on the dendritic membrane was calculated. This was achieved by comparing the proportion of the surface of dendritic membranes involved in the formation of symmetric or asymmetric synapses to the percentages of gold particles associated with striatal and subthalamic-like synapses (Fig. 2B). A randomly selected sample of 20 dendrites in GPe and 20 dendrites in GPi were photographed and measured. In total, these dendrites accounted for approximately 2 mm of dendritic membrane from which data was collected. Since no significant difference was found between GPe and GPi, the percentages of dendritic membrane involved in synaptic specializations in both pallidal segments were pooled (Fig. 2B). These measurements revealed that ~85% of the membrane of pallidal dendrites does not contribute to synaptic specializations whereas ~13% contribute to symmetric synapses and ~2% contribute to asym-

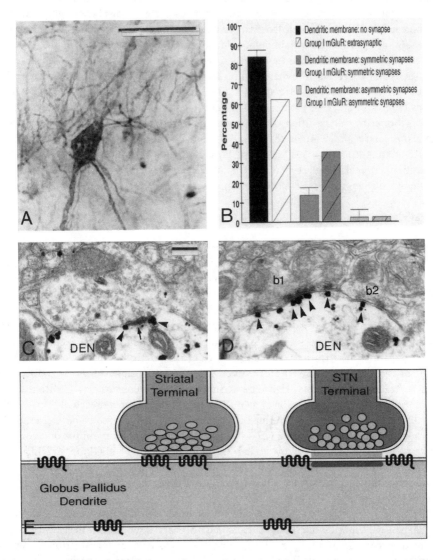

Figure 2. Group I mGluR in the monkey globus pallidus. (A) mGluR1 α-immunoreactive neuron in GPi. (B) Histogram comparing the percentage (Mean ± SD) of dendritic membrane of GPe and GPi neurons not contributing to any synaptic junctions *(no synapse)* or involved in the formation of symmetric and asymmetric synapses, with the percentages of gold particle labeling for group I mGluRs on pallidal dendrites. The proportion of gold particles at both types of synapses or located extrasynaptically was estimated from a total of 1337 membrane-bound gold particles in 250 dendrites randomly selected in mGluR1α and mGluR5-immunostained GPe and GPi. (C) mGluR5 immunoreactivity perisynaptic (arrowhead) to an asymmetric post-synaptic specialization (arrow) in GPe. (D) mGluR5 immunolabeling in the main body of axodendritic symmetric synapses (arrowheads) established by two striatal-like boutons (b1, b2) in GPe. (E) Summary of the pattern of distribution of group I mGluR immunoreactivity in relation to striatal and subthalamic terminals in contact with GPe and GPi dendrites. Scale bars: A: 50 μm; C: 0.25 μm (valid for D).

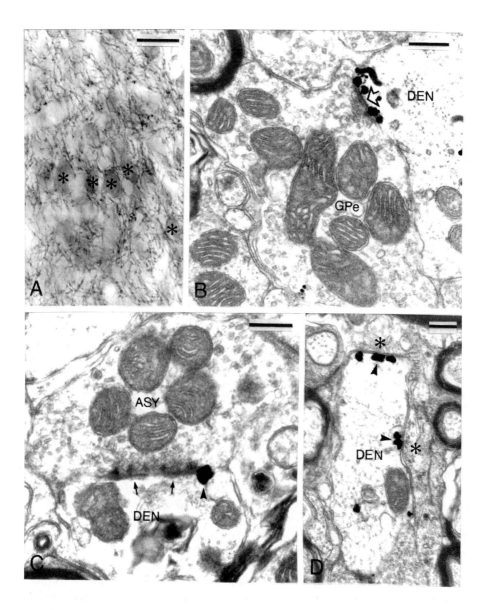

Figure 3. MGluR1α labeling in the monkey subthalamic nucleus. (A) Dense plexus of immunoreactive processes and lightly labeled perikarya (asterisks). (B) mGluR1α immunoreactivity at a symmetric synapse (open arrow) established by a GPe-like bouton. (C) Perisynaptic labeling (arrowhead) associated with an asymmetric axo-dendritic synapse (arrows). (D) mGluR1α synaptic labeling (arrowheads) at "en passant" type axodendritic symmetric synapses established by vesicle-filled axon-like processes. Scale bars: A: 25 μm; B: 0.25 μm; C: 0.25 μm; D: 0.25 μm.

metric synapses (Fig. 2B). Thus, based on a random distribution, a substantially lower percentage of gold particle labeling would be predicted at symmetric striatopallidal synapses than was found for Group I mGluR labeling in both pallidal segments (Fig. 2B).

3.3. Subthalamic nucleus

Although neuronal perikarya were lightly labeled, the neuropil of the subthalamic nucleus (STN) displayed strong immunoreactivity for both group I mGluRs (Fig. 3A). In the electron microscope, mGluR1α immunoreactivity was expressed exclusively in post-synaptic elements. Immunogold particles were commonly found perisynaptic to asymmetric post-synaptic specializations on small dendrites (Fig. 3C). In some cases, gold particles were attached to the main body of "en passant type" symmetric synapses established by vesicle-filed axon-like processes (Fig. 3D) or symmetric synapses formed by large GPe-like terminals (Fig. 3B). Extrasynaptic labeling was also detected for both receptor subtypes.

3.4. Substantia nigra

At the light microscope level, the intensity of group I mGluRs labeling in the substantia nigra compacta (SNc) and ventral tegmental area (VTA) was much higher in monkeys than in rats (compare Fig. 4A with 5A). Whereas the labeling was confined to the lateral SNc in rats, the whole extent of SNc and VTA displayed moderate to strong immunoreactivity in monkeys (Fig. 4A). At the electron microscope level, mGluR1α labeling was found along the plasma membrane of SNc neurons whereas most of the gold labeling for mGluR5 was intracytoplasmic (Fig. 4B-E). The pattern of subsynaptic distribution of mGluR1α immunoreactivity in the SNc resembled that found in other parts of the basal ganglia e.g. many gold particles were extrasynaptic, some were found in the core of symmetric synapses and a few were located perisynaptically to asymmetric post-synaptic specializations (Fig. 4B-C). However, in contrast to other basal ganglia nuclei, mGluR1α immunoreactivity was occasionally found in the main body of asymmetric postsynaptic specializations in monkey SNc neurons (Fig. 4C-D).

In the rat SNr, rich plexuses of intensely stained dendritic processes were frequently seen (Fig. 5B). At the electron microscope level, the intensity of immunogold labeling for mGluR1α and mGluR5 was much higher in SNr than SNc neurons (Fig. 5C-D). However, the relative distribution of membrane-bound versus intracytoplasmic immunogold labeling varied significantly for the two group I mGluR subtypes. Quantitative measurements revealed that most gold particles were attached to the plasma membrane in mGluR1α-immunoreactive dendrites (Fig. 5C), whereas less than 15% of the total gold labeling for mGluR5 was membraneous (Fig. 5D) (Hubert et al., 1999). As was found for the globus pallidus, most of the membrane-bound gold particles in dendrites were extrasynaptic or associated with the main body of symmetric synapses established by striatal-like terminals (Fig. 5C) while the remaining particles were located perisynaptically to asymmetric post-synaptic specializations (Fig. 5C).

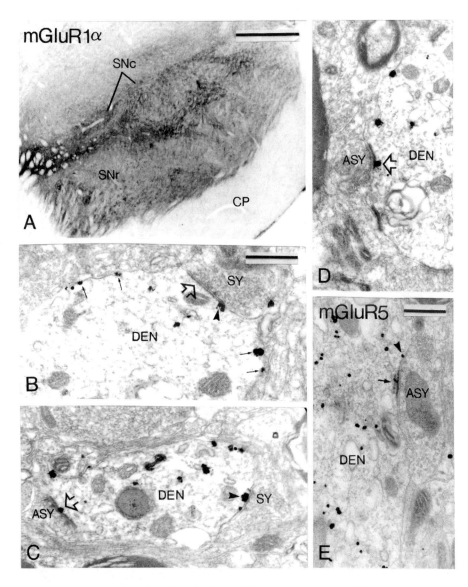

Figure 4. mGluR1α (A-D) and mGluR5 (E) immunolabeling in the monkey SNc. (A) Dense mGluR1α immunostaining of SNc and VTA neurons. (B) Perisynaptic mGluR1α labeling (arrowhead) at a symmetric axo-dendritic synapse (open arrow). Extrasynaptic gold particles are also indicated (arrows). (C) mGluR1α labeling in the main body of an axo-dendritic asymmetric synapse (open arrow) and a symmetric synaptic contact (arrowhead). (D) mGluR1α labeling in the main body of an asymmetric post-synaptic specialization (open arrow). (E) mGluR5 immunoreactivity in a proximal dendrite. Note that a single gold particle is attached to the membrane (arrowhead) not far from an asymmetric synapse (arrow). In contrast to the mGluR1α immunoreactivity which is largely membraneous, most of the mGluR5 immunolabeling is intracytoplasmic. Abbreviations: CP: cerebral peduncle; VTA: ventral tegmental area. Scale bars: A: 0.5 mm; B: 0.5 μm (valid for C,D); E: 0.5 μm.

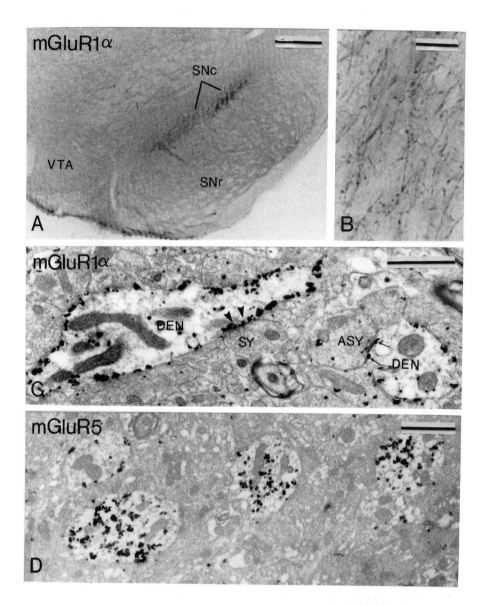

Figure 5. mGluR1α (A-C) and mGluR5 (D) immunolabeling in the rat SNr. (A) Low power view of mGluR1α immunoreactivity in the substantia nigra. Note the low level of immunoreactivity in the medial SNc and VTA in comparison to the monkey (see Fig. 4A). (B) High power view of mGluR1α-immunoreactive processes in the SNr. (C and D) Comparison of the pattern of distribution mGluR1α (C) and mGluR5 (D) immunoreactivity in dendrites of the SNr. Note that most mGluR1a labeling is attached to the plasma membrane whereas the mGluR5 immunoreactivity is intracytoplasmic. The arrowheads in C indicate gold particles in the main body of a symmetric synapse established by a striatal-like terminal whereas the arrows point to two gold particles perisynaptic to an asymmetric axo-dendritic synapse. Scale bars: A: 0.25 mm; B: 50 μm; C: 1 μm; D: 1 μm.

4. SUMMARY AND CONCLUDING REMARKS

The findings presented in this study demonstrate that both group I mGluR subtypes (mGluR1α and mGluR5) are widely distributed and largely expressed postsynaptically throughout the basal ganglia. Three main features characterize their subsynaptic localization: (**1**) They are located perisynaptically or extrasynaptically to asymmetric post-synaptic specializations of putative glutamatergic synapses. Gold particles were never found in the main body of asymmetric synapses, except in the SNc where some asymmetric post-synaptic specializations were labeled. (**2**) Both group I mGluR subtypes were found in the main body of symmetric synapses established by putative GABAergic terminals. This was particularly obvious in the globus pallidus and the SNr where symmetric synapses formed by striatal terminals were decorated with large numbers of gold particles. (**3**) The relative abundance of membrane-bound versus intracytoplasmic labeling for the two group I mGluR subtypes was strikingly different. Whereas mGluR1α immunolabeling was almost exclusively found on the plasma membrane, a large proportion of mGluR5 immunoreactivity was intracytoplasmic. Whether this indicates a differential level of internalization between the two group I mGluR subtypes remains to be established.

Although the perisynaptic localization of Group I mGluRs at asymmetric synapses was expected on the basis of previous findings in the cerebellum and hippocampus (Baude et al., 1993; Nusser et al., 1994; Lujan et al., 1996; Ottersen and Landsend, 1997), the labeling of symmetric striatal synapses was surprising and raises questions about the functions and mechanisms of activation of group I mGluRs at GABAergic synapses. To our knowledge, this is the first evidence of post-synaptic group I mGluRs at putative GABAergic synapses in the CNS. Three potential sources of activation should be considered: (1) extra-synaptic diffusion of glutamate from subthalamic and/or cortical terminals (Asztely et al. 1997; Barbour and Hausser, 1997; Kullmann and Asztely, 1998), (2) release of glutamate from astrocytes (Antanitus, 1998; Araque et al., 1999) and (3) release of glutamate from striatal terminals (Dubinsky, 1989; White et al., 1994). If glutamate, indeed, activates mGluRs at striatofugal synapses, it is likely that the post-synaptic mGluR responses regulate GABA currents in the target neurons (Glaum and Miller, 1993; 1994). Activation of mGluRs can also modulate post-synaptic GABA responses via the reduction of K^+ conductances which leads to an increase of membrane excitability and depression of GABA receptor sensitivity (Glaum and Miller, 1994; Conn and Pin, 1997).

It is worth noting that group I mGluR activation has been shown to mediate both slow IPSPs and EPSPs in midbrain dopamine neurons (Fiorillo and Williams, 1998). Thus, while looking for a functional role for group I mGluRs at GABAergic synapses, it is important to consider that these receptors could have different effects depending on the specific conditions surrounding the activation of these receptors.

5. ACKNOWLEDGMENTS

The authors thank Jean-François Paré and Jeremy Kieval for technical assistance and Frank Kiernan for photography. This research was supported by NIH grants RO1 NS37423-02 and P50 NS38399-01 and a grant from the American Parkinson Disease Association.

6. REFERENCES

Abe T, Sugihara H, Nawa H, Shigemoto R, Mizuno N, and Nakanishi S. Molecular characterization of a novel metabotropic glutamate receptor mGluR5 coupled to inositol phosphate/Ca^{2+} signal transduction. J Biol Chem 1992; 267: 13361-13368.

Antanitus DS. A theory of cortical neuron-astrocyte interaction. Neuroscientist 1998; 4: 154-159.

Araque A, Parpura V, Sanzgiri R, and Haydon PG. Tripartite synapses: glia, the unacknowledged partner. Trends Neurosci 1999; 22: 208-215.

Asztely F, Erdemli G, and Kullmann DM. Extrasynaptic glutamate spillover in the hippocampus: dependence on temperature and the role of active glutamate uptake. Neuron 1997; 18: 281-293.

Barbour B, and Hausser M. Intersynaptic diffusion of neurotransmitter. Trends Neurosci 1997; 20: 377-384.

Baude A, Nusser Z, Roberts JD, Mulvihill E, McIlhinney RA, Somogyi P. The metabotropic glutamate receptor (mGluR 1 α) is concentrated at perisynaptic membrane of neuronal subpopulations as detected by immunogold reaction. Neuron 1993; 11: 771-787.

Berthele A, Laurie DJ, Platzer S, Zieglgänsberger W, Tölle TR, Sommer B. Differential expression of rat and human type I metabotropic glutamate receptor splice variant messenger RNAs. Neuroscience 1998; 85: 733-749.

Bolam JP, Powell JP, Wu J-Y, Smith AD. Glutamate decarboxylase-immunoreactive structures in the rat neostriatum. A correlated light and electron microscopic study including a combination of Golgi-impregnation with immunocytochemistry. J Comp Neurol 1985; 237: 1-20.

Bradley SR, Standaert DG, Rhodes KJ, Rees HD, Testa LM, Levey AI, Conn PJ. Immunohistochemical localization of subtype 4a metabotropic glutamate receptors in the rat and mouse basal ganglia. J Comp Neurol 1999; 407: 33-46.

Calabresi, P, Pisani A, Mercuri NB, Bernardi G. Heterogeneity of metabotropic glutamate receptors in the striatum: electrophysiological evidence. Eur J Neurosci 1993; 5:1370-1377.

Calabresi P, Pisani A, Mercuri NB, Bernardi G. Post-receptor mechanisms underlying striatal long-term depression. J Neurosci 1994; 14: 4871-4881.

Calabresi, P, Pisani A, Mercuri NB, Bernardi G. The corticostriatal projection: from synaptic plasticity to dysfunctions of the basal ganglia. Trends Neurosci 1996; 19:19-24.

Conn PJ, Pin JP. Pharmacology and functions of metabotropic glutamate receptors. Ann Rev Pharmacol & Toxicol, 1997; 37: 205-237.

Cozzi A, Attuci S, Peruginelli F, Luneia R, Pellicciari R, Moroni F. Type 2 metabotropic glutamate (mGlu) receptors tonically inhibit transmitter release in rat caudate nucleus: In vivo studies with (2S,1'S,2'S,3'R)-2-(2'carboxy-3'-phenylcyclopropyl)glycine, a new potent and selective antagonist. Eur J Neurosci 1997; 9: 1350-1355.

Dubinsky JM. Development of inhibitory synapses among striatal neurons in vitro. J Neurosci 1989; 9: 3955-3965.

Fiorillo CD, Williams JT. Glutamate mediates an inhibitory post-synaptic potential in dopamine neurons. Nature, 1998; 394: 78-82.

Freund TF, Powell J, Smith AD. Tyrosine hydroxylase-immunoreactive boutons in synaptic contact with identified striatonigral neurons with particular reference to dendritic spines. Neuroscience 1984; 13: 1189-1215.

Gereau IV RW, Conn PJ. Multiple presynaptic metabotropic glutamate receptors modulate excitatory and inhibitory synaptic transmission in hippocampal area CA1. J Neurosci 1995; 15: 6879-6889.

Glaum SR, and Miller RJ. Activation of metabotropic glutamate receptors produces reciprocal regulation of ionotropic glutamate and GABA responses in the nucleus of the tractus solitarius of the rat. J Neurosci 1993; 13: 1636-1641.

Glaum SR, and Miller RJ. Acute regulation of synaptic transmission by metabotropic glutamate receptors. In: The Metabotropic Glutamate Receptors (Conn PJ and Patel J, eds), 1994; pp 147-172. Totowa, NJ: Humana Press Inc

Hanson JE, Smith Y. Group I metabotropic glutamate receptors at GABAergic synapses in monkeys. J Neurosci 1999; 19:6488-6496.

Houamed KM, Kuijper JL, Gilbert TL, Haldeman BA, O'hara PJ, Mulvihill ER, Almers W, Hagen FS. Cloning, expression, and gene structure of a G-protein-coupled glutamate receptor from rat brain. Science 1991; 252: 1318-1321.

Hubert GW, Paquet M, and Smith Y. Differential subcellular localization of mGluR1 α and mGluR5 in the rat and monkey substantia nigra. J. Neurosci 2001; 21: 1838-1847.

Kerner JA, Standaert DG, Penney Jr JB, Young AB, Landwehrmeyer GB. Expression of group one metabotropic glutamate receptor subunit mRNAs in neurochemically identified neurons in the rat neostriatum, neocortex, and hippocampus. Mol Brain Res 1997; 48: 259-269.

Kinoshita A, Shigemoto R, Ohishi H, Van der Putten H, Mizuno N. Immunohistochemical localization of metabotropic glutamate receptors, mGluR7a and mGluR7b, in the central nervous system of the adult rat and mouse: A light and electron microscopic study. J Comp Neurol 1998; 393: 332-352.

Kosinki CM, Standaert DG, Testa CM, Penney Jr JB, Young AB. Expression of metabotropic glutamate receptor 1 isoforms in the substantia nigra pars compacta of the rat. Neuroscience 1998; 86: 783-798.

Kullmann DM, and Asztely F. Extrasynaptic glutamate spillover in the hippocampus: evidence and implications. Trends Neurosci 1998; 21: 8-14.

Lovinger, D. Trans- l-aminocyclopentane-1,3-dicarboxylic acid (t-ACPD) decreases synaptic excitation in rat striatal slices through a presynaptic action. Neurosci Lett, 1991; 129:17-21

Lovinger, D, McCool, BA. Metabotropic glutamate receptor- mediated presynaptic depression at corticostriatal synapses involves mGLuR2 or 3. J Neurophysiol 1995; 73:1076-1083

Lujan R, Nusser Z, Roberts JD, Shigemoto R, Somogyi P. Perisynaptic location of metabotropic glutamate receptors mGluR1 and mGluR5 on dendrites and dendritic spines in the rat hippocampus. Eur J Neurosci 1996; 8: 1488-1500.

Manzoni O, Bockaert J. Metabotropic glutamate receptors inhibiting excitatory synapses in the CA1 area of rat hippocampus. Eur J Neurosci 1995; 7: 2518-2523.

Manzoni O, Michel J-M, Bockaert J. Metabotropic glutamate receptors in the rat nucleus accumbens. Eur J Neurosci 1997; 9: 1514-1523.

Martin LJ, Blackstone CD, Huganir RL, Price DL. Cellular localization of a metabotropic glutamate receptor in rat brain. Neuron 1992; 9: 259-270.

Masu M, Tanabe Y, Tsuchida K, Shigemoto R, Nakanishi S. Sequence and expression of a metabotropic glutamate receptor. Nature 1991; 349: 760-765.

Minakami R, Katsuki F, Sugiyama H. A variant of metabotropic glutamate receptor subtype 5: an evolutionally conserved insertion with no termination codon. Biochem Biophys Res Commun 1993; 194: 622-627.

Nakanishi S. Metabotropic glutamate receptors: synaptic transmission, modulation, and plasticity. Neuron, 1994; 13: 1031-1037.

Nicoletti F, Meek JL, Iadarola MJ, Chuang DM, Roth BL, Costa E. Coupling of inositol phospholipid metabolism with excitatory amino acid recognition sites in rat hippocampus. J Neurochem 1986a; 46: 40-46.

Nicoletti F, Wroblewski JT, Novelli A, Alho H, Guidotti A, Costa E. Their activation of inositol phospholipid metabolism as a signal-transduction system for excitatory amino acids in primary cultures of cerebellar granule cells. J Neurosci 1986b; 6: 1905-1911.

Nusser Z, Mulvihill E, Streit P, Somogyi P. Subsynaptic segregation of metabotropic and ionotropic glutamate receptors as revealed by immunogold localization. Neuroscience 1994; 61: 421-427.

Ottersen OP, Landsend AS. Organization of glutamate receptors at the synapse. Eur J Neurosci 1997; 9: 2219-2224.

Parpura V, Basarsky TA, Liu F, Jeftinija S, and Haydon PG. Glutamate-mediated astrocyte-neuron signaling. Nature 1994; 369: 744-747.

Pearce B, Albrecht J, Morrow C, Murphy S. Astrocyte glutamate receptor activation promotes inositol phospholipid turnover and calcium flux. Neurosci Lett 1986; 72: 335-340.

Petralia RS, Wang YX, Singh S, Wu C, Sh i L, Wei J, Wenthold RJ. A monoclonal antibody shows discrete cellular and subcellular localizations of mGluR1α metabotropic glutamate receptors. J Chem Neuroanat 1997; 13: 77-93.

Pin JP, Duvoisin R. The metabotropic glutamate receptors: structure and functions. Neuropharmacol 1995; 34: 1-26.

Pisani A, Calabresi P, Centonze D, Bernardi G. Activation of group III metabotropic glutamate receptors depresses glutamatergic transmission at corticostriatal synapses. Neuropharmacol 1997; 36: 845-851.

Rainnie DG, Holmes KH, Shinnick-Gallagher P. Activation of postsynaptic metabotropic glutamate receptors by trans-ACPD hyperpolarizes neurons of the basolateral amygdala. J Neurosci 1994; 14: 7208-7220.

Sladeczek F, Pin J-P, Récasens M, Bockaert J, Weiss S. Glutamate stimulates inositol phosphate formation in striatal neurones. Nature 1985; 211: 182-185.

Smith Y, Bennett BD, Bolam JP, Parent A, Sadikot AF. Synaptic relationships between dopaminergic afferents and cortical or thalamic input in the sensorimotor territory of the striatum in monkey. J Comp Neurol, 1994; 344: 1-19.

Smith Y, Shink E, Bevin MD, Bolam JP. Synaptology of the direct and indirect striatofugal pathways. Neuroscience 1998; 86: 353-387.

Testa CM, Standaert DG, Young AB, Penney JB, Jr. Metabotropic glutamate receptor mRNA expression in the basal ganglia of the rat. J Neurosci 1994; 14: 3005-3018.

Testa CM, Friberg IK, Weiss SW, Standaert DG. Immunohistochemical localization of metabotropic glutamate receptors mGluR1α and mGluR2/3 in the rat basal ganglia. J Comp Neurol 1998; 390: 5-19.

White LE, Hodges HD, Carries KM, Price JL, and Dubinsky JM. Colocalization of excitatory and inhibitory neurotransmitter markers in striatal projection neurons in the rat. J Comp Neurol 1994; 339: 328-340.

LOCALIZATION AND PHYSIOLOGICAL ROLES OF METABOTROPIC GLUTAMATE RECEPTORS IN THE INDIRECT PATHWAY

Michael J. Marino, Stefania R. Bradley, Hazar Awad, Marion Wittmann, and P. Jeffrey Conn*

1. INTRODUCTION

Parkinson's disease (PD) is a common neurodegenerative disorder characterized by disabling motor impairments including tremor, rigidity, and bradykinesia. The primary pathological change giving rise to the symptoms of Parkinson's disease is loss of dopaminergic neurons in the substantia nigra pars compacta that modulate the function of neurons in the striatum and other nuclei in the basal ganglia (BG) motor circuit (Fig. 1). Currently, the most effective pharmacological agents for treatment of PD include levodopa (L-DOPA), the immediate precursor of dopamine, and other drugs that replace the lost dopaminergic modulation of BG function.[1] Unfortunately, dopamine replacement therapy ultimately fails in most patients due to loss of efficacy with progression of the disease and severe motor and psychiatric side effects.[2] Because of this, a great deal of effort has been focused on developing new approaches for treatment of PD.

The primary input nucleus of the basal ganglia is the striatum (caudate, putamen, and nucleus accumbens), which receives dense innervation from the cortex and subcortical structures. The primary output nuclei of the basal ganglia are the substantia nigra pars reticulata (SNr) and the entopeduncular nucleus (EPN) which send GABAergic projections to the thalamus. The current model of cortical information flow through the basal ganglia states that the striatum projects to these output nuclei both directly, and indirectly through the globus pallidus and subthalamic nucleus (STN).[3,4] The direct pathway provides a GABAergic inhibition of the SNr/EPN, while the projection to globus pallidus relieves a GABAergic inhibition of STN, resulting in a glutamatergic excitation of SNr/EPN. A delicate balance between the inhibition of the output nuclei by the direct pathway, and excitation by

---

* M. J. Marino, S.R. Bradley, H. Awad, M. Wittmann, and P.J. Conn, Emory University School of Medicine Department of Pharmacology Atlanta, GA 30322

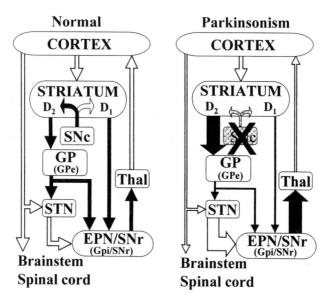

Figure 1. A model of how the Parkinson's-related loss of dopamine neurons in the SNc impacts information flow throughout the basal ganglia. Note the increase of glutamatergic transmission at the STN-SNr synapse. Inhibitory connections are depicted by black arrows, excitatory transmission depicted by white arrows. Figure modified from reference 3.

the indirect pathway is believed to be crucial for control of movement, and any imbalance in this system underlies the pathophysiology of movement disorders.

Recent studies reveal that loss of nigro-striatal dopamine neurons results in a series of neurophysiological changes that lead to over activity of the indirect pathway, resulting in a pathological excitation of the STN. Increased activity of STN neurons leads to an increase in glutamate release at STN synapses onto GABAergic projection neurons in the output nuclei. This glutamate-mediated over excitation of BG output ultimately produces the motor impairment characteristic of PD.[5] Discovery of the pivotal role of increased activity in the indirect pathway in PD has led to a major focus on surgical approaches for treatment. For instance, lesions or high frequency stimulation of the STN provides a therapeutic benefit to PD patients.[6] In addition, pallidotomy, a surgical lesion of the GP, produces similar therapeutic effects by reversing the impact of increased activity of STN neurons.[7,8] Development of these highly effective neurosurgical approaches provides a major advance in our understanding of the pathophysiology of PD. However, surgical approaches are not widely available to Parkinson's patients. Due to their invasive nature, high cost, and considerable expertise required, such treatment is reserved for patients that are refractory to dopamimetic therapy.

An alternative to surgical approaches to reducing the increased excitation of basal ganglia output nuclei in PD patients would be to employ pharmacological agents that counteract the effects of overactivation of the STN neurons by reducing transmission through the indirect pathway. One approach would be to target metabotropic glutamate receptors (mGluRs). Eight mGluR subtypes have been cloned (designated mGluR1-mGluR8) from mammalian brain. These mGluRs are classified into three major groups based on sequence homologies, coupling to second messenger systems, and selectivities for various agonists. Group I mGluRs, which include mGluR1 and mGluR5, couple primarily to increases in phosphoinositide hydrolysis. Group II mGluRs (mGluR2 and mGluR3), and group III mGluRs (mGluR4, 6, 7,and 8) couple to inhibition of adenylyl cyclase. The mGluRs are widely distributed throughout the central nervous system and play important roles in regulating cell excitability and synaptic transmission (for review see [9,10]). One of the primary functions of the mGluRs is a role as presynaptic receptors involved in reducing transmission at glutamatergic synapses. The mGluRs also serve as heteroreceptors involved in reducing GABA release at inhibitory synapses. Finally, postsynaptically localized mGluRs often play an important role in regulating neuronal excitability and in regulating currents through ionotropic glutamate receptors. If mGluRs play these roles in basal ganglia, particularly in the indirect pathway, members of this receptor family may provide an exciting new target for drugs that could be useful for the treatment of PD, as well as other disorders of BG function. In this chapter we will describe our current understanding of mGluR distribution and function in the indirect pathway. Unless otherwise noted, all results presented are from studies of rat basal ganglia.

2. THE STRIATO-PALLIDAL SYNAPSE

The indirect pathway arises from the striatal enkephalinergic medium aspiny neurons.[11,12]. These GABAergic neurons project to cells in the GP, forming the first synapses in the indirect pathway. Striatal neurons express mRNA for group I, II and III mGluRs.[13] Of these, the group III mGluRs, mGluR4 and mGluR7, have been localized to presynaptic striatopallidal terminals using both confocal and electron microscopy.[14-16] Neurons in the GP express mRNA for mGluR1 and 5,[13] and are immunoreactive for mGluR7.[15,16] Postsynaptic localization has been demonstrated for mGluR1,[17] and mGluR7.[15,16] In addition, mGluR5 has been localized to postsynaptic sites at primate striato-pallidal synapses.[18]

To date there have been no studies on the function of the mGluRs at the striato-pallidal synapse. However, the receptor localization raises some interesting possibilities. While the primary input to the pallidum is GABAergic, there is some sparse glutamatergic input from the STN.[19] Therefore, activation of the STN could directly excite pallidal neurons by actions on postsynaptic ionotropic and metabotropic glutamate receptors, and disinhibit pallidal neurons by actions on presynaptic mGluRs modulating GABA release. The resulting excitation of the GP would in turn inhibit the STN. This inhibitory feedback loop may play a role in regulating the balance of activity through the indirect pathway under normal conditions. However, in the case of PD, the sparse glutamatergic input may be insufficient to maintain this feedback control. The potential therapeutic value of restoring balance at this site will be determined by future studies on the role of group III mGluRs in modulating transmission at this synapse.

3. THE PALLIDO-SUBTHALAMIC SYNAPSE

In contrast to the striato-pallidal synapse, relatively little is known about the distribution of mGluRs at the pallido-subthalamic synapse. The projection neurons of the GP express mRNA for mGluR1 and 5, and the glutamatergic projection neurons of the STN express mGluR1, 2, 3, and 5 mRNA.[13] Recently, the group I mGluRs have been postsynaptically localized to dendrites of STN neurons at both symmetric and asymmetric synapses.[20,21]

Activation of group I mGluRs induces a robust depolarization of STN neurons,[20,21] Interestingly, this depolarization is blocked by the mGluR5-selective antagonist MPEP, but not by the mGluR1-selective antagonist CPCCOEt, indicating that only one of the group I mGluRs (mGluR5) localized at this synapse mediates the direct depolarization of these neurons. A role for the mGluR1 found at postsynaptic sites in the STN remains to be determined. In addition to directly depolarizing the STN neurons, group I mGluR activation also has been demonstrated to increase the frequency of STN burst firing[20,21,22]. Since the switch from single spike activity to a burst-firing mode is one of the characteristics of parkinsonian states in animal models[23-25] and parkinsonian patients,[26,27] this effect may play a key role in the neuropathology of this disease.

4. THE SUBTHALAMO-NIGRAL SYNAPSE

Glutamatergic projections from the STN to the BG output nuclei constitute the final synapse in the indirect pathway. To date, the only study of mGluRs in the EPN has been an *in situ* study, and the results closely parallel findings in the SNr.[13] Therefore, we will focus on studies of the STN-SNr synapse. Neurons in the STN express mRNA for mGluR1, 2, 3, and 5, and the SNr GABAergic neurons express mRNA for mGluR1, 3, and 5.[13] Immunocytochemical studies have demonstrated presynaptic localization of mGluR2/3[28] and 7[15,16] at asymmetric synapses in the SNr. The presence of mGluR2 is of particular interest because it exhibits a rather restricted distribution in the BG. In addition to the STN, the only other BG cells found to express mGluR2 are the striatal cholinergic interneurons.[13] Therefore, compounds selective for mGluR2 would be expected to exhibit relatively few side effects. The group I mGluRs have been found postsynaptically localized at symmetric and asymmetric synapses in the SNr.[17,29]

Several recent studies have provided a great deal of information on the physiological roles mGluRs play in regulating the STN-SNr synapse. Both group II and group III receptors have been shown to inhibit glutamatergic transmission at this synapse.[28,30] In accord with the immunocytochemical studies, the pharmacology and physiology of this inhibition is consistent with actions on presynaptic mGluR2/3 and 7.[28] Activation of group I mGluRs produces a robust direct depolarization of SNr GABAergic neurons [29]. This effect is blocked by the mGluR1-selective antagonist CPCCOEt, but not by the mGluR5-selective MPEP. Therefore, in contrast to the effect of group I mGluR agonists in the STN, this effect appears to be mediated solely by mGluR1. Interestingly, stimulation of glutamatergic afferents in the SNr at frequencies consistent with the normal firing rate of STN neurons induces an mGluR-mediated slow EPSP which is completely blocked by CPCCOEt.[29] This indicates that postsynaptic mGluR1 may play an important role in tonic regulation of basal ganglia output.

Since increased activity in the STN is believed to play a key role in the pathophysiology of PD,[5] the STN-SNr synapse is a logical site to target pharmacological interventions. The findings that the group II mGluRs are effective at decreasing transmission at this synapse, and exhibit a somewhat restricted distribution, indicate that these receptors could provide an ideal target for the development of antiparkinsonian compounds. Consistent with this, recent studies have demonstrated that the systemic injection of the highly selective group II mGluR agonist LY354740 decreases haloperidol-induced muscle rigidity[31] and catalepsy[30] in a rat model of PD.

5. METABOTROPIC GLUTAMATE RECEPTORS IN OTHER BASAL GANGLIA REGIONS

While this review has focused on the indirect pathway, it should be noted that mGluRs are expressed throughout the BG and have functional relevance at multiple sites. For example, input to the BG at the cortico-striatal synapse is modulated both presynaptically by group II and III mGluRs[14-17] and postsynaptically by group I mGluRs.[17,32] The main effect of the presynaptic mGluRs is to reduce the cortical input to the striatum.[33-35] Activation of the postsynaptic group I mGluRs produces a direct excitation of the indirect pathway.[36,37] Interestingly, mGluR5 has been found to exclusively colocalize with enkephalin in striatal medium aspiny neurons[32] indicating that the selective activation of the indirect pathway may be mediated by this receptor.

The group III mGluRs mGluR4, and 7 have been localized to presynaptic symmetric striato-nigral terminals.[14-16] Activation of group I and III mGluRs decreases inhibitory transmission in the SNr,[38] demonstrating that the mGluRs also play a role in modulating the direct pathway. In the case of the group I mGluRs, this is of particular interest since, as discussed above, mGluR1 has been demonstrated to directly activate SNr neurons. This direct excitation coupled with a group I-mediated disinhibition suggests that group I receptor activation could dramatically increase SNr output. Therefore, in addition to the relevance for PD, compounds selective for the group I mGluRs may hold therapeutic relevance for disorders involving alteration of activity through the direct pathway such as Huntington's disease, Tourette's syndrome, and epilepsy.

Finally, all three groups of mGluRs have been shown to modulate glutamatergic transmission in the substantia nigra pars compacta.[39] This finding is of particular interest since glutamate release in the SNc is hypothesised to play a role in the degeneration of the nigrostriatal dopamine system. While the source of the excitatory afferents regulated by mGluRs in SNc was not defined in these studies, it is likely that these EPSCs are mediated in part by activity at STN terminals. These data raise the exciting possibility that group II mGluR agonists have potential not only for reducing the symptoms of established PD, but could also slow progression of PD. Future studies will be needed to clearly define the role of increased STN activity in contributing to progression of the disorder and to rigorously define the mGluR subtypes involved in regulating transmission at STN-SNc synapses.

In summary, the mGluRs are expressed throughout the indirect pathway and selectively modulate synaptic transmission and cell excitability at each synapse in the pathway (Table 1). Studies of this family of receptors not only provides insight to BG function, but holds promise for the development of therapeutic compounds for the treatment of movement disorders.

Table 1. Summary of distribution and physiological effects of mG1uRs in the indirect pathway. Numbers are indicated for mGluR subtypes detected at the mRNA or protein level. The mRNA columns refer to mRNA expression in the neurons of origin for the presynaptic terminals, and the target neurons for the postsynaptic terminals. See text for references.

PRESYNAPTIC LOCALIZATION AND EFFECTS

Synapse	mRNA (Presynaptic cells)	Protein (Presynaptic Terminal)	Physiological Effect
Striato-pallidal	1, 3, 4, 5	4, 7	?
Pallido-subthalamic	1, 5	?	?
Subthalamo-nigral	1, 2, 3, 5	2/3, 4, 7	Decrease Glutamate Release

POSTSYNAPTIC LOCALIZATION AND EFFECTS

Synapse	mRNA (Postsynaptic cells)	Protein (Postsynaptic Terminal)	Physiological Effect
Striato-pallidal	1, 5	1, 5, 7	?
Pallido-subthalamic	1, 2, 3, 5	1, 5	Direct Depolarization
Subthalamo-nigral	1, 3, 5	1, 5	Direct Depolarization

6. REFERENCES

1. Poewe, W.H. & Granata, R. Movement Disorders: Neurological Principals and Practice. Watts,R.L. (ed.), pp. 201-219 (McGraw-Hill, New York, 1997).
2. Poewe, W.H., Lees, A.J. & Stern, G.M. Low-dose L-dopa therapy in Parkinson's disease: a 6-year follow-up study. *Neurology 36*, 1528-1530 (1986).
3. DeLong, M.R. Primate models of movement disorders of basal ganglia origin. *Trends Neurosci. 13*, 281-285 (1990).
4. Bergman, H., Wichmann, T. & DeLong, M.R. Reversal of experimental parkinsonism by lesions of the subthalamic nucleus. *Science 249*, 1436-1438 (1990).
5. Wichmann, T. & DeLong, M.R. Movement Disorders: Neurological Principals and Practice. Watts,R.L. (ed.), pp. 87-97 (McGraw-Hill, New York,1997).
6. Limousin, P., Pollak, P., Benazzouz, A., Hoffmann, D., Le Bas, J.F., Broussolle, E., Perret, J.E., & Benabid, A.L. Effect of parkinsonian signs and symptoms of bilateral subthalamic nucleus stimulation. *Lancet 345*, 91-95 (1995).
7. Laitinen, L.V., Bergenheim, A.T. & Hariz, M.I. Leksell's posteroventral pallidotomy in the treatment of Parkinson's disease. *J. Neurosurg. 76*, 53-61 (1992).
8. Baron, M.S., Vitek, J.L., Bakay, R.A., Green, J., Kaneoke, Y., Hashimoto, T., Turner, R.S., Woodard, J.L., Cole, S.A., McDonald, W.M., & DeLong, M.R. Treatment of advanced Parkinson's disease by posterior GPi pallidotomy: 1-year results of a pilot study. *Ann. Neurol. 40*, 355-366 (1996).
9. Conn, P.J. & Pin, J.P. Pharmacology and functions of metabotropic glutamate receptors. *Annu. Rev. Pharmacol. Toxicol. 37*, 205-237 (1997).

10. Anwyl, R. Metabotropic glutamate receptors: electrophysiological properties and role in plasticity. *Brain Res. Brain Res. Rev. 29*, 83-120 (1999).
11. Beckstead, R.M. & Kersey, K.S. Immunohistochemical demonstration of differential substance P-, metenkephalin-, and glutamic-acid-decarboxylase-containing cell body and axon distributions in the corpus striatum of the cat. *J. Comp Neurol. 232*, 481-498 (1985).
12. Anderson, K.D. & Reiner, A. Extensive co-occurrence of substance P and dynorphin in striatal projection neurons: an evolutionarily conserved feature of basal ganglia organization. *J. Comp Neurol. 295*, 339-369 (1990).
13. Testa, C.M., Standaert, D.G., Young, A.B. & Penney, J.B., Jr. Metabotropic glutamate receptor mRNA expression in the basal ganglia of the rat. *J. Neurosci. 14*, 3005-3018 (1994).
14. Bradley, S.R., Standaert, D.G., Rhodes, K.J., Rees, H.D., Testa, C.M., Levey, A.I., & Conn, P.J. Immunohistochemical localization of subtype 4a metabotropic glutamate receptors in the rat and mouse basal ganglia. *J. Comp Neurol. 407*, 33-46 (1999).
15. Bradley, S.R., Standaert, D.G., Levey, A.I. & Conn, P.J. Distribution of group III mGluRs in rat basal ganglia with subtype-specific antibodies. *Ann. N. Y. Acad. Sci. 868*, 531-534 (1999).
16. Kosinski, C.M., Bradley, S.R., Kerner, J.A., Conn, P.J., Levey, A.I., Landwehrmeyer, G.B., Penney, J.B.,Jr., Young, A.B., Standaert,D.G. Expression of metabotropic glutamate receptor 7 mRNA and protein in the rat basal ganglia. *J. Comp Neurol .415*, 266-84 (1999).
17. Testa, C.M., Friberg, I.K., Weiss, S.W. & Standaert, D.G. Immunohistochemical localization of metabotropic glutamate receptors mGluRla and mGluR2/3 in the rat basal ganglia. *J Comp Neurol. 390*, 5-19 (1998).
18. Hanson, J.E. & Smith, Y. Group I metabotropic glutamate receptors at GABAergic synapses in monkeys. *J Neurosci. 19*, 6488-6496 (1999).
19. Shink, E. & Smith, Y. Differential synaptic innervation of neurons in the internal and external segments of the globus pallidus by the GABA- and glutamate-containing terminals in the squirrel monkey. *J. Comp Neurol. 358*, 119-141 (1995).
20. Awad, H. & Conn, P.J. Physiological actions of metabotropic glutamate receptors in neurons of the subthalamic nucleus. Neuropharmacology 38, A2.(1999).
21. Awad, H. & Conn, P.J. Regulation of neurons of the subthalamic nucleus by metabotropic glutamate receptors. *Society For Neuroscience Abstracts 25*, 176.15 (1999).
22. Beurrier, C., Congar, P., Bioulac, B. & Hammond, C. Subthalamic nucleus neurons switch from single-spike activity to burst-firing mode. *J. Neurosci. 19*, 599-609 (1999).
23. Hollerman, J.R. & Grace, A.A. Subthalamic nucleus cell firing in the 6-OHDA-treated rat: basal activity and response to haloperidol. *Brain Res. 590*, 291-299 (1992).
24. Bergman,H., Wichmann,T., Karmon,B. & DeLong,M.R. The primate subthalamic nucleus. II. Neuronal activity in the MPTP model of parkinsonism. *J. Neurophysiol. 72*, 507-520 (1994).
25. Hassani, O.K., Mouroux, M. & Feger, J. Increased subthalamic neuronal activity after nigral dopaminergic lesion independent of disinhibition via the globus pallidus. *Neuroscience 72*, 105-115 (1996).
26. Benazzouz, A. *et al.* Single unit recordings of subthalamic nucleus and pars reticulata of substantia nigra in akineto-rigid parkinsonism. Society For Neuroscience Abstracts 22, 91.18. (1996).
27. Rodriguez, M.C. *et al.* Characteristics of neuronal activity in the subthalamic nucleus and substantia nigra pars reticulata in Parkinson's disease. Society For Neuroscience Abstracts 23, 183.6. (1997).
28. Bradley, S.R., Marino, M.J., Wittmann M, Rouse, S.T., Levey, A.I., Conn, P.J. Physiological roles of presynaptically localized type 2,3 and 7 metabotropic glutamate receptors in rat basal ganglia. *Society For Neuroscience Abstracts 25*, 176.16 (1999).
29. Marino, M.J., Bradley, S.R., Wittmann, M. & Conn, P.J. Direct excitation of GABAergic projection neurons of the rat substantia nigra pars reticulata by activation of the mGluR1 metabotropic glutamate receptor. *Society For Neuroscience Abstracts 25*, 176.17 (1999).
30. Marino, M.J., Bradley, S.R., Wittmann, M., Rouse, S.T. & Levey, A.I. Potential antiparkinsonian actions on metabotropic glutamate receptors in the substantianigra pars compacta. Neuropharmacology 38, A28. (1999).
31. Konieczny, J., Ossowska, K., Wolfarth, S. & Pilc, A. LY354740, a group II metabotropic glutamate receptor agonist with potential antiparkinsonian properties in rats. *Naunyn Schmiedebergs Arch. Pharmacol. 358*, 500-502 (1998).

32. Testa, C.M., Standaert, D.G., Landwehrmeyer, B., Penney, J.B., Jr. & Young, A.B. Differential expression of mGluR5 metabotropic glutamate receptor mRNA by rat striatal neurons. *J. Comp Neurol. 354*, 241-252 (1995).
33. Lovinger, D.M. & McCool, B.A. Metabotropic glutamate receptor-mediated presynaptic depression at corticostriatal synapses involves mGLuR2 or 3. *J. Neurophysiol. 73*, 1076-1083 (1995).
34. East, ST, Hill, M.P. & Brotchie, J.M. Metabotropic glutamate receptor agonists inhibit endogenous glutamate release from rat striatal synaptosomes. *Eur. J. Pharmacol. 277*, 117-121 (1995).
35. Pisani, A., Calabresi, P., Centonze, D. & Bernardi, G. Activation of group III metabotropic glutamate receptors depresses glutamatergic transmission at corticostriatal synapse. *Neuropharmacology 36*, 845-851 (1997).
36. Kearney, J.A., Frey, K.A. & Albin, R.L. Metabotropic glutamate agonist-induced rotation: a pharmacological, FOS immunohistochemical, and [14C]-2-deoxyglucose autoradiographic study. *J. Neurosci. 17*, 4415-4425 (1997).
37. Kaatz, K.W. & Albin, R.L. Intrastriatal and intrasubthalamic stimulation of metabotropic glutamate receptors: a behavioral and Fos immunohistochemical study. *Neuroscience 66*, 55-65 (1995).
38. Wittmann, M., Marino, M.J., Bradley, S.R. & Conn, P.J. GABAergic Inhibition of Substantia Nigra Pars Reticulata Projection Neurons is Modulated by Metabotropic Glutamate Receptors. Society For Neuroscience Abstracts 25, 176.18.(1999).
39. Wigmore, M.A. & Lacey, M.G. Metabotropic glutamate receptors depress glutamate-mediated synaptic input to rat midbrain dopamine neurones in vitro. *Br. J. Pharmacol. 123*, 667-674 (1998).

NICOTINE AFFECTS STRIATAL GLUTAMATERGIC FUNCTION IN 6-OHDA LESIONED RATS

Charles K. Meshul, Cynthia Allen, and Tom. S. Kay*

1. INTRODUCTION

Parkinson's disease is a progressive disorder that is characterized by degeneration of the dopamine containing neurons located within the midbrain (substantia nigra pars compacta). There is also substantial loss of dopamine within nerve terminals located within the striatum which originate from those dopamine neurons. Current therapy involves administration of the precursor to dopamine, namely *l*-dopa. This chemical is taken up into the brain and then converted to dopamine. Although replacement of dopamine is effective over the first few years, other movement disorders are associated with long-term *l*-dopa therapy[9]. The *l*-dopa induced dyskinesias limit the usefulness of this type of therapy.

There is suggestive evidence that smokers have a decreased risk of developing Parkinson's disease[18,34]. This is consistent with recent animal studies in which nicotine was reported to counteract the effects of a lesion of the nigrostriatal pathway in terms of either dopamine cell loss in the midbrain, striatal glucose utilization, striatal dopamine levels or changes in dopamine D-2 receptors[10,11,13,23]. However, this neuroprotective effect of nicotine has been challenged in the MPTP mouse model of Parkinson's disease.[6]

Although loss of dopamine is the major neurochemical deficit in Parkinson's disease, other neurotransmitters within the striatum may also be altered. There is a major projection from the cortex to the striatum [14]. The corticostriatal pathway uses the excitatory neurotransmitter, glutamate, and dopamine is known to modulate the activity of glutamatergic synapses[35]. We recently reported changes in glutamatergic function within the striatum of rats following a unilateral lesion of the pathway from the substantia nigra to the striatum with the neurotoxin, 6-hydroxydopamine (6-OHDA)[16]. Using *in vivo* microdialysis, we reported that 3 months following the lesion, there was a decrease in the extracellular level of striatal glutamate. This was associated with an increase in the density of nerve terminal glutamate immunolabeling, using immuno-gold electron microscopy.

---

* C.K. Meshul, C. Allen, and T.S. Kay, Research Services (RD-29), Neurocytology Lab, V.A. Medical Center and Departments of Behavioral Neuroscience and Pathology, Oregon Health Sciences University, Portland, OR 97201.

The possible neuroprotection of dopamine neurons by nicotine pretreatment is of interest, however, the efficacy of nicotine in animals previously lesioned with 6-OHDA was evaluated in the current study. Specifically, the effect of nicotine on the glutamatergic corticostriatal pathway was determined in 6-OHDA lesioned animals which had also been administered the dopamine agonist, apomorphine, a model for *l*-dopa induced dyskinesias[24].

2. METHODS

2.1. *In Vivo* Microdialysis Measurement of Extracellular Glutamate in Non-Lesioned Animals

Details of the microdialysis procedure have been published elsewhere [16]. Extracellular glutamate levels were analyzed within the dorsolateral caudate. Non-lesioned animals were injected with nicotine (NIC-subchronic)(0.4 mg/kg as the base, s.q., N = 9) or saline (NIC-acute)(1 ml/kg, s.q., N = 9) for 6 days. Another set of animals was injected with saline and did not undergo microdialysis but were used for the ultrastructural studies only (control group, N = 6). On the day of the microdialysis, the animals were then injected with nicotine (0.4 mg/kg) and samples taken every 15 minutes for the following 2 hours. At the conclusion of the experiment, the animals were perfused with glutaraldehyde fixative (see below) and tissue taken to either verify probe placement or cut and processed for electron microscopy. The mean probe recovery was $7.5 \pm 1.2\%$.

2.2 Unilateral Lesion of Nigrostriatal Pathway

Details of this method have been previously described [16]. Following a lesion of the left medial forebrain bundle, all rats were then tested for contralateral turning 2-3 weeks following the lesion with 0.05 mg/kg of apomorphine (Sigma Chem. Co.). Only those animals showing robust contralateral turning (> 10 turns/min) that were initially treated with 6-OHDA were used in subsequent experiments[21]. None of the sham lesioned animals rotated[20].

One week after testing the 6-OHDA treated animals with apomorphine (baseline determination), the following experimental groups were used:

1. sham lesion: inject with normal saline (1 ml/kg/day) (N = 7)(SAL).
2. 6-OHDA lesion: inject with normal saline (1 ml/kg/day) (N = 7)(6-OH/SAL).
3. 6-OHDA lesion: inject with nicotine (0.4 mg/kg)(N = 7) (6-OH/NIC).
4. 6-OHDA lesion: inject with apomorphine (0.05 mg/kg)(N = 7)(6-OH/APO).
5. 6-OHDA lesion: inject with nicotine first (0.4 mg/kg) followed 15 minutes later by apomorphine (0.05 mg/kg)(N = 7)(6-OH/APO&NIC).

2.3 Electron Microscopy/Immunocytochemistry

Details of the electron microscopic procedure for processing tissue and quantitative ultrastructural immunolabeling have been previously described[16]. One day after the last injection of drug for the 6OHDA/APO/NIC experiments or after the last dialysate sample collected (see above), the rats were anesthetized, perfused transcardially with fixative, the dorsolateral striatum dissected and embedded in plastic. Thin sections were cut and analyzed for the number of asymmetrical synapses containing a perforated or non-perforated postsynaptic density (PSD) per field of view[16]. Other thin sections were exposed to the glutamate primary antibody (1:250,000; non-affinity purified, rabbit polyclonal; Arnel, Brooklyn, N.Y) and then to a secondary antibody conjugated to 10 nm gold particles. The number of gold particles per nerve terminal associated with an asymmetrical synapse containing a perforated or non-perforated PSD was counted and the area of the nerve terminal and spine determined using Image Pro Plus software (Media Cybernetics, Silver Springs, MD, Version 3.01). Glutamate containing nerve terminals were typically photographed making a synaptic contact on a dendritic spine, indicating that they most likely originated from the cortex. Since we had established that the vast majority of the glutamate pool was associated with synaptic vesicles and that less than 10% of the glutamate was found within the cytoplasmic pool, only the vesicular pool was analyzed [16]. In addition, the mitochondrial pool of glutamate, as an index of background labeling, was determined within each nerve terminal. The data were analyzed using a one-way ANOVA followed by a post-hoc analysis using Perit'z F-test for comparison of multiple means[16].

2.4 Contralateral Rotations

Details of the procedure for apomorphine-induced contralateral rotations have been previously published[17]. The baseline number of contralateral turns was determined 2-3 weeks after the 6-OHDA lesion. One day after the final drug injection (NIC, APO, APO&NIC, or SAL) the number of contralateral rotations was measured for 3 minutes after an apomorphine injection (challenge). The mean number of contralateral rotations/min was then determined for each experimental group. The change in contralateral rotations between the challenge and baseline levels was determined as a percent of baseline. An overall mean percent was calculated and the data were analyzed using a one-way ANOVA followed by a post-hoc analysis using Perit'z F-test for comparison of multiple means[16].

3. RESULTS

In non-lesioned animals, subchronic treatment with nicotine for 6 days resulted in a significant decrease in the basal extracellular level of striatal glutamate compared to the saline group (Figure 1). Following the collection of baseline samples, nicotine was then injected into both the subchronic and saline groups and samples collected over the next 2 hours. Although there was a small increase in the overall level of extracellular glutamate following nicotine administration in both treatment groups, this change was not statistically significant compared to the baseline (data not shown).

After the last glutamate sample was collected following the acute injection of nicotine, the animals were perfused with fixative and the dorsolateral striatum processed for glutamate immunolabeling. There was a significant increase in the density of nerve terminal glutamate immunolabeling associated with synapses making an asymmetrical synaptic contact in the subchronic nicotine treated group compared to the group given either an acute injection of nicotine or the control group (Figure 2). In the animals given an acute injection of nicotine,

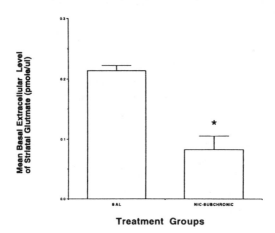

Figure 1. Basal extracellular level of striatal glutamate (picomoles/îl) in non-lesioned rats following 6 days of subchronic nicotine (0.4 mg/kg) or saline treatment (SAL) as measured by *in vivo* microdialysis. There was a significant decrease in the extracellular glutamate level following 6 days of drug treatment compared to the saline group. Values are means ± SEM.
* - $p < .05$ compared to the saline group using the Student's t-test.

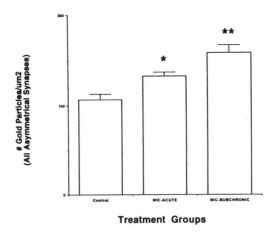

Figure 2. Density of striatal nerve terminal glutamate immunoreactivity associated with synapses making an asymmetrical synaptic contact following an acute or subchronic (7 days) administration of nicotine (0.4 mg/kg) in non-lesioned rats. The acute and subchronic groups underwent microdialysis (see Fig. 1) while the control group did not and was injected with saline for 7 days. Two hours after the challenge dose with nicotine, the animals were perfused with fixative and processed for glutamate immunolabeling. Values are means ± SEM.
* - $p < .05$ compared to the control, as determined by Peritz' f-test for comparison of multiple means.
** - $p < .05$ compared to the acute nicotine and control group, as determined by Peritz' f-test for comparison of multiple means.

there was a small but significant increase in the density of nerve terminal glutamate immunolabeling compared to the control group. Although there was no change in the extracellular level of glutamate following the acute injection of nicotine as measured by *in vivo* microdialysis, the technique of ultrastructural immunocytochemistry has greater resolution and was able to detect a small change in nerve terminal glutamate density. As a further measure of the specificity of the glutamate immunolabeling procedure, the density of glutamate immunolabeling of the mitochondrial pool associated with asymmetrical nerve terminals (i.e. background metabolic pool labeling) was analyzed. There were no differences in glutamate immunolabeling within the mitochondrial pool between any of the groups (data not shown).

As an animal model for *l*-dopa-induced dyskinesias, subchronic treatment with the nonspecific dopamine agonist, apomorphine, results in an increase in the number of contralateral rotations (ie sensitization)[24]. The role of the corticostriatal pathway in apomorphine-induced sensitization and whether nicotine treatment can effect the development of this sensitization was then determined. The nigrostriatal dopamine pathway was first injected with 6-OHDA or vehicle (sham control), then two weeks later, the 6-OHDA injected animals were tested for success of the lesion with apomorphine (0.05 mg/kg) and the number of contralateral rotations determined (baseline level). Only those animals showing robust rotations (> 10/min) were used for the subsequent study. Two weeks later guide cannulae were implanted and 1 week later, animals received either daily treatment with apomorphine

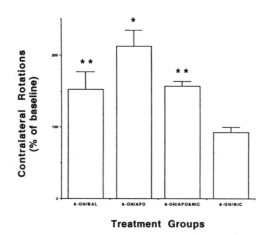

Figure 3. Three weeks after the injection of 6-OHDA or saline into the medial forebrain bundle (MFB), the animals were tested for the success of the lesion by administration of the dopamine agonist, apomorphine (0.05 mg/kg). The number of contralateral rotations was counted for 3 minutes, starting 15 minutes after the apomorphine injection. The next day, the animals were injected with either saline (6-OH/SAL), apomorphine (6-OH/APO), nicotine (6-OH/NIC), or nicotine followed 30 minutes later by apomorphine (6-OH/APO&NIC). This treatment schedule continued for the next 6 days. On day 7, all the animals were injected with just apomorphine and the number of contralateral rotations was determined as mentioned above. The group injected with saline into the MFB did not rotate and that data is not included in this figure.
*- $p < .05$ compared to 6-OH/APO & NIC and 6-OH/NIC groups, as determined by Peritz' f-test for comparison of multiple means.
** - $p < .05$ compared to the 6-OH/NIC group, as determined by Peritz' f-test for comparison of multiple means.

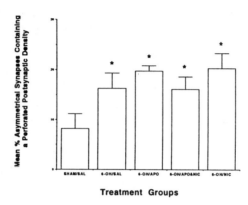

Figure 4: Effect of nicotine and apomorphine treatment on the mean percentage of striatal asymmetrical synapses containing a perforated PSD in animals injected with 6-OHDA or saline into the MFB. In all the 6-OHDA lesioned groups, there was an increase in the mean percentage of asymmetrical synapses containing a perforated PSD compared to the saline group (SHAM/SAL).
* - $p < .05$ compared to the saline group (SHAM/SAL), as determined by Peritz' f-test for comparison of multiple means.

(0.05 mg/kg, s.q.), nicotine (0.4 mg/kg, s.q.), nicotine + apomorphine, or saline (1 ml/kg). The sham group only received vehicle. Six days after drug treatment, all animals were injected with apomorphine (0.05 mg/kg) and the number of contralateral rotations determined (challenge level). Figure 3 shows that apomorphine treatment in 6-OH lesioned rats (6-OH/APO) resulted in a significant increase (sensitization) in the number of contralateral rotations compared to its baseline level. There was over a 100% increase in the number of apomorphine-induced contralateral turns. Saline treatment alone (60H/SAL), followed by a challenge dose of apomorphine, lead to a 50% increase in the number of contralateral turns. This apomorphine-induced sensitization was blocked by the co-administration of nicotine (6-OH/APO/NIC). Nicotine treatment alone (6-OH/NIC) resulted in a significant decrease in the number of contralateral rotations compared to the other three groups.

As an indication of changes in glutamate nerve terminal function [16], the mean percentage of striatal asymmetrical synapses containing a perforated postsynaptic density was determined. All 6-OHDA lesioned groups, regardless of which drug was administered, showed a significant increase in the mean percentage of synapses associated with a perforated PSD (Figure 4). In addition, we find a significant decrease in the density of nerve terminal glutamate immunolabeling in the apomorphine sensitized group (6-OH/APO) compared to the saline treated group (6-OH/SAL)(Figure 5). There was a further decrease in glutamate immunoreactivity in the apomorphine sensitized group that was treated with nicotine (6-OH/APO/NIC). It appears that apomorphine treatment alone results in an increase in striatal glutamate neurotransmitter release (decrease in density). However, blockade of increased contralateral rotations (sensitization) by nicotine (Figure 3) continues to result in a decrease in striatal

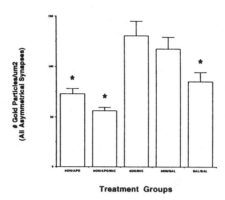

Figure 5: Same groups as in Figure 4 except that the density of nerve terminal glutamate immunolabeling from the dorsolateral striatum was determined.
*- $p < .05$ compared to 6OH/NIC and 6OH/SAL groups, as determined by Peritz' f-test for comparison of multiple means.

glutamate immunolabeling (6-OH/APO/NIC) compared to the apomorphine group (6OH/APO).

4. DISCUSSION

Subchronic nicotine administration results in a decrease in the basal extracellular level of striatal glutamate (Fig. 1) and this is associated, in the same group of animals, with an increase in the density of nerve terminal glutamate immunoreactivity (Fig. 2). This suggests that a decrease in neurotransmitter release results in an accumulation of neurotransmitter presynaptically, which is consistent with our previous findings[16]. In addition, we report that in 6-OHDA lesioned rats, nicotine treatment blocks the increase in contralateral rotations due to subchronic apomorphine administration (Fig. 3). Blockade of this behavioral sensitization did not result in any changes in the increase in the mean percentage of striatal asymmetric synapses containing a perforated PSD (Fig. 4) or the decrease in the density of nerve terminal glutamate immunoreactivity in the 6-OH/APO/NIC group (Figure 5).

Subchronic treatment with continuous nicotine administration at the time of a mechanical lesion of the nigrostriatal pathway partially protected substantia nigra pars compacta dopamine neurons from degenerating[10]. This interesting observation was the basis for the current study in which the nigrostriatal pathway was completely lesioned with 6-OHDA and then several weeks later, nicotine treatment was initiated. This protocol was used to determine the consequences of nicotine treatment in rats in which the nigrostriatal pathway was previously lesioned, similar to patients first diagnosed with Parkinson's disease. The decrease in the basal levels of striatal glutamate following subchronic administration of nicotine (Fig. 1) suggests that this drug may have an antiglutamatergic effect. There has been recent interest in the role of the glutamatergic system in Parkinson's disease and the role of glutamate antagonist drugs to treat this movement disorder[31].

In addition, the current therapeutic use of *l*-dopa to treat Parkinson's disease is associated with the development of dyskinesias[9]. In the current study, in order to mimic the *l*-dopa induced dyskinesias, 6-OHDA lesioned rats were administered the dopamine agonist, apomorphine. This leads to an increase in the number of contralateral rotations (i.e. sensitization), a model for *l*-dopa induced dyskinesias. Co-administration of nicotine blocked the development of apomorphine-induced sensitization. In addition, following 6 days of nicotine treatment alone in 6-OHDA rats, acute administration of apomorphine then resulted in a significant decrease in the number of contralateral turns compared to the subchronic apomorphine group. This suggests that nicotine may be useful in the treatment of *l*-dopa induced dyskinesias. In agreement with this hypothesis, there have been two recent reports of the successful use of an anticholinergic drug to counteract the behavior deficits in two different models of Parkinson's disease; reserpine in the mouse and MPTP in monkeys [15,30].

The concept of an antiglutamatergic effect of nicotine may seem counterintuitive since there have been several reports showing that acute nicotine treatment results in an increase in either glutamate release or in excitability [26]. For example, the nicotine-induced increase in *c-fos* expression is dependent on stimulation of the N-methy*l*-D-aspartate subtype of glutamate receptor [12]. It has been reported that direct injection of nicotine through the microprobe results in an increase in the extracellular level of glutamate [8,32,33]. However, these studies involved acute injection of nicotine and the current study used subchronic administration of the compound. As a working hypothesis, we are in the process of assessing the role of dopamine following nicotine administration to control the release of glutamate, since it is known that nicotine can stimulate the release of dopamine[5] and that dopamine agonists can block the release of neuronal glutamate[35].

The initial aim of this study was to investigate the interactions between nicotine and glutamate within the striatum. As a behavioral tool, the dopaminergic nigrostriatal pathway was lesioned and the effect of nicotine on apomorphine-induced contralateral rotations was assessed. Although nicotine was able to prevent the apomorphine-induced sensitization, this blockade did not translate into a reversal of the nerve terminal glutamatergic changes found after apomorphine treatment (Fig. 5). Although there is evidence that apomorphine administration results in an increase in the extracellular level of striatal glutamate [27] and an associated decrease in the density of nerve terminal glutamate immunolabeling (Fig. 5), blockade of the behavioral sensitization by nicotine did not affect the morphology of striatal glutamate synapses (Fig. 4 and 5). However, the nicotine/apomorphine treated animals continued to show contralateral rotation; only the sensitization was blocked. Therefore, it is likely that the corticostriatal pathway is still involved in apomorphine-induced rotations, since a lesion of the motor cortex (but see Crossman et al.[4]) and injection of lidocaine into the denervated striatum significantly reduced the number of rotations following apomorphine or *l*-dopa treatment[3,19]. However, the decrease in contralateral rotations due to nicotine treatment may involve other basal ganglia structures, such as the substantia nigra pars reticulata (SN-PR)[22,28,29].

The direct and indirect basal ganglia pathways converge at the level of the SN-PR[1,7]. The SN-PR receives an excitatory glutamatergic input from the subthalamic nucleus (indirect pathway) and an inhibitory GABA projection from the striatum (direct pathway). It is possible that the interaction between these two pathways may be important in controlling the inhibitory input to the motor thalamus. This would affect the activity of the thalamo-cortico-striatal pathway. In support of the hypothesis, a lesion of the subthalamic nucleus results in a significant reduction in the number of apomorphine-induced contralateral rotations [2,25]. Changes in glutamate and GABA levels within the SN-PR may be detectable with the current quantitative immunocytochemical techniques and these studies are currently ongoing in the lab.

5. ACKNOWLEDGMENTS

Supported by the Dept. of Veterans Affairs Merit Review Program and the Smokeless Tobacco Research Council.

6. REFERENCES

1. R.L. Albin, A.B. Young, and J.B. Penney, The functional anatomy of basal ganglia disorders, *Trends Neurosci.* 12:366 (1989).
2. P. Burbaud, C. Gross, A. Benazzouz, M. Coussemacq, and B. Bioulac, Reduction of apomorphine-induced rotational behaviour by subthalamic lesion in 6-OHDA lesioned rats is associated with a normalization of firing rate and discharge pattern of pars reticulata neurons, *Exp. Brain Res.* 105:48 (1995).
3. M.A. Cenci and A. Bjorklund, Transection of corticostriatal afferents reduces amphetamine- and apomorphine-induced striatal Fos expression and turning behaviour in unilaterally 6hydroxydopamine-lesioned rats, *Eur. J. Neurosci.*, 5:1062 (1993).
4. A.R. Crossman, M.A. Sambrook, S.W. Gergies, and P. Slater, The neurological basis of motor asymmetry following unilateral 6-hydroxydopamine brains lesions in the rat: the effect of motor decortication, *J. Neurol. Sci.*, 34:407 (1977).
5. G. DiChiara and A. Imperato, Drugs abused by humans preferentially increase synaptic dopamine concentrations in the mesolimbic dopamine system of freely moving rats, *Proc. Natl. Acad. Sci. USA* 85:5274(1988).
6. B. Ferger, C. Spratt, C.D. Earl, P. Teismann, W.H. Oertel, and K. Kuschinsky, Effects of nicotine on hydroxyl free radical formation in vitro and on MPTP-induced neurotoxicity in vivo, *Arch. Pharmacol.* 358:351 (1998).
7. C.R. Gerfen, T.M. Engber, L.C. Mahan, Z. Susel, T.N. Chase, F.J. Monsma Jr., and D.R. Sibley, D_1 and D_2 dopamine receptor-regulated gene expression of striatonigral and striatopallidal neurons, *Science,* 250:1429 (1990).
8. Y. Gioanni, C. Rougeot, P.B.S. Clarke, C. Lepouse, A.M. Thierry, and C. Vidal, Nicotinic receptors in the rat prefrontal cortex: increase in glutamate release and facilitation of mediodorsal thalamo-cortical transmission, *Eur. J. Neurosci.* 11:18 (1999).
9. J.J. Hagan, D.N. Middlemiss, P.C. Sharpe, and G.H. Poste, Parkinson's disease: prospects for improved therapy, *Trends Pharmacol. Sci.*, 18:156 (1997).
10. A.M. Janson, K. Fuxe, L.F. Agnati, I. Kitayama, A. Harfstrand, K. Andersson, Goldstein M., Chronic nicotine treatment counteracts the disappearance of tyrosine-hydroxylase-immunoreactive nerve cell bodies, dendrites and terminals in the mesostriatal dopamine system of the male rat after partial hemitransection, *Brain Res.*, 455:332 (1988).
11. A.M. Janson, P.B. Hedlund, K. Fuxe, G. von Euler, Chronic nicotine treatment counteracts dopamine D_2 receptor upregulation induced by a partial meso-diencephalic hemitransection in the rat, *Brain Res.*, 655:25-32 (1994).
12. H. Kiba and A. Jayaraman, Nicotine induced *c-fos* expression in the striatum is mediated mostly by dopmaine D_1 receptor and is dependent on NMDA stimulation, *Molec. Br. Res.*, 23:1 (1994).

13. R. Maggio, M. Riva, F. Vaglini, F. Fornai, R. Molteni, M. Armogida, G. Racagni, and G.U. Corsini, Nicotine prevents experimental Parkinsonism in rodents and induces striatal increase of neurotrophic factors, *J. Neurochem.* 71:2439 (1998).
14. A.J. McGeorge and R.L.M. Faull, The organization of the projection from the cerebral cortex to the striatum in the rat, *Neurosci.* 29:503 (1989).
15. F. Menzaghi, K.T. Whelan, V.B. Risbrough, T.S. Rao, and G.K. Lloyd, Interactions between a novel cholinergic ion channel agonist, SIB-1765F and L-dopa in the reserpine model of Parkinson's disease in rats, *J. Pharm. Exper. Therapeutics* 280:393 (1997).
16. C.K. Meshul, N. Emre, C.M. Nakamura, C. Allen, M.K. Donohue, and J.F. Buckman, Time-dependent changes in striatal glutamate synapses following a 6-hydroxydopamine lesion, *Neuroscience,* 88:1-16 (1999).
17. C.K. Meshul and C. Allen, Haloperidol reverses the changes in striatal glutamatergic immunolabeling following a 6-OHDA lesion. *Synapse,* 36:129 (2000).
18. D.M. Morens, A. Grandinetti, D. Reed, L.R. White, and G.W. Ross, Cigarette smoking and protection from Parkinson's disease, *Neurol.* 45:1041 (1995).
19. A. Mura, J. Feldon, and M. Mintz, Reevaluation of the striatal role in the expression of turning behavior in the rat model of Parkinson's disease, *Brain Res.,* 808:48 (1998).
20. K. Neve, M.R. Kozlowski and J.F. Marshall, Plasticity of neostriatal dopamine receptors after nigrostriatal injury: relationship to recovery of sensorimotor function and behavioral supersensitivity, *Brain Res.* 244:33 (1992).
21. S.J. O'Dell and J.F. Marshall, Chronic L-dopa alters striatal NMDA receptors in rats with dopaminergic injury, *NeuroRep.* 7:2457-2461 (1996).
22. D. Orosz and J.P. Bennett, Simultaneous microdialysis in striatum and substantia nigra suggests that the nigra is a major site of action of L-dihyroxyphenyalanine in the "Hemiparkinsonian" rat, *Exp. Neurol.,* 115:388 (1992).
23. Ch. Owman, K. Fuxe, A.M. Janson, J. Kahrstrom, Chronic nicotine treatment eliminates asymmetry in striatal glucose utilization following unilateral transection of the mesostriatal dopamine pathway in rats, *Neurosci Lett.,* 102:279 (1989).
24. S.M. Papa and T.N. Chase, Levodopa-induced dyskinesias improved by a glutamate antagonist in Parkinsonian monkeys, *Ann Neurol* 39:574 (1996).
25. B. Piallat and A. Benazzouz, Subthalamic nucleus lesion in rats prevents dopaminergic nigral neuron degeneration after striatal 6-OHDA injection: behavioural and immunhistochemical studies, *Exp. Brain Res.,* 105:48 (1996).
26. K.A. Radcliffe and J.A. Dani, Nicotinic stimulation produces multiple forms of increased glutamatergic synaptic transmission, *J. Neurosci.* 18:7075 (1998).
27. M.S. Reid, M. Herrera-Marschitz, J. Kehr, and U. Ungerstedt, Striatal dopamine and glutamate release: effects of intranigral injections of substance P, *Acta Physiol Scand* 140:527 (1990).
28. G.S. Robertson and H.A. Robertson, Evidence that the substantia nigra is a site of action for L-DOPA, *Neurosci. Lett.,* 89:204 (1988).
29. G.S. Robertson and H.A. Robertson, Evidence that L-Dopa-induced rotational behavior is dependent on both striatal and nigral mechanisms, *J. Neurosci.,* 9:3326 (1989).
30. J.S. Schneider, A. Pope-Coleman, B.S. Van Velson, F. Menzaghi, and G.K. Lloyd, Effects of SIB -1508Y, a novel neuronal nicotinic acetylcholine receptor agonist, on motor behavior and Parkinsonian Monkeys, *Movement Dis.,* 13:637 (1998).
31. M.S. Starr, Glutamate/dopamine D_1/D_2 balance in the basal ganglia and its relevance to Parkinson's disease. *Synapse,* 19:264 (1995).
32. E. Toth, H. Sershen, A. Hashim, E.S. Vizi, and A. Lajtha, Effect of nicotine on extracellular levels of neurotransmitters assessed by microdialysis in various brain regions: role of glutamic acid, *Neurochem Res.,* 17:265 (1992).
33. E. Toth, E. Vizi, and A. Lajtha, Effect of nicotine on levels of extracellular amino acids in regions of the rat brain in vivo, *Neuropharm.,* 32:827 (1993).
34. C. Tzourio, W.A. Rocca, M.M. Breteler, M. Baldereschi, J.-F. Dartigues, S. Lopez-Pousa, J.-M. Manubens-Bertran, and A. Alperovitech, Smoking and Parkinson's disease, *Neurol.* 49:1267 (1997).
35. B.K. Yamamoto and S. Davy, Dopaminergic modulation of glutamate release in striatum as measured by microdialysis, *J. Neurochem.,* 58:1736 (1992).

SUBCELLULAR AND SUBSYNAPTIC LOCALIZATION OF GLUTAMATE TRANSPORTERS IN THE MONKEY BASAL GANGLIA

Ali Charara, Maryse Paquet, Jeffrey D. Rothstein and Yoland Smith*

1. INTRODUCTION

Glutamate is the main excitatory neurotransmitter in the central nervous system. In addition to its essential role as mediator of rapid signaling, this excitatory amino acid also contributes to brain damage observed in acute insults (e.g. head trauma) and some chronic neurodegenerative disorders (e.g. Parkinson's disease) (Choi, 1992; Fornai et al., 1997). Therefore, removal of glutamate from the synaptic cleft plays an important role not only in the dynamics of glutamate receptor activation but also in the maintenance of subtoxic levels of glutamate. This requirement is handled by the family of high-affinity glutamate transporters (Amara, 1992).

Molecular cloning studies have now identified five different subtypes of glutamate transporters: glutamate transporter 1 (GLT-1), glutamate aspartate transporter (GLAST), excitatory amino acid carrier 1 (EAAC1), as well as excitatory amino acid transporters 4 (EAAT4) and 5 (EAAT5) (Kanai and Hediger, 1992; Pines et al., 1992; Storck et al., 1992; Arriza et al., 1994, 1997; Fairman et al., 1995). GLAST and GLT-1 are primarily found in glial cells with differences in their respective cellular and regional contents (Rothstein et al., 1994; Chaudhry et al., 1995; Lehre et al., 1995; Schmitt et al., 1996; Brooks-Kayal et al., 1998), whereas EAAC 1 is predominantly localized to neurons and widely distributed throughout the brain (Rothstein et al., 1994; Conti et al., 1998). EAAT4 is abundant only in the cerebellum (Fairman et al., 1995; Nagao et al., 1997), while EAAT5 is restricted to the retina (Arriza et al., 1997). The importance of these transporters in limiting the diffusion of glutamate and, therefore its excitotoxicity, was suggested by their alterations in brain diseases such as amyotrophic lateral sclerosis (Rothstein et al., 1995), seizure (Miller et al., 1997) as well as Alzheimer's disease (Li et al., 1997) and Huntington's disease (Arzberger et al., 1997). Furthermore, recent knock-out ex-

---

* A. Charara, Department of Neuroscience, University of Pittsburgh, Pittsburgh, PA, USA. M. Paquet, and Y. Smith, Yerkes Regional Primate Research Center and Department of Neurology, Emory University, Atlanta, GA, USA. J.D. Rothstein, Department of Neurology, Johns Hopkins University, Baltimore, MD, USA.

periments of the genes encoding these transporters have demonstrated the major role of astroglial transport in maintaining low extracellular glutamate level and preventing chronic neurotoxicity (Rothstein et al., 1996).

Very little is known about the role of GLT-1, GLAST and EAAC1 in regulating the glutamatergic neurotransmission in the basal ganglia. Apart from the striatum where the subcellular distribution of glutamate transporters has been investigated in rats (Rothstein et al., 1994), most information on their localization in the basal ganglia is limited to *in situ* hybridization and light microscope immunohistochemical studies (Rothstein et al., 1994; Torp et al., 1994; Lehre et al., 1995; Velaz-Faircloth et al., 1996). In this review, we present some of our recent findings on the subcellular and subsynaptic localization of the three main subtypes of glutamate transporters in the monkey basal ganglia.

2. MATERIALS AND METHODS

Five adult rhesus monkeys and two Sprague-Dawley rats were deeply anesthetized with an overdose of pentobarbital and perfused transcardially with cold oxygenated Ringer solution, followed by fixative containing 4% paraformaldehyde and 0.1% glutaraldehyde in phosphate buffer (PB; 0.1 M, pH 7.4). The brains were cut into 60 μm-thick sections with a vibrating microtome in the frontal plane. Sections were collected in cold phosphate-buffered saline (PBS; 0.01 M, pH 7.4) and treated with sodium borohydride (1% in PBS). The sections were then placed in a cryoprotectant solution and frozen at -80°C for 20 min., thawed and washed in PBS before being processed for immunocytochemistry.

2.1 Antisera

Polyclonal antibodies were generated against the protein-conjugated synthetic peptides corresponding to the C-terminal regions of GLT-1, GLAST and EAAC1 as described previously (Rothstein et al., 1994). The crude antisera were affinity-purified prior to their use in immunocytochemistry at a concentration of 0.2 μg/ml for GLAST, 0.06 μg/ml for EAAC1 and 0.17 ig/ml for GLT-1 (Rothstein et al., 1994). Two other commercially available polyclonal antibodies raised against GLT-1 (dilution 1:6000) and GLAST (dilution 1:6000) and one EAAC1 monoclonal antibody (1 μg/ml) were also used in the present study (Chemicon International, Temecula, CA). The mouse tyrosine hydroxylase (TH) antibody used for double labeling experiments in the substantia nigra was purchased form INCSTAR (Stillwater, MN, USA) and used at 1:1000 dilution. The post-embedding for glutamate and GABA was carried out with rabbit antisera raised against glutamate (Arnel Products, New York, NY; Dilution 1:10,000) or GABA (1:10,000). The specificity of these antibodies is described in detail elsewhere (Hodgson et al., 1985; Hepler et al., 1988).

2.2 Single Immunostaining for GLT-1, GLAST, EAAC1

Both pre-embedding immunoperoxidase and silver-intensified immunogold techniques were used in the present study. Details of immunostaining protocols for light and electron microscopy are described elsewhere (Hanson and Smith, 1999). Briefly, sections processed for immunoperoxidase were stained with diaminobenzidine (DAB) using the avidin-biotin-peroxidase method (ABC, Vetcor Labs, Burlingame, CA) whereas sections processed for immunogold were incubated with 1.4 nm gold-conjugated secondary antibodies and then silver-intensified for 5-10 min with HQ Silver Kit (Nanoprobes, Stonybrook, NY). As controls, sections were incubated in solutions from which the primary antisera were omitted. After immunostaining and processing for electron microscopy, ultrathin sections of various basal ganglia nuclei were prepared as described in previous study (Hanson and Smith, 1999).

2.3 Double Immunostaining for TH/EAAC1 and GLAST/GLT-1

Some sections through the substantia nigra-ventral tegmental area complex were processed for the localization of TH and EAAC1 or GLAST and GLT-1. In those sections, TH and GLAST were revealed first with DAB using the ABC technique whereas EAAC1 and GLT-1 were localized with the pre-embedding immunogold method. As controls, the TH or the glutamate transporters antibodies were omitted in turn from the incubation solutions while the rest of the procedure remained the same.

2.4 Post-embedding Immunocytochemistry for Glutamate and GABA

Some ultrathin sections from the substantia nigra, immunostained for GLT-1 with DAB, were also processed for post-embedding glutamate and GABA immunocytochemistry according to the protocol described in a previous study (Skink et al., 1995). Control experiments that were carried out in this material are described in the same study (Skink et al., 1995).

2.5 Analysis of Material

Transverse sections containing basal ganglia were taken form each of the five monkeys for both light and electron microscope analyses. Rat striatal tissue was only used for GLT-1 immunostaining. To analyze the relationships between gold particles and post-synaptic specializations, micrographs of dendrites in the substantia nigra were taken at 25000X and 31500X from the surface of the TH/EAAC1 double immunostained sections where the labeling was optimal. The gold particles were then counted and pooled in six categories according to their localization (intracellular, extrasynaptic on the membrane, perisynaptic to asymmetric synapses, perisynaptic to symmetric synapses, synaptic in the active zone of asymmetric synapses and in the active zone of symmetric synapses).

3. RESULTS

3.1 Striatum

The pattern of labeling for each transporter was homogeneous and relatively similar throughout the caudate nucleus and putamen. There was no patchy distribution of immunostaining. Overall, the striatum displayed strong GLT-1 and EAAC1 immunoreactivity (IR), whereas GLAST immunostaining was much more discrete. GLT-1 IR was found in glial cell processes that formed a dense uniform network among non-immunoreactive dendrites and perikarya (Fig. 1A). GLAST IR was confined to thin glial processes, whereas EAAC1 IR

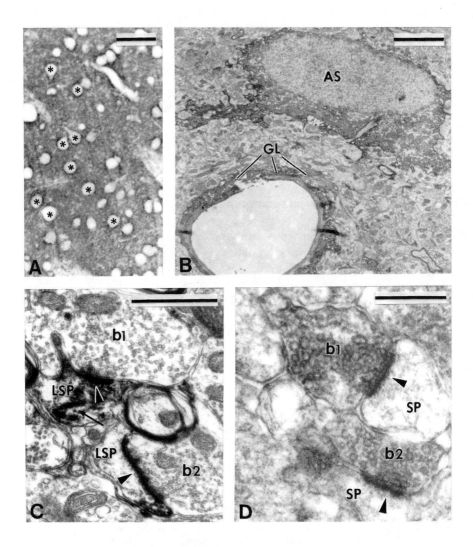

Figure 1. GLT-1 immunoreactivity in the striatum. (A) GLT-1-immunoreactive processes ensheat numerous unlabeled medium-sized neurons (asterisks) in the monkey putamen. (B) Low power electron micrograph of a GLT-1-immunoreactive astrocyte (AS) and a GLT-1-immunoreactive process (GL) surrounding a capillary in the monkey putamen. (C) Two unlabeled terminal boutons (b1, b2) form asymmetric synapses (arrowheads) with two dendritic spines (LSP) immunolabeled for GLT-1 in the monkey putamen. Note that the peroxidase product is preferentially associated with the post-synaptic densities and the spine apparatus (arrow). (D) shows two GLT-1-immunoreactive boutons (b1, b2) forming asymmetric synapses (arroweads) with two non-immunoreactive dendritic spines (SP) in the rat striatum. Scale bars (A) 50μm, (B) 3μm, (C) 0.5μm, (D) 0.3μm.

was found in numerous large- and medium-sized neurons throughout the putamen and caudate nucleus (Fig. 2A). Double immunostaining experiments showed that EAAC1 IR was localized in numerous calbindin-immunoreactive projection neurons as well as in somatostatin- and parvalbumin-containing interneurones (Smith et al., 1998).

Ultrastructural analysis revealed species differences in the distribution of GLT-1 IR. Although this transporter was mostly expressed in glial cells in rats and monkeys (Fig. 1B), a differential pattern of expression in neuronal elements was found in the two species. In monkeys, GLT-1 immunostaining was expressed in the head of large dendritic spines where it was preferentially associated with the post-synaptic densities of asymmetric synapses and the spine apparatus (Fig. 1C) whereas in rats, GLT-1 IR was found in terminal boutons forming asymmetric axospinous synapses but not in dendritic spines (Fig. 1D). GLAST IR was much more discrete than GLT-1 labeling and confined to thin glial processes and heads of small dendritic spines (Fig. 2C,D). Consistent with light microscopic observations, EAAC1 IR was found in large- and small-sized perikarya with smooth or indented nuclei and dendritic processes of various sizes, where it was predominantly associated with microtubules (Fig. 2B). In addition, EAAC 1 IR was encountered in the head of dendritic spines where it displayed a pattern of staining similar to that of GLT-1, i.e. the IR was associated with the post-synaptic densities of asymmetric synapses and the spine apparatus (Fig. 2B).

3.2 Globus Pallidus

At the light microscopic level, strong GLT-1 IR was confined to processes surrounding unlabeled neurons and dendrites, while GLAST IR was less abundant and expressed in very thin processes around non-immunoreactive cells in both the internal (GPi) and external (GPe) segments of the globus pallidus. On the other hand, EAAC1 IR was expressed in neuronal cell bodies and proximal dendrites in both GPe and GPi.

Ultrastructural analysis revealed that the pattern of staining for each transporter was the same in GPe and GPi. GLT-1 IR was selectively found in glial cells from which arose long processes that formed a tight sheath around axon terminals in contact with dendrites and neuronal cell bodies (Fig. 3A). A remarkable feature was that the labeled processes did not extend between terminals to reach the synaptic cleft, except for a subpopulation of boutons that formed asymmetric synapses (Fig. 3B). It is worth noting that other boutons also forming axo-dendritic asymmetric synapses, were devoid of GLT1 IR at their synaptic junction. GLAST-immunoreactive processes were thinner and much less abundant than GLT-1-immunoreactive elements (Fig. 3C). EAAC1 IR was associated not only with microtubules and asymmetric post-synaptic densities, but also with small preterminal axonal segments (Fig. 3D-E).

3.3 Subthalamic Nucleus

Overall, the pattern of staining for the three transporters in the STN resembled that described for GPe and GPi. GLT-1 and GLAST IR was confined to glial cell bodies and processes, whereas EAAC IR was found in neuronal perikarya and dendrites.

At the electron microscope level, prominent immunoperoxidase labeling for GLT-1 was found in glial cell bodies and processes that ensheated asymmetric axo-dendritic and axo-axonic synapses (Fig. 4A), whereas GLAST IR was light and found in thin glial pro-

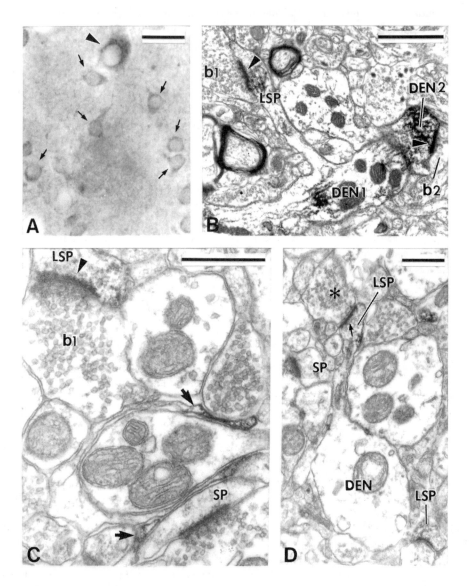

Figure 2. EAAC1 and GLAST immunoreactivity in the monkey striatum. (A) EAAC1 is present in medium-sized neuronal perikarya (arrows) and large-sized cell bodies (arrowhead) and dendrites in the putamen. (B) A EAAC1-immunoreactive (ir) dendritic spine (LSP) emerges from a labeled dendrite (DENT), and forms an asymmetric synaptic contact (arrowhead) with a non-ir bouton (b1). Another labeled dendrite (DEN2) forms an asymmetric synapse (arrowhead) with a non-ir bouton (b2) in the same field. The peroxidase product is associated with the post-synaptic densities of asymmetric synapses, the spine apparatus and microtubules. (C) GLAST immunoreactivity is expressed in thin glial processes (arrows), and a dendritic spine (LSP) which forms an asymmetric synaptic contact (arrowhead) with an unlabeled bouton (b1) in the putamen. A non-ir dendritic spine (SP) is also present in the same field. (D) A GLAST-ir dendritic spine (LSP) which emerges from a non-ir dendrite (DEN), forms an asymmetric synapse (small arrow) with a non-ir terminal (*). A labeled (LSP) and an unlabeled (SP) dendritic spines are seen in the same field. Scale bars (A) 25μm, (B) 1μm, (C) 0.5μm, (D) 0.5μm.

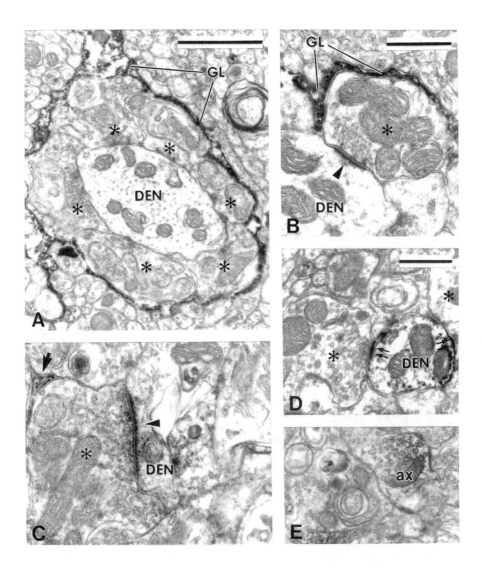

Figure 3. GLT-1, GLAST, and EAAC1 immunoreactivity in the monkey globus pallidus. (A) GLT-1 immunoreactivity (GL) is expressed in glial cell processes that form a tight sheat around axon terminals (*) in contact with a dendrite (DEN) in the external segment of the globus pallidus (GPe). (B) A GLT-1-immunoreactive glial process ensheats a subthalamic-like terminal (*) forming an asymmetric synapse (arrowhead) with a dendrite (DEN) in the GPe. (C) GLAST immunoreactivity (arrow) is present in thin glial cell processes in the vicinity of a bouton (*) forming an asymmetric axo-dendritic synapse (arrowhead) in the GPe. (D) A EAAC1-immunoreactive dendrite (DEN) forming synaptic contact (double arrows) with two terminal boutons (*) in the GPe. (E) EAAC1 immunoreactivity is occasionally found in small preterminal axonal segments (ax) in the GPe. Scale bars (A) 1μm, (B) 0.5μm (also valid in C), (D) 0.5μm (also valid in E).

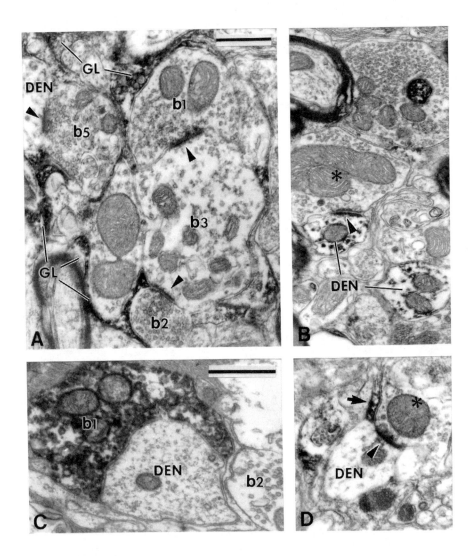

Figure 4. GLT-1, EAAC1 and GLAST immunoreactivity in the monkey subthalamic nucleus. (A) GLT-1-immunoreactive glial cell processes (GL) ensheat terminal boutons (b1, b2, b3, b5) forming asymmetric axo-axonic and axo-dendritic synapses (arrowheads). (B) shows two EAAC1-immunoreactive dendrites (DEN). Note that one of these dendrites forms an asymmetric synaptic contact (arrowhead) with an unlabeled bouton (*). (C) A EAAC1-immunoreactive terminal bouton (b1) is apposed to a dendrite (DEN) in the vicinity of a non-immunoreactive terminal (b2). (D) shows a thin GLAST-immunoreactive process (small arrow) close to an asymmetric axo-dendritic synapse (arrowhead). Scale bars (A) 0.5μm (also valid in B and D), (C) 0.5μm.

cesses which, in many cases, were close to asymmetric synapses (Fig. 4D). EAAC1 IR was mostly expressed by cell bodies, dendrites and the post-synaptic densities of asymmetric axo-dendritic synapses (Fig. 4B). A few EAAC1-labeled terminal boutons were seen but none of them formed clear synaptic contact (Fig. 4C).

3.4 Substantia nigra/ventral tegmental area

At the light microscopic level, GLT-1 IR was found in processes that surrounded non-immunoreactive neurons in the substantia nigra pars compacta (SNc), substantia nigra pars reticulata (SNr) and ventral tegmental area (VTA). GLAST IR was less abundant and confined to very thin processes around unlabeled neuronal perikarya. In sections labeled for EAAC1, moderate IR was found in the SNr in comparison to an intense immunolabeling in cell bodies and proximal dendrites of neurons in the dorsal tier of the SNc (SNc-d) and VTA. A subpopulation of neurons in the ventral tier of the SNc (SNc-v) that merged with the SNr also displayed strong EAAC1 IR (Fig. 6A).

Ultrastructural analysis revealed that GLT-1 and GLAST IR was exclusively found in glial elements. In both the SNc and SNr, GLT-1-immunoreactive processes ensheated neuronal cell bodies, dendrites as well as glutamate-immunoreactive axon terminals (Fig. 5A), while GLAST IR was much more discrete and found in thin processes around neuronal cell bodies and dendrites (Fig. 5B-C). Preliminary results showed that both GLAST and GLT-1 are co-localized in many glial cell processes throughout the SNc (Fig. 5B-C). EAAC1 IR was expressed in both postsynaptic and presynaptic elements including preterminal axonal segments and terminal boutons forming symmetric synapses (Fig. 6B-D). In sections labeled for EAAC 1 and TH, EAAC1 IR was found in both TH-immunopositive (SNc) and TH-immunonegative elements (SNr) (Fig. 6E-F). However, the relative distribution of gold particles varied between the two structures. Although most gold particles were intracytoplasmic in both populations of neurons, quantitative measurements revealed that there was more gold particles attached to the plasma membrane in TH-immunonegative than TH-immunopositive dendrites (Figs. 6E-F, 7). Most of the membrane-bound gold particles were extrasynaptic though occasional synaptic labeling was encountered (Fig. 6F, 7). In such cases, gold particles were located perisynaptically to asymmetric synapses or associated with the main body of symmetric synapses established by striatal-like terminals (Figs. 6E-F; 7). In addition, a few gold particles were also seen at the active zone of asymmetric synapses in the SNc or perisynaptic to symmetric axo-dendritic synapses in the SNr (Fig.7).

4. SUMMARY AND CONCLUDING REMARKS

Our study demonstrates that the three glutamate transporters, GLT-1, GLAST, and EAAC1 are widely distributed throughout the basal ganglia. Overall, GLT-1 and EAAC1 are the most abundant in all basal ganglia nuclei while GLAST is much more restricted. In the monkey striatum, GLT-1 and GLAST, known as glial transporters, are also expressed in dendritic spines, whereas EAAC1 is found exclusively in neuronal elements, including dendrites and spines. Interestingly, both GLT-1 and EAAC1 IR are found to be preferentially associated with the spine apparatus and the post-synaptic densities of asymmetric axo-spinous synapses. Dendritic spines are known as the main targets of glutamatergic afferents in the

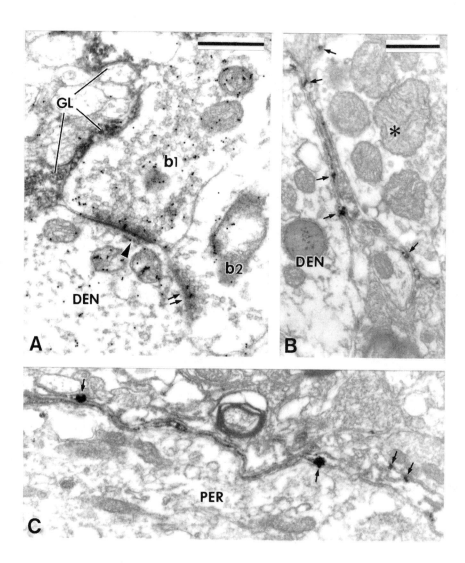

Figure 5. GLT-1 and GLAST immunoreactivity in the monkey substantia nigra. (A) GLT-1-immunoreactive glial cell process (GL) surrounding a glutamate-enriched bouton (b1) forming an asymmetric synapse (arrowhead) with a dendrite (DEN) in the SNc. Note the presence of a bouton (b2), with a low level of glutamate immunoreactivity, that forms a symmetric synapse (small arrows) with the same dendrite. (B) shows a glial cell process double immunolabeled for both GLAST (peroxidase reaction product) and GLT-1 (gold particles/small arrows) which extends between a dendrite (DEN) and a terminal bouton (*) in the SNc. (C) GLAST (peroxidase product) and GLT-1 (gold particles/small arrows) immunoreactivity are expressed in a glial cell process surrounding a neuronal perikaryon (PER) in the SNc. Scale bars (A) 0.5μm; (B) 0.5μm (also valid in C).

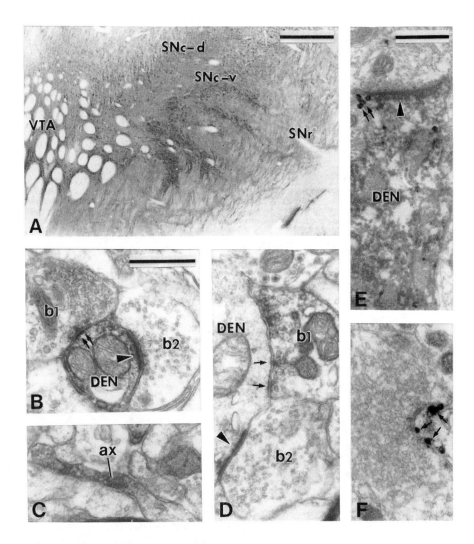

Figure 6. EAAC1 immunoreactivity in the monkey substantia nigra. (A) EAAC1 immunoreactivity is expressed in neurons of the dorsal tier of the substantia nigra pars compacta (SNc-d), ventral tegmental area (VTA) and a subpopulation of neurons in the ventral tier of the substantia nigra pars compacta (SNc-v). A moderate level of immunoreactivity is found in the substantia nigra pars reticulata (SNr) neurons. (B) A EAAC1-immunoreactive terminal bouton (b1) forms a symmetric synapse (double arrows) with a labeled dendrite (DEN) in the SNc. Note the presence of a non-immunoreactive bouton (b2) forming an asymmetric synapse (arrowhead) with same dendrite. (C) shows a EAAC1-positive preterminal axonal segment (ax) in the SNc. (D) A EAAC1-labeled bouton (b1) forms a symmetric synapse (small arrows) with a nonimmunoreactive dendrite (DEN) in the SNc. An unlabeled bouton (b2) forms an asymmetric synaptic contact with the same dendrite. (E) EAAC1 immunoreactivity (immunogold) is located perisynaptically (double arrows) to an asymmetric synapse (arrowhead) on a dendrite (DEN) immunostained for tyrosine hydroxylase in the SNc. (F) EAAC1 immunoreactivity (small arrows) is associated with the main body of a symmetric axo-dendritic synapse. Scale bars (A) 1mm, (B) 0.5μm (also valid in C, D and F), (E) 0.5 μm.

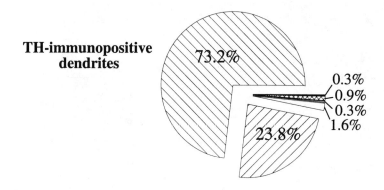
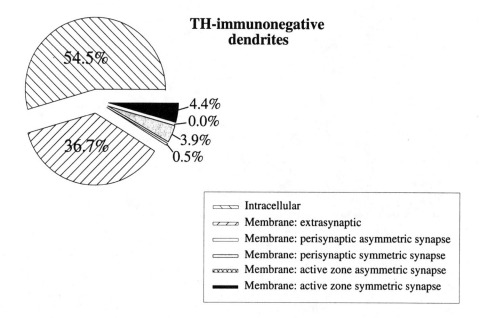

Figure 7. Pie chart showing the relative distribution of gold particles for EAAC1 on TH-immunopositive and TH-immunonegative dendrites in the monkey substantia nigra. A total of 100 TH-positive and 68 TH-negative dendrites were examined.

striatum (Smith and Bolam, 1990). Both *N*-methyl-D-aspartate (NMDA) and α-amino-3-hydroxy-5-methylisoxazole-4-propionate (AMPA) glutamate receptor subunits have been localized in the main body of asymmetric post-synaptic specializations (Bernard et al., 1997; Bernard and Bolam, 1998). Furthermore, AMPA receptors have been associated with the spine apparatus in the hippocampus (Nusser et al., 1998). Altogether, these data suggest that glutamate transporters, in addition to their role in clearing glutamate from the synaptic cleft, and AMPA receptors may also play a role in synaptic plasticity (see Nusser et al., 1998).

The main feature that characterize glutamate transporters localization in the globus pallidus is that GLT-1-immunoreactive processes ensheat axon terminals in contact with dendrites but do not reach the synaptic cleft except in the case of a subpopulation of boutons forming asymmetric axo-dendritic synapses. Since GLT-1 mediates most of the clearance of extracellular glutamate (Robinson, 1999), our results suggest that glutamate released from axon terminals in the globus pallidus can spillover from synapses and activate neighboring glutamate receptors located peri- and extra-synaptically to asymmetric synapses or synaptically at GABAergic synapses (Clarke and Bolam, 1998; Hanson and Smith, 1999). The importance of glutamate spillover in the activation of distant glutamate receptors was recently shown in the hippocampus and the cerebellum (Barbour and Hausser, 1997; Kullmann and Asztely, 1998; Dzubay and Jahr, 1999).

In the midbrain, strong level of EAAC1 IR is associated with SNc-d, VTA and a subpopulation of SNc-v neurons. Although the exact role of EAAC1 in dopaminergic cells remains to be established, such compartmental organization suggests that glutamate transporters might underlie the differential sensitivity of dopaminergic neurons to degeneration in Parkinson's disease. Furthermore, recent evidence that some nigrostriatal neurons might release glutamate as a neurotransmitter (Sulzer et al., 1998) should be kept in mind while looking for the functional significance of these data.

In conclusion, glutamate transporters are abundant in the basal ganglia and, therefore, probably play major roles in modulating glutamatergic transmission. An abnormal regulation of these transporters could participate in the excitotoxic death of striatal and nigral neurons in Huntington's and Parkinson's diseases respectively.

5. ACKNOWLEDGMENTS

The authors thank Jean-François Paré and Jeremy Kieval for technical assistance and Frank Kiernan for photography. Thanks are also due to Dr. Peter Somogyi for the generous gift of his GABA antiserum. This research was supported by NIH grants RO1 NS37423-02 and P50 NS38399-01 and a grant from the American Parkinson Disease Association. Ali Charara holds a fellowship from the FCAR.

6. REFERENCES

Amara SG. Neurotransmitter transporters: a tale of two families. Nature 1992; 360: 420-421.
Arriza JL, Fairman WA, Wadiche JI, Murdoch GH, Kavanaugh MP, Amara SG. Functional comparison of three glutamate transporter subtypes cloned from human motor cortex. J Neurosci 1994: 14: 5559-5569.

Arriza JL, Eliasof S, Kavanaugh MP, Amara SG. Excitatory amino acid transporter 5, a retinal glutamate transporter coupled to a chloride conductance. Proc Natl Acad Sci USA 1997; 94: 4155-4160.
Arzberger T, Krampft K, Leimgruber S, Weindl A. Changes of NMDA receptor subunit (NR1, NR2B) and glutamate transporter (GLT-1) mRNA expression in Huntington's disease-An *in situ* hybridization study. J Neuropathol Exp Neurol 1997; 56: 440-454.
Barbour B, Hausser M. Intersynaptic diffusion of neurotransmitter. Trends Neurosci 1997; 20: 377-384.
Bernard V, Somogyi P, Bolam JP. Cellular, subcellular, and subsynaptic distribution of AMPA-type glutamate receptor subunits in the neostriatum of the rat. J Neurosci 1997; 17: 819-833.
Bernard V, Bolam JP. Subcellular and subsynaptic distribution of the NR1 subunit of the NMDA receptor in the neostriatum and globus pallidus of the rat: co-localization at synapses with the GluR2/3 subunit of the AMPA receptor. Eur J Neurosci 1998; 10: 3721-3736.
Brooks-Kayal AR, Munir M, Jin H, Robinson MB. The glutamate transporter, GLT-1, is expressed in cultured hippocampal neurons. Neurochem Int 1998; 33: 95-100.
Chaudhry FA, Lehre KP, Campagne MVL, Ottersen OP, Danbolt NC, Storm-Mathisen J. Glutamate transporters in glial plasma membranes: highly differentiated localizations revealed by quantitative ultrastructural immunocytochemistry. Neuron 1995; 15: 711-720.
Choi DW. Excitotoxic cell death. J. Neurobiol 1992; 23: 1261-1276.
Clarke NP, Bolam JP. Distribution of glutamate receptor subunits at neurochemically characterized synapses in the entopeduncular nucleus and subthalamic nucleus of the rat. J Comp Neurol 1998; 397: 403-420.
Conti F, Debiasi S, Minelli A, Rothstein JD, Melone M. EAAC1, a high-affinity glutamate transporter, is localized to astrocytes and Gabanergic neurons besides pyramidal cells in the rat cerebral cortex. Cereb Cortex 1998; 8: 108-116.
Dzubay JA, Jahr CE. The concentration of synaptically released glutamate outside of the climbing fiber-Purkinje cell synaptic cleft. J Neurosci 1999; 5265-5274.
Fairman WA, Vandenberg RJ, Arriza JL, Kavanaugh MP, Amara SG. An excitatory amino-acid transporter with properties of a ligand-gated chloride channel. Nature 1995; 375: 599-603.
Fornai F, Vaglini F, Maggio R, Bonuccelli U, Corsini GU. Species differences in the role of excitatory amino acids in experimental parkinsonism. Neurosci Biol Rev 1997; 21: 401-415.
Hanson JE, Smith Y. Group I metabotropic glutamate receptors at GABAergic synapses in monkeys. J Neurosci 1999; 19: 6488-6496.
Hepler JR, Toomin CS, McCarthy KD, Conti F, Battaglia G, Rustioni A, Petrusz P. Characterization of antisera to glutamate and aspartate. J Histochem Cytochem 1988; 36: 13-22.
Hodgson AJ, Penke B, Erdi A, Chubb IW, Somogyi P. Antisera to γ-aminobutyric acid. 1. Production and characterization using a new model system. J Histochem Cytochem 1985; 33: 229-239.
Kanai Y, Hediger MA. Primary structure and functional characterization of a high affinity glutamate transporter. Nature 1992; 360: 467-471.
Kullmann DM, Asztely F. Extrasynaptic glutamate spillover in the hippocampus: evidence and implications. Trends Neurosci 1998; 21: 8-14.
Lehre KP, Levy LM, Ottersen OP, Storm-Mathisen J, Danbolt NC. Differential expression of two glial glutamate transporters in the rat brain: quantitative and immunocytochemical observations. J Neuroscience 1995; 15: 1835-1853.
Li S, Mallory M, Alford M, Tanaka S, Masliah E. Glutamate transporter alterations in Alzheimer's disease are possibly associated with abnormal APP expression. J Neuropathol Exp Neurol 1997; 56: 901-911.
Miller HP, Levey AI, Rothstein JD, Tzingounis AV, Conn PJ. Alterations in glutamate transporter protein levels in kindling-induced epilepsy. J Neurochem 1997; 68: 1564-1570.
Nagao S, Kwak S, Kanazawa I. EAAT4, a glutamate transporter with properties of a chloride channel, is predominantly in Purkinje cell dendrites, and forms parasagittal compartments in rat cerebellum. Neuroscience 1997; 78: 929-933.
Nusser Z, Lujan R, Laube G, Roberts JDB, Molnar E, Somogyi P. Cell type and pathway dependence of synaptic AMPA receptor number and variability in the hippocampus. Neuron 1998; 545-559.
Pines G, Danbolt NC, Bjoras M, Zhang Y, Bendahan A, Eide L, Koepsell H, Storm-Mathisen J, Seeberg E, Kanner BI. Cloning and expression of a rat brain L-glutamate transporter. Nature 1992; 360: 464-467.
Robinson MB. The family of sodium-dependent glutamate transporters: a focus on the GLT-I/EAAT2 subtype. Neurochem Int 1999; 33: 479-491.
Rothstein JD, Martin L, Levey AI, Dykes-Hoberg M, Jin L, Wu D, Nash N, Kuncl RW. Localization of neuronal and glial glutamate transporters. Neuron 1994; 13: 713-725.

Rothstein JD, Van Kammen M, Levey AI, Martin L, Kuncl RW. Selective loss of glial glutamate transporter GLT-1 in amyotrophic lateral sclerosis. Ann Neurol 1995; 38: 73-84.

Rothstein JD, Dykes-Hoberg M, Pardo CA, Bristol LA, Jin L, Kuncl RW, Kanai Y, Hediger MA, Wang Y, Schielke JP, Welty DF. Knockout of glutamate transporters reveals a major role for astroglial transport in excitotoxicity and clearance of glutamate. Neuron 1996; 16: 675-686.

Schmitt A, Asan E, Püschel B, Jöns Th, Kugler P. Expression of the glutamate transporter GLT-1 in neural cells of the rat central nervous system: non-radioactive *in situ* hybridization and comparative immunocytochemistry. Neuroscience 1996; 71: 989-1004.

Shink E, Smith Y, Differential synaptic innervation of neurons in the internal and external segments of the globus pallidus by the GABA- and glutamate-containing terminals in the squirrel monkey. J Comp Neurol 1995; 358: 119-141.

Smith AD, Bolam JP. The neural network of the basal ganglia as revealed by the study of synaptic connections of identified neurons. Trends Neurosci 1990; 13: 259-265.

Smith Y, Charara A, Rothstein JD, Freedman LJ. Glutamate transporters in the striatopallidal complex and the subthalamic nucleus (STN) in monkeys. Soc Neurosci Abstr 1998; 24: 2068.

Storck T, Schulte S, Hofmann K, Stoffel W. Structure, expression, and functional analysis of a Na^+-dependent glutamate/aspartate transporter from rat brain. Proc Natl Acad Sci USA 1992; 10955-10959.

Sulzer D, Joyce MP, Lin L, Geldwert D, Haber SN, Hattori T, Rayport S. Dopamine neurons make glutamatergic synapses in vitro. J Neurosci 1998; 18: 4588-4602.

Torp R, Danbolt NC, Babaie E, Bjoras M, Seeberg E, Storm-Mathisen J, Ottersen OP. Differential expression of two glial glutamate transporters in the rat brain: an *in situ* hybridization study. Eur J Neurosci 1994; 6: 936-942.

Velaz-Faircloth M, McGraw TS, Malandro MS, Fremeau RT, Kilberg MS, Anderson KJ. Characterization and distribution of the neuronal glutamate transporter EAAC1 in rat brain. Am J Physiol 1996; 270: C67-C75.

GLUTAMATE-DOPAMINE INTERACTIONS IN STRIATUM AND NUCLEUS ACCUMBENS OF THE CONSCIOUS RAT DURING AGING

F. Mora, A. Del Arco, and G. Segovia*

Abstract: In this chapter we review the interaction of glutamate and dopamine in striatum and nucleus accumbens of the conscious rat during aging. We studied the effects of increasing concentrations of endogenous glutamate (obtained by blocking its reuptake) on the extracellular concentrations of dopamine in striatum and nucleus accumbens in the young rat. It was found that increasing concentrations of glutamate correlated significantly with increasing concentrations of dopamine in both structures. The increase of dopamine were significantly reduced after blockade of NMDA and AMPA/kainate glutamate receptors, suggesting that the increase of dopamine was mediated by glutamate. The interaction glutamate/dopamine showed a significant age-related decrease in nucleus accumbens but not in striatum. It is suggested that the interaction glutamate-dopamine represents a balanced input to the GABA neuron in the striatum and nucleus accumbens and that during aging this balance is disrupted.

1. GLUTAMATE-DOPAMINE INTERACTIONS IN STRIATUM AND NUCLEUS ACCUMBENS

In recent years, the interaction of glutamate and dopamine in striatum and nucleus accumbens has received much attention due, at least in part, to our present knowledge on the connectivity and neurochemistry of this area of the brain (12, 23, 39). In striatum, glutamatergic inputs from the cortex synapse on GABA neurons. Also synaptic dopaminergic inputs from the substantia nigra impinge upon GABA neurons in this area of the brain. In the nucleus accumbens a similar synaptic arrangement is found. Thus, both glutamatergic and dopaminergic inputs synapse on GABA neurons. In the nucleus accumbens, however, the glutamatergic terminals arise from cell bodies located in amygdala, hyppocampus and

---

* F. Mora, A. Del Arco, and G. Segovia, Department of Physiology, Faculty of Medicine, University Complutense of Madrid, Ciudad Universitaria, 28040 Madrid, Spain.

prefrontal cortex while dopaminergic inputs arise from the ventral tegmental area of the mesencephalon. Despite the fact of glutamatergic receptors being located on dopaminergic terminals and dopamine receptors on glutamatergic presynaptic endings (21), no direct connections seem to exist between glutamatergic and dopaminergic terminals in striatum and nucleus accumbens (2, 10, 38). This suggest that a volumetric type of interaction between dopamine and glutamate exists within the striatum and nucleus accumbens (9). The interaction between glutamate and dopamine and also GABA in neostriatum and nucleus accumbens seem relevant because of its implication in neurodegenerative diseases and also aging (4, 34).

In striatum and nucleus accumbens several studies *in vivo* have reported the release of dopamine induced by glutamate agonists (for review see (25)). Thus, local infusions of the glutamate agonist N-methyl-D-aspartate (NMDA) increases extracellular dopamine (14, 16, 26, 35), and also intrastriatal infusion of an amino-3-hydroxy-5-methylisoxazole-4-propionate (AMPA)/kainate-receptor agonist increases extracellular dopamine (16, 29, 40).

We recently have developed a novel approach to study the effects of endogenous glutamate on different neurotransmitters in the brain of the awake rat (8, 33, 34, 36). This approach was to increase the endogenous concentrations of glutamate and see whether such an increase produced any effect on the extracellular concentrations of dopamine (33, 34, 36). For that, endogenous extracellular glutamate was selectively increased in striatum and nucleus accumbens by perfusing locally, through microdialysis cannulae, the selective glutamate uptake inhibitor L-trans-pyrrolidine-3,4-dicarboxylic acid (PDC). PDC is a transportable glutamate analogue that produces a selective and potent inhibition of glutamate uptake without interfering with ionotropic or metabotropic glutamate receptor binding (3, 42) or producing neurotoxic damage *in vivo* in doses up to 100 mM (20, 27).

Using PDC, we found a positive significant correlation between increases of glutamate and increases of dopamine in both striatum and nucleus accumbens (Fig. 1). The correlation glutamate-dopamine was not only significant but independent of the doses of PDC used to increase the endogenous glutamate (34, 37).

The action of glutamate receptors on dopamine was shown by a series of experiments in which perfusing the striatum and nucleus accumbens with specific glutamate receptor blockers, the NMDA-receptor antagonist 3-[R-2-carboxypiperazin-4-yl]-propyl-1-phosphonic acid (CPP) and the AMPA/kainate receptor antagonist (6,7-dinitroquinoxaline-2,3-dione) (DNQX), reduced significantly the increases of dopamine produced by glutamate (33, 36). These results suggest that the effects of endogenous glutamate on dopamine in striatum and nucleus accumbens was mediated, at least in part, by both NMDA and AMPA/kainate receptors.

It is of interest the fact that despite of a lack of direct glutamate-dopamine synapses in striatum and accumbens a significant correlation between increases of endogenous glutamate and endogenous dopamine was found. Of relevance in this context is the report in which an extrasynaptic, volumetric type of interaction betweeen glutamate and dopamine has been suggested in the striatum (18) and also that glutamate released from synapses diffuses to perisynaptic zones to act on glutamate receptors (9, 17, 31, 32).

Several studies have shown that endogenous glutamate can diffuse to act on receptors located extrasynaptically (1, 32). This action of glutamate has been shown to be of significance in different areas of the brain, particularly the hippocampus (17, 31). Complementary to that, we have recently suggested the possibility of increases of extracellular glutamate (as

STRIATUM

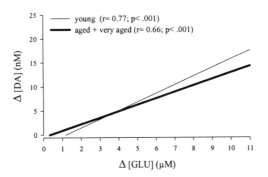

N. ACCUMBENS

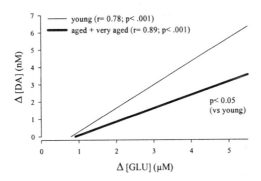

Figure 1. Effects of aging on the correlations between increases of dialysate dopamine and glutamate in striatum and nucleus accumbens of the awake rat.

detected by microdialysis) having their origin in the neuron-glia network in the brain (9). Therefore the actions of extracellular glutamate on dopamine could be attributed to an extrasynaptic/volumetric diffusion of glutamate through the extracellular space (24).

2. GLUTAMATE-DOPAMINE INTERACTIONS IN STRIATUM AND NUCLEUS ACCUMBENS DURING AGING

We have recently investigated the effects of aging [groups of rats of 2-3 months (young), 12-14 months (middle age), 27-32 months (aged) and 37 months (very aged)] on the actions of increasing concentrations of endogenous glutamate on the extracellular concentrations of dopamine in striatum and nucleus accumbens of the conscious rat using microdialysis. We have found that there are age-related changes in the effects of endogenous glutamate on extracellular dopamine and that these changes are specific to the area of the brain studied (see below).

We found no differences between aged and very aged rats when compared to young rats in striatum (Fig. 1). In contrast, the increases of extracellular concentrations of dopamine produced by glutamate in nucleus accumbens were lower in middle-aged and very aged rats than in young rats (Fig. 1). When the ratio dopamine to glutamate in all groups of animals was studied, an age-related decrease in the nucleus accumbens but not in striatum was found (Fig. 2). These findings suggest that the changes in the glutamate- dopamine interactions during the normal process of aging are regionally specific and that the effects of aging in the basal ganglia exhibits a dorsal-to-ventral pattern of effects with changes in the most ventral parts (nucleus accumbens) and no changes in the dorsal ones (dorsal striatum).

Several other reports have indicated a similar dorsal-to-ventral gradients of effects produced by aging in the basal ganglia (7, 11). Thus, it has been reported that the effects of amphetamine or a D2-receptor agonist, quinpirole on Fos expression in the basal ganglia, exhibits a dorsal-to-ventral pattern, with changes in nucleus accumbens and no changes in dorsal striatum (7). Also, the amount of release of dopamine, after high potassium stimulation, decreases in nucleus accumbens but less in dorsal striatum of aged rats when compared

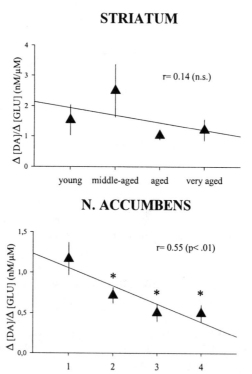

Figure 2. Effects of aging on the ratios of increases of dialysate dopamine to increases of glutamate in striatum and nucleus accumbens of the awake rat. The dopamine to glutamate ratio was calculated for the maximum increases of glutamate produced by PDC (4 mM).

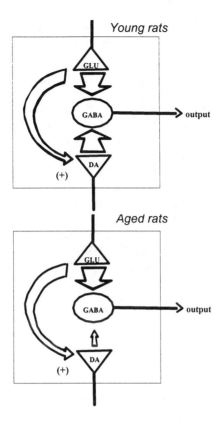

Figure 3. Schematic diagram showing the changes in endogenous interactions of glutamate and dopamine during aging in the nucleus accumbens of the awake rat.

to young rats (11). Other studies using microdialysis have reported a lack of effects of aging on the chemically stimulated release of dopamine in dorsal striatum (15, 30).

The effects of aging on the actions of endogenous glutamate on extracellular dopamine in striatum and nucleus accumbens may be a consequence of a differential decrease in the number or in the activity of the glutamatergic receptors that mediate the effects of glutamate. In fact, different studies have shown a decrease in the density of NMDA receptors in striatum and nucleus accumbens (19, 28) and also a reduction of the responses mediated by this type of receptor in striatum (6).

3. ON THE EFFECTS OF AGING ON THE FUNCTIONAL INTERACTION OF GLUTAMATE AND DOPAMINE IN STRIATUM AND NUCLEUS ACCUMBENS

The age-related changes of the interaction of glutamate and dopamine found in our studies are relevant in order to understand the alterations of motor behavior that occur in the normal process of aging (5, 41). Morevoer, our studies provide evidence for a differential effect of aging on these two neurotransmitters depending of the area in the basal ganglia studied. Thus, only the ventral striatum (nucleus accumbens) but not dorsal striatum may account for motor deficits based on the glutamate-dopamine interactions in the basal ganglia.

We have indicated that the interactions between glutamate and dopamine in the basal ganglia of the young rat could indicate a balanced input of both neurotransmitters to the GABA neuron in this area of the brain. As an effect of aging in the nucleus accumbens we found that glutamate is less potent to increase the extracellular concentrations of dopamine and that this decrease is negatively correlated with age (Fig. 3). This would be indicative of a disruption of this balanced input (glutamate-dopamine) to the GABA neuron which consequences could be the changes in the spontaneous locomotor activity found in rats during aging (7, 13).

It has been proposed that the mesolimbic dopamine system which terminals are located in the nucleus accumbens could play the role of an interface between motivation (limbic system) and action (neostriatum) (22). That is, the translation in neuronal coding from motivational to porposeful motor behavior. If that were the case, the findings reported in this review for the nucleus accumbens and ocurring during aging could be indicative of a deficit not of the mechanisms of motor plan and execution but to the coupling between motivation and the proper motor execution.

4. ACKNOWLEDGEMENT

This research has been suppored by grants DGICYT PM96-0046; CM99-08.5/0031/99; CM2000-08.5/00/70.1/2000 and DGINV-SAF-2000/0112.

5. REFERENCES

1. Barbour, B., Häusser, M.A., 1997, Intersynaptic diffusion of neurotransmitter, *Trends Neurosci.*, 20: 377-384.
2. Bouyer, J.J., Park, D.H., Joh, T.H., Pickel, V.M., 1984, Chemical and structural analysis of the relation between cortical inputs and tyrosine hydroxylase-containing terminals in rat neostriatum, *Brain Res.*, 302: 267-275.
3. Bridges, R.J., Stanley, M.S., Anderson, M.W., Cotman, C.W., Camberlin, A.R., 1991, Conformationally defined neurotransmitter analogues. Selective inhibition of glutamate uptake by one pyrrolidine-2,4-dicarboxylate diastereomer, *J.Med.Chem.*, 34: 717-725.
4. Carlsson, M., Carlsson, A., 1990, Interactions between glutamatergic and monoaminergic systems within the basal ganglia - implications for schizophrenia and Parkinson's disease, *Trends Neurosci.*, 13: 272-276.
5. Carlsson, M., Svensson, A., Carlsson, A., 1992, Interactions between excitatory amino acids, catecholamines and acetylcholine in the basal ganglia. In: *Excitatory amino acids*, (R.P. Simon, ed.) Thieme, New York: pp 189-194.
6. Cepeda, C., Li, Z., Levine, M.S., 1996, Aging reduces neostriatal responsiveness to N-methyl-D-aspartate and dopamine: an *in vitro* electrophysiological study, *Neuroscience*, 73: 733-750.
7. Crawford, C.A., Levine, M.S., 1997, Dopaminergic function in the neostriatum and nucleus accumbens of young and aged Fisher 344 rats, *Neurobiol.Aging*, 18: 57-66.
8. Del Arco, A., Mora, F., 1999, Effects of endogenous glutamate on extracellular concentrations of GABA, dopamine and dopamine metabolites in the prefrontal cortex of the freely moving rat: involvement of NMDA and AMPA/kainate receptors, *Neurochem.Res.*, 24: 1027-1035.

9. Del Arco, A., Segovia, G., Fuxe, K., Mora, F., 2002, Changes of dialysate concentrations of glutamate and GABA in the brain: an index of volume transmission mediated actions?, *J.Neurochem.*, (in press).
10. Desce, J.M., Godeheu, G., Galli, T., Artaud, F., Cheramy, A., Glowinski, J., 1992, L-glutamate-evoked release of dopamine from synaptosomes of the rat striatum: involvement of AMPA and N-methyl-D-aspartate receptors, *Neuroscience*, 47: 333-339.
11. Friedemann, M.N., Gerhardt, G.A., 1992, Regional effects of aging on dopaminergic function in the Fisher-344 rat, *Neurobiol.Aging*, 13: 325-332.
12. Groenewegen, H.J., Berendse, H.W., Meredith, G.E., Haber, S.N., Voorn, P., Wolters, J.G., Lohman, A.S.M., 1991, Functional anatomy of the ventral, limbic system-innervated striatum. In: *The mesolimbic dopamine system: from motivation to action*, (P. Willner, J. Scheel-Krüger, eds.) John Wiley & Sons, Chichester: pp 19-59.
13. Huang, R.-L., Wang, C.-T., Tai, M.-Y., Tsai, Y.-F., Peng, M.-T., 1995, Effects of age on dopamine release in the nucleus accumbens and amphetamine-induced locomotor activity in rats, *Neurosci.Lett.*, 200: 61-64.
14. Imperato, A., Honoré, T., Jensen, L.H., 1990, Dopamine release in the nucleus caudatus and in the nucleus accumbens is under glutamatergic control through non-NMDA receptors: a study in freely-moving rats, *Brain Res.*, 530: 223-228.
15. Kametani, H., Iijima, S., Spangler, E.L., Ingram, D.K., Joseph, J.A., 1995, In vivo assessment of striatal dopamine release in the aged male Fischer 344 rat, *Neurobiol.Aging*, 16: 639-646.
16. Kendrick, K.M., Guevara-Guzman, R., De la Riva, C., Christensen, J., Ostergaard, K., Emson, P.C., 1996, NMDA and kainate-evoked release of nitric oxide and classical transmitters in the rat striatum: in vivo evidence that nitric oxide may play a neuroprotective role, *Eur.J.Neurosci.*, 8: 2619-2634.
17. Kullmann, D.M., Erdemli, G., Asztely, F., 1996, LTP of AMPA and NMDA receptor-mediated signals: evidence for presynaptic expression and extrasynaptic glutamate spill-over, *Neuron*, 17: 461-474.
18. Lannes, B., Micheletti, G., 1994, Glutamate-dopamine balance in the striatum: pre- and post-synaptic interactions. In: *The basal ganglia IV. New ideas and data on structure and function*, (G. Percheron, J.S. McKenzie, J. Feger, eds.) Plenum Press, New York: pp 475-489.
19. Magnusson, K.R., Cotman, C.W., 1993, Effects of aging on NMDA and MK801 binding sites in mice, *Brain Res.*, 604: 334-337.
20. Massieu, L., Morales-Villagrán, A., Tapia, R., 1995, Accumulation of extracellular glutamate by inhibition of its uptake is not sufficient for inducing neuronal damage: an in vivo microdialysis study, *J.Neurochem.*, 64: 2262-2272.
21. Maura, G., Giardi, A., Raiteri, M., 1988, Release-regulation D-2 dopamine receptors are located on striatal glutamatergic nerve terminals, *J.Pharmacol.Exp.Ther.*, 247: 680-684.
22. Mogenson, G.J., Jones, D.L., Yim, C.Y., 1982, From motivation to action: fuctional interface between the limbic system and the motor system, *Prog.Neurobiol.*, 14: 69-97.
23. Mora, F., Porras, A., 1994, Interactions of dopamine, excitatory amino acids and inhibitory amino acids in the basal ganglia of the conscious rat. In: *The basal ganglia IV. New ideas and data on structure and function*, (G. Percheron, J.S. McKenzie, J. Féger, eds.) Plenum Press, New York: pp 441-447.
24. Mora, F., Segovia, G., Del Arco, A., 1999, Endogenous glutamate-dopamine-GABA interactions in specific circuits of the brain of the awake animal. In: *Recent research developments in neurochemistry*, (S.G. Pandalai, ed.) Research Signpost, Trivandrum: pp 171-178.
25. Morari, M., Marti, M., Sbrenna, S., Fuxe, K., Bianchi, C., Beani, L., 1998, Reciprocal dopamine-glutamate modulation of release in the basal ganglia, *Neurochem.Int.*, 33: 383-397.
26. Morari, M., O'Connor, W.T., Ungerstedt, U., Fuxe, K., 1993, N-methyl-D-aspartic acid differentially regulates extracellular dopamine, GABA, and glutamate levels in the dorsolateral neostriatum of the halothane-anesthetized rat: an *in vivo* microdialysis study, *J.Neurochem.*, 60: 1884-1893.
27. Obrenovitch, T.P., Urenjak, J., Zilkha, E., 1996, Evidence disputing the link between seizure activity and high extracellular glutamate, *J.Neurochem.*, 66: 2446-2454.
28. Ossowska, K., Wolfarth, S., Schulze, G., Wardas, J., Pietraszek, M., Lorenc-Koci, E., Smialowska, M., Coper, H., 2001, Decline in motor functions in aging is related to the loss of NMDA receptors, *Brain Res.*, 907: 71-83.
29. Patel, D.R., Young, A.M.J., Croucher, M.J., 2001, Presynaptic alpha-amino-3-hydroxy-5-methyl-4-isoxazolepropionate receptor-mediated stimulation of glutamate and GABA release in the rat striatum in vivo: a dual-label microdialysis study, *Neuroscience*, 102: 101-111.

30. Santiago, M., Machado, A., Cano, J., 1993, Effects of age and dopamine agonists and antagonists on striatal dopamine release in the rat: an *in vivo* microdialysis study, *Mech.Ageing Dev.*, 67: 261-267.
31. Scanziani, M., 2002, Competing on the edge, *Trends Neurosci.*, 25: 282-283.
32. Scanziani, M., Salin, P.A., Vogt, K.E., Malenka, R.C., Nicoll, R.A., 1997, Use-dependent increases in glutamate concentration activate presynaptic metabotropic glutamate receptors, *Nature*, 385: 630-634.
33. Segovia, G., Del Arco, A., Mora, F., 1997, Endogenous glutamate increases extracellular concentrations of dopamine, GABA, and taurine through NMDA and AMPA/kainate receptors in striatum of the freely moving rat: a microdialysis study, *J.Neurochem.*, 69: 1476-1483.
34. Segovia, G., Del Arco, A., Mora, F., 1999, Effects of aging on the interaction between glutamate, dopamine and GABA in striatum and nucleus accumbens of the awake rat, *J.Neurochem.*, 73: 2063-2072.
35. Segovia, G., Mora, F., 1998, Role of nitric oxide in modulating the release of dopamine, glutamate, and GABA in striatum of the freely moving rat, *Brain Res.Bull.*, 45: 275-279.
36. Segovia, G., Mora, F., 2001, Involvement of NMDA and AMPA/kainate receptors in the effects of endogenous glutamate on extracellular concentrations of dopamine and GABA in the nucleus accumbens of the awake rat, *Brain Res.Bull.*, 54: 153-157.
37. Segovia, G., Porras, A., Del Arco, A., Mora, F., 2001, Glutamatergic neurotransmission in aging: a critical perspective, *Mech.Ageing Dev.*, 122: 1-29.
38. Sesack, S.R., Pickel, V.M., 1992, Prefrontal cortical efferents in the rat synapse on unlabeled neuronal targets of catecholamine terminals in the nucleus accumbens septi and dopamine neurons in the ventral tegmental area, *J.Comp.Neurol.*, 320: 145-160.
39. Smith, A.D., Bolam, J.P., 1990, The neural network of the basal ganglia as revealed by the study of synaptic connections of identified neurones, *Trends Neurosci.*, 13: 259-265.
40. Smolders, I., Sarre, S., Vanhaesendonck, C., Ebinger, G., Michotte, Y., 1996, Extracellular striatal dopamine and glutamate after decortication and kainate receptor stimulation, as measured by microdialysis, *J.Neurochem.*, 66: 2373-2380.
41. Svensson, L., Zhang, J., Johannessen, K., Engel, J.A., 1994, Effect of local infusion of glutamate analogues into the nucleus accumbens of rats: an electrochemical and behavioural study, *Brain Res.*, 643: 155-161.
42. Thomsen, C., Hansen, L., Suzdak, P.D., 1994, L-glutamate uptake inhibitors may stimulate phosphoinositide hydrolysis in baby hamster kidney cells expressing mGluR1a via heteroexchange with L-glutamate without direct activation of mGluR1a, *J.Neurochem.*, 63: 2038-2047.

THE IMMUNOHISTOCHEMICAL LOCALISATION OF $GABA_A$ RECEPTOR SUBUNITS IN THE HUMAN STRIATUM

H.J. Waldvogel, W.M.C. van Roon-Mom, and R.L.M. Faull*

Keywords: striosome, matrix, GABA.

Abstract: The regional localisation of gamma-aminobutyric acid$_A$ ($GABA_A$) receptor subunits was investigated in the human striatum using immunohistochemical staining for five $GABA_A$ receptor subunits (α_1, α_2, α_3, $\beta_{2,3}$ and γ_2) and other neurochemical markers. The results demonstrated that $GABA_A$ receptors in the striatum showed considerable subunit heterogeneity in their regional distribution and cellular localisation. High densities of $GABA_A$ receptors in the striosome compartment contained the α_2, α_3, $\beta_{2,3}$ and γ_2 subunits and lower densities of receptors in the matrix compartment contained the α_1, α_2, α_3, $\beta_{2,3}$ and γ_2 subunits. These results show that the subunit composition of $GABA_A$ receptors display considerable regional variation in the human striatum.

1. INTRODUCTION

Gamma-aminobutyric acid (GABA) plays a critical role in the functioning of the mammalian basal ganglia. GABA is the major inhibitory neurotransmitter in the basal ganglia and exerts its effects mainly via $GABA_A$ receptors which are localised principally on the postsynaptic membranes of GABAergic synaptic terminals.

Molecular biological studies have shown that the $GABA_A$ receptor is a pentameric complex consisting of five subunits assembled from a range of different classes of subunits, α_{1-6}, β_{1-3}, γ_{1-3}, $\rho_{1,2}$, δ, and ϵ subunit (22). There is evidence for considerable heterogeneity of the subunit composition of $GABA_A$ receptors at both the regional and cellular levels in the mammalian brain (10, 11, 12 27). In the basal ganglia, previous studies indicate the presence of a variety of $GABA_A$ receptor subunit configurations (20, 28).

---

* H.J. Waldvogel, W.M.C. van Roon-Mom, and R.L.M. Faull, Department of Anatomy with Radiology, Faculty of Medical and Health Science, University of Auckland, Auckland, New Zealand.

The Basal Ganglia VI
Edited by Graybiel *et al.*, Kluwer Academic/Plenum Publishers, 2002

It is now well established that, at the regional level, the mammalian striatum can be divided into two major interdigitating neurochemical compartments, which consist of a larger matrix compartment, and a smaller patch or striosome compartment (16). Most of the known neurotransmitters and their receptors in the striatum, as well as a variety of other neuroactive substances, are cytoarchitecturally arranged according to this striosome/matrix compartmental organisation. The striosome and matrix compartments are delineated by high concentrations of a range of neurochemicals (13 18, 25) including $GABA_A$ receptors (9). Studies in the baboon and human caudate putamen show that the regional distribution of $GABA_A$ receptors conforms to the complex striosome and matrix compartmental organisation (8, 9, 26).

In the present study, in order to gain a better understanding of GABAergic mechanisms in the human basal ganglia, we have undertaken an investigation of the localisation of $GABA_A$ receptor subunits using immunohistochemical techniques at the light microscope level. We have investigated the overall regional distribution of the various $GABA_A$ receptor subunits in the human striatum and observed whether they conform to the striosome and matrix compartmental organisation of the striatum.

2. MATERIALS AND METHODS

For this study the basal ganglia from 15 post mortem human brains were obtained from the New Zealand Neurological Foundation Human Brain Bank (Department of Anatomy with Radiology, University of Auckland). The protocols used in these studies were approved by the University of Auckland Human Subject Ethics Committee. The tissue used in this study was from cases with no history of neurological disease and showed no neurological abnormalities. The brains were fixed by perfusion through the basilar and carotid arteries, with 15% formalin in 0.1M phosphate buffer pH 7.4. Blocks of the basal ganglia were then dissected out and post-fixed for 24 hours. These blocks were cryo-protected in 30% sucrose in 0.1M phosphate buffer and sectioned on a freezing microtome at a thickness of 50-70µm. The sections were collected in PBS with 0.1% sodium azide (PBS-azide).

A range of monoclonal and polyclonal antibodies were used to label $GABA_A$ receptor subunits in the human basal ganglia as follows. Two monoclonal antibodies specific for the α_1 (bd24) and the $\beta_{2,3}$ (bd17) subunits of the $GABA_A$ receptor (6, 17), and polyclonal antibodies against the α_2, α_3, and γ_2 subunits were raised in guinea pigs (3, 12). Other antibodies used were against enkephalin (Seralab Ltd, UK). All antibodies were dissolved in immunobuffer consisting of 1% goat serum in PBS with 0.2% Triton-X and 0.4% Thimerosol.

In order to investigate whether the patterns of distribution of the $GABA_A$ receptor subunits within the caudate-putamen followed the neurochemically defined striosome and matrix compartments, sections adjacent to those labelled with the various $GABA_A$ receptor subunits were labelled with antibodies to enkephalin which define the striosome compartment of the striatum.

Sections of the basal ganglia were processed free-floating in tissue culture wells using standard immunohistochemical procedures. Sections were washed in PBS and 0.2% Triton-X (PBS-triton) and then incubated for 20 minutes in 50% methanol and 1% H_2O_2 and washed again. The sections were then incubated in primary antibodies for 2-3 days on a shaker at

4°C, the primary antibodies were washed off and then the sections were incubated overnight in the appropriate species specific biotinylated secondary antibodies. The secondary antibodies were washed off and the sections incubated for 4 hours at room temperature in streptavidin-conjugated HRP complex. The sections were then reacted in 0.05% DAB with 0.4% nickel ammonium sulphate and 0.01% H_2O_2 in 0.1M phosphate buffer pH7.4 for 15-30 minutes to produce a blue-black reaction product. The sections were washed in PBS, mounted on chrome-alum coated slides, dehydrated through a graded alcohol series to xylene, and coverslipped. In addition, some sections were processed as control sections to determine non-specific staining.

3. RESULTS

3.1 Immunohistochemical localisation of $GABA_A$ receptor subunits in the striatum.

The regional immunohistochemical labelling patterns of the $GABA_A$ receptor subunits α_1, α_2, α_3, $\beta_{2,3}$, and γ_2, and their relation to the neurochemical marker enkephalin in the human striatum are illustrated in Figures 1 and 2. The immunohistochemical labelling for the various $GABA_A$ receptor subunits showed a marked heterogeneity in their regional staining patterns throughout the striatum.

In order to investigate whether the heterogeneous pattern of immunoreactivity for the various $GABA_A$ receptor subunits corresponded to the striosome/patch and matrix compartments of the striatum, the $GABA_A$ receptor subunit staining pattern was compared to the pattern of staining in adjacent sections that were stained with enkephalin which delineates the striosome/matrix compartments in the dorsal striatum (25)

α_1 *subunit*. The regional localisation of α_1 subunit immunoreactivity showed a marked heterogeneous distribution throughout the striatum (Fig. 1B). The heterogeneity of labelling was evident as patches of low intensity of immunoreactivity that were of various shapes and sizes in the dorsal and ventral striatum (arrows, Fig. 1B, 2A). The patches were surrounded by a background matrix of higher densities of labelling. In the dorsal striatum, the patches of low α_1 subunit immunoreactivity corresponded closely to the regions of high enkephalin immunoreactivity (arrows, Figs. 1A, 2B) which delineated the striosome compartment in adjacent sections while the surrounding region of relatively high densities of α_1 subunit immunoreactivity corresponded with the regions of relatively low densities of enkephalin immunoreactivity which delineated the matrix compartment. In the ventral striatum, the patches of low α_1 subunit immunoreactivity (double arrow, Fig. 2B) corresponded to the patches of low enkephalin (double arrow, Fig. 1A) immunoreactivity in adjacent sections.

α_2 *subunit*. α_2 subunit immunoreactivity showed high intensities of labelling throughout the dorsal striatum (Figs. 1C, 2C), but showed especially intense labelling in patches (arrows, Figs. 1C, 2C) which could be identified as striosomes on adjacent sections stained for enkephalin (arrows, Figs. 1A, 2B); the intense labelling of the α_2 subunit corresponded to the regions of intense enkephalin immunoreactivity on adjacent sections. Individual α_2-rich striosomes also appear to be surrounded by an α_2-poor annular zone intervening between the intensely immunoreactive striosome and the moderately high intensity of labelling of the surrounding matrix compartment (Figs. 1C, 2C). In the ventral striatum, the low density

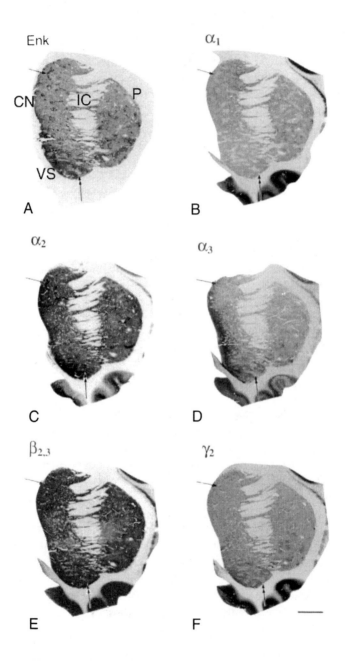

Figure 1. Immunohistochemical localisation of Enkephalin (A), and the $GABA_A$ receptor subunits α_1 (B), α_2 (C), α_3 (D), $\beta_{2,3}$ (E) and γ_2 (F) in the human caudate-putamen. CN: caudate nucleus; IC: internal capsule; P: putamen; VS ventral striatum. scalebar = 0.5cm

patches of α_2 subunit immunoreactivity (double arrow, Fig. 1C) corresponded to the low density patches of enkephalin, α_1, α_3, $\beta_{2,3}$ and γ_2 immunoreactivity (double arrows, Figs. 1A, B, D-F)

α_3 subunit. α_3 subunit immunoreactivity was moderately low throughout the striatum but did show some heterogeneity of staining (Fig. 1D). In the dorsal caudate nucleus patches of higher intensities of α_3 subunit labelling were evident (arrow, Fig. 1D) and these appear to correspond with striosomes labelled with other markers on adjacent sections (Fig. 1). In the ventral striatum, α_3-poor patches were scattered within regions of high α_3 subunit immunoreactivity (double arrow, Fig. 1D); these patches in the ventral striatum appeared to be in register with the patches of low intensities of staining for the α_1, α_2, $\beta_{2,3}$ and γ_2 subunits (double arrows, Fig. 1)

$\beta_{2,3}$ subunit. $\beta_{2,3}$ subunit immunoreactivity showed high intensities of labelling throughout the striatum (Figs. 1E). In the dorsal striatum, patches of intense $\beta_{2,3}$ subunit labelling could be distinguished (arrows, Figs. 1E) which aligned with striosomes as identified by markers on adjacent sections (arrows, Fig. 1). In addition, a very pale staining annular region was evident intervening between the striosome and matrix regions (arrow Fig. 1), and this was most obvious in the dorsal regions of the caudate-putamen. By contrast, in the ventral striatum, patches of lower $\beta_{2,3}$ subunit immunoreactivity were evident (double arrow, Fig. 1E) which were in register with poorly stained patches of enkephalin (double arrows, Fig. 1A), and low immunoreactive patches of α_1, α_2, α_3 and γ_2 subunits (double arrows, Fig. 1).

γ_2 subunit γ_2 subunit immunoreactivity showed moderate intensities throughout the striatum (Fig. 1F). The caudate nucleus and putamen showed relatively homogeneous labelling; however, small regions of higher intensity of γ_2 labelling which corresponded to

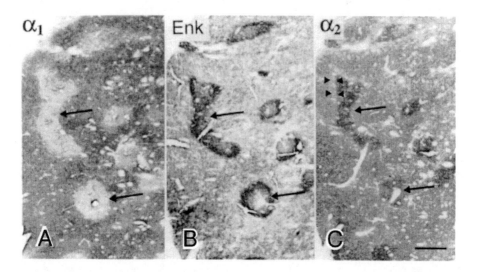

Figure 2. Light micrographs showing a series of adjacent sections of the caudate nucleus (arrowed in Figure. 1A, B, C) comparing the staining patterns of α_1 subunit (A), enkephalin (B) and α_2 subunit (C). Note the regions of low α_1 subunit corresponds with the regions of high enkephalin and α_2 subunit (arrows). The arrowheads in C indicate the annular zone of low α_2 subunit which intervenes between the striosome and matrix compartments. Scalebar =1mm.

striosomes were evident in the dorsal region of the caudate nucleus and putamen (arrows, Fig. 1F), and these regions were often delineated by a surrounding immunonegative annular zone (arrows; Fig. 1F). In the ventral striatum γ_2 subunit-poor patches (double arrow, Fig. 1F) were located within a more densely stained background matrix region; these γ_2-poor patches corresponded to regions of low immunoreactivity for, enkephalin, and for the subunits α_1, α_2, α_3 and $\beta_{2,3}$ (Fig. 1).

4. DISCUSSION

The compartmental distribution of $GABA_A$ receptors shown by previous ligand binding studies (9) is confirmed and considerably extended by the results of these immunohistochemical studies showing that the distribution of $GABA_A$ receptor subunits also follows the striosome/matrix compartmental organisation. As illustrated in Figures 1 and 2, the striosome compartment of the caudate nucleus and putamen showed very distinctive intense immunoreactivity for the α_2, α_3, $\beta_{2,3}$ and γ_2 subunits of the $GABA_A$ receptor, whereas the matrix compartment showed a less intense, moderate immunoreactivity for the α_1, α_2, α_3, $\beta_{2,3}$ and γ_2 subunits. These findings suggest a different molecular configuration for the $GABA_A$ receptors in these two major neurochemical compartments of the human striatum – that is, the α_1 subunit of the $GABA_A$ receptor is only present in the matrix compartment, whereas the α_2, α_3, $\beta_{2,3}$ and γ_2 subunits are present in both compartments, although they are present in the highest concentrations in the striosome compartment.

Molecular biological studies have shown that $GABA_A$ receptors are heteropentameric chloride channels comprised of a range of different classes of subunits which include α_{1-6}, β_{1-3}, γ_{1-3}, δ, $\rho_{1,2}$ and ε subunits (1, 22). These subunits can be assembled into a variety of $GABA_A$ receptor complexes with different subunit configurations which show characteristic pharmacological profiles and ligand binding properties (21, 23, 28). The results of previous pharmacological studies show that the $\alpha_1 \beta_x \gamma_2$ (where β_x is any β subunit) subunit configuration displays the characteristics of type I $GABA_A$ receptors, and that substituting the α_1 subunit for other α subunits such as α_2 or α_3 displays the characteristics of type II $GABA_A$ receptors (24). Our present immunohistochemical results on the subunit configuration of $GABA_A$ receptors in the human dorsal and ventral striatum have shown that the $GABA_A$ receptors in the matrix have a molecular configuration which includes α_{1-3}, $\beta_{2,3}$, γ_2, whereas the $GABA_A$ receptors in the striosomes have a subunit composition comprising α_{2-3}, $\beta_{2,3}$ γ_2. These results are in agreement with our previous findings (9) showing that type I receptors are only found in the matrix and that type II receptors are found in the highest concentrations in the striosomes, but are also present in the matrix.

In addition, the striosome and matrix compartments in the dorsal caudate-putamen complex are separated by a small, intervening annular region which is immunonegative for all of the $GABA_A$ receptor subunits investigated in the present study. These findings suggest that this annular region contains very low levels of $GABA_A$ receptors and is therefore functionally distinct from the striosome and matrix compartments. This conclusion is supported by other studies showing that this annular or perisomal zone is distinguished by high levels of neurotensin receptors (7), and by high concentrations of substance P and substance P receptors (19).

Numerous previous studies on the mammalian striatum have shown that the striosome and matrix compartments are neurochemically distinct. In particular, this same pattern of neurochemical heterogeneity of the striosome/matrix compartments has been confirmed in other studies on the primate and human striatum; for example the striosome compartment is especially well delineated by enkephalin or substance P (2, 25, 26) and the matrix compartment by high concentrations of acetylcholinesterase, calbindin and parvalbumin (15, 18). Furthermore, this compartmentalisation of the striatum has been further substantiated by findings that the cytoarchitecture (14) and input-output relations (5, 13, 16) are also organised according to the striosome and matrix organisation. In particular, connectional studies in the rat (4) suggest that the striosome and matrix compartments may, through their separate and distinctive pattern of afferent and efferent connections, represent functionally separate input-output striatal processing channels - the striosomes representing a 'limbic' processing channel and the matrix a 'sensorimotor' processing channel. In particular, our results showing a heterogeneity of $GABA_A$ receptor subtypes in the two compartments supports the concept that the striosome and matrix compartments represent functionally distinct compartments of the human caudate-putamen complex, and that through the different putative molecular configurations of the $GABA_A$ receptors in the two compartments, GABA may exert quite different effects in the modulation of 'limbic' striosomal and 'sensorimotor' matrix processing.

5. ACKNOWLEDGEMENTS

This research was supported by the New Zealand Health Research Council and The New Zealand Neurological Foundation.

6. REFERENCES

1. Barnard EA, Darlison MG, Seeburg P (1987) Molecular biology of the $GABA_A$ receptor: the receptor/channel superfamily. TINS *10*: 502-509
2. Beach TG, McGeer EG (1984) The distribution of substance P in the primate basal ganglia:an immunohistochemical study of baboon and human brain. Neuroscience *13*: 29-52
3. Benke D, Fritschy JM, Trzeciak A, Bannwarth W, Mohler H (1994) Distribution, prevalence, and drug binding profile of gamma-aminobutyric acid type A receptor subtypes differing in the beta-subunit variant. J. Biol. Chem. *269*: 27100-7
4. Donoghue JP, Herkenham M (1986) Neostriatal projections from individual cortical fields conform to histochemically distinct striatal compartments in the rat. Brain Res. *365*: 397-403
5. Elben F and Graybiel AM (1995) Highly restricted origin of prefrontal cortical inputs to striosomes in the macaque monkey. J. Neuroscience 15: 5999-6013
6. Ewert M, Shivers BD, Luddens H, Mohler H, Seeburg PH (1990) Subunit selectivity and epitope characterisation of mAbs directed against the GABAA/benzodiazepine receptor. T J.Cell Biol. *110*: 2043-2048
7. Faull RLM, Dragunow M, Villiger JW (1989) The distribution of neurotensin receptors and acetylcholinesterase in the human caudate nucleus: evidence for the existence of a third neurochemical compartment. Brain Res. *488*: 381-6
8. Faull RLM, Villiger JW (1986) Heterogeneous distribution of benzodiazepine receptors in the human striatum: a quantitative autoradiographic study comparing the pattern of receptor labelling with the distribution of acetylcholinesterase staining. Brain. Res. *381*: 153-158

9. Faull RLM, Villiger JW (1988) Multiple benzodiazepine receptors in the human basal ganglia: a detailed pharmacological and anatomical study. Neuroscience. *24*: 433-451
10. Fritschy JM, Benke D, Mertens S, Oertel WH, Bachi T, Mohler H (1992) Five subtypes of type A gamma-aminobutyric acid receptors identified in neurons by double and triple immunofluorescence staining with subunit-specific antibodies. Proc. Natl. Acad. Sci.USA *89*: 6726-30
11. Fritschy JM, Mohler H (1995) $GABA_A$ receptor heterogeneity in the adult rat Brian: differential regional and cellular distribution of seven major subunits. J. Comp. Neurol. *359*: 154-194
12. Gao B, Hornung J-P, Fritschy J-M, (1995) Identification of distinct $GABA_A$-receptor subtypes in cholinergic and parvalbumin positive neurons of the rat and marmoset medium septum-diagonal band complex. Neuroscience. *65*: 101-117
13. Gerfen CR, Herkenham M, Thibault J (1987) The neostriatal mosaic: II. Patch- and matrix-directed mesostriatal dopaminergic and non-dopaminergic systems. J. Neuroscience *7*: 3915-34
14. Goldman-Rakic PS (1982) Cytoarchitectonic heterogeneity of the primate striatum: subdivision into island and matrix cellular compartments. J. Comp. Neurol. *205*: 398-413
15. Graybiel AM, Ragsdale CW, Jr. (1978) Histochemically distinct compartments in the striatum of human, monkcys, and cat demonstrated by acetylthiocholinesterase staining. Proc. Natl. Acad. Sci. USA *75*: 5723-6
16. Graybiel AM, Ragsdale CW,Jr, Yoneoka ES, Elde RP (1981) An immunohistochemical study of enkephalins and other neuropeptides in the striatum of the cat with evidence that the opiate peptides are arranged to form mosaic patterns in register with the striosomal compartments visible by acetylcholinesterase staining. Neuroscience. *6*: 377-397
17. Haring P, Stahli C, Schoch P, Takacs B, Staehelin T, Mohler H (1985) Monoclonal antibodies reveal structural homogeneity of g-aminobutyric acid/benzodiazepine receptor in different brain areas. Proc. Natl. Acad. Sci. USA *82*: 4837-4841
18. Holt DJ, Graybiel AM, Saper CB (1997) Neurochemical architecture of the human striatum. J. Comp. Neurol. *384*: 1-25
19. Jakab RL, Hazrati LN, Goldman Rakic P (1996) Distribution and neurochemical character of substance P receptor (SPR)-immunoreactive striatal neurons of the macaque monkey - accumulation of SP fibers and SPR neurons and dendrites in striocapsules encircling striosomes. J. Comp. Neurol. *369*: 137-149
20. Kultas-Ilinsky K, Leontiev V, Whiting PJ (1998) Expression of 10 $GABA_A$ receptor subunit messenger RNAs in the motor-related thalamic nuclei and basal ganglia of Macaca mulatta studied with in situ hybridization histochemistry. Neuroscience *85*: 179-204
21. Luddens H, Wisden W (1991) Function and pharmacology of multiple $GABA_A$ receptor subunits. TIPS *12*: 49-51
22. Morrow AL (1995) Regulation of $GABA_A$ receptor function and gene expression in the central nervous system. Int. Rev. Neurobiol. *38*: 1-41
23. Olsen RW, Tobin AJ (1990) Molecular biology of $GABA_A$ receptors. FASEB J *4*: 1469-1480
24. Pritchett DB, Luddens H, Seeburg PH (1989) Type 1 and type 11 $GABA_A$-benzodiazepine receptors produced in transfected cells. Science. *245*: 1389-1392
25. Waldvogel HJ, Faull RLM (1993) Compartmentalization of parvalbumin immunoreactivity in the human striatum. Brain. Res. *610*: 311-316
26. Waldvogel HJ, Fritschy JM, Mohler H, Faull RLM (1998) $GABA_A$ Receptors in the Primate Basal Ganglia - an Autoradiographic and a Light and Electron Microscopic Immunohistochemical Study of the a_1 and $b_{2,3}$ Subunits in the Baboon Brain. J. Comp. Neurol. *397*: 297-325
27. Wisden W, Laurie DJ, Monyer H, Seeburg PH (1992) The distribution of 13 $GABA_A$ receptor subunit mRNAs in the rat brain. I. Telencephalon, diencephalon, mesencephalon. J. Neuroscience. *12*: 1040-1062
28. Wisden W, Seeburg PH (1992) $GABA_A$ receptor channels: from subunits to functional entities. Curr. Opin. Neurobiol. *2*: 263-269

SYNAPTIC LOCALIZATION OF GABA_A RECEPTOR SUBUNITS IN THE BASAL GANGLIA OF THE RAT

F. Fujiyama, J-M. Fritschy, F.A. Stephenson and J.P. Bolam*

1. INTRODUCTION

The inhibitory amino acid, γ-aminobutyric acid (GABA), plays a critical role in the neuronal networks of the basal ganglia. Most of the major classes of neurons in the basal ganglia utilize GABA as a neurotransmitter. Thus neurons of the globus pallidus (GP), and basal ganglia output neurons in the entopeduncular nucleus (EP) and the substantia nigra pars reticulata (SNr) are GABAergic (Smith and Bolam, 1990). Within the striatum spiny projection neurons are GABAergic (Smith and Bolam, 1990) as are populations of GABAergic interneurons (Bolam et al., 1983, 1985) that express different calcium binding proteins (Cowan et al., 1990; Kita et al., 1990; Kubota et al., 1993; Clarke and Bolam, 1997) or synthesize nitric oxide (Kubota et al., 1993). Furthermore, the striatum receives GABAergic afferents from the globus pallidus (Bevan et al., 1998; Smith et al., 1998) and possibly the substantia nigra (van der Kooy et al., 1981).

In view of the widespread distribution of GABA and its receptors and its critical role in basal ganglia function it is important to characterise the position and composition of GABA receptors in relation to the synaptic circuitry of the basal ganglia. In this study we examine the localization of subunits of the GABA_A receptor in the basal ganglia of the rat. The primary objectives were threefold. First, to determine the sub-cellular localization of subunits of the GABA_A receptor, in particular to determine their spatial relationship to synaptic specializations. Secondly, to attempt to characterize the axon terminals presynaptic to the GABA_A receptor-positive synapses by GABA_A immunolabelling. Thirdly, to determine whether different subunits of the GABA_A receptor that have been shown to be co-expressed at the cellular level are co-expressed at individual synapses.

These issues were addressed by using the post-embedding immunogold technique on freeze-substituted, Lowicryl-embedded tissue from rat basal ganglia. Ultrathin sections of striatum, SNr, SNc, EP, GP and subthalamic nucleus (STN) were collected on coated single

---

* F. Fujiyama and J.P. Bolam, MRC Anatomical Neuropharmacology Unit, Dept. of Pharmacology, Mansfield Road, Oxford, UK. J-M. Fritschy, Institute of Pharmacology, University of Zurich, Switzerland. F.A. Stephenson, School of Pharmacy, University of London, UK.

slot or mesh grids. The sections were then immunolabelled essentially as described previously (Baude et al 1993; Bernard et al 1997; Clarke et al 1998; Nusser et al 1998) using the antibodies directed against subunits of the GABA$_A$ receptor, against GABA itself or mixtures of antibodies followed by the appropriate secondary antibodies conjugated to colloidal gold of different diameters (5-20 nm, British BioCell Int.). Primary antibody was used at the following dilutions: anti-GABA (1:5000, Somogyi et al., 1985a,b; Hodgson et al., 1985), anti-GABA$_A$ α 1 subunit (11 μg/ml, Jones et al., 1997), anti-GABA$_A$ β2/3 subunits (10 μg/ml, Haring et al., 1985), anti-GABA$_A$ γ2 subunit (10 μg/ml, Benke et al., 1996).

2. LOCALIZATION OF IMMUNOGOLD LABELLING FOR GABA$_A$ RECEPTOR SUBUNITS

To determine the distribution of GABA$_A$ receptor subunits, the immunogold particles coding for the β2/3 subunits were analysed in the striatum by systematic examination of adjacent photomicrographs at a final magnification of 33,000 (366 gold particles; 138 μm^2) or by systematic scans of sections on mesh grids (2760 gold particles; approximate area: 8980 μm^2). Although the intracellular labelling accounted for 54.4% of the gold particles, they most commonly occurred as single gold particles and only rarely were associations of two or more particles observed. Forty four percent of gold particles were associated with membranes i.e. either touching the membrane or within one diameter of it. Of the membrane-associated particles, 55.5% were localized at symmetrical synapses where gold particles lined up along the synaptic specialization (Figure 1A, B). Immunopositive asymmetrical synapses were not observed. Although quantitative analyses have not yet been performed, it was evident that there is also a selective association of β2/3 subunits of the GABA$_A$ receptor at symmetrical synapses in the other regions of the basal ganglia examined (Figures 2 & 3).

3. IS THE GABA$_A$ RECEPTOR LOCATED AT SYMMETRICAL SYNAPSES ESTABLISHED BY GABA-POSITIVE TERMINALS?

In order to determine the association between GABAergic terminals and β2/3 subunits adjacent photomicrographs were examined or systematic, nonoverlapping, scans of sections on mesh grids were made. A synapse was considered positive by the presence of two or more immunogold particles along the synaptic membranes. Each receptor-positive synapse and the level of immunogold labelling for GABA in the presynaptic bouton was assessed. A bouton was considered to be GABA-negative or to have low levels of GABA if the number of gold particles was two or less. In the striatum, systematic scans of double-labelled sections revealed that most (59.3%) of the boutons forming synapses that were positive for the β2/3 subunits of the GABA$_A$ receptor were also positive for GABA. The boutons were of variable size and made symmetrical synaptic contacts with dendrites, spines and perikarya. They contained from 0-4 mitochondria and sometimes formed synapses with more than one structure in the same plane. In addition to the GABA-positive boutons, 40.7% of receptor-positive synapses were formed by boutons that possessed low or undetectable levels of GABA. These synaptic boutons were identified in sections in which structures that

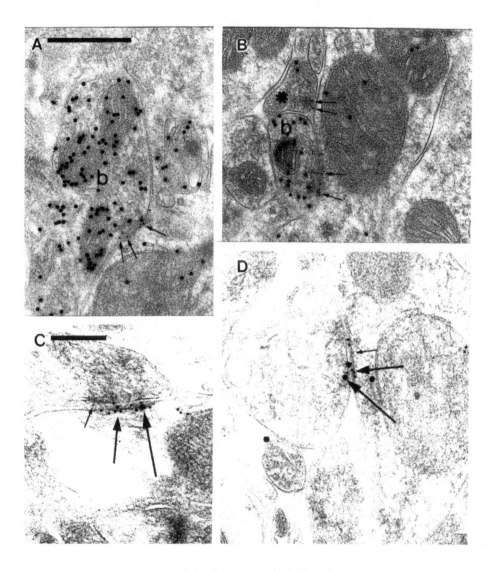

Figure 1. Localization of subunits of the GABA$_A$ receptor in the striatum.
A & B. GABA-positive boutons (b), identified by the accumulation of large immunogold particles (20 nm), form symmetrical synapses with dendrites. The synapses are positive for the β2/3 subunits of the GABA$_A$ receptor as identified by the presence of the small immunogold particles (10 nm; arrows). In **B** a second bouton (asterisk) forms a receptor-positive synapse but only possesses low levels of GABA immunolabelling.
C & D. Co-localization of different subunits of the GABA$_A$ receptor at symmetrical axodendritic synapses in the striatum. α1 subunits are identified by the 20 nm gold particles (large arrows), β2/3 subunits by the 10 nm gold particles (medium arrows) and γ2 subunits by the 5 nm gold particles (small arrows).
Scales: A & B: 0.5 μm ; C & D: 0.2 μm.

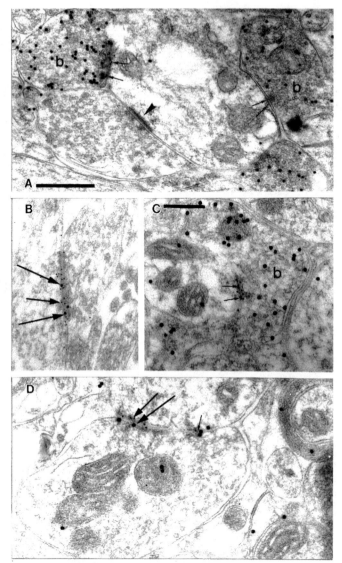

Figure 2. Localization of subunits of the GABA_A receptor in the entopeduncular nucleus and subthalamic nucleus.
A. GABA-positive boutons (b; probably derived from the striatum) in the EP, identified by the accumulation of large immunogold particles (20 nm), forming symmetrical axodendritic synapses. The synapses are positive for the β2/3 subunits of the GABA_A receptor as identified by the presence of the small immunogold particles (10 nm; arrows). Note the large, immunonegative bouton (probably derived from the STN) forming an asymmetric synapse (arrowhead) that does not possess receptor immunolabelling. **B.** A bouton in the EP forming a synapse that is positive for αl subunits (identified by the 20 nm gold particles; large arrow) and the β2/3 subunits (10 nm gold particles; medium arrow) of the GABA_A receptor. **C.** A GABA-positive bouton in the STN that is probably derived from the GP forms a symmetrical synapse that is positive for the β2/3 subunits of the GABA_A receptor (arrows). **D.** Co-localization of different subunits of the GABA_A receptor at a symmetrical axodendritic synapse in the STN. αl subunits are identified by the 20 nm gold particles (large arrows), β2/3 subunits by the 10 nm gold particles (medium arrows) and γ2 subunits by the 5 nm gold particles (small arrows).
Scales: A & B: 0.5 μm ; C & D: 0.2 μm.

were strongly labelled for GABA were identified in the close vicinity. Postsynaptic targets included spines, dendrites and perikarya.

In the GP, STN, EP, SNr and SNc the majority of boutons forming receptors positive synapses were positive for GABA (Figure 2 & 3). In the GP, EP, SNr and SNc the majority of GABA-positive boutons forming the receptor-positive synapses had morphological characteristics of boutons derived from the striatum. A smaller proportion possessed the characteristics of terminals derived from pallidal neurons (Smith et al 1998). In the STN most of them had the characteristics of pallidal terminals. Quantitative analysis in the EP, SNr and SNc revealed that, unlike the striatum, only 9.3%, 4.1% and 3.8% of boutons forming receptor-positive synapses respectively, possessed low or undetectable levels of GABA.

There are several possible explanations for the presence of the second population of axon terminals that formed $GABA_A$ receptor positive synapses i.e. those possessing low or undetectable levels of GABA in the striatum. First, it is possible that the low or undetectable levels of GABA are a technical artifact caused by the failure to maintain the antigenicity for GABA within those boutons. This however is unlikely as other boutons that were strongly positive for GABA were found in the vicinity of the boutons with the low levels of GABA, furthermore, in the EP, SNr and SNc much smaller proportions of boutons with low or undetectable levels of GABA formed receptor-positive synapses. A second possibility is that the terminals with low or undetectable levels of GABA truly represent a population of terminals that are GABAergic but are at the lower end of the spectrum in terms of their content of GABA and are simply below the level of detection in this tissue. Different populations of GABAergic terminals in the striatum may be derived from neurons that express different levels of GABA (Bolam et al., 1985; Kita and Kitai, 1988; Kawaguchi et al., 1995; Smith et al., 1998). Thus, in the striatum the terminals of GABA interneurons and the terminals of globus pallidus neurons are those terminals with high levels of GABA whereas the terminals with low or undetectable levels of GABA are the local axon terminals of the spiny projection neurons. A third possibility is that the terminals are indeed non-GABAergic, but form synapses that are positive for GABA receptor subunits. Mismatches between the putative transmitter of synaptic terminals and the receptor located within the synapse have been reported for GABA and for other receptors in the subthalamic nucleus (Clarke and Bolam, 1998) and cerebellum (Nusser et al., 1998) and there are many populations of non-GABAergic terminals in the striatum that form symmetrical. The possibility of the presence of $GABA_A$ receptors at synapses formed by these classes of terminals cannot, as yet, be excluded.

4. WHAT COMPOSITION OF $GABA_A$ RECEPTOR SUBUNITS IS EXPRESSED IN THE BASAL GANGLIA?

Quantification of the proportion of receptor-positive synapses that express immunoreactivity for $\alpha 1$, $\beta 2/3$ and $\gamma 2$ subunits was performed by photographing every immunopositive synapse in well preserved strips of an ultrathin section. Micrographs were printed at a final magnification of 33,000. The gold particles for $\alpha 1$, $\beta 2/3$ and $\gamma 2$ subunits were counted on 50 synapses. The proportion of receptor-positive synapses expressing the different subunits were calculated.

Sections immunolabelled to reveal $\alpha 1$ and $\gamma 2$ subunits of the $GABA_A$ receptor revealed labelling for both subunits although the labelling was not as robust as that obtained with the

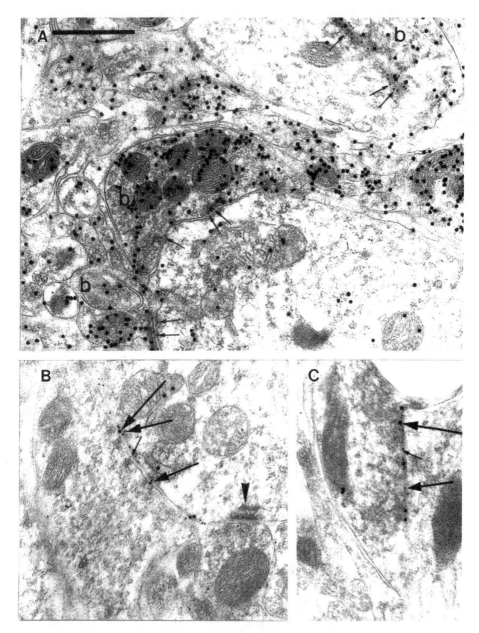

Figure 3. Localization of subunits of the GABAa receptor in the SN.
A. Symmetrical axodendritic synapses in SNc immunolabelled for the β2/3 subunits of the GABAa receptor identified by the presence of the small immunogold particles (10 run; arrows). The presynaptic boutons (b) are immunolabelled for GABA (large immunogold particles; 20 nm). **B & C.** Colocalization of different subunits of the GABAa receptor at symmetrical axodendritic synapses in the SNc **(B)** and SNr **(C)**. α1 subunits are identified by the 20 nm gold particles (large arrows), β2/3 subunits by the 10 nm gold particles (medium arrows) and γ2 subunits by the 5 nm gold particles (small arrows). Note the asymmetric synapse in **B** (arrowhead) that does not display receptor labelling. Scale: 0.5 pm.

antibodies against the β2/3 subunits. Immunogold particles were observed both on membranes and at intracellular sites but the most prominent labelling, in the form of groups of immunogold particles, occurred at symmetrical synapses. Labelling for each of the subunits was observed at symmetrical synapses in each of the regions and involved spines, perikarya, and, most frequently, dendritic shafts.

Triple-labelling for the α1, β2/3 and γ2 subunits with three different sizes of gold particles revealed the co-localization of all three subunits the GABA$_A$ receptor subunits at individual symmetrical synapses (Figures 1-3). Systematic analysis of receptor-positive synapses in EP, GP and SNr revealed similar patterns of labelling. Thus, 63-76%, 82-92% and 12-28% were positive for α1, β2/3 and γ2 subunits respectively. In contrast, in the striatum the α1 subunit is less commonly expressed and systematic analysis of receptor-positive synapses revealed that 42%, 92% and 26% were positive for α1, β2/3 and γ2, subunits respectively.

This finding corroborates previous radioligand-binding, *in situ* hybridization and immunohistochemical studies indicating the co-localization of GABA$_A$ receptor subunits in the striatum at the regional and cellular levels (Fritschy and Mohler, 1995; Caruncho et al., 1996, 1997; McKeman and Whiting, 1996; Waldvogel et al., 1997, 1998; Riedel et al., 1998). In fact the α1, β2/3 and γ2 receptor subunit configuration has been proposed as the most common for GABA$_A$ receptors in the mammalian brain. From pre-embedding immunocytochemical and *in situ* hybridisation studies (Fritschy and Mohler, 1995; Caruncho et al., 1996, 1997; McKernan and Whiting, 1996; Waldvogel et al., 1997, 1998; Riedel et al., 1998) it is evident that the α1 subunit in not expressed, or expressed at low levels, by medium spiny neurons but is expressed by small populations of striatal neurons which include GABA interneurons and a large type of projection neuron (Caruncho et al., 1996; Waldvogel et al., 1997, 1998). It is thus likely that the synapses that were positive for the α1 subunits with or without the β2/3 and γ2 subunits are formed by these latter populations of neurons. Those synapses that were negative for the α1 subunits may express some other α subunit, possibly α2, and may thus represent the synapses of spiny neurons. It is thus evident that there are differences in the GABA$_A$ receptor subunit profiles at synapses on medium sized projection spiny neurons and those on interneurons in the striatum.

5. CONCLUSIONS

The present findings demonstrate the precise localization of subunits of the GABA$_A$ receptor in relation to symmetrical synaptic specializations in the rat basal ganglia. The main conclusion that we can draw from this study is that GABA$_A$ receptors are primarily located at symmetrical synapses. The boutons that form the receptor-positive synapses generally possess high levels of GABA. In the striatum however, the boutons are heterogeneous with respect to their morphology and neurochemistry. Finally, different subunits of the GABA$_A$ receptor co-localize at the level of *individual* synapses in each of the regions of the basal ganglia studied. Thus GABA transmission mediated by GABA$_A$ receptors in the basal ganglia is likely to occur primarily at symmetrical synapses and, at some synapses, it is mediated by receptors possessing α1, β2/3 and γ2 subunits. Experiments are in progress to identify the origin of the synaptic boutons involved in these circuits.

6. ACKNOWLEDGEMENTS

The authors thank Caroline Francis, Paul Jays, Liz Norman and David Roberts for technical assistance. The work was funded by the Medical Research Council, U.K. F.F. was supported by the Japanese Ministry of Education and Saga Medical School. F.A.S. is supported by an Medical Research Council ROPA award.

7. REFERENCES

Baude, A., Z. Nusser, J.D.B. Roberts, E. Mulvihill, R.A.J. Mcilhinney and P. Somogyi (1993) The metabotropic glutamate receptor (mGluR1a) is concentrated at perisynaptic membrane of neuronal subpopulations as detected by immunogold reaction. Neuron 11:771-787.

Benke, D., M. Honer, C. Michel and H. Mohler (1996) GABAA receptor subtypes differentiated by g subunit variants: prevalence, pharmacology and subunit architecture. Neuropharmacol. 35:1413-1422.

Bernard, V., P. Somogyi, and J P. Bolam (1997) Cellular, subcellular and subsynaptic distribution of AMPA-type glutamate receptor subunits in the neostriatum of the rat. J. Neurosci. 17, 819-833.

Bevan, M.D., P.A.C. Booth, S.A. Eaton and J.P. Bolam (1998) Selective innervation of neostriatal interneurons by a subclass of neuron in the globus pallidus of the rat. J. Neurosci. 18:9438-9452.

Bolam, J.P. and B. Bennett (1995) The microcircuitry of the neostriatum. In M. Ariano, and D.J. Surmeier (ed): Molecular and Cellular Mechanisms of Neostriatal Functions. Austin, TX: R. G. Landes Company, pp. 1-19.

Bolam, J.P., D.J. Clarke, A.D. Smith and P. Somogyi (1983) A type of aspiny neuron in the rat neostriatum accumulates (^{3}H) γ-aminobutyric acid: combination of Golgi-staining, autoradiography and electron microscopy. J. Comp. Neurol. 213:121-134.

Bolam, J.P., J.F. Powell, J.-Y. Wu and A.D. Smith (1985) Glutamate decarboxylase-immunoreactive structures in the rat neostriatum. A correlated light and electron microscopic study including a combination of Golgi-impregnation with immunocytochemistry. J. Comp. Neurol. 237:1-20.

Caruncho, H.J., I. Liste and J.L. Labandeira-Garcia (1996) GABAA receptor α1-subunit-immunopositive neurons in the rat striatum. Brain Res. 722:185-189.

Caruncho, H.J., I. Liste, G. Rozas, E. López-Martín, M.J. Guerra and J.L. Labandeira-García (1997) Time course of striatal, pallidal and thalamic α1, α2 and β2/3 GABAA receptor subunit changes induced by unilateral 6-OHDA lesion of the nigrostriatal pathway. Mol. Brain Res. 48:243-250.

Clarke, N.P. and J.P. Bolam (1997) Colocalization of neurotransmitters in the basal ganglia of the rat. Brit. J. Pharmacol. 120:281 P.

Clarke, N.P. and J.P. Bolam (1998) Distribution of glutamate receptor subunits at neurochemically characterized synapses in the entopeduncular nucleus and subthalamic nucleus of the rat. J. Comp. Neurol. 397:403-420.

Cowan, R.L., C.J. Wilson, P.C. Emson and C.W. Heizmann (1990) Parvalbumin-containing GABAergic interneurons in the rat neostriatum. J. Comp. Neurol. 302:197-205.

Fritschy, J.-M. and H. Mohler (1995) GABAA-receptor heterogeneity in the adult rat brain: Differential regional and cellular distribution of seven major subunits. J. Comp. Neurol. 359:154-194.

Hanley, J.J. and J.P. Bolam (1997) Synaptology of the nigrostriatal projection in relation to the compartmental organization of the neostriatum in the rat. Neuroscience 81:353-370.

Håring, P., C. Stahli, P. Schoch, B. Takacs, T. Staehelin and H. Mohler (1985) Monclonal antibodies reveal structural homogeneity of GABAA/benzodiazepine receptors in different brain areas. Proc. Natl. Acad. Sci. USA 82:4837-4841.

Hodgson, A.J., B. Penke, A. Erdei, I.W. Chubb and P. Somogyi (1985) Antisera to γ-aminobutyric acid. I. Production and characterization using a new model system. J. Histochem. Cytochem. 33:229-39.

Jones, A., E.R. Korpi, R.M. McKernan, R. Pelz, Z. Nusser, R. Mäkelä, J.R. Mellor, S. Pollard, S. Bahn, F.A. Stephenson, A.D. Randall, W. Sieghart, P. Somogyi, A.J.H. Smith and W. Wisden (1997) Ligand-gated ion channel subunit partnerships: GABAA receptor α6 subunit gene inactivation inhibits δ subunit expression. J. Neurosci. 17:1350-1362.

Kawaguchi, Y., C.J. Wilson, S.J. Augood and P.C. Emson (1995) Striatal interneurones: chemical, physiological and morphological characterization. Trends Neurosci. 18:527-535.

Kita, H. and S.T. Kitai (1988) Glutamate decarboxylase immunoreactive neurons in rat neostriatum: Their morphological types and populations. Brain Res. 447:346-352.

Kita, H., T. Kosaka and C. W. Heizmann (1990) Parvalbumin-immunoreactive neurons in the rat neostriatum: a light and electron microscopic study. Brain Res. 536:1-15.

Kubota, Y., S. Mikawa and Y. Kawaguchi (1993) Neostriatal GABAergic interneurones contain NOS, calretinin or parvalbumin. Neuroreport 5:205208.

McKernan, R.M. and P.J. Whiting (1996) Which GABAA-receptor subtypes really occur in the brain? Trends Neurosci. 19:139-143.

Nusser, Z., W. Sieghart and P. Somogyi (1998) Segregation of different GABAA receptors to synaptic and extrasynaptic membranes of cerebellar granule cells. J. Neurosci. 18:1693-1703.

Riedel, A., W. Hartig, J.M. Fritschy, G. Bruckner, U. Seifert and K. Brauer (1998) Comparison of the rat dorsal and ventral striatopallidal system - A study using the GABAA-receptor $\alpha 1$-subunit and parvalbumin immunolabeling. Exp. Brain Res. 121:215-221.

Smith, A.D. and J.P. Bolam (1990) The neural network of the basal ganglia as revealed by the study of synaptic connections of identified neurones. Trends Neurosci. 13:259-265.

Smith, Y., M.D. Bevan, E. Shink and J.P. Bolam (1998) Microcircuitry of the direct and indirect pathways of the basal ganglia. Neuroscience 86:353-387.

Somogyi, P. and A.J. Hodgson (1985a) Antisera to g-aminobutyric acid. III. Demonstration of GABA in Golgi-impregnated neurons and in conventional electron microscopic sections of cat striate cortex. J. Histochem. Cytochem. 33:249-57.

Somogyi, P., A.J. Hodgson, I.W. Chubb, B. Penke and A. Erdei (1985b) Antisera to γ aminobutyric acid. II. Immunocytochemical application to the central nervous system. J. Histochem. Cytochem. 33:240-8.

Van der Kooy, D., D.V. Coscina and T. Hattori (1981) Is there a non-dopaminergic nigrostriatal pathway? Neuroscience 6:345-357.

Waldvogel, H.J., J.M. Fritschy, H. Mohler and R.L.M. Fault (1998) GABAA receptors in the primate basal ganglia: An autoradiographic and a light and electron microscopic immunohistochemical study of the $\alpha 1$ and $\beta 2/3$ subunits in the baboon brain. J. Comp. Neurol. 397:297-325.

Waldvogel, H.J., Y. Kubota, S.C. Trevallyan, Y. Kawaguchi, J.-M. Fritschy, H. Mohler and R.L.M. Fault (1997) The morphological and chemical characteristics of striatal neurons immunoreactive for the $\alpha 1$-subunit of the GABAA receptor in the rat. Neuroscience 80:775-792.

AFFERENT CONTROL OF NIGRAL DOPAMINERGIC NEURONS

The role of GABAergic inputs

James M. Tepper, Pau Celada, Yuji Iribe and Carlos A. Paladini*

Key words: substantia nigra, firing pattern, burst firing, pars reticulata, globus pallidus, subthalamic nucleus, autocorrelogram, disinhibition

1. INTRODUCTION

In *in vivo* extracellular recordings from anesthetized adult rats, midbrain dopaminergic neurons fire spontaneously at low rates, averaging around 4 spikes/second (Bunney et al., 1973; Guyenet and Aghajanian, 1978; Deniau et al., 1978; Tepper et al., 1982). Under these conditions, the neurons exhibit 3 distinct patterns or modes of firing. The first is that of pacemaker-like firing, characterized by very regular interspike intervals (Wilson et al., 1977; Tepper et al., 1995). The second, and most common pattern of activity *in vivo* is a random, or occasional mode (Wilson et al., 1977), characterized by long post-firing inhibition which rises smoothly into a flat autocorrelation function indicating that the remaining interspike intervals are distributed randomly, best characterized by a Poisson distribution. The third, least common mode of firing, is burst firing, in which the neurons exhibit stereotyped bursts of 2-8 action potentials in which the first intraburst interspike interval is around 60 ms, followed by progressively increasing interspike intervals and progressively decreasing spike amplitudes (Grace and Bunney, 1984). The bursts are not usually rhythmic or continuous, and occur embedded in background random single-spike firing activity. Both in anesthetized and unanesthetized rats (Freeman et al., 1985), dopaminergic neurons often switch between different firing modes, and the three firing patterns are best thought of as a continuum, with the pacemaker-like firing on one end and burst firing on the other. However, *in vitro*, the burst pattern and the random pattern are not seen. Instead, virtually all dopaminergic neurons fire in the pacemaker mode (Grace, 1987; Kang and Kitai, 1993). The absence of the burst and

---

* J.M. Tepper, P. Celada, Y. Iribe and C.A. Paladini, Center for Molecular and Behavioral Neuroscience, Rutgers University, Newark, NJ 07102.

random firing modes *in vitro* suggests very strongly that the different firing patterns of dopaminergic neurons are primarily controlled by afferent input. However, precisely which afferents are responsible for modulating the firing pattern of dopaminergic neurons has remained unclear.

Several studies have suggested that glutamatergic afferents, specifically those acting at NMDA receptors, may be responsible for the burst firing mode (see Overton and Clark, 1997 for review). However, the predominant input to dopaminergic neurons is GABAergic (Ribak et al., 1976), yet the nature of the GABAergic modulation of dopaminergic neurons is only poorly understood. In this chapter we review some of the literature on the afferent control of dopaminergic neurons, focusing on the GABAergic afferents, and describe some of our recent attempts at understanding the role of GABAergic inputs in the control of the firing pattern of nigral dopaminergic neurons.

2. GABAERGIC AFFERENTS TO SUBSTANTIA NIGRA

The best characterized GABAergic inputs to nigral dopaminergic neurons arise from the striatum and the globus pallidus (GP, Grofová, 1975; Smith and Bolam, 1990; Somogyi et al., 1981). Pharmacological studies of the striatal input have led to some contradictory results. Dopaminergic neurons express both $GABA_A$ and $GABA_B$ receptors, and respond to agonists of each with hyperpolarization (Lacey, 1993). Whereas an early *in vivo* study showed that striatal-evoked inhibition of nigral dopaminergic neurons was blocked by systemic administration of the $GABA_A$ receptor antagonist, picrotoxin (Grace and Bunney, 1985), a subsequent *in vitro* study showed the presence of $GABA_B$ IPSPs presumably originating from striatum in VTA dopaminergic neurons (Cameron and Williams, 1993). The pharmacology of the pallidal input has not been studied previously.

When recorded extracellularly *in vivo* with multi-barrel pipettes allowing local pressure application of drugs, both striatal and pallidal stimulation-evoked inhibition could be completely blocked by bicuculline or picrotoxin. In contrast, the selective $GABA_B$ antagonists, 2-OH saclofen and CGP55845A, were ineffective at blocking the evoked inhibition. However, in many cases the $GABA_B$ antagonists produced an *increase* in the evoked inhibition, and in some cases even revealed an inhibition that could not be seen in the absence of $GABA_B$ receptor blockade (Paladini et al., 1999). In previous *in vitro* studies some of the GABAergic afferents to substantia nigra dopaminergic neurons have been shown to possess presynaptic $GABA_B$ autoreceptors that inhibit stimulus-evoked GABA release and reduce the size of $GABA_A$-mediated IPSPs or IPSCs (e.g., Haüsser and Yung, 1994; Shen and Johnson, 1997).

Taken together, these data indicate that *in vivo* the postsynaptic effects of striatal and pallidal afferents to nigral dopaminergic neurons are mediated predominantly or exclusively by $GABA_A$ receptors. The role of the postsynaptic $GABA_B$ receptors remains unclear. They may be activated by extra-striatal, extra-pallidal inputs, or they may respond to GABA overflow resulting from sustained and synchronous activation of striatal and/or pallidal inputs that are difficult to elicit under *in vivo* experimental conditions (see Paladini et al., 1999 for detailed discussion). In addition, both striatal and pallidal afferents to dopaminergic neurons possess presynaptic $GABA_B$ autoreceptors that are functional *in vivo* under "normal" physiological conditions where there is sufficient GABA tone to activate these receptors thereby suppressing GABA release. In the presence of $GABA_B$ antagonists, striatal and pallidal terminal autoreceptors are disinhibited, calcium- and activity-dependent GABA release is increased thereby facilitating striatal and pallidal-evoked inhibition.

3. THERE IS AN IMPORTANT GABAERGIC INPUT TO NIGRAL DOPAMINERGIC NEURONS ORIGINATING FROM THE AXON COLLATERALS OF PARS RETICULATA PROJECTION NEURONS

Although the substantia nigra is heavily innervated from striatum and GP, it was recognized relatively early on that there might be an extra-striatal, extra-pallidal source of GABAergic innervation to the dopaminergic neurons. Intracellular recording and staining of GABAergic pars reticulata projection neurons *in vivo* revealed that their axons emitted local collaterals that arborized both in the pars compacta and the pars reticulata (Grofová et al., 1982). Large kainic acid lesions of pars reticulata reduced GAD activity in pars compacta by 30-40%, while not significantly altering GAD activity in pars reticulata. In contrast, transections anterior to substantia nigra which eliminated striatal and pallidal GABAergic inputs reduced GAD activity in pars reticulata by up to 90% while reducing GAD activity in pars compacta by only 65% (Grofová and Fonnum, 1982; see also Nitsch and Risenberg, 1988). These data suggest that while most of the GABAergic innervation of pars reticulata originates from striatum and GP, a significant proportion of the GABAergic innervation of pars compacta originates in the pars reticulata.

Early electrophysiological evidence also suggested the existence of a pars reticulata GABAergic innervation of dopaminergic neurons. Simultaneous extracellular recordings of pars compacta dopaminergic neurons and unidentified GABAergic neurons in pars reticulata showed that the two cell types fired reciprocally with one another (Grace et al., 1980). This intriguing finding was consistent with an inhibitory effect of a GABAergic pars reticulata neuron on pars compacta dopaminergic neurons. However, the dendrites of dopaminergic neurons release dopamine (Cheramy et al., 1981) and extend deeply into pars reticulata (Juraska et al., 1977; Tepper et al., 1987), and one could not rule out the possibility that the reciprocal firing obtained was due to dopaminergic inhibition of pars reticulata neurons (Timmerman and Abercrombie, 1996) rather than from a GABAergic inhibition of the dopaminergic neuron. A subsequent *in vitro* study showed that stimulation of pars reticulata in slices taken from rats transected anterior to substantia nigra several days earlier to allow striatonigral and pallidonigral afferents to degenerate resulted in IPSPs in pars compacta dopaminergic neurons (Hajüs and Greenfield, 1994). However, like the Grace et al. (1980) experiment, the identification of the pars reticulata neuron as an interneuron or a projection neuron could not be determined from these studies.

A monosynaptic GABAergic inhibition of pars compacta dopaminergic neurons by GABAergic pars reticulata projection neurons was subsequently demonstrated by stimulating the thalamus or tectum while recording from identified dopaminergic neurons *in vivo* (Tepper et al., 1995; Paladini et al., 1999). Because GABAergic pars reticulata output neurons but not dopaminergic neurons project to thalamus and/or superior colliculus and because there are no reciprocal projections from these regions back to substantia nigra, stimulation of thalamus or tectum could be used to antidromically activate pars reticulata projection neurons selectively. Antidromic spikes in reticulata projection neurons resulted in a reliable, short latency inhibition of dopaminergic neurons. The onset of inhibition was not significantly longer than the average antidromic latency of the pars reticulata neurons suggesting that the orthodromic inhibition of the dopaminergic neurons was monosynaptic (Tepper et al., 1995). Like the inhibition evoked by stimulation of striatum or GP, the inhibition of dopaminergic neurons evoked by selective activation of pars reticulata projection neurons was blocked completely by the $GABA_A$ antagonists, picrotoxin or bicuculline (Tepper et al., 1995; Paladini et al., 1999). Also similar to the case with striatal and pallidal afferents, local application of the $GABA_B$ antagonists 2-OH saclofen or CGP55845A never abolished the inhibition; in about 50% of the cases application of 2-OH saclofen or CGP55845A *increased* the duration and/or the magnitude of the inhibition, indicating the presence of functional inhibitory presynaptic $GABA_B$ receptors on the pars reticulata axon collateral terminals, as is also the case with striatal and pallidal terminals (Paladini et al., 1999).

4. ANATOMY OF THE RETICULATA-COMPACTA INTERACTION

The morphological substrates of the thalamic and tectal-evoked inhibition were identified with *in vivo* intracellular recording and filling of antidromically identified pars reticulata

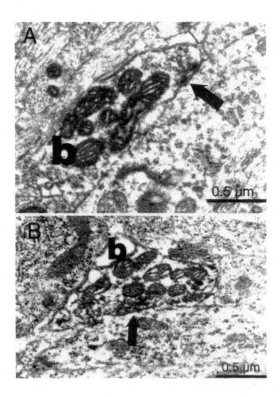

Figure 1. A. Labeled bouton from antidromically identified nigrothalamic neuron intracellularly labeled with biocytin *in vivo* makes symmetrical synapse (arrow) on presumed dopaminergic soma in pars compacta. **B.** Another labeled bouton makes a symmetrical synapse onto immunocytochemically identified proximal dopaminergic dendrite in pars compacta.

neurons (Damlama et al., 1993). As noted previously (Deniau et al., 1982; Grofová et al., 1982), the axons of reticulata projection neurons issued local collaterals that arborized within both pars reticulata and pars compacta. The local collaterals were studded with large and small varicosities that were distributed at irregular intervals. In some cases, collaterals exhibited long stretches without varicosities followed by several varicosities in close proximity to one another.

At the electron microscopic level, the varicosities proved to be synaptic boutons. They were large, approximately 1-2 μm in diameter, contained 3-8 mitochondria and were loosely packed with pleomorphic vesicles, as shown in Figure 1. In pars compacta, labeled boutons were seen to form symmetric synapses with the cell bodies and proximal dendrites of dopaminergic neurons. In one case, a labeled axon was observed to form a pericellular basket around the soma and proximal dendrites of a dopaminergic neuron, similar to that reported for pallidal inputs to dopaminergic neurons (Smith and Bolam, 1990).

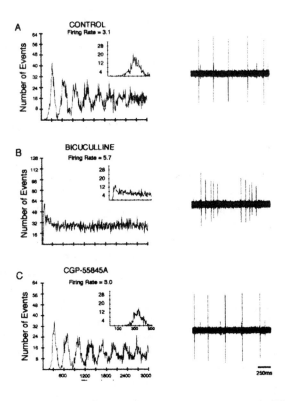

Figure 2. Effects of local administration of GABA antagonists on firing pattern of dopaminergic neuron recorded *in vivo*. **A.** Pre-drug, the neuron fires in a pacemaker pattern. **B.** Bicuculline causes dramatic shift to burst firing. **C.** After recovery from bicuculline, CGP55845A increases the regularity of firing over that of control (note the number of peaks in the autocorrelogram). From Paladini and Tepper, 1999 with permission.

5. GABAERGIC INPUT SUPPRESSES BURST FIRING IN DOPAMINERGIC NEURONS IN VIVO

How does $GABA_A$-mediated inhibition modulate the firing pattern of nigral dopaminergic neurons? When $GABA_A$ antagonists, were applied locally, all dopaminergic neurons, regardless of baseline firing rate or pattern, immediately switched to a burst firing pattern, as characterized by the shape of the autocorrelation histogram, an increase in the coefficient of variation of the interspike interval (CV), an increase in the percentage of spikes fired in bursts and the mean number of spikes per burst (Tepper et al., 1995; Paladini and Tepper, 1999), as illustrated for one representative neuron in Figure 2.

The burst firing was not due to increased firing rate; although bicuculline significantly increased the mean firing rate (from about 4.6 to 5.8 spikes/sec), neither picrotoxin nor gabazine caused a significant increase in firing rate. None of the measures of burst activity (except for burst duration) was significantly correlated with either baseline firing rate or the change in firing rate after $GABA_A$ receptor blockade (Paladini and Tepper, 1999). Local application of 2-OH saclofen or CGP55845A, never exerted these effects, but instead shifted the firing pattern away from burst firing towards more regular, pacemaker firing. This effect was presumably due to blockade of presynaptic $GABA_B$ autoreceptors on the terminals of the GABAergic inputs resulting in increased GABA release and increased activation of postsynaptic $GABA_A$ receptors (see above). These data indicate that GABAergic input to dopaminergic neurons, acting via $GABA_A$ receptors, serves to suppress burst firing *in vivo*.

6. ORIGIN OF THE GABAERGIC INPUT THAT SUPPRESSES BURST FIRING IN DOPAMINERGIC NEURONS

What is the source of the GABAergic input that suppresses burst firing in dopaminergic neurons *in vivo*? Among striatum, GP and the pars reticulata, the GP and the pars reticulata are the more likely because unlike striatal efferents, pallidal and nigral GABAergic neurons fire tonically at a high rate. To examine the involvement of the GP in the GABAergic control of dopaminergic neuron firing pattern, local infusions of muscimol or bicuculline were made into the GP in order to decrease or increase the firing rate of pallidal output neurons respectively, while recording the effects on the firing pattern of substantia nigra dopaminergic neurons in anesthetized rats (Celada et al., 1999). Pallidal inhibition led to a powerful regularizing effect on the firing pattern of dopaminergic neurons; pallidal inhibition produced significant decreases in the CV, the percentage of spikes fired in bursts and the mean number of spikes/burst, and an increase in the number of peaks in the autocorrelogram. Increases in pallidal activity led to precisely the opposite effects measured by all of these parameters (Figure 3). The changes in firing pattern were associated with unexpectedly modest and anomalous changes in firing rate; pallidal inhibition produced a 17% *decrease* in spontaneous firing rate whereas pallidal excitation led to a 40% *increase* in firing rate.

These effects on burst firing and firing rate were opposite to what would be expected for manipulating a monosynaptic GABAergic pathway from GP to the dopaminergic neurons, and suggested that the effects seen were indirect, mediated by a second inhibitory neuron interposed between GP and the dopaminergic neuron. Examination of the response of pars reticulata GABAergic neurons (some of which were identified antidromically as nigrothalamic

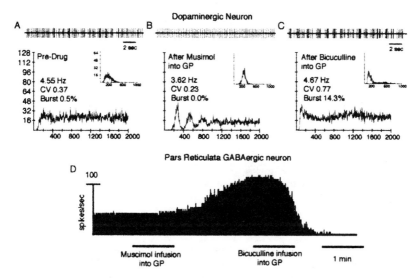

Figure 3. Effects of decreasing and increasing pallidal activity on the firing pattern of a pars compacta dopaminergic neuron and a pars reticulata neuron. **A.** Spike train, autocorrelogram and first order interval histogram of atypical dopaminergic neuron showing random firing under control conditions. **B.** After muscimol-induced inhibition of GP the firing rate decreased and the firing pattern shifted to the pacemaker mode. **C.** Subsequent infusion of bicuculline into GP increased the firing rate of the neuron and not only reversed the pacemaker effect of the prior muscimol infusion but shifted the dopaminergic neuron into the burst firing mode. **D.** Ratemeter showing effects of the same manipulations of GP firing rate on an unidentified pars reticulata GABAergic neuron. Modified from Celada et al., 1999 with permission.

neurons) to manipulation of pallidal activity were consistent with this. Unlike the dopaminergic neurons, the firing rates of reticulata GABAergic neurons were dramatically altered by increases or decreases in pallidal activity in a manner consistent with the changes being due to alteration of a monosynaptic GABAergic pathway. Increasing pallidal activity led to a complete cessation of spontaneous activity in many neurons, while decreases in pallidal activity led to increases in firing rate of over 100% (Celada et al., 1999). These data were interpreted to mean that pars reticulata GABAergic neurons (certainly the projection neurons and possibly also interneurons) comprise a principal source of the GABAergic tone that suppresses burst firing in dopaminergic neurons, and that these neurons are themselves under the control of GABAergic efferents from GP. Although it remains to be demonstrated, presumably similar effects would obtain following manipulation of striatal output.

7. INTERACTION OF SUBTHALAMIC AFFERENTS WITH INTRINSIC GABAERGIC NEURONS

The data described above point toward an important role of intrinsic GABAergic circuitry in controlling the activity of nigral dopaminergic neurons. Might this circuitry also modulate the effects of excitatory afferents? The response of dopaminergic neurons to stimulation of the subthalamic nucleus (STN) is complex. Although the output of the nucleus is strictly glutamatergic, responses recorded in dopaminergic neurons after subthalamic stimulation are a mixture of excitation, inhibition followed by excitation, or inhibition (Chergui et al., 1994; Hammond et al., 1978; Robledo and Féger, 1990; Smith and Grace, 1992), although pars reticulata GABAergic neurons almost always respond with excitation (Nakanishi et al., 1987; Robledo and Féger, 1990).

Synaptic potentials were recorded intracellularly from pars compacta dopaminergic neurons in parasagittal slices in response to stimulation of the STN as illustrated in Figure 4 (Iribe et al., 1999). STN-evoked depolarizing synaptic responses in dopaminergic neurons reversed at approximately -31 mV, intermediate between the expected reversal potential for an EPSP and an IPSP. Blockade of $GABA_A$ receptors with bicuculline caused a positive shift in the reversal potential to near 0 mV, suggesting that STN stimulation evoked a near simultaneous EPSP and IPSP. Both synaptic responses were blocked by application of the glutamate receptor antagonist, CNQX, indicating that the IPSP could not be monosynaptic and required glutamatergic excitation of a GABAergic neuron in substantia nigra.

The confounding influence of inhibitory fibers of passage from GP and/or striatum by STN stimulation was eliminated by unilaterally transecting striatonigral and pallidonigral fibers three days prior to recording. The reversal potential of STN-evoked synaptic responses in dopaminergic neurons in slices from transected animals was approximately -30 mV. Bath application of bicuculline shifted the reversal potential to ~ 5 mV as it did in intact animals, suggesting that the source of the IPSP was within substantia nigra.

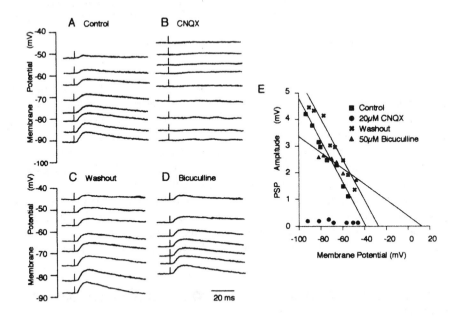

Figure 4. The IPSP component of the STN-evoked DPSP is polysynaptic. Under control conditions STN stimulation produced a DPSP with a reversal potential of -38.8 mV (A, E). Addition of CNQX completely abolished the DPSP (B,E). After a one hour wash, the DPSP returned and still exhibited a hyper-polarized reversal potential (C,E). Subsequent application of bicuculline shifted the reversal potential in the positive direction to 12.6 mV (D,E). Traces in A-D are the average of 4 single sweeps. Modified from Iribe et al., (1999) with permission.

These data indicate that electrical stimulation of the STN elicits a mixed EPSP-IPSP in nigral dopaminergic neurons due to the co-activation of an excitatory monosynaptic and an inhibitory polysynaptic connection between the STN and the dopaminergic neurons of substantia nigra pars compacta. The EPSP arises from a direct monosynaptic excitatory glutamatergic input from the STN. The IPSP arises polysynaptically, most likely through STN-evoked excitation of GABAergic neurons in substantia nigra pars reticulata which produces feed-forward $GABA_A$-mediated inhibition of dopaminergic neurons through inhibitory intranigral axon collaterals. This is likely the reason that so many of the previous attempts to examine subthalamic effects on dopaminergic neurons reported mixed excitatory and inhibitory responses.

8. SUMMARY AND CONCLUSIONS

The activity of nigral dopaminergic neurons is significantly modulated by GABAergic afferents. In addition to striatum and GP, GABAergic synapses arising from the axon collaterals of pars reticulata GABAergic projection neurons (and perhaps also interneurons) exert a powerful modulation of the firing pattern of dopaminergic neurons. All three of these GABAergic inputs appear to exert their effects predominantly or exclusively through $GABA_A$ receptors, and GABA release from all three is subject to modulation from inhibitory presynaptic $GABA_B$ autoreceptors. Pharmacological blockade of $GABA_A$ receptors, or reduction in GABAergic tone caused by the inhibition of firing of pars reticulata GABAergic neurons produces a dramatic shift in dopaminergic neuron firing pattern to the burst mode, whereas increases in firing of reticulata neurons leads to pacemaker-like firing and a reduction in bursting. The ionic mechanisms underlying the burst firing unmasked by interruption of the tonic GABAergic input remain to be determined. Nevertheless, the firing pattern of dopaminergic neurons *in vivo* is controlled to an important extent by disinhibition exerted by changes in the activity of the pars reticulata GABAergic neurons, which in turn, is modulated by other GABAergic inputs from GP and striatum.

9. ACKNOWLEDGEMENTS

Thanks to Paul Bolam for taking the electron micrographs in Figure 1 and to Elizabeth Abercrombie for comments on the manuscript. Supported by NS-34865.

10. REFERENCES

Bunney, B.S., Walters, J.R., Roth, R.H. & Aghajanian, G.K. (1973) Dopaminergic neurons: effect of antipsychotic drugs and amphetamine on single cell activity. *J. Pharmacol. and Exp.Therapeut.185:* 560-571.

Cameron D.L., & Williams, J.T. (1993) Dopamine D1 receptors facilitate transmitter release. *Nature (Lond.) 366:*344-347.

Celada, P., Paladini C.A., & Tepper, J.M. (1999) GABAergic control of rat substantia nigra dopaminergic neurons: Role of globus pallidus and substantia nigra pars reticulata. *Neuroscience 89*:813-825.

Cheramy, A., Leviel, V. & Glowinski, J. (1981) Dendritic release of dopamine in the substantia nigra. *Nature 289*: 537-542.

Chergui, K., Akaoka, H., Charlety, P.J., Saunier, C.F., Buda, M., and Chouvet, G. (1994)Subthalamic nucleus modulates burst firing of nigral dopamine neurones via NMDA receptors. *Neuroreport 5*: 1185-1188.

Damlama, M., Bolam, J.P., & Tepper J.M. (1993) Axon collaterals of pars reticulata projection neurons synapse on pars compacta neurons. *Soc. Neurosci. Abstr. 19*:1432.

Deniau, J.M., Hammond, C., Riszk, A. & Feger, J. (1978) Electrophysiological properties of identified output neurons of the rat substantia nigra (pars compacta and pars reticula): evidence for the existence of branched neurons. *Exp. Brain Res. 32*: 409-422.

Deniau, J.M., Kitai, S.T., Donoghue, J.P., & Grofova, I. (1982) Neuronal interactions in the substantia nigra pars reticulata through axon collaterals of the projection neurons. *Exp. Brain Res. 47*:105-113.

Freeman, A.S., Meltzer, L.T. & Bunney, B.S. (1985). Firing properties of substantia nigra dopaminergic neurons in freely moving rats. *Life Sciences 36*:1983-1994.

Grace, A.A. (1987) The regulation of dopamine neuron activity as determined by in vivo and in vitro intracellular recordings. In: Chiodo, L.A. & Freeman, A.S. (Ed.) *Neurophysiology of dopaminergic systems-current status and clinical perspectives*. Lakeshore Publishing Company, Grosse Pointe, pp. 1-66.

Grace, A.A. & Bunney, B.S. (1984) The control of firing pattern in the nigral dopamine neurons: Burst firing. *J. Neurosci. 4*: 2877-2890.

Grace, A.A., & Bunney, B.S. (1985) Opposing effects of striatonigral feedback pathways on midbrain dopaminergic cell activity. *Brain Res. 333*:271-284.

Grace, A.A., Hommer, D.W. & Bunney, B.S. (1980) Peripheral and striatal influences on nigral dopamine cells: Mediations by reticulata neurons. *Brain Res. Bull. 5 (Suppl 2)*: 105-109.

Grofovi, I. (1975) Identification of striatal and pallidal neurons projecting to substantia nigra. An experimental study by means of retrograde transport of horseradish peroxidase. *Brain Res. 91*:286-291.

Grofová, I., Deniau, J.M. & Kitai, S.T. (1982) Morphology of the substantia nigra pars reticulata projection neurons intracellularly labeled with HRP. *J. Comp. Neurol. 208*: 352-368.

Grofová, I., & Fonnum, F. (1982) Extrinsic and intrinsic origin of GAD in pars compacta of the rat substantia nigra. *Soc. Neurosci. Abstr. 8*:961

Guyenet, P.G. & Aghajanian, G.K. (1978) Antidromic identification of dopaminergic and other output neurons of the rat substantia nigra. *Brain Res.* 150: 69-84.

Hajós M. and Greenfield S.A. (1994) Synaptic connections between pars compacta and pars reticulata neurones: Electrophysiological evidence for functional modules within the substantia nigra. *Brain Res. 660*: 216-224.

Hammond, C., Deniau, J.M., Rizk, A., and Féger, J. (1978) Electrophysiological demonstration of an excitatory subthalamonigral pathway in the rat. *Brain Res.* 151: 235-244.

Haüsser M.A. and Yung W.H. (1994) Inhibitory synaptic potentials in guinea-pig substantia nigra dopamine neurones in vitro. *J. Physiol. (Loud) 479*: 401-422.

Iribe, Y., Moore, K., Pang, K.C. & Tepper, J.M. (1999) Subthalamic stimulation-induced synaptic responses in nigral dopaminergic neurons in vitro. *(J. Neurophysiol., in press)*.

Juraska, J.M., Wilson, C.J., & Groves, P.M. (1977) The substantia nigra of the rat: A Golgi study. *J. Comp. Neurol. 172*:585-599.

Kang, Y. & Kitai, S.T. (1993) Calcium spike underlying rhythmic firing in the dopaminergic neurons of the rat substantia nigra. *Neurosci. Res. 18*: 195-207.

Lacey, M.G. (1993) Neurotransmitter receptors and ionic conductances regulating the activity of neurones in substantia nigra pars compacta and ventral tegmental area. In: G.W. Arbuthnott and P.C. Emson (Eds.) *Chemical Signalling in the Basal Ganglia, Progress in Brain Research, Volume 99*, pp. 251-276.

Nakanishi, H., Kita, H., and Kitai, S.T. (1987) Intracellular study of rat substantia nigra pars reticulata neurons in an in vitro slice preparation: electrical membrane properties and response characteristics to subthalamic stimulation. *Brain Res.* 437: 45-55.

Overton, P. & Clark, D. (1997) Burst firing in midbrain dopaminergic neurons. *Brain Res. Rev. 25*: 312334

Paladini, C.A., Celada, P., & Tepper, J.M. (1999) Striatal, pallidal, and pars reticulata evoked inhibition of nigrostriatal dopaminergic neurons is mediated by $GABA_A$ receptors in vivo. *Neuroscience 89*:799-812.

Paladini, C.A., & Tepper, J.M. (1999) $GABA_A$ and $GABA_B$ antagonists differentially affect the firing pattern of substantia nigra dopaminergic neurons in vivo. *Synapse 32*:165-176.

Ribak, C.E., Vaughn, J.E., Saito, K., Barber, R. & Roberts, E. (1976). Immunocytochemical localization of glutamate decarboxylase in rat substantia nigra. *Brain Res.* *116*:287-298.

Robledo, P., and Féger, J. (1990) Excitatory influence of rat subthalamic nucleus to substantia nigra pars reticulata and the pallidal complex: electrophysiological data. *Brain Res.* 518: 47-54.

Shen K.Z. and Johnson S.W. (1997) Presynaptic $GABA_B$ and adenosine Al receptors regulate synaptic transmission to rat substantia nigra reticulata neurones. *J Physiol. (Lond)* *505*: 153-163.

Smith, I.D., and Grace, A.A. role of the subthalamic nucleus in the regulation of nigral dopamine neuron activity. *Synapse* 12: 287-303, 1992.

Smith, Y., & Bolam, J.P. (1990) The output neurones and the dopaminergic neurones of the substantia nigra receive a GABA-containing input from the globus pallidus in the rat. *J. Comp. Neurol.* *296*:47-64.

Somogyi P. Bolam J.P. Totterdell S. and Smith A.D. (1981) Monosynaptic input from the nucleus accumbens-ventral striatum region to retrogradely labeled nigrostriatal neurones. *Brain Res.* *217*:245-263.

Tepper, J.M., Martin, L.P. & Anderson, DR. (1995) GABA*A* receptor-mediated inhibition of nigrostriatal dopaminergic neurons by pars reticulata projection neurons. *J. Neurosci.* *15*:3092-3103.

Tepper, J.M., Sawyer, S.F., & Groves, P.M. (1987) Electrophysiologically identified nigral dopaminergic neurons intracellularly labeled with HRP: Light microscopic analysis. J. *Neuroscience.* 7:2794-2806.

Tepper, J.M., Nakamura, S., Spanis, CW., Squire, L.R., Young, S.J., & Groves, P.M. (1982) Subsensitivity of catecholaminergic neurons to direct acting agonists after single or repeated electroconvulsive shock. *Biol. Psychiat.* *17*:1059-1070.

Timmerman, W. & Abercrombie, E.D. (1996) Amphetamine-induced release of dendritic dopamine in substantia nigra pars reticulata: D1-mediated behavioral and electrophysiological effects. *Synapse* 23: 280-291.

Wilson, C.J., Young, S.J. & Groves, P.M. (1977). Statistical properties of neuronal spike trains in the substantia nigra: Cell types and their interactions. *Brain Res. 136*: 243-260.

THE ROLE OF STRIATAL ADENOSINE A_{2A} RECEPTORS IN MOTOR CONTROL OF RATS

Wolfgang Hauber, Jens Nagel, Partic Neuscheler, Michael Koch[*]

1. INTRODUCTION

It is well known that adenosine and ATP play a central role in the energy metabolism of all organisms. However, adenosine is also an important endogenous modulator which is involved in the regulation of numerous physiological processes. It is released in part from metabolically active cells by facilitated diffusion. In addition, adenosine is generated in the extracellular space by rapid and quantitative ectoenzymatic degradation of released ATP.[1]

In the central nervous system adenosine is no classical neurotransmitter, because it is not stored in synaptic vesicles or released in quanta. Extracellular adenosine which is rapidly removed by reuptake mechanisms or by adenosine deaminase degradation modulates neuronal activity through distinct cell-surface receptors coupled to G-proteins. They are termed as A_1, A_{2A}, A_{2B} and A_3 receptors which have a differential distribution in the central nervous system. A_1 receptors are widely distributed in the brain, with high densities in the hippocampus, cerebellum and neocortical areas.[2] In contrast, A_{2A} receptors have a much more restricted distribution with the striatum showing the highest density of A_{2A} receptors in the brain.[3]

The striatum is the main input structure of the basal ganglia and receives glutamatergic afferents from the cerebral cortex and the thalamus as well as a dopaminergic projection from the substantia nigra pars compacta (Fig.1). The information conveyed by the glutamatergic and dopaminergic inputs to the striatum are transmitted through a direct and an indirect pathway to the basal ganglia output nuclei, in rats the entopeduncular nucleus and the substantia nigra pars reticulata, and from there to the thalamus and to the brainstem. These two pathways have opposite effects on the neuronal activity of both output structures, which - in turn - exert tonic inhibitory effects on motor activity.

A_{2A} receptors are co-expressed with dopamine D_2 receptors and are mainly restricted to striatal neurons projecting to the globus pallidus. These striatopallidal neurons are part of the indirect pathway. In contrast, striatonigral and striatoentopeduncular neurons constituting

---

[*] W. Hauber, J. Nagel, and P. Neuscheler, Department of Animal Physiology, University of Stuttgart, D-70550 Stuttgart; M. Koch, Department Neuropharmacology, University of Bremen, D-28334 Bremen, Germany.

The Basal Ganglia VI
Edited by Graybiel *et al.*, Kluwer Academic/Plenum Publishers, 2002

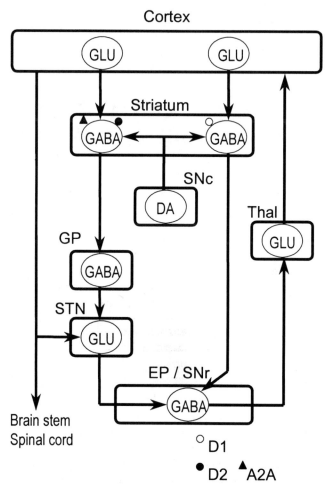

Figure 1. The localization of adenosine A_{2A} and dopamine D_1 and D_2 receptors on striatal projection neurons. D_2 and A_{2A} receptor are mainly co-localized on GABAergic striatopallidal neurons. EP: Entopeduncular nucleus; GP: Globus pallidus; STN: Subthalamic nucleus; SNr: substantia nigra pars reticulata; SNc: Substantia nigra pars compacta, Thal: Thalamus; GLU: Glutamate, DA: Dopamine.

the direct pathway to the output structures are regulated predominantly by A_1 and dopamine D_1 receptors (see [4] for a recent review).

In the work presented here we summarize our recent studies on the role of adenosine to modulate motor behaviour of rats. In particular, we focus on striatal interactions between adenosine, dopamine and glutamate. It is well known that a blockade of D_2 receptor mediated transmission in the caudate-putamen, the dorsal part of the striatum, induces in rodents massive problems to initiate and to execute spontaneous motor behaviour, an impairment termed akinesia. The degree of akinesia can be easily measured in the catalepsy test by placing a rat in an abnormal body position, for example placing both forelegs on an elevated bar or placing the animal on a vertical grid. Akinetic animals can not leave such a body position rapidly and the latency to initiate spontaneous motor behaviour is used as a measure for the degree of akinesia.

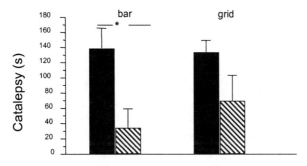
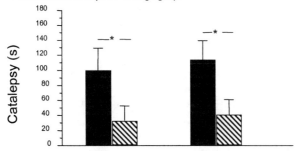

Figure 2. Intracaudate infusion of the A_{2A} antagonist MSX-3 reversed akinesia induced by systemic pretreatment with the D_1 antagonist SCH23390 or the D_2 antagonist raclopride (from Ref. [8]). Akinesia was measured in rats using the bar and grid catalepsy test.

2. RESULTS AND DISCUSSION

Caffeine as well as theophylline are among the most frequently consumed psychoactive drugs world-wide. They exert their effects through an unselective blockade of adenosine A_1 and A_2 receptors. Systemic co-administration of theophylline potently reversed akinesia induced by systemic[5] or striatal[6] D_2 receptor blockade. 8-(3-Chlorostyryl)caffeine (CSC), a selective A_{2A} receptor antagonist, produced similar anti-akinetic effects.[6] Blockade of striatal D_1 receptors by microinfusion of SCH23390 also induced strong akinesia which was dose-dependently reversed by systemic co-administration of theophylline, and to a markedly lesser extent by the A_{2A} antagonist CSC.[6] Although it is likely that motor stimulant effects induced by a systemic adenosine A_{2A} receptor blockade are mediated by the caudate-putamen, direct evidence for this hypothesis is lacking, because selective adenosine receptor antagonists have a very low water-solubility which seriously limits their use for *in vivo* studies with intracerebral infusions. Using a novel water-soluble compound MSX-3,[7] we were now for the first time able to directly investigate motor effects induced by an intrastriatal A_{2A} receptor blockade. We found that striatal blockade of A_{2A} receptors by MSX-3 produced a moderate motor stimulation measured by an enhanced sniffing activity.[8] Furthermore, akinesia induced by a systemic[8] or intrastriatal (Hauber and Neuscheler, unpublished observations) D_1 or D_2 receptor blockade was potently reversed by simultaneous A_{2A} receptor blockade in the caudate-putamen (Fig. 2).

On the other hand, we found that a stimulation of A_{2A} receptors in the caudate-putamen induced a pronounced akinesia[9] which was as severe as after D_2 receptor blockade or dopamine depletion. Therefore, using intracerebral microinfusions we were able to confirm a key role of striatal A_{2A} receptors in control of motor behaviour suggested by previous studies using systemic drug administration.

In view of the selective receptor expression, motor effects of an A_{2A} receptor blockade or stimulation are most probably mediated by striatopallidal neurons (Fig.1). Although A_{2A} and D_1 receptors are not located on the same striatal efferent neurons, antagonistic effects of the direct and indirect pathways at the level of the output structures might explain the reversal of D_1 receptor mediated akinesia by a striatal A_{2A} receptor blockade. In contrast, the reversal of D_2 receptor mediated akinesia could be due to direct antagonistic interactions between D_2 and A_{2A} receptors on striatopallidal neurons. From studies at the cellular level there is compelling evidence that striatal A_{2A} receptors can directly interact with D_2 receptors in the membrane and regulate their binding and signal transduction characteristics.[10] Accordingly, stimulation of striatal A_{2A} receptors was found to decrease the affinity of D_2 receptors.[11] Recent evidence suggests that stimulation of striatal A_{2A} receptors suppressed GABA-mediated inhibitory postsynaptic currents.[12] Thus, apart from interacting with D_2 receptors, another major action of striatal A_{2A} receptors might be the inhibition of GABA release from recurrent collaterals of projection neurons.[13] Anatomical data do not clearly show whether A_{2A} receptors are localized predominantly on dendrites of striatal projection neurones or local axon collaterals.[14] However, *in vivo* microdialysis data showing that striatal GABA release was not altered by A_{2A} receptor ligands are at variance with this notion (see [15] for review). Thus the cellular localisation and precise mechanism of action of striatal A_{2A} receptor is still unclear in detail. This issue needs to be clarified, because it is important for the question to which extent effects of adenosine are dependent or independent of dopamine D_2 receptor transduction mechanisms.

Striatopallidal neurons which represent a major neuronal substrate of the motor effects of A_{2A} receptor ligands also receive strong glutamatergic inputs from the cortex and thalamus. Given that the akinetic effects induced by striatal A_{2A} receptor stimulation are mediated by an increased neuronal activity of striatopallidal neurons, a simultaneous blockade of excitatory glutamatergic input might reduce the increased activity of striatopallidal neurons and antagonize akinesia caused by an A_{2A} receptor stimulation. Indeed, a blockade of glutamate receptors of the N-methyl-D-aspartate (NMDA) receptor subtype reduced A_{2A} receptor mediated akinesia. Dizocilpine, an uncompetitive antagonist as well as CGP37849, a competitive antagonist at NMDA receptors were found to potently reverse akinesia induced by intrastriatal infusion of the adenosine receptor agonist CGS21680.[9] Overall, our data show that in the caudate-putamen A_{2A} receptors, D_2 receptors, and NMDA receptors contribute to motor control probably by regulating the activity of striatopallidal neurons.

These data imply that A_{2A} receptor antagonists might be useful for the treatment of Parkinson's disease.[13] There is considerable support in keeping with this notion, as KF17837, another selective A_{2A} receptor antagonist, counteracted akinesia induced by dopamine receptor blockade and dopamine depletion. In addition, this A_{2A} antagonist acted synergistically with the dopamine precursor L-DOPA under conditions both of reserpine-induced dopamine depletion and haloperidol–induced dopamine receptor blockade.[16] Furthermore, A_{2A} receptor antagonists potentiated rotation induced by D_1 or D_2 agonists in rats with unilateral nigrostriatal lesions.[17] A_{2A} receptor antagonists have been also shown to improve the motor disability of marmosets pretreated with the selective nigral toxin MPTP.[18] Importantly, no evidence of dyskinesia was reported after treatment with A_{2A} receptor antagonists. Finally, in Parkinsonian patients it has been shown that unselective adenosine receptor antagonists caused measurable improvements in both subjective and objective scores of the disease.[19] Thus, A_{2A} receptor antagonists might have a potential as anti-Parkinson drugs either given alone or as supplements to L-DOPA therapy.[13]

Adenosine also plays a role opposite to dopamine in the nucleus accumbens, part of the ventral subdivision of the striatum. Comparable to the caudate-putamen, A_{2A} receptor stimulation in the nucleus accumbens produced similar effects as an intra-accumbal D_2 receptor blockade and inhibited locomotor activity in mice[20] and rats.[21] The nucleus accumbens represents an important interface between limbic and motor structures. It is involved in the conversion of motivation into action,[22] but also in control of sensorimotor gating mechanisms.[23] We investigated a role of adenosine in regulating the prepulse inhibition of acoustic startle responses which is considered to represent a sensorimotor gating mechanism. Briefly, prepulse inhibition refers to the phenomenon that the amplitude of a startle response is reduced by prior presentation of a weak prepulse which precedes the startling pulse. It is well known that prepulse inhibition is disrupted by a stimulation of dopaminergic system, in particular within the nucleus accumbens. In the initial experiments, we found that an unselective blockade of adenosine receptors by theophylline produced a similar reduction of prepulse inhibition as did apomorphine, while stimulation of A_1 receptors by the selective agonist N^6-cyclopentyladenosine (CPA) even enhanced prepulse inhibition.[24] We have not yet specified the neural substrate of this effect, but we suppose that it might be mediated by a presynaptic, A_1 receptor-dependent control of dopaminergic terminals in nucleus accumbens. Finally, stimulation of A_{2A} receptors in the nucleus accumbens reversed apomorphine-induced disruption of prepulse inhibition.[25] Collectively, these data point to a role A_{2A} receptors in the nucleus accumbens in control not only of locomotor activity but also of sensorimotor gating.

Interestingly, in schizophrenic patients there is a disruption of sensorimotor gating processes. Therefore the loss of prepulse inhibition in rats with hyperactive dopamine system is considered to be a useful model for studying the neurobiology of impaired sensorimotor gating in schizophrenic patients.[26] Of course, this idea implies a role of the central dopaminergic system in the pathophysiology of schizophrenia. This hypothesis is partly based on the assumption that a blockade of D_2 receptors in the nucleus accumbens mediates the antipsychotic action of neuroleptics or, at least, their therapeutic effect on positive symptoms in schizophrenia and that a D_2 receptor blockade in the caudate-putamen mediates their extrapyramidal side effects (see [27] for a recent review). Needless to say that the dopamine hypothesis of schizophrenia is still a matter of debate as dysfunction of numerous other transmitter systems has been implicated in the pathophysiology of schizophrenia as well. Furthermore, there is also evidence supporting the cerebral cortex and cortical dopaminergic mechanisms as a critical site for the mechanisms underlying the action of antipsychotics.[28] Bearing these caveats in mind, our finding that A_{2A} receptor agonists reversed apomorphine-induced disruption of prepulse inhibition adds further evidence to the idea that that A_{2A} agonists might have a potential in the treatment of schizophrenia.[27]

In conclusion, adenosine is an important neuronal mediator which plays a fundamental role to modulate motor behaviour in the basal ganglia. The present studies reveal some important interactions between adenosine, dopamine and glutamate as well as intrinsic basal ganglia pathways which might mediate motor effects induced by a stimulation or a blockade of A_{2A} receptors. Our findings on interactions between dopamine and adenosine support the view that adenosine A_{2A} receptor ligands might be potentially useful therapeutics for treatment of basal ganglia disorders involving a dopaminergic dysfunction as Parkinson's disease and perhaps also schizophrenia.

3. ACKNOWLEDGEMENTS

The excellent technical assistance of Stefan Nitschke is gratefully acknowledged.

4. REFERENCES

1. Linden, J., *Purinergic systems*, in *Basic Neurochemistry*, G.J. Siegel, *et al.*, Editors. 1994, Raven Press: New York. p. 401-416.
2. Rivkees, S.A., S.L. Price, and F.C. Zhou, Immunohistochemical detection of A1 receptors in rat brain with emphasis on localization in the hippocampal formation, cerebral cortex, cerebellum and basal ganglia, *Brain Res.*, 677 (1995) 193-203.
3. Ongini, E. and B.B. Fredholm, Pharmacology of adenosine A2a receptors, *Trends Pharmacol. Sci.*, 17 (1996) 364-372.
4. Ferré, S., B.B. Fredholm, M. Morelli, P. Popoli, and K. Fuxe, Adenosine-dopamine receptor-receptor interactions as an integrative mechanism in the basal ganglia, *Trends Neurosci.*, 20 (1997) 482-487.
5. Casas, M., S. Ferre, T. Guix, and F. Jane, Theophylline reverses haloperidol-induced catalepsy in the rat, *Biol. Psychiatr.*, 24 (1988) 642.
6. Nagel, J. and W. Hauber, *Adenosine differentially modulates motor behaviour mediated by striatal dopamine D1 and D2 receptors*, in *New Neuroethology on the move*, N. Elsner and R. Wehner, Editors. 1998, Thieme: Stuttgart. p. 68.
7. Müller, C.E., R. Sauer, Y. Maurish, R. Huertas, F. Fülle, K.-N. Klotz, J. Nagel, and W. Hauber, A2A-selective adensoine receptor antagonists: Development of water-soluble prodrugs and new tritiated radioligands, *Drug Dev. Res.*, 45 (1998) 190-197.
8. Hauber, W., J. Nagel, R. Sauer, and C.E. Müller, Motor effects induced by a blockade of adenosine A2A receptors in the caudate-putamen, *Neuroreport*, 9 (1998) 1803-1806.
9. Hauber, W. and M. Münkle, Stimulation of adenosine A2a receptors in the rat striatum induces catalepsy that is reversed by antagonists of N-methy-D-aspartate receptors, *Neurosci. Lett.*, 196 (1995) 205-208.
10. Fuxe, K., S. Ferre, M. Zoli, and L.F. Agnati, Integrated events in central dopamine transmission as analyzed at multiple levels. Evidence for intramembrane adenosine A(2A) dopamine D-2 and adenosine A(1) dopamine D-1 receptor interactions in the basal ganglia, *Brain Res. Rev.*, 26 (1998) 258-273.
11. Ferré, S., G. Von Euler, B. Johansson, B.B. Fredholm, and K. Fuxe, Stimulation of high-affinity adenosine A2 receptors decreases the affinity of dopamine D2 receptors in rat striatal membranes, *Proc. Natl. Acad. Sci. USA*, 88 (1991) 7238-7241.
12. Mori, A., T. Shindou, M. Ichimura, H. Nonaka, and H. Kase, The role of adenosine A2a receptors in regulating GABAergic synaptic transmission in striatal medium spiny neurons, *J. Neurosci.*, 16 (1996) 605-611.
13. Richardson, P.J., H. Kase, and P. Jenner, Adenosine A2A receptor antagonists as new agents for the treatment of Parkinson's disease, *Trends Pharmacol. Sci.*, 18 (1997) 338-344.

14. Rosin, D.I., A. Robeva, R.I. Woodard, P.G. Guyenet, and J. Linden, Immunohistochemical localization of adenosine A2A receptors in the rat central nervous system, *J. Comp. Neurol.*, 401 (1998) 163-186.
15. Fredholm, B.B. and P. Svenningsson, Striatal adenosine receptors- where are they? What do they do?, *Trends Pharmacol. Sci.*, 19 (1998) 46-47.
16. Kanda, T., S. Shiozaki, J. Shimada, F. Suzuki, and J. Nakamura, KF17837: a novel selective adenosine A(2A) receptor antagonist with anticataleptic activity, *Eur. J. Pharmacol.*, 256 (1994) 263.
17. Morelli, M., S. Fenu, A. Pinna, and G. Di Chiara, Adenosine A(2) receptors interact negatively with dopamine D-1 and D-2 receptors in unilaterally 6-hydroxydopamine-lesioned rats, *Eur. J. Pharmacol.*, 251 (1994) 21-25.
18. Kanda, T., T. Tashiro, Y. Kuwana, and P. Jenner, Adenosine A(2A) receptors modify motor function in MPTP-treated common marmosets, *Neuroreport*, 9 (1998) 2857-2860.
19. Mally, J. and T.W. Stone, The effect of theophylline on parkinson symptoms, *J. Pharm. Pharmacol.*, 46 (1994) 515-518.
20. Barraco, R.A., K.A. Martens, M. Parizon, and H.J. Normile, Role of adenosine A2a receptors in the nucleus accumbens, *Prog. Neuropsychopharm. Biol. Psychiatr.*, 18 (1994) 545-549.
21. Hauber, W. and M. Münkle, Motor depressant effects mediated by dopamine D2 and adenosine A2a receptors in the nucleus accumbens and the caudate-putamen, *Eur. J. Pharmacol.*, 323 (1997) 127-131.
22. Groenewegen, H.J., C.I. Wright, and A.V.J. Beijer, *The nucleus accumbens: Gateway for limbic structures to reach the motor system?*, in *Emotional Motor System*, G. Holstege, R. Bandler, and C.B. Saper, Editors. 1996, Elsevier Science Pub.: 1000 AE Amsterdam. p. 485-511.
23. Koch, M., The neurobiology of startle., *Progr. Neurobiol.* 56 (1999) 1-22.
24. Koch, M. and W. Hauber, Regulation of sensorimotor gating by interactions of dopamine and adenosine in the rat, *Behav. Pharmacol.*, 9 (1998) 23-29.
25. Hauber, W. and M. Koch, Adenosine A2a receptors in the nucleus accumbens modulate prepulse inhibition of the startle response, *Neuroreport*, 8 (1997) 1515-1518.
26. Swerdlow, N.R., D.L. Braff, N. Taaid, and M.A. Geyer, Assessing the validity of an animal model of deficient sensorimotor gating in Schizophrenic patients, *Arch. Gen. Psychiatry*, 51 (1994) 139-154.
27. Ferré, S., Adenosine-dopamine interactions in the ventral striatum. Implications for the treatment of schizophrenia, *Psychopharmacology*, 133 (1997) 107-120.
28. Lidow, M.S., G.V. Williams, and P.S. Goldman-Rakic, The cerebral cortex: a case for a common site of action of antipsychotics., *Trends. Pharmacol. Sci.*, 19 (1998) 136-140.

A NOVEL NEUROMODULATOR IN BASAL GANGLIA

M. Clara Sañudo-Peña*

1. INTRODUCTION

Cannabinoids are a family of compounds that naturally occur in the marihuana plant *(Cannabis sativa)* and produce their effects through recently cloned G-coupled cannabinoid receptors (Matsuda et al. 1990; Munro et al. 1993). Of the myriad of physiological actions of cannabinoids, the motor effects seem to be mediated by the neural CB1 cannabinoid receptor (Rinaldi-Carmona et al. 1994, Compton et al. 1996). In general, activation of CB1 receptors by cannabinoid agonists inhibits neurotransmission through various mechanisms: by inhibition of adenylate cyclase thus decreasing the cellular levels of cAMP (Howlett et al 1995), by activation of K+ channels thus decreasing the excitability of the cell (Deadwyler et al. 1993), and by inhibition of N/Q Ca^{2+} channels thus inhibiting neurotransmitter release (Mackie et al 1992).

2. MOTOR ACTIONS OF CANNABINOIDS IN THE BASAL GANGLIA

A complex pattern of opposite effects on movement is observed when cannabinoids are administered into basal ganglia nuclei. Cannabinoids activated movement when microinjected into the striatum (Sañudo-Peña et al. 1998a) or the substantia nigra reticulata (Sañudo-Peña et al. 1996). The opposite action on movement, inhibition, was obtained when they were microinjected into the globus pallidus (Sañudo-Peña and Walker 1998) or the subthalamic nucleus (Miller et al. 1998). Similar complexity of cannabinoid effects on movement within each single basal ganglia nucleus is observed (Sañudo-Peña et al. 1996, 1998a,b,c; Sañudo-Peña and Walker 1997, 1998, see below).

2.1. Distribution of CB1 Cannabinoid Receptors in the Basal Ganglia

The basal ganglia constitutes one neural substrate for the well-known motor actions of cannabinoids (Romero et al. 1995) since it expresses the highest levels of CB 1 receptors in

---

* M.C. Sañudo-Peña, The Alan M. Schrier Research Laboratory, Department of Psychology, Brown University, Providence, RI 02912, USA.

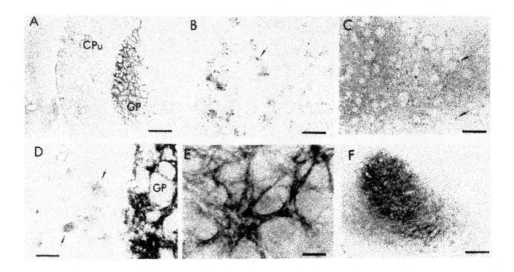

Figure 1. A) Coronal section of the rat brain at the septo-fimbrial level with a very densely stained GP, while the CPU and the cortex are moderately stained. In the CPU and the GP, there is a lateral-to-medial density gradient, the lateral region being the more densely labeled. B) In the CPU, there are numerous elliptical moderately stained CB1 immunoreactive neurons, 10-15 μm in long axis, with scant punctuated cytoplasm, and unindented unstained nuclei. They are in the size range and shape typical of medium-sized spiny neurons. The number of these neurons is higher in the rostral and lateral part of the CPU. C)In the medial part of the rostral CPU, there are numerous intensely stained immunoreactive fiber bundles coursing medially and caudally into the GP D) As they approach the GP, the bundles group together and become larger. E) In the GP, a dense fine unbeaded CB1-like immunoreactive nerve fiber meshwork is traversed by the large immunonegative fascicles. Besides the GP, other projection areas of the CPU, such as the entopeduncular nucleus, and the SNr where the CB1-like inunoreactivity is shown as fine dots due to cross section of the projection fibers (F), also show intense CB 1-like immunoreactivity. In some parasagital sections (not shown), an almost continuous band of CB1 immunoreactivity can be seen from the CPU through the GP, entopeduncular nucleus to the SNr. The immunoreactivity in these three target areas occurred in unbeaded fine axons; no immunoreactive neurons were found in these areas (21). Scale bars=500μm, (A); 200μm, (C); 501μm (B-F).

the brain (Herkenham et al. 1991a,b; Tsou et al. 1998)(Fig. 1) as well as the enzyme inolved in the degradation of endogenous cannabinoids (Tsou et al. 1999). In contrast, the levels of CB 1 receptors are low in brainstem which may explain the low toxicity of cannabinoid agonists, an attractive quality for putative therapeutic uses. The striatum (CPU) and subthalamic nucleus (Sth), input structures of the basal ganglia and major inhibitory and excitatory sources, respectively (Graybiel 1990), express mRNA for the cannabinoid receptor (Mailleux and Vanderhaegen 1992). Both output nuclei, the globus pallidus (GP) and substantia nigra pars reticulata (SNr), contain the brain's highest levels of CB 1 receptors which are located on input terminals from the striatum and subthalamic nucleus (Herkenham et al. 1991 a; Sañudo-Peña and Walker 1997). The cells in the GP and SNr are tonically active as is the excitatory input they receive from the subthalamic nucleus (Robledo and Feger 1991) and serve to excite or inhibit the production of movement, respectively. On the other hand, the striatal input to the output nuclei is mainly silent (Wilson 1993).

2.2. Motor Effects of Cannabinoids in the SNr

Turning correlates well with the cellular activation or inhibition of basal ganglia nuclei. Cellular activation in the SNr leads to inhibition of movement when it occurs bilaterally and ipsilateral turning when it occurs unilaterally. Conversely, inhibition of the SNr increases movement when the treatment is bilateral and produces contralateral turning when the treatment is unilateral. Intranigral administration of the cannabinoid agonist CP55,940 produced contralateral rotation as compared to vehicle (Fig. 2). At the time this study was done (Sañudo-Peña et al. 1996), the only proposed mechanism of cannabinoid action in the substantia nigra was the inhibition of GABA release from striatonigral terminals (Miller et al. 1995). However, this mechanism cannot account for the observed effect because any treatment that disinhibits SNr neurons would inhibit movement, but the result obtained was the opposite. The contralateral turning observed implies that the cannabinoid agonist somehow inhibits SNr neurons.

The location of mRNA for the cannabinoid receptor in the Sth is consistent with a possible expression of the receptor at the terminals in the SNr. This suggested an alternative site of action for the cannabinoid agonist within the SNr, because an inhibitory action of cannabinoids on the release of the excitatory transmitter glutamate from subthalamonigral terminals would inhibit SNr neurons and produce contralateral turning. Further support for this possibility stems from the fact that while the striatonigral input to the substantia nigra reticulata is mainly silent, the subthalamonigral input is tonically active. Therefore, the pri-

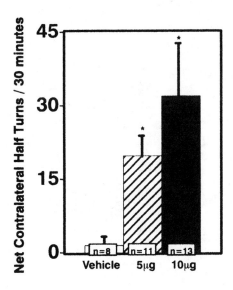

Figure 2. Effect of unilateral microinjections of various doses of the cannabinoid agonist CP 55,940 into the substantia nigra pars reticulata on turning behavior; * significantly different from vehicle group, p< 0.05).

mary action of a cannabinoid under basal conditions would be at the active subthalamic nucleus input.

2.2.1 Cannabinoid Actions at Subthalamonigral Terminals

We tested the possibility of an effect of cannabinoids on the subthalamic influence in the SNr using electrophysiological methods (Sañudo-Peña and Walker 1997). In these experiments an electrode was placed in the SNr to obtain single cell extracellular recordings in this structure while stimulating the Sth. A knife cut was performed rostral to the location of the Sth to disrupt the striatonigral pathway, an assumed site of action of cannabinoids in the SNr (Fig.3).

The firing rate of SNr neurons was increased by stimulation of the Sth (Fig. 4A). The cannabinoid agonist WIN 55,212-2 blocked the increased firing in SNr neurons produced by stimulation of the subthalamus (Fig. 4B,C). It is clear that this effect was mediated by cannabinoid receptors, because it was blocked by the cannabinoid antagonist SR141716A (Fig. 4C). These results indicate that cannabinoids dampen the excitatory input to the SNr from the subthalamic nucleus.

Cannabinoids also inhibit the actions of dopamine D2 agonists in the substantia nigra (Fig.5). Administration of the dopamine D2 agonist quinpirole into the SNr induced ipsilateral rotation. Presumably, this occurs by inhibition of dopamine neurons through a D2 autoreceptor mechanism at the dendrites of dopamine neurons which invade the SNr. The cannabinoid agonist reversed this ipsilateral rotation and further induced contralateral rotation similar to that produced by the cannabinoid agonist alone, again apparently showing a cannabinoid action at the very active subthalamonigral input, which as noted above, would be the basis for the cannabinoid actions under basal conditions, while the striatonigral input is mainly silent (Sañudo-Peña et al. 1996).

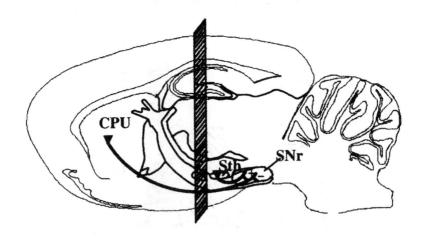

Figure 3. Schematic of the preparation employed to study the action of cannabinoids at the subthalamic input to the SNr. Drawing of the rat brain in sagittal section showing the known inputs to the SNr that contain cannabinoid receptors and the level at which the knife cut was performed (hatched vertical plane). The striatonigral pathway is inhibitory, whereas the subthalamonigral input excitatory.

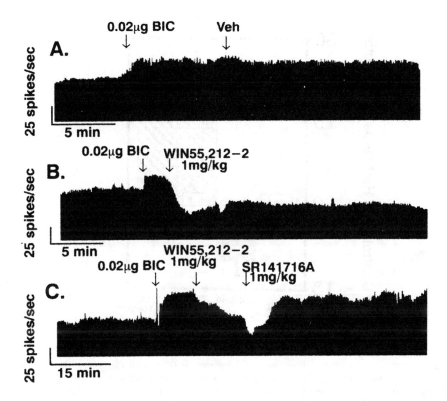

Figure 4. Firing rate histograms obtained from three different neurons as follows: a) Microinjection of bicuculline (0.01-0.02 μg/0.2-0.4μl) in the Sth caused long lasting excitation of SNr neurons; b) WIN55,212-2 (1mg/kg, i.v.) significantly attenuated this effect; c) SR141716A (1mg/kg, i.v.) antagonized the effect of WIN55,212-2. The short-lasting decrease in the firing rate apparent immediately following administration of the antagonist could not be attributed to the drug because similar shortlasting decreases were seen in other animals in the absence of the antagonist.

2.2.2. Cannabinoid Actions at Striatonigral Terminals

Cannabinoids also control the striatonigral pathway by presynaptic inhibition. Intranigral administration of the dopamine D1 agonist SKF82958 induced contralateral rotation (Fig.6A), an effect known to be due to the stimulation of GABA release from striatonigral terminals where dopamine D1 receptors are located. The coadministration of the cannabinoid agonist together with the dopamine D1 agonist partially reversed the contralateral rotation induced by the dopamine D1 agonist, due to the opposite effects of cannabinoid and dopamine D1 receptors on striatonigral terminals (Sañudo-Peña et al. 1996).

Supporting data is found in an electrophysiological study (Miller et al. 1995) (Fig. 6B) in which the cannabinoid agonist WIN 55,212-2 blocked the inhibition of the firing of SNr neurons produced by electrical stimulation of the CPU, while no effect was observed after administration of the inactive enantiomer WIN 55,212-3.

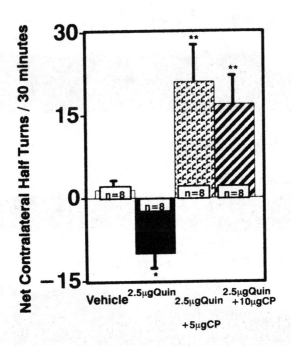

Figure 5. Effect of unilateral microinjections of the dopamine D2 agonist Quinpirole alone or together with the cannabinoid agonist CP 55,940 into the substantia nigra pars reticulata on turning behavior ** significantly different from the rest of the groups*, "significantly different from the vehicle group and the group treated with quinpirole). The cannabinoid reversed the ipsilateral turning induced by the dopamine D2 agonist.

2.2.3. Simultaneous Action of Cannabinoids on Subthalamic and Striatal Terminals at the SNr

The coadministration of the cannabinoid agonist together with the dopamine D1 agonist only partially reversed the contralateral rotation induced by the dopamine D1 agonist. Some contralateral rotation remained, which was similar in magnitude to that induced by the cannabinoid alone (Fig. 6A). This suggests a dual effect of cannabinoids, first on the striatonigral pathway where they counteract the effect of the dopamine D1 agonist releasing GABA from striatonigral terminals, and second in inhibiting the tonically active subthalamonigral input reflecting the basal effect of cannabinoids.

2.3. Motor Effects of Cannabinoids in the GP: Actions at Subthalamopallidal and Striatopallidal Terminals

As was the case for the SNr, it appears that the output neurons of the globus pallidus are inhibited by local administration of a cannabinoid which in this structure leads to ipsilateral rotation (Fig. 7A)(Sañudo-Peña and Walker 1998).

A NOVEL NEUROMODULATOR IN BASAL GANGLIA

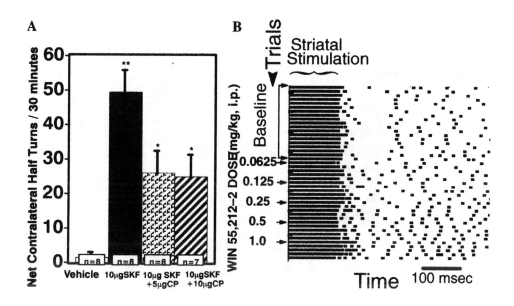

Figure 6. A. Effect of unilateral microinjections of the D_1 dopamine receptor agonist SKF82958 alone or together with the cannabinoid agonist CP 55,940 into the substantia nigra pars reticulata on turning behavior (* significantly different from the vehicle group and the group treated with SKF 82958, **significantly different from the rest of the groups, $p< 0.05$). The cannabinoid significantly reduced the contralateral turning induced by the dopamine D_1 agonist. **B.** Rasterplot showing the effect of WIN 55,212-2 on the inhibition of a substantia nigra pars reticulata neuron evoked by electrical stimulation of the striatum. Each square represents the time of occurrence of a single action potential. Each row represents one stimulation trial. Striatal stimulation produced a brief inhibition of activity in the substantia nigra pars reticulata which was more pronounced during the first 100 ms following stimulation. This is seen as the lack of action potentials (squares) in the top part of the record (period prior to drug administration). WIN 55,212-2 injections began following the 20th trial (cumulative doses of 0.0625-1.0 mg/kg) and were repeated every fifth trial. Note the reversal of the inhibition following injection of WIN 55,212-2.

The inhibition of pallidal neurons by the cannabinoid and in turn the induction of ipsilateral turning is consistent with the finding that cannabinoids enhance the catalepsy produced by pallidal microinjections of GABA agonists (Pertwee and Wickens 1991) and a study showing loss of cannabinoid receptors in the globus pallidus of Huntington's disease patients in early stages of the disease (Richfield and Herkenham 1994). A loss of inhibition of the pallidal output neurons is associated with the chorea observed in early stages of the disease (Albin et al. 1989). These findings are consistent in indicating that cannabinoids inhibit pallidal output neurons and that this would be expected to result in the ipsilateral turning observed. As noted above, inhibition of the excitatory input from subthalamic nucleus terminals appears to account for the turning behavior produced by microinjections of cannabinoids in the substantia nigra (Sañudo-Peña and Walker 1997). A similar mechanism may

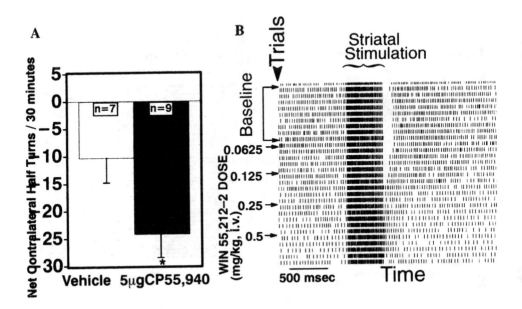

Figure 7. A. Effect of unilateral microinjections of the cannabinoid agonist CP 55,940 into the globus pallidus on turning behavior; * significantly different from vehicle group, p< 0.05). B. Rasterplot showing the effect of WIN55,212-2 on the inhibition of a GP neuron evoked by electrical stimulation of the CPU. Each tick represents the time of occurrence of a single action potential. Each row represents one stimulation trial. CPU stimulation produced a brief inhibition of activity in the GP which was more pronounced during the first 100 ms following stimulation. WIN 55,212-2 injections began following the 20th trial (cumulative doses of 0.0625-0.5 mg/kg) and were repeated every fifth trial. Note the reversal of the inhibition following injection of WIN55,212 -2.

account for the actions of cannabinoids in the globus pallidus, since it also receives a major excitatory input from the subthalamic nucleus. In support of this possibility, glutamate antagonists enhanced the catalepsy induced by cannabinoids (Kinoshita et al. 1994) and induced ipsilateral rotation when injected into the globus pallidus (Yamaguchi et al. 1986). Therefore, the site of action of cannabinoids in the globus pallidus under basal conditions seems to be the same as that observed in the substantia nigra reticulata: the tonically active subthalamic input. In a study very similar to that reported above for the substantia nigra reticulata, cannabinoids administered in the globus pallidus blocked the inhibitory action of striatal stimulation on the activity of globus pallidus neurons (Fig. 7B)(Miller et al. 1996). It appears that cannabinoids thus modulate both the striatopallidal and subthalamopallidal pathways.

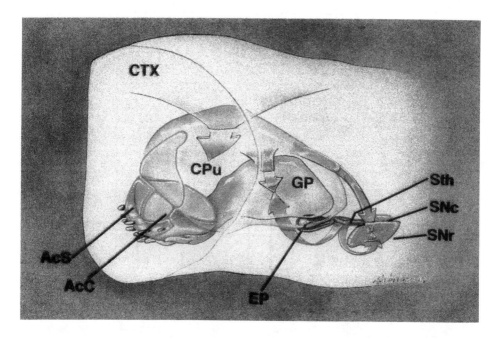

Figure 8. Model of cannabinoid action in the output nuclei of the basal ganglia. The inhibitory striatal (CPU) input is not tonically active (light arrows) while the opposite is true for the excitatory subthalamic (Sth) input (dark arrows). Cannabinoids act on both inputs and the effect on motor function would depend upon the level of activity of each one.

2.4. Modulatory Action of Cannabinoids in Basal Ganglia Output Nuclei

In conclusion, cannabinoids show similar modulatory actions at both output nuclei of the basal ganglia (GP and SNr), where they act upon both the major excitatory and the major inhibitory input (Fig.8). The entopeduncular nucleus shares a similar input arrangement to the GP and SNr. Thus, similar cannabinoid actions would be expected in this nucleus. The effect on motor function would thus depend upon which input is active at the time, and thus would serve to return the system toward basal levels of activity.

3. SYSTEMIC ACTIONS OF CANNABINOIDS ON MOVEMENT

When a cannabinoid is administered systemically it decreases movement at very low doses in an autoreceptor like effect followed by a dose dependent stimulating effect on activity that is interrupted by the appearance of rigidity and catalepsy (Fig. 9) (Sañudo-Peña

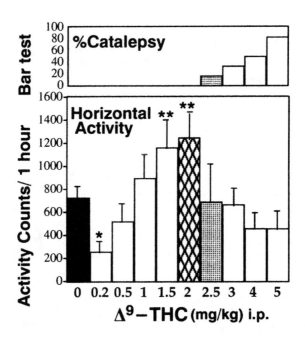

Figure 9. Dose-curve of systemic administration of Δ^9-THC effects on horizontal activity in rats. Note the increase in activity with relatively low doses (1-2 mg/kg) of the cannabinoid agonist. (* significantly different from the rest of the groups except the ones receiving 4 or 5mg/kg of Δ^9-THC, $p < 0.05$, **significant ly different from the rest of the groups except the one receiving 1mg/kg of Δ^9 THC > $p < 0.05$).

et al. 1999). We propose that cannabinoids have an activational role in movement that is overridden at higher doses by the major modulatory actions of these compounds counteracting opposing systems. Recently, a study of CB1 receptor knockout mice revealed a decrease in the general activity of these animals suporting an activational role of CB1 receptors on movement (Zimmer et al. 1999).

If we compare the systemic actions of cannabinoids with what we observed within basal ganglia nuclei there appears to be a close parallel. For instance, in the SNr, the subthalamic input is tonically active and thus would be the primary determinant of the action of a cannabinoid agonist. As mentioned above the inhibitory action on neurotransmitter release from these terminals by a cannabinoid would induce movement. As the dose of the cannabinoid is increased the secondary action of cannabinoids blocking the GABAergic phasically active striatal transmission would produce an opposite inhibitory effect on movement. The simultaneous increase and decrease in motor output may be expressed as rigidity and catalepsy.

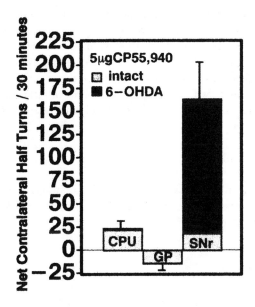

Figure 10. Effect of unilateral microinjections of 5 µg of the cannabinoid agonist CP 55,940 into the striatum (n=15), globus pallidus (n=12, or substantia nigra reticulata (n=13), on turning bahavior in rats with unilateral 6-OHDA dopaminergic lesions of the nigrostriatal pathway (dark shading). The three groups are significantly different from their respective vehicle group (origin line), p < 0.05. The standard error of the mean for the vehicle control group in each structure is; CPU: ±6.85 (n=7), GP: ±5.79 (n=8), SNr: ±8.21 (n=13). The levels of turning of intact rats after unilateral microinjections of 5 µg of the cannabinoid agonist CP 55,940 into the striatum, globus pallidus, or substantia nigra reticulata are represented for the interest of comparison (light shading).

4. MOTOR EFFECTS OF CANNABINOIDS IN A RAT MODEL OF PARKINSON'S DISEASE

Finally, the actions of cannabinoids were studied in the 6-hydroxydopamine rat model of Parkinson's disease (Fig. 10) (Sañudo-Peña et al. 1998). Animals with unilateral lesions of the dopaminergic innervation of the basal ganglia were compared to intact animals. The administration of a cannabinoid agonist into the GP of lesioned animals induced the same relative amount of turning and in the same direction as in intact animals. However, in the SNr, the contralateral turning induced by the cannabinoid agonist increased over nine-fold in the lesioned compared to the intact animals. This result is in accordance with a cannabinoid action at the subthalamonigral pathway, because after 6-OHDA lesions, stimulation of the subthalamic nucleus leads to a greatly exaggerated excitatory response in the substantia nigra reticulata, but not in other areas innervated by the subthalamic nucleus, such as the globus pallidus (Robledo and Feger 1991). Our data also rule out a dopaminergic involvement on the motor effects of cannabinoids in the basal ganglia, which is in accordance with a lack of cannabinoid markers on dopaminergic neurons (Herkenham et al. 1991b; Mailleux and Vanderhaegen 1992; Tsou et al. 1998) and suggests a therapy other than lesions (DeLong 1990) to specifically counteract Sth hyperactivity.

Furthermore, as mentioned above, cannabinoid agonists oppose the action of dopaminergic drugs acting on both main families of dopamine receptors in the SNr in intact animals, but this effect does not occur in animals with dopamine lesions (Sañudo-Peña et al.). Therefore, we hypothesize that the inhibitory action of cannabinoids on subthalamonigral terminals may be significant for movement disorders that stem from overactivity of the subthalamonigral pathway, which includes Parkinson's disease and suggests the potential for a coadjunctive treatment with cannabinoids and dopamine agonists.

5. ACKNOWLEDGEMENT

I would like to thank N.I.H. for the financial support (DA12999) and for the gift of Δ^9-THC, Sanofi Recherche (Montpellier, France) for the gift of SR141716A, and Pfizer, (Groton, CT) for the generous gift of CP55,940.

6. REFERENCES

Albin R.L., Young A.B., Penney J.B. 1989. The functional anatomy of basal ganglia disorders, Trends Neurosci, 12: 366-375.

Compton D.R., Aceto M.D., Lowe J., Martin BR. 1996. In vivo characterization of a specific cannabinoid receptor antagonist (SR141716A): inhibition of D9-tetrahydrocannabinol-induced responses and apparent agonist activity. J.Pharm.Exp.Ther. 277:586-594.

Deadwyler S.A., Hampson R.E., Bennett B.A., Edwards T.A., Mu J., Pacheco M.A., Ward S.J., Childers S.R. 1993. Cannabinoids modulate potassium current in culture hippocampal neurons. Receptors Channels, 1: 121-134.

DeLong MR. Primate models of movement disorders of basal ganglia origin. 1990. Trends Neurosci., 13: 281-285.

Graybiel A.M. Neurotransmitters and neuromodulators in the basal ganglia. 1990. Trends Neurosci., 13: 244-253.

Herkenham M., A. B. Lynn, B.R. de Costa, E. K Richfield. 1991a. Neuronal localization of cannabinoid receptors in the basal ganglia of the rat. Brain Res., 547: 267-274.

Herkenham M., A.B. Lynn, M.R. Johnson, L.S. Melvin, B.R. de Costa, K.C. Rice. 1991b. Characterization and localization of a cannabinoid receptor in rat brain: a quantitative in vitro autoradiographic study. Journal of Neuroscience, 11: 563-583.

Howlett A.C. Pharmacology of cammabinoid receptors. 1995. Ann. Rev. Pharmacol. Toxicol., 35: 607-634.

Kinoshita H, Hasegawa T, Kameyama T, Yamamoto I, Nabeshima T. 1994. Competitive NMDA antagonists enhance the catalepsy induced by delta 9-tetrahydrocannabinol in mice. Neurosci. Lett., 174: 101-104.

Mackie K., Hille B. 1992. Cannabinoid inhibit N-type calcium channels in neuroblastoma-glioma cells. Proc. Nail. Acad. Sci., 89: 3825-3829.

Mailleux P., J. J. Vanderhaeghen. 1992. Distribution of neuronal cannabinoid receptor in the adult rat brain: A comparative receptor binding radioautography and in situ hybridization histochemistry. Neurosci., 48: 655-688.

Matsuda L.A., Lolait SJ., Brownstein M., Young A., Boner T.I. 1990. Structure of a cannabinoid receptor and functional expression of the cloned cDNA, Nature, 346: 561-564.

Miller A., Sañudo-Peña M.C., Walker J.M. 1998. Ipsilateral turning behavior induced by unilateral microinjections of a cannabinoid into the rat subthalamic nucleus. Brain Res., 793: 7-11.

Miller A., J. M. Walker. 1995. Effects of a cannabinoid on spontaneous and evoked neuronal activity in the substantia nigra pars reticulata. Eur. J. Pharmacol., 279: 179-185.

Miller A., J. M. Walker. 1996. Electrophysiological effects of a cannabinoid on neural activity in the globus pallidus. Eur. J. Pharmacol., 304: 29-35.

Munro S., Thomas K.L., Abu-Shaar M. 1993. Molecular characterization of a peripheral receptor for cannabinoids, Nature, 365: 61-65.

Pertwee R.G. Wickens A.P 1991. Enhancement by chlordiazepoxide of catalepsy induced in rats by intravenous or intrapallidal injections of enantiomeric cannabinoids, Neuropharmacology, 30: 237-244.

Richfield E.K., Herkenham M. 1994. Selective vulnerability in Huntington's disease: preferential loss of cannabinoid receptors in lateral globus pallidus.Ann. Neurol., 36: 577-84.

Rinaldi-Carmona M, Barth F, Heaulme M, Shire D, Calandra B, Congy C, Martinez S, Maruani J, Neliat G, Caput D, et al. 1994. SR141716A, a potent and selective antagonist of the brain cannabinoid receptor. FEBS Lett., 350: 240-244.

Robledo E, Feger J. 1991. Acute monoaminergic depletion in the rat potentiates the excitatory effect of the subthalamic nucleus in the substantia nigra pars reticulata but not in the pallidal complex. J. Neural. Transm., 86: 115-126.

Romero J., de Miguel R., García-Palomero E., Fenández-Ruiz J.J., Ramos J.A. 1995. Time-course of the effects of anandamide, the putative endogenous cannabinoid receptor ligand, on extrapyramidal function. Brain Res., 694: 223-232.

Sañudo-Peña M.C., Patrick S.L., Patrick R.L., Walker J.M. 1996. Effects of intranigral cannabinoids on rotational behavior in rats: interactions with the dopaminergic system, Neurosci. Lett. 206: 21-24.

Sañudo-Peña M.C., Walker J.M. 1997. Role of the subthalamic nucleus on cannabinoid actions in the substantia nigra of the rat, J. Neurophysiol, 77: 1635-1638.

Sañudo-Peña M.C., Walker J.M. 1998. Effects of intrapallidal cannabinoids on rotational behavior in rats: interactions with the dopaminergic system, Synapse 28: 27-32.

Sañudo-Peña M.C., Force M., Tsou K., Miller A.S., Walker J.M. 1998a. Effects of intrastriatal cannabinoids on rotational behavior in rats: interactions with the dopaminergic system, Synapse 30: 221-226.

Sañudo-Peña, M.C., Saundra L. Patrick, Robert L. Patrick and J. Michael Walker. 1998b. Cannabinoid control of movement in the basal ganglia in an animal model of parkinson disease. Neuroscience Letters, 248: 171-174.

Sañudo-Peña M.C., J.M. Walker. 1998. A novel neurotransmitter system involved in the control of motor behavior by the basal ganglia. Ann. NY Acad. Sci., 860 "Neuronal mechanisms for generating locomotor activity", 475-479.

Sañudo-Peña M. C., K. Tsou, J.M. Walker. 1999a. Motor Actions of cannabinoids in the basal ganglia output nuclei. Life Sciences special issue on Endogenous Cannabinoids, 65: 703-713.

Sañudo-Peña M.C., Fride E. 2002. Marijuana and movement disorders. In Biology of Marijuana: From Gene to Behavior. Ed. by E.S. Onaivi, Integra Software Services, Pondicherry, India, 205-233.

Sañudo-Peña M.C., J. Romero, J.J. Fernandez Ruiz, J. M. Walker. Activational role of cannabinoids on movement. European Journal of Pharmacology, 391 (2000) 269-74.

Tsou K., Brown S., Mackie K., Sañudo-Peña M.C., Walker J.M. 1998. Immunohistochemical distribution of cannabinoid CB1 receptors in the rat central nervous system. Neuroscience, 83: 393-411.

Tsou K., M. I. Nogueron, S. Muthian, M.C. Sañudo-Peña, C. J. Hillard, D. G. Deutsch, and J. M. Walker, 1998b, Fatty acid amide hydrolase that degrades anandamide is located preferentially in large neurons in the rat central nervous system as revealed by immunohistochemistry. Neuroscience Letters, 254,137-140.

Wilson C.J. 1993. The generation of natural firing patterns in neostriatal neurons. In GW Arbuthnott, PC Empson (Eds.) Chemical signalling in the basal ganglia, Progr. Brain Res., 99: 277-298.

Yamaguchi K., Nabeshima T., Kameyama T. 1986. Role of dopaminergic and GABAergic mechanisms in discrete brain areas in phencyclidine-induced locomotor stimulation and turning behavior. J. Pharmacobiodyn., 9: 975-98.

Zimmer A, Zimmer AM, Hohmann AG, Herkenham M, Bonner TI., 1999, Increased mortality, hypoactivity, and hypoalgesia in cannabinoid CBI receptorknockout mice.Proc Natl Acad Sci U S A, 96, 5780-5

Index

Adenosine: *see* Striatal adenosine receptors and motor control (in rat)
Afferent innervation, role in organotypic cultures, striatal patch/matrix organization development, 388–390
Age level, visuo-spatial attention, P3 component, 68–71
Aging: *see* Glutamate-dopamine interactions during aging (in rat)
Akinesia: *see* Antipsychotic drug-induced rigidity
Anterograde tracing, of corticostriatal afferents, striatal patch/matrix organization development, 386–387
Antipsychotic drug-induced rigidity, 519–525
    discussed, 523–524
    methods, 520–521
    overview, 519–520
    results, 521–523
Antisera
    glutamate transporter localization (in monkey), 600
    subsynaptic localization of group I mGlu receptors, 568–569
Aperiodic behavior, tremor, 53–55
Apomorphine
    dopamine-induced rotation and firing rate in BG output ganglia, 539
    globus pallidus pathophysiology, 18–19
Apomorphine responses, 23–31
    discussed, 28–29
    methods, 25
    overview, 23–25
    results, 26–28
Arm movements: *see* Bimanual repetitive arm movements; Movement inhibition and next sensory state prediction; Pedunculopontine tegmental nucleus blockade (effects on arm movements in monkey)

Axon protection, glial cell line-derived neurotrophic factor (GDNF) effects (rodent), 121–123

Barrel cortex neuron classification, 399–409
    discussed, 406–407
    methods, 401–402
    overview, 399–401
    results, 402–405
Basal ganglia
    epileptic seizure control (animal models), 172–173; *see also* Epileptic seizure control (animal models)
    interspecies variations, 139
    multiunit activity of basal ganglia (monkey), 97–106; *see also* Multiunit activity of basal ganglia (monkey)
    stereotactic surgery and, 3–14; *see also* Stereotactic surgery
Basal ganglia reward system: *see* Cognitive decision processes
Basal ganglia thalamic distribution, 455–462
    discussed, 460–461
    methods, 456
    overview, 455
    results, 456–460
        cerebellothalamic/basal ganglia distribution compared, 459–460
        cerebellothalamic distribution, 459
        nigrothalamic distribution, 456–458
        pallidothalamic distribution, 458–459
Behavior: *see* Noradrenaline and motivated behaviors; Tonically active neurons and predictable sensory events (monkey striatum); Tonically active neurons and switch in behavioral set (monkey striatum)

Behavioral analysis
  multiunit activity of basal ganglia (monkey), 99
  pedunculopontine tegmental nucleus blockade
    (effects on arm movements in monkey), 152
  procedural learning (dopamine influence), 313, 314
  subthalamic nucleus lesions (in hemiparkinsonian
    marmoset), 140–141, 143–145, 146–147
Behavioral task: *see* Sequential stages of behavioral
    task (putamental neurons in monkey)
Bilateral posteroventral pallidal stimulation, clinical
    observations, 6
Bilateral subthalamic nucleus (STN) stimulation,
    clinical observations, 7
Bimanual repetitive arm movements, 43–50
  apparatus and tasks, 45
  data analysis, 45–46
  discussed, 48
  overview, 43–44
  results, 47
    levodopa effects, 47
    pallidotomy effects, 47
  subjects, 44–45
Border cells, white matter laminae and, globus
    pallidus pathophysiology, 16–18

CAG repeat; *see also* Neuronal dysfunction (mouse)
  neuronal dysfunction (mouse), 107–108
  neuronal intranuclear inclusions, 84–85
Cannabinoids, 661–673
  basal ganglia output nuclei, 669
  globus pallidus, 666–669
  Parkinson's disease (rat model), 671–672
  receptor distribution in basal ganglia, 661–663
  subthalamic nucleus, 663–666
  systemic actions, 669–671
Catecholaminergic neurons, dopamine depleting
    lesions and, 477–487; *see also* Dopamine
    depleting lesions
Cell body protection, glial cell line-derived
    neurotrophic factor (GDNF) effects (rodent),
    121–123
Centralized selection architecture, selection problem,
    259
Cerebellothalamic distribution: *see* Basal ganglia
    thalamic distribution
Cerebellum-linked thalamo-cortical paths: *see*
    Thalamo-cortical paths (pallidum and
    cerebellum-linked)
C-Fos
  MAPkinase role in cortically-induced immediate
    early gene (IEG) induction, 560–563
  transynaptic induction of, striatal patch/matrix
    organization development, 387–388
Cholinergic interneurons: *see* Striatal cholinergic
    interneurons (dopamine actions)

Cingulate motor areas: *see* Corticostriatal projections
    from cingulate motor areas
Classically conditioned task, tonically active neurons
    and predictable sensory events (monkey
    striatum), 351–352
Clozapine: *see* Antipsychotic drug-induced rigidity
Cocaine self-administration, 547–555
  accumbal mechanisms, 552–553
  firing patterns, 549–552
    phasic changes, 550–552
    tonic firing rate, 549–550
  methods, 548–549
  overview, 547–548
Cognition, visuo-spatial attention, 61–81; *see also*
    Visuo-spatial attention
Cognitive activation studies, striatal function in
    obsessive compulsive disorder, 89–90
Cognitive decision processes, 303–310
  asymmetrical basal ganglia reward system, 308–309
  decision theory, 304–308
  discussion, 309
  overview, 303
Cognitive neuroscience, striatal function in obsessive
    compulsive disorder, 90
Coincidence detection
  information processing in cortico-striato-nigral
    circuits (in rat), 204–207
  striatal regions, dopamine and ensemble coding,
    241–242
Contralateral rotation, nicotine and striatal
    glutamatergic function (in 6-OHDA lesioned
    rats), 591
Control of spiking: *see* Spiking control in striatal
    and pallidal neurons
Corticostriatal afferents, anterograde tracing of,
    striatal patch/matrix organization
    development, 386–387
Corticostriatal circuity, striatal function in OCD,
    88–89
Corticostriatal projections from cingulate motor
    areas, 419–428; *see also* Striatal patch/matrix
    organization development
  discussed, 425–426
  materials and methods, 420–422
  overview, 419–420
  results, 422–425
Corticostriatal stimulation: *see* MAPkinase role in
    cortically-induced immediate early gene
    (IEG) induction
Cortico-striato-nigral circuits (in rat), information
    processing in, 199–207; *see also* Information
    processing in cortico-striato-nigral circuits (in
    rat)
Cortico-subthalamo-pallidal hyperdirect pathway (in
    monkey), 217–223

INDEX 677

Cortico-subthalamo-pallidal hyperdirect pathway (in monkey) (cont.)
   discussed, 220–222
   methods, 218–220
   overview, 217–218
   results, 220

Decision processes: see Cognitive decision processes
Dendritic changes: see Medium spiny neuron dendritic changes (in weaver mouse striatum)
Difference late positivity, visuo-spatial attention, ERP components, 71–73
Difference negativity, visuo-spatial attention, ERP components, 71
Distributed selection architecture, selection problem, 258–259
Dopamine depleting lesions, 477–487
   discussed, 482–484
   methods, 478–479
   overview, 477–478
   results, 479–482
Dopamine effects: see Antipsychotic drug-induced rigidity; Multisecond oscillations in STN (dopamine effects); Noradrenaline and motivated behaviors; Procedural learning (dopamine influence); Selection (dopamine effects); Striatal cholinergic interneurons (dopamine actions)
Dopamine and ensemble coding, 237–244
   nucleus accumbens, 238–241
      ensemble coding, 238
      up and down states, 239–241
   overview, 237
   striatal regions, coincidence detection in, 241–242
Dopamine-glutamate interactions: see Glutamate-dopamine interactions during aging (in rat)
Dopamine-induced rotation and firing rate in BG output ganglia, 537–545
   discussed, 540–544
   methods, 538–539
   overview, 537–538
   results, 539–540
Dopamine receptor (D_1 and D_2) differential modulation, 527–535
   discussed, 533–534
   methods, 528–530
   overview, 527–528
   results, 530–533
Dopamine receptor-deficient (D_2) mutant mice, 507–518
   discussion, 514–515
   functional D_2 protein expression, 511–512
   overview, 507–511
   presynaptic DA function, 512–514
Dopamine receptor stimulation effects, 465–476

Dopamine receptor stimulation effects (cont.)
   discussed, 473–474
   firing pattern changes, 469–471
   firing rate changes, 466–469
   multisecond oscillations, 471–473
   overview, 465–466
Dopaminergic agonist effects on pallidal neurons, 499–505
   discussed, 504
   methods, 500–501
   overview, 499–500
   results, 501–503
Dopaminergic innervation of frontal lobe, differential reductions, in monkey: see MPTP effects on dopaminergic innervation of frontal lobe (in monkey)
Dopaminergic neurons, 641–651; see also Gamma-aminobutyric acid (GABA) receptor control of dopaminergic neurons
Dorsal prefrontal loop, basal ganglia function, 3
Drug addiction: see Cocaine self-administration

Electroencephalography (EEG), visuo-spatial attention, 64–65
Electron microscopy, nicotine and striatal glutamatergic function (in 6-OHDA lesioned rats), 590–591
Electrophysiological recordings
   dopaminergic agonist effects on pallidal neurons, 500–501
   multiunit activity of basal ganglia (monkey), 99–100
   organotypic co-culture of SN, PPN, and STN, 439, 443–446
   surround inhibition, 192–193
Enkephalin-expressing cells (rodent globus pallidus), 411–417
   discussed, 415–416
   materials and methods, 412–413
   overview, 411–412
   results, 413–415
Ensemble coding: see Dopamine and ensemble coding
Epileptic seizure control (animal models), 169–178
   basal ganglia role, 172–173
   discussed, 174–175
   initiation and control circuits, 170–171
   nigro-collicular projection, 174
   overview, 169–170
   substantia nigra role, 171
Event-related potentials, visuo-spatial attention, 61–81; see also Visuo-spatial attention
Excitatory cortical inputs: see Cortico-subthalamo-pallidal hyperdirect pathway (in monkey)
Experimental Huntington's disease, neuronal dysfunction (mouse), 107–115

Experimental Parkinsonism
  cannabinoids, 671–672
  glial cell line-derived neurotrophic factor (GDNF) effects on nigrostriatal dopamine neurons (rodent), 117–130; *see also* Glial cell line-derived neurotrophic factor (GDNF) effects (rodent)
  multiunit activity of basal ganglia (monkey), 97–106; *see also* Multiunit activity of basal ganglia (monkey)
External globus pallidus (GPe), 33–41
  discussed, 39
  methods and subjects, 33–37
  overview, 33
  recordings, 16
  results, 37–39

Feature-integration theory, visuo-spatial attention, 62
Feed-back/feed-forward, surround inhibition, 194–195
Fos (C-fos), transynaptic induction of, striatal patch/matrix organization development, 387–388; *see also* C-Fos
Fos immunoreactivity, dopamine-induced rotation and firing rate in BG output ganglia, 540
Functional imaging, stereotactic surgery, 7–10; *see also* Stereotactic surgery

Gamma-aminobutyric acid$_A$ (GABA$_A$) localization (in human), 623–630
  discussed, 628–629
  materials and methods, 624–625
  overview, 623–624
  results, 625–628
Gamma-aminobutyric acid$_A$ (GABA$_A$) localization (in rat), 631–639
  discussed, 637
  expression in basal ganglia, 635–637
  GABA-positive terminals, 632–635
  immunogold particles, 632
  overview, 631–632
Gamma-aminobutyric acid (GABA) receptor control of dopaminergic neurons, 641–651
  burst firing suppression, 646–647
  overview, 641–642
  pars reticulata, 643–644
  pars reticulata-compacta interactions, 644–645
  substantia nigra, 642–643
  subthalamic afferents, 647–649
Genetics, Parkinson's disease, 131–132
Glial cell line-derived neurotrophic factor (GDNF) effects (rodent), 117–130
  clinical considerations, 127
  functional effects, 125–126
  intrastriatal 6-OHDA lesion model, 118–120

Glial cell line-derived neurotrophic factor (GDNF) effects (rodent) (*cont.*)
  morphological and biochemical effects, 121–125
  cell body and axon protection, 121–123
  regeneration, 123–125
  overview, 117–118
Globus pallidus; *see also* Enkephalin-expressing cells (rodent globus pallidus); External globus pallidus (GPe)
  apomorphine responses in, 23–31; *see also* Apomorphine responses
  basal ganglia function, 3
  cannabinoids, 666–669
  glutamate transporter localization (in monkey), 603
Globus pallidus pars externalis (GPe)
  dopaminergic agonist effects on pallidal neurons, 501–503
  multiunit activity of basal ganglia (monkey), 97–106; *see also* Multiunit activity of basal ganglia (monkey)
Globus pallidus pars internalis (GPi)
  dopaminergic agonist effects on pallidal neurons, 501
  multiunit activity of basal ganglia (monkey), 97–106; *see also* Multiunit activity of basal ganglia (monkey)
Globus pallidus pathophysiology, 15–21
  apomorphine studies, 18–19
  improvement mechanisms, 20
  overview, 15–16
  recordings, 16–18
    GPe, 16
    GPi, 18
    optic tract and internal capsule, 18
    white matter laminae and border cells, 16–18
  surgery, 19–20
Glutamate
  dopamine receptor (D_1 and D_2) differential modulation, 527–535; *see also* Dopamine receptor (D_1 and D_2) differential modulation
  subsynaptic localization of group I mGlu receptors, 567–580; *see also* Subsynaptic localization of group I mGlu receptors
Glutamate-dopamine interactions during aging (in rat), 615–622
  overview, 615
  striatum and nucleus accumbens, 615–617
  striatum and nucleus accumbens (aging), 617–620
Glutamatergic function, striatal: *see* Nicotine and striatal glutamatergic function (in 6-OHDA lesioned rats)
Glutamate transporter localization (in monkey), 599–613
  discussed, 607–611

INDEX

Glutamate transporter localization (in monkey) (*cont.*)
  materials and methods, 600–601
  overview, 599–600
  results, 601–607
    globus pallidus, 603
    striatum, 601–603
    substantia nigra/ventral tegmental area, 607
    subthalamic nucleus, 603–607
Grooming
  neuronal correlates of, natural action sequences, 281–284
  syntactical, natural action sequences, 280–281

High responders to novelty, noradrenaline and motivated behaviors, 327–330
Histology
  multiunit activity of basal ganglia (monkey), 100
  neuronal dysfunction (mouse), comparative histology of two models, 110–111
  pedunculopontine tegmental nucleus blockade (effects on arm movements in monkey), 153
  procedural learning (dopamine influence), 313–314
  subthalamic nucleus lesions (in hemiparkinsonian marmoset), 141–143
Huntington's disease; *see also* Neuronal dysfunction (mouse); Neuronal intranuclear inclusions
  experimental, neuronal dysfunction (mouse), 107–115
  neuronal intranuclear inclusions in neostriatal striosomes, 83–86

Immediate early gene (IEG) induction: *see* MAPkinase role in cortically-induced immediate early gene (IEG) induction
Immunocytochemical localization: *see* Tyrosine hydroxylase localization in human nigral-striatal system
Immunocytochemistry; *see also* Dopamine depleting lesions
  glutamate transporter localization (in monkey), 601
  nicotine and striatal glutamatergic function (in 6-OHDA lesioned rats), 590–591
  subsynaptic localization of group I mGlu receptors, 568–569
Immunogold particles, gamma-aminobutyric acid$_A$ (GABA$_A$) localization (in rat), 632
Immunohistochemistry
  gamma-aminobutyric acid$_A$ (GABA$_A$) localization, 623–630; *see also* Gamma-aminobutyric acid$_A$ (GABA$_A$) localization (in human)
  multiunit activity of basal ganglia (monkey), 100
  organotypic co-culture of SN, PPN, and STN, 439
Immunostaining, glutamate transporter localization (in monkey), 600–601

Implicit learning, striatal function in OCD, cortico-striatal circuity, 88–89
Indirect pathway: *see* Metabotropic glutamate receptor localization and function
Information processing in cortico-striato-nigral circuits (in rat), 199–207
  anatomical channeling, 199–204
  coincidence detection, 204–207
  overview, 199
Instrumental task, tonically active neurons and predictable sensory events (monkey striatum), 350–351
Internal capsule, optic tract and, globus pallidus pathophysiology, 18
Iontophoresis, dopamine receptor (D$_1$ and D$_2$) differential modulation, 527–535; *see also* Dopamine receptor (D$_1$ and D$_2$) differential modulation

Learning, 297–302
  discussion, 301–302
  methods, 297–298
  overview, 297
  results, 299–301
  striatal function in OCD, cortico-striatal circuity, 88–89
Levodopa, bimanual repetitive arm movements, 43–50; *see also* Bimanual repetitive arm movements
Limb movements: *see* Movement inhibition and next sensory state prediction
Localization; *see also* Gamma-aminobutyric acid$_A$ (GABA$_A$) localization (in human); Gamma-aminobutyric acid$_A$ (GABA$_A$) localization (in rat)
  immunocytochemical: *see* Tyrosine hydroxylase localization in human nigral-striatal system
  metabotropic glutamate receptor localization and function, 581–588; *see also* Metabotropic glutamate receptor localization and function
  subsynaptic localization of group I mGlu receptors, 567–580; *see also* Subsynaptic localization of group I mGlu receptors
Looped circuits, basal ganglia function, 3
Low responders to novelty, noradrenaline and motivated behaviors, 327–330

Magnetic resonance imaging, striatal function in obsessive compulsive disorder, 89–90
MAPkinase role in cortically-induced immediate early gene (IEG) induction, 559–566
  c-Fos induction pattern, 560–562
  overview, 559–560
  signaling pathways, 563–565
  striatal neurons expressing c-Fos, 562–563

Medium spiny neuron dendritic changes (in weaver mouse striatum), 379–384
  discussed, 382–383
  methods, 379–381
  overview, 379
  results, 381–382
Memory, striatal function in OCD, cortico-striatal circuitry, 88–89
Mesolimbic noradrenaline: *see* Noradrenaline and motivated behaviors
Metabolic glutamate receptors: *see* Subsynaptic localization of group I mGlu receptors
Metabotropic glutamate receptor localization and function, 581–588
  multiple sites, 585–586
  overview, 581–583
  pallido-subthalamic synapse, 584
  striato-pallidal synapse, 583
  subthalamo-nigral synapse, 584–585
1-Methyl-4-phenyl-1,2,3,6-tetrahydropyridine (MPTP): *see* MPTP effects on dopaminergic innervation of frontal lobe (in monkey)
Microdialysis studies, noradrenaline and motivated behaviors, 327–330
Molecular alterations, in striatal neurons, neuronal dysfunction (mouse), 111–113
Motivated behavior: *see* Noradrenaline and motivated behaviors
Motor control: *see* Striatal adenosine receptors and motor control (in rat)
Motor loop, basal ganglia function, 3
Motor thalamus: *see* Basal ganglia thalamic distribution
Movement: *see* Natural action sequences
Movement inhibition and next sensory state prediction, 267–277
  discussion, 274–275
  functional pathway, 270–271
  model results summarized, 271–274
  movement inhibition, 268
  next sensory state information, 268–270
  overview, 267
MPTP dosing, procedural learning (dopamine influence), 312–313
MPTP effects on dopaminergic innervation of frontal lobe (in monkey), 159–167
  discussed, 163–165
  materials and methods, 160
  overview, 159–160
  results, 160–163
Multisecond oscillations, dopamine receptor stimulation effects, 471–473
Multisecond oscillations in STN (dopamine effects), 245–254
  bursting, oscillations and synchrony, 246–248

Multisecond oscillations in STN (dopamine effects) (*cont.*)
  discussed, 250–252
  dopamine receptor stimulation effects, 249–250
  firing rates, 246
  multisecond oscillations, 248–249
  overview, 245–246
Multiunit activity of basal ganglia (monkey), 97–106
  discussed, 102–103
  experimental procedures, 98–100
    animals, 98
    behavioral assessment, 99
    electrophysiological recordings, 99–100
    immunohistochemistry and histology, 100
    protocol, 99
    stereotaxic surgery, 98–99
  overview, 97–98
  results, 100–101
Muscle rigidity: *see* Antipsychotic drug-induced rigidity
Musimol microinjection, pedunculopontine tegmental nucleus blockade (effects on arm movements in monkey), 153, 155

Natural action sequences, 279–287
  discussion, 284–285
  neuronal correlates of grooming movements, 281–284
  overview, 279–280
  syntactical grooming sequences, 280–281
Neostriatal neuron spiking, 225–235
  nomenclature, 225–226
  overview, 225
  PANS, 228–229
  research in, 226–228
  STANS, 232–233
  TANS *in vitro*, 229–232
Neostriatal striosomes, in neuronal intranuclear inclusions, 83–86; *see also* Neuronal intranuclear inclusions
Neural networks (tremor)
  example, 56–57
  periodic and aperiodic behavior in, 53–55
Neuronal dysfunction (mouse), 107–115
  comparative histology of two models, 110–111
  molecular alterations in striatal neurons, 111–113
  NMDA receptors, neuronal sensitivity to stimulation of, 108–110
  overview, 107–108
Neuronal intranuclear inclusions, 83–86
  discussed, 85–86
  methods, 84
  overview, 83–84
  results, 84–85
Next sensory state prediction: *see* Movement inhibition and next sensory state prediction

INDEX

Nicotine and striatal glutamatergic function (in 6-OHDA lesioned rats), 589–598
  discussed, 595–597
  methods, 590–591
  overview, 589–590
  results, 591–595
Nigral-striatal system: *see* Tyrosine hydroxylase localization in human nigral-striatal system
Nigro-collicular projection, epileptic seizure control (animal models), 174
Nigrostriatal dopamine neurons, glial cell line-derived neurotrophic factor (GDNF) effects on (rodent), 117–130; *see also* Glial cell line-derived neurotrophic factor (GDNF) effects (rodent)
Nigrostriatal pathway lesions, nicotine and striatal glutamatergic function (in 6-OHDA lesioned rats), 590
Nigrothalamic distribution, basal ganglia thalamic distribution, 456–458
NMDA injection, transynaptic induction of c-fos, striatal patch/matrix organization development, 387–388
NMDA receptors, neuronal sensitivity to stimulation of, neuronal dysfunction (mouse), 108–110; *see also* Neuronal dysfunction (mouse)
Noradrenaline and motivated behaviors, 323–333
  discussion, 330–331
  microdialysis studies, 327–330
  overview, 323–324
  pharmaco-behavioral studies, 325–327
  relevance of, 324–325
Nucleus accumbens; *see also* Cocaine self-administration; Glutamate-dopamine interactions during aging (in rat)
  cocaine self-administration, 547–555
  dopamine and ensemble coding, 238–241
    ensemble coding, 238
    up and down states, 239–241

Obsessive compulsive disorder, striatal function in, 87–93; *see also* Striatal function in obsessive compulsive disorder
Optic tract, internal capsule and, globus pallidus pathophysiology, 18
Organotypic co-culture of SN, PPN, and STN, 437–454
  discussed, 446–450
    afferent inputs, 449–450
    cellular organization, 446–447
    DA and non-DA neurons, morphology and physiology, 448–449
    projection patterns, 447–448
  methods, 438–439
  overview, 437–438

Organotypic co-culture of SN, PPN, and STN (*cont.*)
  results, 439–446
    electrophysiology, 443–446
    MPT-explant, 440–441
    STN-explant, 440
    VM-explant, 441–443

Pallidal neurons: *see* Cortico-subthalamo-pallidal hyperdirect pathway (in monkey); Dopaminergic agonist effects on pallidal neurons; Spiking control in striatal and pallidal neurons
Pallidal stimulation
  bilateral posteroventral, clinical observations, 6
  functional imaging, 10
  globus pallidus pathophysiology, 19–20
Pallido-subthalamic synapse, metabotropic glutamate receptor localization and function, 584
Pallidothalamic distribution, basal ganglia thalamic distribution, 458–459
Pallidotomy
  bimanual repetitive arm movements, 43–50; *see also* Bimanual repetitive arm movements
  external globus pallidus (GPe) neuronal activity, 33–41; *see also* External globus pallidus (GPe)
  functional imaging, 7–10
  globus pallidus pathophysiology, 19–20
  unilateral posteroventral, clinical observations, 5–6
Pallidum, subsynaptic localization of group I mGlu receptors, 571–574
Pallidum-linked thalamo-cortical paths: *see* Thalamo-cortical paths (pallidum and cerebellum-linked)
PANS: *see* Neostriatal neuron spiking
Parkinson's disease; *see also* Nicotine and striatal glutamatergic function (in 6-OHDA lesioned rats); Tremor
  apomorphine responses, 23–31; *see also* Apomorphine responses
  basal ganglia function, 3–4; *see also* Stereotactic surgery
  bimanual repetitive arm movements, 43–50; *see also* Bimanual repetitive arm movements
  cannabinoids, rat model, 671–672
  causes of, 131–132
  dopaminergic innervation of frontal lobe, differential reductions, in monkey, 159–167; *see also* MPTP effects on dopaminergic innervation of frontal lobe (in monkey)
  experimental
    glial cell line-derived neurotrophic factor (GDNF) effects on nigrostriatal dopamine neurons (rodent), 117–130; *see also* Glial cell line-derived neurotrophic factor (GDNF) effects (rodent)

Parkinson's disease (cont.)
  experimental (cont.)
    multiunit activity of basal ganglia (monkey), 97–106; see also Multiunit activity of basal ganglia (monkey)
    external globus pallidus (GPe) neuronal activity, 33–41; see also External globus pallidus (GPe)
    globus pallidus pathophysiology, 15–21; see also Globus pallidus pathophysiology
    metabotropic glutamate receptor localization and function, 581–588; see also Metabotropic glutamate receptor localization and function
    MPTP model, 132–133
    procedural learning (dopamine influence), 311–321; see also Procedural learning (dopamine influence)
    subthalamic nucleus lesions (in hemiparkinsonian marmoset), 139–149; see also Subthalamic nucleus lesions (in hemiparkinsonian marmoset)
    surround inhibition (in monkey), 183–184
    tremor in, 51–60
    vesicular monoamine transporter (VMAT2) role, 131–137
    visuo-spatial attention in, 61–81; see also Visuo-spatial attention
    VMAT2 neuroprotection, 133–134
  Pars reticulata
    cannabinoids, 663–666
    gamma-aminobutyric acid (GABA) receptor control of dopaminergic neurons, 643–646
Patch/matrix organization markers, 393–398; see also Striatal patch/matrix organization development
  discussed, 395–397
  methods, 394
  overview, 393–394
  results, 394–395
Pedunculopontine nucleus: see Organotypic coculture of SN, PPN, and STN
Pedunculopontine tegmental nucleus blockade (effects on arm movements in monkey), 151–158
  discussed, 156–157
  materials and methods, 152–153
    behavioral analysis, 152
    histology, 153
    musimol microinjection, 153
    neuronal activity recordings, 152–153
    surgical, 152
  overview, 151–152
  results, 154–156
Periodic behavior, tremor, 53–55
Phasically active neurons (PANS): see Neostriatal neuron spiking

Positron emission tomography
  globus pallidus pathophysiology, 20
  pallidal stimulation, 10
  pallidotomy, 7–10
  subthalamic stimulation, 10
Predictable sensory events: see Tonically active neurons and predictable sensory events (monkey striatum)
Primary antisera: see Antisera
Procedural learning (dopamine influence), 311–321
  discussion, 317–318
  methods, 312–314
    behavioral assessment, 313
    histology, 313–314
    MPTP dosing, 312–313
    Rotating Object Retrieval puzzle, 312
  overview, 311–312
  results, 314–316
    anatomy, 314–316
    behavior, 314
Projection patterns, organotypic co-culture of SN, PPN, and STN, 447–448
P3 event-related potentials, visuo-spatial attention, 61–81; see also Visuo-spatial attention
Putamental neurons (monkey): see Sequential stages of behavioral task (putamental neurons in monkey)

Raclopride: see Antipsychotic drug-induced rigidity
Reaction time task, tonically active neurons and predictable sensory events (monkey striatum), 350
Reciprocal inhibition, selection problem, 258–259
Regeneration, glial cell line-derived neurotrophic factor (GDNF) effects (rodent), 123–125
Repetitive arm movements: see Bimanual repetitive arm movements
Rotating Object Retrieval puzzle, procedural learning (dopamine influence), 312, 314
Rotational behavior: see Dopamine-induced rotation and firing rate in BG output ganglia

Saccade generation, movement inhibition and next sensory state prediction, 271–272
Salience, selection problem, 258
Seizures: see Epileptic seizure control (animal models)
Selection (dopamine effects), 257–266
  core function, 260–262
  overview, 257
  problem of, 258–259
  short-latency dopamine response, 262–264
Sequential arm movements, movement inhibition and next sensory state prediction, 272–274

Sequential stages of behavioral task (putamental neurons in monkey), 289–296
  discussion, 293–294
  methods, 289–291
  overview, 289
  results, 291–293
Serial reaction time test (SRT), striatal function in OCD, 89–90
Short-latency dopamine response, selection (dopamine effects), 262–264
Single-unit activity, dopamine receptor (D_1 and D_2) differential modulation, 527–535; see also Dopamine receptor (D_1 and D_2) differential modulation
Sometimes tonically active neurons (STANS): see Neostriatal neuron spiking
Spiking: see Neostriatal neuron spiking
Spiking control in striatal and pallidal neurons, 209–216
  discussed, 214–215
  methods, 210
  overview, 209–210
  results, 210–214
Spiny neuron dendritic changes (in weaver mouse striatum): see Medium spiny neuron dendritic changes (in weaver mouse striatum)
Spiny projection neurons, 188–192; see also Dopamine and ensemble coding; Neostriatal neuron spiking
STANS: see Neostriatal neuron spiking
Stereotactic surgery, 3–14
  apomorphine responses in, 23–31; see also Apomorphine responses
  clinical observations, 5–7
    bilateral posteroventral pallidal stimulation, 6
    bilateral subthalamic nucleus (STN) stimulation, 7
    unilateral posteroventral pallidotomy, 5–6
  dopaminergic agonist effects on pallidal neurons, 500
  external globus pallidus (GPe) neuronal activity, 33–41; see also External globus pallidus (GPe)
  functional imaging, 7–10
    pallidal stimulation, 10
    pallidotomy, 7–10
    subthalamic stimulation, 10
  multiunit activity of basal ganglia (monkey), 98–99
  overview, 3–5
  synthesis, 11–12
Striatal adenosine receptors and motor control (in rat), 653–659
  discussed, 655–658
  overview, 653–655

Striatal cholinergic interneurons (dopamine actions), 489–497
  discussed, 494–495
  methods, 490
  overview, 489
  results, 490–494
Striatal function in obsessive compulsive disorder, 87–93
  cognitive activation studies, 89–90
  cognitive neuroscience, 90
  cortico-striatal circuitry, 88–89
  neurobiological models of OCD, 88
  overview, 87
  phenomenology of OCD, 87–88
  research directions, 91
Striatal glutamatergic function: see Nicotine and striatal glutamatergic function (in 6-OHDA lesioned rats)
Striatal neurons
  expressing c-Fos, MAPkinase role in cortically-induced immediate early gene (IEG) induction, 562–563
  spiking control in: see Spiking control in striatal and pallidal neurons
Striatal patch/matrix organization development, 385–391
  afferent innervation role in organotypic cultures, 388–390
  anterograde tracing of corticostriatal afferents, 386–387
  overview, 385–386
  transynaptic induction of c-fos, 387–388
Striatal regions
  dopamine and ensemble coding, coincidence detection in, 241–242
  dopamine receptor (D_1 and D_2) differential modulation, 527–535; see also Dopamine receptor (D_1 and D_2) differential modulation
Striato-nigral afferents: see Substantia nigra pars compacta neuron distribution and striato-nigral afferents (3D modeling)
Striato-nigral system: see Tyrosine hydroxylase localization in human nigral-striatal system
Striato-pallidal synapse, metabotropic glutamate receptor localization and function, 583
Striatum; see also Glutamate-dopamine interactions during aging (in rat)
  gamma-aminobutyric acid$_A$ (GABA$_A$) localization, 625–628
  glutamate transporter localization (in monkey), 601–603
  subsynaptic localization of group I mGlu receptors, 569–571
Subcellular/subsynaptic glutamate transporter localization: see Glutamate transporter localization (in monkey)

Substantia nigra; *see also* Organotypic co-culture of SN, PPN, and STN
  epileptic seizure control (animal models), 171
  gamma-aminobutyric acid (GABA) receptor control of dopaminergic neurons, 642–643
  glutamate transporter localization (in monkey), 607
  subsynaptic localization of group I mGlu receptors, 574–576
Substantia nigra pars compacta neuron distribution and striato-nigral afferents (3D modeling), 359–367
  contribution of, 364–365
  functional considerations, 363
  overview, 359–360
  perspectives on, 365
  three dimensional organization, 360–363
Subsynaptic localization of group I mGlu receptors, 567–580
  materials and methods, 568–569
  overview, 567–568
  results, 569–576
    pallidum, 571–574
    striatum, 569–571
    substantia nigra, 574–576
    subthalamic nucleus, 574
Subthalamic nucleus
  cannabinoids, 663–666
  gamma-aminobutyric acid (GABA) receptor control of dopaminergic neurons, 647–649
  glutamate transporter localization (in monkey), 603–607
  multisecond oscillations in (dopamine effects), 245–254; *see also* Multisecond oscillations in STN (dopamine effects); Organotypic co-culture of SN, PPN, and STN
  multiunit activity of basal ganglia (monkey), 97–106; *see also* Multiunit activity of basal ganglia (monkey)
  subsynaptic localization of group I mGlu receptors, 574
Subthalamic nucleus lesions (in hemiparkinsonian marmoset), 139–149
  discussed, 145–147
  methods, 140–141
    analysis, 141
    behavioral analysis, 140–141
    histology, 141
    surgery, 140
  overview, 139–140
  results, 141–145
    behavioral analysis, 143–145
    histology, 141–143
    post-operative observations, 143
Subthalamic stimulation, functional imaging, 10

Subthalamo-nigral synapse, metabotropic glutamate receptor localization and function, 584–585
Subthalamus, apomorphine responses in, 23–31; *see also* Apomorphine responses
Supplementary motor area, basal ganglia function, 3
Surround inhibition, 187–197
  anatomical basis, 187–188
  electrophysiological tests, 192–193
  feed-back versus feed-forward, 194–195
  macroscopic level, 193–194
  overview, 187
  spiny projection neurons, 188–192
Surround inhibition (in monkey), 181–186
  mechanisms and sites, 184–185
  normal subjects, 182–183
  overview, 181
  parkinsonism subjects, 183–184
Switch in behavioral set: *see* Tonically active neurons and switch in behavioral set (monkey striatum)
Syntactical grooming sequences, natural action sequences, 280–281

TANS: *see* Neostriatal neuron spiking
Task demand effects, visuo-spatial attention, P3 component, 66–68
Thalamo-cortical paths (pallidum and cerebellum-linked), 429–436
  discussed, 433–435
  materials and methods, 430–431
  overview, 429
  results, 431–432
Thalamus
  motor: *see* Basal ganglia thalamic distribution
  ventrolateral, basal ganglia function, 3
Three-dimensional modeling: *see* Substantia nigra pars compacta neuron distribution and striato-nigral afferents (3D modeling)
Tonically active neurons and predictable sensory events (monkey striatum), 347–355
  discussion, 353–354
  methods, 348–349
  overview, 347–348
  results, 350–352
    classically conditioned task, 351–352
    free rewards, 352
    instrumental task, 350–351
    reaction time task, 350
Tonically active neurons and switch in behavioral set (monkey striatum), 335–346
  discussion, 341–345
  materials and methods, 336–337
  overview, 335
  results, 337–341

INDEX   685

Tonically active neurons (TANS): *see* Neostriatal neuron spiking
Transynaptic induction, of c-fos, striatal patch/matrix organization development, 387–388
Tremor, 51–60
  discussed, 58–59
  example, 56–57
  modeling approach, 52–53
  overview, 51–52
  periodic and aperiodic behavior, 53–55
  physiological plausibility, 57–58
Tyrosine hydroxylase localization in human nigral-striatal system, 369–378
  discussed, 374–376
  methods, 370–372
  overview, 369–370
  results, 372–374

Unilateral posteroventral pallidotomy, clinical observations, 5–6

Ventral tegmental area, glutamate transporter localization (in monkey), 607
Ventrolateral thalamus, basal ganglia function, 3

Vesicular monoamine transporter (in Parkinson's disease)
  MPTP model, 132–133
  PD causes, 131–132
  VMAT2 neuroprotection, 133–134
Visuo-spatial attention, 61–81
  data analysis, 65
  discussed, 73–76
    P3 amplitude modulations, 74–75
    P3 latency prolongation, 75–76
    SV and VO tasks, 73–74
  EEG recording and averaging, 64–65
  overview, 61–63
  results, 65–73
    behavior measurements, 65–66
    ERP components, 71–73
    P3 component, 66–71
  stimuli and procedures, 64
  subjects, 63
  tables, 76–79

Weaver mouse striatum: *see* Medium spiny neuron dendritic changes (in weaver mouse striatum)
White matter laminae, border cells and, globus pallidus pathophysiology, 16–18